Alcohol and Temperance in Modern History

An International Encyclopedia

Advisory Board

Alcohol and Temperance in Modern History

An International Encyclopedia

Volume II: M–Z

Jack S. Blocker, Jr., David M. Fahey, and Ian R. Tyrrell, Editors

ABC CLIO

Santa Barbara, California Denver, Colorado Oxford, England

Library of Congress Cataloging-in-Publication Data

Blocker, Jack S.

Alcohol and temperance in modern history : an international encyclopedia / Jack S. Blocker, Jr., David M. Fahey, and Ian R. Tyrrell.

p. cm.

Includes bibliographical references and index.

ISBN 1-57607-833-7 (hardcover : alk. paper) — ISBN 1-57607-834-5 (e-book)

1. Alcohol—Encyclopedias. 2. Drinking of alcoholic beverages—Encyclopedias. 3. Alcoholism—Encyclopedias. 4. Temperance—Encyclopedias. 5. Alcoholic beverage industry—Encyclopedias. I. Fahey, David M. II. Tyrrell, Ian R. II. Title.

HV5017.B56 2003

362.292'03—dc21

8679

06 05 04 03 0 9 8 7 6 5 4 3 2 1

This book is also available on the World Wide Web as an e-book. Visit abc-clio.com for details.

ABC-CLIO, Inc.

130 Cremona Drive, P.O. Box 1911

Santa Barbara, California 93116-1911

This book is printed on acid-free paper ♾.

Manufactured in the United States of America

Contents

Alcohol and Temperance in Modern History

Volume II

Volume II

M

Mackenzie, Alexander (1822–1892)

Alexander Mackenzie (1822–1892), a stonemason, temperance supporter, and politician, was born on 28 January 1822 in Logierait, Perthshire, Scotland, to Alexander Mackenzie and Mary Stewart Fleming, and died on 17 April 1892 in Toronto, Ontario, Canada. As Canada's second prime minister, he oversaw the passage of the Canada Temperance Act (1878), Canada's first federal Prohibition law.

Mackenzie immigrated to Canada in 1842, staying five years in Kingston, Ontario, before joining his six brothers in Sarnia, Ontario. During his early years in Sarnia, Mackenzie established himself as a building contractor of some note. He prized the pragmatic and eschewed the frivolous. A Baptist convert from Presbyterianism, he practiced a sincere Christian piety infused with a strong sense of self-discipline. Politically, he held reform sympathies, and thus began his involvement with those politicians who coalesced into Canada's Liberal Party.

In keeping with his religious and secular principles, Mackenzie viewed drunkenness as both sinful and unproductive behavior. Upon his arrival in Sarnia, he became active in local temperance affairs, sitting as an executive member of the Port Sarnia Temperance Society. Most notably, he supervised the construction of Sarnia's first temperance hall in 1854. As a member of the Port Sarnia Temperance Society, Mackenzie worked closely with fellow townsmen Malcolm Cameron, who introduced the first bill of prohibition in the provincial legislature, and Alexander Vidal, who later became the inaugural president of the Dominion Alliance for the Total Suppression of the Liquor Traffic. All three men were also involved with the Sons of Temperance, an organization that had found its way to Sarnia in the early 1850s. Despite their shared interest in temperance reform, a rift in Sarnia's political scene, in which Cameron successfully sued Mackenzie for libel, brought about the demise of the Port Sarnia Temperance Society and the local Sons of Temperance lodge.

Mackenzie moved on to politics, gaining a seat in the provincial legislature in 1861. With Canadian Confederation six years later, he successfully jumped to the federal arena, where his political skills saw him elected as the federal Liberals' first official leader. In November 1873, with the fall of Sir John A. Macdonald's scandal-ridden government, Mackenzie and his party assumed power and held it until electoral defeat in 1878. With a fellow traveler in office, Canada's temperance lobby immediately pressed the new prime minister for Dominion-wide prohibition. Although personally sympathetic to the cause, Mackenzie rebuffed these advances, believing that a significant segment of the Canadian population would not tolerate such a law. He later accepted the Dominion Alliance's proposal for a statute of local-option prohibition, a concept that appealed to his egalitarian sensibilities as it put the choice for prohibition directly in the hands of electors. Of course, it also afforded him the middle ground in dealing with a contentious issue. The prohibitory law received assent as the Canada Temperance Act on 10 May 1878. Along with establishing the secret ballot and creating the Supreme Court of Canada, passing the Canada Temperance Act ranks among Alexander Mackenzie's most significant public accomplishments.

Glen C. Phillips

See also: Canada Temperance Act; Dominion Alliance for the Total Suppression of the Liquor Traffic; Sons of Temperance

References

National Archives of Canada. Alexander Mackenzie Papers (MG 26, B).

Riddell, Russell. "Community Leadership in Sarnia: The Evolution of An Elite, 1830 to 1865" (M.A. thesis, University of Western Ontario, 1991).

Sarnia Observer. Selected issues, 1853–1860.

Thomson, Dale C. *Alexander Mackenzie: Clear Grit.* Toronto: Macmillan, 1960.

Maguire, Robert (1826–1890)

Robert Maguire (1826–1890), a chief founder of the Church of England Total Abstinence Society in 1862, edited its journals and those of its successor, the Church of England Temperance Society (CETS), established in 1873. An evangelical, anti–Roman Catholic polemicist, the St. James, Clerkenwell, London, rector was led to teetotalism by his parish council, which was unusual in being controlled by working people. A National Temperance League member, he helped organize the London Coffee House Conference that led to the formation of the church teetotal society. Maguire wrote *Temperance Landmarks: A Narrative of the Work and the Workers* (1880) and many polemical discourses and pamphlets.

Maguire took the pledge in November 1860 only after a crisis of conscience: Earlier he had "regard[ed] the teetotalism of a working man as a good part of a good character," but, like other clergymen, he had rejected total abstinence, believing that alcohol was beneficial to health. After abstaining, Maguire believed that his health improved. Working-class teetotalers helped Maguire administer a parish teetotal society in 1861. A shoemaker, Thomas Bouffler, who had urged the society's initiation, became its first registrar.

Maguire and the four other evangelicals who initiated the Church of England Total Abstinence Society (Julia Wightman, teetotal clergymen Stopford Ram, Francis Close, and Henry Ellison, who introduced teetotalism to Anglican parishes in the mid-1850s) sanctioned the teetotalism of working-class parishioners and overcame fears for health and reputation. With their American-inspired movement, associated with radicals, chartists, skeptics, secularists, Nonconformists, and Roman Catholics, they risked violating Anglican theology and social deference. Updating evangelicalism, they responded to the masses at the parish level rather than imposing control from above, yet they insisted on ministerial direction of parish teetotalism. Like other middle-class teetotalers, and unlike the working-class abstainers, many of these evangelical clergymen saw excessive drinking as a cause of poverty, rather than the reverse.

Maguire and other teetotal clergymen exaggerated how Anglican temperance could help revive the established church's parochial system. Most bishops ignored their disrespectful commands to espouse teetotalism and join the church teetotal society. Some accused Maguire of favoring teetotalism above the church. As editor of the *Church of England Temperance Magazine,* Maguire sympathized with Nonconformist teetotalers and accused Anglican bishops of idolatrous veneration of alcohol.

Maguire suppressed his anti-Romanism to appear at temperance meetings with Roman Catholics, such as Archbishop Henry Edward Manning, as well as with Nonconformists. Demonstrating the church's social utility, ecumenical cooperation in antidrink movements was said to have political value in combating disestablishment; this inspired establishment sanction of the CETS in 1873. Maguire refused to link his parish teetotal society with the new dual-basis CETS, with its sections for teetotalers and moderate drinkers, although he edited the new *Church of England Temperance Chronicle.*

The British Library and Lambeth Palace Library hold Maguire's writings.

Stephanie Olsen and Gerald Wayne Olsen

See also: Christian Socialism; Church of England Temperance Society (CETS); Close, Francis; Cruikshank, George; Ellison, Henry; Evangelical Temperance (United Kingdom); Moule, Handley Carr Glyn; National Temperance League (NTL); Wightman, Julia Bainbrigge

References

Olsen, G. W. "From Parish to Palace: Working-Class Influences on Anglican Temperance Movements, 1835–1914" in *Journal of Ecclesiastical History.* 40, no. 2 (April 1989): 239–252.

———. *Pub and Parish: Drink and Temperance in the Church of England, 1829–1875.* Unpublished book manuscript.

Rooke, Thomas, ed. *Clerical Experiences of Total Abstinence.* London: Church of England and Ireland Temperance Reformation Society, 1867.

Mailloux, Father Alexis (1801–1877)

Father Alexis Mailloux (1801–1877) was an austere Catholic priest who in 1851 inherited the mantle of leadership of the French Canadian temperance campaign after the flamboyant Father Charles Chiniquy was discredited by his sexual misadventures. Father Mailloux resolutely rejected pleasures of the flesh and encouraged his followers to do the same. They were to embrace an ascetic way of life by joining the Société de la Croix, which instructed them to hang a large black cross on a prominent wall of their home as a reminder of their pledge. Though Father Mailloux's Société de la Croix did not arouse the same enthusiasm as Father Chiniquy's patriotic temperance crusade had, it became the leading French Canadian temperance society in the third quarter of the nineteenth century. Mailloux succeeded in enrolling large numbers of new parishes and members between 1847 and 1877, doing his part to foster the sobriety that persisted in many parts of rural Quebec well into the twentieth century.

As the chief theorist of French Canadian temperance in the 1850s and 1860s, Mailloux's abhorrence of worldly vanities contrasted with the Anglo-American stress on social progress, individual self-help, and empowerment of female victims of intemperance. (He did, however, share their interest in humane childrearing, and his 1851 parenting manual enjoyed a century of use.) Capable of sermons so frightening that hearers feared entering his confessional, Mailloux kept the hellfires burning in Quebec while English Canadians began theorizing heavens on earth through prohibition, urban renewal, and female suffrage.

As a full-time itinerant preacher from 1847, Father Mailloux became the leading spokesman for the Société de la Croix (founded by his colleague Father Edouard Quertier in 1842). This particular brand of temperance was authoritarian and patriarchal, with the parish priest officiating over its meetings, masses, and ceremonies. If the priest decided that a man was worthy to join, his whole family was automatically enrolled. Far from stressing temperance as a key to progress and prosperity, Mailloux's 1850 *La Croix présentée aux membres de la Société de tempérance* presented sobriety as a rural people's refuge from the nineteenth century's "revolutions, insubordination, pride, luxury, irreligion, perverted passions for pleasure, worldly joys and well-being."

The leading temperance publicist in French Canada in the third quarter of the nineteenth century, Mailloux succeeded in enrolling sixty-three Quebec diocese parishes by 1850. Although some balked at joining this somber society, eventually all the parishes of Canada East were enrolled. In the early twentieth century, branches were still appearing, though the Société de la Croix no longer held preeminence among a diverse group of Catholic temperance societies. Perhaps the heyday of Mailloux's success was in 1855, when Montreal's Bishop Bourget proudly announced that the French Canadians had become a nondrinking people. However, drinking soon resurfaced in urban areas, and Quebec's per capita consumption in the second half of the nineteenth century tended to outrun other provinces.

The sectarian nature of the Société de la Croix had a chilling effect on cooperation with the predominantly Protestant English Canadian prohibitionists. Still, it helped consolidate the social leadership of parish priests in one of the most Catholic cultures in the world, arguably forestalling an endangered minority's absorption into the North American mainstream in temperance as in other matters.

Jan Noel

See also: Alcohol, Consumption of, per Capita (Canada); Chiniquy, Father Charles Paschal Telesphore

References

Gagnon, Serge. "Mailloux, Alexis" in *Dictionary of Canadian Biography,* vol. 10, 488–489. Toronto: University of Toronto Press, 1972.

Noel, Jan. *Canada Dry: Temperance Crusades since Confederation.* Toronto: University of Toronto Press, 1995.

Maine Law

In 1851, Maine became the first state to outlaw the manufacture and sale of alcoholic beverages. In the next four years, twelve states and territories followed Maine's example and adopted prohibitory legislation. The Maine Law revealed the growing disillusionment of antebellum reformers with the effectiveness of moral suasion. This shift in mood also affected the abolition and antigambling movements; a desire for governmental interference, particularly at the state level, began to replace the emphasis on individual will. In the less confident United States of the 1850s, many reformers believed that only the state acting against private property in the name of the general welfare could allow the individual to resist temptation. Although unsuccessful in the short term, the Maine Law and its cousins began the process that would eventually culminate in the Eighteenth Amendment.

The Second Great Awakening had convinced many Americans that previously acceptable behaviors such as drinking, gambling, and slave-holding were actually evil influences that weakened society by destroying individuals. As early as 1833, the American Temperance Union passed a controversial resolution advocating legislation to support the prohibitionist position. In 1838, Massachusetts banned the sale of distilled spirits in amounts of fewer than 15 U.S. gallons, and within the next decade, states as diverse as New York and Mississippi passed some sort of restriction on alcohol sales. In 1847, the U.S. Supreme Court boosted the prohibitionist cause in three cases known as the *License* cases. The Court, although speaking in six separate opinions, unanimously rejected the arguments of Daniel Webster and Rufus Choate and ruled that state governments could, in fact, constitutionally deny licenses to sell distilled spirits.

In Maine, a legislative committee headed by James Appleton had decided in 1837 that the complete prohibition of alcohol was the best means of improving the state's licensing laws. Neal Dow (1804–1897), a wealthy merchant and ex-Quaker who had adopted temperance in 1827, organized the Maine Temperance Union in 1838. In 1842, the city of Portland voted by a two-to-one margin (943 to 498) to stop the sale of alcohol. In 1846, Dow spearheaded a drive that collected almost 4,000 signatures on a 59-foot petition that convinced the Maine legislature to pass a law forbidding the sale of intoxicating beverages in lots of fewer than 28 U.S. gallons. Both the city and state laws were widely evaded by citizens, tavern owners, and government officials.

In early 1851, Dow became mayor of Portland. Using extensive propaganda and intensive lobbying, he prodded Maine's legislature into passing the Maine Law on 31 May. Although antebellum temperance is often portrayed as a Whig Party issue, Maine's Democratic-controlled legislature passed the measure by two-thirds majorities in both chambers. The Maine Law, signed by Democratic Governor John Hubbard on 2 June, banned the manufacture, sale, and keeping-for-sale of intoxicating liquors. It also gave the state government the right of search and seizure on complaint of three residents (eliminating the need for informers) and allowed for the confiscation of illegal supplies. A requirement of jail for third offenders reduced the discretion of prosecutors and judges in liquor cases. The Maine Law did, however, allow the sale of alcohol for medical or mechanical purposes and permitted personal usage and importation from other states.

As mayor of Portland from 1851 to 1859, Dow struggled to keep the city free of alcohol. If the Maine Law was the forerunner of the Volstead Act, Maine residents also pioneered many of the forms of subterfuge and evasion that made prohibitory legislation difficult to enforce. The Maine Law was repealed in 1856 but reenacted two years later after a landslide referendum; in 1884, it was added to Maine's constitution as the state's Fifth Amendment. The controversy over the Maine Law probably inspired Timothy Shay Arthur's best-selling *Ten Nights in a Barroom, and What I Saw There* (1854), the most famous American temperance work.

The Maine Law achieved nationwide fame and many states rushed to copy it. By 1855, all of New England was "dry," as was New York, Michigan, Indiana, Iowa, Delaware, and the Minnesota and Nebraska territories; voters in Pennsylvania and Illinois only narrowly defeated proposed "Maine Laws" in referenda. The prohibitory laws of the 1850s reflected the alarm of middle-class Protestants made uneasy by the social and cultural changes brought by capitalism and the Industrial Revolution. They also felt threatened by waves of Irish and German immigration in the 1840s and 1850s; both nationalities came

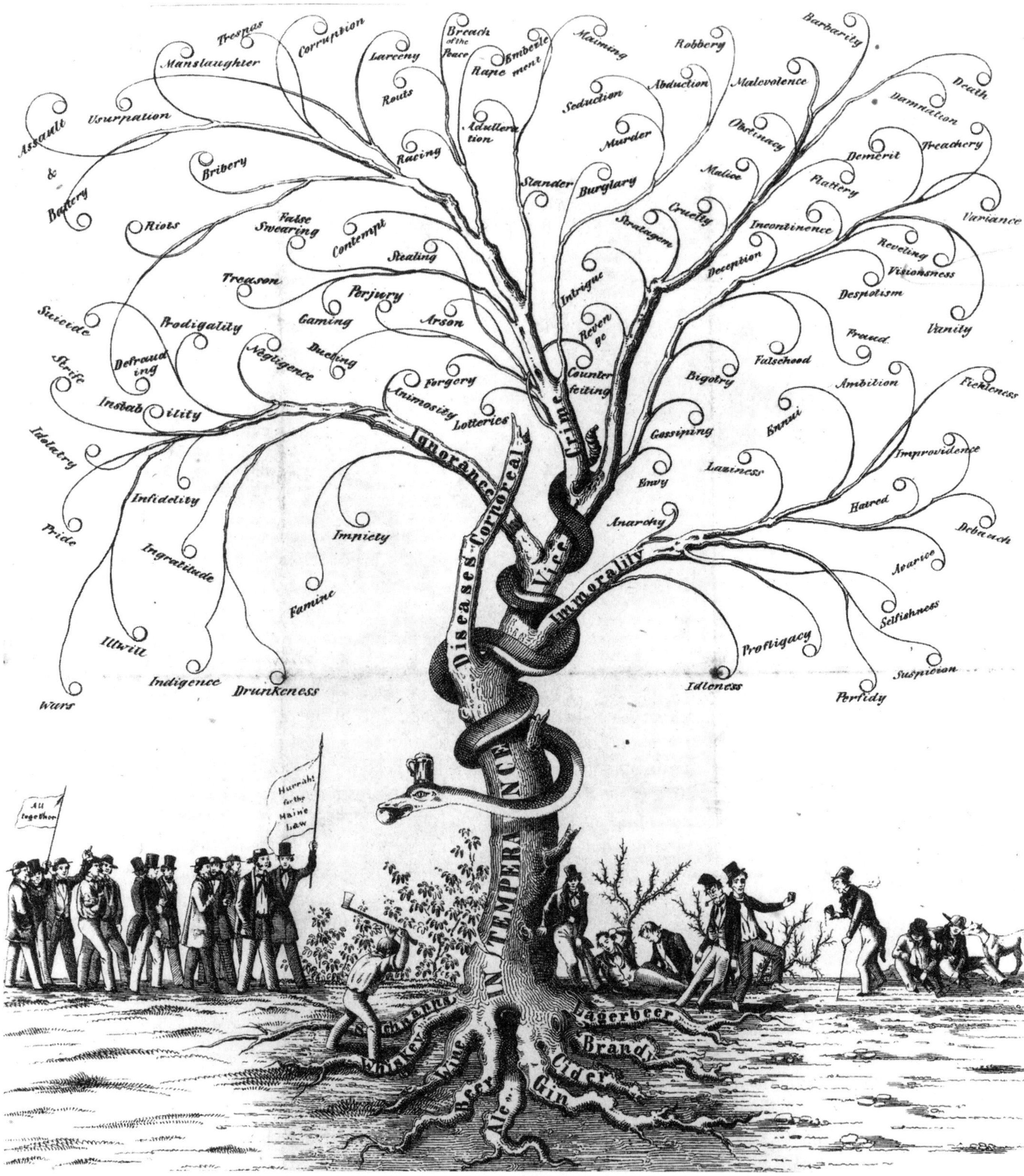

A print issued by A. D. Fillmore in 1855 expounding on the benefits of temperance and the evils of drink (Library of Congress)

from cultures in which drinking was an integral part of social culture. Most prohibition laws were repealed, however, when immigrant and other antireform groups discovered the political rewards of an antitemperance coalition. By the end of the Civil War, "Maine Laws" had virtually disappeared outside New England, either unenforced, overturned, or struck down by state courts as unconstitutional.

Jon Sterngass

See also: Dow, Neal; Fifteen Gallon Law; State Prohibition (United States)

References

Byrne, Frank. *Prophet of Prohibition: Neal Dow and His Crusade.* Madison: State Historical Society of Wisconsin, 1961.

Dow, Neal. *The Reminiscences of Neal Dow: Recollections of Eighty Years.* Portland, ME: Evening Express, 1898.

Pegram, Thomas. *Battling Demon Rum: The Struggle for a Dry America, 1800–1933.* Chicago: Ivan Dee, 1998.

Tyrrell, Ian. *Sobering Up: From Temperance to Prohibition in Antebellum America, 1800–1860.* Westport, CT: Greenwood, 1979.

Malins, Joseph (1844–1926)

Joseph Malins (1844–1926) introduced Good Templary to England and became a leader of the Independent Order of Good Templars (IOGT), one of the most influential temperance organizations in the late nineteenth century. In 1876, he became a key figure in the split within the IOGT on the question of whether to uphold its long-standing tradition of universalism and so allow African Americans to become members everywhere despite the objections of local whites in the American South.

Born in Worcester, Malins attended school for only two years before being forced to go to work to help with financial problems brought upon his family by his father's drunkenness. In 1860, one month after his father died from the effects of a lifetime of heavy drinking, he took a personal vow of total abstinence.

Malins eventually sought out others who had taken a similar pledge. In 1862, he joined an Anglican organization, the St. Thomas's Bow Street Total Abstinence Society. When that organization failed two years later, he joined the Hope Street Temperance Society. His involvement with the latter came to an end when Malins, who by livelihood was a furniture and decorative painter, immigrated to Philadelphia. It was there that he learned about Good Templary. He joined a west Philadelphia lodge. Because of his wife's illness, Malins decided to return to England. Before his departure he received authorization to organize the IOGT in England.

In 1868, Malins organized the Columbia Lodge in Birmingham. Only eleven individuals were in attendance, including several family members. He later gained a few additional followers from the now defunct Hope Street society as additional charter members. At first things did not go well for the English Templars. They operated at a disadvantage compared to other English fraternal organizations of the day in that they did not offer a mutual insurance plan. At one point, the Columbia Lodge was located over a horse stable with a terrible odor; Malins wearied other officers by knocking on their doors at six o'clock in the morning to teach them in their kitchens about Templar rules and rituals. Malins, a rather prickly man, soon found himself the target of a plan to oust him as leader. Fortunately for Malins, the American officials of the IOGT backed him against the dissidents.

Despite the early setbacks Malins proved to be an adept organizer as well as an extremely hard worker who put in twelve-hour days for the cause. By 1870, he had organized the requisite number of local lodges for the IOGT to allow the establishment of the Grand Lodge of England. Still in his twenties, Malins was elected grand worthy chief templar, the highest position within a grand lodge and an office he would hold until he retired in 1914. Fairly soon, the position allowed Malins to enjoy a comfortable middle-class life with a salary that at its peak reached £500, a significant income for a man who several years earlier had to pawn his overcoat to pay for the printing of Templar pamphlets. Besides his wife, the former Lucy Ellen Jones, Malins's family eventually included five children, all of whom were lifelong abstainers, perhaps not surprising for a family in which the eldest son was named Templar.

Following the establishment of the grand lodge, English Templary underwent a period of rapid expansion. By 1874, there were more than 200,000 members meeting in more than 2,000 lodges. During its first fifteen years, the English Grand Lodge gave the teetotal pledge to more than 1.3 million individuals. At the same time, Malins became active in Templar activities outside England. He established the first Welsh lodge and in 1873 sent deputies to form lodges in France, Portugal, South Africa, Bermuda, Belgium, and India. In 1880, he went to Sweden to establish the first Scandinavian grand lodge. In recognition of his work, at the 1874 session of the Right Worthy Grand Lodge (RWGL), the international governing body of the IOGT, Malins was elected right worthy grand councilor, the second highest office in the international organization.

Soon after achieving this office Malins became involved in a bitter controversy. Universalism—allowing women to be full members, for example—was a hallmark of Good Templary from its early days in the 1850s in New York State, and it was an aspect of Good Templary that Malins held dear. After the Civil War, however, the IOGT began to expand rapidly in the states of the former Confederacy, and in the southern grand lodges white supremacy and segregation were substituted for the concept of universalism. Malins's chief opponent in this dispute was John J. Hickman, the leader of the Grand Lodge of Kentucky, the oldest and largest of the southern grand lodges. Hickman refused to grant charters to black Kentuckians and instead used money from the Kentucky Grand Lodge to pay the expenses for the creation of a separate black temperance organization, not affiliated with the IOGT, to be known as the United Order of True Reformers. In 1876, while meeting in Kentucky, the RWGL rejected Malins's policy of authorizing Good Templars from outside the American South to organize African American lodges in the former slave states. Most Templars believed that this solution would have violated jurisdictional rights. At this point, Malins decided there was no other choice but to organize a rival RWGL of the World. Malins did not admit that he had seceded from the RWGL. He argued instead that the original RWGL lost its legitimacy by breaking with the tradition of universalism. His enemies claimed that his break with the old RWGL had a different motive: fear that the international organization might insist on the division of the very large English Grand Lodge into provincial grand lodges that would be analogous state grand lodges in America. The dispute became so identified with the personalities of Malins and Hickman that contemporaries labeled the two factions "Malinsites" and "Hickmanites."

Approximately 10 percent of English Templars joined the Hickmanites and formed their own English Grand Lodge. Most of these individuals, such as their leader, Dr. F. R. Lees, rejected the racism of the southerners but were motivated by an intense dislike of Malins's autocratic leadership style. Malins tried to organize black lodges in the American South for the RWGL of the World but was not very successful. The Templar split ended in 1887 when the new leader of the RWGL, John Finch, convinced Malins to accept a compromise that essentially allowed segregation to continue. Some English Templars were not pleased with the terms, feeling that expediency had won out over the ideal of racial equality. For Malins, however, ending the split was the more important task because it had been highly damaging to English Templary. Membership had fallen to fewer than 66,000 during the conflict, and Malins was personally aggravated by the presence in England of a Hickmanite Grand Lodge. He also was strongly committed to Templar internationalism. He eventually was elected the head of the reunited Templar international organization.

Malins was involved with nearly all the national temperance organizations in Great Britain. He served for many years as a vice president of the United Kingdom Alliance and helped found the National Temperance Federation, an umbrella organization of prohibitionist temperance groups. Malins was also a member of the executive committees of the National Commercial Temperance League and the National Free Church Council. Despite his meager formal education, he wrote numerous pamphlets on temperance issues as well as a large body of temperance poetry. For a time he served as the editor of the *International Good Templar,* and in 1888 he founded a temperance press agency that distributed temperance propaganda freely to temperance, religious, and other groups. In politics he was a Liberal Unionist and in religion a Methodist.

When Malins retired as leader of the Grand Lodge of England on the eve of World War I, the organization was clearly in decline with fewer than 45,000 members. Nevertheless, few contemporaries could match his overall contribution to the IOGT and to the temperance cause. To honor him, the English Grand Lodge created a new office, patriarch templar. He served the international order as well as the English Templars. Malins estimated that his travels on behalf of the IOGT added up to more than 700,000 miles. When he died, he was planning to attend the 1927 meeting of the Templar international organization in Philadelphia, at which time his sixtieth anniversary in the IOGT was to be celebrated.

Kenneth Pearl

See also: Good Templars (IOGT); Hoyle, William; Lees, Frederic Richard; National Temperance Federation (NTF); United Kingdom Alliance (UKA); World Prohibition Federation

References

Fahey, David M. *Temperance and Racism: John Bull, Johnny Reb and the Good Templars.* Lexington: University of Kentucky Press, 1996.

Malins, Joseph (son). *The Life of Joseph Malins: Patriarch Templar, Citizen and Temperance Reformer.* Birmingham: Templar Press, 1932.

Malt Distillers' Association of Scotland (MDA)

The Malt Distillers' Association (MDA) of Scotland was previously known as the Pot Still Malt Distillers' Association (PSMDA), which in turn developed from the North of Scotland Malt Distillers' Association (NSMDA). Today it provides an advisory service to its members.

The introduction of the continuous Coffey still for grain whiskey production in 1830 and an 1860 law that allowed the blending of spirits from different distilleries had a major impact on the Scotch whiskey industry. Output of the cheaper grain whiskey had overtaken that of malt by the 1870s, and the rapid growth of blended whiskey left the malt distillers fearing their livelihood to be under attack. They decided collective action was required but only managed to form regional organizations. In 1874, the NSMDA was formed with thirty-six member distilleries.

The aims of the NSMDA were to increase the interchange of ideas, to press for reform of excise laws, and to establish a more equitable trading relationship with blenders. There were early successes when proposed increases in excise duty were rejected on three occasions between 1888 and 1915. The association's grievance with blenders was that the distillers provided warehousing without charge, but its inability to speak for all malt distillers proved to be a stumbling block in establishing a standard rental.

The definition of whiskey was another major concern. The NSMDA view was that only malt whiskey merited this description. In 1905, the successful prosecution of a London publican for selling blended whiskey that was not of the quality demanded appeared to be a victory, although the NSMDA had not been directly involved. The grain distillers responded by achieving the establishment of a royal commission to investigate, among other things, the use of the word "whisky" (its British spelling) as a trade description. NSMDA submitted its views, but the commission's 1909 report found that "whisky" could be described as a spirit distilled from a mash of cereal grains, thus including grain whiskey.

During World War I malt distilling ceased for eighteen months, and maintaining the financial stability of the trade brought the NSMDA into cooperation with grain distillers and blenders. After the war, NSMDA managed to curb overproduction by bringing about reduced output in the period 1925–1927 by as much as 25 percent per annum. Implementing restrictions highlighted the need for a single malt distillers' organization, and in 1927 NSMDA was replaced by PSMDA, which covered the whole of Scotland. The new organization was active in regulating output and lobbying about duty levels and the importation of barley. The number of independent malt distillers declined, however, and the Distillers Company Limited, which owned fifty-one malt distilleries, took the lead in many areas that had been the sole province of the association.

During World War II there were again restrictions on output, and the PSMDA had to implement the decision to limit the number of operating distilleries to twenty-two in 1942–1943. In the postwar period the Scotch Whisky Association, originally established to protect the interests of exporters, became recognized as the main industry body. The PSMDA reexamined its role; it entered into wage negotiations with trade unions on behalf of its members, and it continued to recommend prices until 1956. It also began to provide the expertise that its smaller members demanded, covering such topics as effluent, transportation, valuation, and rating.

In 1972, the association's name was altered to MDA. It has continued its dual role as a pressure group and provider of advisory services, mainly in the fields of health and safety, customs and excise, and environmental issues.

Frank O. Robson

See also: Distillers Company Limited (DCL); Scotland

References

Morrice, Philip. *The Schweppes Guide to Scotch.* Alphabooks, 1983.

Weir, Ronald B. *The History of the Malt Distillers' Association of Scotland.* Elgin, Scotland: Malt Distillers' Association, 1974.

Maltsters' Association of Great Britain (MAGB)

The Maltsters' Association of Great Britain (MAGB) is the trade association of the malting industry in the United Kingdom. Close connection with the brewing and distilling industries has meant that malting has long suffered government attention, both directly and indirectly. The association came into existence in the nineteenth century with the specific aim of ameliorating the restrictive regulations then applied to malting practice, and it continues to promote and defend the industry.

Taxation of malt became a major source of public revenue in eighteenth-century Britain. Measures to avoid fraud and evasion by maltsters led to legislation that prescribed every step in malt production. These regulations hindered the technological development of malting and became a considerable nuisance even to the most law-abiding maltsters. Matters came to a head in 1827 with the introduction of a new Malt Act that not only listed 101 penalties for transgressors but also further complicated already labyrinthine regulations. On 3 December 1827, about thirty maltsters met in London to form an Association of Maltsters of the United Kingdom. Soon the new body had about 1,800 members. A joint committee comprising association members and surveyors general of excise was established to negotiate reform of taxation. By 1830, nearly two-thirds of the regulations and penalties had been repealed, with further reform following to minimize vexation to the maltsters. The association had proven its worth. Apart from this early success, no record of the association's activities remains, and it is believed to have dissolved in 1880 following repeal of the malt tax.

The need for a malting industry voice was again recognized during World War I. Fears of the repercussions of proposed state purchase of the drinks industry by David Lloyd George's government and prohibitive wartime restrictions on malting led to resuscitation of the association as the MAGB in 1917. Within a year, the association had 226 members, with all but a handful of the smallest malting companies joining. Nationalization was avoided, but the malting industry mirrored the brewing industry in shedding companies in the difficult trading conditions of the interwar years. Membership of the MAGB had dropped to 91 by 1939. The association developed as a progressive force in malting during and after World War II. Its primary role remained as a protector of its member companies' interests, but it also promoted change in the industry, backing mechanization and the improvement in working conditions of a notoriously poorly paid labor force. Routine activities embraced noncompetitive aspects of barley quality and supply, scrutiny of legislation, food safety issues, liaison with government and trade bodies, and, from 1981, vocational examinations. Until 1968, membership of the association was restricted to sales maltsters. It was then opened to brewer and distiller maltsters who malted exclusively for their own use, but this measure could not prevent long-term decline in membership as the drinks industry became increasingly concentrated. By 2002, the MAGB represented the producers of over 98 percent of the malt made in the United Kingdom, the world's third largest malting nation, exporting around 30 percent of its production, but had only fourteen members.

Raymond G. Anderson

See also: Beer; Lloyd George, David

References

Clark, Christine. *The British Malting Industry since 1830.* London: Hambledon Press, 1998.

"Maltsters' Association of Great Britain" in *Brewers' Guardian Centenary Edition* (1971): 151–152.

Murrell, Ivor. "The Role of the MAGB with the UK and European Malting Industry" in *Brewing Room Book, 2001–2003,* edited by R. H. B. Beach, 34–37. Ipswich: Pauls Malt, 2001.

Mann, Marty (1904–1980)

Marty Mann (1904–1980) was one of America's greatest public health care reformers. In 1944, she founded and directed what became known during her lifetime as the National Council on Alcoholism (NCA; now the National Council on Alcoholism and Drug Dependence, NCADD). Mann's goal was to reduce the pervasive stigma attached to alcoholism so that alcoholics would seek help. Her ambitious method was to educate whole communities, both in the general public and within systems of medicine, health care, business and industry, government, law and the penal system, social work, schools, the military, and the clergy and churches, about the facts of alcoholism in order to mobilize and develop community resources on behalf of alcoholics. Her message was that alcoholism is a disease and the alcoholic is a sick person, that the alcoholic can be helped and is worth helping, and that alcoholism is a public health problem and therefore a public responsibility.

A gifted journalist and publicist, Mann reached enormous numbers of people by constant and extensive writing on her subject for journals, magazines, and newspapers. Also a charismatic speaker, she averaged more than 200 talks a year during her career. Numerous radio interviews, and later TV, helped spread the word. She tirelessly mentored individuals and organizations who then carried the message of hope to their particular lay and professional communities. The national NCADD office and local affiliates continue the work.

By the time Mann died in 1980, the general public of the United States and other countries were aware of the term *alcoholic* and viewed alcoholism as a primary disease, not a sin, moral failing, or form of insanity. These were the concepts that Mann had promulgated, and their widespread acceptance today is largely her achievement.

Mann was also the first woman to achieve long-term sobriety in Alcoholics Anonymous (AA). Destitute from several years of acute alcoholism, she began attending AA meetings in 1939, four years after that organization's founding. Her leadership in the NCA and her promotion of AA as a successful program where alcoholics could find help opened the way for other women and many men to seek recovery. The composition and membership of AA were forever changed by her example and her efforts.

Mann was born in Chicago, Illinois, on 15 October 1904, the eldest child of very wealthy parents. She was married briefly in 1925 to John Blakemore of New Orleans. After her divorce, she resumed her maiden name but retained the title "Mrs." When her father lost his money in 1929, she went to work. In about 1930 her drinking suddenly escalated into uncontrollable compulsion and she lost everything. Convinced she must be insane, Mann was admitted in 1938 to New York's Bellevue Hospital as a charity patient, and six months later to Blythewood Sanitarium in Greenwich, Connecticut. Her physician at Blythewood was Dr. Harry Tiebout, who introduced her to AA. There were no other women in AA at the time, so Bill Wilson, AA's cofounder, became her mentor, champion, and later colleague.

In the early 1940s, she met Priscilla Peck, who became the longtime art editor and later art director of *Vogue* magazine. Mann and Peck were devoted, loving partners for nearly forty years, until Mann's death. They both were active and influential in helping lesbian and gay alcoholics to seek recovery.

Yale University's Center of Alcohol Studies sponsored and subsidized NCA's early years. When Yale withdrew in 1948, NCA nearly went under, but it was rescued in 1954 by the largesse of R. Brinkley Smithers, heir to an IBM fortune. NCA stabilized and began to flourish.

Mann wrote two books that were influential contributions to public understanding of alcoholism. *Marty Mann's New Primer on Alcoholism* was published in 1958, and *Marty Mann Answers Your Questions about Drinking and Alcoholism* in 1970.

Because of Mann's and NCA's grassroots educational efforts, landmark legislation was enacted in 1970 that for the first time established federal responsibility for treatment of alcoholics. The bill, introduced and sponsored by Iowa Senator Harold Hughes, created the National Institute on Alcohol Abuse and Alcoholism (NIAAA). Treatment programs proliferated throughout the country.

In addition, the American Medical Association (AMA) proclaimed in 1967 that alcoholism is indeed a complex disease. Mann and the NCA created the climate in which an AMA specialty in addictive medicine was formed. Most medical schools initiated courses in alcoholism as a disease. Thousands of hospitals began to admit and treat alcoholics instead of turning them away. Business and industry established employee assistance programs to address the issue of alcoholism.

In her later years, Mann received many honors and was recognized internationally as a pioneer and leading force in the alcoholism movement. She died of a stroke on 22 July 1980.

Sally Brown

See also: Alcoholics Anonymous (AA); Alcoholism; Christopher D. Smithers Foundation; Employee Assistance Programs (EAPs); Hughes, Harold Everett; National Council on Alcoholism and Drug Dependence (NCADD); National Institute on Alcohol Abuse and Alcoholism (NIAAA); Wilson, William Griffith; Yale Center of Alcohol Studies

References

Brown, Sally, and David R. Brown. *A Biography of Mrs. Marty Mann: The First Lady of Alcoholics Anonymous.* Center City, MN: Hazelden, 2001.

Kutrz, Ernest. *Not-God: A History of Alcoholics Anonymous.* Expanded ed. Center City, MN: Hazelden, 1991.

White, William L. *Slaying the Dragon: The History of Addiction Treatment and Recovery in America.* Bloomington, IL: Chestnut Health Systems/Lighthouse Institute, 1998.

"Women Suffer, Too" in *Alcoholics Anonymous: The Story of How Many Thousands of Men and Women Have Recovered from Alcoholism,* 4th ed., 200–207. New York: Alcoholics Anonymous World Services, Inc., 2001.

Manning, Henry Edward (1808–1892)

Henry Edward Manning (1808–1892) was one of the most prominent Roman Catholic temperance advocates in the nineteenth century. An English archbishop, Manning established the League of the Cross, a Catholic total abstinence society, in 1873. He was also a leading spokesman for the United Kingdom Alliance (UKA), a heavily Protestant, prohibitionist organization.

Educated at Oxford, Manning took holy orders in the Anglican Church in 1832 and married Caroline Sargent, a clergyman's daughter, the following year. Widowed in 1837, he never married again. He strongly supported the Oxford Movement, which called for the incorporation of Catholic teachings and practices into Anglicanism. Like John Henry Newman and some of the movement's other leaders, Manning grew increasingly dissatisfied with Anglicanism. In 1851, he converted to Catholicism and was ordained a priest later that year. In 1865, Pope Pius IX appointed him archbishop of Westminster, which made him the highest-ranking prelate in English Catholicism. Ten years later, the pope made him a cardinal.

Soon after becoming an archbishop, Manning became interested in the temperance movement. Visiting his mostly Irish, mostly impoverished parishioners in the London slums, he became convinced that the evil of intemperance was responsible for a host of other vices. In 1867, he initiated the Truce of St. Patrick in an effort to curb the drunkenness associated with that holiday. Those Catholics who refrained from drinking from the 16th through the 18th of March were granted special indulgences. When the UKA's leaders learned of this initiative they sent delegates to meet with him. In October 1867, he appeared on the platform at the UKA's annual meeting strongly endorsing the Permissive Bill, which would have authorized voters to impose prohibition in their communities, and in 1868, he appeared before a committee of the House of Commons to support a Sunday closing proposal. Until the end of his life, Manning championed any and all measures that would make alcoholic beverages less available. Unlike some temperance advocates, however, he did not believe that temperance legislation on its own would solve the problems of the working class in England. He also wanted the government to provide better housing and recreation facilities for the poor.

In the 1870s, Manning's temperance enthusiasms took him in another direction. Having taken the pledge himself in 1872, he established the League of the Cross in the following year on a nationwide basis. He said he hoped that the league would carry on the work that the Irish temperance advocate Father Theobald Mathew had begun in the 1840s. In an effort to avoid the political questions that so occupied the UKA, Manning drew on the model of the Salvation Army when organizing his new society. Like the Salvation Army, the league had leaders with military-style ranks, and its members regularly marched and carried banners in military-style parades. In its early years, the league expanded quickly. By 1876, it had 58,000 pledged members in twenty-two chapters throughout the country. Its strongholds were London and Liverpool, where Father James Nugent had been working tirelessly to promote the pledge. For a time, it even had chapters in Australia, India, and Ireland.

Although the league received the pope's blessing, not all English Catholics were so enthusiastic. Many "old Catholics"—those wealthy families who had stayed loyal to Rome through the Reformation—considered Manning a fanatic. Among those opposed to Manning's temperance efforts was a fellow convert and cardinal, John Henry Newman, who declared that he was not sure whether England had too many pubs or too few. Comments such as these did not deter Manning in the least. Right up until his death, he called on the English people to support the UKA's prohibitionist proposals and urged his fellow Catholics to take the pledge and enroll in the league. After his death, however, no English bishop took up the cause, and the league soon began to decline.

John F. Quinn

See also: Mathew, Father Theobald; Nugent, James; St. Patrick's Day; United Kingdom Alliance (UKA)

References

Dingle, A. E., and Brian Harrison. "Cardinal Manning As Temperance Reformer." *Historical Journal.* 12 (1969): 485–510.
Gray, Robert. *Cardinal Manning.* New York: St. Martin's, 1985.
McClelland, Vincent. *Cardinal Manning: His Public Life and Influence, 1865–1892.* London: Oxford University Press, 1962.

Marsh, John (1788–1868)

John Marsh (1788–1868), a Congregational minister and secretary of the American Temperance Union from 1837 to 1866, was a leading figure in the temperance movement in the United States for more than thirty years. A lecturer, writer, editor, and activist, Marsh advocated teetotalism and emphasized individual self-control and the immorality of drinking. He edited the *Journal of the American Temperance Union* from its first issue in 1837 until 1866 and wrote numerous sermons, books, and tracts devoted to temperance. Throughout his life, Marsh argued that temperance was inseparable from evangelical Christianity and that only spiritual salvation could ensure sobriety.

Born in Wethersfield, Connecticut, on 12 April 1788, Marsh graduated from Yale University in 1804 and served as pastor of the Congregational Church in Haddam, Connecticut, from 1818 to 1832. Marsh became increasingly devoted to temperance reform, acting as secretary of the Connecticut Temperance Society from 1829 to 1832. His sermon "Putnam and the Wolf" (1829), which compared intemperance to a "ferocious beast" that "bred putrefaction and death among the holy," circulated widely as a tract, with more than 150,000 copies in print. Following an invitation from the American Temperance Society, Marsh moved to Philadelphia in 1833 to become secretary of the Pennsylvania Temperance Society, a post he held until 1837.

In 1836, Marsh served as secretary of the National Temperance Convention in Saratoga, New York, which saw the reorganization of the American Temperance Society as the American Temperance Union and its official adoption of a teetotal position. The union appointed Marsh corresponding secretary and editor of its new organ, *The Journal of the American Temperance Union,* first published in January 1837. Marsh held these posts until the dissolution of the American Temperance Union in 1866. For those three decades, Marsh played a prominent role in antialcohol reform. He considered the Washingtonians "low" and insufficiently pious, charging that the spirit of their meetings too closely resembled that of the saloon and in the early 1840s helped lead an attack that contributed to the group's demise. He similarly criticized the Order of the Sons of Temperance, deeming their fraternal rites and paraphernalia distractions from the true spiritual and political goals of temperance. In the 1850s and 1860s, Marsh supported prohibition and no-license campaigns. During the Civil War, he corresponded with Abraham Lincoln about temperance issues and organized an antiliquor propaganda campaign for the Union armies.

Marsh moved to New York City in 1838 and lived there until his death on 5 August 1868. In 1839, he founded the *Youth's Temperance Advocate,* a religious newspaper for children that he edited until 1865. Marsh's books and tracts included *The Bow of Promise* (1840), *Hannah Hawkins: The Reformed Drunkard's Daughter* (1843), *Temperance Hymn Book* (1844), *Temperance Anecdotes* (c. 1848), *The Temperance Battle Not Man's but God's* (1853), and his autobiography *Temperance Recollections* (1866). He also edited an important collection of American Temperance Union reports entitled *Permanent Temperance Documents* (1836). He served as delegate to the World's Temperance Conventions in London in 1846 and in New York in 1853, as well as to the North American Temperance Convention in Chicago in 1857.

Joshua Paddison

See also: American Temperance Society (ATS); Sons of Temperance; Washingtonians

References

Lender, Mark Edward. "Marsh, John," in *Dictionary of American Temperance Biography,* 323–325. Westport, CT: Greenwood, 1984.

Marsh, John. *Temperance Recollections: Labors, Defeats, Triumphs: An Autobiography.* New York: Charles Scribner, 1866.

Tyrrell, Ian R. *Sobering Up: From Temperance to Prohibition in Antebellum America, 1800–1860.* Westport, CT: Greenwood, 1979.

Martha Washington Societies

Martha Washington Societies emerged as the female branch of the male-dominated Washingtonian temperance movement. Begun in Baltimore in 1840 by a group of reformed drunkards, the Washingtonian movement aimed at not just preventing intemperance but also redeeming those who were already inebriates. The movement, which was primarily run by working-class men, spread quickly to other cities, claiming some 200,000 members by 1841. In New York, the first city outside of Baltimore to welcome the Washingtonians, women as well as men became interested in aiding intemperate men and their families. To facilitate these efforts, women formed Martha Washington Societies as auxiliaries to the male groups. These societies spread rapidly, especially in the Northeast and Midwest. New York City experienced the greatest growth, with more than forty societies and 6,000 members by 1842.

Martha Washington groups incorporated the functions of both temperance societies and charitable associations. Martha Washingtonians held fairs and bazaars to raise money for charity and temperance work, provided financial assistance to the families of drunkards, and supported the efforts of the male Washingtonian groups, often attending their meetings in large numbers. Members took a pledge of total abstinence from all intoxicating liquors, worked to reform drunkards, and attempted to prevent further intemperance.

Strategies and Tactics

Like their male Washingtonian counterparts, the Marthas endorsed moral suasion as the primary means of reform. In their case, they relied on a particular kind of suasion, known in the nineteenth century as female influence. Women's unique access to men's hearts and consciences as wives, mothers, sisters, and daughters, they believed, gave women a prominent and powerful role in reform. The Marthas worked directly with male drunkards, arguing that the society of sober, respectable women sympathetic to the plight of the reformed man would buttress his determination to avoid alcohol. At the same time, the Marthas' concern for the wives and children of drunkards reflected their understanding of intemperance as a primarily male vice that injured families through the abusive behavior of fathers and husbands. As the male Washingtonians provided a support network for intemperate men, Martha Washingtonians endeavored to provide similar assistance to their families. Recognizing that the families of drunkards often suffered from poverty and unemployment, the Marthas provided clothing so that men could look for work and thereby feed their families. In addition to raising money for clothing, food, and other essentials needed by households impoverished by alcohol, Washingtonian women also visited families to encourage, support, and advise the wives and children of drunkards.

Distinguishing Features of the Work

Three aspects of Martha Washingtonianism distinguished it from previous women's efforts at temperance reform. First, working-class women dominated the movement to a much greater extent than in other female reform efforts. This class dimension shaped the Marthas' treatment of the wives of drunkards. Like other temperance groups, Martha Washington Societies exhorted inebriates to sign the pledge of total abstinence and encouraged their wives and families to persevere, behave morally, and support newly sober men. But unlike the middle- and upper-class women who took the lead in other female reform societies, the Marthas identified and sympathized with the often working-class recipients of their efforts. They did not attempt to impose unworkable middle-class standards of domestic behavior on the impoverished households of their charges. Rather, they evolved working-class standards of feminine deportment and domestic order that better fit the situations of the families with whom they were working. Moreover, the Martha Washingtonians opposed the condescending and destructive practice, common in many cities, of requiring destitute women to enter poor houses, often giving up their children in the process, to receive relief. Instead of taking the unfortunate victims of intemperance from their homes to reform them, the Marthas attempted to reach them at home and thereby bring about change in the home itself.

Second, more than previous temperance organizations, Martha Washington Societies highlighted the problem of female intemperance. Though the Marthas continued to view intemperance as a male vice, unlike the more genteel middle-class temperance societies they did not seek to deemphasize or deny the existence of female drunkards. Familiar with the realities of working-class life, the Marthas understood that women might accede to the urgings of their intemperate husbands to join them in drinking, or sometimes turn to liquor themselves to alleviate the misery of their situation. Believing that the efforts of their own sex could restore intemperate women to usefulness and happiness, they sought to minister to the needs of female as well as male drunkards. Though their primary focus remained reforming drunken men and helping their families, the Martha Washingtonians were the first temperance society to take seriously the problem of female drunkenness.

Third, the Marthas increased dramatically the scope of women's reform efforts. In earlier temperance societies, women had occupied a decidedly auxiliary and subordinate role. Men undertook all major initiatives and formulated significant policies. By contrast, women took the lead in the Martha Washington Societies. Women established relatively independent organizations, raised funds and decided how they would be spent, and engaged in much face-to-face work with drunkards. To a far greater degree than middle-class female reformers, Martha Washingtonians possessed personal experience, or gained first-hand knowledge, of the ravages of alcohol abuse on families. This intimate understanding of the problem, coupled with their preference for visiting the homes of aid recipients and potential converts, took them beyond the domestic sphere and into the realm of social activism. Some Martha Washington Societies engaged in other forms of activism, encouraging notoriously intemperate volunteer fire companies to take the pledge, and presenting banners to those that did at public ceremonies. Other Washingtonian women went further still, establishing temperance newspapers such as the *Olive Plant and Ladies' Temperance Advocate* (New York) and the *New York Washingtonian and Ladies' Literary Pearl* to facilitate communication among members and spread the movement's message. The female editors of these newspapers articulated a distinct and independent women's voice within the larger temperance movement. By the late 1840s, both the male and female branches of the Washingtonian movement were in decline, but Martha Washingtonianism had influenced temperance activism and provided a precedent for active female participation in social reform.

Scott C. Martin

See also: Moral Suasion; Washingtonians

References

Alexander, Ruth M. "'We Are Engaged as a Band of Sisters': Class and Domesticity in the Washingtonian Temperance Movement, 1840–1850" in *Journal of American History.* 75 (December 1988): 763–785.

Dannenbaum, Jed. "The Origins of Temperance Activism and Militancy among American Women" in *Journal of Social History.* 15 (Winter 1981): 235–252.

Johnson, Lorenzo Dow. *Martha Washingtonianism, or, A History of the Ladies' Temperance Benevolent Societies.* New York: Saxton and Miles, 1843.

Tyrrell, Ian. "Women and Temperance in Antebellum America, 1830–1860" in *Civil War History.* 28 (June 1982): 128–152.

Massachusetts Society for the Suppression of Intemperance (MSSI)

The Massachusetts Society for the Suppression of Intemperance (MSSI) was the first statewide temperance society in the United States. Begun in 1813, the MSSI had auxiliary societies in 41 Massachusetts towns by 1815. This pioneering organization relied on the tactics of personal influence and public appeals, but it also sought the enforcement of laws against drunkenness and unlicensed liquor sales. Its activities and effects were modest. When a new surge of temperance enthusiasm began in 1826, the MSSI had been lifeless for several years.

The General Association of Massachusetts Congregational Churches appointed a committee on intemperance in 1811 and the following year approved a proposal to create a temperance society. The first meeting was in Boston in 1813. The new society intentionally remained small, with at most several hundred men. It sought prestigious members, and its ranks soon included many well-known ministers, merchants, and lawyers. The religious preference was orthodox Congregationalism, although approximately 40 percent of the members were Unitarian. Nearly everyone belonged to the Federalist Party, and several members held high positions in state government.

This group of distinguished men joined the MSSI for several reasons over and beyond their religious convictions. They deplored the sharp rise in liquor consumption in postrevolutionary America, when cheap whiskey from western states poured into eastern markets. The MSSI blamed intemperance, not the economic turmoil before and during the War of 1812, for onerous poor relief expenditures. But the MSSI also saw intemperance as part of a pervasive deterioration in the virtue of a citizenry that had foolishly voted Federalists out of office, disestablished religion, and challenged the colonial era deference that had insured respect for the elite. The changes throughout American society made these reformers uneasy. They thought intemperance was symptomatic of a moral and civic turpitude that they lamented and resisted. It was especially worrisome that the mass of laborers were willing to drink heavily rather than imitate the decorous manners and virtuous piety of the upper strata of society.

The Bostonians attracted a select band of prominent men, but the MSSI auxiliaries recruited a larger share of the local population. In the town of Concord, for example, 123 of the 318 men on the 1814 tax list joined the MSSI. As in the Boston branch, the Concord auxiliary drew church members from several denominations. Another similarity was the appeal to gentlemen of property. Twenty-two of the 30 wealthiest men in Concord joined the MSSI, and the entire membership paid, on average, more than twice the tax exacted of nonmembers. For a third similarity, the purchases recorded in a ledger book from a Concord store make clear that drinkers joined (the MSSI never forbade moderate drinking).

Yet the Concord case modifies the view of the MSSI as a quest to recapture the power once exercised by an elite now in precipitous decline. Comparisons over time of tax records indicate that MSSI members achieved greater upward economic mobility (both inter- and intra-generational) than did other Concordians. The Concord MSSI members also held a disproportionate share of town offices, and that share rose, not fell, from 1780 to 1814. The evidence from Concord suggests the broad appeal of temperance in its earliest years. It was not solely a reaction by displaced elites against a world they no longer controlled.

MSSI members did not set out to reclaim drunkards. Most evaluations of the chances of success were bleak. The parent society never sought inebriates as members, and its publications rarely noted cases of reformation. Some spokesmen wanted employers to withhold liquor rations from laborers, and they advised the auxiliaries to try to coax drunkards to sobriety, but there is little evidence that the local societies did much in that regard. The Scituate auxiliary, for instance, voted to make a list of intemperate townspeople. Rather than approach them directly, however, the society asked the selectmen to admonish the drunkards.

Beyond admonishment was the law, and temperance societies periodically urged town officials to arrest drunkards and unlicensed sellers. If the selectmen knew that respectable citizens wanted diligent law enforcement, they would not avoid their duties, the MSSI believed. The temperance auxiliaries did more than scold the selectmen. In several towns, they helped with law enforcement, and in others they petitioned for a reduction in the number of liquor licenses granted. In Scituate, they drafted a list of conditions for license applicants to sign, and later they appointed a committee to gather evidence of illegal liquor sales. The local exertions fell far short of the extensive activity undertaken by prohibitionists in the 1850s and later, but there was widespread local interest in better enforcement of the laws.

In its brief life, the MSSI failed to achieve its goals. The amount of liquor consumed remained as high as it had been. Retailers usually refused to cooperate, and selectmen continued to approve license applications. Within a few years, the MSSI became less and less active. In the 1818 annual report, the secretary reported information for only 6 of the 41 auxiliaries.

The MSSI suffered from several handicaps. It never had a vigorous organizational center with a charismatic leader, weekly newspaper, and other methods to mobilize widespread support. Its tight association with the Federalists hurt, as did the loss of urgency once the economic woes inflicted by the War of 1812 eased. The tolerance of moderate drinking undercut its ability to castigate liquor as harmful and sinful. Even so, the MSSI had introduced important tactics that would shape later temperance initiatives: the antipathy for drunkards, the pursuit of legal remedies, and the recruitment of socially prominent members. Future temperance crusaders would refine or alter those strategies, but they all wrestled with the enduring issues first explored by the MSSI.

Robert L. Hampel

See also: American Temperance Society (ATS)

References

Barnett, Redmond. "From Philanthropy to Reform: Poverty, Drunkenness, and the Social Order in Massachusetts, 1780–1825" (Ph.D. dissertation, Harvard University, 1973).

Hampel, Robert L. *Temperance and Prohibition in Massachusetts, 1813–1852.* Ann Arbor, MI: UMI Research Press, 1982.

Rorabaugh, William J. *The Alcoholic Republic: An American Tradition.* New York: Oxford University Press, 1979.

Tyrrell, Ian R. *Sobering Up: From Temperance to Prohibition in Antebellum America, 1800–1860.* Westport, CT: Greenwood, 1979.

Mathew, Father Theobald (1790–1856)

Father Theobald Mathew (1790–1856) was the most influential nineteenth-century Irish advocate of temperance. His antidrink crusade had a major impact during the 1840s, not only in Ireland itself but in Britain and the United States as well.

Mathew was born in October 1790 into a minor branch of an aristocratic Catholic family in County Tipperary in the rural

Father Theobald Mathew (1790–1856) (Library of Congress)

southwest of Ireland. In 1808, he was sent to train for the priesthood at the national seminary in Maynooth, County Kildare, but during his first year he was forced to leave after getting caught in a student prank. Mathew then joined the small Capuchin Order, and in 1814 he was ordained and almost immediately posted to Cork city. There he served for nearly twenty-five years, rising to be head of his order. Although he helped establish churches and schools and was widely respected by all classes and creeds, before 1838 his career was relatively obscure.

Cork was a major center of the brewing and distilling industries and the temperance societies established in the city from the early 1830s, mainly by Quaker and Unitarian merchants and Anglican clergy, had little impact. A total abstinence society that appeared in 1835 had some success in recruiting working-class Catholics, although the leadership remained largely in the hands of middle-class Protestants. Mathew's decision to become president of this society in April 1838 appears to have resulted from an appeal for support by Catholic members who feared they were becoming targets for Protestant proselytizers.

For the next eighteen months, Mathew preached teetotalism in Cork, attracting larger and larger audiences, so that by November 1839 his supporters were claiming he had administered 66,000 pledges. The late 1830s were years of bad weather, poor harvests, growing poverty, and threats of famine in Ireland. Against this background, Mathew's promotion of abstinence as a panacea for the country's many economic woes and as a means of social betterment had an obvious appeal. It was only during the winter of 1839, however, that his crusade began to attract national attention. In November and December of that year he conducted missions in the cities of Limerick and Waterford. Extraordinary numbers of people eagerly took the pledge: contemporary estimates suggested as many as 150,000 in Limerick and around 80,000 in Waterford. Although these figures are probably inflated, there is little doubt that tens of thousands flocked from the south and west of Ireland to hear Mathew's antidrink message.

Mathew took his crusade further afield in 1840, in particular to Dublin in March and November, although his support always remained strongest in the southwest of Ireland. By the spring of 1841 he was claiming a membership of more than 5 million; in that same year, a national census calculated the total Irish population at a little over 8 million. The reliability of Mathew's figure is by no means assured, but it is clear that during a period of just three years he had convinced millions—perhaps as many as half the Irish population—to accept total abstinence.

Historians have struggled to explain the remarkable appeal of this antidrink crusade in Ireland, where previously temperance had made little headway. Historian H. F. Kearney (1979) characterized Mathew as an "apostle of modernisation" and argued that his crusade was a medium for spreading urban, middle-class values of self-restraint, thrift, literacy, and respectability to rural Ireland. Another author, Elizabeth Malcolm (1986) disagreed, stressing the messianic and millenarian aspects of the crusade. She noted the widespread belief that Mathew could perform miracles, especially heal the sick, and pointed out that similar powers were invested in the medals and cards that he distributed generously. Paul Townend (2002) argued that previous writers had put too much emphasis on Mathew's personal appeal. He saw the crusade as an indication of the Irish people's disillusionment with their existing spiritual and political leaders and of their desire for national regeneration.

Traditionally, the Great Famine of 1845–1849 had been seen as marking the end of Mathew's crusade, but recent studies have established that the movement peaked in the early 1840s and was already showing signs of decline before 1845. Historian Colm Kerrigan (1992) pointed out that by 1841–1842 Mathew had probably reached all those in Ireland likely to accept his message. The rise of Daniel O'Connell's campaign for repeal of the Act of Union in 1842–1843 also distracted popular attention from teetotalism. Mathew, wanting to avoid involvement in politics, curtailed his missions in Ireland during these years and instead spent time preaching total abstinence in Britain. After short visits to England and Scotland in 1841–1842, he spent two months there on a major crusade in 1843. Whereas teetotalism had made significant strides in Britain during the 1830s, this success had occurred mainly among Protestants. As in Ireland, however, Mathew for the first time convinced large numbers of working-class Catholics to embrace the antidrink cause; in doing so, he laid the basis for a Catholic temperance movement in Britain.

The waning of the crusade, the heavy debts he had incurred in promoting it, and repeated criticism of his methods, even from within his own church, demoralized Mathew, and from 1845 he concentrated on relieving famine distress in Cork. His health deteriorated, and in 1848 he suffered a stroke. Nevertheless, in 1849–1851 he conducted a long crusade in the United States. There, however, he quickly became embroiled in a bitter controversy with abolitionists when he refused to condemn slavery. He argued that such a condemnation would prevent him from conducting a temperance mission in the South, but his refusal to publicly back abolition dismayed many of his liberal supporters, not only in the United States, but in Ireland as well.

His last five years, until his death in December 1856, were spent in Ireland in retirement due to poor health. Family members reported that he was deeply depressed at what he considered the failure of his temperance work.

Elizabeth Malcolm

See also: Ireland

References

Bretherton, George. "The Battle between Carnival and Lent: Temperance and Repeal in the Making of Modern Ireland" in *Histoire sociale.* 27 (1994): 295–320.

Kearney, H. F. "Father Mathew: Apostle of Modernisation" in *Studies in Irish History Presented to R. Dudley Edwards,* edited by Art Cosgrove and Donal McCartney, 164–175. Dublin: University College, 1979.

Kerrigan, Colm. *Father Mathew and the Irish Temperance Movement, 1838–49.* Cork: Cork University Press, 1992.

Maguire, John F. *Father Mathew: A Biography.* London: Longman, Roberts and Green, 1863.

Malcolm, Elizabeth. *"Ireland Sober, Ireland Free": Drink and Temperance in Nineteenth-Century Ireland.* Dublin: Gill and Macmillan, 1986; Syracuse: Syracuse University Press, 1986.

Quinn, John F. *Father Mathew's Crusade: Temperance in Nineteenth-Century Ireland and Irish America.* Amherst: University of Massachusetts Press, 2002.

Townend, Paul A. *Father Mathew: Temperance and Irish Identity.* Dublin: Irish Academic Press, 2002.

McBride, Francis Scott (1872–1955)

Francis Scott McBride (1872–1955), United Presbyterian minister and national superintendent of the Anti-Saloon League of America for almost three decades, was described in the *Christian Advocate* in 1924 as "large in frame, an ex-star in football. Scotch in blood, he has the physical and moral tenacity of his forebears." In the same year, Bishop Thomas Nicholson, a former national president of the Anti-Saloon League, praised McBride's tact, courage, and energy, and portrayed him as effective alike on the platform and behind the scenes. Born in Carroll County, Ohio, on 29 July 1872, he attended Muskingum College and Allegheny Theological Seminary. In 1901, he was ordained as a minister in the United Presbyterian Church. He served churches in Kittanning, Pennsylvania (1901–1909), and Monmouth, Illinois (1909–1911). During these years, he was also a member of the state board of directors of the Anti-Saloon League of Pennsylvania.

In 1911, McBride left the active ministry and became superintendent of the Springfield District for the Anti-Saloon League of Illinois. One year later, he was promoted to state superintendent, a position he would hold until June 1924. His twelve years in office were marked by a great deal of success. Though the league was understaffed and faced a well-financed liquor lobby based in Chicago, he was able to organize and manage temperance forces to achieve victories in a number of local elections and gain control of the state legislature. He received national recognition among the temperance forces for his part when Illinois became the twenty-sixth state to ratify the Eighteenth Amendment and then passed a more stringent state code than the federal Volstead Act.

In 1924, the Anti-Saloon League of America was at a crossroads as to what course it should take after Purley Baker, its national superintendent, died. Some of the leaders, led by Ernest Cherrington, founder of the World League Against Alcoholism, felt that the league should redouble its efforts to educate the public about the dangers of alcohol. Others, led by Wayne Wheeler, general counsel and legislative superintendent, felt that law enforcement should be the main concern. Wheeler put forth the Illinois superintendent as his candidate and, at a special Indianapolis meeting on 9 April, engineered McBride's victory in the election for the league's national superintendent. During his first four years in office, McBride maintained the league offices in Westerville, Ohio. Then, to counter stiff opposition from the Association Against the Prohibition Amendment (AAPA), which was using many of the same tactics as the Anti-Saloon League, McBride moved his offices to Washington, D.C.

McBride's time in office was not as successful as many had hoped it would be. He was not able to heal the divisions within the organization, and by the late 1920s the league had become more defensive as it defended Prohibition. By the time National Prohibition ended on 5 December 1933, the league had become a mere shadow of the crusading organization it had once been. McBride nevertheless continued to press temperance issues as national superintendent into the 1940s. He died on 23 April 1955 and is buried in Westerville, Ohio.

T. Jason Soderstrum

See also: Anti-Saloon League of America (ASLA); Cherrington, Ernest Hurst; National Prohibition (United States); Wheeler, Wayne Bidwell

References

Kerr, Kathel Austin. *Organized for Prohibition: A New History of the Anti-Saloon League.* New Haven: Yale University Press, 1985.

McBride, Francis Scott, and John J. Lentz. *Radio Addresses by Hon. John J. Lentz and Dr. Francis Scott McBride.* Westerville, OH: American Issue Publishing, 1928.

McCarthy, Raymond Gerald (1901–1964)

Raymond Gerald McCarthy (1901–1964) was an American educator who shocked 1950s America by estimating that there were several million alcoholics in the United States. Born in Brockton, Massachusetts, on 30 April 1901, he attended Boston College and Harvard University. Upon graduating in 1928, he taught high school at various institutions in New England. During this time he received his master's degree in education in 1939 from Harvard. Five years later, McCarthy joined the faculty at Yale University as an instructor of health education. He would remain in this position for the next decade before moving to the position of associate professor at the Yale Center of Alcohol Studies. In 1944, he was appointed the first staff member of the Yale Plan Clinics, an institution established to study alcoholism in a clinical setting, and in 1948 he became associate director of the Yale Summer School of Alcohol Studies. The center moved its facilities to Rutgers University in New Brunswick, New Jersey, in 1962 and was renamed the Rutgers Center of Alcohol Studies. In the same year, McCarthy was rewarded by being named a professor of education in the center and promoted to the position of executive director of the summer school. He would remain in these posts until his death on 25 June 1964.

McCarthy contributed to several professional groups, including serving as a consultant to the community services branch of the National Institute of Mental Health, and as a member of the committee for the study of links between disease and alcoholism for the American Public Health Association. He was also president of both the North American Association of Alcoholism Programs and the Association for the Advancement of Instruction about Alcoholism and Narcotics. From March 1961 until shortly before his death, he was director of the division of alcoholism of the New York State Department of Mental Hygiene. He was also director of the alcoholism program for the Massachusetts State Department of Public Health from 1961 to 1962. His publishing credits include *Alcohol and Social Responsibility, a New Educational Approach* (1949), *Teen-agers and Alcohol* (1956), *Drinking and Intoxication* (1959), and *Alcohol Education for the Classroom and Community* (1964). He served as associate editor of the *Quarterly Journal of Studies on Alcohol.*

Although most of McCarthy's writing focused on education, especially among the young, he was a strong proponent of the concept of alcoholism as a disease. For him alcoholism was an illness with genetic, psychosocial, and environmental factors. This illness concept had a great effect on those who studied with him at the Yale Summer School of Alcohol Studies, especially among those representing the religious denominations. In 1953, he estimated that 68 million people consumed alcohol in the United States. Of that group, between 3 million and 4 million were "in difficulty associated with that use" ("Prof. Raymond McCarthy Dies; Rutgers Expert on Alcoholism," 29). He also stressed that intelligence had nothing to do with alcoholism. He proposed that a voluntary organization of citizens had greater power to change this situation than specialists in the field. McCarthy influenced thousands of people who studied in his classrooms and read his books.

T. Jason Soderstrum

See also: Alcoholism; *Journal of Studies on Alcohol;* Rutgers Center of Alcohol Studies; Yale Center of Alcohol Studies

References

Bacon, Selden D. "Raymond G. McCarthy, 1901–1964," in *Quarterly Journal of Studies on Alcohol.* 25 (1964): 413–416.

McCarthy, Raymond G. *Alcohol Education for Classroom and Community: A Source Book for Educators.* New York: McGraw Hill, 1964.

———. *Drinking and Intoxication: Selected Readings in Social Attitudes and Controls.* New Haven: Yale Center of Alcohol Studies, 1959.

McCarthy, Raymond G., and Edgar M. Douglas. *Alcohol and Social Responsibility: A New Educational Approach.* New York: Thomas Y. Crowell, 1949.

"Prof. Raymond McCarthy Dies; Rutgers Expert on Alcoholism" in *New York Times.* 26 June 1964, 29.

McClung, Nellie Letitia (1873–1951)

Nellie Letitia McClung (1873–1951) was Canada's best-known feminist and temperance leader. An ardent member of the Woman's Christian Temperance Union (WCTU), leading suffragist, spell-binding lecturer, and best-selling author, whose nineteen volumes championed temperance and women's rights, she helped usher in franchise and prohibition victories during World War I. Her book of essays, *In Times Like These,* first published in 1915 and still in print, remains a superb expression of first-wave feminism in North America.

Born Helen Letitia Mooney in Grey County, Ontario, Canada, on 21 October 1873, she and her Methodist Scots-Irish family soon joined the land rush to the Canadian prairies that displaced Native peoples. On a homestead near Souris, Manitoba, she did not attend school until age ten, but she always insisted that farm life was best. An evangelical Protestant and enthusiastic schoolteacher (1890–1896), she joined the provincial WCTU in order to educate the public to alcohol's dangers. In 1896, she married pharmacist Robert Wesley McClung, whose mother Annie was a leader of the WCTU. This marriage, which according to custom required her to leave teaching, produced five children.

McClung's observation that women and children were alcohol's particular victims confirmed her conclusion that the world treated her sex unfairly. Like most WCTU activists, she anticipated that women's suffrage would banish the bar and protect the family. Moving to Winnipeg in 1911, she joined the Canadian Women's Press Club and the Political Equality League. In 1914, she won accolades as the misogynous female premier in the city's suffrage Mock Parliament. Well known for the best-seller *Sowing Seeds in Danny* (1908), McClung used humor to stir audiences. Hard-fought campaigns brought prohibition and federal and provincial franchises for most women during World War I. The first did not survive, but suffrage gradually expanded to include all Canadian women by 1961.

McClung moved to Alberta, where she was elected as a Liberal member of the legislature in the 1921 campaign won by the United Farmers. She maintained a nonpartisan commitment to good government and attributed her 1926 defeat to "liquor interests" opposed to prohibition. An ecumenical Methodist, she joined the United Church of Canada in 1924 and campaigned for the right of women to become ministers. In 1929, with other women activists, nicknamed "The Famous Five," she successfully petitioned the Judicial Committee of the Privy Council of Great Britain for recognition of women as persons in the Constitution. She became the first female member of the board of governors of the Canadian Broadcasting Corporation in 1936 and a Canadian delegate to the League of Nations in 1938. She died in Victoria, British Columbia, on 1 September 1951.

Veronica Strong-Boag

See also: Woman's Christian Temperance Union (WCTU) (Dominion of Canada)

References

Fiamengo, Janice. "A Legacy of Ambivalence: Responses to Nellie McClung" in *Rethinking Canada: The Promise of Women's History,* edited by Veronica Strong-Boag, Mona Gleason, and Adele Perry, 149–163. Toronto: Oxford University Press, 2002.

Savage, Candace. *Our Nell: A Scrapbook Biography of Nellie McClung.* Saskatoon, Saskatchewan: Western Producer, 1979.

Warne, Randi. *Literature As Pulpit: The Christian Social Activism of Nellie L. McClung.* Waterloo, Ontario: Wilfrid Laurier, 1993.

McEwan, William (1827–1913)

William McEwan (1827–1913), a leading Scottish brewer, was born at Alloa in 1827, the son of John McEwan, a ship owner. Alloa had a thriving port and industries including malting, brewing, and distilling. After completing his schooling, McEwan spent the first eight years of his career working in a succession of clerical posts in Alloa, Glasgow, and Huddersfield, where he gained commercial experience and furthered his education. In 1851, he began five years of technical and management training with his uncle, John Jeffrey, proprietor of the Heriot Brewery, Edinburgh. His apprenticeship was completed by 1856, and using family and borrowed capital, he established his own business at the Fountain Brewery, Edinburgh, near the terminus of a canal and adjacent to the railway.

Although expanding markets worked in his favor, McEwan was remarkably successful, and within four years his turnover was £40,000 per annum. Taking advantage of cheaper transport, he penetrated markets in the Glasgow conurbation and was so successful that the area soon accounted for more than 50 percent of the firm's trade. McEwan opened up other markets locally, established a foothold in the north of England, and developed a profitable export trade to the empire. McEwan's personal success as a businessman can be gauged by the fact that the firm's turnover by 1885 was £92,000, and when registered in 1889 had a nominal capital of £1 million.

A Gladstonian Liberal, McEwan turned to politics in 1886, becoming a member of Parliament (M.P.) for Central Edinburgh, a seat he subsequently held for fourteen years. Although a highly regarded constituency M.P., who even managed to gain support from the temperance lobby, McEwan remained a backbencher, though active on drink trade issues and Irish home rule. At the general election in 1895, he was returned unopposed. He is said to have declined a title, remarking, "No, I would rather be first in my own order, than be at the tail end of another" (Donnachie 1986–1990, 2: 44), indicating his personal pride as a self-made businessman. He was made a privy councillor in 1907.

Apart from his association with the firm and product that still bears his name, William McEwan was a notable philanthropist who made gifts to the nation and the city of Edinburgh. He augmented a valuable art collection, presenting paintings to the National Gallery of Scotland, and gifted the McEwan Hall to the University of Edinburgh in 1897. McEwan was made an honorary doctor of laws and presented with the freedom of the city of Edinburgh. He was a deputy lieutenant of Edinburgh for thirty years. His last home, Polesden Lacey, Surrey, was ultimately bequeathed to the National Trust by his daughter, Margaret Greville, a noted socialite and friend of royalty.

McEwan died in London in 1913 at age eighty-six. His estate was valued at £1.5 million, an enormous sum for the time, representing his holdings in the business and industrial, railway, and mining stock internationally. McEwan was undoubtedly one of the most successful brewers of his generation; by good fortune, he entered the industry when it was experiencing a period of rapid expansion. One obituary described him as "a shrewd, hard headed, hard working businessman . . . one of the merchant princes of Scotland" who built up his firm from "small beginnings to huge dimensions and world wide reputation" (Donnachie 1986–1990, 2: 4)—a suitable tribute to McEwan's enterprise and energy.

Ian Donnachie

See also: Beer; Scotland

References

Donnachie, Ian. *A History of the Brewing Industry in Scotland.* 2d ed. Edinburgh: John Donald, 1998.

———. "William McEwan" in *Dictionary of Scottish Business Biography,* edited by Anthony Slaven and Sydney Checkland. 2 vols. Aberdeen: University of Aberdeen Press, 1986–1990.

Medicine, Alcohol as

For much of the nineteenth century, alcohol was popularly believed to have medicinal qualities: It stimulated and supported the system, prevented fevers and infectious diseases, and furnished the stamina necessary for hard physical labor. Many practitioners considered the effects of alcohol to be not only benign in small doses, but even therapeutically beneficial in stimulating the "vital powers" essential to the natural healing processes. From the perspective of the majority of physicians and prospective patients, the use of alcohol as a stimulant was an acceptable replacement for the heroic therapeutics based upon bleeding, purging, and opiates. As the temperance movement gathered steam after mid-century, however, physicians often found themselves in the unenviable position of opposing prohibitionists.

From the 1820s, the use of alcohol in therapeutics became increasingly popular in Great Britain, France, and North America. Its acceptance was part of a complex interaction among a shift in medical paradigms, patient preferences, and professional competition. For centuries, the treatment model of choice for Western physicians had been heroics. Based upon the ancient Galenic typologies of humors, physicians sought to redress the imbalances of body fluids and temperatures thought to be responsible for most diseases. The diseased organism was considered to be "sthenic," that is, afflicted by the overstimulation of normal bodily functions, which required depletive therapies. Through bleeding the patient of significant quantities of blood, producing excessive salivation through the administration of calomel (a mercury derivative), purging the digestive system with strong laxatives, diuretics, and emetics, or stopping up the bowel with opiates, physicians demonstrated their prowess with the immediate and dramatic results of their therapies.

By the 1830s, however, a growing number of practitioners and their patients were seeing or experiencing the negative results of these practices, particularly when injudiciously administered by improperly educated physicians. Families were especially repulsed by such practices when administered to children. In the spirit of democratization that swept the United States in the 1830s and 1840s, known as the Jacksonian era, state legislatures began stripping the medical profession of its licensing privileges and opening up the profession to a host of "irregulars," including homeopaths, naturopaths, and hydropaths. The most damaging to the regular profession in the short term was the work of Samuel Thomson, an irregular physician who published his botanical remedies in a guide for the average household. The popularity of this work signaled a revolution in self-help and mistrust in medical orthodoxy. Thomson's manual and the many that soon followed were based upon gentle, nonintrusive remedies and were favored by domestic healers, usually mothers, who lived in rural areas far from physicians or who refused to submit to heroic therapies.

This development created a crisis in orthodox medicine, one that was also generated by internal transformations. By mid-century, an increasing number of physicians were abandoning, or at least reducing, heroic therapeutics in favor of more benign remedies such as beverage alcohol. Yet the profession took pains to state that the change had not been driven by the loss of patient revenue. By the 1850s and 1860s, physicians averred that their therapeutics were changing because the nature of illness itself had altered. Diseases, like everything else in the nineteenth century, had been transformed by industrialization. Faced with the external overstimulation of new technologies and communications and changes in work and social roles, the modern diseased state had become "asthenic," or enfeebling, and therefore required stimulation.

Furthermore, in a nod to the irregulars, the new medical orthodoxy stated that the body had great natural restorative powers, which should be supported rather than interfered with. In 1860, distinguished American physician and author Oliver Wendell Holmes promoted the healing power of nature as an alternative to excessive drugging, leeching, and cupping. The physician's role was to support and strengthen the enfeebled body so that nature could work its healing course. Such support included prescriptions of rest, moderate exercise, nutritious diet (beef tea was a favorite), and judicious amounts of stimulants. Quinine, iron tonics, and cinchona bark were possibilities, but by the 1860s beverage alcohol, which until well into the twentieth century was considered as a stimulant rather than a depressant, was the stimulant of choice. Whiskey, brandy, and wine were the most common types of alcohol administered.

During the American Civil War, the foundation of most prescriptions for military personnel was whiskey. Alcohol was a specific for the treatment of infectious diseases such as typhoid fever, dysentery, and pneumonia. Pneumonia patients at mid-century were prescribed 8 to 12 ounces of spirits a day for several days. In the 1870s and 1880s, when the medical technology to measure temperature became available, physicians became preoccupied with precisely quantifying temperature levels in the progress of disease. Beverage alcohol was considered useful in lowering fever.

Just as heroic treatments had been discredited and disfavored by patients owing to their injudicious use, by the 1870s the overly free use of stimulants also was decried by alternative healers, temperance advocates, and many regular physicians themselves. This situation led to a number of anomalies. The self-help mentality, which originated with the irregulars, soon was exploited by the patent- or proprietary-medicine industry, which used the tools of the emerging mass media, such as blanket newspaper and billboard advertising, to create many of the millionaires of the new industrial age. Many, if not most, of the innocuously named concoctions then available, such as Hood's Sarsaparilla Cures or Paine's Celery Compound, contained significant quantities of alcohol as well as opiates, cocaine, strychnine, arsenic, and/or chloral hydrate. The contents of these potions were closely guarded secrets. Ironically, patent medicines were favored by women and the clergy, two groups

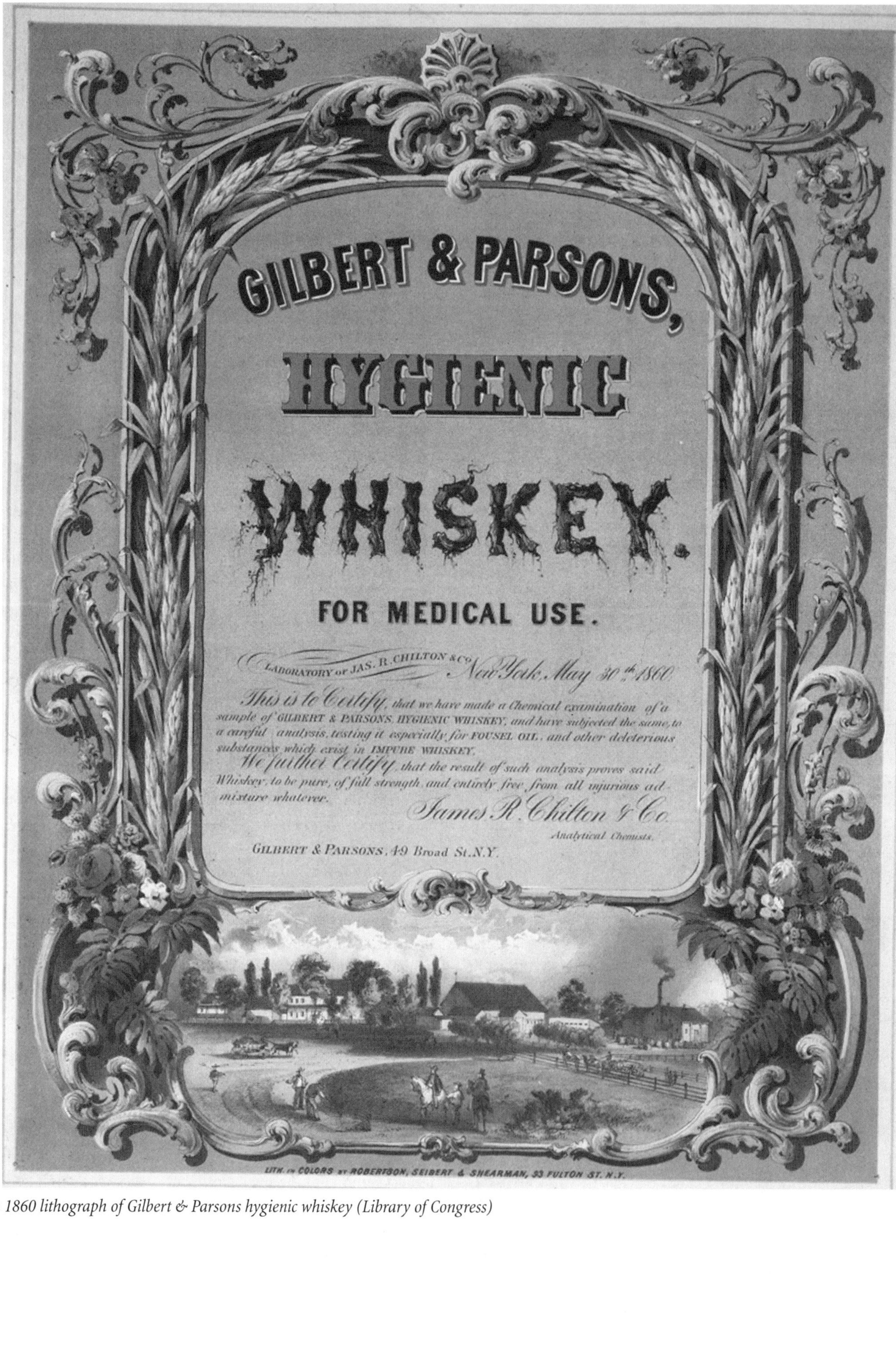

An 1860 lithograph of Gilbert & Parsons hygienic whiskey (Library of Congress)

that constituted the core of the temperance movement. Physicians were quick to note the absurdity of prohibitionists who regularly partook of tonics with higher alcohol content than straight whiskey, but the public dismissed their criticisms as attempts to monopolize medical care.

The physicians' own practice of using liquor in therapeutics faced sustained attack from temperance advocates, and by the end of the nineteenth century, the bromides and salicylates replaced beverage alcohol for the most part. Nor was alcohol limited to general medical practice. Asylum physicians in the late nineteenth century also used spirits in the medical treatment of the insane. This issue was raised in the Ontario legislature in Canada in 1877, and asylum superintendents (employees of the provincial government) were forced to defend their treatment choices. Henry Landor of the London Asylum for the Insane believed that alcohol restored the patients' vital powers, and Daniel Clark of the Toronto Asylum considered alcohol an essential medication less pernicious than its alternatives: chloral hydrate, opium, or morphine. By the end of the century, however, asylum superintendents were removing alcohol from their shelves and instead formulating the concept of inebriety as a mental disease.

Cheryl Krasnick Warsh

References

Berman, Alex. "The Thomsonian Movement and Its Relation to American Pharmacy and Medicine" in *Bulletin of the History of Medicine.* 25, no. 5 (1951): 405–428.

"Government Promotion of Inebriety" in *Maritime Medical News.* 15, no. 2 (1903): 66.

Warner, John Harley. *The Therapeutic Revolution: Medical Practice, Knowledge and Identity in America 1820–1885.* Cambridge, MA: Harvard University Press, 1986.

Young, James H. *The Toadstool Millionaires.* Princeton: Princeton University Press, 1961.

Methodist Board of Temperance, Prohibition, and Public Morals (MBTP)

The Methodist Board of Temperance, Prohibition, and Public Morals (MBTP) served as an official agency of the northern Methodist Episcopal Church during the Prohibition era in the United States. The board's charter stated that its mission was to promote "the voluntary total abstinence from all intoxicants and narcotics by the members of the Church, . . . and to secure the speedy enactment of statutory and constitutional laws prohibiting the traffic in alcoholic liquors." The dual emphasis on education and political activism, combined with effective leadership and ample funding, led the MBTP to become one of the best-known prohibition lobbies in the nation. During the 1920s, it was second only to the Anti-Saloon League of America in the public eye. And as with the league, it was this presence in the public arena that led to charges that the MBTP's tactics violated the separation of church and state.

Origins

The MBTP had its origins in the permanent Committee on Temperance and Prohibition founded by the northern Methodist Episcopal Church in 1888. The church's 1904 quadrennial General Conference recognized the need for a dedicated temperance organization and established the Methodist Temperance Society with headquarters in Chicago. The society campaigned for statewide prohibition in Oklahoma, and the 1908 General Conference apportioned $25,000 to support the fledgling organization. In 1910, the society's board of managers appointed Clarence True Wilson as general secretary, a position he would hold for the next quarter century. An ordained Methodist minister, Wilson had begun his prohibition career in the 1890s and had served as president of the Oregon Anti-Saloon League from 1908 to 1910. Upon assuming his post in the Methodist group he devoted himself to building up the organization's position within the church and advancing the cause of prohibition. His organizational and oratorical talents were major factors in the society's growth. The 1912 General Conference approved the society's relocation to Topeka, Kansas, and increased its annual allocation to $50,000. Over the next four years the society expanded its activities by appointing additional secretaries for research, extension services, and work among African Americans. It participated in twenty-seven state prohibition campaigns and established two publications—the *Clipsheet,* a weekly press bulletin issued to the nation's newspapers and pastors, and *The Voice,* a monthly journal. The society also published and distributed temperance tracts across the country. Wilson made several cross-country speaking tours and participated in the 1915 "Flying Squadron" campaign, in which prohibition speakers conducted marathon speaking engagements from coast to coast.

The Washington Years

The 1916 General Conference voted to rename the society and relocate its headquarters to Washington, D.C. These developments reflected the governing body's conviction that the hour for a national referendum on prohibition was drawing near and that "the influence of the Church Temperance Society is needed in the nation's capital, where many bills are pending upon which our representatives need to hear officially from the Methodist Episcopal Church" (*Journal of the General Conference* 1916: 681–682).

Over the next decade, the MBTP established itself as the major Protestant denominational presence on Capitol Hill. In 1917, its board of managers purchased a building lot across First Street from the Capitol Grounds and directly opposite the future Supreme Court complex. Dedication of the Methodist Building in 1924 secured the Methodist lobby's presence in the heart of federal Washington. The membership of the MBTP's board of managers and advisory members during this period reveals its growing influence in the national arena. The list included North Carolina Senator Josephus Daniels, California

Senator Charles Randall, dime-store magnate S. S. Kresge, and New York retailer J. C. Penney. These political and financial connections facilitated the MBTP's rise to prominence.

More important, however, was the board's campaign to shape public opinion. Despite the assertion of its general secretary that it was nothing more than an educational enterprise, the MBTP was dedicated to grassroots mobilization. Wilson's contact with legislators (some of whom rented offices in the Methodist Building) and frequent appearances before the press were seconded by the *Clipsheet* and *The Voice.* Both publications worked to keep church congregations abreast of legislative developments, and Wilson frequently inferred in his speeches that the MBTP, as the official agency of the Methodist Episcopal Church, represented a constituency of some 15 million people. The *Clipsheet,* edited by Deets Pickett and widely distributed to the nation's newspapers and Protestant ministers, was especially effective at transmitting the board's viewpoints. Writing in the *North American Review,* one columnist later compared Wilson's impact to that of the Anti-Saloon League's Wayne Wheeler: "Though Wheeler may have beaten the politicians at the Capital, it was Wilson who beat the tom-toms among the people" (Tucker 1930: 129).

Criticism of Wilson and the board increased following Wilson's appearance before the Senate hearings on National Prohibition in April 1926. In his testimony, Wilson reviewed the material and spiritual benefits of Prohibition before proceeding to accuse immigrants of attempting to sabotage the effort. Whether his statements reflected the view of the board or the Methodist Episcopal Church as a whole remains an open question. *The Voice* expressed the same combination of evangelical optimism and nativist pessimism throughout the 1920s. Its columns reveal that the board's agenda included issues—such as immigration, movies, dancing, cigarette smoking, and Catholicism—outside the purview, strictly speaking, of temperance and prohibition. These concerns were, however, proof of the board's self-appointed role as guardian for Methodism's conservative wing. As the 1920s progressed, the MBTP's aggressive defense of conservative moral standards brought it into increasing conflict with the modern American mainstream and progressive elements within the church. Several commentators noted the board's insistence on regulating personal conduct while it remained silent over the more serious issue of Methodism's involvement with the Ku Klux Klan. In the late 1920s, the MBTP participated in another congressional inquiry, the Senate Committee on Lobbies, to determine whether it had wielded undue influence with legislators. Despite receiving a clean bill of health, the board's proselytizing during the 1928 presidential campaign led to further controversy, and the General Conference journals indicate that politicians formerly affiliated with the MBTP abandoned the organization around this time.

Following the stock market crash in 1929, the MBTP gradually faded from the national scene as contributions dried up and the nation's attention was drawn to other matters. Following revocation of the Eighteenth Amendment in 1933, the 1936 General Conference amended the board's constitution, vesting the power to elect the general secretary with itself, and restructuring the MBTP's board of managers to ensure a more representative body. Wilson stepped down for reasons of health and was succeeded by former Anti-Saloon League head Ernest Cherrington. Henceforth, the MBTP emphasized temperance education and curtailed its political activism as mainstream Methodism shifted its historical emphasis on moral perfectionism for a more social orientation. In 1960, the board was merged into the Board of Christian Social Concerns, which became the United Methodist Church General Board of Church and Society, still housed in the Methodist Building in Washington, D.C., in 2002.

C. Wyatt Evans

See also: Anti-Saloon League of America (ASLA); Cherrington, Ernest Hurst; Federal Regulation of Alcohol before 1920; National Prohibition (United States)

References

Journal of the General Conference of the Methodist Episcopal Church. New York and Cincinnati: Methodist Book Concern, 1904–1936.

Merz, Charles. "The Methodist Lobby" in *New Republic.* 13 October 1926: 213–215.

"The Methodist Board of Temperance, Prohibition and Public Morals: An Editorial" in *Christian Century.* 24 December 1930: 1582–1585.

National Prohibition Law. Hearings before the Subcommittee of the Committee of the Judiciary, U.S. Senate, 69th Cong., 1st sess., vol. 2. Washington DC: Government Printing Office, 1926.

Pezet, Washington. "The Temporal Power of Evangelism: The Methodists in National Politics" in *Forum.* October 1926: 481–491.

Tucker, Ray T. "Prophet of Prohibition" in *North American Review.* August 1930: 129–136.

United Methodist Church General Board of Church and Society. http://www.umc-gbcs.org.

United Methodist Church General Commission on Archives and History (GCAH). http://www.gcah.org.

Wilson, Clarence True. "Methodist Rights in Politics" in *Forum.* November 1926: 668–681.

Mexico

Mexico is a vast and varied country, geographically as well as culturally. This diversity is reflected in alcohol-drinking habits and customs. Native rural traditions persist, living side by side with the cosmopolitan ways of urban regions.

Mexican Preferences in Alcohol

From pre-Columbian times to the present, more than forty-two regional beverages have been identified in Mexico, fruit liquors being the most common. Called *moscos* or *ponches* in some places, the ones obtained from cacti are predominant—in North Mexico, *sotol* (Chihuahua) and *bacanora* (Sonora); in the Southeast, *posch* (Chiapas) and *xtabentún* (Yucatan). *Mezcal,* from Oaxaca, and *tequila,* a name that can be applied only to the drink that comes from a region in Jalisco, are the ones of international renown. There still are a number of traditional domestic beverages to which chemical substances are added to rapidly increase their power and toxicity—such as *ocoxochitl,* made of herbs and 96 proof alcohol, which is very cheap to produce clandestinely (Natera 1987).

Pulque deserves special attention: Made from the fermented sap of the maguey, known as *aguamiel,* it was a powerful and sacred beverage in pre-Columbian times. Its alcoholic content varies between 3 and 6 proof, although it has been reported to be up to as much as 12 proof. People drink it in very arid places as a substitute for water when the latter is not available. By the eighteenth century, it had lost much of its original association with divinity in ritual practices. Nonetheless, it is still being drunk, mainly in Central Mexico, and is currently a trendy drink among the young of the middle and upper classes, competing with beer. There is ongoing research about the process of production, storage, and distribution of pulque as well as about the psychosocial and health effects of pulque on the general population. Researchers have attempted to demonstrate a link between pulque drinking and the high mortality rate in Mexico due to hepatic cirrhosis. It is well documented that in rural Mexico pulque and other spirituous liquors have been used as currency to pay the peasants their day's wages (Bunzel 1940). At the present time, pulque represents 3.1 percent of the total per capita consumption of ethanol.

Distilled beverages and wine were introduced to Mexico in colonial times, stimulating an important development in the economic relationship between New Spain and the Spanish crown. During the eighteenth century, the vice-regal government tried to grow grapes; ultimately, this experiment failed in the nineteenth century owing to social changes in the country. Wine consumption then diminished noticeably, and it was not until the second half of the twentieth century that it increased again, mainly among women. In the 1970s, a quarter liter of wine was drunk each year per capita; by 1990, consumption had reached almost 2 liters (0.5 U.S. gallon) per year.

Beer has been known in Mexico since 1580. Introduction of drinkable water and electricity in the twentieth century stimulated the creation of diverse industries in different parts of the country, including the production of alcoholic beverages, and most of all beer.

The 1994 Income and Spending Poll and local studies show that in every sector of the population preference is regulated by acquisitive power. Wine is the drink of choice of people with a high economic status, whereas pulque and various types of hard liquor *(aguardientes)* rule at the other extreme. This same poll shows that, in 1994, 84 percent of alcoholic consumption was of beer; wine followed this lead at 5 percent, distilled beverages garnered 4 percent, and fermented drinks, 1 percent. It is predicted that by 2010 tequila will be number one among the distilled beverages.

Traditions and Practices

Alcohol drinking in Mexico has never been homogeneous among men and women, despite the fact that in pre-Columbian societies women had the right to become inebriated. Patterns of consumption have been closely linked to culture, and especially to values and customs. Social connotations of drunkenness are very different for men and for women: In males it is widely accepted, whereas in females it is strongly rejected. In Mexico, alcohol functions as a social integrator, and it is common to hear that "without alcohol there's no party." In rural zones, alcohol drinking is strongly associated with celebration of the saints' many religious festivities; at these events, women are allowed to drink (and even some who are abstemious do).

M. E. Medina-Mora (2000) has observed that there is no well-defined norm in Mexican attitudes about drinking. On the one hand, many Mexicans feel that getting drunk "is a harmless way of having fun," or that "it is good for oneself to get drunk once in a while"; on the other, it is common to hear them attribute negative qualities to alcohol, such as that "it brings out the worst in people." There is no clear definition available for what constitutes "drinking" or "drinking excessively" in Mexican culture, and generally there is little information about how much alcohol is incapacitating for executing concrete actions.

Males generally do not drink with their wives, even when they drink at home, but she is the one serving the drinks, waiting on his friends, and bearing the consequences of his drinking (violence, illnesses, and the like). It is a Mexican tradition that women must refuse when offered a drink, unlike men, who are compelled to accept (Natera 1987).

In the last third of the twentieth century, behavior patterns in Mexico surrounding alcohol changed as a result of globalization. By the early twenty-first century, abstinence among young people was becoming less common every day, and women were starting to drink at an earlier age. Adolescent girls now drink almost as much as boys; among adults, females adopt a moderate pattern, while males tend to drink more. After age fifty, the decrease in consumption is lower in women than in men; some women start drinking only then, and the ones who already did drink tend to drink more.

Epidemiological investigations of drinking patterns begun in 1964 have focused on quantity and frequency of consumption. Medina-Mora (2000) reported that Mexicans' most typical way of drinking is explosive: One always drinks to get drunk. Qualitative studies conducted in rural zones by Natera and collaborators have verified this excessive way of drinking in expressions such as "one must drink fully, or else it will not even feel like drinking" (Natera 1987). When one drinks, it must be enough to get drunk, and afterward one goes to church to "swear" not to drink anymore for some period of time. Quantity is more important than frequency. For women, drunken males are undesirable but have to be tolerated.

It is common practice among Mexicans to get drunk on weekends. In the past, this caused a high rate of absenteeism at work on Monday, or "Holy Monday," as it was called. Drinking usually started on Saturday night and ended on Sunday; Mondays were reserved for recovering from hangovers. Since the labor week was reduced to five days in the 1970s, drinking has started on Friday night and continued on Saturday; Sunday is left for resting, so Holy Mondays have tended to disappear.

The legal age for public drinking in Mexico is eighteen, but there are minors who drink alcohol at parties, in bars, and at home. The last national poll (1998) reported a trend toward heavier consumption among women and adolescents. In 1988, it was recorded that 63 percent of women were abstemious, and by 1998 the percentage had diminished to 53 percent. Nine percent of women between ages eighteen and sixty-five have five drinks or more on each occasion, and 1 percent has dependence symptoms.

The percentage of the population that is alcohol-dependent (9.2 percent of the males, 1.5 percent of the females) is low in comparison with the United States, Canada, Spain, and France, but consequences of acute intoxication in moderate drinkers are greater in Mexico, particularly in terms of accidents and acts of violence, which are among the first ten causes of death nationwide. Abuse in alcohol consumption represents 9 percent of the total incidence of illness in Mexico. The mortality rate due to cirrhosis of the liver for each 100,000 inhabitants is 48.6, or 72.5 for men and 21.8 for women. Of the 29 percent of urban women who suffer physical violence by their spouses nationwide, in 60 percent of the cases alcohol was involved (Secretaría de Salud 1998).

Production, Sales, and Consumption

Traditionally, many alcoholic beverages have been produced clandestinely in Mexico or are produced by small enterprises not subject to official control—presently, only companies producing more than 25,000 liters (6,605 U.S. gal.) per year are regulated. Moreover, alcohol is often homemade for private consumption, particularly in rural zones. For this reason, plus the fact that there is a high percentage of abstinence, the consumption rate per capita does not reflect the reality of alcohol distribution, although it is useful for knowing the tendencies. For instance, consumption rate per capita in 1970 was at 2.06 liters (0.5 U.S. gal.); by 1996, it was calculated at 5.5 liters (1.45 U.S. gal.) in the population above age fifteen. In fact, it has been ascertained that 60 percent of the production of alcoholic beverages is under no sanitary or fiscal control, is sold in recycled bottles with labels similar to those of patented beverages, or is homemade for private or regional consumption. In addition, 96 proof alcohol, widely available and very cheap, is drunk by *teporochos* (homeless winos) and the lower social classes. Alcoholic drink lacking sanitary and quality controls has caused great harm in the population: In 1999, more than 100 deaths due to adulterated alcohol were reported in one peasant village.

In the second half of the eighteenth century, a well-organized elite that successfully influenced government decisions controlled the pulque market. Cultural guidelines of pulque shops, characteristic from their origins well into the twentieth century, were thus defined to include live music, dancing, spicy food, games of chance, and yet other attractions to stimulate higher alcohol consumption (Lozano 1998).

During the 300 years of Spanish rule, production of mezcal and other hard liquors was forbidden in the greater part of New Spain on the argument that they were extremely harmful to health (Sahagún 1979; Soberón 1992). Manufacture and commercialization of cane liquor became legal in 1796. Production was carried out in an informal, artisan-like manner and had regional variations. Cane liquor, known today as rum, was closely related to the development of sugar cane haciendas and is considered a traditional drink nationwide.

Beer production developed rapidly at the end of the nineteenth century, and at one point there were as many as twenty-nine registered beer companies. Mexican industries came to export beer to the value of as much as 637 billion pesos at that time. Beer is the most consumed beverage by teenagers and men and women among all social strata, and it frequently reaches the farthest corners of the country more efficiently than water. Many beliefs about the alleged health benefits of beer and pulque persist because of their status as a food in popular perceptions.

Prohibitions have taken place at different times in Mexico. In the eighteenth century, pulque was forbidden because it was thought to be a cause of violence and a danger to the interests of the Spanish crown and the mestizos in charge of its production, sale, and distribution. Prohibition lasted five years, until the authorities decided to sell the beverage again in an "orderly and open" manner (Corcuera 1991).

Jesuits founded pulque haciendas in the middle of the eighteenth century. This was the period of splendor for pulque production and circulation (Soberón 1992). Pulque shops became very profitable, and by the end of the century the production and sale of pulque was one of the most prosperous activities in New Spain. From the period of the War of Independence to the end of the nineteenth century, the pulque industry flourished once more. In 1909, the Pulque Retailing Company was established. By then, there were already 989 pulque shops in existence in Mexico City (Soberón 1998).

While the Volstead Act was in effect in the United States, control of consumption in Mexico was attempted by means of extra taxes on alcoholic beverages. The taxes were not applied uniformly to all beverages, however, which excited strong protest; mezcal, tequila, and every kind of spirituous beverage was excluded, but beer was not, in spite of being "scientifically" considered to be harmless ("Hay mucha" 1930; "Todo el país" 1930).

The feminist movement in the 1980s eliminated restrictions that had surrounded the sale of pulque. During the nineteenth century and into the twentieth, pulque was sold in homes, with women (and especially widows) being in charge of sales. In spite of this, *cantinas* and *pulquerías* displayed signs prohibiting entry to women, children, and men in uniform. Pulque was sold to women only in a contiguous room, through a small window. This changed with the feminist movement; nevertheless, the almost null presence of women in these places is still a fact.

In the last decade of the twentieth century, alcoholic beverages mixed with sparkling water and sweet soft drinks with an approximate alcohol content of 2 percent have been successfully incorporated into the Mexican beverage market. These drinks are bottled in pleasant presentations that attract the young and women. At present, pulque and other regional beverages of low cost and limited commercial distribution, known as *aguardientes* (strong liquors), are consumed mainly in households that lack public services, running water, and telephones and that may have dirt floors (INEGI 1994). These are consumed mostly in rural zones and among the urban population that inhabit poor quarters.

The production of alcohol and of the necessary sugars for producing alcoholic beverages is still controlled by the state. This is an outgrowth of the 1940s, when the alcoholic beverage industry grew at a very fast pace with an oligopolistic character. Almost 90 percent of the production of beer, wines, and distillates and the necessary production materials was concentrated in a few firms, there being big profits for the state (through taxes) and for the sellers. Discussion continues about whether or not this arrangement constitutes a disguised monopoly.

Movements for Alcohol Control and against Alcohol Abuse

Current movements promoting moderation in Mexico stem from antecedents that existed from the eighteenth century on. For example, various codices and writings by sixteenth-century Spanish chroniclers report that when someone got drunk outside the limits of ritual practice, he was punished by having his head shaved publicly, by flogging, and even by stoning if he repeated the offense. Yet, these rules were not applied equally to everyone (Taylor 1987). Old people, women who had just given birth, and victorious warriors were all exempted from punishment. In pre-Columbian times, rules on moderation laid emphasis on who was allowed to drink alcohol and when, rather than on quantity, so those who drank could do it without shame. These regulations disappeared with the Spanish conquest.

Although the Indians were "soft and tame" in Spanish eyes, they also were prone to drunkenness, which in turn generated other vices. When colonization came and the ways of the Indian changed, Native peoples had to be convinced of the goodness of the new rules, moderation included. Indians kept on drinking until they fell, and the church acted as a coercive force: "If you get drunk"—the church threatened—"you'll go to hell." In the eighteenth century, missionaries stopped trying to convince the Indians not to drink, but instead of asking whether they themselves were the ones who had failed to transmit the moral message, they attributed to the Indians a natural incapacity to understand, something the Indians took advantage of, indulging in drinking as they pleased (Corcuera 1996).

During the nineteenth century, control was exercised by imposing a limited timetable on the wine shops (they had to close at 1 P.M. on Sundays and holidays, and were fined in case of disobedience) and through increases in customs costs. Vending hours were extended in 1833, but in 1856 sales were restricted in specific urban areas. Stricter sanitary control measures were imposed in 1871.

During Porfirio Diaz's mandate (which spanned the last third of the nineteenth century and the first decade of the twentieth), "scientific" knowledge prevailed; alcohol abuse was explained as a part of the national decadence that had to be fought publicly. Alcoholism was characterized as a "social pathology" associated with criminality and violence and cataloged as genetically transmitted. It was considered contrary to accepted custom, to the spirit of progress, and to prevailing morality. This view was useful for legitimating social control, supporting ethnic movements, strengthening social class divisions, and establishing the difference between "decent people" and "degenerates." The most important thing for Diaz's dictatorship was to maintain obedience and production through a national project closely modeled on the modernizing progress of the United States and Europe.

Governmental reaction to the problem of alcoholism has changed over the past two centuries. In the nineteenth century and the beginning of the twentieth, alcoholism was referred to as a moral problem and described as a vice. Ambition, deception, and lack of education were identified as causes. Since the last third of the twentieth century, the notion of alcoholism as a sickness became general, as did the belief that alcoholism can be prevented and that patterns of consumption can be changed in the direction of moderation. Campaigns in the media and in schools focus on strengthening protective actions (such as informing people that to prevent high concentrations of alcohol in the blood, one has to eat and let some time pass between drinks). The main objectives include bolstering values and self-esteem in children, developing healthier lifestyles in the general population, teaching people how to avoid illness, increasing awareness of risk factors, and promoting moderation in the consumption of the adult population.

There has not been an official policy on prevention through damage reduction, but some experimental programs directed toward young people (such as the "designated driver" program) have been established to help them avoid associated complications—car accidents, mostly. These are being put into practice with positive results. In one study, 3 percent of adolescents twenty-two to twenty-seven years of age revealed they have more than five drinks per event, which is why efforts have focused on decreasing their consumption and providing information about the hazards of drinking and driving (Medina-Mora 2002).

Clear internal regulations on the use and abuse of alcohol and other drugs are encouraged in the workplace to sensitize the captive population to the quantity of alcohol they consume and the risks of excess. In prevention campaigns, the association of alcohol with the use of other drugs, and the finding that tobacco and alcohol use at a young age was the entrance door to illegal drugs, were an important alarm.

Family as an informal control of alcohol abuse and drunkenness has been a protective factor and one of the most useful resources for society. The "swearing," which consists of a written promise to the Virgin, made before a priest, to stop drinking for long periods, is a deeply rooted popular custom among Catholics. The promise is usually kept, sometimes for several years, but success in having the strength to remain abstinent paradoxically requires people to get drunk to celebrate.

The alcohol industry, pressured by the state during the 1990s, has promoted alcoholism prevention campaigns. Publicity regulation has also been a state resource to encourage moderate consumption. Mexico was one of the first countries on the North American continent in which bottle labels were used to warn consumers about the consequences of excessive drinking. Moreover, obligation to clearly display a precautionary legend has been extended to all alcohol publicity (be it in video or printed). The effectiveness of this measure has not been sufficiently evaluated.

By putting aside the concept of vice and turning to more public health–oriented views, official campaigns have contributed to modifying negative attitudes toward excessive consumers. This change in perspectives has encouraged a better social perception of alcoholics, making it possible for them to seek treatment without fear of stigma.

Alcoholism as a social problem is currently confronted through the design of production, distribution, and commercialization policies to protect people's health and to support those negatively affected by their drinking habit so that they can have access to treatment. There are also hot lines to access relevant information. Enforcing the law, which requires the consumer to prove legal age in shops and bars, not selling more alcohol to people who are drunk, and prohibiting consumption promotions are also a part of preventive policies.

Institutional Answers

For many years, only alcohol dependence was dealt with and not the problems associated with it. It was not until 1980 that some attention was given to other consumption-related problems. Alcoholics Anonymous (AA) has had a large impact on the alcoholism problem in Mexico. It is best known as a source of recovery and is increasingly recognized among health professionals.

AA was introduced to Mexico in 1940 by a group of U.S. citizens. As only English was spoken, prejudice made it at first hard for the groups to be accepted. Nonetheless, from 1969 on, AA spread, forming groups composed of all the economic strata, men and women whose common goal was to avoid drinking alcohol and to improve and keep their own health as well as that of their families. By 1997, there were already 13,200 groups, a number that put Mexico in second place worldwide. Half of these groups are concentrated in Mexico City, each with an average of 19 members, 87 percent men and 13 percent women. Fifty percent are between thirty and fifty years old, and only 20 percent are under thirty. The group AA 24 Hours, which started its activities in 1975, already by 1997 consisted of 150 groups and eight farms, and other AA 24 Hours groups with the modality of intensive care include 360 groups and fifteen farms. The latter are for poor people with severe alcoholism problems (Rosovsky 1998). There are also private models of treatment, based on the Minnesota model developed by the Hazelden Foundation.

Several sanitation laws in Mexico affect the alcohol trade. The 1959 Sanitary Code includes clauses restricting the opening of new alcohol shops to sites more than 300 meters away from schools, churches, and workplaces. The Sanitary Ordinance of 1963 stipulates methods of manufacture and establishes legal and illegal operations relative to alcohol. After various revisions to the Mexican Code of Health, the measures the state must implement to fight alcoholism in the country were specified in the 1970s. It was reiterated that alcoholism is an illness and that its care is the responsibility of the Public Health Bureau.

Nowadays there are very clear regulations regarding alcohol in Mexican legislation on health, transit, labor, and sanitation. Banning of alcohol sales on holidays, a measure enforced to varying degrees for the past seventy years, is today the subject of strong controversy and could be eliminated if legislators are willing to trust the maturity of the population. Currently, the "dry law" is only enforced during local and federal elections to guarantee order.

Guillermina Natera

See also: Comité Nacional de Lucha contra el Alcoholismo; Dirección Anti-Alcohólica; Franco, *Ing.* Luis G.; Hazelden Foundation; Portes Gil, Emilio

References

Berruecos, L. "Bebidas y licores de hoy" in *Bebidas Nacionales: México Desconocido.* 18 (1994): 50–84.

Bunzel, Ruth. "The Role of Alcoholism in Two Central American Cultures" in *Psychiatry.* 3 (1940): 361–387.

Corcuera de Mancera, S. *El fraile, el indio y el pulque: Evangelización y embriaguez en la Nueva España, 1523–1548.* México: Fondo de Cultura Económica, 1991.

———. *Entre gula y templanza: Un aspecto de la historia mexicana.* México: Fondo de Cultura Económica, 1996.

"Hay mucha excitación en el estado de Tamaulipas a causa de un impuesto ilegal" in *Excelsior.* 2 February 1930.

Instituto Nacional de Estadística, Geografía e Informática (INEGI). *Encuesta nacional de ingresos y gastos (internal report).* México: INEGI, 1994.

Lozano, Teresa. "De Fuego y de Maguey: Mezcal" in *Beber de Tierra Generosa: Historia de las Bebidas Alcoholicas en México,* edited by M. E. Medina-Mora and L. Echeverría, vol. 1, 110–126. México: Fundación de Investigaciones Sociales, A.C., 1998.

———. "Del Chinguirito al Ron" in *Beber de Tierra Generosa.* México: Fundación de Investigaciones Sociales, A.C., 1998.

Medina-Mora, M. E., G. Borges, and J. Villatoro. "The Measurement of Drinking Patterns and Consequences in Mexico" in *Journal of Substance Abuse.* 12 (2000): 183–196.

Medina-Mora, M. E., G. Natera, and G. Bourges. "Alcoholismo y Abuso de Bebidas Alcoholicas" in *Observatorio Mexicano en Tobaco, Alcohol y Otras Dragos,* 15–26. México: Consejo Nacional Contra las Adicciones, 2002.

Natera, G. "El consumo de alcohol en zonas rurales de México" in *Salud Mental.* 10, no. 4 (1987): 59–66.

———. "Mexico" in *International Handbook on Alcohol and Culture,* edited by D. B. Heath, 179–189. Westport, CT: Greenwood, 1995.

Piccato, P. "The Discourse about Alcoholism and Criminality in México City, 1890–1917" (M.A. thesis, University of Texas at Austin, 1993).

Rosovsky, H. "Alcoholics Anonymous in Mexico: A Strong but Fragmented Movement" in *Diversity in Unity: Studies of Alcoholics Anonymous in Eight Societies.* NAD Publication No. 3, 1998: 165–184.

Sahagún, Fray Bernardino de. *Historia General de las Cosas de la Neuva España.* México: AGN, 1979.

Secretaría de Salud, Dirección General de Epidemiología, Instituto National de Psiquiatría. *Encuesta Nacional de Adicciones Consejo Nacional Contra las Adicciones.* México, 1988 and 1998.

Soberón, Arturo, and Miguel Angel Vázques. "El consumo de pulque en la ciudad de México, 1750–1800" (M.A. thesis, Universidad Aútonoma Nacional de México, 1992).

———. "Elixir milenario: el pulque" in *Beber de tierra generosa: Historia de las bebidas alcohólicas en México,* vol. 1, edited by M. E. Medina-Mora and L. Echeverría, 26–49. México: Fundación de Investigaciones Sociales, A.C, 1998.

Taylor, W. *Drinking, Homicide and Rebellion in Colonial Mexican Villages.* Palo Alto: Stanford University Press, 1987.

"Todo el país secunda la política de la presidencia sobre el alcoholismo" in *Excelsior.* 19 April 1930.

Microbreweries (Japan)

Microbreweries (*jibiru*) in Japan became possible in 1994, when the minimum output required for a beer production license was lowered from 2,000 kiloliters (c. 528,000 U.S. gallons) a year to 60 (c. 16,000 U.S.). Two local breweries obtained licenses in December 1994, and in February 1995, the first Japanese microbrewery, Echigo Beer, was born with a brewpub. It was operated by the Uehara Sake Winery in Niigata. At the same time, *happoshu,* sparkling low-malt liquor, became legal.

The beer market in Japan had been dominated by four major companies, Asahi, Kirin, Sapporo, and Suntory, which had been producing mainly light pilsners. Until 1995, the only beers available to the Japanese public were those put out by these companies, which all tasted similar. After the founding of small-scale microbreweries, however, Japan, the world's fourth-biggest beer-producing country, saw a microbrewery boom. The number of microbreweries steadily increased from 17 in 1995 to 75 in 1996, dramatically to 182 in 1997, 240 in 1998, and 304 in 2000. The number of brands reached more than 1,000 in 2000.

In the first year of microbrewery operation, Ryoji Oda founded the Japan Craft Beer Association (JCBA) to popularize and promote micro beers. The country's first major microbrew festival, held in Tokyo in 1995, featured twelve local beers from around Japan. JCBA runs beer-tasting seminars and in 2000 launched the Beer Taster Organization (BTO) to provide certification to qualified beer tasters. JCBA and BTO organize the "Japan Beer Cup" and the "International Beer Competition."

With the new industry booming, the Japan Brewers Association (JBA) was established in 1999 for the stabilization of local beer-brewing operations through exchange of information and improvement of technology and management. The JBA has also held the annual national microbrewery festival. As of July 2002, the JBA consisted of 162 companies including 32 sponsors.

The majority of the Japanese public, however, has not developed a taste for microbrewery beers with their richer and more complex flavors than the major beers, which contain zero beer yeast. After the boom reached its peak in 2000, the number of microbreweries began to decrease, reaching 258 in March 2001. The Japanese public had been too accustomed to the mass-produced pilsners marketed by the big four. Home brewing had long been illegal, which meant that widespread appreciation of handmade beer was hard to foster. Furthermore, many of Japan's microbreweries were financed to attract and stimulate a local economy, but not necessarily to make more delicious beer. Though deregulation enabled venture businesses outside the existing liquor industry to participate in the new market, for those small-scale microbrewers management could be challenging. Although some of these companies have been successful, even the biggest microbrewer, Ginga Kogen Beer, filed bankruptcy in June 2002.

In the continuing recession, microbreweries are suffering from the high costs of ingredients, paying more for them than the mainstream beer companies, yet they pay the same rate of tax, as high as 45 percent. The inexpensive and therefore increasingly popular *happoshu,* with a 40 percent share in the beer market in 2002, benefits from a lower tax rate. This is hurting microbreweries. The JBA asked the National Tax Agency in 2001 to reduce the tax on microbreweries in order to protect the industry and allow the Japanese the choice of enjoying nonmainstream beers.

Fusako Ogata

See also: Japan; Microbreweries (United States, United Kingdom, and Canada)

References

Harrell, Bryan. "Tokyo's First Craft Beer Festival." http://celebrator.com/9602/Harrell-JapFest.html.

Hozumi, Tadahiko. *Jibiru Sanka* (Cheers to microbreweries). Tokyo: Ken-Yu Kan, 1998.

Japan Brewers Association. http://beer.gr.jp/.

Yoshihara, Akino. "Raising a Glass to the Micro-Beer Boom" in *The Daily Yomiuri.* 22 November, 2000. http://beertaster.org/dailyy001122.html.

Microbreweries (United States, United Kingdom, and Canada)

Microbrewing is small-scale production of beer in major industrial societies where large brewers can brew up to 100 million barrels (31 U.S. gallons per barrel; 1 UK barrel contains 36 imperial gallons [43.23 U.S. gallons]) a year. Microbrewing in Great Britain, Canada, and the United States began as a response to the ever-increasing concentration of their national brew markets. Microbrewing is one segment of the craft beer industry, which is itself a subsection of the beer industry. Craft brewing has four segments: microbreweries, brewpubs, contract brewers, and regional breweries. A microbrewery produces fewer than 15,000 barrels a year.

The United States

For over a decade beginning in the 1980s, microbreweries were hot commodities with an ever-growing market share. Craft beer appealed perfectly to baby-boom generation Americans with its standards of quality and high pricing similar to imports. The strength of the trend gave microbreweries a 2 percent market share by 1996. In 2001, U.S. craft breweries produced 6.2 million barrels.

Most microbreweries are quite small and operate on the local level. In a survey of Portland, Oregon, breweries, for example, the leading brewer, Deschutes Brewery, had a barrellage of 54,965. The twenty-fifth largest brewery produced 701 barrels. Deschutes Brewery represents 2.2 percent of the Oregon beer market. Due to its level of production it is actually a regional brewer because it has a capacity of between 15,000 and 2 million barrels. Deschutes also has a pub.

Brewpubs, or restaurants connected to breweries, sell over 50 percent of their beer on-site in a restaurant physically connected to the brewery. As of 2000, there were 1,023 brewpubs in

Paul Camusi (left) and Ken Grossman (right), owners of Sierra Nevada Brewing Company (Ed Kashi/Corbis)

the United States and 3,000 microbrew labels. The joining of dining and brewing represented a change in post-Prohibition drinking legislation, which usually prohibited such facilities.

Contract breweries are companies that use the facilities of a preexisting brewery to produce their beer. Boston Beer Company, producer of Sam Adams beer, is the best-known contract brewer in the United States. Although Boston Beer's corporate headquarters are in Boston, Sam Adams is primarily brewed at the Pittsburgh Beer brewery in Pennsylvania. As of 1997, Boston Beer was the nation's seventh largest brewer, with an output of 1,352,000 barrels. Anheuser-Busch produced 100 million barrels in 2001.

Several organizations represent small breweries in the United States. The oldest is the Brewer's Association of America (BAA), a trade association for small brewers and allied industries. The main legislative goal of the BAA is to maintain the tax differential favoring small brewers. Smaller breweries pay $7 a barrel rather than the full tax of $18 a barrel that the big companies pay. To be a full member of the BAA a company must brew fewer than 2 million barrels annually. The large brewers and importers can be associate members, however. There are also allied industry memberships and wholesaler memberships. Anheuser-Busch does not belong to the BAA, but Boston Beer does. Malting companies, printers, and glass manufacturers are some of the allied industries involved in the BAA.

Canada

In September 2000, over seventy Canadian small brewers joined together to request a tax reduction. The industry employs more than 3,000 Canadians and pays $21 billion in taxes. From 1984 to 1994, Canadians consumed an average of 2 billion liters (529 million U.S. gallons) of beer per year. As of 1994, there were more than twice as many breweries in Canada as there were in 1982. All Canadian brewers, large or small, pay about $2.30 in excise tax on a 24-bottle case of beer. Although costs are greater for microbreweries, they pay the same rate as the behemoths of Labatt and Molson, which represent over 90 percent of the Canadian market.

The United Kingdom

Underlying late-twentieth-century changes in brewing in Canada and the United States was beer drinkers' quest for quality and selectivity. North American consumers sought a change from the uniformity and homogeneity of big brewing. Although consumers in Great Britain had the same desires for their beer, the changes in the British brewing industry took a somewhat different shape. The key element in the recent transformation of English beer has been the Campaign for Real Ale (CAMRA). CAMRA is a consumer group seeking to maintain standards for brewers and publicans. Prior to its founding in 1971, no new ale breweries had been created in Britain for fifty years. There are now 300 new brewers in England. The two growth areas for British brewing are premium lagers and real ales. Real ale is beer stored in casks containing live yeast, which produces a secondary fermentation. Kegged lager, in contrast, is inert and can be stored for long periods of time and shipped long distances. CAMRA also promotes changes in legislation and fairer tax policies for small brewers as well as publishing guides to good beer.

Microbrewers in the United Kingdom have had an organization to represent them exclusively since 1980. The Society of Independent Brewers (SIBA) works for legislative changes in both Great Britain and the European Union. As in Canada, the SIBA is seeking a reduced tax rate for small brewers.

Amy Mittelman

See also: Anheuser-Busch; Campaign for Real Ale (CAMRA); Microbreweries (Japan)

References

"Annual Craft Brewing Growth Statistics for 2001." Institute for Brewing Studies website. http://www.beertown.org/PR/pdf/fact_sheet.pdf.

Brewers Association of America website. http://207.8.155.214/.

Campaign for Real Ale website. http://www.camra.org.

"The Craft Brewing Explosion" in *Modern Brewery Age.* 46, no. 21 (22 May 1995): 6–10.

Erickson, Jack. *Star Spangled Beer.* Reston, VA: RedBrick Press, 1987.

Hancock, Jay. "Beneath the Foam a Mania Goes Flat" in *Baltimore Sun.* 9 June 2002.

Lehndorff, John. "From Beer to Eternity: Belly Up to Taste a Few of the 1,200 Brews" in *Rocky Mountain News* (Denver, CO). 5 October 2000, 7D.

Pigg, Susan. "70 Microbrewers Join Forces to Make Case for Excise-Tax Cut" in *Toronto Star.* 20 September 2000.

Society of Independent Brewers website. http://www.siba.co.uk/index.htm.

"Steven Beaumont on Beer and Canada." World of Beer website. http://worldofbeer.com/bg1.html.

"Top 10 U.S. Brewers in 1997." American Brewery History Page website. http://www.beerhistory.com/library/holdings/top10.shtml.

"Top 25 Microbreweries" in *Business Journal-Portland.* 18, no. 51 (December 2001): 43.

Miller Brewing Company

After many years of success as first a local, then a national brewery under family control, the Miller Brewing Company passed through a period as part of a major conglomerate, then became absorbed into an international brewing operation. Miller Brewing Company traces its origins to Frederick J. Miller's purchase of the Plank Road Brewery in Milwaukee in 1855. Miller was born in Riedlingen, Germany, on 24 November 1824 and learned brewing from an uncle who had a brewery in Nancy, France. He operated the royal brewery of the Hohenzollerns in Sigmaringen, Germany, before immigrating to the United States in 1850. He traveled around the country for a time before deciding that Milwaukee offered him the best opportunity to pursue his trade.

Lorenz and Charles Best, sons of Milwaukee brewer Jacob Best, had established the Plank Road Brewery, with an extensive cave system for aging its beer, in 1850. Lorenz died in 1853, and Charles was not interested in pursuing the business. Frederick Miller purchased the brewery property and equipment for $3,500. The new brewery was known as Fred Miller's Plank Road Brewery. In addition to the brewery, Miller operated a beer garden. He brewed 300 barrels of beer during his first year. By the 1880s, Miller was brewing 80,000 barrels a year. Miller not only brewed a quality product but also pursued innovation. He was among the first brewers to establish a bottling department and offer bottled as well as draft beer, for example. When Miller died in 1888, his brewery was firmly established as a major firm.

Upon his death the company was organized as a stock company known as the Fred Miller Brewing Company. Miller's oldest son, Ernest, took over management of the company. Under his leadership the brewery continued to increase production and innovate in brewing technology while establishing the brand names and trademarks associated with Miller beer today. In 1906, a modern refrigerated building replaced the cave system under the brewery for aging beer. In that same year the firm sponsored a contest to name its premium beer, which became known as Miller High Life, with the trademark tag line—"the Champagne of Bottled Beers." A year later the "girl in the moon" logo first appeared. When Ernest died in 1922, his brother, Frederick A. Miller, became head of the firm.

The second Fred Miller picked up the challenge of leading the company through National Prohibition. Unlike many breweries, which closed forever, Miller continued in business, making near beer, malt syrup, carbonated soft drinks, and a malt-based health tonic. The firm survived. When repeal took effect thirteen years later, Miller shipped a case of High Life to the White House for President Roosevelt and offered free beer as part of Milwaukee's celebration. Repeal was quickly followed by another challenge to the brewing industry, however, World War II. The war effort led to the diversion of raw materials and caused a reduction in production. Much of the beer that was produced was for the armed services. Although the company survived these two crises, by the end of the war it had only one "label," or type of beer, Miller High Life. At various times during its history, the brewery had produced around thirty-five labels. In 1947, Frederick C. Miller succeeded his uncle Frederick A. as head of the firm.

Miller grew rapidly in the postwar economy, and production soon reached 2 million barrels a year. When Frederick C. Miller died in 1954, leadership of the firm passed out of the family, ending an era. Miller was not the only firm caught up in these changes. Many brewing families were passing control to professional managers or leaving what was rapidly becoming a national industry. The number of local and regional breweries, already greatly reduced by Prohibition, was further contracted by increased competition as some brewers sought to market their beer nationally. Miller was one of the traditional brewing companies that decided to expand to the national level.

Miller acquired its former Milwaukee competitor, A. Gettelman, in 1961. In 1966, the company acquired two breweries outside Milwaukee when it bought the Carling brewery in Fort Worth, Texas, and a brewery in Azusa, California, from the General Brewing Company. W. R. Grace Company purchased 53 percent of the Miller family's stock in the firm that same year, and three years later the Philip Morris Company bought the balance of the family's stock and a controlling interest in the firm. Miller Brewing was now part of a large, multiproduct corporation with a powerful and effective marketing operation. Miller ranked eighth in the nation in sales in 1968, by 1977 it had risen to fifth, and in 1977 it ranked second, a position it still holds. This growth came from additional acquisitions and expansion of brewing capacity, aggressive marketing of Miller brands, and new Miller labels and product lines to adapt to changing market dynamics and customer interests.

Miller embarked on a major expansion program in 1972, acquiring several brands either outright or through licensing agreements and buying or building additional breweries. Today Miller beers are brewed in Washington State, California, Texas, North Carolina, Georgia, and Ohio as well as in Milwaukee and Chippewa Falls, Wisconsin. The company also introduced Miller Lite, which established a new niche in the beer market for lower calorie beer, in 1972. Lite benefited from the marketing experience of Philip Morris and several very successful and memorable advertising campaigns. "Everything you've ever wanted in a beer and less" was followed by the "Less filling, tastes great" campaign. Lite beer has become a major market niche, and Miller Lite remains the dominant label. In 1986, the company's tradition of technological innovation led to Miller Genuine Draft, produced with a cold filtration process that allows shelf life without pasteurization. Miller Genuine Draft Lite was introduced in 1988. Another part of the expansion involved repositioning Miller High Life from the premium market—as a "country club" beer—to one with more appeal to the average man and woman. "Miller Time" came to represent the end of the workday. As the beer market became stratified, the firm had a label for each niche—Lowenbrau for the premium level, High Life and Miller Genuine Draft, each with a Lite, for the mid range, Lite for that lucrative niche, Milwaukee's Best, Meister Brau, and Milwaukee's Best Lite for the lower price market, and Olde English 800 Malt Liquor and Mickey's for the malt liquor market.

As the demand grew for craft beers made in microbreweries and brewpubs, Miller again adjusted to changing trends, acquiring the Leinenkugel Brewing Company of Chippewa Falls, Wisconsin, in 1988 to produce a broad range of beers for the craft brew market. The Plank Road Brewery was reactivated to brew and market Red Dog, Southpaw Light, and Ice House for the same reason. With the firm's size and marketing and distribution network, its beers are very competitive price-wise in the craft beer niche. During the 1990s and the first years of the twenty-first century, Miller has developed a number of products in the flavored malt beverage area. Miller Brewing remains solidly number two in the U.S. beer industry, with overseas sales increasing since the 1986 establishment of Miller Brewing International. In 2002, Miller was purchased from Philip Morris by South African Breweries to become part of the new company SABMiller plc.

William H. Mulligan Jr.

See also: South African Breweries (SAB)

References

Apps, Jerry. *Breweries of Wisconsin.* Madison: University of Wisconsin Press, 1992.

Baron, Stanley. *Brewed in America: The History of Beer and Ale in the United States.* Boston: Little, Brown, 1962.

Greer, Douglas F. "The Causes and Consequences of Concentration in the US Brewing Industry" in *Quarterly Review of Economics and Business.* 21, no. 4 (1981): 87–106.

McGahan, A. M. "The Emergence of the National Brewing Oligopoly: Competition in the American Market, 1933–58" in *Business History Review.* 65, no. 2 (1991): 229–284.

Miller Brewing Company. "About Miller Brewing Company." http://www.millerbrewing.com/aboutMiller/aboutMain.asp.

Molson Incorporated

Molson Inc., one of Canada's most dominant brewers and ranking among the fifteen largest brewing companies in the world, is North America's oldest operating brewer. Molson traces its beginnings to the brewery established by Thomas Loid in Montreal, Quebec, in 1782. The next year, nineteen-year-old John Molson, an emigrant from Lincolnshire, England, joined Loid in partnership. In 1785, Molson bought the brewery outright. He turned out his first batch of beer in 1786, a year that Molson Inc. upholds as its official founding. Molson's brewery prospered amidst a city swelling with British and Irish immigrants. The brewer also made inroads into Montreal's French-speaking community. In 1816, John Molson, then quite wealthy, took his sons, John Jr., Thomas, and William, into partnership under the name John Molson and Sons. The patriarch then gradually withdrew from active business affairs. When he died on 11 January 1836, he left a legacy of success not only in brewing but also in steamboat shipping, railway building, warehousing, hotel-keeping, banking, and philanthropy. Although his descendants also pursued other commercial and charitable interests, brewing remained the Molsons' chief enterprise, except during three decades from the mid-1830s, when they ran a large whiskey distillery in conjunction with their brewery. The brewery remained under Molson control for the rest of the nineteenth century and into the twentieth through a series of partnerships involving John Molson's sons and grandsons.

Until the early 1900s, Molson was essentially provincial in outlook, cultivating most of its market in the Montreal area, only halfheartedly making agency forays into Ontario and the Maritime provinces. However, Molson's geographic strategy was not misplaced, since the Montreal market was sufficiently lucrative. Plant expansion and improvement, such as electrification in 1900, accompanied sales growth. After the turn of the century, a time when outside brewers increasingly competed for Montreal sales, Molson pursued a more aggressive marketing plan, pushing its sales reach beyond the Montreal area and concentrating upon brand-driven promotion. In 1903, the brewer introduced a new flagship brand, Molson Export Ale. The new plan immediately set the brewery on an upward course. Annual output reached 1 million imperial gallons (1.2 million U.S. gallons) in 1907 and 2 million imperial gallons (2.4 million U.S. gallons) in 1909. A private joint-stock company, Molson's Brewery Limited, was incorporated two years later. Just prior to World War I, an automatic bottling line enhanced Molson's plant efficiency.

Although war restrictions on beer manufacture were a setback, Quebec's liberal postwar prohibitory regime—provincial prohibition was effectively nonexistent—gave Molson an edge over non-Quebec brewers. Indeed, sales growth encouraged a $2 million brewery expansion during an era when many brewers struggled to stay afloat. U.S. Prohibition also helped Molson, as Americans visited a "wet" Montreal in droves and Montreal beer was smuggled into a "dry" United States by the trainload, sometimes quite literally. Another benefit of Prohibition—a more lasting one—was that the sizable Ontario market came to Molson's permanent attention. In 1928, the Montrealer established Molson's Ontario Limited in Toronto to coordinate sales in Ontario. To support its widening sales geography, Molson hired an advertising agency.

Commitment to Ontario and more sophisticated advertising allowed Molson to compete against E. P. Taylor's brewing conglomerate, the Brewing Corporation of Canada (later Canadian Breweries Limited). By the 1940s, however, Taylor's substantial control of the Canadian beer market convinced the Molsons, like the Labatts, that survival depended on nationwide production. To secure expansion capital, the family reincorporated Molson's Brewery Limited as a public company on 15 February 1945. A few years later, construction began on a new Toronto brewery (opened in 1955). From 1958 to 1962, expansion continued with the acquisition of seven breweries in western Canada and Newfoundland. A brewery in Barrie, Ontario, was acquired in 1974. In response to changing national tastes, the brewer moved into lager brewing. Molson Golden was introduced in 1954, and Molson Canadian, the brewer's flagship lager, came on in 1959. Meanwhile, a new promotional direction was taken in 1957 with company sponsorship of Hockey Night in Canada, Canada's most popular televised sports program. That same year, the Molson family bought the Montreal Canadiens of the National Hockey League. During the 1960s and

1970s, Molson, using profits derived from a growing share in a growing national beer market, diversified into other commercial sectors, notably buying Diversey Corporation, a water technologies company; Beaver Lumber, a building products retailer; and the Montreal Canadiens (purchased from the Molson family in 1978). Concentrated U.S. distribution commenced in 1971. Molson, Labatt, and Carling-O'Keefe (the reworked Canadian Breweries Limited) each accounted for a third, more or less, of Canadian beer sales by the 1970s.

Canada's beer market reached maturity and began to shrink, however, during the 1980s. To maintain market share, Molson sought enhanced sales in specific market segments, such as that for light beer. The same maintenance strategy saw the company conclude a licensing arrangement with Coors, the Colorado brewer. However, Molson's most daring initiative was its merger with Carling-O'Keefe in 1989. The move gave Molson command of nearly 60 percent of the Canadian beer market; Carling-O'Keefe's owners, Elder's IXL (later the Foster's Brewing Group), assumed half ownership of Molson, a share that was sold back to Molson in 1998. The merger saddled Molson with overcapacity, which became quite alarming as a leaner Labatt quickly ate away at Molson's market-share advantage. Consequently, over the 1990s and into the early 2000s, Molson consolidated its breweries from sixteen to seven, divested itself of nonbrewing assets, and increasingly targeted market niches for discount, light, high-alcohol, ice, and dry beers and for fruit-flavored malt beverages. At the same time, the brewer increasingly emphasized the major foreign brands it brewed under license, such as Coors, Miller, Foster's, Heineken, and Corona. The overall plan worked, and with its financial house in order, the cash-rich Molson Inc. (incorporated in 1999) has recently expanded beyond Canada. In 2000, the brewer turned to the world's fourth biggest and fastest growing beer market and bought Brazil's Bavaria Breweries. The company followed this purchase by acquiring another Brazilian brewer, Kaiser. With these acquisitions, Molson is now the world's thirteenth largest brewer.

Glen C. Phillips

See also: Carling Brewery; Labatt Brewing Company Limited; O'Keefe Brewery

References

Denison, Merrill. *The Barley and the Stream: The Molson Story, a Footnote to Canadian History.* Toronto: McClelland and Stewart, 1955.

Dubuc, Alfred. "John Molson" in *Dictionary of Canadian Biography,* vol. 7, *1836–1850.* Toronto: University of Toronto Press, 1988.

Hunter, Douglas. *Molson: The Birth of a Business Empire.* Toronto: Viking, 2001.

Molson, Karen. *The Molsons: Their Lives and Times, 1760–2000.* Willowdale, Ontario: Firefly, 2001.

National Archives of Canada. Molson Archives.

Woods, Shirley E., Jr. *The Molson Saga, 1786–1986.* Toronto: Doubleday Canada, 1986.

Moonshine

"Moonshine" is an American name for a worldwide phenomenon, the illegal sale of liquor home-brewed from traditional agricultural products to avoid payment of taxes. Wherever a government imposes excise taxes on alcohol, there will be moonshine, from Irish *poteen* to Indian *hootch* to Peruvian *pisco* to African palm wine. The liquors are indigenous, traditional products that became subject to taxation only when a state imposed its authority on rural people. Moonshine is only one of many names for the American product, which got its name from having to be made secretly at night, by moonlight. The liquor was also known by the names "white lightnin'," "mountain dew," "pop skull," "bust-head," "white mule," "bumblings," "wildcat," "brush whiskey," "blockade," or "forty rod," and a host of other descriptive epithets in the United States. It is usually associated with the Appalachian South's corn whiskey but also appeared as rum in the Irish waterfront neighborhoods of Brooklyn during the 1860s and 1870s, or potato vodka in New York's lower east side neighborhood.

The Appalachian product, usually corn whiskey but also fruit brandies depending on the region, is the most famous in folksong and story. Mountain dew required several weeks of processing before emerging from the still in liquid form. Traditional whiskey recipes began with sprouting of unground corn by pouring warm water over the ears for about three days, draining through a small hole in the barrel. When sprouts appeared, the corn's starch had converted to sugar and the substance was ready to be dried out and ground into meal. Pouring boiling water over the meal produced "sweet mash," which the distiller allowed to stand for two days. He or she (there were women moonshiners, too) next fermented this mash, which required eight to ten days. Fermentation tested a distiller's skill, for it required maintaining a constant temperature without a thermometer. The distiller judged the degree of fermentation by the sound in the barrel. A sound like rain falling on a roof or pork frying in a pan signified that "sour mash" had reached perfection, the liquid portion or "beer" being alcoholic and "sour enough to make a pig squeal." The distiller then poured the beer into the still, a copper pot with a removable cap attached to a spiral tube, the "worm." This worm was surrounded by a watertight barrel or box through which cold water was kept constantly running, the reason stills had to be located conveniently to flowing water. The moonshiner built a fire under the still, often set into a stone fireplace enclosing the pot to prevent loss of heat. The heat caused the alcohol in the beer to vaporize and then condense into liquid while passing through the cold worm. The liquor, at this stage impure "singlings," drained into a receiving tub or barrel. The singlings were then passed through the still again with the fire at a lower temperature, a process again requiring skill to avoid problems: If the "doubling" process were too brief, weak and sour spirits would result; if too long, pure alcohol developed. The goal was a clear whiskey, ready for immediate consumption without aging.

This is a simple traditional recipe. Twentieth-century refinements included a "thump keg" or doubler placed between the still and the worm, through which the vapor passed with a great buildup of heat, resulting in a second distillation that avoided "doubling." Distillers also saved time by using sugar instead of sprouted corn to begin fermentation, an innovation disdained by old-timers who said the result was not true "corn juice." Unscrupulous distillers adulterated weak whiskey with buckeye tree pods to give a good "bead," bubbles that appeared when the liquor was shaken, an indication of proof. In the Prohibition era of the 1920s, moonshiners used automobile radiators as worms, which added a poisonous flavor of lead solder to their product. Indeed, there is a story of one community outside of Atlanta that inadvertently poisoned urban customers and understandably never regained its reputation for good white lightnin'.

Homebrewed liquor was a craft, with fine and poor craftsmen. Accounts of its taste range from appreciation to disgust; of its effects from mild warmth to passing out. When moonshining was "turning from a little business into a big business," as writer Horace Kephart put it in the early twentieth century, quality suffered in favor of quick profits. Stills with hundreds of gallons of capacity dwarfed the 5- or 10-gallon copper pots of old-timers.

All sorts of people became moonshiners, and for varied motives. Some found selling liquor without the tax (at about half the cost of taxed spirits in the later nineteenth century) a way to supplement income, especially during times of agricultural depression when it was difficult to market corn. Others became full-time "blockaders" (who ran the blockade of tax collectors), shipping their liquor long distances and relying on urban or rural customers living within or outside the mountains. They sold to "blind tiger" saloons in towns as well as to lumberjacks, miners, and railroad and textile mill workers, who were the vanguard of social and economic change in the southern mountains in the late nineteenth and early twentieth centuries. Moonshiners, men and women, black and white, sometimes well-known local characters famed for the quality of their product or as daring leaders of clans of illicit distillers, were often entrepreneurial risk-takers realizing the possibilities of expanding markets.

Moonshine would not have entered American history if American officials had not followed British precedent and levied excise taxes on liquor. The conflict between small farmers' assertions of a "right to make a little licker" and the government's efforts to collect taxes created folk heroes in the traditional American mode: free citizens battling an overpowering government, quintessential individualists in a nation that never liked to pay taxes. The first confrontation over liquor taxes was the western Pennsylvania Whiskey Rebellion of 1794. Many of the settlers of the region were people of Scots-Irish ancestry who had brought traditional whiskey making with them. As part of Alexander Hamilton's program to establish sources of revenue for the new U.S. government, Congress enacted an excise on liquor in 1791. Many farmers in backwoods areas expressed their discontent, but there was no serious confrontation until 1794. Officers appointed to collect the new tax were harassed, much as the British stamp-tax collectors had been before the Revolution; indeed, the language of protest against the tax sounded the same. Vigilante groups organized to drive the tax collectors out of the region, prompting President George Washington to call for a force of 15,000 militia to quell the insurrection. Accompanied part of the way by Hamilton, the troops dispersed the rebels without deaths on either side. The rebellion was crushed and two of its leaders convicted of treason, and the government merely asserted its authority without provoking the people by renewing its efforts to collect the excise, which was abolished under President Thomas Jefferson.

Renewed temporarily during the War of 1812, the excise did not weigh heavily on small producers. During the Civil War, Congress levied numerous excise taxes on a multitude of products and transactions to help finance the war effort. Most of these were repealed after Union victory in 1865, but the liquor and tobacco taxes remain to this day. Although some reformers argued that taxation would reduce drinking and smoking, that prediction soon proved untrue, and temperance advocates criticized the government for relying on revenue from immoral and dangerous products. Levied on small and large distillers throughout the country, the liquor excise developed into the federal government's second-largest source of revenue after the tariff. The Bureau of Internal Revenue, created to collect the taxes, did not get off to a good start, becoming entangled in the notorious Whiskey Ring scandal of the early 1870s. Large distillers connived with local revenue officials to evade taxation in exchange for bribes, with the taint of scandal reaching up to the highest officials in Washington. Secretary of the Treasury Benjamin Bristow organized a series of simultaneous raids on distilleries to seize untaxed barrels of whiskey as evidence. This tactic broke the ring and led to prosecution of several distillers but few officials. Bristow's zeal cost him his job, but large distillers soon became allies of the government when threatened with loss of profits from the competition of thousands of small moonshiners, particularly in the mountain South.

Individually, small evaders did not cost much in lost profits or federal revenue, but collectively, they represented substantial loss to both tax-paying distillers and the government in the regions in which they were active. Gradually a bipartisan consensus developed in favor of the internal revenue system. Though created by Republicans during the war, Democrats, after years of shrill denunciation of the tyranny of the liquor tax, discovered when they gained control of the executive branch (1885–1889, 1893–1897) that the excise was too useful to do without. Even in the South, finding that revenue officer appointments proved useful as patronage sources, the Democrats muted their defense of moonshiners and attacks on the tax.

After the period of inefficiency and scandal, the Bureau of Internal Revenue was in need of an energetic leader to restore

A moonshine still confiscated by the Bureau of Internal Revenue, photographed at the Treasury Department, 1920s (Library of Congress)

its reputation and stop the drain of government income. President Ulysses S. Grant appointed Green B. Raum, lawyer and Civil War veteran, commissioner of the bureau in 1876. Raum convinced the cabinet and Congress that southern moonshining was a major problem, then launched a full-scale attack; he was able to leave office proclaiming that he had solved it. He emphasized coordinated sweeps through moonshiners' territory to catch violators in the act, as well as traditional reliance on informers' reports of hidden stills. At first his campaigns throughout the mountain South generated heightened resistance. Moonshiners received warning of impending raids and hid their stills. Ambushes and battles led to casualties on both sides. Local juries refused to convict moonshiners, and hostile state officials did their best to frustrate Raum's efforts. Raiding parties normally included deputy revenue collectors ("revenuers"), who seized or destroyed stills, and deputy U.S. marshals, who made arrests. Raum combined a sustained attack on moonshiners with a policy of leniency toward first offenders and encouragement of mountaineers to operate small tax-paying distilleries. At first the outcome of his campaign was uncertain, but by the time he left office in 1883, he could point to decreased violence, increased revenues, and greater sympathy from local citizens, mainly town dwellers, suggested by newspaper editorials and increased conviction rates.

Sporadic guerrilla warfare continued under Raum's successors, with revenuers permanently ensconced in the mountains and moonshiners refusing to give up completely. The next period of major confrontation was during the period of national depression from 1893 to 1897, when farmers became desperate for the cash that could be earned from mountain dew, especially in light of the fact that an increasing number of southern counties and some states were enacting total prohibition of alcohol sales. Both these factors contributed to the commercialization of moonshining as the only source of liquor in some areas. Additionally, Congress raised the liquor tax to compensate for declining revenue from all sources. "Whitecap" vigilantes appeared in some states to support moonshiners; they were effective in intimidating local officials and punishing detested informers, who received $5 for each still they reported. Federal prosecutors did succeed in breaking up the whitecap movement and by the early 1900s had managed again to contain, if not entirely eliminate, violent resistance.

The advance of prohibition through the South enlisted state officials against moonshiners but sometimes created opponents: The revenuers only wanted to collect taxes, indeed depended on the existence of distilleries, whereas state officials were enforcing prohibition of all liquor sales. The nation did not learn the lessons of local and state prohibition, however, and when National Prohibition went into effect in 1920, moonshining and bootlegging, the sale of illegal liquor, became a nationwide phenomenon supervised by mobsters. The era also witnessed the use of automobiles in chases, said to be the origins of stock-car racing.

Moonshining did not disappear with repeal of Prohibition in 1933; people still preferred to buy cheap liquor, and many may have developed a taste for raw corn whiskey. Though some people say that descendants of moonshiners have gone into the more profitable business of growing marijuana, they remain active to this day. Federal authorities have sporadically cracked down on illicit distillers, their most recent effort being a combined federal and state raid in March 2000 on Rocky Mount, Virginia, which claimed to be the "moonshine capital of the world." Federal revenuers (now members of the Bureau of Alcohol, Tobacco, and Firearms) and state officials netted three culprits and gave agents the old-fashioned thrill of chopping up stills with an axe.

At various times moonshine is defined as an official problem, but the people who distill untaxed home brew persist as a symbol of American individualism. Moonshine's mythic status also endures. The latest news on the moonshine front as of May 2002 was that a young West Virginia entrepreneur was making and selling legal moonshine, hoping to capture the market of people looking for something new. He claimed to be the first marketer of such a product, but ten years before, the author of this article bought a legal mason-jar bottled product called "Georgia Moon." He still has most of it, as for most people raw corn whiskey is not something to be drunk in large quantities.

Wilbur R. Miller

See also: Bureau of Alcohol, Tobacco, and Firearms (BATF); Palm Wine; Taxation of Liquor (United States); Whiskey Rebellion; Whiskies

References

Carr, Jess. *The Second Oldest Profession: An Informal History of Moonshining in America.* Englewood Cliffs, NJ: Prentice-Hall, 1972.

Kellner, Esther. *Moonshine: Its History and Folklore.* Indianapolis, IN: Bobbs-Merrill, 1971.

Kephart, Horace. *Our Southern Highlanders.* New York: Outing Publishing, 1913.

Miller, Wilbur R. *Revenuers and Moonshiners: Enforcing Federal Liquor Law in the Mountain South, 1865–1900.* Chapel Hill: University of North Carolina Press, 1991.

"North Carolina Moonshine: A Survey of Moonshine Culture, 1900–1930." www.ibiblio.org/Moonshine.

Moral Suasion

Moral suasion refers to the primary tactic used by British and American temperance reformers from the late eighteenth to the mid-nineteenth century. Stemming from a larger tradition of voluntarism in Anglo-American reform, moral suasion assumed that bad or undesirable behavior could be changed by appealing to a person's conscience. Through rational argument, sympathetic appeal, or moral entreaty, suasionists advocating reforms from abolitionism to sabbatarianism hoped to convince wrongdoers of their error, thereby setting them on the correct path. At times, moral suasion also included arguments based on self-interest: Proper behavior produced a better life in material as well as moral terms. Nowhere were these tendencies more apparent than in the battle against intemperance.

Temperance advocates employed a variety of arguments to persuade drinkers to moderate their alcohol use or abstain entirely, though rational arguments occupied a central place in their use of moral suasion. Beginning in the late eighteenth century, physicians, following Benjamin Rush's lead, reminded drinkers of the dire physical consequences of alcohol abuse. Political economists condemned the wastefulness and uselessness of the drink trade, both for nations and individuals. Clerics warned drinkers that alcohol threatened their eternal welfare, reminding them of the biblical proscription against drunkards entering the kingdom of heaven (1 Cor. 6:10, for example). Other reformers painted heartrending pictures of the violence, neglect, and abuse visited upon the wives and children of drunkards. Because women were believed to wield a unique and powerful moral influence over men proceeding from their relationships as sisters, wives, and mothers, female temperance reformers played a major role in urging intemperate men to reform. In short, moral suasionists implored drinkers to mend their ways, arguing that temperance held moral, financial, and emotional benefits.

Moral suasion represented the opposite pole from legal coercion in the reform continuum. In general, advocates of moral suasion at first resisted efforts to impose legal sanctions on the sale or use of alcohol. Since many temperance advocates viewed drunkenness as an individual sin or moral failing, they believed that persuasion, rather than coercion, offered the best means of motivating inebriates to change their behavior. Moreover, prohibiting or regulating the use of alcohol, opponents of coercion feared, would undermine the authority and limit the rightful prerogatives of heads of households to govern their families. Still, the inability of moral suasion to eradicate intemperance or its attendant ills had, by the 1850s, caused many of its advocates to reconsider their opposition to more forceful measures. "Moral suasion is moral balderdash," one American temperance journal remarked in 1852 (Ginzberg 1986: 601).

Faced with indisputable evidence that moral means alone could not conquer drunkenness, temperance advocates looked to government at various levels to pass legislation outlawing the sale, possession, or use of alcohol. In the United States, many state governments, beginning with Maine in 1851, enacted prohibitory legislation, which set the tone for future attempts to control intemperance. In Great Britain, the newly formed United Kingdom Alliance advocated outlawing trade in alcoholic beverages in 1853. Though temperance advocates continued to endorse moral suasion as a useful means of reform, antiliquor forces had, by the late nineteenth century, begun to embrace legal coercion as their primary tactic.

Scott C. Martin

See also: Maine Law; Martha Washington Societies; United Kingdom Alliance (UKA)

References

Ginzberg, Lori. "'Moral Suasion Is Moral Balderdash': Women, Politics, and Social Activism in the 1850s" in *Journal of American History.* 73 (December 1986): 601–622.

Harrison, Brian. *Drink and the Victorians: The Temperance Question in England, 1815–1872.* Pittsburgh: University of Pittsburgh Press, 1971.

Mothers Against Drunk Driving (MADD)

Mothers Against Drunk Driving (MADD) is a nonprofit organization with the goal of stopping drunk driving and supporting the victims of drunk driving accidents. Candy Lightner founded the organization in 1980 in Sacramento, California, after a drunk driver killed her thirteen-year-old daughter. The driver had a blood-alcohol content (BAC) of 0.20 percent and five previous drunk driving convictions, the latest just two days before the accident in which he killed Cari Lightner. He pleaded guilty and received two years in prison. Lightner was furious at the lenient sentence and quickly began organizing for stricter laws and penalties. MADD's campaigns for changes in the legal framework governing drinking and driving met considerable success during the 1980s and 1990s. In recent years, its mission has included preventing underage drinking.

Lightner's charisma and dramatic personal story helped make her campaign an immediate success. In the late 1970s and early 1980s, the death toll on American roadways from drunk driving was substantial, penalties were lenient, and there was little public awareness of the problem. The extent of the concern among those most directly affected ensured that MADD was only the most successful of several anti–drunk driving organizations that were established at this time. MADD had a special appeal as the "voice of the victim" and received tremendous support from the many victims of accidents and their families, especially bereaved parents. By the end of 1982, MADD had 100 chapters. In 1983, NBC produced a made-for-television movie called *MADD: The Candy Lightner Story.* A month later, a poll revealed that 84 percent of Americans had heard about MADD. By 1985, there were 320 chapters across the United States, and by 2001 the number had grown to more than 600 chapters in the United States plus affiliates in Guam, Canada, and Puerto Rico. Although MADD is one of the most successful nonprofit organizations in the United States and has more chapters than ever before, it is currently finding it more difficult to raise funds, and in recent years critics have charged that it has become overly zealous in its approach.

From the beginning, MADD paid careful attention to the rights of the victim, and it has played an important role in the victim's rights movement. It puts considerable effort into supporting victims through a victim's services office, a wide variety of brochures about grieving, and a magazine called *MADDvocate.* Local MADD chapters provide victim advocates who aid victims with criminal justice matters and help them make links with counseling personnel. They also provide support groups for people who have been injured in or bereaved in a drunk driving accident. Since 1998, MADD's website has featured a Victim's Tribute, a weekly story about someone who died in a drunk driving accident. MADD strongly supports the right of victims of alcohol-related traffic crashes to seek financial compensation from establishments and servers who have provided alcohol to people who are already intoxicated or to minors. It has also put considerable effort into developing Victim Impact Panels, which give victims of drunk driving accidents the opportunity to share their stories with first-time and second-time Driving While Intoxicated (DWI) offenders. MADD believes that these panels have mental health benefits for victims and help to reduce recidivism.

In the legislative arena, MADD met with enormous success in its early years. Between 1981 and 1985, all fifty states strengthened their laws against drinking and driving, and the number of states requiring mandatory jail sentences for first offenders convicted of DWI doubled. In 1984, Congress, under pressure from MADD, passed a law requiring states to increase their minimum drinking age to twenty-one by 1987 or lose a percentage of their federal highway grants. In 1984, only twelve states had minimum drinking ages of twenty-one. By the end of the year, twelve additional states had raised their drinking age, and by the early 1990s nearly all states had minimum drinking ages of twenty-one.

More recently, MADD scored a substantial victory when President Bill Clinton in 2000 signed a bill requiring states to pass a law allowing a maximum 0.08 percent BAC for all drivers. States not complying by 1 October 2003 would lose 2 percent of their federal highway construction funds, and additional penalties would be phased in over time. States that have not passed such legislation by 1 October 2006 will lose 8 percent of their highway construction funds. At the time of this legislation, the majority of states had a maximum allowable BAC of 0.10 percent. This measure attracted significant opposition from the beverage industry, which argued that the bill unfairly targeted

Mothers Against Drunk Driving buttons (Catherine Karnow/Corbis)

social drinkers. MADD continues to lobby for the confiscation of vehicles and license plates of habitual impaired drivers, mandatory confinement for repeat offenders, no reduction in charges, and ignition interlock devices.

In 1996, MADD announced a new focus on underage drinking and driving. MADD is lobbying for "Zero Tolerance" laws that prohibit youth under twenty-one years of age from driving with any measurable amount of alcohol in their blood system. MADD has repeatedly declared that "alcohol" is America's number one youth drug problem. It has lobbied the Office of National Drug Control Policy to put some of its drug education money into alcohol education. Its programs directed at youth, including a series of hard-hitting public service announcements, emphasize that alcohol use leads to sexually transmitted diseases, shortened life spans, date rape, and obesity. MADD encourages parents not to serve alcohol to their underage children at home. It also supports graduated licensing programs, which would impose additional restrictions on young drivers. It has called upon the alcoholic beverage industry to avoid any advertising that might appeal to youth, such as using celebrities, athletes, or cartoon characters, and to avoid the sponsorship of events with youth appeal, such as sporting events.

Public education has come to assume a greater importance in MADD's programs. At a Candlelight Vigil in 1986, attendees were asked to take the red ribbons that surrounded their vigil candles and "tie them on" their automobiles as a reminder not to drink and drive. This is called the "Tie It on for Safety" campaign. Red ribbons are distributed at schools, offices, and churches, and participants are encouraged to attach them to their cars to raise awareness of drunk driving, especially during the high-risk holiday season. MADD has also popularized the concept of the "designated driver," asking that whenever people's celebrations involve alcohol, a designated driver should be appointed who does not drink. The public awareness program is focused around the four deadliest holidays of the year—Labor Day, the December holiday season, Memorial Day, and St. Patrick's Day. There is also a poster campaign, targeted at alcohol-serving establishments, to encourage adults to choose a designated driver. Also, in the public education arena, MADD has endorsed the concept of requiring warning labels on alcoholic beverages stating that alcohol will impair skills necessary for the operation of motor vehicles or heavy machinery.

MADD's programs place the blame for drunk driving on individuals and stress the need for self-regulation by drinkers and by the alcohol industry. MADD has never addressed the role of structural factors such as suburban sprawl, extensive highway construction, and lack of public transport in making drunk driving accidents possible. In recent years, with its new emphasis on preventing alcohol use by teenagers and its campaign to have maximum allowable BAC lowered, it is moving increasingly in a neo-prohibitionist direction, garnering new enemies, not just from the beverage industry, but also from civil libertarians. MADD launched one of the most successful alcohol reform movements in American history, but it remains to be seen if its current program will gather the same support as its earlier initiatives.

Catherine Carstairs

See also: Drunk Driving Laws (United States); Students Against Destructive Decisions (SADD)

References

Ameda, Phyllis Jean. "Temperance for a New Age: The Crusade against Drunk Driving, 1980–1997" (M.A. thesis, California State University, Fresno, 1998).

MADD website. http://www.madd.org.

Reinarman, Craig. "The Social Construction of an Alcohol Problem: The Case of Mothers Against Drunk Drivers and Social Control in the 1980s" in *Theory and Society.* 17 (1998): 91–120.

Weed, Frank. "The Victim-Activist Role in the Anti–Drunk Driving Movement" in *Sociological Quarterly.* 31 (Summer 1990): 459–473.

Moule, Handley Carr Glyn (1841–1920)

Handley Carr Glyn Moule (1841–1920), a Cambridge evangelical and bishop of Durham, supported teetotalism for individualistic, evangelical, and laissez-faire reasons, largely to obviate state intervention. His father, Henry Moule (1801–1880), vicar of Fordington, Dorchester, Dorset, inventor and writer, was converted to teetotalism in the late 1830s by working people as he overcame fears of performing his pastoral duties without alcohol. Dorchester was one of the 1,496 Welsh and southern English rural communities where parson and squire had banned drink. But by 1864, Henry Moule was one of only seven priests in those communities identified as a total abstainer. He was subject not only to rural patriarchal influences but to metropolitan teetotal pressures. Handley Moule, later his father's curate at Dorchester, was raised as a teetotaler along with his seven brothers.

A Cambridge alumnus and the evangelical Ridley Hall's first principal, Handley Moule advocated temperance as promoting British individualism, as contrasted with European statism and the collectivism of his Anglo-Catholic and Broad Church teetotal colleagues. Moule's teetotalism sought individual self-improvement by simply rejecting the sin of drunkenness. As Cambridge United Temperance Council president, Moule told students: "There may be one man in a hundred for whom some kind of stimulant is a necessity; before you touch it, make quite sure you are the hundredth man." Cambridge evangelicals inspired temperance campaigns at Mission Houses in urban slums.

Despite his evangelicalism, as bishop of Durham (1901–1920) Handley Moule continued the temperance programs of his Anglo-Catholic predecessors in that northern seafaring and mining community. Already a vice president of the Church of England Temperance Society (CETS), in 1902 Handley Moule joined the teetotal insurance society, the Independent Order of Rechabites, to advocate "temperance of body and discipline of habit . . . and thrift, which itself promotes independence and develops liberty of action in a hundred ways" (Harford and Macdonald, 236–237). Individual liberty was England's response to Germany's early government welfare, where the "community has become almost the deity of the individual, fatal to the higher interest of the community. For it dulls the individual conscience, and the collective conscience after all, is but the summing up of individual convictions" (Harford and Macdonald, 237–238).

Contradicting Christian Socialists, Moule's temperance principles entrusted social progress to the personal behavior of the poor, making unnecessary extensive governmental regulation beyond limiting the drink interest's power. Yet, with Archbishop Randall Davidson of Canterbury and a slight majority of bishops, including Christian Socialists, Moule promoted the Liberal government's 1908 Licensing Bill as a matter of conscience over commerce. The Anglican establishment's drink investment proved stronger, and the bill was devastated in the House of Lords. By 1914, Moule's evangelical individualism had been largely surpassed in the CETS by Broad Church and Anglo-Catholic socialism.

Stephanie Olsen and Gerald Wayne Olsen

See also: Christian Socialism; Church of England Temperance Society (CETS); Conservative Party (United Kingdom); Davidson, Randall; Ellison, Henry; Evangelical Temperance (United Kingdom); Gore, Charles; Green, T. H.; Liberal Party (United Kingdom); Rechabite Friendly Society; Temple, Frederick; Wilberforce, Basil and Wilberforce, Ernest

References

Harford, J. B., and F. C. Macdonald. *Handley Carr Glyn Moule, Bishop of Durham: Biography.* London: Hodder and Stoughton, 1922.

Olsen, G. W. *Drink and the British Establishment: The Church of England Temperance Society, 1873–1914.* Unpublished book manuscript.

———. *Pub and Parish: Drink and Temperance in the Church of England, 1829–1875.* Unpublished book manuscript.

Peake, A. S. "Moule, Henry" in *Dictionary of National Biography, Supplement, 1912–1921,* 390–391. London: Oxford University Press.

Murphy, Francis (1836–1907)

Francis Murphy (1836–1907) founded the Gospel Temperance Movement in the United States in 1873. Never a believer in legislative prohibition of alcohol, Murphy refused to attack dealers and instead relied on the persuasive power of reformed alcoholics to convince others of the evils of drink. Speakers recruited and trained by Murphy fanned out across North America in the 1870s to spread his message.

Born on 24 April 1836 to a poor widowed Roman Catholic mother in County Wexford, Ireland, Murphy came to the United States at the age of sixteen. After landing in New York City, he quickly drank away all of his money and possessions. He then moved to Quebec, Canada, for a fresh start but lost his hotel job for intemperance. Murphy next found work on an upstate New York farm before enlisting in the Union Army upon the outbreak of the Civil War and serving until 1865. Married at eighteen, he became sober for the sake of his wife. Taking his wife and children from New York to Portland, Maine, Murphy opened a hotel. Soon he began drinking again and, in a drunken brawl, he fell down the stairs of his saloon with a customer who died of his injuries. Murphy was tried for murder but acquitted.

In 1869, he declared bankruptcy, and shortly thereafter he was jailed for habitual drunkenness. Convinced of the error of his ways by a Methodist evangelist, Cyrus Sturdevant, who visited the prisoners, Murphy held his first prayer meeting in the Portland jail. Although he remained behind bars, his wife and six children were evicted from their home. With his family starving, Murphy pleaded for his release and became a free man on 30 October 1870. His wife succumbed to typhoid three weeks later.

On 3 April 1873, Murphy spoke before an audience at City Hall in Portland. It was the first of a series of lectures that he delivered in the city. A tall man with gray hair and an enormous black mustache, he held audiences spellbound with the story of his fall and redemption. Murphy aimed to save alcoholics, teach people to trust a higher power and support other Christians, utilize the influence of women upon men, and arouse a sense of self-respect among drinkers. To achieve these goals, he relied on the public testimony of those who signed his pledge: "I do pledge my word and honor, God helping me, to abstain from all intoxicating liquors as a beverage and that I will, by all honorable means, encourage others to abstain." Typically, the former drinking companions of the men who took the pledge would turn out to listen to them speak about the sufferings of their families and then sign the pledge themselves. The movement progressed quietly in Maine, New Hampshire, Iowa, Illinois, and New York until 1876, when Murphy received an invitation to speak in Pittsburgh. During his visit, 40,000 people signed his pledge. The Murphy Movement, also known as the Blue Ribbon Movement for the way its followers adorned themselves, remained strong throughout the 1870s before losing ground to coercive movements. Murphy died on 30 June 1907 in Los Angeles following a long illness.

Caryn E. Neumann

See also: Blue Ribbon Movement; Reformed Men's Clubs

References

Ferris, George T., and Thurlow Weed Brown. *Minnie Hermon; or the Curse of Rum: A Tale for the Times Embracing also the Life and Work of Francis Murphy and Dr. Henry A. Reynolds.* New York: Henry S. Goodspeed, 1878.

Murphy, Rebecca Fisher. *Memoirs of Francis Murphy, the Great Temperance Apostle, by his Wife.* Long Beach, CA: Graves and Aersey, 1908.

Music

See Drinking Songs (United States); English Songs, Representations of Drinking in (1600–1900); Music, Popular, Representations of Drinking in; Music Halls

Music, Popular, Representations of Drinking in

Stories about alcohol have been part of all cultures and heritages for ages. The German composer Carl Orff (1895–1982) adapted his famous cantata *Carmina Burana* (created in 1937) from old profane songs from the Middle Ages; the lyrics honored the ephemeral joys of life, free loving, and drinking in the tavern: "They drink to the Pope and the King alike. All drink without restraint" (*In taberna quando sumus,* from *Carmina Burana*). This long tradition confirms that the celebration of elation through drinking alcohol has been present in art and music for many centuries. Moreover, not only has alcohol been mentioned in some operas, classical music, or old folk songs, it is also an important element of popular culture in the traditional music from many Western countries. There are traditional drinking songs in many languages: German beer drinking songs ("Ein Prosit"), Irish drinking songs, and French songs about wine that even find an inspiring echo in French Canada. Now and then, these songs are designed to be a soft form of propaganda, or at least a legitimation to incite the one who hesitates to drink to do so. In every such song, the message can be expressed by something like: "Let's sing along and drink together! Do as you're told and say like me!"

If drinking is present and justified in many songs, temperance seems to be absent, indeed to be almost a taboo in popular music. Parody appears to be the only way to talk about the habit of not drinking. On a recent CD rerelease titled *Golden Age of Comedy,* famous comedian W. C. Fields (1879–1946) does a few satirical monologues about drinking, prohibition, and temperance. After more than half a century, we can listen to two corrosive pieces that make fun of agents preaching temperance: "Fields Drank a Glass Of Water" and "The Temperance Lecture," two of the many sketches that he invented during the 1930s.

The first recordings in modern times featuring drinking themes were probably those made in the United States during the 1920s. One of the oldest examples (still available) of a recording of an American song about drinking addiction is possibly by a blues singer from Nashville, Tennessee, named Leroy Carr (1905–1935), who recorded in 1930 his song "Sloppy Drunk Blues." Just a few years later, this talented singer and piano player sadly died of alcoholism at only thirty years of age. His song became a blues standard that was also recorded by Sonny Boy Williamson (1947), Jimmy Rogers (1954), and in the 1960s by other bluesmen such as Sleepy John Estes (1899–1977) and Big Joe Williams. On the same topic, Estes also recorded "Liquor Store Blues" in the late 1930s.

Recordings from that period are becoming easier to find these days, thanks to some new rereleases on CD available through specialized stores. One of the first such collections was issued around 1990 by Sony Music; it featured more than twenty artists in a style that was labeled Race Records. The new title for that special collection was *Booze and the Blues* (1996); it was released in the Legacy Roots 'n' Blues Series, with historical notes on the CD booklet.

The African American artists represented on that eclectic CD sing about a wide range of drinking and addiction themes representing common points of view that were in the air before 1950. The titles are revealing. Artists and titles include: Barbecue Bob ("Me and My Whiskey"), Memphis Minnie ("Drunken Barrel House Blues"), Kid Prince Moore ("Bug Juice Blues"), Lewis Black ("Corn Liquor Blues"), Walter Vinscon's Mississippi Sheiks ("Bootleggers' Blues"), and the Memphis Jug Band ("Rukus Juice and Chittlin'"). There are also duets by Charley Jordan and Peetie Wheatstraw ("Cherry Wine Woman"), Charlie Spand and Big Bill Broonzy ("Rock and Rye"), Peetie Wheatstraw and Casey Bill Weldon ("More Good Whiskey Blues"), Josh White and Walter Roland ("Pigmeat and Whiskey Blues"), Merline Johnson and Lucille Bogan, with pianist Walter Roland ("Drinking Blues"), and Rosetta Howard with Willie Dixon's Big Three Trio ("When I Been Drinking"). Other songs and artists include Bumble Bee Slim ("I Keep on Drinking to Drive My Blues Away"), Little Bill Gaither ("Moonshine by the Keg"), Merline Johnson ("Bad Whiskey Blues"), Rev. W. M. Mosley ("Drinking Shine"), Jack Newman ("Blackberry Wine"), Sloppy Henry ("Canned Heat Blues"), The State Street Swingers ("You Drink Too Much"), Casey Bill Weldon ("Give Me Another Shot"), Peetie Wheatstraw ("More Good Whiskey Blues"), and Joshua White ("Pigmeat and Whiskey Blues").

Of course, this selection does not include every blues song on the matter. One of the famous blues artists of that period, Big Bill Broonzy (1893–1958), composed one of his best songs ever, entitled "Good Liquor Gonna Carry Me Down," after National Prohibition. It's the story of a man who could get everything from life, including a charming girl, but chooses instead to get drunk. The boogie-type melody is set to a vivid rhythm. Big Bill Broonzy's lyrics are funny and light:

> But I just keep on-a drinkin', yeah, I keep on drinkin'
> Yeah, I just keep on drinkin', till good liquor carry me down. . . .

The same year, Broonzy recorded in Chicago another song with the same melody and chords, changing the words to avoid any mention of alcohol, to produce "Mountain Blues" (1935).

Between 1950 and 1954, another blues singer named Amos Milburn (1926–1980) released at least six songs about drinking. This talented piano player was an exceptional entertainer. He was a unique artist, a fine blues singer and composer who created a great number of rhythm-and-blues songs related to alcoholism. Among those is the famous "One Bourbon, One Scotch and One Beer," which was redone by many artists, including John Lee Hooker. The lyrics are sung in a natural way:

> Since my baby's been gone everything is lost,
> I'm on this kick and I can't get off.

"Let Me Go Home, Whisky" was another Milburn song about alcohol dependence and addiction. Here, alcohol is represented almost as a real character that has taken the drinker's soul as a prisoner: "Whisky in the morning; whisky in the night." The drinker is not single or on his own, nor does he suffer pain, unhappy love, or any other problem that would supply a reason for his drinking: His baby asks him not to be drunk again but he drinks nonetheless! On the same theme, "Bad, Bad, Whisky" features an alcoholic who lost his happy home because of his drinking habits. On an opposite theme, "Good, Good Whisky" is about the promise of apparent relief that alcohol will give. The drinking saga goes on with "Milk and Water," about the thin possibility of becoming sober ("I won't be on this kick for long"). "Vicious, Vicious Vodka" introduces an alternative to all the Milburn songs dealing unanimously with whisky. Sadly, Amos Milburn had a tragic destiny: He became an alcoholic by the mid-1950s, and that partly ended his by-then successful career.

American music produced many other standards about drinking, such as "Pink Champagne," a trademark for bluesman and jazzman Jimmy Rushing. "Drinking Wine Spoo-Dee-O-Dee" was recorded by countless artists, including bluesman Lightnin' Hopkins in 1961 (on his LP entitled *Blues in My Bottle*), and was revisited by Jerry Lee Lewis in 1973. Other artists in different countries have recorded songs about alcoholism through the years.

Alcoholism in French Songs

Numerous songs have been written about drinking in France, a country famous for its wine production and consumption. Only a few examples by the most famous artists will be mentioned. Charles Aznavour's first recording in France was a light song composed with his friend Pierre Roche entitled "J'ai bu" (1948), which means "I have been drinking." Although he composed very few songs about drinking, Aznavour tells here in a funny mode the story of a husband who understands that his wife has been cheating on him for a long time. He sings: "I'm all barmen's friend, and the enemy of all the street lamps and reverberators." The opposition between sensual feelings and drinking is often stated: "I forgot your caresses; I feel better when I am drunk." That story was used again with more melodramatic lyrics in one of Aznavour's best songs, "Tu te laisses aller" (1960), which was translated into English with the title "You've let yourself go." Aznavour even recorded a new version of that song in duet with Liza Minnelli. The story is interesting although quite misogynous: The husband explains that he drinks just to have the strength to tell his wife everything he really thinks about her. He declares the worst possible things and almost everything a good husband should never say. He concludes: "As you are now, you look like your mother." Almost forty years after his first recording, Aznavour made another song, more tragic, not really about drinking but rather about being alcoholic (even though he never mentions in the song the word "alcoholic"). The title is "Je bois," or "I Drink," and the reason follows: "to give me the illusion that I exist" (1987). The strongest parts of that aggressive song say that this man drinks because his union is childless; when he drinks, he irrigates his own cirrhosis. Luckily, the lyrics are not autobiographical.

Another interesting thing to notice is that French artists who recorded songs about drinking often used jazz rhythms. This was the case not only with Aznavour's first song but also with Boris Vian's famous parody of a washed-up alcoholic entitled "Je bois" (1956), which could be translated "I am used to drink." Boris Vian was a complete artist, a novelist, trumpet player, and poet. In his song, drinking is linked with an unfaithful wife and leads to the idea of suicide. With humor, Vian says he drinks bad wine and recognizes that it is disgusting, but he does it whenever he has free time, to avoid suicide. The song was also recorded by singer/actor Serge Reggiani during the early 1960s.

On his first record, French singer Claude Nougaro composed a comic jazz song where he imitates a drunken man singing (and also on stage and in the promotional film for the song). The title is "Je suis sous" (1962). In French, these words make a double entendre: "I am drunk," but also "I am under," as in, "I am under your balcony." The song says: "Take me back and I will never drink again; I will change": a promise from an alcoholic.

In Belgium during the 1960s, Jacques Brel composed two sad songs about drinking: "Jef" and "Amsterdam," a famous song that was translated by Mort Shuman and recorded by many English artists, and even by the young David Bowie in 1973. "Amsterdam" is about drinking sailors, prostitutes, and unfaithful women. In "Jef," two friends get drunk because the desperate one has lost the woman he loved; his friend promises fun, travels, and projects, even though he doesn't have much money. Drinking is presented as the solution, an escape, not a problem. The singer is ashamed not because his friend is drunk, but because he cries in the street.

Canadian Examples

French-Canadian folk music has created some original examples of drinking songs. The best is "Prendre un verre de bière mon minou" (Take a glass of beer my friend), which celebrates the joys of being drunk all the time. The words are almost comic: *Saoul avant-hier, saoul hier au-soir, saoul à soir encore et puis saoul tout le temps* ("Drunk the day before yesterday, drunk yesterday night, drunk tonight, drunk all the time"). The song appears on many different LPs. In 1930, folk artist Ovila Légaré composed and recorded a satirical song entitled "Chapleau fait son jour de l'an" (Mr. Chapleau celebrates the New Year). This joyous song was about a real character, the premier of the province of Québec, Joseph-Adolphe Chapleau (1840–1898), who had a reputation as a heavy drinker.

Female alcoholism was quite unusual in songs; Québec's rock singer Luce Dufault nonetheless made a hit in 1996 with the song "Soirs de scotch" that was composed for her by two men (words by novelist Christian Mistral and music by singer Dan Brigas). Although the lyrics deal with liquor addiction, the song was often played on airwaves and the videoclip was aired on Canadian television.

> Les soirs de scotch m'enchantent.
> Je sais pas si je te l'ai dit,
> L'ivresse est caressante,
> Suave et chaude comme un nid.
>
> (The nights when I drink scotch enchant me.
> I don't know if I told you,
> Drunkenness is like a caress,
> Bland and warn as a nest.)

Contemporary Examples from England

English pop music has created a few songs about temperance. A few weeks after leaving the Beatles, John Lennon recorded in one of his first solo singles a song about the desperate need of alcohol. Titled "Cold Turkey" (1969), the song was about Lennon's own addiction to hard drugs. It was a raw, heavy song with an obsessive melody, and a lot of screaming, by a new group he called the Plastic Ono Band.

The British pop group the Kinks invented a satiric song entitled "Alcohol" (1971) in which alcohol was compared to a demon ("Here's the story of a sinner / Used to be a winner / Who enjoyed a life of prominence and position"). The LP was entitled *Muswell Hillbillies* (1971).

David Bowie recorded a strange and erratic version of Bertolt Brecht and Kurt Weil's "Moon of Alabama" (from *Threepenny Opera*) in 1980; it has been a standard recorded by many artists and rock groups (even the Doors did a version around 1969). The refrain repeats that because mother is dead, we all must have whiskey.

Recent Examples in the United States

During the 1970s, singer/actor Tom Waits created a character, a lonely but spiritual drinker with a jazzy voice, who became the persona for his songs about drinking, bars, bad women, and all types of alcohol: wine, beer, "gin and vermouth," as he mentions in "Warm Beer and Cold Women." In this song, Waits sings about the lack of communication that makes him get drunk: "I just don't fit in every joint I stumbled into tonight. . . . I'll be drinkin' to forget you." But in the refrain, he adds: "And the drinks are on me tonight." His best album is a live recording made in California, *Nighthawks at the Diner* (1975), which recreates the atmosphere of a smoky bar; the listener can hear the noise of the glasses. It features ten songs with long introductions by Tom Waits in his unique voice that resembles Louis Armstrong's. In concert, Waits often seems to be half-drunk: He talks slowly, tells funny stories and weird jokes, and always smokes on stage, but he is in control, backed by excellent musicians. His later records are less interesting, perhaps because his voice got tired. In the song "Better Off without a Wife," Waits expresses his friendship with his comrades in a universe of sympathetic barmen and snack-bar waitresses. He sings, "The bar stools are on fire" in a song about strange things happening in a bar, "The Piano Has Been Drinking (Not Me)," from the album *Small Change* (1977).

Conclusion

Countless songs mention drinking, alcohol, and habits of drinking. This brief panorama is far from complete, but it proves that important artists representing different styles and countries have created (or adapted) songs on alcohol-related themes. Many of these songs are constructed as an attempt to explain the fundamental question, "Why am I drinking?" The words "I drink because . . ." are followed by some reason, which often happens to be a woman. She might not love him anymore, or the other way round. If women are easily accused, very few characters say they drink because they can't help it, or just because they like it. Curiously, no song of lasting popularity has ever expressed the advantages of temperance or the joys of not being drunk. Even Albert Collins's song "I Ain't Drunk" (composed by Jim Liggins) gives all the possible reasons to let him drink with the hope of never being drunk. Taken from the CD *Live 92–93*, the song says frankly not to bother the singer and let him drink: "You all oughta mind your own business."

Is there a song that could be considered a masterpiece about alcoholism? If so, it is probably Charles Aznavour's "Je Bois" (1987), because it synthesizes all the major themes and explains with many moving details the distress and the justification for an unhappy man who cannot live without drinking. Some drinking songs are sad ones.

Yves Laberge

See also: Drinking Songs (United States); English Songs, Representations of Drinking in (1600–1900)

References

Aznavour, Charles. *Charles Aznavour au Carnegie Hall.* CD, EMI (France), 2003.

———. *Je Bois.* CD, Trema (France), 1989 (1987).

———. *20 Great Songs in English.* CD. Sony Music International/Globetrotter, 1999.

Booze and the Blues. CD. Legacy Roots 'n' Blues Series. Columbia, 67002.22, 1996.

Brel, Jacques. *Master Series Jacques Brel.* CD. Polygram 816458, 1990.

Broonzy, Bill. *The Young Big Bill Broonzy.* LP. Yahoo L-1011, 1928–1935.

Cohn, Lawrence, ed. *Nothing but the Blues: The Music and the Musicians.* New York and Paris: Abbeville Press, 1994.

Collins, Albert. *Live 92/93.* CD. Virgin 40658, 1995.

Dufault, Luce. *Luce Dufault de Luce Dufault.* CD. Arpège AMCD-94, 1996.

Fields, W. C. *Golden Age of Comedy.* CD. Pulse 50XLF, 2000. (From mono recordings made in the 1930s and early 1940s.)

Herzatz, Gérard. *La grande encyclopédie du blues.* Paris: Fayard, 1997.

Hopkins, Lightnin'. *Blues in my Bottle.* CD. Original Blues Classics (Fantasy) series. Prestige/Bluesville OBCCD-506–2 (BV-1045), 1990 [1961].

The Kinks. *Muswell Hillbillies.* CD. Velvet Records, 1988 [1971].

Légaré, Ovila. *Ovila Légaré: 12 chansons.* CD. Montréal, Collection Portraits. Fonovox/Radio-Canada/DEP, PORT2–1052, 1997 [1930].

Milburn, Amos. *The Best of Amos Milburn: Down the Road Apiece.* CD. EMI Records USA, 7243 8 27229 2 2, 1993. (Also available in Europe on Mosaic Records.)

Nougaro, Claude. *Les talents du siècle. Claude Nougaro.* CD. Polygram, 1998.

Plastic Ono Band. *Cold Turkey.* 45-rpm single record. Apple 1001 (UK); Apple 1814 (North America), 1969.

Rushing, Jimmy. *Rushing Lullabies.* CD. Sony/Legacy Records 65118, 1997 [1959].

Tremblay, Gynette, and Simonne Voyer. "Quadrille, Cotillon, Reel, Brandy" in *Cap-aux-Diamants: Revue d'histoire du Québec.* No. 67 (2001): 38–45.

Vian, Boris. *Boris Vian chante Boris Vian.* CD. Polygram 845 912–2, 1991.

Waits, Tom. *Nighthawks at the Diner.* CD. Elektra Entertainment, 2008, 1989 [1975].

———. *Small Change.* CD. Elektra/Asylum, 1990 [1976].

Music Halls

Music halls provided popular working-class British stage entertainment during the nineteenth and early twentieth centuries. They were originally located in public houses as a means of entertaining patrons while they consumed the landlord's liquors. An evening of music hall entertainment consisted of "turns," short performances by individual artistes who performed a cosmopolitan mix of songs, dances, comedy, acrobatic and circus routines, and short skits. Though always centered on London, music hall shows became the first international entertainment industry, and its most popular artistes were the first international celebrities.

Music hall entertainment began in 1854 when Charles Morton, who would become known nostalgically as the "father of the halls," built a large room with a stage adjacent to his South London pub, the Canterbury Arms. Morton named this new structure the Canterbury Music Hall, a grand title that other such venues would imitate. The Canterbury's design reflected its primary purpose; the floor was flat with tables and benches arranged in long rows, so that waiters delivering drinks could pass easily through the crowd. Many other publicans built similar rooms in London and provincial cities as businessmen realized that bigger profits could be made by inducing patrons to drink for longer periods of time. Not surprisingly, many of the most popular songs and skits in the earliest days of the music hall phenomenon celebrated prodigious feats of drinking or the misadventures of inebriates.

Until about 1890, an ostentatiously dressed "chairman" sat at the foot of the stage to oversee music hall entertainment, introducing the turns, emphasizing the risqué aspects of songs, and encouraging the audience to drink. Chairmen embodied many of the patrons' aspirations by consuming huge quantities of alcohol—often bought for them by audience members—and bantering on familiar terms with the biggest stars. The most famous chairmen—generally the flashiest dressers and heaviest drinkers—became celebrities in their own rights; only the most favored patrons, those who spent most freely on drink, would be invited to sit at the chairman's table.

The music hall's close links between the bottle and the stage were evident. Artistes were expected to treat their fans liberally to drinks in the halls and at pubs. This relationship was cemented by George Leybourne, a singer known as the Great Mackney, whose 1868 smash hit "Champagne Charlie" made him the music hall forum's first great star. Leybourne worked under a contract that called for him to always appear in public drinking champagne and dressed as the rakish dandy he played on stage. Champagne Charlie, who was always ready for a spree, appeared in London just as the British government lowered import duties on champagne, making the tipple available to the British middle classes for the first time. A small army of Mackney imitators, known as "swells" for their supposed love of drinking and carousing, sang about a variety of alcoholic products in thinly veiled advertisements.

British immigrants took this style of performance across Europe and around the world, and by the 1870s, there were halls throughout the empire. At the same time, music hall performers forged transatlantic connections with American vaudeville, thereby creating the first global entertainment industry. The most popular artistes were stars in North America and the British Empire.

By the 1890s, a combination of moral indignation and commercialization began changing an industry that had been built on the drinks trade. First, management syndicates acquired chains of halls. These men perceived that they could generate greater revenues by catering to the growing, increasingly affluent middle classes. Though references to alcohol remained popular on stage—for instance, Charlie Chaplin shot to fame just before World War I playing an inebriated aristocrat—this intense capitalization fundamentally changed music hall's relationship with drinking. Syndicates built lavish halls in each of the major British cities and signed artistes to exclusive contracts. Bars—often reputed to be the haunt of prostitutes—and drinking were confined to specific enclosed areas at the back of the auditorium. Chairmen disappeared, and tip-up style theater seats replaced benches and tables as a more formal theatrical style of presentation was adopted. In addition to traditional music hall routines, patrons now saw a salubrious blend of dance, opera, and classical music in what was renamed "variety" entertainment. Even so, the biggest halls sat atop a hierarchy of local venues that often still resembled the raucous drinking rooms of Morton's day. In many towns and poorer neighborhoods, the music hall remained staunchly working class and centered on the consumption of alcohol.

Though the physical links between the halls and the bottle were less apparent at the start of the twentieth century than they had been fifty years before, the toll exacted on performers remained high. In a profession that idealized the life of the drunk, Mackney, the Great Vance, Dan Leno, Jennie Hill, Marie Lloyd, and many other of the music hall's biggest stars succumbed to alcohol-related illnesses.

The music hall did not survive World War I. Cinema replaced live theater as the most popular form of entertainment in Britain, and U.S. troops brought jazz rhythms to Europe. The United States was now the center of world entertainment. Nevertheless, music hall influences remained strong thanks to British performers such as Chaplin and Stan Laurel who immigrated to Hollywood and the turns style of performance that heavily influenced the variety programs that dominated the early days of radio and television.

Andrew Horrall

See also: Pubs and Beerhouses (United Kingdom)

References

Bailey, Peter, ed. *Music Hall: The Business of Pleasure.* Milton Keynes, UK: Open University Press, 1986.

Bratton, J. S., ed. *Music Hall: Performance and Style.* Milton Keynes, UK: Open University Press, 1986.

Kift, Dagmar. *The Victorian Music Hall: Culture, Class, and Conflict.* Cambridge and New York: Cambridge University Press, 1996.

MacInnes, Colin. *Sweet Saturday Night.* London: MacGibbon and Kee, 1967.

N

Nation, Carry Amelia Moore (1846–1911)

Carry Nation (1846–1911) developed a radical technique to protest the sale of liquor that brought her national attention. Born Carrie Amelia Moore in Kentucky, in early adulthood she was active locally in religious and reform circles in the various communities in which she lived, particularly after her marriage to her second husband David Nation, a Disciples of Christ minister. Her violent attacks on drugstores and saloons in Kansas and elsewhere, which she called "hatchetation," established her reputation as a controversial celebrity. In her notoriety she became an emblem for fanatical prohibition.

Voters in Kansas had adopted a prohibition amendment to the state constitution in 1880, but lack of enforcement and an 1890 Supreme Court decision allowing for interstate sale of liquor in original packaging had rendered the measure ineffective.

Carry Nation (1846–1911) (Hulton/Archive)

Throughout the 1890s, political prohibitionists and the Kansas Woman's Christian Temperance Union (WCTU) had struggled to maintain the ban, but by 1900 saloons were operating openly in many towns and druggists sold alcohol with impunity. Local officials often profited from fines levied and bribes paid by saloon keepers. Carry Nation, by that year the middle-aged president of the local WCTU chapter at Kiowa, Kansas, began her crusade by attacking a local drugstore where she correctly suspected the proprietor of selling liquor. In later attacks at Wichita, Enterprise, and Topeka, Kansas, Nation would enter saloons, often accompanied by like-minded "smashers," and wreak havoc, destroying not only stores of liquor but also mirrors, glassware, and pictures that she considered lewd. Frequently arrested and jailed, she lambasted judges and local law enforcement officials for their alliances with illegal saloons. Her outrageous behavior and her abrasive personality alienated her from many Kansas prohibitionists but gained her considerable attention from the national press and a devoted following. The publicity surrounding her attacks also contributed to the failure of efforts to resubmit the prohibition amendment to Kansas voters, an action that might have led to its repeal. The Kansas legislature also passed in 1901 legislation that made it much easier to prosecute suspected saloon keepers and to confiscate their property.

Following her success in Kansas, Nation began traveling, accepting speaking invitations that occasionally climaxed in the smashing of a local saloon. As her fame grew, the circus-like atmosphere of her appearances often overshadowed the political reform. Saloon keepers would invite her to smash their bars in the hope that the ensuing publicity would make up in sales the losses in fixtures and inventory. Nevertheless, Nation continued to lobby for temperance and, later, antitobacco legislation as she toured.

Toward the end of her career, she interspersed engagements at temperance rallies with appearances on vaudeville stages, even performing in productions of *Ten Nights in a Barroom.* She moved to Oklahoma and began publishing *The Hatchet* in 1905 in support of prohibition in that territory. Two years later, she moved to Washington, D.C., where she continued her public appearances and lobbying. In 1909, she retired to Arkansas and built "Hatchet Hall," which served as her home, a school, and a boarding house for otherwise homeless women. She died in 1911.

James D. Ivy

References

Asbury, Herbert. *Carry Nation.* New York: Alfred A. Knopf, 1929.

Bader, Robert Smith. *Prohibition in Kansas: A History.* Lawrence: University Press of Kansas, 1986.

Grace, Fran. *Carry A. Nation: Retelling the Life.* Bloomington: Indiana University Press, 2001.

National British Women's Temperance Association (NBWTA)

The National British Women's Temperance Association (NBWTA) was formed during the annual meeting of the British Women's Temperance Association (BWTA) in 1893 when the majority of the executive committee walked out and established the Women's Total Abstinence Union (WTAU). The NBWTA continued under this name until 1925.

Difficult Times for Temperance Women

The split within the British women's temperance movement arose from a number of issues related to the objectives, organization, and connections of the BWTA. Reformers within BWTA wanted to include suffrage as an objective, as they believed that votes for women would be beneficial to the temperance cause and that the "Do Everything" policy of the WCTU and its leader Frances Willard (1839–1898) would enhance their chances of influencing the government. The members of the rival WTAU disagreed.

BWTA leader Lady Henry Somerset (1851–1921) had campaigned hard for the hearts and minds of the membership, but the ordeal had been stressful for her. After the secession of the executive committee members, she traveled with Willard to Switzerland to recuperate. On her return, Somerset produced a plan of action. Somerset was disadvantaged by the loss of *Wings,* the journal of the BWTA, and of capable and politically aware women who had defected to the new WTAU. Without a journal, the NBWTA had lost its main channel of communication. The search for a new journal proved to be difficult. Eva McLaren proposed that the *Women's Signal* (1894–1899) be adopted as the official organ. This suggestion did not satisfy all members; some complained that the journal did not provide enough space for Christian articles. Somerset financed the *Women's Signal* for a short period, but disputes occurred between Somerset and the editor, Florence Fenwick Miller (1854–1935), over the amount of space that should be given to temperance branches and their work.

Somerset radically restructured the NBWTA administration, introducing departments headed by superintendents. The superintendent of an expanded suffrage department encouraged members to fight for the vote in order to advance the temperance cause. Unfortunately, commitment to the new organizational structure varied; even Manchester with its extensive network of branches found it difficult to be unanimous. Some branches became anxious over the amount of work that the departments generated. Many members believed that some departments were inappropriate for a temperance society.

Errors of Judgment

All these domestic problems paled in significance when Willard, and the NBWTA by association with her, became embroiled in a debate in 1894 over lynching in the southern United States. Willard refused to condemn lynching when challenged by Florence Balgarnie (1857–1928) and by a black American, Ida B. Wells (1862–1931). In the resulting controversy, Somerset dismissed Balgarnie from her position as an NBWTA superintendent.

Later, in 1897, Somerset made a further grave error of judgment when, in her capacity as president of the NBWTA and joint vice president of the World's WCTU British committee, she replied in the *Times* to a letter from Lord George Hamilton on the subject of the Cantonment Acts in India. In her letter Somerset accepted regulated prostitution. This position outraged a number of women identified with the fight against prostitution, such as Josephine Butler, as well as the anti–contagious diseases societies. The controversy damaged Somerset's position as superintendent of Social Purity for the World's WCTU. She offered her resignation as president of the NBWTA but was reinstated. In protest, Fenwick Miller took over complete control of the *Women's Signal* and banned Somerset from its premises.

By 1896, Somerset's views were at a variance with those of the temperance movement as a whole. She moved toward the high license (costly fees intended to reduce the number of drinksellers) and the Gothenburg scheme of disinterested management and away from prohibition. Her testimony to the Royal Commission on Liquor Licensing Laws, which asked her more than 640 questions, was extensive. Her response to Sir Algernon West's question whether "she wished to make all the population teetotallers" caused consternation in the NBWTA. In answering West's question, she said that she had no wish to interfere with those who chose to take alcohol in moderate quantities.

Politics of Temperance and the Liberal Party

Somerset's successor, Rosalind Howard, Countess of Carlisle (1845–1921), was a staunch supporter of prohibition, an activist in the Women's Liberal Federation, and a supporter of women's rights. She felt that Somerset had become a "loose cannon" as far as temperance politics were concerned. It is true that Somerset was not in tune with the NBWTA membership. The NBWTA felt it had been pushed into accepting her policy of high license but would not go a step further on her path. In the judgment of most members, the Gothenburg experiment in Scandinavia had failed; restrictions on the sale of spirits there had simply increased the sale of beer.

When Carlisle challenged Somerset in 1903, she was elected as the new president. She immediately began to dismantle Somerset's department structure and to stamp her own policies on the membership, managing the association with the help of a small group of women whose politics she considered sound. Her ties with the Liberal Party allowed the NBWTA to gain political clout. This cooperation between the association and the party was a natural alliance. Most leading temperance women were middle-class Liberal partisans, although their support for women's suffrage divided them from the Liberal Party leadership.

Under Carlisle's presidency, the NBWTA developed lobbying links to Parliament. With a membership of more than 100,000, she was able to command respect in the temperance movement. Politically active NBWTA members had participated in local government; several of them had been elected to school boards, which gave them a platform from which they could support education for temperance.

The visit of the American reformer Mary Hunt to the United Kingdom in 1903 encouraged attempts to introduce what was called Scientific Temperance Instruction (STI). Impressed by Hunt's rhetoric, Carlisle became convinced that STI should be part of NBWTA strategy. Carlisle was not alone in supporting STI; more than 14,000 medical practitioners petitioned for the introduction of such a curriculum. Although there were limited successes in introducing STI, Carlisle's plan for a fully developed course of lessons on alcohol, hygiene, and related topics never became a reality.

Small Victories

The NBWTA and other temperance reformers rejoiced in a few steps toward a teetotal country. Bessbrook, Armagh, and Toxteth near Liverpool kept their people free from the temptation of public houses. In fact, the people hadn't voted for this protection. Bessbrook was a new town whose founders banned pawnshops and public houses.

By the beginning of the twentieth century, the beleaguered Liberal Party had shifted its focus from temperance to other issues. Although Carlisle still sought temperance reform through the Liberal Party, some members of the rival Women's Total Abstinence Union were impressed by temperance socialists and their teetotalism.

The outbreak of World War I brought dramatic changes in British society. In the early days, temperance women provided nonalcoholic refreshment for the troops. Despite such patriotic efforts, the temperance cause continued to decline, and by the end of the war, temperance was no longer an issue in national politics. Shortly after the war ended, Parliament granted women over the age of thirty the vote for elections to the House of Commons and the right to stand for election. Taking advantage of their right to vote, temperance women fought for the retention of wartime restrictions on the sale of alcoholic drink. The first woman to take a seat in Parliament was Nancy Witcher Astor (1879–1964), a Conservative, in 1919. She was a teetotaler who had been born in the United States. Margaret Wintringham (1879–1955), the second woman to be elected to Parliament, in 1921, was a Liberal who had been the vice president of the Lincolnshire Total Abstinence Association. Astor and Wintringham worked together on a bill to prohibit the sale of alcohol to those under the age of eighteen.

When Carlisle died in 1921, she was succeeded as NBWTA president by one of her daughters, Lady Cecilia Roberts (1869–1947), who in turn was succeeded by the veteran temperance reformer Agnes Slack (1857–1946). In 1926, the NBWTA and the Women's Total Abstinence Union reunited, forming the National British Women's Total Abstinence Union. In 1993, it was renamed the White Ribbon Association.

Margaret Barrow

See also: Astor, Lady Nancy Wicher Langhorne; British Women's Temperance Association (BWTA); Carlisle, Countess of (Rosalind Howard); Conservative Party (United Kingdom); Heath, H. Cecil; Hunt, Mary Hannah Hanchett; Labour Party (United Kingdom); Liberal Party (United Kingdom); Slack, Agnes Elizabeth; Somerset, Isabella Caroline (Lady Henry Somerset); Willard, Frances Elizabeth Caroline; Woman's Christian Temperance Union (WCTU) (United States); Women's Total Abstinence Union (WTAU); World's Woman's Christian Temperance Union (WWCTU)

References

Barrow, Margaret. "Temperate Feminist: The British Women's Temperance Association" (Ph.D. dissertation, University of Manchester, 1999).

A Century of Service, 1876–1976. Birmingham: National British Women's Total Abstinence Union, 1976.

Henley, Lady Dorothy. *Rosalind Carlisle, Countess of Carlisle.* London: Hogarth, 1958.

Roberts, Charles. *The Radical Countess: The History of the Life of Rosalind, Countess of Carlisle.* Carlisle: Steele Brothers, 1962.

Snowden, Philip. *Socialism and the Question of Drinking.* London: Independent Labour Party, 1908.

National Council on Alcoholism and Drug Dependence (NCADD)

For more than fifty years, the National Council on Alcoholism and Drug Dependence (NCADD) has served as the leading voluntary public policy advocacy group in the United States on alcoholism-related issues. In 1942, Dwight Anderson, representing the Research Council on Problems of Alcohol, called for a "new public health movement" to change the way America perceived alcoholism and the alcoholic. That movement began two years later, when Marty Mann founded the National Committee for Education on Alcoholism (NCEA). Mann drew her inspiration for NCEA from the success of other public health organizations that had been formed to support education, research, and treatment for tuberculosis, cancer, and heart disease.

Mann was aided in her initial efforts to organize NCEA by E. M. Jellinek of Yale University, Ruth Bangs of the *New York Herald Tribune,* Austin McCormick of the Osborne Association, Dwight Anderson of the New York State Medical Society, and Dr. Ruth Fox, a psychiatrist specializing in alcoholism. Jellinek was so taken by Mann's vision that he offered the Yale Center of Alcohol Studies' sponsorship of NCEA. The new organization began operations in April 1944, with an annual budget of $13,000.

Mann established local NCEA branches around the country in order to pursue NCEA's five-point program: (1) educating local communities about alcoholism; (2) establishing local alcohol information and referral centers; (3) enlisting the aid of local hospitals in detoxifying alcoholics; (4) establishing clinics for the diagnosis and treatment of alcoholism; and (5) establishing "rest centers" for the long-term care of alcoholics. These local chapters were staffed by volunteers from among recovered alcoholics and their family members, as well as physicians, clergy, and others concerned about the problem of alcoholism.

The primary means of changing public attitudes and public policies was through instilling in the American culture a set of "kinetic" ideas formulated by Anderson and refined by Mann. The five ideas that were at the heart of NCEA's educational campaigns were: (1) Alcoholism is a disease; (2) The alcoholic, therefore, is a sick person; (3) The alcoholic can be helped; (4) The alcoholic is worth helping; (5) Alcoholism is our No. 4 public health problem, and our public responsibility.

NCEA, along with the Research Council on Problems of Alcohol, Alcoholics Anonymous, and the Yale Center of Alcohol Studies, forged a "modern alcoholism movement" that affirmed hope for recovery from alcoholism, called for a new scientific approach to the study and resolution of alcohol problems, and laid the foundation for a national network of community-based alcoholism prevention and treatment programs.

Coming of Age (1950–1980)

In 1950, NCEA separated from Yale to become an independent organization. This separation was facilitated by the financial support of philanthropist R. Brinkley Smithers and the Christopher D. Smithers Foundation. Signaling this new independent status was the board's decision to change the name of the organization to the National Committee on Alcoholism (1950) and later to the National Council on Alcoholism (NCA) (1956). Through Mann's visits to communities across the country, the number of local NCEA/NCA affiliates grew from fourteen in 1946 to fifty-three in 1956.

NCA focused on educating the public through its publishing activities and its influence upon the press, radio, television, and motion pictures throughout the 1950s and 1960s. Major initiatives included lobbying to change policy positions on alcoholism by major medical and public health organizations, hosting conferences with themes on alcoholism for helping professionals, and lobbying for federal legislation for funding to support local alcoholism education and treatment efforts. As the number of local NCA affiliates reached eighty, NCA successfully pushed the passage of landmark federal legislation (such as the 1970 Hughes Act) that provided federal funding for local alcoholism treatment programs.

During the 1970s, NCA helped expand alcoholism treatment and occupational alcoholism programs and provided organizational sponsorship of the American Society of Addiction Medicine. NCA also intensified its educational campaigns to destigmatize alcoholism through its media blitzes, seeding television shows such as *Maude* and *All in the Family* with information on alcoholism, and hosted professional education and training events. The boldest of such activities was its 1976 "Operation Understanding." In this highly publicized event, fifty-two prominent Americans proclaimed their recovery from alcoholism. Another significant change within NCA in the 1970s was the decision not to continue to allow alcohol industry representatives on the NCA board.

Retrenchment and Renewal (1980–2002)

Marty Mann served as NCA's director for twenty-four years and continued to influence NCA's advocacy vision until her death on 22 July 1980. Her speaking and writing exerted a profound influence on American attitudes toward alcoholism, and her political savvy and spellbinding oratory coaxed many state and federal legislators into formulating public health responses to the problem of alcoholism. Following Mann's death, NCA experienced a period of financial struggle. The Smithers Foundation provided crucial financial support to NCADD during this period.

NCA changed its name to the National Council on Alcoholism and Drug Dependence (NCADD) in 1990, bowing to the growing integration of the alcoholism and drug addiction fields. The debate over the name change signaled a larger uncertainty regarding NCA's future direction as an organization. Many local NCADD affiliates had taken on the responsibility of providing local treatment services—services that progressively dwarfed their public education and public policy activities. As a result of these changes, relationships between NCADD and its affiliates weakened. By 2000, the number of local NCADD affiliates, which had peaked at more than 230 in the early 1980s, dropped to below ninety.

Today, NCADD is undergoing a renewal process. The board has recommitted NCADD to its historical public education and public policy mission and to rebuilding the grassroots affiliate relationships that have contributed to the organization's greatest successes. NCADD continues its media campaigns, taking leadership roles in Alcohol Awareness Month and National Recovery Month activities. It reaches Americans through cable television programs, its newsletters and publications, and its website, which reports more than 10,000 visits per month. NCADD also continues its public policy advocacy activities by supporting parity in insurance for addiction treatment and by serving as a watchdog over the marketing practices of the alcohol industry.

Perhaps the best measure of NCADD's effect on American perceptions about alcoholism is the successful inculcation of its core ideas. In 1943, the year before NCADD's founding, only 6 percent of Americans surveyed believed alcoholism was a disease. Fifty years later, 87 percent of Americans accepted this view.

William L. White

See also: Alcoholics Anonymous (AA); Alcoholism; Christopher D. Smithers Foundation; Jellinek, Elvin Morton; Mann, Marty

References

"For 50 Years, The Voice of Americans Fighting Alcoholism." National Council on Alcoholism and Drug Dependence (NCADD) website. http://www.ncadd.org.

Johnson, Bruce Holley. "The Alcoholism Movement in America: A Study in Cultural Innovation" (Ph.D. dissertation, University of Illinois at Urbana, 1973).

Mann, Marty. "Formation of a National Committee for Education on Alcoholism" in *Quarterly Journal of Studies on Alcohol.* 5, no. 2 (1944): 354.

White, William L. *Slaying the Dragon: The History of Addiction Treatment and Recovery in America.* Bloomington, IL: Chestnut Health Systems, 1998.

National Institute on Alcohol Abuse and Alcoholism (NIAAA)

The U.S. National Institute on Alcohol Abuse and Alcoholism (NIAAA) is the primary source of funding for research on all aspects of alcohol abuse and alcoholism. It is one of nineteen institutes making up the National Institutes of Health based in Bethesda, Maryland. Its most recent fiscal-year budget was $384 million, of which $317 million was dedicated to the support of research grants and contracts and $9.5 million was set aside for the support of research training grants. The remainder of the budget covers administrative and communication activities.

Structure and Functions

In 2003, NIAAA supported a wide range of research via approximately 1,000 grants. Although most of the studies are centered around a biomedical definition of alcohol dependence, it would be inappropriate to characterize the institute's research portfolio as heavily skewed toward biomedical research. Many studies involve multidisciplinary models and researchers from multiple backgrounds.

In addition to administrative divisions and offices, NIAAA's extramural research program is divided into three divisions and eight branches. The Division of Clinical and Prevention Research includes branches on health services, prevention, and treatment. The Division of Basic Research has branches on biomedical research, genetics and proteomics, and neuroscience and behavior. The Division of Biometry and Epidemiology has biometry and epidemiology branches.

NIAAA also supports a large intramural research program through its Division of Intramural Clinical and Biological Research. This work is almost exclusively oriented toward biomedical research and includes seven laboratories devoted respectively to molecular and cellular neurobiology, molecular physiology, neurogenetics, membrane biochemistry and biophysics, clinical studies, integrative neuroscience, and physiologic studies.

The institute supports extramural research through a variety of mechanisms. Most common are "R01" grants, which can be awarded from one to five years for any amount of funding, and may be renewed indefinitely. Applications for such funding follow a common format. Review of applications occurs within specialized committees composed of appointed research experts. Review procedures are precise. Numerous other specialized funding mechanisms exist, including a variety of opportunities for students and scholars to obtain funding to support both beginning and advanced training in the techniques of alcohol research.

NIAAA supports fifteen national alcohol-research centers that are funded through different procedures, each of which is focused on a narrow set of research questions, ranging from the molecular and genetic levels to prevention and social epidemiology. In addition to supporting an administrative core, each center grant typically supports three to five integrated projects that each resemble an R01 grant project. The institute also supports thirty-one research training centers, with a similar diversity of topical foci, many of which are associated with the alcohol research centers.

Early History

NIAAA was officially established on 31 December 1970 with the signing of congressional legislation by President Richard M. Nixon. The legislation, known as the Hughes Act, was the culmination of lobbying by alcoholism interests and hearings conducted by Senator Harold Hughes of Iowa, the first admitted recovering alcoholic to serve in Congress. President Nixon, who was committed to a policy of reducing the size of the federal government, was initially opposed to the bill, but he was convinced to sign by several major political supporters who were also constituents of the alcoholism community. The NIAAA was the first U.S. federal agency exclusively committed to alcohol issues since the abolition of the Prohibition Bureau with repeal of National Prohibition in 1933.

Establishment of the NIAAA culminated an organized campaign that began with the founding of the National Committee for Education on Alcoholism in 1943, soon renamed the National Council on Alcoholism (NCA). This organization had strong ties both to Alcoholics Anonymous (AA) and to the Yale Center of Alcohol Studies, representing, respectively, a clear-cut approach to treatment and scientific research rather than reliance on moral guidance. NCA's goal was to build social, cultural, and political orientations toward alcoholism as a treatable illness deserving a legitimate share of medical attention on par with other illnesses. Further, NCA promoted the concept that alcoholics should receive the same opportunities as those impacted by other illnesses and argued that there was no foundation for the stigma typically ascribed to recovering alcoholics.

Important commitment came with passage of a position statement by the American Medical Association in 1955 asserting that alcoholism was an illness that could be successfully treated. Efforts to decriminalize public intoxication and other alcohol-linked minor crimes pressed forward during the 1960s, which finally culminated in the passage of the Uniform Alcoholism and Intoxication Act in 1976 (with considerable behind-the-scenes motivation emanating from NIAAA). Several major leaders of business, industry, and philanthropy, especially Thomas Pike and R. Brinkley Smithers, pressed for the establishment of a federal agency. These various interests were pulled together by the presence of Senator Hughes, who was able to generate the enabling legislation, conduct nationwide hearings, and then work among his congressional colleagues for its passage.

The new NIAAA sought to define its goals beyond the stereotyped vision of the alcoholic as a socially isolated public inebriate. One of its early promotions centered on "Project 95," which drew attention to the 95 percent of the alcohol-troubled population that was *not* on Skid Row but was dispersed throughout the "respectable" segments of the working, middle, and upper social classes.

Thus, NIAAA was launched with a broad mandate to address prevention, treatment, and research issues related to both alcohol abuse and alcoholism. From 1971 to 1981, it wielded almost total influence on the growth and direction of alcohol-related programming and supported most of the alcohol-related research being conducted. Its mandated responsibilities during this period included administration of formula grant funds to the states (which constitute a very large proportion of the total federal dollars spent to support treatment and prevention activities) as well as the development of innovative approaches to treatment and prevention. Most of the latter was accomplished through the funding and monitoring of competitive demonstration project grants in organizations throughout the nation. These projects were oriented toward finding new mechanisms for prevention and treatment among such diverse special populations as American Indians, drinking drivers, women, employed persons, the poor and homeless, and youth.

During this period, NIAAA was a very large and complex organization and was often visible in the mass media. Its administrators came under criticism for providing grant funds to build constituency groups that would in turn lobby at the local, state, and federal levels for increased allocations for alcohol-related programming. It was also charged that constituency representatives came to have undue influence in the allocation of grant funds. Later, the institute came under criticism for advocating "responsible drinking," which was viewed as an indirect message to promote alcohol consumption. Finally, a firestorm of criticism erupted when an institute-sponsored research report indicated that a small number of persons treated for alcoholism had successfully resumed normal drinking following treatment.

Recent History

In 1981, NIAAA was dramatically transformed when actions by the Reagan administration reduced the agency's mandate to dealing only with research and research training. The authorities to administer formula grant funds to the states and to fund treatment programs and demonstration projects were removed and ultimately transferred to a new organization, the Substance Abuse and Mental Health Services Administration. Authority within this new agency is highly decentralized, and drug and alcohol issues are usually considered together as policies and programs are developed. The centralized authority and magnetism for constituency development represented by the NIAAA of the 1970s was not to return. With this change, the constituency that could battle for attention and resources for alcohol issues has dwindled dramatically. Although a recent director of NIAAA held out hope for rallying these constituencies around alcohol-research issues, creating a significantly potent link between the public and science seems unlikely.

NIAAA's organizational home has shifted several times since its founding. Prior to its founding, the National Institute of Mental Health (NIMH) had housed a small office that supported a number of alcohol-related research studies. For a brief period after its founding, NIAAA was housed within NIMH. After it became an independent institute, it was folded into the larger Alcohol, Drug Abuse and Mental Health Administration (ADAMHA), which included the three institutes represented in its title. The National Institute on Drug Abuse (NIDA) had been founded in 1973 and has been structured in a manner very similar to NIAAA. All three institutes became exclusively focused on research with the changes of 1981, and in 1993 all three were added to the National Institutes of Health with the dismantling of ADAMHA.

The move to NIH has created pressure for NIAAA and its sister institutes to be almost completely focused on science. This shift was to be demonstrated by commitment to a biomedical orientation and to research studies that are isomorphic with other forms of medical research supported throughout NIH. Thus the NIAAA supported a massive effort called Project MATCH, an eight-year multisite study of patients' responses to different treatment approaches. This nationwide clinical trial, one of the largest ever centered on behavioral therapy, involved 1,726 patients at over thirty institutions and treatment agencies. The trial employed three individually delivered treatments with widely differing philosophies and practices: (1) twelve-step facilitation therapy to help patients affiliate with Alcoholics Anonymous; (2) cognitive-behavioral therapy to induce coping skills to prevent relapse; and (3) motivational enhancement therapy to increase commitment to behavioral change. Assignments to treatment regimens were based on research that predicted who would respond best to each treatment. Carefully designed and repeated outcome evaluations produced results that were surprising and disappointing to many. They showed no significant differences in outcomes across the three regimens and suggested that triaging clients to individual therapy is not necessary for treatment success.

Other examples of biomedical research include a large multisite collaborative study of the genetics of alcoholism, which has continued over many years. Recently, the institute has supported a new project called COMBINE. This is a multicenter, randomized clinical trial to evaluate combinations of three interventions for treating alcohol dependence. Two of the interventions consist of pharmacological treatment with naltrexone or acamprosate. The third intervention is a comprehensive behavioral therapy including such components as motivational enhancement therapy, cognitive behavioral therapy, and referral to self-help groups, including AA. All three interventions include support for compliance to medications and reduction in drinking.

Attention to behavioral research has not been minimal, however. Research to develop effective interventions to deal with drinking among children and teenagers has been widely supported. During the 1990s, NIAAA supported many research studies, publications, and meetings centered on abusive drinking by college students. Following a report developed over three years by the NIAAA-supported Task Force on College Drinking, the institute committed an additional $8 million over the most recent two fiscal years to the issue of college drinking. The institute also has supported extensive studies of interventions for reducing accidents, injuries, and death associated with drinking and driving.

Over its history, NIAAA has had seven directors, all but one of whom served for relatively brief periods, and all but one of whom have been male physicians. Morris Chafetz, M.D., was the founding director, serving from 1971–1975, followed by Ernest Noble, M.D., Ph.D., 1976–1978; John DeLuca, M.A., 1979–1981; William E. Mayer, M.D., 1982–1983; Robert G. Niven, M.D., 1983–1985; Enoch Gordis, M.D., 1986–2001; and Ting-Kai Li, M.D., 2002–present.

The NIAAA has a comprehensive and easily navigable website. In addition to extensive organizational information about NIAAA, the website is the home of "ETOH," an international bibliographic database covering research and practice associated with alcohol abuse and alcoholism and including both complete literature citations and abstracts. There is substantial current information on the website, including findings from new research and numerous on-line publications for researchers, practitioners, and the general public that can be readily downloaded. The website also features extensive information on available research funding and thorough directions about how to apply for funds.

The NIAAA has a congressional mandate to publish a report that summarizes the current state of the nation's problems with alcohol abuse and alcoholism and the latest findings from research that has been supported by the institute. This is the *Alcohol and Health Report,* a very extensive volume, which most recently appeared in its tenth edition in June 2000. The institute also publishes a journal, *Alcohol Research and Health,* which is mainly composed of research articles summarizing the current state of research findings and pressing research issues in specialty areas. This journal, published since 1973, is written at a level to be useful to both researchers and educated laypersons.

The institute also publishes brief summaries of research findings and practical applications in its Alcohol Alert series. Finally, the institute occasionally publishes research monographs and summaries of research conferences. All of these materials are available in full-text format for downloading at the website.

Paul M. Roman

See also: Alcoholics Anonymous (AA); Alcoholism; Chafetz, Morris Edward; Hughes, Harold Everett; National Council on Alcoholism and Drug Dependence (NCADD); Prohibition Bureau; Rand Report; Yale Center of Alcohol Studies

References

"Conversation with Enoch Gordis," Journal Interview 58 in *Addiction.* 97, no. 2 (2002): 141–151.

Hertzman, Marc. "The NIAAA: Past, Problems, and Prospects" in *Contemporary Drug Problems.* 5, no. 1 (1976): 45–56.

Hewitt, Brenda. "The Creation of the National Institute on Alcohol Abuse and Alcoholism" in *Alcohol, Health and Research World.* 19, no. 1 (1995): 12–16.

National Institute on Alcohol Abuse and Alcoholism (NIAAA) website. http://www.niaaa.nih.gov.

Roman, Paul, and Terry Blum. "Notes on the New Epidemiology of Alcoholism in the USA" in *Journal of Drug Issues.* 17, no. 4 (1987): 321–332.

Wiener, Carolyn. *The Politics of Alcoholism: Building an Arena around a Social Problem.* New Brunswick, NJ: Transaction Books, 1981.

National Prohibition (United States)

National Prohibition, which by means of the Eighteenth Amendment to the Constitution banned the manufacture, sale, transportation, and importation of beverage alcohol in the United States as of 16 January 1920, marked the high point of a generation-long effort to dismantle the U.S. liquor industry and the culture of drinking it supported. Although Prohibition reduced alcohol consumption and helped root out the "old-time" saloon that temperance advocates had long opposed, popular enthusiasm for National Prohibition declined for several reasons, including widespread defiance of the liquor ban (especially in cities); the surprisingly stringent definition of illegal intoxicating beverages contained in the Volstead Act, the legislation that implemented the Eighteenth Amendment; the underfunded, inept, and venal enforcement of the Volstead Act by federal agents; and lax support for Prohibition from Congress and state governments. After thirteen years of National Prohibition, ratification of the Twenty-First Amendment on 5 December 1933 repealed the Eighteenth Amendment and ended the "dry reform."

The Coming of National Prohibition

The drive for a Prohibition amendment to the Constitution began in earnest in 1913 following a string of state prohibition laws and the passage that year of the Webb-Kenyon Act, a federal statute that barred interstate shipments of liquor that violated the laws of dry states. Dry strategists realized that state prohibition was an impossibility in the area from Massachusetts to Maryland and in other "wet" strongholds across the nation. (Twenty-one states remained wet when Congress passed the Eighteenth Amendment in 1917.) Yet drys were encouraged by the possibilities of federal-state cooperation exemplified by Webb-Kenyon. National constitutional Prohibition was fashioned to outflank the wet states with national authority while safeguarding the states' rights sensibilities of many drys by stipulating that the states and the national government would concurrently enforce Prohibition through legislation of their own choosing.

Relentless lobbying by the Anti-Saloon League of America (ASLA) and the Woman's Christian Temperance Union (WCTU) combined with the crisis atmosphere of U.S. entry into World War I to accelerate congressional passage of the Prohibition amendment. Wartime grain restrictions led Congress to shut down the distilling industry, and President Woodrow Wilson cut grain allotments to breweries and ordered the alcohol content of beer reduced to 2.75 percent. That same month, on 18 December 1917 Congress passed the Prohibition amendment and sent it to state legislatures for ratification. The Eighteenth Amendment was formally ratified on 16 January 1919 and slated to go into effect one year later. Eventually, forty-six of the forty-eight states ratified the amendment. Congress refused to wait for ratification, however, and enacted the Wartime Prohibition Act on 21 November 1918 (after World War I had ended), which barred the sale of intoxicants beginning in July 1919. National Prohibition was thus in place before it officially commenced.

Critical to the enforcement of National Prohibition and its public reception was the Volstead Act, which more specifically defined the features of the National Prohibition law. Congress passed the Volstead Act, officially known as the National Prohibition Act, over Wilson's veto on 27 October 1919, after the Eighteenth Amendment had been formally ratified. The Volstead Act extended and reinforced the Wartime Prohibition Act (which prompted Wilson's veto), laid out the procedures for regulating the production of industrial alcohol, and, most significantly, set the rules for National Prohibition. The Eighteenth Amendment had banned the production, sale, and movement of intoxicating liquors but had not precisely defined that term. Many urban and working-class drinkers hoped low alcohol beer, such as that produced during World War I, would be legal under Prohibition. The Volstead Act, using an Internal Revenue Service (IRS) standard for taxing alcoholic beverages, defined intoxicating drinks as those with an alcohol content of 0.5 percent by volume or more. This standard eliminated virtually all alcoholic drinks from production and sale for beverage purposes.

Enforcement of the law was assigned to the commissioner of internal revenue in the Treasury Department. Agents were empowered to declare as public nuisances buildings, vehicles, and other property that were used to manufacture, move, sell, or store illegal alcohol and to seize, sell, or close them for up to one year. Violators of the law could be fined $1,000 and imprisoned for up to six months; later alterations stipulated that repeat offenders could receive maximum fines of $10,000 and five years' incarceration. Alcohol for medical and sacramental uses required permits. There were loopholes for ordinary citizens. Private residences could not be searched unless there was evidence of unlawful liquor sales within the dwelling. Home manufacture of "nonintoxicating" ciders and fruit juices was permitted, which led to loud complaints that rural folk were unmolested in the enjoyment of hard ciders and fruit wines while urbanites could not produce home brew. Finally, legal stocks of liquor acquired before Prohibition could be served to family members and "bona fide guests" without violating the law.

Enforcement and Resistance

Rigid standards, state and local resistance, and poor administration of federal enforcement mechanisms plagued National Prohibition from the outset. Hoping to achieve a more modest local definition of prohibition, legislatures in New York, New Jersey, and Massachusetts passed laws in 1920 authorizing the sale of light wines and low alcohol beer; the Supreme Court crushed those challenges to the Volstead Act standard that same year. Although the doctrine of concurrent enforcement allowed states to adopt more stringent standards than federal law mandated, it did not permit looser regulations. As more than a dozen states had refused to pass state prohibition laws at any time, reliance on concurrent enforcement proved to be a significant weakness in the Prohibition enforcement network. Every state except Maryland had passed a state-level companion to the Volstead Act creating local enforcement mechanisms. Few, however, invested sufficient funds to sustain effective enforcement; twenty-eight states set aside no money at all for Prohibition enforcement in 1927, for example. Beginning with New York in 1923, a handful of states repealed their Prohibition enforcement acts. Local police were sporadic in their efforts to crack down on the illegal traffic in alcohol, and some urban juries proved reluctant to convict their fellow citizens for behavior they deemed insufficiently criminal to merit prosecution. In large cities, webs of underground drinking clubs, dubbed speakeasies, operated more or less openly, supplied by criminal syndicates who battled one another over control of the illegal liquor business. Complaints of a Prohibition crime wave, the extent of which historians debate, became more prominent in the late 1920s, especially after the St. Valentine's Day massacre in Chicago between rival bootlegging gangs, and politicians such as New York Governor Al Smith renewed calls for the modification of the Volstead Act.

It was clear that federal officials would be responsible for the vast bulk of Prohibition enforcement. But here, too, there were debilitating weaknesses. From its formation in October 1919 until March 1927, the Prohibition Unit, the federal agency empowered to enforce National Prohibition under the IRS commissioner, underwent three complete reorganizations and numerous personnel changes, producing "seven years of chaos," according to an official report (National Commission on Law Observance 1931: 197). Not until 1927 did agents in the renamed Prohibition Bureau come under civil service; its absence before then allowed for political interference in the agency and poor-quality officers. Skimpy congressional appropriations for Prohibition enforcement—fluctuating between $6 million and $10 million annually from 1921 through 1926—meant that agents were few in number (slightly more than 3,000 at any time) and ill-paid, the latter an important deficiency given the possibilities of corruption attached to the work. Between 1920 and 1929, one out of every twelve Prohibition enforcement agents, 8.5 percent of the force, was fired for illegal or unauthorized acts. Nevertheless, Prohibition agents made thousands of arrests, clogging a court system that did not expand to meet the new demand. Overwhelmed judges avoided backlogs by assigning small fines to those pleading guilty to violating the liquor laws. Although aided by the Coast Guard and Customs Service in countering smuggling, the Prohibition Bureau was incapable of carrying out the massive task of Prohibition enforcement.

Prohibition advocates and Congress were complicit in these shortcomings. Wayne Wheeler, general counsel of the ASLA, had pushed for IRS stewardship of Prohibition and exemption from civil service for agents, thinking that the ASLA would exert greater political pressure on enforcement under such an arrangement. Dry leaders also were reluctant to criticize federal Prohibition enforcement or to demand higher appropriations to support more efficient measures, fearing that acknowledging the need for more resources would stimulate greater opposition to National Prohibition. For its part, Congress was content to leave the volatile Prohibition issue alone, using the Prohibition Unit as a source of patronage and making minimal appropriations for additional enforcement.

Achievement and Repeal

For all its obvious flaws, National Prohibition did reshape the American culture of drinking. Americans drank less than they had before Prohibition. Large sections of the country were effectively dry during the 1920s, and even in cities alcoholic drinks during Prohibition were more expensive (in some cases up 500 percent in price). Consumption estimates indicate that Americans drank less alcohol after Prohibition than had been the case before its implementation. Between 1911 and 1915, Americans fifteen years old and above consumed an estimated 2.56 U.S. gallons (9.7 liters) of absolute alcohol (the alcohol contained in drinks) per capita; in 1934, after Prohibition had ended, the comparable figure was 0.97 U.S. gallon (3.7 liters). Moreover, Americans drank in a different context during and after Prohibition. Drinking in pre-Prohibition America was most often a male, largely working-class activity that took place in saloons. Prohibition drinking in speakeasies was more likely

Two men of the U.S. Internal Revenue Bureau carry packages of confiscated liquor, 1920s. (Library of Congress)

to be in mixed gender groups of middle-class people in a restaurant or club atmosphere.

In its final years, National Prohibition enforcement was better organized. Civil service improved the quality of Prohibition Bureau agents, and, in 1930, the bureau was finally moved from the Treasury Department to the Department of Justice. In 1929, President Herbert Hoover announced his intention to improve Prohibition enforcement and commissioned a thorough study of National Prohibition. The 1931 report of the National Commission on Law Observance and Enforcement, popularly known as the Wickersham Commission, made plain the disastrous history of Prohibition enforcement to that point, badly damaging confidence in the enterprise. By that time, the immensity of the Great Depression had demolished Hoover's credibility and made the continuation of a policy that closed down an entire industry and punished one of the small pleasures of life seem excessive. The Association Against the Prohibition Amendment, an effective business group dedicated to the repeal of the Eighteenth Amendment, made headway in the late 1920s. It was joined in 1929 by the Women's Organization for National Prohibition Reform, a group that rebutted the WCTU's claim that American women stood unanimously behind Prohibition. In 1928, Democratic presidential nominee Al Smith had announced his support for modification of the Prohibition laws; in 1932, the Democratic Party and its candidate, Franklin D. Roosevelt, stood for repeal. With a speed that astonished those who considered the Eighteenth Amendment unassailable, Hoover lost to Roosevelt, Congress passed a repeal amendment in early 1933, and Roosevelt, after taking office, revised the Volstead Act to allow 3.2 percent alcohol beer. By December, the Twenty-First Amendment was ratified and National Prohibition was cast aside.

Thomas R. Pegram

See also: Alcohol, Consumption of, per Capita (United States); Anti-Saloon League of America (ASLA); Association Against the Prohibition Amendment (AAPA); Democratic Party (United States); Federal Regulation of Alcohol before 1920 (United States); Prohibition Bureau; Republican Party (United States); Smith, Albert Emanuel; State Prohibition (United States); Wheeler, Wayne Bidwell; Woman's Christian Temperance Union (WCTU) (United States); Women's Organization for National Prohibition Reform (WONPR)

References

Hamm, Richard H. *Shaping the 18th Amendment: Temperance Reform, Legal Culture, and the Polity, 1880–1920.* Chapel Hill: University of North Carolina Press, 1995.

Kyvig, David E. *Repealing National Prohibition.* 2d ed. Kent, OH: Kent State University Press, 2000.

Merz, Charles. *The Dry Decade.* New York: Doubleday, Doran, 1931.

National Commission on Law Observance and Enforcement. *Enforcement of the Prohibition Laws,* vol. 2. Washington, DC: U.S. Government Printing Office, 1931.

Pegram, Thomas R. *Battling Demon Rum: The Struggle for a Dry America, 1800–1933.* Chicago: Ivan R. Dee, 1998.

Sinclair, Andrew. *Prohibition: The Era of Excess.* New York: Harper & Row, 1962.

National Temperance Federation (NTF)

The National Temperance Federation (NTF) was founded in Britain in 1883 with the avowed purpose of bringing about unity among the numerous temperance pressure groups. The idea was to mobilize support behind a common denominator of legislative proposals upon which all societies could agree. In practice this objective proved impossible to achieve, and the NTF was throughout its existence a narrower body dominated by the more radical prohibitionists. The first president and a leading figure in the NTF was W. S. Caine, a member of Parliament (M.P.) who headed the organization until his death in 1903. Joseph Malins, leader of the Good Templars in England, chaired the executive committee from 1884 until his death in 1926.

By the 1880s, many temperance leaders were becoming frustrated at the lack of legislative progress. The prohibitionist bodies, in particular the United Kingdom Alliance (UKA), dominated pressure group policies. The UKA was not interested in schemes for licensing reform and only wanted the local veto. It had failed to offer substantial backing when William Gladstone's government introduced a licensing bill in 1871. Since that date, various other bodies had developed, including the Church of England Temperance Society, which sought various restrictions on the sale of alcohol, other than local veto, and also the replacement of the magistrates as the licensing authority. By the 1880s, a rash of licensing reform schemes were being put forward allowing for reduction in licenses or a combination of local veto with other options and reforms.

In 1883, an initiative came from the British Temperance League (BTL), a body with close ties to the UKA, to found a federation for promoting action "on the points upon which they are agreed . . . such common action to extend . . . only so far as there is common agreement." W. S. Caine, who had been elected a Liberal M.P. in 1880 and who later served as a junior minister in the Admiralty (1884–1885), was elected chairman of the new NTF. Caine was a leading figure in the BTL and also a vice president of the UKA. The new federation proved largely ineffective in promoting temperance unity, given the continued existence of jealousies among the various societies and pressure groups. Another problem was that, although the idea of giving local communities a voice in licensing matters found favor among most temperance reformers, the interpretations placed upon this "local option" varied widely. The NTF proved more effective as a negative force to oppose unwelcome government initiatives, notably in 1888 and 1890 when the Unionist government attempted a license reduction scheme on terms that were seen to favor the drink trade, notably financial compensation. The NTF provided a useful parliamentary focus of a cross-party kind, particularly in the period after the Liberal split of 1886 when Liberal Unionist M.P.s, including Caine, supported the Conservative-dominated administration of Lord Salisbury. By 1904, the NTF was completely dominated by the prohibitionists, who by this time were in bitter dispute with supporters of the Gothenburg principles of disinterested management.

Some NTF records can be found at Alliance House, London, headquarters of the United Kingdom Temperance Alliance.

John Greenaway

See also: British Temperance League (BTL); Caine, William Sproston; Conservative Party (United Kingdom); Gothenburg Schemes/Disinterested Management; Liberal Party (United Kingdom); Local Option; Malins, Joseph; United Kingdom Alliance (UKA)

References

Cherrington, Ernest Hurst, et al., eds. *Standard Encyclopedia of the Alcohol Problem.* 6 vols. Westerville, OH: American Issue Publishing Co., 1925–1930: 1860–1861.

Dingle, A. E. *The Campaign for Prohibition in Victorian England: The United Kingdom Alliance, 1872–1895.* London: Croom Helm, 1980.

Newton, John. *W. S. Caine, M.P.: A Biography.* London: James Nisbet, 1907.

National Temperance League (NTL)

The National Temperance League (NTL), organized in 1856, was a London-based society that worked to create a public opinion in favor of total abstinence. It was formed by the amalgamation of the National Temperance Society (organized in 1842) and the London Temperance League (1851), both of them moral suasion societies. The National Temperance Society, in turn, had been formed after the dissolution in 1842 of other metropolitan organizations: the New British and Foreign Temperance Society and the British and Foreign Society for the Suppression of Intemperance, which had been organized in 1839 in a dispute over whether a teetotal pledge should include a commitment not to provide alcoholic beverages to others.

This genealogy helps explain the NTL's approach and the support it received. The organization was moderate, practical-minded, and for many years relatively well funded. Quakers were prominent in the leadership and among financial supporters. For instance, Samuel Bowly (1802–1884), a wealthy Quaker, was one of the early NTL presidents. Although the NTL supported restrictionist legislation, its focus was on education. In this, the NTL contrasted with organizations that made legislation their priority. Yet many prominent NTL officers and members were also active in other societies that emphasized legislation, even prohibition. For instance, Sir Walter C. Trevelyan, Bt. (1797–1879), was the first president of both the United Kingdom Alliance and the NTL; he broke with the league in 1861 when the NTL declined to amalgamate with the alliance.

The NTL hoped to persuade influential groups to support teetotalism. It aided in the formation of temperance organizations for the Church of England and Nonconformist denominations, distributing to clergymen 10,000 copies of Julia Wightman's *Haste to the Rescue!* (1858). It courted physicians and helped organize a teetotal society of medical practitioners, sponsored meetings of teetotal mayors, and helped plan national and international temperance conferences. It also served humbler people. For instance, it worked to organize temperance societies among soldiers and sailors and was active in promoting temperance instruction in the schools. Encouraging women to participate in temperance reform, it had a women's branch as early as 1860, long before there was a national women's temperance society. As a moral suasion organization, the NTL (and its predecessor the London Temperance League) helped sponsor lecture tours by the Anglo-American reformed drunkard John B. Gough. It also published a number of magazines such as the *Weekly Record* (later the *Temperance Record*), the *Temperance Mirror,* and the *National Temperance Quarterly.*

Well organized, the NTL began with the Scottish-born bookseller and publisher William Tweedie (1821–1874) as one of its honorary secretaries. Another Scot, John Rae, became its paid secretary in 1861. He served until 1899; his son, John Turner Rae, succeeded him, remaining in the position until 1929.

Despite the NTL's bureaucratic efficiency, in the 1890s, alarmed by a drop in funds, dissident members criticized the leadership as oligarchic and incompetent. Although at the turn of the century the NTL negotiated for a merger with the provincial British Temperance League (BTL), nothing came of the idea at this time.

Finally, in 1952 the NTL merged, apparently as a junior partner, with the BTL to create the British National Temperance League. The Livesey Library at the University of Central Lancashire has the proceedings of the NTL's executive and general committees for 1856–1928, and the diary of Samuel Sims (1826–1891), an NTL agent in the 1880s.

David M. Fahey

See also: Army Temperance Association; British Medical Temperance Association (BMTA); British Temperance League (BTL); Gough, John Bartholomew; Horsley, Sir Victor; Moral Suasion; Rae, John Turner; Robinson, Sarah; Temple, Frederick; United Kingdom Alliance (UKA); Weston, Agnes; Whittaker, Thomas

References

Burns, (James) Dawson. *Temperance History: A Consecutive Narrative of the Rise, Development, and Extension of the Temperance Reform.* London: National Temperance Publication Depot, 1889–1891.

Cherrington, Ernest Hurst, et al., eds. *Standard Encyclopedia of the Alcohol Problem.* 6 vols. Westerville, OH: American Issue Publishing, 1925–1930: 1861–1862.

Gourlay, William. *"National Temperance": A Jubilee Biography of the National Temperance League, Instituted 1856.* With an Introduction by John Turner Rae. London: Richard J. James, 1906.

Harrison, Brian. *Drink and the Victorians: The Temperance Question in England, 1815–1872.* London: Faber & Faber, 1971; 2d ed., Keele: Keele University Press, 1994.

Shiman, Lilian Lewis. *Crusade against Drink in Victorian England.* New York: St. Martin's, 1988.

National Temperance Society and Publication House (NTSPH)

The National Temperance Society and Publication House (NTSPH), founded in 1865, produced more antiliquor propaganda than any other nineteenth-century organization—more than a billion pages in its fifty years of existence—distributed through a far-reaching network of churches, fraternal lodges, and local temperance groups. The NTSPH, which replaced the often sectarian American Temperance Union, sought to provide a "common center and ground of union" for all factions of the U.S. temperance movement. It published two monthly periodicals (*National Temperance Advocate* and *Youth's Temperance Banner,* with a combined circulation of about 600,000 in the 1870s) and more than 2,000 books, tracts, flyers, and broadsides promoting abstinence. By blanketing the nation with literature, the NTSPH helped sustain the temperance movement through some of its bleakest years.

The NTSPH was established in August 1865 at the National Temperance Convention at Saratoga, New York. Noting the lack of a "continuous, systematic effort" to publish and distribute temperance literature, Pennsylvania attorney James Black successfully petitioned the convention to launch the society in New York City. New York businessman William Earl Dodge was named president, a position he held until his death in 1883. John Newton Stearns served as publications agent and editor of the society's periodicals until 1895. Vice presidents included many of the leading male American temperance activists of the Gilded Age, such as William Buckingham, William H. Burleigh, Samuel Cary, Edward Delavan, Neal Dow, John B. Gough, Horace Greeley, Oliver Otis Howard, John Marsh, Gerrit Smith, and Reuben Walworth.

The society's publishing efforts were vast and diverse. *The National Temperance Advocate,* published until 1897, replaced the defunct *Journal of the American Temperance Union* as the nation's leading temperance periodical. *The Youth's Temperance*

Banner, a four-page newspaper for children, attained a monthly circulation of almost 150,000 by 1876. The society's book publishing efforts included textbooks, encyclopedias, sermons, speeches, medical treatises, anthologies of essays, theology, political manifestoes, almanacs, hymnals, sheet music, memoirs, novels, short stories, children's books, and poetry. It produced hundreds of millions of pages of temperance-related tracts, pamphlets, broadsheets, flyers, manuals, certificates, curricular materials, concert exercises, and pictorial cards.

The NTSPH solicited manuscripts from clergymen, doctors, politicians, and activists, both male and female. In addition to works by its almost 100 vice presidents, the society published books and tracts by T. S. Arthur, Charles Jewett, Howard Crosby, and dozens of others. Its women authors included Julia B. Ballard, Mary Dwinell Chellis, Josephine Pollard, Effie Raymonds, Ella Wheeler, and Mrs. J. McNair Wright.

The NTSPH also promoted temperance in a host of other ways. It organized annual national conventions and, in 1876, an International Temperance Conference in Philadelphia. It sent deputations to numerous state and local temperance conventions. It held public meetings, receptions, and "parlor conferences" in New York and other cities. The society urged ministers to preach on "the Christian duty of total abstinence" on the last Sunday of each year. It led nationwide petition campaigns for prohibition and no-license legislation and for the creation of an independent federal commission to carry out a thorough investigation of the liquor industry. In 1873, a bill supported by the society that would have created such a commission was passed by the U.S. Senate but died in the House of Representatives.

Unlike the earlier American Temperance Union, which had competed with other organizations for control of the temperance movement, the NTSPH won support from a range of antiliquor groups, including the Sons of Temperance, the Independent Order of Good Templars, and the Woman's Christian Temperance Union. The society was adamantly nondenominational, nonsectarian, and nonpartisan. Despite the generality of its appeal, the NTSPH was frequently in debt because it sold most of its publications at cost and often donated materials to "needy and destitute" churches and temperance organizations. The society survived through the support of philanthropists, including Andrew Carnegie, Cornelius Vanderbilt, James H. Kellogg, and John D. Rockefeller, Sr. Donations to the society declined after 1897, and in 1915 it merged with the Federal Council of Churches of Christ in America. Philadelphia retailer John Wanamaker was the society's last president.

Joshua Paddison

See also: American Temperance Society (ATS); Delavan, Edward Cornelius; Dow, Neal; Good Templars (IOGT); Gough, John Bartholomew; Marsh, John; Sons of Temperance; Temperance Commission of the Federal Council of Churches; Woman's Christian Temperance Union (WCTU) (United States)

References

Centennial Temperance Volume: A Memorial of the International Temperance Conference, Held in Philadelphia, June 1876. New York: National Temperance Society and Publication House, 1877.

Lender, Mark Edward. *Dictionary of American Temperance Biography.* Westport, CT: Greenwood, 1984.

Rumbarger, John J. *Profits, Power, and Prohibition: Alcohol Reform and the Industrializing of America, 1800–1930.* Albany: State University of New York Press, 1989.

National Trade Defence Association (NTDA)

The National Trade Defence Association (NTDA) began in 1888 but did not take this name until 1900. Originally, it was called the National Trade Defence Fund. J. Danvers Power, secretary of the Country Brewers' Society, and H. Cosmo Bonsor, a London brewer and Unionist member of Parliament, played a major role in its organization. Eventually, there was a parallel Scottish association.

At a time when English brewers lacked a national trade society, the fund tried to unite them for political purposes. As its name implied, the fund tried to marshal the money needed to present the case of the drink trade before the public. Soon it began to work to elect friends and defeat enemies in parliamentary elections. The fund at first recruited only brewers and distillers but quickly extended a collective membership to organizations of retailers. Publicans were in a position to influence working-class voters, and much of the money came from the big London brewers.

Conflict swirled around the fund in the early 1890s. Midland and northern brewers who resented the fund's oligarchic leadership created the General Association of the Licensed Trade, broader in membership and more ambitious in its organizational structure. After a few months, it collapsed for want of money. The fund made concessions to its critics, enlarging its governing committee and subsidizing the appointment of district agents on a permanent basis.

Historian David W. Gutzke has argued that it was these full-time, salaried agents who were behind the fund's success. The Midland Counties Federated Brewers' Association maintained autonomy, however, publishing in Birmingham the *Licensed Trade News,* edited by E. Lawrence Levy, and the fund left direct supervision of elections in metropolitan London to the Licensed Victuallers' Central Protection Society.

All was not harmony. The fund and then the NTDA often were at odds with the organization of provincial publicans, the Licensed Victuallers' National Defence League. The league complained that the NTDA undercut it and starved it of funds. The league disliked NTDA concessions on behalf of trade unity to off-license holders, especially to licensed grocers. Few northern brewers bothered to subscribe to NTDA funds. In the

Edwardian era, the NTDA remained heavily dependent on London brewers. Despite the wealth of the drink trade, the NTDA seldom had enough money to implement its political strategy fully.

During the magisterial reduction campaign of 1903, the NTDA purchased the rights and obligations of the Licenses Insurance Corporation and Guarantee Fund, Limited, which had insured the Farnham license holders against nonrenewal, and suppressed the application for appeal. Apparently with the memory of the 1891 case of *Sharp v. Wakefield* in mind, the NTDA feared that an appeal to the House of Lords might end in a general affirmation of the powers of the licensing magistrates to deny license renewals for the purpose of reducing the number of licenses in a locality.

After 1910, parliamentary elections mattered little for the drink trade and the NTDA became a trade good-will organization. Eventually, in 1956, it was renamed the National Trade Development Association. It was dissolved in 1975.

David M. Fahey

See also: Brewers' Society (BS); Licensed Victuallers' National Defence League (LVNDL)

References

Fahey, David M. "Brewers, Publicans, and Working-Class Drinkers: Pressure Group Politics in Late Victorian and Edwardian England" in *Histoire sociale.* 3 (1980): 85–103.

Gutzke, David W. *Protecting the Pub: Brewers and Publicans against Temperance.* Woodbridge, Suffolk: Boydell for the Royal Historical Society; Wolfeboro, NH: Boydell & Brewer, 1989.

National Woman's Christian Temperance Union

See Woman's Christian Temperance Union (United States)

Native Americans: Drinking Patterns and Temperance Reform

Perhaps no other group in history has received more attention for its alcohol use than North American Indians. Embedded in the discourse about the role of alcohol in Native societies is the assumption that Indian drinking was and is qualitatively and quantitatively different from that of non-Indians. From the time of first contact, European observers defined American Indians as "addicted" to alcohol. Despite this rhetoric, it is important not to accept the judgments of these observers at face value. In looking at alcohol use, misuse, and control, it is important to question cultural biases and to examine the historical contexts in which alcohol consumption, alcohol control, and alcohol rhetoric occurred.

For some Native groups, alcohol became an outlet for numbing the experience of cultural dislocation, epidemic disease, and European conquest. Still, for many of these same groups, excessive alcohol use began decades after first contact with Europeans and alcohol. For other Native peoples, access to alcohol in the first decades of colonialism was quite limited and consumption rare. In short, there was and is no single style of drinking for Indians generically—rather, the experience of alcohol was and is one that is culturally and historically specific.

Alcohol found its way into Native societies through the European trade. Some Indian groups did have fermented beverages, such as cactus and persimmon wine and corn beer, prior to contact. However, the alcoholic content of these beverages probably averaged somewhere between 2 and 5 percent, compared to the 80- and 100-proof distilled liquors introduced with European contact.

Although the trade evolved slowly over the seventeenth and eighteenth centuries, wherever European colonists met Native Americans, some alcohol usually traded hands. In many interior villages, the alcohol that accompanied the trade was initially part of the gift giving that established friendly relationships between traders and consumers. Among the Ojibway of the Great Lakes, for example, alcohol was often exchanged as food. The European trader would supply a keg in exchange for Native items such as maple sugar and corn, with Ojibway women playing a critical role in the exchange. But there is every indication that alcohol was in short supply in most of the North American interior during the colonial era and even into the nineteenth century.

Along the borderlands, where Indian peoples and Europeans conducted trade on a wider scale, however, alcohol was a medium of exchange that had cultural as well as economic repercussions for colonial societies and Native peoples. One Dutch colonist in New Netherlands claimed that selling alcohol to Indians was both custom and necessity for colonists who needed Indian corn. Many European colonists, long used to drinking themselves, saw little reason for denying access to Indians. Even colonial officials recognized the value of alcohol in lubricating social relations between Indians and non-Indians.

By the middle of the eighteenth century, however, the trade in alcohol became the primary way in which the British gained the edge over the French in the Indian trade. British American colonists in New England, using sugar and molasses imported from the West Indies, illegally produced rum, which then made its way to the interior tribes. There were an estimated 140 distilleries in British North America by 1770, producing almost 5 million imperial gallons (6 million U.S. gallons) of rum annually. Over time, the trade moved from gift exchange to a profitable commodity for European traders.

How Native peoples received and used alcohol within their societies varied considerably and also changed over time. Prior to European contact, some Indian people—the Zunis, Yumas, and Apaches of the Southwest, for example—used wine for secular purposes. Others, including the Tohono O'odham and the Pima, ingested alcohol for medicinal and spiritual reasons. Many early European observers maintained that Indians be-

A wood engraving from Frank Leslie's Illustrated Newspaper, *1871, depicts a European colonist trader offering a bottle of alcohol to a Native American on the plains. (Hulton/Archive)*

lieved that they gained power when drunk. According to one eighteenth-century colonist in the Carolinas, Indian people there used alcohol "as solemnly . . . as if it were part of their religion," preferring to drink only at night and only in amounts sufficient to stupefy them (Lawson 1967 [1709]: 202). Similarly, a French observer near Montreal noted that when the Indian people there drank, they would "experience a new sort of elation that promptly and effectively achieved the end of taking them out of themselves" (quoted in Carpenter 1959: 148). According to these same observers, neither the Carolina nor the Montreal Indians condemned the behavior of those who were drunk.

An American observer of an early-nineteenth-century Ojibway gathering witnessed someone sprinkling a tobacco sacrifice into a pot of heating rum before consumption, which suggested that the Ojibway ritually incorporated alcohol into their belief system. Other observers also reported that the Ojibway routinely used alcohol in mourning practices. Although these same outsiders also noted the excessive use of alcohol by some Ojibway, there were rules that regulated these rituals. For example, Ojibway adoptee John Tanner told of his mother's excessive drinking in the aftermath of her husband's death. Yet, when Tanner also began to drink to excess, his mother admonished him, claiming that his drunkenness was a sign of his "ingratitude to her." Tanner's role as her son was to console and care for his grieving mother, making sure that *her* alcohol use did not spiral out of control (James 1940: 65–66).

As some Native societies came under increasing pressure from colonialism, disease, and conquest, excessive alcohol use had a devastating effect on both individuals and societies. Those villages closest to Euro-American settlements and hence the trade were most affected. Further, Indian peoples' "drunken comportment" shocked and even horrified European observers, especially missionaries. Often the closest non-Indian observers of drinking "frolics," as they called them, missionaries bemoaned the "devil-like" demeanor of drunken Indians that made their villages "open hells." Native sexuality, already more open than that of European colonists, seemed absolutely depraved when combined with alcohol.

By far the most devastating impact from the Native perspective, however, was the violence that accompanied excessive alcohol consumption. One French observer reported: "A husband will kill his wife, and those women who get as drunk as the men knife their husbands and their children" (quoted in Mancall 1995: 89). Other, more common incidents included an increase in quarrelling and domestic violence between husbands and wives, parents and children, and brothers. Conflicts with European colonists also seem to have increased when one or both parties had consumed alcohol. Evidence suggests that colonists on occasion may have supplied liquor to Native peoples with the express purpose of inciting violence as an excuse for assaulting them. Whatever the reasons, where alcohol-related violence wreaked havoc, the lives most often lost were those of Native people. Combined with the deadly epidemics that swept through Native villages in the eighteenth and nineteenth centuries, alcohol brought about declines in health and higher rates of mortality than might have otherwise been present. The full extent of alcohol's impact in early and untimely deaths among North American Indians cannot be known.

Many scholars have suggested that alcohol was an important lubricant in treaty negotiations and in individual land sales across time. Perhaps the most pernicious use of liquor occurred when European and American diplomats conducted official treaty negotiations with Indian tribes. For example, in 1820, Andrew Jackson and other U.S. diplomats met with the Choctaws to produce what became known as the Treaty of Doak's Stand. The United States had the Choctaws in a vulnerable position and they knew it. Jackson warned the Choctaws that failure to trade most of their remaining lands east of the Mississippi for land west of the Mississippi would result in starvation. Utilizing part of the $20,000 budget he had received for conducting the negotiations, Jackson arranged for free access to unlimited liquor for all the Choctaw Indians both before and during the negotiations. The Choctaws eventually traded 6 million acres of land in Mississippi for 13 million acres of land west of the Mississippi River. It may be that the results would have been the same had alcohol not been present. The Choctaws were given no choice about ceding land—only the exact terms were negotiated. The Treaty of Doak's Stand was not the first, nor would it be the last, time that European and U.S. negotiators lubricated treaty negotiations with liquor.

Over time, with greater adaptation to colonialism, drinking patterns in many Indian societies began to reflect more clearly those of their non-Indian neighbors. Some current scholars suggest that studies of Indian drinking should focus on "multiracial regional cultures" to assess how alcohol use occurs in a geographical area rather than focusing exclusively on Indian patterns of drinking. Close analysis of Indian drinking in historical context may indeed be the key to understanding actual drinking patterns and behavior. Such focus might more clearly reveal the North American experience with alcohol, not just the "Indian" experience.

Despite the obvious variety of Indian drinking patterns, most colonial governments thought alcohol was enough of a problem for Native peoples that they passed laws forbidding liquor sales to Indians. In New England in particular, laws against the Indian liquor traffic focused on the need for social order, rather than any attempts at reforming drinking behavior. Enforcement itself was often sporadic. By the eighteenth century, however, many Native peoples recognized the deleterious effects that alcohol was having on their societies and made internal attempts to control alcohol use. One common route was that of spiritual revitalization movements, which often focused on the economic and cultural impact of excessive alcohol consumption. Indeed, Native prophets were in the forefront in promoting total abstinence from alcohol (as opposed to temperate use). To be sure, Native peoples did not form a pan-Indian organization, and the Native prophets' warning against liquor had multiple meanings. Often the language of temperance and abstinence became associated with all things European or all things British, as in the case of the Great Lakes Uprising known as Pontiac's Rebellion in 1763. Over time, the Native prophets' rejection of alcohol intensified, and by the time Tenskwatawa—the Shawnee Prophet—emerged in the Ohio Valley in the early 1800s, Native revitalization was solidly identified with abstinence.

Although many historians have assessed the Native abstinence message as being Christian-influenced, when abstinence among revitalized Indian groups was first enunciated, it was not a Christian concept. The first Euro-American organization to advocate for total abstinence from alcohol (for non-Indians) was the American Temperance Society, founded in 1826. It is most likely that Native peoples developed their philosophies about alcohol independently of imposed colonial or borrowed Christian ideology.

The first federal laws prohibiting alcohol sales to Native peoples in the new United States came in 1802 after a plea by the Miami leader Little Turtle to President Thomas Jefferson. Initially, the discretionary power to regulate the Indian liquor traffic was reserved to the president. By the 1830s, Congress had taken the situation partially into its own hands in invoking its constitutional power to regulate commerce with the Indians. Most often, when Indian nations signed treaties ceding land to the United States, the treaty contained a clause forbidding the sale of distilled spirits both on the ceded lands and on the lands retained by the Native group. For the most part, local and federal law-enforcement officials ignored clauses prohibiting the sale of liquor on lands that had passed out of Indian hands, and enforcement of liquor laws on remaining Indian lands was sporadic. Congress passed a number of federal laws against liquor sales to Indians in the nineteenth century; its most comprehensive piece of legislation came in 1892. This law targeted non-Indians and sought to eradicate the sales and trade itself. Federal legislation ultimately worked to stereotype the Indian drinker and marginalized Native peoples socially in areas where non-Indians and Indians interacted regularly. Thus, the multiracial regional cultures that might have defined drinking patterns for a specific area became somewhat skewed under the weight of discriminatory legislation that targeted Indian peoples as problem drinkers and criminalized non-Indians who sold or provided alcohol to Native peoples.

For a time, National Prohibition turned federal attention away from Native alcohol consumption, as all Americans became subject to a prohibition that previously had applied only to Native Americans. The Indian Bureau's Special Office for the Suppression of Liquor Among the Indians was terminated, and, in 1924, all American Indians born in the United States became U.S. citizens. The end of National Prohibition in 1934 also saw a shift in Indian policy with the "Indian New Deal" of Commissioner of Indian Affairs John Collier. However, alcohol remained an issue for some Indian peoples in the second half of the twentieth century. In the 1940s and 1950s, many reservation-based governments took the lead in prohibiting alcohol sales and consumption on the reservations.

Cultural dislocation and economic marginalization in Indian America continue, and alcohol may sometimes serve in mitigating the effects of internal colonialism. Anthropologist Nancy Lurie suggested that Indian drinking served as a kind of "on-going protest demonstration" against U.S. colonialism (Lurie 1971: 311). Other scholars have suggested that illegal alcohol consumption created a "culture of excitement" for some reservation-based Indians (Weibel-Orlando 1985: 201). Some epidemiologists have begun to question both the generalized approach of the national agencies charged with studying alcohol in American Indian cultures and statistics about Native alcohol consumption that seem overwhelming. Sociologist Philip May has contended that "an analysis of current mortality data . . . will reveal that many current myths are either false or, at best, half truths" and declared that the perception of American Indians as biophysically incapable of metabolizing alcohol "has virtually no basis in fact" (May 1994: 121–122). Similarly a twenty-five-year longitudinal study by Stephen Kunitz and Jerrold Levy of drinking patterns among Navajo men found that many Navajo drinkers, "especially those in the more traditional sample, were able to stop or severely curtail their drinking by the time they were middle-aged," a pattern that is unique both in comparison with other Native groups and with the general population of the United States (Kunitz and Levy 1994: 226). Finally, even as much attention focuses on the drinking problem of Native peoples, some 53 percent of all Native Americans abstain from alcohol use. Further, a number of Native groups, both on-reservation and off, have developed recovery strategies that speak to Native American needs, cultures, and experiences.

Kathryn A. Abbott

See also: Alcohol, Consumption of, by Indices (United States); Colonization, European, and Drinking Behavior among Indigenous Peoples (Portuguese America); Federal Regulation of Alcohol before 1920 (United States); First Nations, the Indian Act, and Prohibition (Canada); Handsome Lake (Ga-nya-di-yoh); Resorts; South America; Tenskwatawa

References

Abbott, Kathryn A. "Alcohol and the Anishinaabeg of Minnesota in the Early Twentieth Century" in *The Western Historical Quarterly.* 30, no. 1 (1999): 25–43.

———. "Liquor Law in Minnesota Indian Country in the Early Twentieth Century" in *Legal Studies Forum.* 25, nos. 3 and 4 (2001): 567–585.

Carpenter, Edmund S. "Alcohol in the Iroquois Dream Quest" in *American Journal of Psychiatry.* 116, no. 1 (1959): 148–151.

James, Edwin, ed. *An Indian Captivity (1789–1822): John Tanner's Narrative of His Captivity among the Ottawa and Ojibwa Indians.* Occasional Papers Reprint Series no. 20, pt. 1. San Francisco: California State Library, 1940.

Kunitz, Stephen J., M.D., and Jerrold Levy. *Drinking Careers: A Twenty-Five-Year Study of Three Navajo Populations.* New Haven, CT: Yale University Press, 1994.

Lawson, John. *A New Voyage to Carolina,* edited by H. T. Lefler. Chapel Hill: University of North Carolina Press, 1967 [1709].

Lurie, Nancy Oestreich. "The World's Oldest On-Going Protest Demonstration: North American Indian Drinking Patterns" in *Pacific Historical Review.* 40, no. 3 (1971): 311–332.

Mancall, Peter C. *Deadly Medicine: Indians and Alcohol in Early America.* Ithaca: Cornell University Press, 1995.

May, Philip A. "The Epidemiology of Alcohol Abuse among American Indians: The Mythical and Real Properties" in *American Indian Culture and Research Journal.* 28, no. 2 (1994): 121–143.

Unrau, William E. *White Man's Wicked Water: The Alcohol Trade and Prohibition in Indian Country, 1802–1892.* Lawrence: University Press of Kansas, 1996.

Weibel-Orlando, Joan. "Indians, Ethnicity and Alcohol: Contrasting Perceptions of the Ethnic Self and Alcohol Use" in *The American Experience with Alcohol: Contrasting Cultural Perspectives,* edited by Linda A. Bennett and Genevieve M. Ames, 201–226. New York: Plenum Press, 1985.

The Netherlands

Alcohol use has been common in the Netherlands for centuries. Since the second half of the 1700s, strong alcoholic beverages have been within economic reach for everyone. From the beginning of the nineteenth century, alcohol has been freely available. Between 1960 and 1975, per capita alcohol use tripled, increasing from 2.6 liters (0.7 U.S. gal.) to 8.7 liters (2.3 U.S. gal) of pure alcohol annually. Nowadays, the per capita consumption fluctuates around 8 liters (2.1 U.S. gal.) of pure alcohol each year (National Drug Monitor 2002). This rate of consumption is moderate compared to that of most other European countries. Most Dutch people drink moderately, consuming one or two glasses of alcoholic beverages on a daily basis. Almost one in ten Dutch adults experiences alcohol-related problems, and about one in ten people with alcohol-related problems gets help.

Over the centuries, the temperance movement in the Netherlands responded to this free availability of alcohol and its attendant problems. The movement experienced first great popularity, then a decline. The peak of temperance popularity occurred in the first decades of the twentieth century, partly in response to the rising consumption of liquor, but also reflecting Calvinist influence from the churches. At that time, temperance advocates scored many successes and had considerable political influence.

Developments in Alcohol Use since 1700

In 1778, the Dutch professor Mathias van Geuns wrote about the ready availability of alcoholic beverages in the Netherlands, stressing "the virtuousness" of "[very easy] to obtain best foods like well prepared national beer" (Geuns 1778). He also expressed concerns with regard to apparent drunkenness, however. In the years that followed, strong alcoholic drink became more popular and beer became less so. Around 1880, the per capita consumption in the Netherlands was 7 liters (1.8 U.S. gal.) of pure alcohol a year. Thanks to the temperance movement, the consumption rate dropped rapidly, to about 2 liters (0.5 U.S. gal.) of pure alcohol per capita in 1950.

After 1950, however, rapid changes again took place. The tripling in the consumption rate between 1960 and 1975 may have been due to better sales and marketing techniques. Or, perhaps the Dutch were simply catching up to the levels of consumption among their European neighbors. Also, alcoholic beverages became even more easily available, both financially and practically. More and more population subgroups began to drink in more varied environments. Up to 1940, most drinking took place in pubs. After 1940, drinking at home became more common. After 1960, people still drank in the pub and at home, but also in other surroundings, such as canteens of sport clubs, community centers, and other places (Garretsen 2001). No major changes in the per capita consumption rate have taken place since 1980, when the figure reached around 8 liters of pure alcohol per capita.

The Role of the Temperance Movement

In the first half of the nineteenth century, Dutch citizens organized themselves in the fight against strong alcoholic beverages. In 1804, the Maatschappij tot Nut van 't Algemeen (Society of Public Welfare) published a little book entitled *Het morgenslokjen'* (The morning drink), an educational document to prevent excessive and problematic drinking. Other activities followed. The first national temperance group, the Dutch Association for Abolition of Spirits, was founded in 1842. This was one of the most important temperance organizations in the Netherlands until after World War II. In 1875, the Volksbond (Peoples Union) was established to promote moderation. This early antispirits activity waned temporarily around 1880 as the use of alcohol increased and interest in the temperance movement declined.

The end of the nineteenth century was characterized by a new wave of temperance work, with a growth in the number of temperance organizations, their activities, and their members. This shift reached its climax in the first decades of the twentieth century. At this time, temperance advocates scored many successes and had considerable political influence. Important leaders of the temperance forces included Dr. K. H. Bouman and the temperance advocate Th. W. van der Woude. Under temperance influence, the first Consultation Bureau for Alcoholism was founded in 1909 to treat alcohol problems, and the per capita consumption dropped dramatically. Then behavior and attitudes changed again, becoming more relaxed in the post–World War II period of prosperity. Nowadays, the temperance movement in the Netherlands is very small indeed. An extensive overview of the temperance movement in the Netherlands is given by J. C. van der Stel (1995).

Present Drinking Habits

Most Dutch people drink alcohol at least occasionally. A recent survey shows that about 90 percent of the adult population has drunk alcohol at least once; almost three-quarters had drunk alcohol recently (National Drug Monitor [NDM] 2001).

The Netherlands cannot be characterized as a country with one predominant type of beverage preference; it does not have a real beer or wine culture (Garretsen and van de Goor 1995). Beer, wine, and sprits are drunk in relatively large quantities. In 2000, the per capita consumption of beer was 4.1 liters (1.1 U.S. gal.) of pure alcohol; of wine, 2.4 liters (0.6 U.S. gal.); and of spirits, 1.7 liters (0.4 U.S. gal.). Among high-school students beer is the most popular drink, followed by liquors, shooters, wine, and *alcopops* (fruit drinks containing alcohol). Among young people, periodic excessive drinking is prevalent, especially on weekends. Men drink more than women, and heavy drinking patterns are more prevalent among men, with 22 percent of men drinking, at least once a week, six glasses of alcohol or more, compared to 5 percent of women. For people aged eighteen to twenty-four, these figures are 46 percent and 16 percent, respectively, and for people aged twelve to seventeen, 14 percent and 5 percent, respectively (NDM 2001).

Some decades ago, rather strong regional differences influenced by religion and degree of urbanization could be seen in the Netherlands. In the south, originally the Catholic part of the country, people drank more than in the Calvinistic north. These differences are now decreasing due to secularization and other broad societal developments.

The relationship between alcohol and social economic status is complex in the Netherlands. Both teetotalers and heavy drinkers are found more among the lower socioeconomic groups. Ethnic minorities in the Netherlands belong more often to the lower economic classes and come mainly from former colonies (Suriname and the Dutch Antilles) and from Turkey and Morocco. In general, these population subgroups drink less than the general Dutch population (for religious reasons, mainly), but heavy drinking and problem drinking does occur, especially among youth.

The attitude toward heavy drinking in the Netherlands can be described as ambivalent. Most people believe that some alcohol consumption is fine, and at some social events alcohol use is encouraged, but problematic use is far less acceptable. Individual freedom is important in the Netherlands, and in most things people are free to do as they wish as long as they do not bother others.

A horse and cart deliver barrels of beer to a public house at Spiegel Gracht, central Amsterdam, 1954. (Huton/Archive)

Problem Drinking

From the eighteenth century onward, alcohol problems became more prevalent in Dutch society, and in the first half of the twentieth century, the concept of alcoholism became known. Alcoholism was seen more and more as a disease. The first *asiel voor drankzuchtigen* (shelter for alcoholics) was established in 1891. Yet the first extensive surveys in the Netherlands on problem drinking were not conducted until 1980 and 1981 in the city of Rotterdam in the western part of the country and in the province of Limburg in the south. These studies were repeated in 1989 in Limburg and in 1994 in Rotterdam. Problem drinking was defined as excessive alcohol consumption connected with somatic, psychological, or social problems for the problem drinker or for others. The prevalence rates varied at between 7 percent and 11 percent (Bongers et al. 1997). In 1999, surveys were held in three relatively large Dutch cities; in these studies, problem drinking was defined as having three or more alcohol-related problems in the year before the survey. Between 3 and 8 percent of women, and 10 to 16 percent of men, were found to be problem drinkers. Another survey in 1996 revealed that 8.2

percent of the Dutch population aged eighteen and sixty-four met the diagnostic criteria for alcohol dependency or alcohol misuse. Most people did not get professional help for alcohol-related problems. It was estimated in 2001 that about one out of ten people with an alcohol-related problem got help (NDM 2002).

Prevention and Governmental Policy

Prevention activities may be divided into two types, those directed toward the demand for alcohol, such as education and information, and those related to the supply of alcohol.

With regard to education and information, the temperance movement played an important role until World War II. The oldest education document is probably *Het morgenslokjen* (The morning drink) from 1804, mentioned above. In the first decades after World War II, alcohol education was not a top priority in the Netherlands. In the 1980s, more attention was given to alcohol and problem drinking, both in research and in governmental policy and prevention.

Mass media campaigns have been running since 1986 as part of the Alcohol Education and Prevention Project of the National Institute for Health Promotion and Disease Prevention. Other projects were initiated in cooperation with regional institutions. The government supports these education programs as well as specific activities with regard to, for example, drunk driving and alcohol in the workplace.

In general, these education programs are not specifically aimed at reducing or preventing alcohol consumption, since education in itself is generally insufficient to effect behavioral change; however, they are useful in spreading information about the risks related to alcohol use and about the possibilities for treatment. Education may also support other activities that, in general, have little public support, for example, a tax increase. Thus, education serves to legitimize other programs that are more directed toward gaining measurable results.

To make headway against the second type of prevention activity, the availability of alcohol, temperance advocates fought for measures and laws to limit easy access. This effort led, among other things, to the "Drankwet" (Alcohol law) of 1881. Nowadays, Dutch laws with regard to alcohol are not very strict—the Netherlands has a "moderate policy focused on moderation." The newest developments relate, among other things, to a ban on the sale of alcohol in motorway (highway or freeway) gas stations and in other nonfood stores. However, it has been argued that too little attention is paid to alcohol compared with other drugs and that legislation and regulations are insufficiently evaluated (Garretsen 2001).

Producers, Prevention Organizations, and Research Institutes

The Netherlands is mainly a beer- and spirits-producing country; wine is of less importance. Widely known producers include Heineken and Grolsch (beer) and Bols (spirits). Producers and importers take part in the Stichting Verantwoord Alcoholgebruik (STIVA; Foundation for Responsible Alcohol Use), a foundation aimed at stimulating responsible alcohol use. Self-regulation (for instance, with regard to advertisements) is one of the key activities.

The Algemene Nederlandse Drankbestrijders Organisatie (ANDO, Dutch Temperance Society), founded in 1962, still exists, but nowadays the temperance movement is very small indeed. The central coordination point for nationwide prevention campaigns is today to be found in the National Institute for Health Promotion and Disease Prevention in the city of Woerden. Sixteen local and regional support bases are active. About forty centers deliver specialized addiction care, among them sixteen bureaus, with some 130 establishments throughout the country, that are descended from the bureaus founded in 1909.

There are several major research institutes in the Netherlands that focus on addiction research, including the Instituut voor Onderzoek naar leefwijzen en Verslavin (IVO, Addiction Research Institute) based at the universities of Rotterdam (Erasmus University), Maastricht, and Tilburg, the Amsterdam Institute for Addiction Research based at the University of Amsterdam, and the Trimbos Institute, based in Utrecht.

H. F. L. Garretsen, I. M. B. Bongers, and L. A. M. van de Goor

References

Bongers, I. M. B., J. A. M. van Oers, L. A. M. van de Goor, and H. F. L. Garretsen. "Alcohol Use and Problem Drinking: Prevalences in the General Rotterdam Population" in *Substance Use and Misuse.* 32 (1997): 1491–1521.

Garretsen, H. F. L. "Dutch Alcohol Policy Developments: The Last Decades and Present State of Affairs" in *Medicine and Law.* 20 (2001): 301–311.

Garretsen, H. F. L., and L. A. M. van de Goor. "The Netherlands" in *International Handbook on Alcohol and Culture,* edited by D. B. Heath, 190–200. London: Greenwood, 1995.

Geuns, M. van. *De handhaving van gezondheid an leven der ingezentenen onzes vaderlands, een zeer belangrijk en noodzakelijk deel der zorge hunner overledenen* (The maintenance of health and life in the inhabitants of our country, a very important and necessary part of the care their deads). Geldersche Hoogeschoole, Harderwijk, 1778.

National Drug Monitor (NDM). *2001 Annual Report.* Utrecht: Trimbos Institute, 2001.

Stel, J. C. van der. "Drinken, drank en dronkenschap: Vijf eeuwen drankbestrijding en alcoholhulpverlening in Nederland" (Drinking, drink and intoxication: Five centuries of temperance movements and alcohol treatment in the Netherlands) (dissertation, University of Utrecht). Hilversum: Uitgeverij Verloren, 1995.

New Year's Eve

Although the societal significance of the holiday has waxed and waned over the course of time, the advent of the New Year has been accompanied by the consumption of alcoholic beverages and celebratory drinking in Western cultures at least as far back as the Roman era. The nineteenth and twentieth centuries have seen a steady increase in New Year's Eve drinking for drinking's sake, and entering the twenty-first century, increasing concern has been expressed for the effects of celebratory drinking to intoxication on public health and safety apart from moral or ethical considerations.

By 1800, it had become tradition to stay awake past midnight to see the New Year in, the moment ushered in with loud noises such as the ringing of church bells or setting off of fireworks. Public and private drunkenness, with attendant boisterousness, often marked by discharging firearms, was commonplace on New Year's Eve during the nineteenth century. In towns and urban settings, groups of men moved from house to house or tavern to tavern expecting to be invited in for drinks. In more polite society, more of a dinner-party atmosphere would prevail; nevertheless, the serving of alcohol would still be typical.

By the turn of the twentieth century, drinking was widely considered a required ritual of the New Year's Eve celebration, whether toasting the New Year with champagne in a private home or nightclub or raising a glass with the crowd at a neighborhood pub. Many individuals who otherwise abstain from drink indulge on 31 December. Holiday drinking is seen as a show of the conviviality and abandon associated with a mass celebration. The advent of radio and television broadcasts of simultaneous celebrations from around the country and the world encourages a sense of group bonding and solidarity, of being part of a larger event. The public celebration is perhaps taken to its extreme in Scotland, with the still-common tradition of "Hogmanay," once a celebration on the day before New Year's Eve, but which often lasts a day or more into the New Year.

Ironically, these traditions often result in the practice of a second New Year's tradition, that of the resolution, one of the most common being the resolve to stop drinking. Although public drunkenness has always carried with it incidents of disturbance and aggressive behavior, the twentieth century's increase in urban and overall population and the preponderance of automobiles to travel to and from celebration sites have exacerbated the pressures upon civil controls. By the end of the century, many large urban areas required the mobilization of extra law-enforcement officers on New Year's Eve. Roadblocks serving as blood-alcohol-level checkpoints were common, as was overcrowding of jails and temporary holding cells for those arrested on alcohol-related charges. Public order and safety on the holiday had become a major concern and strain on public resources.

Anecdotal evidence entering the twenty-first century suggests that these concerns were leading to a gradual shift from public to private celebration of New Year's Eve, the major public observances of the arrival of the year 2000 notwithstanding.

John P. Hundley

See also: St. Patrick's Day

References

Bacon, Seldon D. "Alcohol and Complex Society" in *Society, Culture, and Drinking Patterns,* edited by David J. Pittman and Charles R. Snyder, 78–100. Carbondale and Edwardsville: Southern Illinois University Press, 1973.

Blackburn, Bonnie, and Leofranc Holford-Strevens. *The Oxford Book of Days.* Oxford and New York: Oxford University Press, 2000.

Breathnach, Sarah Ban. *Mrs. Sharp's Traditions.* New York: Simon and Schuster, 1990.

"Liquor Licensing Deregulation: Consultation on New Year's Eve Licensing Hours." January 1999. Institute of Alcohol Studies website. http://www.ias.org.uk/licensinghours.htm.

New Zealand

For much of its modern history alcohol has constituted a morally and socially suspect commodity in New Zealand; it has also been a highly lucrative one, yielding healthy income for business and government. A strong temperance and prohibition movement flourished in the late nineteenth and early twentieth centuries. Although never achieving the ultimate goal, its legacy was a highly restrictive regime of production, sale, and consumption for almost a century. At its height, from 1917 to 1967, early closing had the vast majority of licensed premises closing their doors at 6 P.M., and alcohol consumption at any hour was largely quarantined from all other forms of socializing.

Although drinking came to be defined early as a moral question, and therefore an issue for conscience rather than party affiliation, alcohol has played an important—if contradictory—part in New Zealand's political history. The Woman's Christian Temperance Union (WCTU) took up the campaign for women's suffrage and achieved success. In 1893, New Zealand became the first nation-state in which women won the vote. Moreover, liquor law and debate has been one of the areas in which formal distinction was made between the rights and status of indigenous Maori and Pakeha (white, or non-Maori) citizens. As the legal restriction and earlier punitive social attitudes to alcohol have relaxed over the past three decades, a predominantly masculine, volume-oriented, beer-drinking culture has given way to a more diverse and integrated drinking culture in which wine is increasingly popular. An emergent viticulture industry has supported the trend, producing wines for domestic consumption and a successful, if still comparatively small, export industry.

Colonial Practices and Attitudes

The history of alcohol consumption and of the attitudes surrounding it begins with the voyages of European exploration in the 1760s and the establishment of sealing, whaling, and trading communities on the New Zealand coast, customarily made up of young workingmen partial to bouts of drinking. Sacramental use began with the Church Missionary Society's arrival at the Bay of Islands in 1814. The Wesleyans followed soon after, but it was not until the 1830s that the French and Irish Catholic missions brought a less stringent set of attitudes to both secular and sacred uses of alcohol; they were also responsible for some of the first grapevine plantings.

With the beginning of larger scale European colonization in the 1840s, the production and consumption of alcohol became both more popular and problematic. The first temperance organizations date from this period. Levels of drunkenness were high and social order was not always easily maintained, especially among the gold diggers in the 1860s and among enclaves of settlers inclined to engage in episodes of heavy drinking, especially after the physical exertions of "clearing the bush."

Only in the 1870s did campaigns for subduing the excesses of drink and drinkers begin to have an impact in public life. A settler population drawn largely from British dissenting and evangelical Protestant backgrounds increasingly supported a vocal movement for restraint on the availability of alcohol. In William Fox, the movement found a powerful early parliamentary voice. The first comprehensive regulation and control of the liquor trade was achieved through the 1881 Licensing Act. Intended as a political compromise, in fact it set the key terms for restriction for almost a century: the requirement for anyone selling alcohol to have a license; a strict limit on licenses; a restriction on selling hours, places, and conditions; and a close regulation of the trade.

Antialcohol Societies and Their Impact

Instead of quenching the demand for reform, the 1881 legislation only marked an early point from which the prohibition movement grew in size and force through the 1880s and 1890s. Local branches of the WCTU were formed in the wake of Mary Leavitt's mission to New Zealand in 1885, and a national body quickly came into being. A key link between the campaign for temperance and agitation for the women's vote was forged when Kate Sheppard became national superintendent of the franchise and legislative departments in the WCTU in 1887. The WCTU provided a key organizational vehicle for a nationwide franchise campaign in which the temperance lobby was an important but not the sole component. The New Zealand Alliance for the Suppression and Prohibition of the Liquor Traffic (NZA), formed in 1886 (and its influential paper, *The Prohibitionist*), was the other leader of the campaign. Although the core of support for the prohibition and temperance movement came from Protestant churches, leaders included the agnostic Robert Stout and a number of left-wing political leaders. The opposition was similarly a coalition of interests, including representatives of "the trade" and prominent politicians, such as the aptly named Henry Fish and long-serving Liberal Premier Richard John Seddon, a former publican.

For the next three decades, the trend was strongly toward increasing restriction and showed every sign of culminating in prohibition. The movement appeared to be in reach of achieving its goal when the popular vote reached high points of 55.82 percent of the vote for prohibition in 1911, and an even higher level in 1919, but a 60 percent majority was necessary and support for prohibition fell short by just over 3,000 votes—those cast by World War I soldiers awaiting return to New Zealand at the conclusion of the war—the so-called "soldiers' mandate." Thereafter, support for prohibition faded gradually, dropping to 30 percent in 1935 and 20 percent in 1963. Nonetheless, the sale and consumption of alcohol remained highly regulated. Licenses were subject to a local option from 1893, and after initially being taken as a wartime measure in 1917, 6 P.M. closing became permanent from 1918 to 1967. This early closing led to one of the most distinctive—and decried—aspects of New Zealand's drinking and wider culture: The so-called "six o'clock swill" made hotels little more than austere, standing-room-only places for as much beer to be drunk in as short a time as possible. Women were excluded, along with seats, food, and even rudimentary comforts. A legal disassociation between drinking and any form of dancing or entertainment was also a product of the prohibition pressure.

Treatment of Maori

A distinctive thread in the pattern of regulation of alcohol in New Zealand was the separate, and discriminatory, provision for Maori under licensing law. One of few areas of legal distinction between Maori and Pakeha, early restrictive legislation was designed to "protect" Maori from the deceitful, exploitive, and unscrupulous activities of traders in liquor. From the earliest measures taken by Governor George Grey in the late 1840s, through the various licensing laws, and until the end of World War II, Maori had limited access to alcohol and were even prohibited from buying it for consumption off licensed premises. There were special restrictions on the sale of alcohol in the King Country region, a large area in the North Island that was the center for one of the major tribal groups, and an area that had been the subject of bitter armed contest in the 1860s. This area remained "dry" for many years.

Post–World War II Relaxation of Restrictions

A rising intolerance for the restrictive and contorted drinking culture developed through the 1950s and especially the 1960s. A popular referendum held in 1967 opened the door to legal liberalization. A majority of the population voted to extend opening hours to 10 P.M. Growing sophistication in eating and drinking tastes resulted in greater demand for a range of drinks and in more choices regarding when and where to consume them. Through the 1970s and 1980s, broader social change saw a substantial expansion of a local restaurant culture that had not previously existed, especially in the form of the popular "BYO" (bring your own) restaurants where patrons were encouraged to bring their own bottles of wine to drink with meals. The erosion of the regular five-day workweek in the face of extensions in working and trading hours, along with the development of tourism as a key industry, added impetus to the liberalization of the liquor laws.

The single most significant measure was the 1989 Sale of Liquor Act, which radically altered the basis of regulation from restricting availability to a focus on responsible management, thereby finally repudiating the principle established in the 1881 legislation, which attempted to control consumption by controlling availability. The number of outlets where alcohol could be purchased multiplied considerably, making cafes, wine bars, and a wide array of restaurants all highly popular parts of an expanded eating and drinking culture; licenses for outlets selling alcohol for home use also increased. More controversially, a further extension of the law took place in 1999 lowering the drinking age to eighteen (the voting age), removing the last restrictions on Sunday trading and allowing for the sale of beer as well as wine in supermarkets. Although levels of control in hours, outlets, and venues were all liberalized, government interest was maintained in taxation levels, always relatively high and a source of irritation to the industry.

Alcohol Production

The handful of vineyards in existence in the 1960s generally made small quantities of largely unremarkable table and fortified wines for local consumption. Since then viticulture has flourished and a number of highly successful wine districts have become established, the best of which are producing internationally competitive wines. Marlborough sauvignon blanc, Hawke's Bay chardonnays, and labels such as Cloudy Bay, Church Road, and Montana now claim places on wine lists around the world. High quality pinot noir from the smaller areas of Martinborough and Central Otago have more recently joined these ranks.

Beer production has long been dominated by two big brewers, Dominion Breweries (DB) and Lion Nathan. Throughout much of the twentieth century, the breweries owned the vast majority of hotels operating as licensed premises as well as the majority of off-licenses—a virtual duopoly existed. With liberalization, the stranglehold of breweries over licensed premises has declined. A strong link exists between the major beer brands and sponsorships of men's national sports. Steinlager has for many years been a major brand promoted by the All Black rugby team and more recently has become one of the principal sponsors of the successful America's Cup team. DB has sponsored the Warriors Rugby League Club, which has played in the National Rugby League competition from 1995, as well as a number of rugby sides, and the national men's cricket team. Alcohol sponsorship of sport has survived while tobacco sponsorship has been outlawed. The interlinking of beer, sport, and masculinity is a theme running strongly through New Zealand's history and culture—in particular through the popular sports of rugby football and horse racing, with sports club drinking a key weekend site for socializing. In the past decade, the theme has also been exploited, ironically, in beer promotions.

Although mergers and distribution contracts have largely stymied the maintenance of truly independent regional brands, some labels with strong regional origins have survived and continue to be marketed with local cultural identities. Originally a Dunedin company, Speights maintains a strong association with the city and the Otago region in the south, being promoted under the banner "pride of the south." "Southern Man," the archetypal Speights drinker, a creation of 1990s advertising promotion for the brand, has come to have a following and cultural recognition well beyond the original billboard. He has his own song, which is sung at the local rugby ground, and more generally has come to be the tag for an admired, if somewhat extreme form of a longer standing cultural figure, "man alone." "Southern men" are tough men who show loyalty to their mates (in this context only men) above all other qualities, ruggedness, and a link with the land (through their work and in their attire) and who display a suspicion about city dwellers, suit-wearers, sophistication, women, and non–beer drinkers.

Alcohol and Temperance in New Zealand Literature

A culture in which alcohol has long been subject to high levels of punitive control has, not surprisingly, bred an abundance of fictional characters with problematic relations with alcohol. Temperance novels were produced in abundance through the prohibition era. Twentieth-century literature has produced figures ranging from Jane Mander's Tom Roland (in *The Story of a New Zealand River,* 1920), to poet and dramatist James K. Baxter's Barney Flanagan ("Lament for Barney Flanagan," 1954, and many more) and Alan Duff's more recent "Jack the Muss" in the novel *Once Were Warriors* (1990, and as featured in the 1994 film).

Modern Consumption Patterns

Total alcohol consumption by New Zealanders has been falling since the early 1980s, with beer declining in the face of rising quantities of wine, though beer remains the dominant drink. Ready-mixed spirits drinks (alcopops), introduced in 1995, have risen sharply in popularity (from 1 to 6.8 percent); whether the growth will be sustained remains to be seen. As of 2001, consumption patterns for the three categories of alcohol were: beer, 47 percent; wine, 30 percent; and spirits, 22 percent. Gender is the strongest and most consistent predictor of drink preferences and consumption patterns. Men make up the majority of alcohol drinkers and account for around two-thirds of the total volume of alcohol consumed, though the proportion drunk by women has been increasing in recent years. There is some evidence that alcohol use also varies according to ethnicity. On a recent international table of comparison, New Zealand stands twenty-third overall in terms of volume of alcohol consumed per capita, rating lower in spirits drinking (at thirty-third), higher in beer drinking (thirteenth, between the Netherlands and Finland), and on a par, at twenty-third, in wine drinking.

Whether this is a source of pride or lament can still draw debate. Claims and counterclaims as to the commercial, social, and moral value of alcohol continue, especially at times of proposed changes in the law, but they are no longer led by the NZA and the WCTU, though both organizations still exist. The more significant voices are now those of the major advocacy bodies: the industry, the Alcohol and Liquor Advisory Council (ALAC), and some research organizations, particularly those investigating the connections between health and alcohol; road accidents and alcohol; and patterns of youth drinking with associated sexual health issues. Principal industry groups are the Beer, Wine and Spirits Council (established in 1990 and dominated by the two big brewers), the Hospitality Association of New Zealand (originally the Licensed Victuallers Association, established in 1902), the Distilled Spirits Association, and the Wine Institute of New Zealand. ALAC (established in 1976) is a public body funded from liquor sales and is charged with promoting moderation in alcohol use and developing strategies to reduce alcohol problems. The Alcohol and Public Health Research Unit, headed by Professor Sally Casswell, constitutes the major research body. Formerly connected to the University of Auckland, as of July 2002 it has been largely relocated to the Centre for Social and Health Outcomes Research and Evaluation at Massey University.

The battle over alcohol as inherently good or bad is largely over, but skirmishes remain over age, education, levels of advertising, and some areas of responsibility. New Zealanders are more likely to be found arguing over the merits or otherwise of the latest wine releases, or over the strengths and weaknesses of their local sports team as against their rivals (and their associated beer brands), than over the rights and wrongs of alcohol per se. Alcohol has become a source of national pride, with local products (apart from spirits) strongly preferred over imports. Moral connotations have faded, but cultural meanings remain powerful.

Charlotte Macdonald

See also: Leavitt, Mary Greenleaf Clement; Sheppard, Catherine ("Kate") Wilson; World's Woman's Christian Temperance Union (WWCTU)

References

Alcohol and Public Health Research Unit website. http://www.aphru.ac.nz.

"Alcohol Available in NZ, Part 2." Alcohol Liquor Advisory Council (ALAC) website. http://www.alcohol.org.nz.

Bollinger, Conrad. *Grog's Own Country.* Wellington: Price Milburn, 1959.

Dictionary of New Zealand Biography, Vols 1–5, Wellington and Auckland: Department of Internal Affairs/Bridget Williams Books and Auckland University Press, 1990–2000 (also available at http://www.dnzb.govt.nz).

"Facts and Figures." Beer Wine and Spirits Council website. http://www.beerwsc.co.nz.

Grimshaw, Patricia. *Women's Suffrage in New Zealand.* Rev. ed. Auckland: Auckland University Press/Oxford University Press, 1987.

Robinson, Roger, and Nelson Wattie, eds. *Oxford Companion to New Zealand Literature.* Auckland and Melbourne: Oxford University Press, 1998.

Nonconformists (United Kingdom)

Nonconformists, a religious minority in England, participated disproportionately in the temperance movement from its beginnings. Nonconformity traced its roots to the seventeenth-century Puritans. By the nineteenth century, there were a number of Protestant denominations and sects outside the Church of England that came to be known as Nonconformists. (In the eighteenth century they often were known as Dissenters and in the twentieth century as Free Church.) Diverse, they included a variety of Methodists, the Religious Society of Friends (Quakers), Baptists, Congregationalists, Presbyterians, and Unitarians in England, as well as related Welsh, Scottish, and Irish groups.

Early Nonconformist Influence

In the early nineteenth century, when the temperance movement was young, meetings were frequently held outdoors as speakers found it difficult to hire suitable halls. It was the Nonconformists who welcomed temperance speakers to their premises. By 1848, there were 566 temperance ministers, of whom only 5 percent were Anglican clergy. Early temperance workers were almost inevitably Nonconformists. For instance, the teetotal pioneer Joseph Livesey (1797–1884) was a Scotch Baptist, and the temperance author William Richard Baker (1798–1861) was a Congregationalist. The Band of Hope first convened in a Quaker meetinghouse.

Not all Nonconformists were committed to teetotalism, and there were divisions between and within sects. Wesleyans were never unanimous about temperance, unlike the Methodists in the southwest, who founded staunchly teetotal Primitive Methodist chapels. The strength of Nonconformity lay outside the capital. Appropriately, the expansion of the temperance movement illustrated the saying that "what Manchester does today London does tomorrow." The Nonconformists of northwestern England embraced the temperance movement as a potential cure for some of the ills that came from rapid industrialization.

Temperance Supporters

Of all the Nonconformists, the Quakers probably were the most closely identified with the temperance movement. George Fox drew their attention to the problem of drink as early as 1643, and over a century later they were informed at their yearly meetings that temperance and moderation were virtues that proceeded from true religion. In 1857, the Religious Society of Friends strengthened this message. None of this advice was forced upon the individual, but in practice Friends complied. Further "advice" on the question of drink later moved them from earlier moderation to abstinence.

Prominent Quaker temperance reformers included the Sturges of Bristol, the Brights of northwestern England, the Croppers, the Wilsons of Sheffield, and the Cadburys of Birmingham. When Quakers were elected to Parliament as members of the Liberal Party, they became a pressure group that attempted to influence party policy and campaign for temperance legislation. Two of the most prominent Quaker members of Parliament (M.P.s) were the Bright brothers, John (1811–1898) and Jacob (1821–1899). John served as the president of the British Association for the Promotion of Temperance in 1842. The United Kingdom Alliance sought his support, but he opposed prohibition and eventually, for medical reasons, abandoned his own teetotalism. In contrast, his brother Jacob continued to support the movement, and their sister Margaret Bright Lucas became president of the British Women's Temperance Association.

Nonconformists often were teetotalers. They included successful business leaders such as the founder of the travel industry, Thomas Cook (1808–1892), and the publisher John Cassell (1817–1865). Nonconformist preachers such as James Sherman (1796–1862), a Congregational minister in London, used their pulpits to spread the temperance message. When the temperance program shifted from moral suasion to prohibition in the 1850s, the temperance movement and the Nonconformist role became increasingly political.

Nonconformity, Temperance, and the Liberal Party

In the second half of the nineteenth century, temperance became respectable and changed from being a local issue to a matter of national concern. By this time, Nonconformists, gaining political confidence, were ready to demand radical change. In addition, trade unions and the Reform League helped make the working class politically aware. Most important, in 1867 the Second Reform Act gave many workers the right to vote in parliamentary elections.

In this context, the United Kingdom Alliance, founded in 1853, was successful in making legislation the first priority in the temperance movement. Founding members of the alliance included Samuel Pope (1826–1901), a Nonconformist barrister, and Dawson Burns (1828–1909), a Baptist minister who became the metropolitan superintendent for the alliance. In 1860, the alliance appointed James Hayes Raper (1820–1897), who had been reared as a Wesleyan Methodist, as its parliamentary agent. He worked out the so-called Permissive Bill to allow voters to impose prohibition in their localities, a compromise from the earlier demand for statutory prohibition for the entire United Kingdom.

Provincial liberalism was strongly Nonconformist and radical by the 1860s, but Anglicans and moderate Whigs dominated the Liberal Party leadership. The alliance tried to appeal directly to Parliament, but temperance M.P.s lost their seats in the 1865 election. The 1868 election was a turning point: Sir Wilfrid Lawson (1829–1906) was returned to Parliament, as was Benjamin Whitworth (1816–1893), a member of the alliance executive. Lawson introduced the Permissive Bill a number of times before changing his strategy. He then proposed vague resolutions for local option. In the 1880s, most Liberal M.P.s voted for this local-option resolution. Sir William Harcourt acted as an intermediary between the Nonconformists and the Liberal leadership on behalf of legislation for direct local veto, authorization for local referendums to prohibit the sale of alcoholic drink. It was hoped that the Nonconformists would rally behind the Liberal Party to enact such a bill. Harcourt introduced direct local-veto bills on behalf of a Liberal government in 1893 and 1895, but the government let them die in the House of Commons.

By the turn of the century, the Nonconformists had lost much of their self-confidence in the face of growing secularism. Many Nonconformists moved their allegiance from the Liberals to the Labour Party of James Keir Hardie, himself a teetotaler. This evolution reduced the strength of the alliance in particular and of direct local-veto prohibition in general. It was the beginning of the end for the triangle of temperance, Nonconformity, and the Liberal Party.

Margaret Barrow

See also: Band of Hope; Burns, (James) Dawson; Harcourt, Sir William; Labour Party (United Kingdom); Lawson, Sir Wilfrid; Liberal Party (United Kingdom); Livesey, Joseph; Local Option; Lucas, Margaret Bright; Pope, Samuel; Raper, James Hayes; United Kingdom Alliance (UKA)

References

Christian Faith and Practice in the Experience of the Society of Friends. London: London Yearly Meetings of the Religious Society of Friends, 1960.

Fahey, David M. "Drink and the Meaning of Reform in Late Victorian and Edwardian England" in *Cithara.* 13, no. 2 (May 1974): 46–56.

———. "Pressure Groups and the British Liberal Party, 1883–1908" in *Social Science.* 54, no. 2 (Spring 1979): 76–85.

Harrison, Brian. *Dictionary of British Temperance Biography.* Aids to Research, no. 1, *Bulletin Supplement.* Coventry, UK: Society for the Study of Labour History, 1973.

———. *Drink and the Victorians: The Temperance Question in England, 1815–1872.* London: Faber & Faber, 1971; 2d ed., Keele: Keele University Press, 1994.

Hudson, Thomas. *Temperance Pioneers of the West.* London, National Temperance Publication Depot, 1887; 2d ed., 1888.

Winskill, Peter T. *The Temperance Movement and Its Workers: A Record of Social, Moral, Religious, and Political Progress.* 4 vols. London: Blackie, 1891–1892.

———. *Temperance Standard Bearers of the Nineteenth Century: A Biographical and Statistical Temperance Dictionary.* Liverpool: the author, 1897–1898.

Non-Partisan Woman's Christian Temperance Union

The Non-Partisan Woman's Christian Temperance Union, organized in 1889, was a splinter group of the Woman's Christian

Temperance Union (WCTU) in the United States. The WCTU was bitterly divided on the question of whether or not to affiliate with the Prohibition Party, which it did in 1884 under the leadership of Frances Willard. Willard's partisan strategy raised a hailstorm of criticism from some members. Finally, after years of bitter internal fights, Willard's opponents walked out of the WCTU convention in 1889 and formed the new Non-Partisan WCTU. This schism led to a clarification of nonpartisan political strategy and helped set the course for the eventual "dry" victory.

Judith Ellen Foster was the moving spirit behind the Non-Partisan WCTU, although she refused high office in the organization. Foster, one of the founders of the WCTU in 1874 and a prominent officer thereafter, was active in the Republican Party and founded the Women's Republican Association. Dedicated to prohibition, Foster believed that success lay in pursuing a nonpartisan strategy and pressuring politicians of all parties to support the reform. Willard's decision to pursue the affiliation with the Prohibition Party in 1884 not only cost the Republicans votes but also threatened Republican legislators who were sympathetic to the prohibition cause. Foster believed the endorsement had actually set back prohibition because the Prohibition Party was a small group unlikely to win significant offices. She thus fought Willard on the issue and, unsuccessful in persuading the popular Willard to drop her partisanship, in 1889 finally broke with the WCTU. At least 10,000 members left with Foster, among them Annie Wittenmeyer, who had been the first WCTU president.

Foster recognized that successful parties were complex coalitions seeking to win electoral majorities. This situation necessarily entailed compromises, and Foster was willing to accept these. License laws to regulate the liquor trades, for instance, although far short of what Foster wanted, at least demonstrated government power over the liquor businesses. Although for Republican prohibitionists like Foster, compromises were often disappointing, it was worse to throw support to a third party, such as the Prohibition Party, and divide the dry vote; the result was only to elect more "wet" Democrats. Foster believed that it was necessary to mobilize public opinion to lead the parties toward prohibition. Although a loyal Republican, her higher loyalty was to prohibition, and she would happily see a wet Republican defeated by a dry Democrat. Her goal, and the goal of the Non-Partisan WCTU that she inspired, was to promote abstinence from drink; laws and politics were a means toward that end.

Thus, Foster, anguished over partisanship, developed a clear vision of a nonpartisan strategy. This same strategy was eventually adopted by the Anti-Saloon League of America and successfully followed for the enactment of prohibition. When the league emerged in 1895 as a national organization struggling to raise funds and win supporters, the Non-Partisan WCTU lent it critical support. Local Non-Partisan WCTU chapters provided small sums for the league and helped it contact important benefactors. Wittenmeyer served as president of the Non-Partisan WCTU from 1896 to 1898, while also serving as a trustee of the Anti-Saloon League of America.

The Non-Partisan WCTU receded when the WCTU returned to a nonpartisan policy after Willard's death in 1898. Local chapters sometimes continued, eventually doing philanthropic work under different names.

K. Austin Kerr

See also: Anti-Saloon League of America (ASLA); Foster, Judith Ellen Horton; Prohibition Party (United States); Republican Party (United States); Willard, Frances Elizabeth Caroline; Wittenmeyer, Annie Turner; Woman's Christian Temperance Union (WCTU) (United States)

References

Gustafson, Melanie. "Partisan and Nonpartisan: The Political Career of Judith Ellen Foster" in *We Have Come to Stay: American Women and Political Parties,* edited by Melanie Gustafson, Kristie Miller, and Elisabeth I. Perry, 1–12. Albuquerque: University of New Mexico Press, 1999.

Kerr, K. Austin. *Organized for Prohibition: A New History of the Anti-Saloon League.* New Haven, CT: Yale University Press, 1985.

Norway

Early History

Nature provided Norway with a rather barren land and a relatively inhospitable climate, making survival itself a true challenge. Moderation and hard work therefore have been a cornerstone of social life throughout most of its history. Early alcohol use was primarily related to times of celebration and special occasions, and drinks consisted of traditionally home-brewed beer, and occasionally mead. However, festivities could take place over several days and were at times rather intense.

Spirits became more and more common from the 1600s, and consumption rose rapidly. However, given that Norway was a poor country, consumption never rose to the same levels characteristic of its wealthier and more powerful neighbors, Sweden and Denmark. Generally, anyone could distill their own spirits, but with time restrictions were introduced. First, in 1538, clergymen were forbidden to distill spirits on their lands. Later, general prohibitions were introduced at various times in response to crop failures and famine in order to prevent the use of grains and potatoes for spirits production. A permanent prohibition on spirits production was introduced in 1756. At the time, Norway was under Danish rule, and Danish merchants held a monopoly on the import of spirits, a factor that clearly fostered such a prohibition. This law also led to bitterness among Norwegians and widespread home distilling, however, despite extensive control tactics and severe repercussions.

As a result of the major political changes taking place in Europe (the Napoleonic Wars), Norway fell under Swedish control in 1814, although it adopted its own constitution the same year. During the work of creating the new legislation, it was suggested that the right to distill one's own spirits should be grounded in the constitution. This reasoning should be understood against the backdrop of the situation under Danish rule as well as in light of general liberal trends characterizing the economic philosophy of the time. In the end, the right to distill spirits was not adopted in the constitution, but such a right was granted to all those who owned or disposed of land and all citizens in the townships. Although reliable alcohol statistics are not available from this time, there are strong indications that alcohol consumption rose sharply thereafter, that is, during the so-called "liberal period." In 1933, the Trade Committee of the Norwegian parliament made a serious attempt to calculate spirits production in the country and arrived at a figure equivalent to 12 liters (3.2 U.S. gallons) pure alcohol per inhabitant over fifteen years of age. If beer and wine consumption are put into the equation and perhaps a modest amount of smuggled goods, this means nearly 15 liters (4 U.S. gals.) of pure alcohol per adult inhabitant; that is, more than twice today's real consumption (including "unregistered" alcohol).

Some restrictions and high taxes on spirits were being applied by 1827, and in the years that followed these became even more restrictive. New provisions were applied to the size of distilling equipment, and taxes steadily increased. The Production Law of 1848 introduced so many conditions and such heavy taxes that in reality it brought an end to legal distilling. In 1850, only forty legal distilleries remained in the entire country. The industrialization of spirit production was thereby concluded, making it easier to control activities and utilize prices more effectively both in terms of fiscal and alcohol policy. Of course, full control was never possible, and it is likely that illegal distilling resumed during this period, even if a sizable number of control authorities were also enlisted.

The industrialization of spirits production, and in time beer production, made it possible to obtain fairly reliable alcohol statistics as well. The official alcohol statistics from the Norwegian national statistics bureau provide yearly figures for consumption of beer, wine, and spirits from 1851 until the present day. As seen in the table on p. 46, alcohol consumption showed a general downward trend during the last half of the 1800s. This trend likely started some years before 1851, since consumption during this year was estimated to be less than half of that from 1833. The downward trend can hardly be attributed to restrictions alone. In time, there arose a popular reaction to the high level of consumption during the 1820s and 1830s and the obvious problems associated with it. Reactions were clearly expressed in public debate, enabling the introduction of the many restrictions that came into force during this period. Public opinion also manifested itself in the organization of a moderation movement based on the idea that the example set by the good citizen would conquer the habitual drinking of spirits found among lower social classes in the population.

The moderation movement was thus anchored in the upper social classes. It lost some of its influence during the 1850s, since it was regarded as having achieved many of its goals. A far more deep-seated shift in values seems to have taken place among other segments of the population, and in 1859 Asbjørn Kloster, a Quaker, founded the first teetotaler association in Stavanger. This was the beginning of one of the greatest grassroots movements in Norway's history.

At first deeply religious, the movement was strongest in the southern and western areas of the country, where both the Norwegian Church and a long list of parishes had a stronghold and alcohol use was limited. It was not until after 1880 that a strong rise in membership took place across the rest of the country, and this growth continued until World War I. At the turn of the century, the movement claimed 135,000 members, and at its height probably nearly double this number. The movement enjoyed great sympathy across different groups and also had strong political influence, both locally and at the national level.

Prohibition

The goal of the temperance movement was an alcohol-free society. However, there was some disagreement within the movement with respect to how this goal was to be achieved: Should one dry out the country gradually via local referenda, such as was theoretically provided for by a law passed in 1894? Or should one attempt to push through a prohibition that could dry out the country all at once?

When the war broke out in 1914, headway had been gained with the gradual strategy. World conditions now intervened, however, and in 1916 and 1917 a total prohibition on first spirits and later fortified wine and beer (medium and strong) was introduced. These temporary prohibitions were understood and respected by a majority of the population on account of the scarcity of goods and the uncertain state of the world. This success inspired followers in the movement to seek a permanent national prohibition, and the temperance movement pursued this strategy, albeit with some reservations. The government did not manage to come to agreement on the issue and instead suggested that it be presented through a public referendum, a move that received unanimity in the Storting (parliament).

The referendum was held in October 1919 and concerned a permanent prohibition on the sale of both spirits and fortified wine. It resulted in a 62 percent majority in favor of prohibition. However, there were wide geographical variations. In the towns, only 44.5 percent voted in favor, and influential communities such as the press were overwhelmingly against prohibition. Moreover, maintaining the prohibition effectively soon proved to be difficult. The art of home distilling had in no way fallen into decline, and smuggling eventually became extensive. Control measures were intensified, especially with respect to smuggling rings, and tough confrontations arose that led to the loss of human life. Additionally, a number of physicians and veterinarians prescribed large doses of spirits as "medicine" for humans and animals. The prohibition led to political problems related to trade, since France, Spain, and Portugal demanded that Norway accept substantial quantities of both spirits and fortified wines in exchange for their imports of Norwegian seafood

Alcohol Consumption in Norway, per Capita (Liters of Pure Alcohol) (Population Fifteen Years of Age and Older), 1851–2000

	Total	*Spirits*	*Wine*	*Beer*
1851	5.47	4.68	0.10	0.69
1852	5.38	4.54	0.11	0.73
1853	5.31	4.46	0.10	0.75
1854	5.58	4.58	0.11	0.89
1855	6.04	4.93	0.10	1.01
1856	6.34	5.12	0.09	1.13
1857	6.04	4.73	0.06	1.25
1858	5.98	4.56	0.05	1.37
1859	5.70	4.15	0.09	1.46
1860	5.38	3.86	0.06	1.46
1861	4.96	3.83	0.07	1.06
1862	4.59	3.46	0.06	1.07
1863	4.22	3.06	0.07	1.09
1864	4.62	3.41	0.07	1.14
1865	4.64	3.51	0.08	1.05
1866	5.00	3.98	0.10	0.92
1867	4.93	3.88	0.09	0.96
1868	4.95	3.93	0.09	0.93
1869	4.95	3.96	0.10	0.89
1870	4.24	3.21	0.09	0.94
1871	4.91	3.82	0.10	0.99
1872	4.85	3.69	0.12	1.04
1873	5.23	3.81	0.15	1.27
1874	6.34	4.70	0.16	1.48
1875	6.92	4.98	0.16	1.78
1876	6.87	5.07	0.15	1.65
1877	6.56	4.74	0.16	1.66
1878	5.57	3.81	0.13	1.63
1879	4.31	2.79	0.11	1.41
1880	4.31	2.79	0.13	1.39
1881	4.00	2.51	0.14	1.35
1882	4.16	2.67	0.14	1.35
1883	4.21	2.64	0.13	1.44
1884	4.15	2.61	0.13	1.41
1885	4.24	2.68	0.12	1.44
1886	3.68	2.43	0.12	1.13
1887	3.45	2.20	0.11	1.14
1888	3.73	2.28	0.13	1.32
1889	3.89	2.43	0.15	1.31
1890	4.20	2.41	0.17	1.62
1891	4.73	2.67	0.23	1.83
1892	4.44	2.54	0.17	1.73
1893	4.48	2.57	0.14	1.77
1894	4.68	2.77	0.17	1.74
1895	4.38	2.64	0.22	1.52
1896	3.91	1.96	0.55	1.40
1897	3.73	1.59	0.61	1.53
1898	4.10	1.75	0.51	1.84
1899	4.58	2.18	0.42	1.98
1900	4.74	2.41	0.43	1.90
1901	4.54	2.44	0.37	1.73
1902	4.44	2.41	0.41	1.62
1903	3.93	2.27	0.34	1.32
1904	3.62	2.25	0.13	1.24

(continues)

(continued)

	Total	*Spirits*	*Wine*	*Beer*
1905	3.28	1.92	0.13	1.23
1906	3.21	1.92	0.16	1.13
1907	3.34	2.06	0.17	1.11
1908	3.41	2.14	0.17	1.10
1909	3.59	2.24	0.18	1.17
1910	3.84	2.44	0.21	1.19
1911	3.94	2.45	0.21	1.28
1912	4.22	2.63	0.23	1.36
1913	4.27	2.75	0.24	1.28
1914	3.99	2.35	0.27	1.37
1915	3.94	2.36	0.31	1.27
1916	4.50	2.51	0.46	1.53
1917	1.49	–	0.34	1.15
1918	0.61	–	0.31	0.30
1919	1.90	–	0.93	0.97
1920	2.81	–	0.82	1.99
1921	2.99	0.64	0.30	2.05
1922	2.92	0.65	0.24	2.03
1923	3.04	0.83	0.39	1.82
1924	2.84	0.56	0.67	1.61
1925	3.07	0.37	0.96	1.74
1926	2.78	0.36	0.86	1.56
1927	3.03	1.11	0.58	1.34
1928	3.06	1.43	0.43	1.20
1929	2.99	1.41	0.42	1.16
1930	3.03	1.44	0.42	1.17
1931	2.65	1.32	0.38	0.95
1932	2.48	1.19	0.35	0.94
1933	2.35	1.17	0.32	0.86
1934	2.36	1.16	0.33	0.87
1935	2.43	1.21	0.33	0.89
1936	2.58	1.28	0.36	0.94
1937	2.78	1.38	0.40	1.00
1938	2.84	1.43	0.39	1.02
1939	2.85	1.43	0.37	1.05
1940	2.24	1.05	0.27	0.92
1941	1.88	1.08	0.19	0.61
1942	1.10	0.84	0.13	0.13
1943	1.68	1.49	0.06	0.13
1944	1.81	1.65	0.03	0.13
1945	1.68	1.42	0.04	0.22
1946	3.10	2.33	0.12	0.65
1947	3.06	2.11	0.17	0.78
1948	2.98	2.12	0.20	0.66
1949	3.01	1.84	0.18	0.99
1950	2.87	1.61	0.20	1.06
1951	2.76	1.40	0.23	1.13
1952	2.83	1.42	0.25	1.16
1953	2.75	1.35	0.25	1.15
1954	2.88	1.45	0.26	1.17
1955	3.11	1.51	0.25	1.35
1956	3.14	1.52	0.25	1.37
1957	3.15	1.54	0.25	1.36
1958	3.15	1.52	0.23	1.40
1959	3.33	1.61	0.25	1.47
1960	3.45	1.71	0.25	1.49

(continues)

(continued)

	Total	*Spirits*	*Wine*	*Beer*
1961	3.60	1.78	0.25	1.57
1962	3.54	1.76	0.25	1.53
1963	3.66	1.79	0.26	1.61
1964	3.58	1.71	0.27	1.60
1965	3.72	1.76	0.28	1.68
1966	3.85	1.80	0.30	1.75
1967	4.05	1.88	0.33	1.84
1968	4.32	1.95	0.39	1.98
1969	4.49	1.96	0.42	2.11
1970	4.73	2.07	0.44	2.22
1971	4.93	2.10	0.48	2.35
1972	5.11	2.20	0.50	2.41
1973	5.22	2.18	0.53	2.51
1974	5.58	2.40	0.57	2.61
1975	5.63	2.42	0.59	2.62
1976	5.66	2.44	0.58	2.64
1977	5.74	2.49	0.59	2.66
1978	5.17	2.03	0.51	2.63
1979	5.65	2.35	0.64	2.66
1980	5.98	2.46	0.75	2.77
1981	5.32	2.06	0.69	2.57
1982	4.83	1.57	0.57	2.69
1983	4.85	1.63	0.65	2.57
1984	4.98	1.62	0.71	2.65
1985	5.22	1.77	0.82	2.63
1986	5.27	1.61	0.83	2.83
1987	5.38	1.63	0.93	2.82
1988	5.26	1.45	0.97	2.84
1989	5.08	1.33	0.97	2.78
1990	4.99	1.24	0.95	2.81
1991	4.90	1.12	0.93	2.85
1992	4.67	1.00	0.91	2.76
1993	4.55	0.96	0.92	2.67
1994	4.74	1.00	0.99	2.75
1995	4.79	1.00	1.04	2.75
1996	5.04	1.02	1.12	2.88
1997	5.35	1.08	1.28	2.95
1999	5.45	1.05	1.50	2.84
2000	5.66	1.05	1.62	2.93

products. Thus, the prohibition on fortified wine was dropped by 1923.

As seen in the table, the level of registered alcohol consumption was by no means low during these years, and given that unregistered consumption is assumed to have reached record levels, it is hard to characterize the prohibition as a political success. In addition, large groups of otherwise law-abiding citizens became involved in an illegal alcohol trade. The press's negative opinion likely had its effect as well. Therefore, support for prohibition gradually declined, and in 1926 the government proposed that the public once again be consulted. A referendum was held in October of the same year and a majority of nearly 56 percent favored ending prohibition. On 5 April 1927, the Storting adopted a new alcohol law, and from 2 May one could legally purchase spirits in the nine towns that had had spirits sales in 1916.

The Wine and Spirits Monopoly

Up until the time of prohibition, the sale of alcohol had taken place in private licensed shops, often in the form of specialty stores for wine and spirits. Early on, consideration had been given to monopolizing these sales, and in 1871 the local merchants in Kristiansand joined forces and created the first cooperative spirits trade in Norway. The idea was borrowed from the Gothenburg system in Sweden. In Kristiansand, the cooperative spirits trade was in practice a monopoly already from its inception, but in principle all merchants could still obtain a license to sell wine and spirits. In 1894, fifty-one towns had cooperative liquor stores; in the same year, an act was passed under whose provisions the cooperative stores virtually achieved a monopoly on the sale of liquor. During the years that followed, the strategy of the temperance movement was to reduce the number of cooperative liquor stores through local referenda, and this strategy proved highly effective. The number of spirits stores had dropped to only thirteen by 1913.

During prohibition, more specifically in 1922, A/S Vinmonopolet (The Wine Monopoly Limited) was established according to an act of Parliament. The Vinmonopol was from its inception a private limited company under government control. The name of the company was at the time a fairly accurate description of its activity, which mainly involved the import and sale of ordinary table wines, and from 1923 also fortified wines. Under the Alcohol Act of 1927, the Vinmonopol in time also acquired a monopoly on the import and sale of liquor. This activity started already in 1927, although the cooperative liquor stores still existed in many towns. In spite of the widening of the Vinmonopol's activity, its original name, referring only to wine, was retained.

A Royal Decree of 1932 determined that the Vinmonopol had the sole right to import and wholesale liquor, wine, fruit wine, and beer; to retail liquor, wine, and fruit wine; to serve liquor; and to export liquor.

Despite these provisions, the old cooperative stores were allowed to continue to operate up to 1938, when the Storting resolved that they should close down. The Vinmonopol took over the cooperative stores and a complete state monopoly was thereby created in the areas provided for by the Royal Decree. Beer could be bought from licensed grocery stores, but from 1993 the right to retail strong beer (over 4.75 percent alcohol by volume) was added to the sole rights of the monopoly. However, from 1990 the sole right to serve liquor was repealed, and from 1996, as a result of the European Economic Area (EEA) agreement with the European Union (EU), the sole right to export, import, and wholesale was also abolished. From then on the Vinmonopol has become a pure retail monopoly.

Fifty years ago there were 46 Vinmonopol stores in the country. This number gradually grew throughout the postwar period, and in the year 2000 there were 141 stores. A sharp increase in this number took place during the years around the turn of the millennium, and it was decided that by the end of 2002 there would be 184. Opinions about the monopoly arrangement have become more negative during the past decade; in particular, many people would like wine to be sold in grocery stores. Thus, improving availability has been emphasized in order to avoid a loss of legitimacy for the monopoly. The government is attempting to solve the problem not only through opening more stores but also through steadily transforming monopoly outlets into self-service stores.

Drinking Patterns

As indicated in the table, spirits have been the dominant alcoholic beverage in Norway for most of the period since 1851, and were even more dominant during the centuries preceding this date. However, this trend has now changed. Registered levels of spirits consumption reached their height in 1980 and have since dropped by more than 50 percent. Beer consumption has risen during the entire postwar period, and wine consumption has also risen, especially after the 1960s. Registered beer consumption surpassed spirits consumption in 1968, and wine consumption surpassed spirits consumption in 1995, as calculated in pure alcohol. However, one should bear in mind that these figures concern trends in the *registered* consumption. It has often been pointed out that real consumption might look different, and this is at least partially true. Studies conducted by the Norwegian Institute for Alcohol and Drug Research confirm that a considerable volume, probably 25 percent, of the total alcohol consumption in Norway is unregistered. Much of this alcohol consists of home-distilled and smuggled spirits, which are illegal goods. However, an increasing share of unregistered consumption consists of legal goods brought home after international travel (tax-free or border trade). This trend has led to a slight increase in consumption of unregistered (although not necessarily illegal) spirits since 1980. But even if one includes an estimate for this category, there has been a substantial decline in the total level of spirits consumption, likely on the order of 35 percent. Total alcohol consumption (beer, wine, and spirits) has also dropped since 1980, but it has increased again over the past few years, so that it now is quite close to the level of 1980. It is also clear that a significant shift in people's preferences has taken place. This is a trend that is common for most northern European countries and is probably due in large part to an increase in travel and therewith contact with other drinking cultures, particularly the predominantly wine-drinking cultures of the Mediterranean. Moreover, the internationalization of the media and entertainment industry has been an important factor, particularly with respect to young people.

The age limit for purchasing or being served beer and wine in Norway is eighteen, and twenty for purchasing spirits. Nonetheless, it appears that it is not especially difficult for young people below these ages to obtain alcohol, and the average age for first-time consumption of alcohol falls under fifteen. According to a major international study (Hibell et al. 2000), the average Norwegian youth (fifteen or sixteen years old) comes out relatively low compared to other young people in Europe with respect to frequency of drinking occasions. And yet Norwegian youths rate quite high with respect to the quantity of alcohol consumed and frequency of intoxication. A similar situation is found among the adult population: Average alcohol consumption among Norwegians is very low compared with that of adults in other Western countries, but the Norwegians manage to squeeze quite a lot of intoxication out of the amount that is consumed. In this way, Norwegians can be likened to their neighbors in the other Nordic countries. As mentioned previously, this drinking pattern has deep historical roots.

Some have also posited that this drinking pattern is due to the high price of alcohol in these countries: Alcoholic drinks must be used as "effectively" as possible; one cannot afford to "waste" alcohol on daily sipping. Others maintain the opposite position: that Norwegian drinking practices are so unique and harmful that one must have high alcohol taxes as a means of prevention. Regardless of which position is most correct, it is a fact that prices are quite high in Norway, even in relation to neighboring countries. This situation has led to a generally high level of border trade, especially from Sweden, which places constant pressure upon the Norwegian authorities to reduce taxes.

Alcohol policy practices in Norway are currently under pressure. To some extent, a liberalization has already taken place. Some elements of the monopoly structure have been dissolved, and the path has been paved for even greater availability in the form of more stores and self-service solutions. Strong forces within public opinion and political parties would like to see wine sold in grocery stores, and the temperance movement has been greatly weakened, both in terms of its membership numbers and political influence. So far, taxes have only been marginally affected, but the government is coming under great pressure to reduce them, a situation that will likely be aggravated as a result of the fact that Norway's EU neighbors Sweden and Finland must now allow nearly unlimited private import of alcohol from other and much cheaper EU countries as of 2004. Even if Norway is not a member of the EU, and can therefore follow an independent price policy in principle in this area, the price trends in neighboring countries will naturally have an indirect effect. If Norway continues to have a somewhat open border with neighboring countries and the rest of Europe, the tax levels will in all likelihood be reduced in the near future, but they will likely continue to be the highest in Europe.

Sturla Nordlund

See also: Denmark; Gothenburg Schemes/Disinterested Management; Sweden

References

Brun-Gulbrandsen, S. "Våre forfedres alkoholbruk" (Our ancestors' use of alcohol) in *Alkohol I Norge* (Alcohol in Norway), edited by O. Arner, R. Hauge, and O. J. Skog. Oslo: Universitetsforlaget, 1985.

Fuglam, Pen. *Kampen om alkoholen I Norge, 1816–1904.* Oslo: Universietsforlaget, 1972.

Hibell, B., et al. *The 1999 ESPAD Report: Alcohol and Other Drug Use among Students in 30 European Countries.* Stockholm: Swedish Council for Information on Alcohol and Other Drugs (CAN), 2000.

Holder, H., E. Kühlhorn, S. Nordlund, E. Österberg, A. Romelsjö, and T. Ugland. *European Integration and Nordic Alcohol Policies.* Aldershot: Ashgate, 1998.

Horverak, Ø., S. Nordlund, and I. Rossow. *Om sentrale deler av norsk alkoholpolitikk* (On central parts of Norwegian alcohol policy). Report No. 1. Oslo: Norwegian Institute for Alcohol and Drug Research, 2001.

Johansson, L. "Sources of the Nordic Solutions" in *Broken Spirits: Power and Ideas in Nordic Alcohol Control,* edited by P. Sulkunen, C. Sutton, C. Tigerstedt, and K. Warpenius, 17–43. NAD Publication No. 39. Helsinki: Nordic Council for Alcohol and Drug Research, 2000.

Lohiniva, R. J. B. L., ed. *Alcohol and Drugs in Norway 2001.* Oslo: National Institute for Alcohol and Drug Research, 2001.

Mäkelä, P., K. Fonanger, B. Hibell, S. Sabroe, and J. Simpura. "Episodic Heavy Drinking in Four Nordic Countries: A Comparative Survey" in *Addiction.* 96 (2001): 1575–1588.

Nordlund, S. "Norway: A Country Profile" in *State Monopolies and Alcohol Prevention,* edited by T. Korteinen, 157–190. Report No. 181. Helsinki: Social Research Institute for Alcohol Studies, 1989.

Nugent, James (1822–1905)

James Nugent (1822–1905) was a leading figure in the Roman Catholic temperance movement in England. A priest, he organized the Catholic Total Abstinence League of the Cross in Liverpool in 1872. The following year, Archbishop Henry Edward Manning established the League of the Cross on a nationwide basis and Father Nugent affiliated his society to it.

Born in Liverpool, Nugent was first trained at an English seminary and then sent to the English College in Rome to complete his studies. Ordained in 1846, he was assigned to a parish in Liverpool in 1849 and remained in that city for the rest of his life. In addition to his parish responsibilities, he served as a prison chaplain, founded shelters for homeless boys, pregnant women, and unwed mothers, and began a newspaper, the *Catholic Times.* According to his biographer, however, the total abstinence movement was "what he regarded as his greatest work" (Bennett 1949: 108). For Nugent, total abstinence provided a means for Liverpool's Irish working class to prosper and gain acceptance in English society.

Nugent was not the first person to promote temperance among the Catholics of Liverpool. In 1843, Father Theobald Mathew, Ireland's "Apostle of Temperance," spent several days in Liverpool administering the total abstinence pledge to 40,000 people, mostly Irish immigrants. During the Great Famine (1845–1849), Liverpool's Irish population increased, and Mathew returned twice to offer the pledge to the new arrivals and encourage long-standing teetotalers. Although some individuals remained faithful to their pledges, no solid organization ever developed in England, and the movement languished after 1850.

When Nugent began his temperance work, he operated in a different fashion from Mathew. Although Mathew would give the pledge to Protestants as well as Catholics, Nugent only allowed practicing Catholics to join the league. Although just as ecumenically minded as Mathew, he recognized that his organization needed to be closely tied to the Catholic Church in order to continue over the long term. The American-based Catholic Total Abstinence Union (CTAU) and the Irish-based Pioneers of the Sacred Heart adopted the same membership rules as the league.

Nugent was well aware of the attractions pubs offered working-class men and recognized that teetotalers needed a place of their own to gather and socialize. In 1875, he purchased land and built a grand League Hall that could seat 2,000. Meetings were held there each week, and concerts were performed there every other week.

Although stationed continuously in Liverpool, Nugent maintained strong ties with Ireland and Irish America. In founding the league, he was aided by John Denvir, an Irish revolutionary associated with the Fenians. Although not as militant as Denvir, Nugent sympathized with Irish nationalism.

In 1870, Nugent made the first of a dozen trips to the United States to promote the immigration of destitute children to rural America. During his travels, he came to know a number of Irish-born priests and bishops. Among his friends were Archbishop John Ireland of St. Paul and Bishop John Keane of Richmond, who both visited Liverpool in 1886 and delivered temperance lectures at Nugent's request. Nugent came to Baltimore in 1889 to speak at the Catholic Congress marking the centenary of John Carroll's appointment as the first Catholic bishop in the United States. He was an honored guest as well at the CTAU events held in conjunction with the congress.

Nugent was given the title of monsignor by Pope Leo XIII in 1892. At his death in 1905, people of all creeds and classes in Liverpool mourned him. The following year, a bronze statue was erected in Liverpool honoring him as the "Apostle of Temperance" and the "Father of the Poor." Unfortunately, as his biographer noted, the League of the Cross "did not long survive its founder" (Bennett 1913: 113).

John F. Quinn

See also: Manning, Henry Edward; Mathew, Father Theobald; Pioneer Total Abstinence Association of the Sacred Heart

References

Bennett, John. *Father Nugent of Liverpool.* Liverpool: Wood, Westworth, 1949.

Catholic Encyclopedia, vol. 11, 150–151. New York: Appleton, 1913.

O

Oglethorpe, James Edward (1696–1785)

James Edward Oglethorpe (1696–1785) was founder of Georgia, originally a penal colony in the United States. Born the youngest son in a virulently Jacobite family and therefore from a group of political outsiders, Oglethorpe made his career as a junior officer in the British army before attending Corpus Christi, Oxford, and then serving under Eugene of Savoy in the Balkans against the Ottoman Empire. Although he was a participant, the drunken antics of the soldiers disgusted Oglethorpe, as did the politicians he dealt with later as he ran for a seat in Parliament from Hasslemere, Surrey, especially after an election brawl resulted in the death of a servant at his hands. Taking his seat in the House of Commons after a brief incarceration for manslaughter, Oglethorpe threw himself into the cause of prison reform, heading a commission to clean up practices at London's jails, including the sale of offices, the operation of taverns and brothels for the warden's profit, and the appalling conditions of the prisoners.

Seeing the solution to prison conditions and the plight of the city's poor in emigration, Oglethorpe planned a penal colony in North America, without slaves, parceled out in small units of 50 acres to encourage family farming, and without hard liquor (beer and wine, crucial when the water supply was untrustworthy, were to be allowed). The colony's cash crop would be silk. With nineteen other trustees, Oglethorpe gained a royal grant from George II in 1732, and settlement began in 1733. Oglethorpe insisted on honest dealings with the neighboring Creek Indians, refusing to allow the settlers to sell them alcohol, setting fair prices for trade exchanges, and sponsoring the London visit of the chief's son Tomochichis. He was also remarkably tolerant of settlers who were Moravians, Huguenots, and even Spanish-Portuguese Jews.

Oglethorpe's main occupation in Georgia, however, was defending it against the neighboring Spanish at St. Augustine, leading missions against the Spanish during the War of Jenkins' Ear, and fortifying Fort Frederica against a Spanish landing. Though he was a talented military commander, Oglethorpe's insistence on sober behavior and fair dealings with the Native Americans soured his relationship with the settlers, who often decamped for South Carolina to own slaves and import whiskey. Disillusioned, Oglethorpe returned to London in 1743, where he faced both parliamentary inquiry and a court-martial on complaints from unhappy Georgian settlers, and although he was cleared of both charges, suspicions of Jacobite sympathies during the 1745 rebellion finished his career.

Oglethorpe died in Britain on 30 June 1785. In his later years, he raised money to finance the British Museum and continued in his criticism of prison conditions, navy press gangs, the British prosecution of the American Revolution, and the institution of slavery. Once Oglethorpe had left Georgia, the remaining trustees there made significant changes that destroyed Oglethorpe's hopes for the colony. They allowed rum in 1742, slavery in 1750, and quickly changed to a system of plantation agriculture, negating the founder's grand plans for a penal colony without, in his opinion, society's worst vices.

Margaret Sankey

See also: Native Americans: Drinking Patterns and Temperance Reform

References

Garrison, Webb. *Oglethorpe's Folly: The Birth of Georgia.* Lakemont, GA: Copple House Books, 1982.

Reese, Trevor R. *Colonial Georgia: A Study in British Imperial Policy in the Eighteenth Century.* Athens: University of Georgia Press, 1963.

Vaeth, J. Gordon. *The Man Who Founded Georgia.* New York: Crowell-Collier, 1968.

O'Keefe Brewery

Toronto's O'Keefe Brewery was a major player in the Canadian beer market from the late 1880s to the 1960s. In 1861, Eugene O'Keefe, a native of Bandon, County Cork, Ireland; George M. Hawke; and brewer Patrick Cosgrave bought the Victoria Brewery on Toronto's Victoria Street. The thirty-four-year-old O'Keefe was not a trained brewer; however, his prior experience in the grocery and hotel trades and in banking were to serve the brewery well. Of further advantage were O'Keefe's and Cosgrave's networks in Toronto's growing Irish Catholic community. Through sound financial management and aggressive local marketing, the partners quickly boosted the Victoria Brewery's fortunes, doubling output to 50,000 imperial gallons (60,000 U.S. gallons) within two years. Despite this success, Cosgrave left the firm in 1863 for the West Toronto Brewery. Continuing as O'Keefe and Company, O'Keefe and Hawke answered the dominant popular call by brewing traditional English ales and Irish stout.

Their beer received considerable public favor. By the early 1870s, the brewers were annually producing more than 350,000 imperial gallons (420,000 U.S. gallons), mostly for the Toronto market. To meet growing demand, the Victoria Brewery underwent major plant expansion in 1864–1865 and again in 1872. O'Keefe and Hawke also monitored and responded to consumer trends. Such prudence informed their diversification into lager in the mid-1870s. At the time, the Germanic drink, customarily served chilled, was becoming a favorite during Canada's hot and hazy summer months. Finding success, O'Keefe and Company built a self-contained lager plant in 1878. The new facility, then Canada's biggest, came into full production the next year.

A New Strategy

In 1882, the firm was altered when Widmer Hawke replaced his father George. The new partnership continued as O'Keefe and Company. By the early 1880s, prominent brewers outside Toronto, such as London's Carling and Labatt, had made significant inroads into Toronto, Ontario's largest urban center. Heretofore, O'Keefe and Company had largely relied upon sales in Toronto and its environs. In order to meet the challenge posed by non-Toronto brewers, the firm adopted a broader geographic strategy. Inaugurated in the mid-1880s, the marketing plan was based upon a combination of traveling sales staff, railway shipping, and brand-driven advertising. As new bar and agency accounts were secured across Ontario and in western Quebec, beer was shipped out by rail. Since lucrative Toronto sales had accorded the Victoria Brewery substantial capital strength, competing on a larger stage was feasible. Moreover, economies of scale allowed the brewer to be price competitive in distant markets. O'Keefe's advertising campaign was demographically sophisticated and centered upon regional dailies, such as the *London Free Press,* and upon such audience-specific journals as *The Week,* an up-market current affairs magazine; *The Labor Advocate,* a skilled-trades organ; and a variety of ethnic newspapers, including Ontario's major Irish Catholic and Irish Protestant weeklies. The national advertising campaign featured the brewery's flagship labels, O'Keefe and Company's Lager and O'Keefe and Company's Imperial Pale Ale. The latter brand exploited Canada's sense of Britishness, despite the moderate Irish nationalism of the brewery's senior partner.

The overall program worked, and annual production eclipsed 1 million imperial gallons (1.2 million U.S. gallons) in the early 1890s, firmly placing the O'Keefe brewery among Canada's top ten. On 1 September 1891, the O'Keefe Brewery Company of Toronto Limited was incorporated, with Eugene O'Keefe as president. Stock subscription financed a massive plant enlargement the following year. This saw the brewery expand to cover an entire city block. The new company remained faithful to the sales strategy adopted a decade before. In 1896, a bottling plant was constructed in Ottawa. O'Keefe also introduced new brands in the 1890s and early 1900s, including O'Keefe's Liquid Malt Extract, a medicinal distributed through an appointed pharmaceutical wholesaler, and O'Keefe's Invalid Stout. The same period also witnessed the adoption of new technologies, including crown-cap bottles, artificial refrigeration, plant electrification, and motorized delivery vehicles. Reflecting remarkably increased sales, a new brewery, with a 500,000-barrel capacity, was erected in 1911.

The same year, Eugene O'Keefe's only son, Eugene Bailey O'Keefe, died. Dejected, the elder O'Keefe sold his brewery shares to Widmer Hawke and Sir Henry Mill Pellatt, the Canadian financier who built Casa Loma, a lavish mansion that is still a Toronto landmark. Continuing his previous generous support of Toronto's Catholic community—his philanthropy led to his appointment in 1909 as the first Canadian chamberlain to the Pope—O'Keefe used the share proceeds to finance various local Catholic charities and missions. Most notably, he donated $400,000 of his brewery wealth toward the construction of Toronto's St. Augustine's Seminary.

Prohibition and Absorption

The popular Eugene O'Keefe died on 1 October 1913. Three thousand mourners attended his requiem. His brewery outlasted World War I restrictions on beer production and various provincial and federal forms of Canadian prohibition. Unlike most other Canadian breweries, O'Keefe's weathered the "dry" storm not by exporting to the United States but chiefly by manufacturing soft drinks. Indeed, sales of these beverages were brisk enough that the company reincorporated as O'Keefe Beverages Limited in 1922, and it continued to bottle soft drinks long after the repeal of Canadian Prohibition in 1927. Still, the company did not entirely ignore U.S. demand for Canadian beer, and if persistent legend is to be acknowledged, O'Keefe's stoneware ginger beer bottles often served as the clandestine means to distribute beer domestically, despite Canadian prohibitory law.

The company was reorganized as the O'Keefe Brewing Company Limited in 1933. The next year, E. P. Taylor bought O'Keefe through his Brewing Corporation of Canada Limited (later Canadian Breweries Limited). Although reducing O'Keefe's lineup to three brands—Old Vienna Lager, Extra Old Stock Ale, and Double Stout—Taylor used O'Keefe's Victoria Street brewery as a cornerstone of his brewing empire until 1966, when he closed it and transferred its production to a much more modern facility in northwest Toronto. Three years later, Rothman's bought Canadian Breweries Limited, reforming it into Carling-O'Keefe Limited in 1973, and continued to promote Old Vienna Lager as a flagship brand. In 1989, Molson Breweries acquired Carling-O'Keefe, then owned by Foster's, the giant Australian brewing group. Molson presently brews O'Keefe's Old Vienna Lager and O'Keefe's Ale as secondary labels.

Glen C. Phillips

See also: Carling Brewery; Labatt Brewing Company Limited; Molson Incorporated

References

Bowering, Ian. *The Art and Mystery of Brewing in Ontario.* Burnstown, Ontario: General Store Publishing, 1988.

Power, Michael. "Eugene O'Keefe" in *Dictionary of Canadian Biography,* vol. 14, *1911–1920.* Toronto: University of Toronto Press, 1998.
Report Concerning an Alleged Combine in the Manufacture, Distribution and Sale of Beer in Canada. Ottawa: Restrictive Trade Practices Commission of Canada, 1955.
Toronto Globe, 8 June 1895.

Olympia Brewing Company

The Olympia Brewing Company, for most of its life a regional brewer in the U.S. Pacific Northwest, was founded at the end of the nineteenth century under the name Capital Brewing Company. By the 1970s it was competing nationally, but ultimately it failed to hold that position or even to survive as a regional presence. Today its signature brands are produced by its former competitors.

Founder Leopold F. Schmidt was born in Germany and came to the United States as a merchant seaman in 1866. He worked initially as a carpenter in several parts of the country and eventually found his way to Montana. There, he went into the brewing business, first at Deer Lodge and then in Butte, where he and Daniel Garner established the Centennial Brewery in 1876. At some point while in Montana, Schmidt returned to Germany and attended a brewing school in Worms. While in Germany, he met and married his wife. After his return to Montana, Schmidt became a prominent figure in the state's business and political life, serving as a member of the new state's constitutional convention and, in 1894, on its Capital Commission. In the course of his duties with the commission he visited Olympia, Washington, and learned of the artesian springs at Tumwater, which he visited.

Schmidt established the Capital Brewing Company in Tumwater in 1896 to take advantage of the spring after tests determined its water was perfect for brewing beer. The original brewery complex included a four-story brew house, a five-story cellar building, a bottling and keg plant, and an ice factory

A worker stands on a ladder in the Seattle freezing room for the Olympia Brewing Company, ca. 1910. (PEMCO—Webster and Stevens Collection; Museum of History and Industry, Seattle/Corbis)

powered by a nearby waterfall. By the early years of the twentieth century, Capital Brewing had established itself in the Alaska gold fields, had set up breweries in Salem, Oregon, and in Bellingham, Washington, and had purchased the Port Townsend Brewing Company. The expanded firm was renamed Olympia Brewing Company in 1902. Olympia attracted attention with a large Swiss Chalet pavilion at the Lewis and Clark Centennial Exposition in Portland, Oregon. The pavilion was relocated to Priest Point Park in Olympia after the exposition. Leopold Schmidt moved to Bellingham in 1910 and purchased a hotel there, the Byron, which he operated until his death in 1914, after which it became the Leopold Hotel.

When Washington State introduced prohibition in 1916, Olympia established the Acme Brewery in San Francisco. The Tumwater brewery produced fruit drinks, and the Bellingham Bay Brewery, which had been producing 100,000 barrels annually, closed; it did not reopen when Prohibition was repealed in 1933.

In 1934, the Schmidt family, which still controlled Olympia, resumed brewing and built a new brewery in Tumwater, replacing the original buildings. They were successful in establishing Olympia in the beer market in California. The Schmidt family remained in control of the firm and it continued as a strong regional brewer. By the 1970s, Olympia was the ninth largest brewer in the United States.

Olympia made several acquisitions in the 1970s when Robert A. Schmidt, a grandson of Leopold's, was president and chief executive officer of the firm, as did other family-owned, regional breweries, in an attempt to remain competitive with the emerging large national brewers. The marketplace was changing dramatically, and remaining a regional brewer did not appear to be a good long-term strategy. Expansion was seen as the key to survival. Robert Schmidt's nephew, Leopold F. "Rick" Schmidt, became president of the firm in 1974. Olympia bought the Theodore Hamm Brewery of Minneapolis/St. Paul in 1975 after investigating building a brewery in Beloit, Wisconsin. The acquisition allowed expanded production not possible in Tumwater and brought the firm into the Midwestern market. The expansion did not lead to the expected increase in sales, however. In 1976, Olympia acquired the Lone Star Brewing Company of San Antonio. None of these efforts, however, succeeded, and Olympia never emerged as a major player in the increasingly national market.

The firm's expansion led it into legal and regulatory difficulties that plagued it for several years. In 1977, Robert A. Schmidt retired as CEO, a position assumed by his nephew, who had been serving as company president. Early in 1980, Rick Schmidt was caught up in a very public personal scandal and resigned from the firm. His uncle came out of retirement briefly to head the company, and in October 1980, James Senna was chosen as president of Olympia Brewing. Senna, promoted from senior vice president for marketing, had been with the firm for a number of years and was the first non–Schmidt family member to head the firm. Two members of the Schmidt family were appointed as vice presidents, Robert A. Schmidt Jr., for brewery operations, and Michael K. Schmidt for field marketing.

Olympia continued to be plagued by legal problems resulting from its acquisitions, especially that of Lone Star, and the failure of its expansion strategy to generate increased revenue. The larger part of Olympia was acquired by Pabst in 1982; Lone Star Brewing was acquired separately by G. Heileman in the same year. In 1999, the Tumwater brewery was acquired from Pabst by Miller as part of a restructuring of Pabst that ended its active life as a brewing company. Miller had acquired G. Heileman earlier. Olympia's labels, principally Olympia and Olympia Light, are now brewed by Miller on contract for Pabst, which markets them.

William H. Mulligan Jr.

See also: Heileman, G., Brewing Company; Miller Brewing Company; Pabst Brewing Company

References

Baron, Stanley. *Brewed in America: The History of Beer and Ale in the United States.* Boston: Little, Brown, 1962.

Greer, Douglas F. "The Causes and Consequences of Concentration in the U.S. Brewing Industry" in *Quarterly Review of Economics and Business.* 21, no. 4 (1981): 87–106.

McGahan, A. M. "The Emergence of the National Brewing Oligopoly: Competition in the American Market, 1933–58" in *Business History Review.* 65, no. 2 (1991): 229–284.

Ronnenberg, Herman. *Beer and Brewing in the Inland Northwest: 1850–1950.* Moscow: University of Idaho Press, 1993.

Orange Order

See Loyal Orange Institution (Ireland and Britain)

Ottawa Home for Friendless Women

In Canada's capital, the Ottawa Home for Friendless Women rehabilitated drunkards, prostitutes, criminals, and other "fallen women." Unlike the inebriate homes run by the Church of England Temperance Society, attention was not confined to the intemperate. The Ottawa Home embodied world temperance reformers' concerns with causative links between intemperance and sexual immorality. It was operated by the Young Women's Christian Temperance Union (YWCTU), the local chapter of the international organization that was headed in the United States by Frances Willard. The fact that the home was run by young, single women, rather than the usual matrons, was exceptional. For its first eight years, YWCTU member Bertha Wright served as director.

After a year of visiting the local jail, the YWCTU sought a place where women leaving prison could continue their rehabilitation and avoid vice. A house at 412 Wellington Street was rented in November 1887. Fearing that the YWCTU would be unable to control the inmates and that the residence would be a hub of immorality, down the street from Parliament, community leaders discouraged the project. Nonetheless, cleaned and furnished to board fifteen inmates, the home opened on 9 January 1888.

Accommodations at this first location were uncomfortable, as compulsory laundry work required more space than the facility allowed. Bedrooms doubled as work areas, and a high humidity level prevailed throughout. In March 1890, the laundry was moved to a larger building where modern equipment was installed. The two buildings allowed room for thirty-six inmates.

The home accepted inmates regardless of race, national origin, religion, or age. Irish women formed the largest group, but residents also included immigrants and Canadians of English, Scottish, French, American, and German origins, together with some neglected children. Clients came from jail, the street, the train station, and occasionally brothels.

Resolve, steadiness, responsibility, industry, and abstinence characterized the home's curative regime. Residents had to resolve to change their lives. Pregnant women signed a contract to remain in the home for at least a year, care for their infants, and take their children with them when they left. Mothers were accepted so long as they looked after their children, a reminder of their past wrongdoings.

The home was partly supported by the laundry. To obtain government funding, the home separated from the YWCTU in 1891 and became an incorporated charitable institution. The board of managers passed from control by young, single women to middle-class married women. Bertha Wright, however, saw the home through this transition.

The Ottawa Home for Friendless Women exemplified the practical, existential efforts of western women to cure the physical, emotional, and social maladies of more unfortunate "sisters." Despite judgmentalism, their collective experience advanced the understanding of criminality and of alcoholism as a disease, dispelling myths about the impossibilities of reforming "fallen women."

Cynthia Belaskie

See also: Church of England Temperance Society (CETS) Inebriate Homes; Church of England Temperance Society (CETS) Police Court Mission; Willard, Frances Elizabeth Caroline

References

Hunt, Geoffrey, Jenny Mellor, and Janet Turner. "Wretched, Hatless and Miserable Clad: Women and the Inebriate Reformatories from 1900–1913" in *British Journal of Sociology.* 40, no. 2 (1989): 244–270.

Wright, Bertha. *Lights and Shades of Mission Work, 1885–1892.* Ottawa: Free Press, 1892.

P

Pabst Brewing Company

The Pabst Brewing Company traces its origins to Jacob Best Sr., who owned a brewery and winery in Mettenheim, Germany. Jacob had four sons who were trained as brewers. In the 1830s, Jacob Jr. and Charles, the eldest and third sons, respectively, immigrated to the United States and found their way to Milwaukee. The brothers' first business venture was a vinegar factory. When the entire family arrived in Milwaukee they established the Best Brewery with a capacity of 300 barrels annually. The brewery produced ale, porter, and, after 1844, lager as well as corn and rye whiskey. The entire family did not stay in business together very long. Charles returned to making vinegar in 1845, was back in brewing a few years later, and in 1850 he and his younger brother Lorenz set up the Plank Road Brewery, the forerunner of Miller.

In 1853, Jacob Sr. retired from the brewery, leaving it to Jacob Jr. and Phillip, the second eldest. At that point, Best and Company was producing 2,500 barrels a year and shipping beer to Chicago. The brewery was the fourth largest in Milwaukee. Jacob Jr. and Phillip dissolved their partnership in 1859, with Phillip continuing the business. In the very competitive world of brewing in Milwaukee, Best and Company's future was by no means assured. When German immigration to Milwaukee slowed during the Civil War, the firm fared poorly. In 1863, Phillip took into partnership his son-in-law Captain Frederick Pabst. Pabst and his brother-in-law Emil Schandein bought out Phillip in 1865.

Under their management the firm thrived, in part owing to the resumption of German immigration into Milwaukee, but also because of Pabst's business skill. Pabst sought to improve his beer and expand his market. He traveled in Europe to recruit brewers, plowed profits into expansion, and developed an extensive network of sales connections in the United States and Europe. The firm had forty sales offices across the country by 1893. In 1870, the firm bought the Melms Brewery, which was the largest in Milwaukee. Production reached 100,000 barrels in 1872, making Best and Company the second largest brewer in the country. The business, which had been a partnership, was incorporated as the Phillip Best Brewing Company in 1873.

Best's beer won numerous awards at world's fairs and other competitions. In 1882, Pabst had a blue ribbon tied around the neck of bottles of the company's "Select" beer. Pabst also played a leading role in changing the way beer was taxed. Prior to 1890, the excise tax was applied to barreled beer. Beer had to be barreled before bottling, which was an added expense. The new law allowed beer to move through metered pipelines. The company name was changed to Pabst Brewing Company in 1888. In 1895, the company introduced its Pabst Blue Ribbon label. Production that year was nearly a million barrels.

Captain Pabst was very involved in Milwaukee civic life, supporting many important local projects, including the Pabst Theatre, which he built and donated to the city. Throughout his career he emphasized producing quality beer. When he died on New Year's Day in 1904, Pabst was the third largest brewer in the country. His son Gustav succeeded him as head of the firm and served until he resigned in December 1921. His brother Fred returned to the firm to face the challenges posed by National Prohibition.

Pabst introduced a nonalcoholic beer called Pablo in 1916 and set up the Pabst Corporation and Pabst Realty Corporation to operate its nonbeer businesses. In 1923, it acquired the Sheboygan Beverage Company to produce soft drinks. That same year, Fred Pabst sold his cheese business to the company. The brewery's cellars were used to age cheese, and the sales force switched from marketing beer to marketing cheese. The cheese business was quite successful. After repeal it was sold to Kraft.

In 1930, Pabst acquired the Puritan Malt Extract Company of Chicago and in 1932 merged with Premier Malt Products Company. The new firm was known as Premier-Pabst Corporation, until 1938 when it returned to the name Pabst Brewing Company.

Pabst returned to the beer business after repeal and expanded aggressively. The Milwaukee brewery was expanded and modernized, and in 1934 a new brewery was built in Peoria Heights, Illinois. In 1946, the Hoffman Beverage Company in Newark, New Jersey, was acquired, and two years later the Los Angeles Brewing Company. Under the leadership of Harris Perlstein, who had been president of Premier Malt Products, and Fred Pabst, sales grew steadily through the early 1950s. In 1935 Pabst was the first major brewer to market its beer in cans. It was also among the first to advertise nationally on the radio, sponsoring comedian Groucho Marx in 1943. In 1950, Pabst sponsored boxing on CBS-TV. Fred Pabst retired as chairman in 1954. Soon thereafter the fortunes of the firm changed as sales slipped.

In 1958, the Pabst family tried to regain control of the firm and remove Perlstein. Perlstein countered by bringing James Windham, president of Blatz Brewing, in as president. Windham was widely respected in the industry for reviving Blatz.

Pabst building in Milwaukee, Wisconsin, ca. 1900 (Library of Congress)

Pabst acquired Blatz the same year, hoping to use its brewery to increase capacity. This plan was blocked by the Justice Department antitrust division. Pabst lost after eleven years of litigation ending in the Supreme Court. Windham cut the price of Blue Ribbon in some markets, and sales increased steadily through the 1970s. He followed a very conservative financial course and put off modernization and expansion unless it could be paid for with cash. Only one new brewery, in Perry, Georgia, was built. When Windham died in 1977, Pabst had $70 million in cash on deposit.

Pabst faced rising production costs due to increasingly antiquated equipment at the same time it became an attractive target for a takeover due to its huge cash reserve. For seven years, the board resisted a series of takeover attempts. This was costly and disruptive. Sales rose briefly after the acquisition of the Blitz-Weinhard brewery in 1979, but problems plagued the firm. In 1982, Pabst acquired the Olympia Brewing Company. Serious financial problems and declining sales, however, hit the firm hard. In the early 1980s, a number of executives were fired and five plants closed and sold to G. Heileman. The board sold the firm to Paul Kalmanovitz for $63 million in 1984.

Kalmanovitz cut costs drastically. Most seriously, he slashed the advertising and quality control budgets. Sales of Blue Ribbon and Pabst's other beers dropped dramatically. Industry analysts predicted the firm's imminent demise. Kalmanovitz died in 1988, and his successor, Lutz Issleib, tried to revive Pabst with more advertising and modernization of its breweries. What had been a national brand, ranked number three nationally in sales in 1980, had distribution only in the Midwest and West Coast in 1988. Issleib's strategy did not succeed, and Pabst closed its breweries between 1996 and 2001. Its beers are brewed by Miller under contract and marketed by Pabst, now located in San Antonio, Texas.

William H. Mulligan Jr.

See also: Heileman, G., Brewing Company; Miller Brewing Company; Olympia Brewing Company

References

Apps, Jerry. *Breweries of Wisconsin.* Madison: University of Wisconsin Press, 1992.

Baron, Stanley. *Brewed in America: The History of Beer and Ale in the United States.* Boston: Little, Brown, 1962.

Cochran, Thomas C. *The Pabst Brewing Company: History of an American Business.* New York: New York University Press, 1948.

Greer, Douglas F. "The Causes and Consequences of Concentration in the US Brewing Industry" in *Quarterly Review of Economics and Business.* 21 (1981): 87–106.

McGahan, A. M. "The Emergence of the National Brewing Oligopoly: Competition in the American Market, 1933–58" in *Business History Review.* 65 (1991): 229–284.

Pacific Islands

All available evidence supports the position that the inhabitants of the Pacific Islands (including Melanesia, Micronesia, and Polynesia) had no alcoholic beverages before contact with the West. Although exploration of the Pacific by outsiders began with Ferdinand Magellan's circumnavigation of the globe in 1521, most Pacific Islanders did not encounter Europeans and other foreigners until the late 1700s and early 1800s. In cases where these interactions were peaceable, the outsiders often offered a drink of brandy or other distilled beverage to the apparent island leaders. Typically, the leaders spat this out as unpalatable and bitter, and in Tahiti such beverages were named "British kava." As beachcombers began to settle in the islands, or as whalers took "rest and recreation" ashore, they taught the islanders how to ferment and distill alcoholic beverages from such local products as coconut palm sap, oranges, and the sugary root of the *ti* plant *(Cordyline fructicosa).* Such "home brews" continue to be made in rural areas of the Pacific right up to the present.

Not too long after the establishment of European beach communities on those islands with good harbors for sailing vessels, European (and later, American) missionaries appeared on the scene. Actually, the first Christian missionaries in the Pacific were Spanish Jesuits from Manila, who went to Guam in the 1660s and participated in the conversion and transformation of Chamorro society from a proud, independent people to a colonized hybrid of the precontact culture mixed with Filipino and Spanish elements. In spite of this early incursion into the Mariana Islands, most of the Pacific was not missionized until the late eighteenth and nineteenth centuries, and evangelization continues in the more remote parts of Melanesia in the twenty-first century. Although Catholics did not usually actively oppose consumption of alcoholic beverages by islanders, Protestant missionaries spoke out against the "Demon Rum" from early on, reflecting the rise of temperance movements in their home countries during the nineteenth century (Marshall and Marshall 1976). Even today, being alcohol-free remains a potent and important symbol of the Christian life among Protestant believers in Oceania.

Traders, Colonialism, and Prohibition

By the 1870s, traders were resident across much of the Pacific, representing mercantile firms from Germany, Great Britain, New Zealand, Australia, Japan, and the United States. Many of these men set up business in outer island communities away from the scattered port towns, where they swapped tobacco, cloth, guns, and alcohol in exchange for copra, or dried coconut meat, and bêches-de-mer, a type of edible sea cucumber. They often took local wives. It was through these trade links that commercially manufactured distilled alcoholic beverages first made their way into islanders' hands.

Along with the expansion of trading empires, the last quarter of the nineteenth century saw the imposition of colonial rule nearly everywhere in Oceania. With colonial government came colonial laws, among the first of which banned islanders' access to beverage alcohol (Marshall 1980). Legal prohibition of Pacific Islanders' access to alcoholic drinks was universal for approximately a century, with deprohibition serving as one symbolic marker of self-rule and independence movements throughout the islands between the late 1960s and the mid-1980s. In the postcolonial era of the past thirty-five years, legal drinking has been regulated mainly by age restrictions, although some islands have required drinkers to possess a special permit in order to partake of their favorite beverage.

Results of Legalization

With legalization of drinking came a dramatic increase in the types and brands of alcoholic beverages. Imported beer and wines from Australia and New Zealand, beers from Japan and the United States, and distilled spirits from various European countries began to turn up in the larger stores in Pacific port towns. Even so, most of these imports were quite expensive by island income standards, so only the elites could afford to drink them. The situation soon began to change, however, with the establishment of numerous breweries in the islands. In the mid-1950s, large transnational beer corporations began to put down roots in island countries such as Papua New Guinea and Fiji. By the year 2000, beer had come to dominate the alcohol market, being far and away the beverage preferred by island tipplers, and there were eleven breweries spread about Oceania (Marshall 2003). These ranged from South Pacific Brewery in Papua New Guinea to La Brasserie de Tahiti in French Polynesia. Many island brews, such as Samoa's Vailima or Vanuatu's Tusker, make use of traditional symbols of value and authority in their advertising, and many also dress their cans up in colors mimicking those of the national flag. As breweries have been built in the islands, using equipment and techniques imported from the former colonial countries, the price of a drink has been reduced and the consumption of beer has increased. Although home brew is still made and drunk occasionally on some of the more remote islands, or in the interior of the big islands of Melanesia, most Pacific Islanders now drink beer, and cold beer if they can get it in local clubs and taverns.

Research on Alcohol in the Islands

Given the ubiquity of drinking in the islands today, the symbolic importance of the right to drink in the run-up to political independence, and the historical interest in alcohol's introduction to societies that lacked experience with this substance, remarkably little scholarly research has been conducted on the subject. Most such work has been accomplished over the past forty years.

Edwin M. Lemert pioneered modern alcohol studies in Oceania, publishing a series of papers on drinking in Polynesia during the 1960s (1962, 1964a, 1964b). His work in Hawaii, Tahiti, Samoa, and the Cook Islands was followed by a further report on Tahitian drinking (Levy 1966) and by an excellent paper on "Drinking Behavior and Race Relations" in Bougainville, now a part of Papua New Guinea (Ogan 1966). The 1970s witnessed a mini-explosion of published research on alcohol in the islands. More articles appeared on drinking among Polynesians (e.g., Gluckman 1974; Lemert 1976) and Melanesians (e.g., Hocking 1970; Schwartz and Romanucci-Ross 1974), and alcohol's place in Micronesia also came in for attention (e.g., Marshall 1975, 1976; Nason 1975). From the end of the 1970s through 1990, three books appeared that vastly increased the available information on drinking in Chuuk (formerly Truk), Federated States of Micronesia (Marshall 1979; Marshall and Marshall 1990), and in Papua New Guinea (Marshall 1982). Recent years have seen further additions to the literature on Pacific Islander drinking, and among the more notable of these are articles or chapters by Carucci (1987), Dernbach and Marshall (2001), Donner (1994), Nero (1990), Toren (1994), and Wormsley (1987). All signs are that the subject of alcohol production, use and abuse will continue to garner ever more scholarly attention in the years ahead.

Mac Marshall

See also: Anthropological Theories of Drinking and Temperance; Kava

References

Carucci, Laurence M. "*Jekero:* Symbolizing the Transition to Manhood in the Marshall Islands" in *Micronesica.* 20 (1987): 1–17.

Dernbach, Katherine Boris, and Mac Marshall. "Pouring Beer on Troubled Waters: Alcohol and Violence in the Papua New Guinea Highlands" in *Contemporary Drug Problems.* 28, no. 1 (2001): 3–47.

Donner, William W. "Alcohol, Community, and Modernity: The Social Organization of Toddy Drinking in a Polynesian Society" in *Ethnology.* 33, no. 3 (1994): 245–260.

Gluckman, L. K. "Alcohol and the Maori in Historical Perspective" in *New Zealand Medical Journal.* 79 (1974): 553–555.

Hocking, R. B. "Problems Arising from Alcohol in the New Hebrides" in *Medical Journal of Australia.* 2, no. 20 (1970): 908–910.

Lemert, Edwin M. "Alcohol Use in Polynesia" in *Tropical and Geographical Medicine.* 14 (1962): 183–191.

———. "Drinking in Hawaiian Plantation Society" in *Quarterly Journal of Studies on Alcohol.* 25 (1964b): 689–713.

———. "Forms and Pathology of Drinking in Three Polynesian Societies" in *American Anthropologist.* 66 (1964a): 361–374.

———. "Koni, Kona, Kava: Orange-Beer Culture of the Cook Islands" in *Journal of Studies on Alcohol.* 37, no. 5 (1976): 565–585.

Levy, Robert. "Ma'ohi Drinking Patterns in the Society Islands" in *Journal of the Polynesian Society.* 75 (1966): 304–320.

Marshall, Mac. "A History of Prohibition and Liquor Legislation in Papua New Guinea, 1884–1963." IASER Discussion Paper No. 33 (1980): 1–24.

———. "Market Highs: Alcohol, Drugs and the Global Economy in Oceania" in *Globalization and Culture Change in the Pacific Island,* edited by Victoria S. Lockwood. Englewood Cliffs, NJ: Prentice-Hall, 2003.

———. "The Politics of Prohibition on Namoluk Atoll" in *Journal of Studies on Alcohol.* 36, no. 5 (1975): 597–610.

———. "A Review and Appraisal of Alcohol and Kava Studies in Oceania" in *Cross-Cultural Approaches to the Study of Alcohol: An Interdisciplinary Perspective,* edited by Michael W. Everett, Jack O. Waddell and Dwight B. Heath, 103–118. The Hague: Mouton, 1976.

———. *Weekend Warriors: Alcohol in a Micronesian Culture.* Mountain View, CA: Mayfield, 1979.

———, ed. *Through a Glass Darkly: Beer and Modernization in Papua New Guinea.* IASER Monograph No. 18. Boroko: Papua New Guinea Institute of Applied Social and Economic Research, 1982.

Marshall, Mac, and Leslie B. Marshall. "Holy and Unholy Spirits: The Effects of Missionization on Alcohol Use in Eastern Micronesia" in *Journal of Pacific History.* 11, no. 3 (1976): 135–166.

———. *Silent Voices Speak: Women and Prohibition in Truk.* Belmont, CA: Wadsworth, 1990.

Nason, James D. "Sardines and Other Fried Fish: The Consumption of Alcoholic Beverages on a Micronesian Island" in *Journal of Studies on Alcohol.* 36, no. 5 (1975): 611–625.

Nero, Karen L. "The Hidden Pain: Drunkenness and Domestic Violence in Palau" in *Pacific Studies.* 13, no. 3 (1990): 63–92.

Ogan, Eugene. "Drinking Behavior and Race Relations" in *American Anthropologist.* 68 (1966): 181–188.

Schwartz, Theodore, and Lola Romanucci-Ross. "Drinking and Inebriate Behavior in the Admiralty Islands, Melanesia" in *Ethos.* 2, no. 3 (1974): 213–231.

Toren, Christina. "The Drinker As Chief or Rebel: Kava and Alcohol in Fiji" in *Gender, Drink and Drugs,* edited by Maryon McDonald, 153–173. Oxford: Berg, 1994.

Wormsley, William. "Beer and Power in Enga" in *Drugs in Western Pacific Societies: Relations of Substance,* edited by Lamont Lindstrom, 197–217. ASAO Monograph No. 11. Lanham, MD: University Press of America, 1987.

Palm Wine

Palm wine, also known as toddy, is a milky white, slightly effervescent beverage produced from sap collected by tapping various species of the very large family of palm trees *(Palmae).* It is found in many parts of Africa and Asia, and where appropriate palms thrive this is often the most commonly available cheap alcoholic beverage. There seems to be a considerable variation in the alcohol content of palm wines produced from different species.

Production

The coconut palm *(Cocos nucifera)* is tapped on the East African coast and in the southern part of the Indian subcontinent; other species used for palm wines include the date palms *(Phoenix dactylifera)* in the Middle East and North Africa, the palmyra or borassus palm *(Borassus flabellifer)* in southern India and Indonesia, and oil palms *(Elaeis guineensis)* and raphia palms *(Raphia vinifera* or *Raphia hookeri)* in West Africa. The oil palm is apparently not tapped outside Africa. The technique of tapping varies somewhat, but in most cases it involves climbing the palm, cutting a flower-shoot, and collecting the sap that drips out of the cut into a vessel that is left attached to the shoot. Palms may also be tapped through cuts in the stem, either at the top or at the bottom, or simply through pushing the whole tree down and cutting its top to collect all the sap within it. The latter technique is apparently most commonly followed with raphia palms.

However the tree is tapped, contact with naturally occurring lactic acid bacteria and yeasts begins a quite rapid process of fermentation, as the sap contains a considerable proportion (around 10 percent, in the case of oil palms) of fermentable sugar. Newly tapped sap is often blended with fermented palm wine to encourage and offer some control over the process of fermentation. No further intervention in the process is required, and where palms thrive the ambient temperature is usually high enough to allow the process to occur naturally. Palm wine is usually taken without any additive, though among the Kasai of Zaire the use of a root chewed with the wine is reported, and in India the addition of chloral hydrate has been noted; palm wine may also be watered down. Within a few hours of tapping, the sap is mildly alcoholic; fermentation continues for up to seventy-two hours, with the liquid becoming increasingly sour.

There is a striking variance in reported potency of palm wines. Although there has been no systematic work on this subject, it would appear that the wine tapped from oil and raphia palms in West Africa takes longer to ferment, and peaks at a lower alcohol content, than does coconut palm wine. O. Bassir (1962) has suggested average strengths for oil palm wine of only around 2 percent by volume at time of consumption, rising to 4 or 5 percent after two days; other sources suggest higher figures, up to around 8 percent seventy-two hours after tapping. Coconut palm wine seems stronger; twenty-four hours after tapping it may be some 6 or 7 percent, rising to a peak of 8 percent, and after around sixty hours it becomes too acidic to be palatable. Palm wine is also a source of ascorbic acid and B vitamins. In its palatable, alcoholic state the drink is taken whole, without filtering or straining in any way (though a straw is often used to avoid the drowned insects that tend to gather in the beverage).

Rituals and Social Significance

The innocuous nature of the freshly tapped juice and the potency of the beverage a day later, combined with the extreme ease with which palm wine can be produced, make this a distinctive beverage. In Ghana, myths of its origin stress its unexpected potency and ability to render the unwary drinker insensible; for religious groups that nominally avoid alcohol, the distinction between refreshing, newly tapped juice and intoxicating wine may be agreeably indistinct. In Africa generally, communities accustomed to drinking grain beer have sometimes viewed palm wine with a degree of suspicion as unusually strong. Nonetheless, the ease with which this beverage can be made has encouraged the spread of palms to new areas: an effect notable on the Kenya coast, where very large areas were planted in the late nineteenth and early twentieth centuries. Where palm wine is produced from planted trees—whether borassus, coconut, or oil palm—its introduction may challenge established patterns of involvement and ownership in liquor: Unlike grain beers, or honey, it relies on a substantial capital asset, the tree, whose long-term ownership becomes an important economic issue. In Tamil Nadu in southern India, tree ownership has lain with wealthy and high-status groups, but tapping has been the work of low-status men. In Africa, palm wine and its production have been very much male domains, with young men tapping trees owned by elder men; this is in striking contrast to the production of grain beers, which has very often been a women's domain. In West Africa, there has been some small-scale industrial production of pasteurized palm wine.

Tapping of palms has been identified as an undesirable practice by colonial and postcolonial governments, for it reduces the production of exportable items such as copra and palm oils and encourages instead a local informal economy of alcohol consumption. From the point of view of tree owners, however, production of palm wine is often a much more lucrative field than processing the oil or copra. It has been argued that tapping causes long-term damage to trees, but it is not clear that this is the case where tapping is done through the flower-shoot—though of course a tapped flower-shoot will not produce any fruit. Stem tapping is more destructive.

Justin Willis

See also: Africa, Central; India

References

Bassir, O. "Observations on the Fermentation of Palm Wine" in *West African Journal of Biological Chemistry.* 6 (1962): 20–25.

———. "Some Nigerian Wines" in *West African Journal of Biological and Applied Chemistry.* 10 (1968): 42–46.

Hardgrave, R. L., Jr. *The Nadars of Tamilnad: The Political Culture of a Community in Change.* Bombay and Berkeley, CA: Oxford University Press and University of California Press, 1969.

Mosha, D., J. Wangabo, and G. Mhinzi. "African Traditional Brews: How Safe Are They?" in *Food Chemistry.* 57 (1996): 205–209.

Okafor, N. "Traditional Alcoholic Beverages of Africa: Strategies for Scale-up" in *Process Biochemistry.* 25 (1990): 213–220.

Parkin, D. *Palms, Wine and Witnesses: Public Spirit and Private Gain in an African Community.* London: Intertext Books, 1972.

Parker, Margaret Eleanor (1828–1896)

Margaret Eleanor Parker (1828–1896), a convener and founding member of the British Women's Temperance Association (BWTA) in 1876 at Newcastle, was elected to be its first president. Born in Houghton-le-Skerne near Darlington, she was the first-born of a family of nine children. She married Charles Parker in 1851 and moved to Dundee. In 1854, she became converted to the temperance cause after hearing John B. Gough speak on the subject. Although she enforced temperance in her own home, it was not until 1874 that she became active within the movement, participating in a procession of seventy women who delivered a petition of more than 8,000 women's signatures in favor of reducing the number of licensed houses. An active member of the Independent Order of Good Templars (IOGT), she gained the rank of worthy vice templar of the Grand Lodge of Scotland. When she was elected to represent Scotland at the Right Worthy Grand Lodge sessions held in Illinois in 1875, she traveled in good company with "Brothers" George Gladstone, Joseph Malins, Joshua Pollard, Samuel Capper, and Frederic Lees and "Sister" Mary Lees.

It was her first visit to North America, and she was appointed also to be a delegate to the convention of the National Temperance Society in Chicago. It was in Chicago that she made contact with the leaders of the Women's Temperance Crusade that had flourished from late 1873 to 1874. Parker believed that the Crusade was a defining moment for the women's temperance movement in the United States. She made contact with Eliza Stewart, generally known as Mother Stewart, who had been a leader of the Crusade in Ohio. After Stewart expressed an interest in visiting Britain, the National Temperance League invited her to lecture in the United Kingdom.

Parker and Stewart were delegates at the 1876 IOGT meeting in Newcastle that resulted in the founding of the BWTA. Parker became president of the newly formed association. The new BWTA elected delegates to the international women's convention in Philadelphia. Margaret Bright Lucas and Catherine Impey accompanied Parker to the meeting there. Parker was instrumental during this visit in founding the World Woman's Christian Temperance Union (WWCTU), although the international organization did not convene until 1886.

Following the Philadelphia meeting, Parker traveled widely in the eastern seaboard and southern states and became enamored with North America. She admired the countryside, the people, the homes, and the time-saving household appliances. She published an account of her journey as *Six Happy Months amongst the Americans* (1874). By the time she returned to Britain, Clara Lucas Balfour had replaced her as BWTA president. A prolific writer, Balfour was better known to the membership than was Parker.

Like many other temperance women, Parker was a supporter of the campaign for women's suffrage. She also worked to abolish the Contagious Diseases Act, although men in Scotland believed that it was unseemly for women to participate in an agitation related to sexually transmitted diseases. Continuing to work for women's issues, Parker founded in 1881 a horticultural and supply association worked entirely by women. Her ideal was to purchase or lease an estate near some large city where women could engage in the culture of flowers, fruits, and vegetables and keep cows, bees, and poultry. Unfortunately, the response was not enough for the scheme to flourish.

Following this failure, she organized an immigration program in 1888 to recruit suitable women from Britain to fill the need for servants in California. She superintended their transit and arranged for their maintenance until positions had been found for them. This venture was more successful. Until her death in 1896, she remained a member of the BWTA executive committee and retained her links with temperance women in the United States.

Margaret Barrow

See also: British Women's Temperance Association (BWTA); Good Templars (IOGT); Gough, John Bartholomew; Impey, Catherine; Lees, Frederic Richard; Lucas, Margaret Bright; Malins, Joseph; National Temperance League (NTL); Willard, Frances Elizabeth Caroline; Women's Temperance Crusade; World's Woman's Christian Temperance Union (WWCTU)

References

Parker, C. E. *Margaret E. Parker: A Memorial.* Bolton: Tillotson, 1906.

Parker, M. E. *Six Happy Months amongst the Americans.* Glasgow: privately printed, 1876.

Patent Medicines

In the Victorian period, over-the-counter drugs were called "patent medicines." The term included pills that contained narcotics and tonics that contained large quantities of alcohol. The term itself was a misnomer, as few of these products were actually patented. More properly speaking, they were proprietary drugs; most manufacturers were unwilling to reveal their ingredients to competitors through the patenting process. Although examples of patent medicines may be found in the 1600s, patent medicines were mass-produced after about 1860. In the early twentieth century, they were subject to acerbic attacks, both in literature (notably in H. G. Wells's 1908 novel *Tono Bungay*) and in professional medical forums; these critiques ultimately led to government regulation of most medicines in Europe and North America by mid-century.

In the Victorian period, medicines for common complaints such as headache, coughing, indigestion, fatigue, and obesity flooded the market. They came in a variety of forms, including tonics, syrups, pills, wafers, and electrical devices and other appliances. Popular brands, such as Beecham's, Holloway's, and Eno's, became household names around the world. They were promoted in advertisements with lavish illustrations and hyperbolic text. Moreover, medicines were priced to sell. Thus, Beecham's Pills, a relatively expensive medicine, could advertise famously that they were "Worth a Guinea a Box." "Price-cutting," whereby individual retailers offered discounts, was widely repudiated by manufacturers and also by other retailers. The result was the formation of Proprietary Articles Trade Associations (PATAs) in the United States in 1881, in Canada in 1896, and in Britain in 1897. The main aim of these associations was to protect the standardized pricing of commercial patent medicines; they also limited retailers to a profit of 20 percent or less to prevent one retailer from capturing a disproportionate share of a local market through price cutting.

Medicines were sold in a variety of retail outlets. Only about 20 percent of dealers were chemists; the most popular points of purchase were department stores, because the likelihood of substitution was low and because economies of scale permitted lower prices. Sales in the United States around 1904 exceeded $60 million. By 1914, sales in Britain exceeded £5 million a year. Canadian sales ranked third worldwide.

Victorian patent medicines were subject to very little government control. Persons who sold medicines were required to take out a license for each shop location, and all compounded medicines were taxed. But there was no government inspection of patent medicines. Nor were producers required to identify ingredients. In Britain the Pharmacy Act of 1868 required that medicines containing addictive substances such as cocaine, opium, and morphine be labeled "poison,"

Wolcott's Instant Pain Annihilator advertisement claimed to cure a man beset by the demons of catarrh, neuralgia, headache, weak nerves, and toothache. (Library of Congress)

but court cases rarely enforced labeling. In the United States, the Pure Food and Drugs Law of 1906 required manufacturers to state the presence and amount of dangerous substances. Similar legislation was passed in Britain in 1908, in a new Pharmacy Act, and in Canada in 1919, in an amendment to the Patent Medicine Act. It remained legal to advertise cures for incurable diseases. Laissez-faire ideology supported the application of the cautionary "buyer beware" even to medical products and services. Dissatisfied consumers could seek redress in the courts, which resulted in several highly publicized cases, the most famous of which were the British cases of *Carill v. the Carbolic Smoke Ball Company* (1891), *Medical Battery Company v. D. Jeffery* (1892) and *Bile Beans v. Davidson* (1905).

Some overall patterns about composition were revealed in the publication of the formulae of popular branded remedies in the *British Medical Journal* (published from January 1907 to April 1912); in the *Journal of the American Medical Association* (June 1900 to March 1907); in the reports of the Australian Royal Commission on Drugs and Food in 1907 and of the British Select Committee on Patent Medicines in 1914; and in the popular exposés *Secret Remedies and More Secret Remedies,* published by the British Medical Association in 1908 and 1910, and *Nostrums and Quackery,* published on behalf of the American Medical Association beginning in 1911. Indigestion remedies, the largest category of patent medicine, were largely bicarbonate of soda. Liver pills were predominantly aloe and rhubarb derivatives. Obesity cures were mostly citric acid in water. Especially controversial was the content of so-called "female pills," whose iron content was reputed to be an abortifacient. Medicated wines or tonics, such as Hall's Tonic and Wincarnis, attracted the attention of temperance reformers because they often contained as much as 20 percent alcohol, although they usually did not advertise that fact. Similarly, soothing syrups (or gripe waters) for infants, such as Godfrey's Cordial or Mrs. Winslow's Soothing Syrup, often contained morphine or alcohol; in the recommended dose they might not be harmful, but few working-class households had standardized measuring spoons, and overdosing, especially for a child weakened by purging, was a very real danger. Also of concern was Chlorodyne, a cough remedy containing chloroform first marketed in 1856, which was connected with several cases of overdose in the 1880s and 1890s, and Daisy Powders, a headache remedy widely marketed in the late 1890s, which was reputed in the early 1900s to have caused several deaths.

Opposition arose first from the medical profession. Although pharmacists felt compelled to carry patent medicines to prevent customers from taking their business elsewhere, doctors opposed patent medicines as a form of quackery that could be addictive or potentially lethal, mask symptoms, delay vital professional treatment, or offer false cures. The professional argument, evidenced, for example, in the *British Medical Journal,* the *Journal of the American Medical Association,* and *Canada Lancet,* was taken up in the popular press. Edward Bok, editor of *Ladies' Home Journal,* launched a concentrated campaign against patent medicines and their makers in a series of articles beginning in 1903. Comparable series followed, including in the American magazine *Collier's* in 1905 and in the British magazine *Vanity Fair* in 1910. All portrayed patent medicines as fraudulent products marketed by charlatans to the weak and unsuspecting.

World War I delayed legislative reform, but it became illegal to advertise cures for incurable diseases in Canada in 1934 (Food and Drugs Act), in the United States in 1938 (Food, Drug and Cosmetic Law), and in Britain in 1938 (Food and Drugs Act). Makers were finally forced to name ingredients in 1938 in the United States, in 1941 in Britain, and in 1952 in Canada.

Lori Loeb

See also: Archaeological Approaches to Drinking and Temperance; Medicine, Alcohol as; Treatment Institutions

References

Bartrip, P. W. J. *Mirror of Medicine: A History of the British Medical Journal.* Oxford: Clarendon Press, 1990.

Lears, T. Jackson. *Fables of Abundance.* New York: Basic Books, 1994.

Parsinnen, Terry. *Secret Passions, Secret Remedies.* Philadelphia: Institute for Human Issue, 1983.

Porter, Roy. *Health for Sale.* Manchester: Manchester University Press, 1989.

Pernod-Ricard Group

In 1932, Paul Ricard, then twenty-three, decided to industrialize his pastis recipe (beyond the still in his bedroom) and conceived his first advertising campaign. He eventually gave his name to this aniseed aperitif after managing to convince his father that it would meet with wide success among the countrymen of Marseilles and its surroundings, where his father was a wine merchant. He was right: In 1938, Ricard's sales amounted to more than 2.4 million liters (634,000 U.S. gallons). The Ricard company was created the following year. In the late 1990s, *Le Vrai Pastis de Marseille* ("the genuine Marseilles pastis") was the most widely sold French alcohol-containing drink, available in 140 countries, and the first European trademark. Since 1975, the Ricard company has been part of the Pernod-Ricard Group, which is a leader in the alcoholic drinks field. The Pernod-Ricard Group is actually the fifth largest wine and liquor company worldwide, with a turnover of 3.4 billion euros in sales annually (with 21 percent of the sales in France; 42 percent in other European countries and Africa; 27 percent in the Americas; and 10 percent in Asia and the Pacific Ocean countries).

After seven years of prohibition in 1922, the French government allowed the consumption of aniseed drinks, except for absinthe, which remains prohibited. The major liquor producers (Pernod, Duval, Berger), until then deprived of anise marketing, increased the number of trademarks and kept the ingredients and their proportions secret. In the late 1930s, producers were allowed an alcohol content in pastis of 45 percent, which experts believed to be ideal. Traditionally, an aniseed aperitif is obtained from the aromatization of an alcohol by natural extracts from star-shaped anise, also called fennel, or green anise, plus licorice and other natural plant extracts. Most producers use star-shaped anise because they consider it to be the richest in aroma compounds. Anise is supposed to have been used as early as 1500 B.C.E. for taking care of teeth and gums and in treating heart disease.

The success of Paul Ricard's company since the 1930s is due partly to the new recipe he worked out, based mainly on green anise, fennel, and licorice; however, it was his skill in communication, sales, and marketing, rather than the product quality, that took him to the top. He built an extensive network of representatives and traders to cover distributors and producers in southern France and used advertising to foster recognition of the famous Pastis trademark. As early as 1948, the Ricard company associated itself with the Tour de France, becoming the first commercial sponsor of the cycling race. Pickup trucks at the event displayed the brand's colors, and at every daily stop there was musical entertainment provided by the company. In 1970, Paul Ricard opened a racing circuit bearing his name; for decades it has hosted prestigious international motor races with wide media coverage, such as the French Formula 1 Grand Prix, the Bol d'Or, and the French Motorcycle Grand Prix. The Pernot-Ricard Group's sales representatives visit bars, restaurants, and nightclubs on a regular basis to maintain regular customers and win new ones. The company can profile the purchasing habits of typical customers in any super- or hyper-market based on the extensive data it gathers for this purpose.

Since its creation in 1975, the Pernod-Ricard Group has acquired a large number of brands in the wine and liquor field, such as Jameson, Havana Club, Jacob's Creek, Wild Turkey, and Clan Campbell. The company attained a greater international presence in December 2001 when it purchased part of Seagram's spirits and wine business for U.S.$8.15 billion. The main brands purchased represented about 14 million crates and U.S.$1.2 billion turnover in 2000 and include Chivas Regal, Glen Grant, Royal Salute, and Glenlivet whiskeys as well as Martell cognac and Seagram's Extra Dry gin. The acquisition considerably increased Pernod-Ricard's operations in North America, Latin America, India, and Southeast Asia.

Although Paul Ricard produced and sold only one aniseed aperitif in 1932, the group today provides many variations of the recipe that led to the trademark's success. Next to Ricard, *Le Vrai Pastis de Marseille,* trademarks such as Pastis 51, Pernod, Ouzo Mini, Sambucca Ramazzotto, and 8 Hermanos show how a small-scale production with unsure results has developed into an industry giant, taking a larger and larger market share throughout the world.

Bertrand Dargelos

See also: France, Production and Consumption of Alcohol in; Seagram

References

Pernod-Ricard Group website: http://www.pernod-ricard.com.

Quid. Paris: Robert Laffont, 1996.

Ricard, Paul. *La passion de créer (The passion to create).* Paris: Albin Michel, 1983.

Personal Liberty League

In the years between the early 1870s and the end of National Prohibition, alcohol manufacturers and retailers and their supporters banded together into several short-lived organizations called Personal Liberty Leagues, Personal Right and Liberty Associations, or just Liberty Leagues. These groups started in upper Midwestern states such as Illinois, Wisconsin, and Iowa and were usually dominated by German immigrants. Early Personal Liberty Leagues published their constitutions and other documents in both English and German. Although urban leagues in cities such as Chicago and Milwaukee were most successful, they fostered or inspired organizations in smaller cities and towns. By the early twentieth century, the concept had spread eastward.

Personal Liberty Leagues served lobbying, campaigning, and mutual aid functions. They worked to repeal or prevent the passage of antisaloon legislation such as Sunday closing laws, civil-damage laws, high-license laws, and local-option laws. They endorsed candidates with alcohol industry–friendly views. At the same time, members of many leagues tried to protect each other from suits brought under antiliquor laws. In some cases, they shared information with one another about local "inebriates" who might become subjects of civil-damage suits; in others, they collectively paid to defend fellow members faced with lawsuits.

The Personal Liberty Leagues were part of a larger strategy of alcohol purveyors to depict themselves as defenders of traditional American liberties. The leagues argued that legislative efforts to limit citizens' ability to buy and sell alcoholic beverages violated the constitutional freedoms of "life, liberty, and the pursuit of happiness." This rhetorical strategy aroused the ire of temperance reformers, who, in response, frequently wrote at length about why "personal liberty" did not include the right to buy and sell alcoholic beverages. Personal Liberty League opponents condemned them as secret organizations plotting to undermine the legal system through bribery or even violence.

Although Personal Liberty Leagues were primarily industry groups protecting their legal interests, they were part of a larger tradition of German radicalism. Germany had a strong history of politically active trade unions, and the early Personal Liberty Leagues imported some aspects of these associations. The founder of the powerful League for the Protection of Personal Liberty of Wisconsin, G. Hermann Boppe, for instance, was not in the liquor industry but was the editor of the Milwaukee *Freidenker* (Free-thinker).

Twentieth-century Personal Liberty Leagues moved away from their German roots to encompass all opponents of prohibition. A well-orchestrated organizational drive in the 1910s led to the formation of Liberty Leagues in nineteen states. Drawing on the writings of liberal thinkers such as John Stuart Mill, these leagues used lobbying and campaigning to oppose National Prohibition and other antiliquor legislation, which they understood as infringements upon individual freedom. More specialized organizations, such as trade union Liberty Leagues and women's Liberty Leagues, appealed to more specific constituencies. Despite their best efforts, however, the organizers of the leagues were unable to halt the gathering momentum of the prohibitionist crusade.

Elaine Frantz Parsons

See also: Civil Damage Laws; Local Option

References

Constitution and Plan of Organization of the League for the Protection of Personal Liberty of Wisconsin. Milwaukee, WI: Freidenker Publishing Co., 1884.

Duis, Perry. *The Saloon: Public Drinking in Chicago and Boston, 1880–1920.* Urbana: University of Illinois Press, 1983.

Kerr, K. Austin. *Organized for Prohibition: A New History of the Anti-Saloon League.* New Haven, CT: Yale University Press, 1985.

Rose, Kenneth D. *American Women and the Repeal of Prohibition.* New York: New York University Press, 1996.

Physicians and Alcohol (Australia)

The idea of alcoholism as a disease was a significant factor in the Australian response to alcohol dependence from colonial times. Physicians therefore played a large part historically in attempts to deal with alcohol-related problems. A politico-economic perspective that looked particularly to reducing overall consumption as an important way of containing the extent of alcohol abuse became well established by the 1980s. At the close of the twentieth century, the role of treatment and that of physicians as therapists was still important, but only as part of a larger, multifaceted, national approach to alcohol abuse.

Early Concern about Curing Alcoholism

Addressing the Intercolonial Medical Congress of 1889, Patrick Smith stated that a person became an inebriate because of an inherited, neurotic temperament. The inebriate's only chance was early treatment of his or her disease in a retreat. In 1899, leading psychiatrist Frederic Norton Manning admitted that treatment of confirmed inebriates had largely been a failure, whether in hospitals for the insane, in inebriate retreats, through hypnotic suggestion, or in the form of "cures" offered by charlatans or moral reformers. An inebriate institution had to be run under stringent regulations concerning diet, compulsory work, physical drill, and classification of inmates if there was to be hope of success.

Some Australian doctors chose to focus on strengthening the will of the inebriate using hypnosis. Prominent Sydney physician J. M. Creed used hypnotherapy extensively in his practice. Creed described his use of hypnotic suggestion to colleagues at the 1908 Australasian Medical Congress. He recorded his regret at not being able to persuade the authorities in New South Wales to try out his methods. J. W. Springthorpe, a well-known Melbourne physician, published in 1906 an account of the use of hypnotherapy.

The influential South Australian Royal Commission on Inebriety of 1906 concluded that hypnosis and the use of drugs (to create aversion to alcohol) were the two most common contemporary forms of treatment. For those that could afford them, various private institutes offered drug treatments. The Keeley cure, involving injections of bichloride of gold to eliminate the craving for drink, was the best known of a number of such cures.

Optimism about the curability of alcoholics remained fairly strong until about 1920. By the 1930s, however, doubts about the long-term effectiveness of specialist treatment had become widespread. Thus, in 1939, John Bostock, senior psychiatrist at Brisbane Hospital, claimed that suggestion, whether through hypnosis or counseling, was the mainstay of treatment. But he warned that there was always a high risk of relapse. R. S. Ellery, psychiatrist at Melbourne's Mont Park psychiatric hospital in the interwar period, said of alcoholics whom he treated that they were very liable to fall by the wayside as soon as life stresses were encountered again.

The Modern Approach to Alcoholism

Medicine "rediscovered" alcoholism in the 1950s and 1960s at a time of high and rising consumption of alcoholic beverages. New treatment services in general hospitals or special facilities proliferated. Governments provided funds for these new public facilities as well as for community education campaigns. However, by the 1970s and 1980s, experts, many of them physicians, were questioning the established emphasis on treatment and were calling for more preventive action, especially that directed at reducing general consumption of alcohol. Professor Ronald Sackville, chairman of the South Australian Royal Commission into the Non-Medical Use of Drugs, wrote in 1981 of his deep skepticism about the contribution treatment could make to solving the drug problem.

In 1980, the Federal Liberal-National Party government publicly endorsed the idea of a national health strategy to deal with the dangers of alcohol and drug abuse. Although supporting in principle a reduction in consumption of alcohol, it would not commit itself to concrete goals to this end.

The new Labor prime minister, R. J. L. Hawke, promised electors in 1984 to launch a National Campaign against Drug Abuse (NCADA), and NCADA was formally inaugurated in 1985. In the late 1980s, the Commonwealth and the states agreed to adopt a national alcohol policy. Health professionals working in the field of alcohol abuse were unhappy with the concessions the policy made to the alcoholic beverages industry. But they acknowledged that they had won some victories: Governments now accepted that normal consumption as well as heavy drinking could cause harm and that attention to advertising, product availability, and pricing was as important for harm minimization as support for treatment and community education.

In the early 1990s, NCADA became the National Drug Strategy (NDS). Central to the NDS was the concept of harm minimization, now internationally recognized as a master guiding principle of alcohol and other drugs policy. Although not neglecting treatment, the NDS put great stress on prevention.

In 1997, professors Eric Single and Timothy Rohl officially reviewed the work done under the NDS over the previous five years. They found that early and brief treatment was both cost-effective and beneficial if carried out by properly trained and resourced persons. The trend toward more responsible drinking in 1988–1995 was due partly to the preventive efforts of the NDS and partly to factors such as improvements in treatment and widespread marketing of drinks with lower alcohol content.

Milton James Lewis

See also: Australia; Inebriate Institutions (Australia); Keeley Institutes

References

Bostock, John. "Alcoholism and Its Treatment" in *Medical Journal of Australia.* 1 (1939): 137–138.

Creed, J. M. "Hypnotic Suggestion As a Therapeutic Agent" in *Australasian Medical Congress Transactions.* 3 (1908): 266–276.

Ellery, R. S. *The Cow Jumped over the Moon: Private Papers of a Psychiatrist.* Melbourne: Cheshire, 1956.

Lewis, Milton. *A Rum State: Alcohol and State Policy in Australia.* Canberra: Australian Government Publishing Service, 1992.

Manning, F. Norton. "The Immediate and Ultimate Treatment of the Inebriate" in *Australasian Medical Gazette.* 18 (1899): 221–225.

Royal Commission upon the Question of the Treatment of Inebriates Report. *South Australia Parliamentary Papers.* 2, no. 22 (1906).

Single, Eric, and Timothy Rohl. *The National Drug Strategy: Mapping the Future. An Evaluation of the National Drug Strategy, 1993–1997.* Canberra: Australian Government Publishing Service, 1997.

Smith, Patrick. "Inebriety: Its Etiology and Treatment" in *Intercolonial Medical Congress of Australasia Transactions.* (1889): 860–867.

Springthorpe, J. W. "Alcohol As a Beverage" in *Intercolonial Medical Journal of Australasia.* 11 (1906): 289–298.

Pioneer Total Abstinence Association of the Sacred Heart

The Pioneer Total Abstinence Association of the Sacred Heart, an Irish Catholic teetotal organization, was established in 1898 by the Jesuit priest James Cullen. The Pioneers grew rapidly from their modest beginnings to a total membership of some 500,000 by the middle of the twentieth century. The organization maintained an active role in Irish public life by sponsoring annual outings, parades, and anniversary celebrations to popularize total abstinence. In time, the organization expanded beyond Ireland, and although its membership has declined in recent years, it remains an active international association of teetotalers.

Father Cullen founded the Pioneers on 28 December 1898 in Dublin with the help of four influential women teetotalers: Anne Egan, Lizzie Power, Mary Bury, and Mrs. A. M. Sullivan. A preacher on behalf of temperance for many years, Cullen became convinced by the late 1890s that the cause required a fresh start in order to succeed in Irish circumstances. Initially, he envisioned a carefully selected and extremely dedicated organization that would harness the moral and spiritual power of particularly zealous Irish women, who, along with their children, were so often, as Cullen expressed it, the "guiltless victims" of intemperance. By their prayers and their own "heroic offering" of lifelong total abstinence from alcohol (except in the Eucharist and as medicine), the women were supposed to serve as moral "Pioneers" in the trek toward a sober Irish society. By early 1899, Cullen agreed to allow interested men to establish branches of the Pioneers, and by 1900 there were more than 1,200 members in a number of branch organizations in parishes around Ireland.

Influenced by his own diagnosis of the reasons for the collapse of Father Theobald Mathew's enormously successful Cork Total Abstinence Society a half century earlier, Cullen expected that his association's highly centralized administrative structure, careful financial oversight, and close attachment to the authority and structure of the Roman Catholic Church would ensure a more permanent social reform. The energetic Cullen believed that the ranks of the Pioneers ought to include only the most committed and determined. He exercised close control over the organization until his death in 1921.

The Pioneers maintain extremely strict standards, including a temperate lifestyle and a lengthy probationary period before new members are allowed to take the pledge and adopt the Sacred Heart pin. It was to be worn at all times as an outward sign of their commitment to the organization's principles. Members are expected to recite on a daily basis the Pioneer Prayer: "For Thy greater glory and consolation, O Sacred Heart of Jesus, for Thy sake to give good example, to practice self-denial, to make reparation to Thee for the sins of intemperance, and for the conversion of excessive drinkers, I will abstain for life from all intoxicating drinks. Amen."

The movement grew rapidly under Cullen's direction. In 1901, a branch society was established at Maynooth, the training college for most of Ireland's Catholic clergy, and Cullen eagerly anticipated that a new, sober generation of clerics would help lead the Irish people into temperate practices. By 1904, there were nearly 40,000 Pioneers in Ireland and the movement had begun to establish juvenile branches to prevent the development of intemperate habits. Cullen's careful cultivation of clerical support was fruitful; from 1905 onward, active Catholics received indulgences in exchange for faithful Pioneer membership. The organization claimed 280,554 Irish members in close to 300 branches by 1918, with a further 2,000 members in the United States, Australia, South Africa, and Switzerland.

Pioneer ideology has historically preferred prevention to cure. Notwithstanding its endorsement of stricter licensing regulations, Sunday pub closing, and a "dry" St. Patrick's Day, the movement emphasized the spiritual nature of the individual and community struggle against alcohol. In general, as an organization the Pioneers have endorsed vigorous and emotional persuasion and education over legislative compulsion and prohibition. Public demonstrations and regular meetings became the lifeblood of the organization, and large-scale parades and outdoor shows of Pioneer strength were a notable feature of Irish public life for much of the twentieth century. Local Pioneer centers were encouraged to engage in works of practical charity and to support a variety of social alternatives to public houses and drinking customs, including regular excursions and pilgrimages as well as the establishment of temperance halls and programs of temperance dances and socials.

From its beginning, the organization was rooted in the Mathew-ite tradition that closely associated temperance with a broad Irish patriotism and, indirectly, nationalist politics. Members were encouraged to see the society as a patriotic endeavor aimed at elevating the Irish people collectively. The Pioneers attempted to establish centers within other core Irish social organizations, notably the army, the police force, and the Gaelic Athletic Association.

After Cullen's death, the movement continued to expand under the direction of its founder's Jesuit successors. From the 1930s, the movement assumed an ever stronger presence in Irish schools and targeted the younger generation to be the standard-bearers of an increasingly temperate Ireland (or so it was hoped). The Pioneers established special terminal pledges for young recruits to the cause, encouraging youth to take a "Confirmation Pledge" against alcohol until they reached adulthood. The 1940s and 1950s represented the high-water mark of the movement in Ireland. Membership approached 500,000 in nearly 2,000 centers, and the organization's political and social influence reached its apogee under the Reverend Sean McCarron, its leader from 1943 to 1957. Occasional mass meetings, including the occasions of the movement's golden jubilee in 1949, the centenary of Father Mathew's death in 1956, and diamond jubilee celebrations in 1959, were grand public events that attracted tens of thousands of Irish men, women, and children. Pioneer missionaries, for the most part Irish priests, brothers, and nuns, gave the organization a more international presence from these years. Pioneer-affiliated and -inspired organizations were established in over a dozen countries. The movement was strong in England, parts of Africa, and Australia.

In recent decades, the Pioneers in Ireland have continued to be unambiguous, vigorous, and outspoken opponents of the prominent place of alcohol in Irish public life. Although the growing secularization of Irish society has marginalized the organization, its membership has declined, and its public influence has waned, it remains active in Ireland and elsewhere.

Paul Townend

See also: Ireland; Mathew, Father Theobald; Nugent, James; St. Patrick's Day

References

Ferriter, Diarmaid. *A Nation of Extremes: The Pioneers in Twentieth Century Ireland.* Dublin: Irish Academic Press, 1999.

Malcolm, Elizabeth. *Ireland Sober, Ireland Free: Drink and Temperance in Nineteenth-Century Ireland.* Dublin: Gill and Macmillan, 1986; New York: Syracuse University Press, 1986.

McKenna, Lambert. *Life and Work of Rev. James Aloysius Cullen.* London: Longmans, Green, 1924.

Poland

Poland is a central European country located at a crossroads of different cultures in which alcohol has had numerous symbolic meanings inherent in heavy drinking as well as a long-standing temperance tradition. The social history of alcohol has been strongly associated with political and historical developments, and particularly with the struggles for national independence and class emancipation. The most recent eruption of temperance sentiments occurred in the beginning of the 1980s, when both the Solidarity movement and the state competed in antialcohol claims in their fight for moral superiority. Introduction of a free-market economy was symbolically and practically connected with privatization of the alcohol business and liberalization of alcohol control. A high tide of alcohol consumption followed, and temperance sentiments faded away.

Centuries of Rising Drunkenness

From the beginning of Polish statehood in the medieval ages, the kings and the Catholic Church had an exclusive privilege *(propination)* to produce and sell alcoholic beverages. In most other European countries, monarchies became stronger during this period; in Poland, however, a class of noblemen increased in influence and power at the expense of the throne. Propination, among other rights the noblemen acquired, was extended to them by the end of the fifteenth century. For serfs, this rule often implied coercion to drink in their lords' inns. Drunkenness was widely spread across all social strata, including serfs, gentry, and aristocracy as well as among priests and monks, as is reflected in numerous pamphlets and epigraphs of renowned poets of the Renaissance period, such as Mikolaj Rey and Jan Kochanowski.

There is anecdotal evidence to suggest that the eighteenth century was a period in Poland's history when drunkenness, gluttony, and excessive leisure replaced such virtues as patriotism, courage, and industry. It was the century during which Poland lost its political and economic position, and eventually its sovereignty, to three neighboring superpowers. In the last efforts to reform the country and save its independence, the sobering-up of the nation was placed among the most important objectives. Under the last Polish king, Stanislaw August Poniatowski, a new code of laws was drafted in which drunkenness was ranked fifth among the seven most destructive plagues. In a climate of Enlightenment, many thinkers deplored the disastrous impact of alcohol on productivity. In his *Patriotic Letters,* Jozef Wybicki considered drunkenness "the biggest fault. . . . If we succeed to disroot it, productivity would rise two times" (Lukasz 1951: 55).

Partition of Poland among the Austro-Hungarian Empire, Prussia, and Russia revealed weaknesses in Poland's feudal agriculture, which could not compete in international markets. Overproduction of grain was a problem because there was no use for the excess except by distilleries. Alcohol production, already huge, was magnified by introduction of the potato. New methods of distillation increased spirits productivity per hectare several times over. Noblemen used all their powers to force peasants to drink as much as possible. They had to buy certain quantities of spirits each month and additional amounts on such occasions as baptisms, weddings, and funerals. In some estates, up to 50 percent of all revenues came from alcohol. Consumption soared, and in some counties reached more than 30 liters (7.9 U.S. gallons) per capita.

Temperance Initiatives

The peak of consumption was reached in the 1840s. Causal links between drunkenness and a variety of health, social, and economic problems became visible and provided the rationale for temperance initiatives. They were undertaken by grassroots movements as well as by the state. As early as the beginning of the nineteenth century, the Rogue Society was established in Vilnius with the ultimate goal of ameliorating economic and moral conditions in the country under partition. The society defined drunkenness as one of the major obstacles. Its founder, Jakub Szymkiewicz, published a book entitled *Tract of Drunkenness* in 1818 in which he laid out the foundations of medical treatment of alcoholism. In the 1840s, the sobriety movement was initiated by the Catholic Church, and in a few years it had spread across the Prussian and Russian areas of Poland. Among the most famous temperance activists were priests from a Silesian town, Piekary Slaskie, who formed the Society of Temperance in 1844. In less than one month, the society had about 185,000 members. As a result, alcohol consumption and production declined. In 1845, 85 distilleries in Silesia went bankrupt and a further 206 had to stop their production. When in 1851 the pope approved the by-laws of the Sobriety Fraternities, also called the Society of Temperance from Ardent Spirits, the movement took on high priority for the church. The movement aimed at the moral and patriotic sobering-up of the nation, which was about to lose its identity under foreign rule because of alcoholism and drunkenness.

During the same period, state authorities became deeply concerned because of health and public order considerations. The tsar seems to have issued two edicts on alcohol in 1844 and 1848 to address the poor health status of conscripts. New legislation introduced an excise tax on spirits, set minimum prices, restricted the strength of spirits to 47 percent, reduced the number of alcohol outlets, and limited hours of operation of inns.

The Austro-Hungarian Parliament adopted special alcohol legislation for Galicia and Bukovina, mostly inhabited by Polish and Ukrainian populations. Certain restrictions followed riots that broke out in Galicia in 1846. One of the symbols of feudal exploitation was local inns, where serfs or peasants were forced to buy alcohol and to drink away their extremely modest incomes. The inns were often run by a Jewish leaseholder called an *arendarz.* However, the vast majority of profits fueled the landlord, who owned both the alcohol and the inn. During famous riots, revolting groups of peasants in drunken mobs conquered local inns and set fire to the manor houses.

The alcohol question constituted a significant issue in working-class movements from the mid-1850s. Part of the worker's movement, particularly its Communist wing, criticized temperance and charity organizations, which, according to the *Communist Manifesto* (1848), weakened class conflict and delayed revolution. Nevertheless, the alcohol industry was seen by socialist parties as an element in a system of total exploitation of the working class.

All in all, the second part of the nineteenth century witnessed growing awareness of alcohol problems, a rise in temperance sentiments fueled by different concerns, and diminishing levels of consumption. By the eve of World War I, per capita alcohol intake in the territories inhabited by the former peoples of Poland was within a range of 3–4 liters (0.8–1.0 U.S. gallons) per capita.

State Monopoly and Reawakening of Temperance Sentiment

About one year after Poland regained its independence after World War I, a state monopoly on spirits was introduced by governmental decree in 1919 and then confirmed by Parliament in 1924. The legislation implied that the state controlled distillation of spirits. Blended vodkas, however, could be produced by a limited number of private producers. Selling alcoholic beverages required special licenses that were granted to people who met particular moral criteria.

Temperance sentiments still prevailed, however, as reflected by the alcohol law of 1921, which gave local communities the power to introduce local prohibitions if endorsed by popular vote. "Dry" zones were established in about 10 percent of local communities. Even in the capital city of Warsaw, local prohibition was seriously considered. Those provisions were lifted during the Great Depression, when alcohol came back as an important economic question. The restrictive policy was liberalized in order to increase the state's revenues and employment.

Throughout the whole interwar period, alcohol consumption was low by international standards, varying between 1.5 and 2 liters (0.4–0.5 U.S. gallon) per capita. Vodka constituted close to 90 percent of total alcohol intake. In spite of low consumption, a high level of concern about alcohol was maintained. Temperance slogans and movements from the nineteenth century survived and continued to affect public opinion. Their focus was on patriotism and the moral condition of the nation: "Through abstinence to national renaissance and then to self-control and to triumph of soul over a body." Such a moralistic approach led to extreme ideas, including control of alcoholism through eugenics. There were fourteen temperance organizations and sobriety associations. The Polish Pathfinders organization, with its membership in the hundreds of thousands, assumed complete abstinence of all its members.

The Experience of World War II

World War II, which lasted almost five years in Poland, constitutes a separate chapter in the history of alcohol there. In general, spirits were not available except from on-premise outlets. Farmers, who were obliged to supply agricultural products to the occupying Nazi authorities, which were partially paid for by special vodka coupons, had ready access to alcohol. In addition, outstanding employees in urban areas could receive bonuses in the form of vodka.

For years such policy has been interpreted as purposeful induction of the nation toward alcoholism. And yet there is some anecdotal evidence that the rationing system did not cover the demand for spirits. A black market flourished. Moonshining was popular in rural as well as urban areas. Illegal distillation of spirits could achieve such wide prevalence only with social approval. During the German occupation, every act with ambiguous legal status was perceived as an action against the Nazis, who decreed a total ban on moonshining under penalty of death.

Toward the end of the war, the Polish underground Home Army adopted a very repressive attitude toward illegal production of spirits. In 1944, the army was given orders to destroy illegal distilleries. In addition, drunken army members could be sentenced to death. The main argument for such restrictions was that drunken soldiers and a drunken nation were not able to fight for independence.

Alcohol and Temperance in the Socialist Period

After World War II, Poland became a socialist country under the influence of the Soviet Union. Nationalization of major industries and a one-party system constituted a basis for further development of a centrally planned economy. As early as 1944, the state alcohol monopoly was restored. Within a few years, the state fully controlled production, trade, and distribution of alcohol. Extensive industrialization was associated with a mass migration from the countryside to the cities and large industrial centers. The process of ruralization of urban areas had a visible impact on drinking patterns. Rural drinking habits (rare bouts of heavy alcohol intake) were mixed with urban ones (more frequent drinking of small quantities). In effect, drinking became more frequent, and per-session consumption increased. This particular overlap of rural and urban patterns contributed significantly to the increase in total alcohol consumption. Moreover, the new consumption patterns induced a growing popularity of beer and wine. Both began to be consumed in addition to, rather than instead of, traditional vodka. A large increase in women's drinking was also observed, parallel to their emancipation and growing participation in the labor market. In the beginning of the 1960s, women drank seven times less than men. Twenty years later, the ratio was 1:5. Cultural homogenization and emancipation combined with growing incomes resulted in a significant increase of consumption. In about thirty years between 1950 and 1980, the consumption rate increased almost threefold, from 3.0 liters (0.8 U.S. gallon) per capita in 1946 to 8.4 liters (2.2 U.S. gallons) per capita.

Numerous temperance associations having their roots in prewar times were reinvigorated just after the war. Very soon, however, the general trend of centralization and state control of all areas of social life caused these associations to be banned and replaced by a single organization in 1948, the Polish National Anti-Alcohol Committee. The committee was financed from the state budget and its autonomy was very limited. Particularly in the beginning of its existence, the committee promoted a Marxist perspective on alcoholism, which was treated as a relic of capitalism that should disappear in a matured socialism. Heavy drinking and alcoholism were perceived as a moral deficiency of individuals who did not grow to meet new socialist moral standards.

To the confusion of the ideological apparatus, alcoholism did not vanish. On the contrary, it became particularly visible in large industrial plants that were supposed to be a centerpiece of the new system. In the mid-1950s, in the climate of political thaw, a disease concept of alcoholism replaced the previous "relic of capitalism" definition. Two laws, passed in 1956 and 1959, decriminalized public drunkenness and introduced a system of sobering-up stations where seriously inebriated people were placed to recover. The new legislation also laid foundations for specialized alcohol treatment, with detailed provisions for compulsory treatment for those alcohol addicts whose behavior was antisocial. Instead of being imprisoned, thousands of alcoholics were placed in treatment institutions, which had to be secured to prevent escapes of compulsory patients. If the disease concept of alcoholism made its way into the public perception, it was as a dangerous disease. Alcohol treatment was expected to treat, but first of all to isolate, dangerous individuals.

In the 1970s, Poland enjoyed a decade of relative prosperity. Real incomes doubled. Alcohol consumption increased from 5 liters (1.3 U.S. gallons) per capita in 1970 to 8.3 liters (2.2 U.S. gallons) in 1977. Consumption growth was associated with a wave of related problems that could not be absorbed by specialized treatment programs dealing with patients as individuals. Thus, a public-health approach was adopted that shifted attention from individual addicts to population-based strategies, including efforts to reduce mean alcohol consumption. At the same time, alcohol treatment underwent ideological shifts. Heralds of a humanistic psychiatric approach demanded deinstitutionalization of alcoholics and challenged compulsory provisions. A network of clubs of former alcoholics consolidated. Twelve-step ideas found their way into the thinking of both treatment staff and patients.

The Period of Solidarity

The beginning of the 1980s witnessed another threshold in public debate on alcohol issues. The decade began with a wave of strikes that eventually led to the birth of the independent trade union called Solidarity. In the movement's first year and a half, when it was legal, it loudly accused the state authorities of responsibility for many social and economic problems, alcoholism among them. The popular argument was that all members of society, and especially alcoholics, were victims of the state policy aimed at maximization of revenues from alcohol production. In a short time, the alcohol issue reached the status of a political problem. To regain its moral position, the state introduced severe restrictions on alcohol availability. Production of cheap wines was almost stopped, and production of vodka was significantly diminished. The alcohol market broke, as supply could not meet the high demand for vodka, and alcohol rationing had to be imposed. The restrictive policy had numerous side effects, including the emergence of a black market, in which alcohol rations were redistributed according to the demand. Nevertheless, recorded consumption visibly dropped from 8.4 liters (2.2 U.S. gals.) to 6.3 liters (1.7 U.S. gals.) in about one year. This seems to have been a real, not only an apparent, decline, as a number of related problems such as alcoholic psychosis and public drunkenness also decreased by 20 to 60 percent.

Parallel to the immediate measures, negotiations were held to develop comprehensive legislation on alcohol, which was finally adopted by the Parliament in 1982. Well known as the Law on Up-Bringing in Sobriety and Counteracting of Alcoholism, the law laid foundations for a centralized system of control in which the state had to play a crucial role. Nongovernmental organizations, however, were also encouraged to join the state administration in antialcohol action. The consumption of alcohol remained stable at 7 liters (1.85 U.S. gals.) per capita throughout the remainder of the 1980s.

During the time of martial law when Solidarity was banned and during its fight for relegalization, temperance sentiments resurfaced. Antialcohol slogans were used to remind Poles of Solidarity and its symbols. Those who were involved came from Solidarity and were offered strong support from the Catholic Church.

Transformation toward a Free-Market Economy

Temperance sentiments faded away as soon as the multiparty system and a market economy were introduced to Poland in 1989. Alcohol trade and distribution belonged to the sectors that were first to be privatized. Importation, wholesale operations, and retail sale involved hundreds of thousands of new private entrepreneurs. Uncontrolled, rapid privatization has been associated with numerous scandals and crimes, including

Workers attach labels and wrap bottles in a vodka distillery in Poznan, Poland, ca. 1987. (James Marshall/Corbis)

organized-crime activity. Alcohol lobbies exerted a significant impact on alcohol policy and legislation. In less than two years, the centralized system of control was destroyed. Alcohol became available around the clock. The number of outlets increased fivefold, from 30,000 to more than 150,000. Beer, no matter how high its alcohol content, was practically excluded from any control, except a ban on advertisement. Despite the statutory ban, alcohol producers, particularly brewers, invented a wide range of tricks to market their products. Large billboards advertising apparently nonalcoholic beer appeared at every street corner. Young people constituted a primary target. In less than five years, the proportion of beer consumers among teenagers increased from 50 to 75 percent. Overall beer consumption in Poland more than doubled, from about 30 liters (7.9 U.S. gallons) to more than 60 liters (15.9 U.S. gallons) per capita by volume, and almost tripled in terms of pure alcohol, as average alcohol content grew from 4.5 percent to 6 percent.

Unrecorded consumption seems to constitute 30 percent of overall intake, which is estimated as high as 10 liters (2.6 U.S. gallons) of ethanol per capita. Major sources of unrecorded consumption are smuggling and illicit industrial production.

Alcohol control vanished as a symbol of the previous regime. The free market constitutes a symbol of new times and a new order. Temperance movements as much as alcohol policy underwent reorientation. Their focus shifted from alcohol and the general population to underage drinkers and alcohol addicts.

Grażyna Świątkiewicz and Jacek Moskalewicz

References

Friske, Kazimierz, and Robert Sobiech. *Pijaństwo—interpretacja problemu społecznego.* Warszawa: Instytut Wydawniczy Związków Zawodowych, 1984.

Lukasz, Kurdybacha. *Dzieje Kodeksu Andrzeja Zamojskiego.* Warszawa: Czytelnik, 1951.

Moczarski, Kazimierz. *Historia alkoholizmu i walki z nim.* Warszawa: Społeczny Komitet Przeciwalkoholowy, 1983.

Moskalewicz, Jacek. "The Monopolization of the Alcohol Arena by the State" in *Contemporary Drug Problems.* 12 (Spring 1985): 117–128.

Moskalewicz, Jacek, and Antoni Zieliński. "Poland (Drunk as a Pole?)" in *International Handbook on Alcohol and Culture,* edited by Dwight B. Heath, 224–236. London: Greenwood, 1995.

Ostrowska, Teresa. *Pierwszy polski traktat o pijaństwie z 1818 roku.* Warszawa: Społeczny Komitet Przeciwalkoholowy, 1987.

Świątkiewicz, Grażyna. "Regulating Unregulated Markets" in *Addiction.* 92 (1997): 67–72.

———. "Self-Help Abstainer Clubs in Poland" in *Contemporary Drug Problems.* 19 (Winter 1992): 157–176.

Świątkiewicz, Grażyna, and Antoni Zieliński. "Alcoholics Anonymous in Poland" in *Diversity in Unity: Studies of Alcoholics Anonymous in Eight Societies,* edited by Irmgard Eisenbach-Stangl and Pia Rosenqvist, 149–164. NAD Publication No. 33. Helsinki: Nordic Council for Alcohol and Drug Research, 1998

Romaniuk, Marian P. *Sylwetki trzeźwościowych działaczy XIX wieku oraz antologia ich pism.* Warszawa: Państwowa Agencja Rozwiązywania Problemów Alkoholowych, 1994.

Pope, Samuel (1826–1901)

Samuel Pope (1826–1901), British barrister and prohibitionist, was honorary secretary of the United Kingdom Alliance from 1853 until his death. After failure at business, he started his legal career in 1858. He moved from Manchester to London, where he made a small fortune at the parliamentary bar. Hoping to represent the prohibitionist cause in the House of Commons, he was on five occasions an unsuccessful Liberal parliamentary candidate in the 1850s and 1860s. In 1869, he became recorder of Bolton. Pope contributed generously to the alliance as a wealthy subscriber, a clever speaker, and a genial, popular veteran who encouraged compromise and practical policies. He is one of the few Victorian temperance reformers to receive an entry in the *Dictionary of National Biography* (1912) who had neither a parliamentary seat nor a hereditary title; the entry barely mentions his connection with the alliance, however. He was a staunch supporter of the North in the American Civil War. In politics he was a radical Liberal who remained loyal to William Gladstone at the time of the Irish home rule schism. In religion he was a Nonconformist.

According to rumor, late in life Pope drank alcoholic beverages. For instance, when he presented evidence at the Royal Commission on Liquor Licensing Laws on 19 July 1898, the publican commissioner Charles Walker strongly insinuated that Pope was a drinker. If this were true, Pope still could have lawfully held office in the alliance. The United Kingdom Alliance always asserted that it was not a temperance organization, insisting that its members commit themselves to prohibition by local referendums but not asking that they be teetotalers.

David M. Fahey

See also: Gladstone, William Ewart; Liberal Party (United Kingdom); United Kingdom Alliance (UKA)

Reference

"Bedwell, Cyril Edward Alfred" in *Dictionary of National Biography.* Supplement, *1901–1911.* London: Smith, Elder, 1912.

Harrison, Brian. *Dictionary of British Temperance Biography.* Aids to Research, no. 1, *Bulletin Supplement.* Coventry, UK: Society for the Study of Labour History, 1973: 98–99.

Portes Gil, Emilio (1891–1978)

Emilio Portes Gil (1891–1978) was interim president of Mexico from 1928 to 1930 during the postrevolutionary period of the Maximato (ca. 1928–1934), when former president Plutarcho Elías Calles still wielded most executive power. Portes Gil launched the revolutionary antialcohol campaign in 1929 with the formation of the Comité National de Lucha Contra El Alcoholismo (National Committee for the Struggle against Alcoholism).

Although Portes Gil is not remembered as having exercised much executive power during his brief tenure as interim president, the structure of his antialcohol program appears to have been his own creation, based on his previous experience as governor of the state of Tamaulipas. His Comité Nacional included representatives from each federal bureau. Its aim was to reproduce, on a national scale, the antialcohol measures that Portes Gil had undertaken while governor. In Tamaulipas, he had instituted policies to raise public awareness about the dangers of alcoholism with the goal of catalyzing a popular temperance movement. He felt that the failure of Prohibition in the United States was probably due to the inherent difficulty of trying to use law to change customary behavior. Instead, Portes Gil's plan was designed to encourage the emergence of a grassroots temperance movement. The Comité Nacional set about gathering statistics regarding alcohol use and abuse, creating and disseminating antialcohol propaganda, and encouraging the development of a national movement. In Tamaulipas, approximately 100 communities had requested that the state government enact temperance legislation on their behalf as a result of the persuasion campaign.

As it turned out, Portes Gil's Comité Nacional was short-lived; it disappeared in 1932. However, the national temperance campaign continued, particularly within the ministries of industry, agriculture, hygiene and education. By 1933, a new official national antialcohol directorate had been formed. Luis G. Franco became the head of activities of the new Dirección Anti-alcohólica (Anti-Alcohol Directorate). The functions of the directorate were similar to those of the Comité Nacional, although its bureaucratic structure was different. It was no longer a body composed of representatives from each of the executive departments; it now included not only government bureaucrats, such as Isaac Ibarra from the Department of War, but also representatives from nongovernmental entities, such as Salvador Ruiz de Chávez of the Confederation of Chambers of Commerce.

As in the past, the advocacy of temperance continued within the various executive departments in a quasi-autonomous fashion. The Dirección Anti-alcohólica's motto was *Temperancia: Por la Patria y Por la Raza* ("For the Fatherland and for the Race"). Primarily dedicated to propagandizing in favor of temperance, it sponsored activities such as a national contest for the best antialcoholism song. Two hundred pesos were awarded to the individual who produced the best song denouncing the negative effects of alcohol on men, the home, and the nation. The national directorate also promoted parades and public ceremonies championing temperance.

Stephanie Mitchell

See also: Comité Nacional de Lucha contra el Alcoholismo; Dirección Anti-alcohólica; Franco, *Ing.* Luis G.; Mexico

References

Bethel, Leslie. *Mexico since Independence.* Cambridge: Cambridge University Press, 1991.

Gonzalez, Hugo Pedro. *Portesgilismo y alemanismo en Tamaulipas.* Ciudad Victoria, Tamaulipas, Mexico: Universidad Autonoma de Tamaulipas, Instituto de Investigaciones Históricas, 1983.

Portes Gil, Emilio. *Emilio Portes Gil: Un civil en la Revolución Mexicana.* Ciudad Victoria, Tamaulipas, Mexico: Instituto Tamaulipeco de Cultura, Gobierno del Estado de Tamaulipas, 1989.

Portugal

Portugal has a long history of viticulture and belongs to the group of Mediterranean wine-preferring countries with high alcohol consumption. Wines, and especially fortified wines, have been economically important to the country for many centuries (Hurst, Gregory, and Gussman 1997). For decades, Portugal has had an alcohol-control structure with the aim of safeguarding the economic interests involved in wine production and export. Not until very recently have social and public health issues been included in Portuguese alcohol policy discussions. This delay in bringing the issue into public discourse is due to the fact that wine has been defined as an agricultural product and a foodstuff consumed at meals, not as an alcoholic beverage and an intoxicant. Also, informal social alcohol control has been effective.

There has never been a strong temperance movement in Portugal. Moreover, there are no nationwide nongovernmental organizations acting as pressure groups for preventive alcohol policies. There are organizations that are very interested in increasing wine consumption and in promoting moderate drinking, however (Karlsson and Österberg 2002).

A Portuguese woman carrying an armful of straw-covered bottles (Hulton/Archive)

Alcohol Production and Trade

Portugal produces a wide range of alcoholic beverages, wine being the most important one. The variety of wines includes red, white, green, rosé, and sparkling as well as many different fortified wines such as Madeira, Moscatel, Carcavelos, and port, which is Portugal's most famous fortified wine. The Portuguese wine trade increased rapidly at the beginning of the eighteenth century; in 1703, the British government concluded a treaty according to which Portuguese wines were admitted into England on more favorable terms than French and German wines. Madeira and port accounted for almost half of Portugal's wine exports in 1995 (Hurst, Gregory, and Gussman 1997).

There are thirteen regions in Portugal where wines designated as quality wines from a specified region are produced and twenty-eight other regions with wine production. In the 1980s, there were almost 180,000 wine producers in Portugal; half of them were members of wine cooperatives. About 5 percent of the active population was employed in viticulture, and more than a tenth of the population was economically dependent on wine production and trade. Vineyards cover some 380,000 hectares (939,000 acres), or approximately 8 percent of the country's arable agricultural land (ibid.). In 1999, with a wine production of 7.8 million hectoliters (206 million U.S. gallons), Portugal came ninth in the list of wine-producing countries in the world and fifth in the list of wine-producing countries in Europe. The output of wine varies considerably from year to year, however.

There was no brewing of beer on a commercial basis in Portugal until the beginning of the nineteenth century. In 1995, four brewing companies operated a total of eight breweries. The two largest ones were nationalized in 1977 following the 1974 revolution, but they were again privatized in 1990. These two brewing companies produced almost 98 percent of all domestic beer in 1990. Lager accounted for 95 percent of the beer market in 1995. During the past few decades, off-premise consumption of beer has increased, but in 1990 over 60 percent of beer was still consumed in restaurants, cafés, and the like (ibid.).

Domestic distilled spirits such as Bagaceira are derived mostly from grapes, but fruit-based brandies and liquors such as Beirao are also popular. Locally produced gins, rums, and vodkas are also consumed (ibid.). Domestic production of distilled spirits, however, remained at quite a low level throughout the twentieth century.

Alcohol Consumption

Recorded consumption of alcoholic beverages declined in Portugal from the mid-1960s to the year 2000 from about 14 liters to 10.8 liters (3.7 to 2.9 U.S. gallons) of pure alcohol per capita each year. Wine consumption, especially, has declined in the past four decades. At the beginning of the 1950s, as well as in the mid-1960s, the yearly consumption of wine, counted in liters of the product, was more than 110 liters (29.1 U.S. gallons) per capita. At that time, wine accounted for more than 95 percent of the total alcohol consumption. By 2000, consumption of wine had declined to 50 liters (13.2 U.S. gallons) per capita (*World Drink Trends* 2002).

Yearly consumption of distilled spirits was about 0.5 liter (0.1 U.S. gallon) of pure alcohol per capita in the mid-1960s. It increased at the beginning of the 1970s to about 1 liter (0.3 U.S. gallon) per capita. After this increase the consumption of distilled spirits was more or less stable until the mid-1980s, when it rose to 1.8 liters (0.5 U.S. gallon) per capita by 1990. By 2000, the consumption of distilled spirits decreased to 1.4 liters (0.4 U.S. gallon) of pure alcohol per capita (ibid.).

In the early 1950s, beer consumption was very low in Portugal, only a few liters of the product per capita a year. Since then it has been growing almost constantly. In liters of the product, it increased from 5 liters per capita (1.3 U.S. gallons) at the beginning of the 1960s to 65 liters (17.2 U.S. gallons) per capita by 2000. During the past decade beer consumption has, however, remained almost unchanged (ibid.).

Despite the increases in beer and distilled spirits consumption, Portugal is still a wine-drinking country. In 2000, the proportion contributed by wine of the total alcohol consumption was about 55 percent, that of beer a little over 30 percent, and that of distilled spirits a little under 15 percent. Although alcohol consumption has declined in Portugal, the country still takes one of the top positions in the world when it comes both to recorded and also to real total per capita alcohol consumption (ibid.; Leifman 2002). The prevalence of alcohol-related problems roughly follows the trends in per capita alcohol consumption. For instance, mortality from cirrhosis of the liver peaked in Portugal in the 1970s, slightly after the peak in the total per capita alcohol consumption, and has since declined (Hurst, Gregory, and Gussman 1997).

Drinking Habits

Drinking at meals is an essential part of the Portuguese alcohol culture. In fact, in some of the few Portuguese studies on drinking habits, drinking apart from meals has been taken as an indication of possible problematic drinking (Simpura and Karlsson 2001). Between 15 and 20 percent of men and a little less than 10 percent of women report drinking alcoholic beverages at all meals, and another 20 to 30 percent of men and 10 to 20 percent of women report drinking at almost all meals. The attitude that alcoholic beverages are a necessary element of daily nutrition is still widespread.

In the mid-1990s, lifetime abstinence rates in the population over fifteen years of age were about 5 percent for men and over 25 percent for women. However, 60 percent of men and about 20 percent of women were regular drinkers who drank at almost all meals or more frequently. The remaining 35 percent of men and 55 percent of women were infrequent or very infrequent drinkers. The highest drinking frequencies were found in the age group between thirty-five and forty-four. Women's proportion of total alcohol consumption was about 30 percent (ibid.).

Administrative Structure of Alcohol Policies

Although the sale and distribution of alcoholic beverages are quite free in Portugal, production and trade in wine are controlled through government agencies. The Instituto da Vinha e do Vinho is the official agency responsible for controlling the domestic table-wine market from production to retail sale. It oversees quality control of these products and is responsible for the country's denomination-of-origin system. Port wines are controlled by the Instituto do Vinho do Porto (Hurst, Gregory, and Gussman 1997).

The Commission against Alcohol Problems, established under the jurisdiction of the General Directorate of Health in 1977, aims to provide information and foster the interest of the population in alcohol-related issues. It has created multidisciplinary teams in some psychiatric hospitals and, after November 1988, has made explicit policy statements concerning preventive alcohol policies (Karlsson and Österberg 2002). The Portuguese Committee against Alcoholism, however, paved the way for such statements through proposals presented to the minister of health at the end of 1986.

In 1988, three regional centers were set up in accordance with government health regulations in Coimbra, Lisbon, and Oporto. Their main objectives were to prevent and treat alcohol-related diseases, rehabilitate patients, coordinate prevention and treatment activities, and teach, educate, inform, and investigate alcohol-related issues in their respective zones (ibid.).

The first alcohol action plan approved by the regional office of the World Health Organization in 1994 and the European Charter on Alcohol from the 1995 Paris Conference were important background documents in putting alcohol issues on the agenda of the Portuguese health strategy for the years 1998–2002. In this document, preventive alcohol-policy goals and objectives were for the first time officially presented. This process further resulted in a resolution in May 1999 establishing an interministerial commission to draft a proposal for an action plan to reinforce and strengthen implementation of the health strategy concerning alcoholism (ibid.).

In the fall of 2000, a resolution of the Assembly of the Republic recommended that the government should adopt a national alcohol program, including a national campaign alerting the public to the perils of excessive consumption and targeted specifically to pregnant women, children, teenagers, and excessive drinkers. It further recommended that the government consider raising the legal drinking age, regulating the advertising of alcoholic beverages, and limiting the business hours of bars and coffee shops near schools.

Alcohol Control Measures

Producers of alcoholic beverages, as well as alcohol importers and wholesalers, need a license to operate in Portugal. On-premise retailers also need a license, but off-premise retailers do not (ibid.). Licenses for wine producers are provided through the IVV; they are permanent and help to ensure that wine producers follow certain standards and practices. Licenses for on-premise retailers are granted by the Municipal Department for Economic Activities in each city's administrative offices. These licensing rules have remained virtually unchanged for half a century.

There are no special rules limiting hours of operation for retail outlets in Portugal. General business hours have been extended over the past few decades; establishments selling goods and services to the general public can be open from 6 A.M. until midnight every day of the week. Convenience stores as well as coffee shops, beer shops, tea shops, restaurants, and snack bars can stay open until 2 A.M. Clubs, cabarets, dancing places, fado houses (restaurants that feature fado singing), and the like stay open until 4 A.M. (ibid.).

There are no legal age limits on off-premise retail sale of alcoholic beverages in Portugal. For on-premise sale there is a legal age limit of sixteen. However, this rule has not been strictly enforced. Furthermore, youngsters under sixteen years of age are legally allowed to enter on-premise retail sale establishments and drink when accompanied by parents.

At the beginning of 1982, the blood-alcohol content (BAC) limit for drivers was set at 0.08 percent. Those who by gross negligence drove a vehicle with a BAC equal to or higher than 0.12 percent could be fined or sentenced to one year's imprisonment. New legislation in January 1998 made it illegal to drive a vehicle with a BAC equal to or higher than 0.05 percent. Drivers who were caught were fined and their driver's licenses were suspended. A new law passed in May 2001 further decreased the BAC limit to 0.02 percent. Drivers caught with a BAC higher than 0.02 but lower than 0.05 percent are fined. A judge can also suspend the offender's driver's license or, as an alternative, depending on the offender's record, sentence him or her to collaborate in prevention campaigns against drunk driving or to participate in promoting traffic safety (ibid.).

Alcohol advertising was regulated for the first time by law in 1981, when special legislation was passed aimed at eliminating misleading or dishonest advertising, among other things. In 1983, an advertising code was created prohibiting advertising of alcoholic drinks on television and radio between 6 and 10 P.M. This law was revised in 1995; advertising of alcoholic beverages on television and radio is now prohibited between 7 A.M. and 9.30 P.M., and television commercials are not allowed to show alcoholic beverages being consumed. The alcohol industry also follows a self-regulating code that states the conditions under which alcohol advertising is allowed, regardless of the media used. The advertising of beer and distilled spirits is also banned on billboards and in cinemas, and these restrictions are fairly effectively enforced (ibid.).

Portugal has never implemented a nationwide effort to educate children in the public schools about alcohol-related problems and issues. However, a health education program has been worked out, and there are plans to educate teachers and health personnel as one of the measures to reduce alcohol consumption by 10 percent by the year 2002 (ibid.).

Alcohol Taxation

Since 1993, the excise duty on wine has been set at zero euros. The excise duties on distilled spirits and beer are based directly on alcohol content. For intermediate products, the excise duty is set on the basis of hectoliters of the product. As of 2003, alcohol excise duty rates for beer, intermediate products, and distilled spirits in Portugal are somewhat higher than the minimum rates required by the European Union (ibid.). Calculated on the basis of alcohol in the beverages, the total tax burden, including value-added tax (VAT) and excise duties, is highest on distilled spirits. It is nearly twice the tax burden on beer. Alcohol in beer, in turn, is taxed nearly six times more heavily than alcohol in the form of table wine. Nowadays the standard VAT rate of 17 percent is applied on beer, intermediate product, and distilled spirits. For wine the VAT rate is 5 percent.

During the 1990s, the excise duty rate for distilled spirits doubled in real terms, and the excise duty rate for beer increased along with the general price level. In beer prices, the total tax burden, including excise duties and VAT, is about 28 percent. About half of the price of distilled spirits consists of taxes, whereas the tax burden on wine is only 5 percent (Hurst, Gregory, and Gussman 1997).

Thomas Karlsson and Esa Österberg

References

Hurst, Wendy, Ed Gregory, and Thomas Gussman. *International Survey: Alcoholic Beverage Taxation and Control Policies.* 9th ed. Ottawa: Brewers Association of Canada, 1997.

Karlsson, Thomas, and Esa Österberg. "Portugal" in *Alcohol Policies in EU Member States and Norway: A Collection of Country Reports,* edited by Esa Österberg and Thomas Karlsson. Helsinki: Stakes, 2002.

Leifman, Håkan. "Estimations of Unrecorded Alcohol Consumption Levels and Trends in 14 European Countries" in *Nordic Studies on Alcohol and Drugs.* 18, English Supplement (2002): 54–70.

Österberg, Esa, and Thomas Karlsson, eds. *Alcohol Policies in EU Member States and Norway: A Collection of Country Reports.* Helsinki: Stakes, 2002.
Simpura, Jussi, and Thomas Karlsson. *Trends in Drinking Patterns among Adult Population in 15 European Countries, 1950 to 2000: A Collection of Country Reports.* Helsinki: Stakes, 2001.
World Drink Trends. 2002 Edition. Oxfordshire, UK: NTC Publications, 2002.

Presidential Commission on Drunk Driving

On 14 April 1982, U.S. President Ronald Reagan issued Executive Order 12358, which created the Presidential Commission on Drunk Driving (PCDD). Reagan's order was widely understood to be a co-optive response to the 1980 formation of Mothers Against Drunk Driving (MADD), the nongovernmental activist group that had sharply increased public awareness of the human costs of alcohol use by U.S. automobile drivers.

Reagan charged the twenty-eight-member PCDD with gathering expert recommendations for using public authority to:

> (a) heighten public awareness of the seriousness of the drunk driving problem;
> (b) persuade States and communities to attack the drunk driving problem in a more organized and systematic manner, including plans to eliminate bottlenecks in the arrest, trial and sentencing process that impair the effectiveness of many drunk driving laws;
> (c) encourage State and local officials and organizations to accept and use the latest techniques and methods to solve the problem; and
> (d) generate public support for increased enforcement of State and local drunk driving laws. (Reagan 1982)

In its final report to Reagan, the PCDD called drunk driving "a national menace, a national tragedy, and a national disgrace" and made recommendations for reducing it. Among the commission's thirty-nine recommendations was a call for the formation of a private, nonprofit organization to promote the implementation of the PCDD's other thirty-eight recommendations. The proposed permanent entity, the National Commission Against Drunk Driving (NCADD), began operations in January 1984 and still exists.

The NCADD has always been careful not to offend the alcohol and automobile industries, whose top executives have routinely sat on its governing board and committees. Although the PCDD charter called for systematic approaches and the NCADD states its purpose as "developing strategies and programs that show promise in reducing the incidences of driving impaired," PCDD/NCADD strategies and programs have included neither aggressive public education about and treatment of alcohol addiction nor any hint of a thought of examining the contribution of the United States's automobile-centered transport system to drunk driving.

These aversions to basic issues were manifest from the beginning. The PCDD was chaired not by a public-health expert but by U.S. Transportation Secretary John A. Volpe. Before serving Presidents Nixon and Reagan as transportation secretary, Volpe had become a multimillionaire construction contractor during the postwar economic boom of the late 1940s and early 1950s. In 1956, Volpe had been appointed by U.S. President Dwight D. Eisenhower to serve as federal highway administrator. In that capacity, Volpe oversaw the main phase of construction of the vast U.S. interstate automobile highway system. So long as such foxes are guarding the henhouse, there will be a perpetual drunk driving epidemic for the NCADD to nurse.

Michael Dawson

See also: Mothers Against Drunk Driving (MADD)

References

"About the Presidential Commission." 2003. National Commission on Drunk Driving website. http://www.ncadd.com/about/presidential.cfm.
Reagan, Ronald. "Executive Order 12358—Presidential Commission on Drunk Driving." 1982. http://www.reagan.utexas.edu/resource/speeches/1982/41482c.htm.
Reagan, Ronald. "To the American People." 1983. www.ncadd.com/about/reagan_letter.cfm.

Preston Teetotal Monument

The Preston Teetotal Monument stands in the Preston General Cemetery, Lancashire, England. Constructed in a commonplace Victorian Gothic style, it has no obvious architectural merit, but it is the focal point of an unusual phenomenon. It rises up in the midst of the headstones of some of the town's teetotal pioneers, a few paces distant from the ornate tomb of Joseph Livesey, their most famous member. Creating what was almost a cemetery within the cemetery, the planners of the monument seem to have envisaged something like a *campo santo* for the burial and veneration of those who had taken the teetotal pledge when this was a novel and hazardous commitment.

Preston occupies an important place in temperance history. It was there in 1833 that Richard Turner (d. 1846), a stammering ex-inebriate, coined the word "teetotal" when he was describing the extent of his repudiation of drink. It was there, too, that the "seven men of Preston" had met one year earlier to administer what they claimed was the first total abstinence pledge. Such was the proselytizing zeal with which they and their followers spread their message throughout Britain that Preston soon acquired a reputation as the "Bethlehem" of teetotalism. True to the metaphor, converts started to travel there in the hope of drawing inspiration from scenes about which they had heard so much. The Teetotal Monument was erected in 1859 to serve as the centrepiece of a sacred site where these teetotal pilgrims could reflect on the origins and significance of the movement.

The inscription on the monument reads as follows: "Erected by public subscription, A.D. 1859, to commemorate the origin in Preston of total abstinence from all intoxicating liquors." This was a myth of origin that was indebted to local pride rather than strict historical accuracy, and the Preston teetotalers had to work hard to sustain it. John Catterall, the chairman of the planning committee that erected the monument, admitted as much in January 1859. He had been appalled, he said, to discover that people in other towns were claiming the credit for inventing teetotalism. This monument would silence anyone who "cribbled a bit, and tried to prove that Preston was not the cradle of the cause"; it would "reserve the glory for Preston when they were dead and gone" (Tyrrell, forthcoming).

With the passage of time, the headstones of teetotal stalwarts were erected in adjoining burial plots, collectively endorsing the monument's version of teetotal history and reinforcing its message. Some of the epitaphs are simple formulaic statements of names and dates together with brief invocations of familial piety; only supplementary materials such as obituary notices and other references reveal the sort of mentality and experiences that led these people to become teetotalers in life and remain united in death. Other epitaphs are sufficiently informative to be read as historical sources in their own right. They point to the wide range of the Preston teetotalers' reforming aspirations, especially those that identified them with the poor and disadvantaged. Thus Livesey is commemorated for his "philanthropy and usefulness as author and worker, as the pioneer of temperance, the advocate of moral and social reform and the helpful friend and counsellor of the poor." Edward Grubb's tombstone reiterates the message that Preston was "the birth-place of teetotalism" and grandiloquently hails him as "the noblest Roman of them all" because of his dedication to making teetotalism a "world-wide movement." A "brilliant orator" and "philosophical teacher," he had shown an "incorruptible integrity" that rendered him "proof against the blandishments of wealth." John Catterall is honored on his tombstone by "the friends of temperance, orphans & blind institutions."

One of the headstones is particularly striking. It reads as follows:

> Sacred To The Memory Of Mary Graham;
> Who departed this life on the 4th of February 1868,
> in the 90th year of her age.
> She was a devoted friend to the temperance cause,
> & a teetotaler 33 years.
> She refused to take either medicine, or intoxicating
> liquors, up to her death.

An early convert, Mary Graham was well known as "Temperance Mary" and "the Queen of the Teetotallers"; her epitaph is a rare, possibly unique, public artifact that recognizes the importance of women in the temperance movement. At a time when convention frowned on strong-minded women and did not encourage female participation in the public sphere, Mary Graham had embarked on a great public-speaking campaign that took her to Liverpool, Manchester, Blackburn, Bolton, and other places in Lancashire. She continued to give speeches to the Preston teetotalers until very late in life, and her death notice saluted her as "the oldest female temperance lecturer." The reference on her headstone to her reaching old age as an abstainer from medicine as well as drink serves as a reminder of the close relationship between teetotalism and Victorian campaigns for medical reform. Rejecting the standard medical practice of the day, which relied heavily on intoxicating drinks and dangerous drugs, many teetotalers opted for prophylactic approaches to disease. Under Joseph Livesey's leadership, Preston teetotalism embraced water cures, dietary regimens, physical exercise, and other natural therapies.

Like other public memorials created during an era that *The Times* described as one of "monument mania," the Preston Teetotal Monument and the headstones in its vicinity have lost their meaning for the modern passer-by. For the historian of the temperance movement, however, they still have stories to tell: They encapsulate the values and ideas that took the total abstainers to the front rank among radical reformers of their day.

Alex Tyrrell

See also: Livesey, Joseph

References

Barnard, Sylvia, M. *To Prove I'm Not Forgot: Living and Dying in a Victorian City.* Manchester: Manchester University Press, 1990.

Harrison, Brian. *Drink and the Victorians: The Temperance Question in England, 1815–1872.* London: Faber & Faber, 1971; 2d ed., Keele: Keele University Press, 1994.

Tyrrell, Alex. "Preserving the Glory for Preston: The Campo Santo of the Preston Teetotalers" in *Monuments of Radicalism,* edited by P. A. Pickering and Alex Tyrrell. Aldershot: Ashgate, forthcoming.

Processed Sugar Ferments

The availability of various forms of processed or semiprocessed sugar has increased very considerably in some parts of the world over the past century. One consequence of this has been a substantial increase in the use of such sugar in informal-sector and/or illegal liquor production, generally on a small scale. Sugar has been used to produce fermented beverages, to strengthen ferments made from other materials, and to produce a base for distillation.

Black jaggery, a cheap by-product of sugar refining, has for decades been used in East Africa and India in the production of a fermented base that would then be distilled. In East Africa, fully processed sugar is now sometimes used in the production of ferments, either in addition to other more established ingredients, such as modified grains or ripe fruits, or on its own. The use of sugar alone would seem to derive from a longer-standing use of honey to produce mead: In East Africa, it was communities such as the Maasai, who had traditionally been mead-drinkers, who turned first to the use of liquor made solely from sugar. The techniques they used reflected this tradition: Sugar was dissolved in warm water, a dried root that had been steeped in a previous batch of liquor was used as an inoculum, and fermentation followed.

In the past twenty or thirty years, with the growth of urban populations who may not have grain or fruit of their own to ferment, use of sugar alone in this way as the base for a fermented beverage has grown, owing its popularity partly to the rapidity of the process of fermentation and the perceived potency of the product, which is believed to be usually rather stronger than grain beer (the only reported test shows it at 4.9 percent by volume). When sugar alone is used as the basis for fermentation, the beverage may be flavored: Tea leaves are one popular additive, in a perhaps ironic nod to the importance of sugar as a consumption item in that most sober of drinks, tea. Fermentation is often encouraged by the addition of dried baker's yeast; this, combined with the high levels of sugar, makes for rapid fermentation. Limited observation by the author suggests that this beverage is more stable than grain-based beer or various fruit or sap-based beverages, perhaps because it is less microbiologically rich: A shelf life of up to a week is reported by producers.

More widely in Africa, sugar is used to enhance the potency of grain-based beers, honey wines, or sugarcane beer. Beverages made in this way are very much a feature of urban areas; they are made for sale in the market, rather than being intended for any ritual or other noncommercial use.

Justin Willis

See also: Africa, Central

Reference

Nikander, P., et al. "Ingredients and Contaminants of Traditional Alcoholic Beverages in Tanzania" in *Transactions of the Royal Society of Tropical Medicine and Hygiene.* 85 (1991): 133–135.

Prohibition Bureau

The Prohibition Bureau officially came into existence in 1927 when the U.S. Congress established it within the Treasury Department. Generally, however, commentators have used the term to also include earlier administrative units created by the Treasury Department to enforce the National Prohibition (Volstead) Act. Corruption, inefficiency, and lack of cooperation with other enforcement agencies hampered the enforcement efforts of the Prohibition Bureau and its predecessors. Nonetheless, federal Prohibition enforcement had impressive impacts in terms of arrests, property seizures, reduction of per capita consumption of alcohol, changes in legal doctrine, and pressures on the legal system. Despite these achievements, the efforts of the Prohibition Bureau were insufficient to convince the public to retain National Prohibition.

The realities of Prohibition enforcement demanded substantial federal resources. In 1920, the congressional appropriation for Prohibition enforcement totaled slightly more than $3 million, but this figure had grown to nearly $15 million by 1930. Although the agents authorized for enforcing Prohibition were always insufficient for the task, they significantly increased the number of field agents responsible for the enforcement of federal law.

The Volstead Act assigned responsibility for Prohibition enforcement to the commissioner of internal revenue in the Treasury Department rather than to the Department of Justice. The ostensible reason for this decision was the experience of the Treasury Department in collecting liquor taxes prior to Prohibition, especially since the prior taxes were continued as penalties in the Volstead Act. In fact, Wayne Wheeler—the legislative director of the Anti-Saloon League of America—favored using the Treasury Department because he thought that approach would maximize the Anti-Saloon League's influence in the enforcement of the Volstead Act.

The Volstead Act also exempted Prohibition field agents from civil service laws. Representatives of the Anti-Saloon League later defended the civil service exemption as a political necessity. In any event, it subjected Prohibition enforcement to political influence from both supporters and opponents of Prohibition.

The result of exempting Prohibition agents from civil service laws was inefficiency and corruption. Turnover was high throughout the Prohibition era. Between 1920 and 1930, the Prohibition Bureau and its predecessors appointed nearly 18,000 agents for field positions that never numbered more than 2,300; the average turnover in the enforcement branch was almost 40 percent. When the 1931 fiscal year began in July, the federal government had dismissed 1,604 agents—more than 8 percent of the total number appointed—for cause. At least for the 752 agents dismissed prior to 1926, the most common grounds for dismissals for cause were drunkenness and bribery.

Prohibition enforcement underwent frequent reorganizations. The secretary of the treasury created a unit within the Bureau of Internal Revenue to administer the permitting program for industrial alcohol and to enforce the Volstead Act in 1919. The Prohibition Bureau created by Congress in 1927 was transferred from the Treasury Department to the Justice Department in 1930.

The commissioner of internal revenue organized the Prohibition Unit along state lines. Each state had a director of federal Prohibition with responsibility for both permitting and enforcement. In 1925, President Calvin Coolidge separated the permitting and enforcement functions when he appointed General Lincoln C. Andrews, a retired army officer, as assistant secretary of the treasury with responsibility for Prohibition enforcement. General Andrews consolidated enforcement into twenty-four Prohibition districts, using the federal judicial district as the geographical basis for the reorganization.

Under the 1927 statute, the Prohibition Bureau was assigned responsibility for enforcing Prohibition and administering the permit program for industrial alcohol. The statute also subjected Prohibition agents to the requirements of federal civil service laws and provided that the term of office of previously appointed agents would terminate six months after the effective date of the new law.

When Congress transferred the Bureau of Prohibition and its enforcement responsibilities to the Department of Justice in 1930, the oversight responsibility for industrial alcohol remained in the Treasury Department in a newly created Bureau of Industrial Alcohol. This reform accomplished a reorganization that had first been recommended by the secretary of the treasury when Congress was first considering the Volstead Act.

Another obstacle to effective enforcement of National Prohibition was the lack of cooperation among enforcement agencies. At the federal level, the Coast Guard and the Customs Service had important responsibilities for enforcing the Prohibition against importation of intoxicating liquor, but the federal government never established a coordinated program for enforcement. Despite the grant of "concurrent" authority for state enforcement of the Eighteenth Amendment, the states made few efforts specifically aimed at the enforcement of Prohibition laws. Maryland never enacted a state prohibition law, and several states repealed their state laws as unnecessary in light of the Volstead Act.

Despite the deficiencies of Prohibition enforcement, the enforcement record of the Prohibition Bureau and its predecessors was substantial. Between 1921 and 1929, Prohibition agents made 539,759 arrests and seized 45,177 cars. The number of Prohibition cases in federal courts increased from 29,114 in 1921 to 65,960 in 1932. For the period 1921 to 1933, Prohibition cases constituted 64.6 percent of all cases in federal district courts. Most important, the enforcement of National Prohibition was sufficiently effective to produce a substantial reduction in the estimated consumption of alcohol in the United States.

The Prohibition Bureau and its predecessors relied on a variety of intrusive enforcement practices. Federal agents employed electronic surveillance, warrantless searches of automobiles, informers, and the use of poisons in industrial alcohol in their efforts to enforce National Prohibition. In New York City, Prohibition agents operated the Bridge Whist Club for several months and sold liquor to all comers. Using the information the agents gathered at the Bridge Whist Club, they later made arrests.

Prohibition enforcement could also be dangerous. Between 1920 and 1926, forty-five federal Prohibition agents lost their lives. The Treasury Department acknowledged that federal Prohibition agents killed at least eighty-nine individuals, and opponents of Prohibition claimed that the number of private citizens who were killed was actually much higher.

The flood of Prohibition cases affected law and the legal system. It prompted the Supreme Court to address and to alter a number of important points of federal criminal law. Despite relatively mild penalties, the huge number of prosecutions added thousands of inmates to federal prisons during the 1920s.

Ultimately, of course, the Prohibition Bureau failed. Despite the impact of Prohibition enforcement on the amount of alcohol consumed, Prohibition never achieved a satisfactory level of enforcement to convince the people of the United States to retain National Prohibition.

Kenneth M. Murchison

See also: Anti-Saloon League of America (ASLA); National Prohibition (United States); Wheeler, Wayne Bidwell

References

Hamm, Richard F. "Short Euphorias Followed by Long Hangovers: Unintended Consequences of the Eighteenth and Twenty-First Amendments" in *Unintended Consequences of Constitutional Amendments,* edited by David E. Kyvig, 164–199. Athens: University of Georgia Press, 2000.

Merz, Charles. *The Dry Decade.* Garden City, NY: Doubleday, Doran, 1930.

National Commission on Law Observance and Enforcement, *Report on the Enforcement of the Prohibition Laws of the United States.* Washington, DC: Government Printing Office, 1931.

Sawyer, Albert E. "The Enforcement of National Prohibition" in *Annals of the American Academy of Political and Social Science.* 163 (1931): 10–29.

Schmeckebier, Laurence F. *The Bureau of Prohibition: Its History, Activities, and Organization.* Washington, DC: The Brookings Institution, 1929.

Prohibition Party (United States)

The oldest minor party in the United States, the Prohibition Party continues to hold aloft its demand for a legal ban on alcohol after more than a century of effort. Founded in the aftermath of the Civil War, the party initially hoped to replace the Republicans as the party of reform. During the 1880s, its hopes came close to realization. A debilitating and destructive internal struggle, however, leading to a party split in 1896, brought to an end its aspiration to become a force in national politics. Nevertheless, the intraparty conflict formed a vital context for the birth of a new, radically different, and eventually victorious successor, the Anti-Saloon League of America (ASLA). National Prohibition, secured under the leadership of the ASLA, did not meet the test the party had always accepted, that Prohibition, to be successful, needed to be enforced with thoroughness and determination by the party in power, backed by popular will. Therefore, while the collapse of National Prohibition destroyed the ASLA, the Prohibition Party lives on.

Formation and Early History

The Prohibition Party first arose in a period of legislative doldrums and swelling popular support for the prohibition issue. Statewide prohibitory laws, known as Maine Laws, had been adopted by various states during the 1850s, but most of these were soon nullified. After the close of the Civil War, the most prominent fraternal temperance societies, the Sons of

Prohibition Party Presidential Candidates and Votes, 1872–2000

Year	*Candidate*	*Vote*	*States*	*Percentage of National Vote*
1872	James Black	5,588	6	0.1
1876	Green Clay Smith	9,630	18	0.1
1880	Neal Dow	10,364	18	0.1
1884	John P. St. John	150,957	34	1.5
1888	Clinton B. Fisk	250,122	37	2.2
1892	John Bidwell	271,111	41	2.2
1896	Joshua Levering	131,285	41	0.9
1900	John G. Woolley	210,200	40	1.5
1904	Silas C. Swallow	259,163	40	1.9
1908	Eugene W. Chafin	252,704	39	1.7
1912	Eugene W. Chafin	209,644	40	1.4
1916	J. Frank Hanly	220,505	42	1.2
1920	Aaron S. Watkins	189,467	32	0.7
1924	Herman P. Faris	57,551	17	0.2
1928	William F. Varney	20,106	8	0.1
1932	William D. Upshaw	81,869	21	0.2
1936	D. Leigh Colvin	37,661	26	0.1
1940	Roger W. Babson	58,725	29	0.1
1944	Claude A. Watson	74,761	27	0.2
1948	Claude A. Watson	103,343	20	0.2
1952	Stuart Hamblen	72,778	21	0.1
1956	Enoch M. Holtwick	41,397	10	0.1
1960	Rutherford L. Decker	46,220	11	0.1
1964	E. Harold Munn, Sr.	23,267	11	0.0
1968	E. Harold Munn, Sr.	15,121	12	0.0
1972	E. Harold Munn, Sr.	12,818	5	0.0
1976	Benjamin C. Bubar	15,898	14	0.0
1980	Benjamin C. Bubar (Statesman Party)	7,212	13	0.0
1984	Earl F. Dodge	4,242	7	0.0
1988	Earl F. Dodge	8,002	6	0.0
1992	Earl F. Dodge	961	unknown	0.0
1996	Earl F. Dodge	1,298	unknown	0.0
2000	Earl F. Dodge	208	1	0.0

Temperance and the Independent Order of Good Templars, attracted tens of thousands of new members, and this upsurge convinced fraternal-society leaders that the time was ripe for a new prohibitionist initiative. In addition, the new party's founders perceived the dominant Republican Party as the vehicle of a moral issue, antislavery, and they believed that the Civil War and the Thirteenth Amendment marked fulfillment of the party's purpose. As further proof of the party's moral exhaustion, Republican officials in some northern states had been instrumental in gutting antiliquor legislation. Therefore, after sporadic attempts at organizing on the state level, a group of reformers led by John Russell of Michigan issued a call for the founding convention of a new reform party. The resulting meeting in Chicago's Farwell Hall, on 1 September 1869, created the national Prohibition Party. Organization of state parties followed.

Like its major-party rivals, the Prohibition Party operated as a relatively open and democratic institution. Local conventions, open to anyone who professed the party faith, elected delegates to congressional district and state meetings, which selected candidates and formulated platforms. The state convention, in turn, elected delegates to the quadrennial national convention, which set national policy and nominated the presidential and vice-presidential candidates. In addition, the state convention elected members of the national committee, which oversaw the federal campaign and directed party policy between conventions. The national committee chose the national party chairman. Party propaganda was the shared responsibility of a network of newspapers run by independent-minded editors and a corps of equally autonomous roving speakers. Fund-raising was unsystematic, dependent upon the efforts of the lecturers and the philanthropy of wealthy supporters, such as Isaac K. Funk of the Funk & Wagnalls Publishing Company, publishers of the party's national organ during the 1890s, *The Voice.*

During the 1870s, the Prohibition Party failed to convince a significant portion of the electorate to give priority to issues other than Reconstruction in the South. By the early 1880s, however, an increasing number of Republican voters came to accept prohibitionists' argument that the moral passion that had once driven the party of Lincoln was now extinguished. The Prohibition Party gained national attention in the presidential election of 1884, when its nationwide total soared from 10,000 votes to more than 150,000, and its support in New York State was widely believed to have caused the defeat of the Republican presidential candidate for the first time since before the Civil War. Assisting the party's mushroom growth in 1884 was support from the rapidly growing Woman's Christian Temperance Union (WCTU), headed by Frances Willard.

In the aftermath of the 1884 election, Prohibition Party leaders prepared to challenge for national supremacy. The party was well placed to do so, as its heartland lay in the wealthy, highly industrialized, urbanizing, and culturally diverse belt of states running west from New Jersey to Nebraska. Its leaders were well-educated professionals driven by moral fire and evangelical ardor. Their arguments appealed not to sectarian perspectives but to widely shared concerns about poverty, public health, and political corruption, and the party reached out to all who shared those anxieties. Heirs to the abolitionist tradition, they sought to incorporate into their ranks both African Americans and southern whites. For president in 1888, the party nominated Clinton B. Fisk, a former Freedman's Bureau official and founder of Fisk University. Sensitive to the wishes of their WCTU supporters, prohibitionists made their party the leading partisan voice for women's suffrage.

The Split of 1896 and After

Although the Prohibition Party attracted 100,000 more votes in 1888 than in 1884 and the election was extremely close, neither the party nor its main issue influenced the outcome. Frustration stimulated an intense debate within the party that continued for the next eight years. In 1892, that debate produced the party's most radical platform ever and what was to be the largest national vote in its history, an increase of about 20,000 over 1888. The party demanded regulation of corporations, limits on land ownership, an inflationary monetary policy, unrestricted women's suffrage, equal pay for equal work by women and men, and antilynching measures. But four years later, the party split into reformist ("broad gauge") and conservative ("narrow gauge") wings, which crystallized into separate parties for the campaign. This split finally destroyed the party's hopes of becoming an influential force in national politics.

Observing at close range the Prohibition Party's travails were conservative prohibitionists, who responded to the party's growing radicalism by forming a new kind of prohibitionist organization, the ASLA. Rejecting nearly every aspect of the party, these men and women created a pressure group that was nonpartisan, narrowly focused, more professional and bureaucratic, systematic in its fund-raising, incremental in its approach, and directed from the top down. It was funded by equally conservative businessmen, such as John D. Rockefeller, and directed its appeal primarily to evangelical Protestants. Through its negative example, the Prohibition Party inadvertently shaped the future course of the U.S. prohibition movement.

Although leaders of the ASLA would often have reason to hope it would die, the Prohibition Party survived the debacle of 1896. In the years between 1896 and the onset of National Prohibition in 1920, the party acted as gadfly to the ASLA's plow horse, criticizing the newer organization for what it perceived as unnecessary caution in forwarding the cause and, as a result, sometimes forcing the ASLA into premature campaigns. Indeed, this may have been the case with National Prohibition, whose triumph may have resulted from convergence of various short-term factors rather than from a rising tide of broad public support. Party prohibitionists shared with the men and women of the ASLA a moralistic attitude toward law that influenced the utopian goals embodied in the Eighteenth Amendment and the Volstead Act and counterbalanced a more pragmatic approach that had evolved over many years' experience in grafting prohibitory measures onto an unwieldy federal system.

Having always argued that Prohibition without a party in power to enforce it would be nugatory, party prohibitionists found in ratification of the Eighteenth Amendment no cause to dissolve their organization. For the same reason, they were not deterred by National Prohibition's repeal. A shell of its former self, the party has continued to survive, although restrictive ballot laws have confined its presence to a steadily shrinking number of states. In the presidential election of 2000, its candidates appeared on the ballot in only a single state, Colorado. Nevertheless, occasional local victories, both substantive and moral, keep up the spirits of its thinned ranks. On the left wing of the U.S. political spectrum during the 1890s and the early years of the twentieth century, the party has moved well to the right since the 1930s.

Jack S. Blocker Jr.

See also: Anti-Saloon League of America (ASLA); Dow, Neal; Good Templars (IOGT); Intercollegiate Prohibition Association (IPA); National Prohibition (United States); Republican Party (United States); Sons of Temperance; Willard, Frances Elizabeth Caroline; Woman's Christian Temperance Union (WCTU) (United States)

References
Blocker, Jack S., Jr. *American Temperance Movements: Cycles of Reform.* Boston: Twayne, 1989.
———. *Retreat from Reform: The Prohibition Movement in the United States, 1890–1913.* Westport, CT: Greenwood, 1976.
Colvin, D. Leigh. *Prohibition in the United States: A History of the Prohibition Party and of the Prohibition Movement.* New York: George H. Doran, 1926.
Hamm, Richard F. *Shaping the Eighteenth Amendment: Temperance Reform, Legal Culture, and the Polity, 1880–1920.* Chapel Hill: University of North Carolina Press, 1995.
Pegram, Thomas R. *Battling Demon Rum: The Struggle for a Dry America, 1800–1930.* Chicago: Ivan R. Dee, 1998.
Rumbarger, John J. *Profits, Power, and Prohibition: Alcohol Reform and the Industrializing of America, 1800–1930.* Albany: State University of New York Press, 1989.
Storms, Roger C. *Partisan Prophets: A History of the Prohibition Party.* Denver: National Prohibition Foundation, 1972.

Provincial Liquor Boards (Canada)

A provincial liquor board is the public agency that regulates the sale of liquor in government stores and in licensed facilities such as restaurants and bars. Often in the past known as liquor control boards, these organizations are at the heart of provincial government control of liquor. Government control, or monopoly sales, was the product of Canada's experience with prohibition during World War I and into the 1920s. By 1930, eight of nine provinces had abandoned prohibition for government control, and in various forms liquor boards continue to exist.

Support for prohibition began to wane at the end of World War I, and in 1920 the British Columbia government offered the voters two choices: continued prohibition or the return of liquor under strict government control. The voters chose the latter, and in 1921 the government appointed a three-member liquor board to operate a chain of liquor stores across the province. Quebec and Yukon Territory also adopted government control in 1921, but this new form of alcohol regulation was particularly reinforced when Ontario adopted it in 1927. Prince Edward Island, the final holdout, did not abandon prohibition until 1948. The structure and membership of liquor boards varied. In some provinces they reported to the legislature; in others they reported directly to the government of the day. The exception was Quebec, where the board existed as a separate public corporation. A board might consist of as few as one member or as many as five.

One feature that boards shared from the beginning was that they were a source of controversy. Opposition parties claimed that the boards were beholden to the provincial government and were centers of corruption and political patronage. Consumers often complained about high prices and limited selection in government stores. Yet the supporters of prohibition argued that prices were too low and that government stores encouraged consumption. Liquor-board regulation of public drinking also caused much debate. Drinkers chafed at restrictive board policies, such as no standing while drinking and the separation of the sexes in some facilities. Nondrinkers, of course, chastised the liquor boards for what they considered loose and often unenforced drinking regulations. Provincial government leaders tried to deflect criticism away from themselves by emphasizing the independence of liquor boards, even if that independence was more illusory than real.

Particularly since the 1980s, liquor boards have become more open, accountable, and consumer friendly. Those who seek liquor licenses have seen a diminution in arbitrary decisions and more emphasis on codified procedures, written explanations, and the possibility of appeal. Ordinary drinkers in pubs and restaurants have experienced a reduction in drinking restrictions. The biggest changes in retail sales have been a priority on customer service and a more corporate orientation. Only Alberta went so far as to privatize its government liquor stores, however, a process that began in 1993.

Robert A. Campbell

See also: Advertising Regulations (Canada); Alcohol, Consumption of, per Capita (Canada); Licensing (Canada); Provincial Prohibition (Canada); Saloons, Taverns, Pubs, and Beerhouses (Canada)

References
Campbell, Robert A. *Demon Rum or Easy Money: Government Control of Liquor in British Columbia from Prohibition to Privatization.* Ottawa: Carleton University Press [McGill-Queen's University Press], 1991.
Marquis, Greg. "Civilized Drinking: Alcohol and Society in New Brunswick, 1945–1975" in *Journal of the Canadian Historical Association.* 2 (2000): 173–203.

Provincial Prohibition (Canada)

Canadian provinces were found to have the power to enact prohibition legislation in *A. G. Ontario v. A. G. Canada* (*Local Prohibition* Reference) (1896). The case was decided by the Judicial Committee of the Privy Council under the division of legislative powers established in 1867 by the British North America (BNA) Act. Ontario's local-option law of 1890 was upheld even though virtually identical to the federal Canada Temperance Act that had been found constitutional in *Russell v. The Queen* (1882). The broad application of federal responsibility for Canada's peace, order, and good government that decided *Russell* was replaced with emphasis on the importance of provincial autonomy and the need to strictly limit authority to ensure peace, order, and good government to matters of unquestioned national concern. Federal power over trade and commerce was further narrowed. With earlier rulings that found retail and wholesale licensing to be provincial, *Local Prohibition* Reference ensured that the liquor trade would be an essentially provincial concern and that Canada would have a decentralized system of liquor laws. At the same time, the incongruous but abiding authority of the federal government to prohibit would sustain prohibition as a national issue for many more years. On the broader constitutional canvas, *Local Prohibition* Reference gave further Privy Council encouragement to the provincial-rights understanding of Canadian federalism law.

Local Prohibition Reference resulted from agreement between the Ontario and Dominion governments to seek answers to a series of questions designed to resolve once and for all jurisdictional confusion with respect to liquor, and specifically, prohibition. Most were general questions. One asked whether the provinces had the power to prohibit retail sales, another if they had the power to prohibit liquor wholesaling, and another if they had the authority to prohibit manufacturing. One was more specific: It asked about the constitutionality of the local-option prohibition law, virtually identical to the Canada Temperance Act, that Ontario had passed in 1890. On this and most of the general questions, Ontario emerged victorious.

Setting the Stage

The conditions for that victory were so much in place by 1885 that it is now tempting to see it as foreordained, *Russell v. The Queen* notwithstanding. There was wide support among Canadian judges through the 1870s and early 1880s for the existence of a provincial police power, variously located within the constitution's list of specific provincial powers, that concerned "the lives, health, morals or well-being of the community" (*Keete v. McLennan* [1876–1877 11 N.S.R. 5]). Some judges had concluded that this power included a power to prohibit at least retail trading. More significantly, the power had in substance been endorsed by the Privy Council in *Hodge v. The Queen* (1884). It recognized the "good government of taverns" and the preservation in municipalities of "peace and public decency," as advanced by a licensing law that regulated matters such as the playing of billiards in taverns, to be provincial functions. The public policy rationale for prohibition was that it would better advance these same objectives. Provincial authority to prohibit seemed to follow logically. The barrier to that logic had been interference with federal power over trade and commerce, but *Hodge* said that legislation based on the police power had nothing to do with trade and commerce.

All this was not, however, self-evident in the years between *Hodge* and *Local Prohibition* Reference. The principal reason was the ambiguity of *Russell v. The Queen:* It had upheld federal prohibition under peace, order, and good government without disagreeing with the Supreme Court of Canada that it also rested on trade and commerce. The Nova Scotia case of *R. v. McDougall* (1890), a challenge to provincial legislation that prohibited the importing, manufacturing, or wholesaling of liquor without provincial license, provides a good example of the confusion. Among the three judges who thought the legislation unconstitutional, two said it was because wholesaling (both for regulation and prohibition) was within trade and commerce and therefore federal. The other said it was because the provincial police power (encompassing both retail and wholesale) authorized regulation only, not prohibition, because of the holding in *Russell* that the Canada Temperance Act was valid federal legislation. The two dissenting judges also thought the distinction between retail and wholesale obsolete, but then denied that *Russell* placed prohibition beyond the police power. Such confusion was understandable. The Privy Council had upheld the Canada Temperance Act in *Russell* by finding it to be for the "promotion of public order, safety or morals." It had then upheld Ontario's liquor licensing legislation in *Hodge* by saying it was "calculated to preserve, in the municipality, peace and public decency, and repress disorderly and riotous conduct." Both pronouncements were binding on Canadian judges. The *Hodge* explanation, that subjects that were provincial in "one aspect and for one purpose" might be federal "in another aspect and for another purpose," was not viewed as helpful. It was the mere beginning of the aspect doctrine that would be developed in subsequent decades of constitutional jurisprudence and was not yet understood. It was precisely the similarity of purpose of the Ontario licensing and the federal prohibition legislation that seemed to bring *Hodge* and *Russell* into tension, if not contradiction.

Local Prohibition Reference

The division and confusion extended into the Supreme Court of Canada. In 1893, the Court heard argument in *Huson v. South Norwich,* a private challenge to Ontario's local-option law of 1890. Then, in 1894, Ontario premier Oliver Mowat deflected mounting pressure for province-wide prohibition by holding a provincial plebiscite and, when that resulted in a positive majority, by claiming the need for jurisdictional clarity. This produced the *Local Prohibition* Reference, argued in the Supreme Court while the outcome in *Huson* was still pending. Embarrassingly, deep disagreement on the law combined with a change in Court membership for the two cases to produce contradictory outcomes. In *Huson,* the Court split 3–2 in ruling Ontario's local-option law to be valid, but in *Local Prohibition* Reference the Court ruled the same legislation to be invalid, again by a 3–2 vote. The judges who upheld Ontario's legislation (Ontario's Samuel Henry Strong and Quebec's Henri Elzéar Taschereau and Télesphone Fournier) accepted the three propositions of Ontario's argument on the provincial police power: first, that it rested on the provincial power over municipal institutions that had been ignored in *Russell;* second, that it had to be interpreted in light of the functions performed by municipalities at confederation; and third, that preconfederation legislation confirmed that these functions had included retail prohibition. The judges who argued for invalidity (Ontario's John Wellington Gwynne, New Brunswick's George Edwin King, and Nova Scotia's Robert Sedgewick) based exclusive federal control over prohibition on the trade and commerce power and not on peace, order, and good government. Thus, they distanced themselves from *Russell v. The Queen.* They looked to the objectives of confederation, especially the determination to avoid the weaknesses of the central government under the U.S. Constitution, and they did their own review of confederation-era legislation to refute

Ontario's claim that prohibition had been a municipal function before confederation. They recognized the provincial police power but found it limited to functions that were essential or reasonably necessary to effective regulation, which excluded prohibition.

Provincial Victory

In the Privy Council, the decision was written by Lord Watson, who has long been either blamed or credited for starting the Privy Council's "provincial bias." He found Ontario's law to be valid, but for reasons that defy easy summarization. His judgment was more concerned with the general structure of the division of powers than with definitive resolution of the more particular division of responsibility over liquor. He did not purport to follow *Hodge v. The Queen.* Indeed, he treated it as limited to the question of licensing. He also brushed aside the municipal power in which Mowat had placed such confidence. It was merely a power to legislate on the machinery of municipal government. Instead, authority for the law came either from provincial authority over property and civil rights or provincial power over local matters. Watson said little as to why and found it unnecessary to choose definitively between the two options. His language fitted best with the power over local matters, and this is the power he was later said to have applied. The more important question for Watson was the potential for federal authority to undermine the "autonomy of the provinces." The continuing validity of the Canada Temperance Act was confirmed, but jurisdiction over peace, order, and good government would in the future be reserved for matters of unquestionable national interest and importance. In particular, it would not be allowed to override provincial powers, as could the enumerated federal powers. But the most important of these enumerated powers, trade and commerce, was also given further restriction. It would no longer be available as an alternative foundation for the Canada Temperance Act, because it was what the BNA Act said it was, a power to regulate, not to prohibit trade.

The outcome, whatever the precise rationale, meant a provincial prohibition power that was largely concurrent with that of the Dominion. Two spheres of mutually exclusive powers had each been found to authorize local-option laws. Later, the Privy Council attempted to explain this by supposing (to the amusement of generations of Canadian law students) that intemperance had become so excessive by 1878 as to constitute a threat to national existence that necessitated exceptional intervention from the federal government. But the implications of *Local Prohibition* Reference were broader than giving the provinces equal ability to legislate local option. Most of the general questions were also answered largely in Ontario's favor, albeit ambiguously. Thus, it seemed that both the Dominion and the provinces could prohibit locally or comprehensively, and across the full spectrum of the liquor trade, from manufacturing to wholesaling to retailing. Federal prohibitory jurisdiction would be paramount only on direct conflict with provincial legislation, as with the activation of the Canada Temperance Act and the Ontario local-option law in the same community. Strong reservations were only expressed regarding provincial prohibition that would directly affect interprovincial trade. But this limitation was quickly weakened in *A. G. Manitoba v. Manitoba Liquor Licences Holders' Association* (1901), where Manitoba prohibition legislation was upheld, notwithstanding that its provincial scope meant a greater impact on trade across provincial boundaries than had been at issue in *Local Prohibition* Reference.

The Scope of the Provincial Victory

With *Local Prohibition* Reference, the provinces emerged from the jurisdictional battle over liquor with broad and dominant responsibility for the liquor trade. They had exclusive control of liquor licensing as a revenue power and as an instrument for controlling the adverse social consequences of the trade. They had a power to prohibit that was largely equivalent to that of the federal government. The exclusion of intraprovincial trade from the trade and commerce power gave them authority to regulate most aspects of the trade from the economic as well as the social and moral perspective.

In the years since the *Local Prohibition* Reference, Canadian provinces have used this broad mandate over the liquor trade to move from restrictive licensing in the late nineteenth and early twentieth centuries to prohibition in the latter years of World War I, and in the 1920s to the system of government-controlled distribution and restrictive licensing laws (for on-premises consumption) that replaced prohibition, starting with Quebec in 1919 and finishing with Prince Edward Island in 1948. In more recent years, provincial regimes have gone through a more subtle change. Licensing laws have been relaxed and consumer service and convenience and government revenue have displaced social protection as the dominant objectives of government-controlled distribution systems. All of this evolution has happened at the provincial level because of the breadth of provincial jurisdiction over liquor as defined by *Hodge v. The Queen, Local Prohibition* Reference, and the cases on the trade and commerce power, including *Citizens' Insurance v. Parsons* (1881). In contrast, the federal jurisdiction to prohibit has been essentially dormant since the failure of the Prohibition experiment in the 1920s. The other elements of federal authority over liquor have been exceptional and limited. They include liquor as an aspect of federal responsibility for Indians, liquor in relation to interprovincial transportation, situations of intoxication that have criminal law implications, and the application to liquor of generic federal regulation in areas such as importing and exporting, product labeling, and weights and measures.

In the broader framework of Canadian federalism, Canadian provinces achieved much more through the constitutional battles over liquor than control of the liquor trade. Provincial legislatures and governments had been recognized as sovereign on provincial matters, potentially expansive federal authority over peace, order, and good government and over trade and commerce had been greatly truncated, and provincial authority over property and civil rights and over local matters had been combined to give the provinces a broad and expansive jurisdiction over both the economy and society. To many, these outcomes represented the imposition of a decentralized constitution on Canada by foreign judges. The lack of explicit foundation in the BNA Act for the crucial value placed on provincial autonomy (also invoked by Lord Watson in *Liquidators of Maritime Bank* [1893], the definitive statement of provincial domestic sovereignty), and its inconsistency with the nationalist vision of leading Fathers of Confederation, especially John A. MacDonald, has provoked a tradition of Privy Council criticism and (sometimes) a search for the ulterior motive that led to the judicial decentralization of Canadian federalism. One alternative view is that the Privy Council gave effect to the regional pluralism that had to be accommodated if Canada was to survive. Another is that receptiveness to provincial autonomy was inherent in the federalism of the BNA Act. More fundamentally still, it perhaps reflected the growing influence of an understanding of law, called legal liberalism, that portrayed the judicial definition and enforcement of mutually exclusive spheres of autonomy as the function of all law. On this perspective, Mowat's forensic victories, sometimes as counsel and sometimes as instructing solicitor but always as chief strategist, may have owed much to the alignment of the core principle of provincial rights (provincial autonomy) to the legal individualism that was characteristic of the legal mind of the age.

William Lahey

See also: Canada Temperance Act; Provincial Liquor Boards (Canada); Provincial Regulation (Canada); Provincial Stores (Canada)

References

Cairns, Alan. "The Judicial Committee and Its Critics" in *Canadian Journal of Political Science.* 4 (1971): 301–345.

Greenwood, F. Murray. "Lord Watson, Institutional Self-Interest, and the Decentralization of Canadian Federalism in the 1890s" in *University of British Columbia Law Review.* 9 (1974): 244–279.

Hogg, Peter W. *Constitutional Law of Canada,* 4th ed. Toronto: Carswell, 1997.

Risk, R. C. B. "Canadian Courts under the Influence" in *University of Toronto Law Journal.* 40 (1990): 687–737.

———. "Constitutional Scholarship in the Late Nineteenth Century: Making Federalism Work" in *University of Toronto Law Journal.* 46 (1996): 427–457.

Romney, Paul. *Mr. Attorney: The Attorney General for Ontario in Court, Cabinet and Legislature, 1791–1899.* Toronto: University of Toronto Press, 1986.

Russell, Peter H. *Leading Constitutional Decisions.* Ottawa, Canada: Carleton University Press, 1987.

Saywell, John T. *The Lawmakers: Judicial Power and the Shaping of the Canadian Constitution.* Toronto: University of Toronto Press for the Osgoode Society, 2002.

Smart, Reginald G., and Alan C. Ogborne. *Northern Spirits: A Social History of Alcohol in Canada.* Toronto: Addiction Research Foundation, 1996.

Vaughan, Frederick. "Critics of the Judicial Committee of the Privy Council: The New Orthodoxy and an Alternative Explanation" in *Canadian Journal of Political Science.* 19 (1986): 495–519.

Provincial Regulation (Canada)

Regulation of liquor is largely a provincial responsibility under Canadian federalism. This rule was established through a series of division of powers cases decided under the British North America (BNA) Act (1867) between the mid-1870s and 1900 by the Supreme Court of Canada and, more significantly, by the Judicial Committee of the Privy Council, Canada's true court of final appeal until 1949. The immediate question was the division of powers over liquor licensing and prohibition legislation. More fundamental issues were also implicated, including the general relationship between federal and provincial powers, the constitutional status of the provinces, and the relative scope of the heads of power that defined provincial and federal legislative jurisdiction, not only in respect of liquor, but also across the full spectrum of governmental activity. In *Russell v. The Queen* (1882), federal prohibition legislation was ruled constitutional. In *Hodge v. The Queen* (1884) and the *McCarthy Act* Reference (1885), the provinces were held to have exclusive authority over liquor licensing and, by extension, over the full range of issues that licensing legislation addressed, including the number, location, and hours of opening of drinking establishments, the age and gender of patrons, and regulations regarding food and entertainment in drinking establishments. In *A. G. Ontario v. A. G. Canada* (*Local Prohibition* Reference [1896]), provincial prohibition legislation was upheld. More significantly, seemingly grand federal powers over Canada's peace, order, and good government, and the regulation of trade and commerce, had been given restrictive interpretations. Competing heads of provincial power over property and civil rights and local matters had been read expansively. The provinces had been established as an autonomous and coordinate, rather than as a subordinate, level of government. The result was a decentralization of legislative responsibility for the liquor trade that represented a larger decentralization of Canadian federalism.

The Constitutional Framework

The importance of the liquor trade to the politicians who designed Canadian federalism is apparent from section 92(9) of the BNA Act. It gave the provinces authority over "shop, saloon, tavern, auctioneer and other licences" for the raising of revenue for "provincial, local or municipal purposes." Otherwise, the BNA Act was, at best, opaque on the authority to adopt legislation concerning liquor. In addition to the licensing power, the provinces had jurisdiction (via 92[8]) over "municipal institutions." Some thought this meant authority over the traditional (preconfederation) functions of municipalities, such as the regulation of taverns, but others thought it was merely authority over the institutions of municipal government. Provinces also had jurisdiction (via 92[13]) over "property and civil rights."

This might mean only authority for the common law, or it could mean authority over the social and economic interactions to which the common law applied. Finally, the provinces had authority (via 92[16]) over "matters of a merely local or private nature." This could either be confirmation of a general theme of jurisdiction over the minor and unimportant, or confirmation of the continuing entitlement of each province to significant rights of self-government over their distinct affairs.

The federal power usually invoked, at least until *Russell v. The Queen,* was the power to regulate trade and commerce (91[2]). The federal power over criminal law played little role, partly because of the understanding that criminal law comprised only wrongs that were inherently criminal and partly because neither level of government criminalized drinking or drunkenness. Trade and commerce meanwhile was widely understood to include local as well as national trade and trade in specific goods as well as general authority for the economy. Such breadth was often expressed as an essential element of the strong central government that was fundamental to confederation. Some concluded it meant that provincial jurisdiction, however defined, could not include prohibition, as that would destroy a branch of trade within Parliament's authority. But this meant that the federal government could prohibit and interfere with provincial licensing, and that seemed questionable. Why give the provinces such a right if Parliament could abolish the trade that made it meaningful?

Early Decisions

Through the 1870s, Canadian judges attempted to resolve these complexities by reading the language of the constitution in light of their understandings of the purposes of the confederation they had all lived through. They sought to draw jurisdictional boundaries over liquor by giving each level of government the legislative authority that matched the governmental functions assigned to it under the "compact of union"—the political deal that the BNA Act implemented. This meant a federal trade and commerce power that was expansive enough for the building of a national economy and strong enough to protect Canada from what Canadians saw as the excessive (and nearly fatal) decentralization of the U.S. Constitution. But it also meant giving broad effect to what the judges usually called the provincial "power of municipal police." This was described in a Nova Scotia case as the power to regulate "the lives, health, morals or well-being of the community," in a New Brunswick case as the "power of making such regulations . . . as would tend to the preservation of good order and prevention of disorderly conduct, rioting or breaches of the police," and similarly in cases in other provinces. It was sometimes based on the licensing power, sometimes on the authority over municipal institutions, sometimes on the authority over property and civil rights, sometimes on the more general provincial authority over local matters, and sometimes on a combination of these powers. The referenced provincial power was less important than the common jurisdictional theme that, within the common national and significantly economic framework of confederation, the people of each of the provinces had a right of self-government over matters affecting the well-being of their provincial and local communities.

The many cases about liquor legislation in these early years of confederation (roughly 30 of the 125 constitutional cases decided by 1900) highlighted this jurisdictional theme. Whether it was regulatory or prohibitory, the legislation aimed at social and moral reform rather than at economic objectives. It was legislation that profiled the understanding that local control over the life of the local community (and perhaps, more broadly, over social matters) had been one of the organizing principles of the division of powers. This notion perhaps had some influence on the attention paid to the liquor cases by Oliver Mowat, Ontario's Liberal premier, and on his success in advancing the provincial-rights cause through them. The liquor cases created opportunities for ensuring judicial recognition of a core and underlying theme of provincial legislative power, at risk of being overlooked underneath the BNA Act's dry references to saloon licenses, municipal institutions, and property and civil rights.

That does not mean Canadian judges were on Mowat's side. Some rejected the idea that social questions, and intemperance particularly, were local issues. The most important example was Ontario's John Wellington Gwynne of the Supreme Court of Canada. In the case that became *Russell v. The Queen,* he anticipated the Privy Council by characterizing intemperance as a national problem within Parliament's general authority over peace, order, and good government. The Canada Temperance Act was federal because of, not despite, its social reform objectives. Even judges who accepted the jurisdictional dichotomy between a local power over community life and a federal power over the economy disagreed on the implications for jurisdiction over liquor. Provincial courts in Ontario and Nova Scotia held that provinces could impose prohibition. Courts in New Brunswick and in Quebec ruled they could not, due to the interference with trade and commerce. In *Severn v. The Queen* (1878), a decision of the Supreme Court of Canada, Chief Justice William Buell Richards (for the majority) found that provinces had no authority to license brewers, since brewing and wholesaling were "quasi-national" and within trade and commerce. He had, however, been one of the Ontario judges who had ruled in favor of a provincial power to prohibit retailing. His colleague William Johnstone Ritchie, who had earlier ruled against provincial retail prohibition while on the New Brunswick bench, on the basis of an expansive trade power, reached the opposite conclusion from Richards, that the provinces could license brewers. Such disagreements reflected the fact that the Canadian judges generally attached comparable importance to federal and provincial powers. Where they disagreed was on how to make the protection of a strong central government (especially in the economic realm) compatible with the right of provincial legislatures to make important choices for their separate societies.

Russell v. The Queen cast this even-handedness into doubt and created complication and confusion, not because it upheld the Canada Temperance Act, but because of the supporting reasoning. Like Gwynne, the Privy Council endorsed federal prohibition, not as a regulation of trade, but as legislation that promoted "public order, safety or morals." This placed it within the general authority of Parliament for Canada's peace, order, and good government found in the preamble of section 91. Before *Russell,* the recurring question had been whether a power of social regulation, provincial at its core, extended to prohibition. Some judges said yes because regulation and prohibition were different means to the same ends, and others said no, most often because of the interference with the trade and commerce power. Now the question seemed reversed: If the national Parliament could advance temperance by prohibiting, could it not advance it by regulating? Logic said yes, since again, the distinction between regulation and prohibition was one of means only.

Hodge v. The Queen

John A. MacDonald, Canada's Conservative (and founding) prime minister, enthusiastically embraced this logic. The so-called McCarthy Act (1884) required federal licenses of retailers and wholesalers and therefore challenged continuing provincial competency over regulation short of prohibition. MacDonald's particular target was Ontario's Liquor Licensing Act, the so-called Crooks Act of 1876. It had transferred retail licensing from municipalities to commissioners appointed by the provincial government and authorized these commissioners to make regulations on matters such as hours of opening and the playing of billiards in taverns. The patronage opportunities thus created for Ontario's Liberal government, as much as general jurisdictional ambition, were MacDonald's objective in passing the McCarthy Act. Mowat's response was to use a constitutional challenge to the Crooks Act, already in progress, to rehabilitate the argument (post *Russell*) for a power of local police that was part of a general constitutional principle of provincial autonomy over local matters, including liquor regulation. His opening lay in the failure of counsel in *Russell* to stress the provincial authority over municipal institutions, which Mowat portrayed as the true foundation of the police power that had previously been variously supported.

In *Hodge v. The Queen* (1884), the Privy Council agreed with Mowat that the Crooks Act was "in the nature of police or municipal regulation of a merely local character . . . calculated to preserve, in the municipality, peace and public decency, and repress disorderly and riotous conduct." Provincial power over municipal institutions, penalties for the enforcement of provincial laws, and matters of a merely local or private nature were all found to be applicable. Further, the Ontario licensing law did not "interfere with the general regulation of trade and commerce." This showed the intervening influence of *Parsons v. Citizen's Insurance* (1881), in which both the Supreme Court of Canada and the Privy Council had excluded trade within a province from federal jurisdiction and set the stage for later recognition that even the purely economic regulation of the liquor trade was provincial. But it also suggested Privy Council acceptance of the links that Mowat made between provincial jurisdiction, the social objectives of liquor legislation, and the principle of provincial autonomy. In addition to finding no violation of the trade and commerce power, the Privy Council rejected the argument that the Crooks Act was unconstitutional because it delegated regulatory authority to boards of commissioners and therefore contravened the rule against delegation by an assembly (the provincial legislature) that itself held only delegated and not plenary legislative power. The implication was that provincial powers were inferior to federal powers. This the Privy Council rejected by declaring the provincial legislatures to have the same supremacy within provincial jurisdictional boundaries as the Imperial Parliament held generally.

Mowat's triumph in *Hodge* sealed the fate of the McCarthy Act. In 1884, the Supreme Court struck down the act's retail licenses, and in 1885, the Privy Council struck down the act's wholesale licenses as well. This marked the end of the distinction between retail and wholesale regulation that the Canadian courts had used, in *Severn* and in other early cases, to reconcile a wide power of police with an equally strong and effective power over trade and commerce. More significantly, *Hodge* and the *McCarthy Act* Reference (but especially *Hodge*) established that the provincial power over liquor was not limited to licensing for revenue but extended to licensing as an instrument of social regulation.

Mowat's greatest triumph still lay in the future. The foundations were, however, firmly planted by 1885, although this is easier to see now than it would have been then. *Hodge* once again posed the question of the relationship between regulation and prohibition in terms that seemed favorable to the provinces: If they could regulate through licensing to secure "peace and public decency," why could they not pursue the same objective through prohibition? Given the general limitation of trade and commerce in *Parsons v. Citizen's Insurance* and the specific conclusion in *Hodge* that licensing did not concern trade and commerce, interference with that federal power seemed no longer to be an obstacle. There was, however, one major uncertainty: in *Russell,* as the Privy Council had refrained from saying the Canada Temperance Act was not valid under trade and commerce as well as peace, order, and good government. Nevertheless, the *Hodge* repudiation of the argument that the provincial legislatures were inferior bodies implicitly challenged the assumption that federal powers were the important ones and set the stage for further expansion of the provincial domain in later cases, both as regards the power to deal with liquor and in other areas of jurisdictional conflict.

William Lahey

See also: Canada Temperance Act; Provincial Prohibition (Canada)

References

Abel, Albert S. "The Neglected Logic of 91 and 92" in *University of Toronto Law Journal.* 19 (1969): 487–521.

Hogg, Peter W. *Constitutional Law of Canada,* 4th ed. Toronto: Carswell, 1997.

Risk, R. C. B. "Canadian Courts under the Influence" in *University of Toronto Law Journal.* 40 (1990): 687–737.

Romney, Paul. *Mr. Attorney: The Attorney General for Ontario in Court, Cabinet and Legislature, 1791–1899.* Toronto: University of Toronto Press, 1986.

Saywell, John T. *The Lawmakers: Judicial Power and the Shaping of the Canadian Constitution.* Toronto: University of Toronto Press for the Osgoode Society, 2002.

Provincial Stores (Canada)

Operated by a board or agency of the provincial government, provincial liquor stores are the dominant retailer of packaged liquor in Canada. These stores are the legacy of Canada's experience with Prohibition in World War I and the 1920s. Rather than return the sale of liquor to the private sector after Prohibition, all provinces eventually adopted government-controlled liquor stores. In 1993, Alberta became the first, and so far only, province to privatize the retail sale of packaged liquor. Regulated private retailers exist in some other provinces, but the government liquor store remains paramount.

The creation of the government liquor store was the compromise achieved after Prohibition began to lose favor. In 1921, British Columbia, Quebec, and Yukon Territory all opened stores overseen by public liquor boards. From the beginning, liquor boards faced the dilemma of the appeal of liquor revenue versus the responsibility to control alcohol consumption. That problem influenced everything from the prices charged to the ambience of the stores. Particularly in the early decades, government stores made liquor available but did not actively promote its sale. For example, in British Columbia only a tiny sign identified a liquor store and curtained windows blocked the view of the interior. Patrons needed both a liquor permit and patience to make a purchase. The stock was kept behind a counter, and customers had to place written orders and pay in cash before they received their goods. World War II was particularly trying because of reduced supplies, strict rationing, and limited retail hours. The results were long lines and often frayed tempers.

As attitudes toward alcohol became more liberal, especially in the 1960s, liquor stores changed with the times. In 1962, British Columbia opened its first self-serve liquor store and the first to employ female clerks. The store provoked the wrath of the still feisty Woman's Christian Temperance Union (WCTU). Yet self-serve stores became the norm across the country, and liquor boards made customer service a high priority by the 1980s. The emphasis on service was driven both by patron demands and by some pressure to privatize government outlets. Only Alberta went that far, but government stores adopted more private retailing techniques, such as impulse bins by the cash registers, payment by credit card, and in-store tastings. In Nova Scotia, customers can now bottle bulk wine in government stores.

A government liquor store in New Westminster, Canada (Tenth Annual Report of the Liquor Board of British Columbia, 1931)

As of 2002, all provincial liquor boards have websites that range from the austere (Saskatchewan) to the flashy (Quebec and Nova Scotia). These sites promote the stores and the products that they sell. Some sites act as on-line stores. Yet they also have areas devoted to the socially responsible use of alcohol and links to enforcement, medical, and self-help groups. The New Brunswick and Newfoundland sites also warn that only people of legal drinking age can enter their electronic domain. Thus, even in the world of the Internet, government liquor stores still have ties to their origins.

Robert A. Campbell

See also: Advertising Regulations (Canada); Alcohol, Consumption of, per Capita (Canada); Gothenburg Schemes/Disinterested Management; Provincial Regulation (Canada); Saloons, Taverns, Pubs, and Beerhouses (Canada); State Stores (United States)

References

Brownsey, Keith. "Selling the Store: Privatizing Alberta's Liquor Stores" in *Public Administration and Policy: Governing in Challenging Times,* edited by Martin W. Westmacott and Hugh P. Mellon, 117–125. Scarborough, Ontario: Prentice-Hall, 1999.

Campbell, Robert A. "'Profit Was Just a Circumstance': The Evolution of Government Liquor Control in British Columbia, 1920–1988" in *Drink in Canada: Historical Essays,* edited by Cheryl Krasnick Warsh, 172–192. Montreal: McGill-Queen's University Press, 1993.

Liquor Control Board of Ontario. http://www.lcbo.com/index_eng.html.

Psychological Theories of Drinking

Psychological theories of drinking take the view that some underlying psychological factor, such as a personality trait or emotional condition, makes people drink or drink more heavily. Life events, learning experiences, and psychological traumas could also predispose certain people to drink alcohol or to drink it heavily. Such theories do not include generic or physiological factors and focus only on psychological factors. The most popular psychological approaches are based on psychoanalytic or addictive personality theories, dependency theories, power theories, social learning, risk-taking concepts, and alcohol expectancies. The psychoanalytic and addictive personality theories are well established in the field and have a long history. However, the theories based on risk taking, social learning, and expectancies are much newer. Selecting the best and most persuasive of these theories is no easy matter. Many have no strong empirical support but still have many adherents.

Psychoanalytic Theories

Psychoanalysis is a set of theories of human behavior developed by Sigmund Freud and his colleagues starting almost a hundred years ago when they began analyzing patients, some of whom had drinking problems, in their private practices. Psychoanalysts generally believe that mental illness or disturbed behavior is due to suppressed and unconscious conflicts that occurred early in life, that is, largely before the age of about five. Such conflicts are to be brought out by psychoanalytic treatment, which involves analysis of dreams and early childhood experiences.

There are numerous psychoanalytic theories of drinking. For example, Sigmund Freud and several other analysts thought that heavy drinking had its origin in repressed homosexual feelings in the drinker. The alcoholic's common delusion of jealousy about his wife is seen to represent the homosexual fantasy about loving a man. Other analysts agreed and argued that alcohol intoxication was a replacement for sexual behavior and that alcoholics turn from women to alcohol as a substitute.

Other Freudian theories of drinking emphasize that heavy drinkers have a fixation at the oral stage of development. Thus, alcoholics never progressed beyond this important stage of infancy. They desire pleasure from drinking as infants do from the bottle or breast, and they enjoy the feeling of fullness and sleep resulting from heavy drinking. Thus, the bottle of alcohol becomes the substitution for the mother's breast. Later psychoanalytic thinkers, such as Karl Menninger, believed that alcohol addiction can be a form of suicide used to prevent a greater self-destruction resulting from the aggression that in turn stems from ungratified eroticism. Psychoanalytic theories were often derived from the analyst's experiences with a few patients and were not tested in large-scale empirical studies. Although unsupported in empirical research, they remain influential with some thinkers and analysts.

Addictive or Alcoholic Personality Theories

In the early days of the study of drinking behavior, many people believed that an "alcoholic personality" type would emerge. That is, some set of traits, personality characteristics, or emotions would be found that set alcoholics off from other people and bring them into heavy drinking. There were a vast number of theories proposed. Most of them hypothesized some personality defect made better by drinking. For example, P. Schilder (1941) took the view that the alcoholic is insecure and threatened by his parents and community. Alcohol and intoxication seem to improve this situation. Other theories stated that the drinker is introverted and alcohol allows him to be more at ease socially and to relate to others.

Numerous theories have proposed that heavy drinkers are neurotic or inadequate socially or psychologically and that drinking allows them to overcome their neuroses and inhibitions. Escapist theories are also common, that is, that the heavy drinker drinks to get away from some inadequate emotional or personality factor. In general, the empirically based search for the alcoholic personality has been disappointing. Although a vast amount of research has been done on the topic, researchers still have not identified an "alcoholic personality." There have been a large number of studies using personality tests, projective tests, questionnaires, and interviews, but they are difficult to summarize. Many have found that factors such as depression and psychopathy are characteristic of heavy drinkers, but many have not. It is difficult to know whether these factors precede or follow alcoholic drinking. Also, the definition and reliability of measurement-of-personality scales has been called into question. The search for the alcoholic personality has largely ended because the empirical studies have been so inconclusive and the results contradictory.

Anxiety or Tension Reduction

Many theories of drinking argue that people drink to reduce tension, anxiety, or some other unpleasant emotion. Such theories have usually been subsumed under the heading of tension reduction theory (Cappell 1975); they go back to the 1940s and were especially popular in the early days of alcohol research in the 1950s and 1960s. These theories assert that alcohol can reduce tension, anxiety, and fear, act as a sedative for psychological states, ease unpleasant emotions, or relieve any other aversive psychological state, and that people drink alcohol in order to reach an ameliorative state.

The empirical support for the tension reduction theory is mixed. Several experimental studies have shown that alcohol reduces conflict behavior in animals, especially in experimental neurosis and other simple avoidance situations. At least one study with humans has shown that alcohol can restore suppressed behavior in a conflict situation (Vogel-Sprott 1967, 337–341). Some studies have shown that alcohol in small amounts can reduce tension and make people more relaxed in social situations.

Many studies have been made of the effect of alcohol on normal and alcoholic subjects. Some of these studies with college students found a reduction in anxiety after small amounts of alcohol but an increase in anxiety after larger amounts. Also, many studies of alcoholics have found that alcohol given in experimental situations increased tension or anxiety. Several large-scale experimental studies have been made of alcoholics confined in the wards of hospitals. These studies typically show that after large amounts of alcohol (30 oz. of spirits per day for fourteen days), anxiety levels increased, as did depression and aggressiveness. Despite these effects, most alcoholics were strongly motivated to drink more alcohol.

Support for the tension reduction theory of drinking is weak and unlikely to improve given all of the negative findings and the contradictory ones presented so far. The theory is essentially a plausible and commonsense one and may be useful in explaining some social drinking. However, alcohol does create significant negative emotions in alcoholics, and hence the tension reduction theory is mostly rejected now.

Alcohol Expectancy Theories

Expectancy theories have a long history in the psychology of motivations. Such theories state that if people perceive a positive association between an observed or actual behavior and some desired outcome, they will be more likely to engage in that behavior. If the association has a negative outcome, they will avoid that behavior. The application of expectancy theory to the drinking of alcohol is much more recent (Critchlow 1986). The expectancy theory states that although young children do not drink, they have expectancies or guesses about the effect of alcohol; for example, it will make them feel good, fall down, talk more, be more popular, and the like. Alcohol expectancies affect drinking initially, and after children start to drink their drinking affects their expectancies about alcohol. Expectancies and experience with actual effects of alcohol interact and influence each other in a complex manner.

Before starting to drink alcohol, children have mainly negative-effect expectancies about how alcohol affects people. After they drink, however, there is an increase in the more positive alcohol expectancies. Much research shows that young people with more positive alcohol expectancies are likely to drink more alcohol. These expectancies can be modified by the individuals' response to mass media, parents, peers, and educational experiences, all of which can create expectancies about alcohol effect.

Although it is a new and expanding area, expectancy theory is a promising area for explaining drinking behavior. This is especially true because it is a theory about normal, youthful drinking, whereas many psychological theories deal mostly with heavy drinking.

Power Theory

The power theory of alcohol consumption was developed by David McClelland and his colleagues in the 1970s. This theory deals only with male drinking and has nothing to say about female drinking. It does, however, attempt to explain both normal drinking and excessive drinking. The theory claims that male drinking is primarily about gaining power, and that men with more power needs drink more excessively.

McClelland and his colleagues conducted a large number of studies, many of them with students at Harvard University. Several of these studies, using the Thematic Apperception Test (TAT), found that power fantasies increased after high levels of alcohol consumption. Men with more power needs normally felt weak but felt more powerful when they drank. McClelland observed students at cocktail parties and had them do the TAT before and after drinking alcohol. He found no evidence that normal drinking decreased anxiety or led to oral gratification or dependency fantasies. He did find large increases in power fantasies after drinking by men. After drinking, the men had more thoughts of personal power, especially those that could be characterized as selfish or self-centered. Some examples were thoughts of dominance over others, sexual conquest, strength, and athletic achievement. After about six drinks, power fantasies seem to dominate thoughts of male drinkers. Thus, McClelland concluded, men drink primarily to feel stronger and more powerful (McClelland 1972).

Research has not shown increases in power fantasies after drinking alcohol among women, however. Female alcoholics tend to value the maternal role more than do women who are not alcoholics, and women with more masculine identities or weak feminine identities tend to drink more than others.

Much research supports the prediction of power theory; however, it seems to have fallen out of favor. Not much research has been done on it in the past few years.

Dependency Theories

Several scientists have proposed that drinking, and especially heavy drinking, are related to dependency needs. Like power theory, dependency theories are usually applied only to males and not females. Women can usually express their dependency needs and hence do not develop conflicts and guilt over them. In men, however, strong dependency needs develop and are not satisfied sufficiently. Men often want more maternal care but feel guilty at some level about this. They want more dependence but also want to be independent. According to this theory, drinking helps men to maintain a fake image of independence. Alcohol consumption provides a feeling of fullness, warmth, and independence, but the effect is temporary.

Howard Blane (1968) argued that male dependency needs are central in the alcoholic's development. Alcoholics can resolve these unfilled dependency needs in several ways: Some make no effort to become independent of their parents; others may be "counter dependent" and avoid expressing dependency needs; and others fall into a mode of denying dependency needs. Support for this theory is limited and it is most useful in explaining certain types of alcoholic behavior rather than normal drinking.

Learning Theories of Drinking

Learning theories of drinking neglect underlying traits and personality characteristics of drinkers, and any assumed or inferred emotional states, as factors in drinking behavior. Rather, they emphasize the reinforcement that alcohol provides to drinkers. This reinforcement may be a reduction in unpleasant emotions, increased relaxation and comfort, or better social skills. Alcohol use is seen to be similar to any other learned behavior and to depend upon reinforcements, that is, valued or pleasurable outcomes when it is used.

Learning theories have been well described by P. C. Rivers (1994). Alcohol is usually consumed in a social setting by social drinkers. For many drinkers, it facilitates social situations and makes them more relaxed, more talkative, and more at ease with other people. Drinking may be modeled on other people's drinking, such as peers or parents, and hence is a learned, socially based behavior.

Some older learning approaches are based on drive reduction theories and involve classical conditioning. The drives reduced may be related to anxiety, fear, avoidance of pain, and the like. A number of studies with rats show that alcohol can reduce such drives and hence be reinforcing. Reinforced behaviors become more frequent; punished behaviors become less frequent.

More recent learning theories of drinking emphasize social learning. Heavy drinkers are people who, through special modeling experiences with alcoholic parents or others and the reinforcing effects of alcohol, have learned to use alcohol to relieve unpleasant situations in their lives. People learn to drink in cultures, families, and social groups that forbid, permit, or encourage heavy drinking. These intergroup differences play an important role in establishing the level of drinking in each family or social group. They provide varying amounts of heavy drinking models. They also determine how likely alcohol is to be chosen as a reinforcement in dealing with problems or unpleasant emotional states. If people are raised in a culture where heavy drinking is allowed, or a family of that sort, they are much more likely to be heavy drinkers themselves.

Generally, there is much support for learning theories of drinking. They combine cultural, learning, and social perspectives on drinking in their most developed forms. They also continue to be a focus of research on the motivation for drinking.

Conclusion

Although most of the psychological theories about drinking that have been proposed in the past century have relatively little empirical support, some are still influential. No one theory has strong enough empirical support to dominate thinking in the field, however. Expectancy and learning theories have the best support, but much remains to be done to formulate and test these and other theories about drinking.

Reginald G. Smart

See also: Alcoholism; Treatment Institutions

References

Blane, H. T. *The Personality of the Alcoholic: Guises of Dependency.* New York: Harper & Row, 1968.

Cappel, H. "An Evaluation of Tension Models of Alcohol Consumption" in *Research Advances in Alcohol and Drug Problems,* edited by R. J. Gibbins, 177–210. New York: John Wiley, 1975.

Critchlow, B. "The Powers of John Barleycorn: Beliefs about the Effects of Alcohol on Social Behavior" in *American Psychologist.* 41 (1986): 751–764.

McClelland, D. C., W. N. Davis, R. Kalin, and E. Wanner. *The Drinking Man: Alcohol and Human Behavior.* New York: Free Press, 1972.

Menninger, K. A. *Man against Himself.* New York: Harcourt Brace, 1938.

Rivers, P. C. *Alcohol and Human Behavior: Theory, Research and Practice.* Englewood Cliffs, NJ: Prentice-Hall, 1994.

Schilder, P. "The Psychogenesis of Alcoholism" in *Quarterly Journal of Studies on Alcohol.* 4 (1941): 277–292.

Vogel-Sprott, M. "Alcohol Effects on Human Behaviour under Reward and Punishment" in *Psychopharmacologia* 11 (1967): 337–341.

Publicans (England)

Although commonly applied to a variety of posts relating to the management of licensed premises, the term "publican" refers, most simply, to an individual charged with the keeping of a public house. Emerging as distinctive figures at the start of the eighteenth century, publicans ever since have been at the heart of the English licensed trade. After attaining their greatest level of prominence within the newly expanded cities of the Victorian era, they steadily declined in importance to society during the twentieth century. Although less numerous and less powerful than their historic counterparts, publicans continue to be key figures within many communities.

To a significant degree, publicans were the direct descendants of those already engaged in the retailing of drink, particularly alehouse keepers. The historian Peter Clark (1983) identified 1750 to 1830 as the transitional period. Prior to the late seventeenth century, alehouse keepers held a relatively modest position in the communities they served. Tending to trade from a domestic dwelling and often coming from a marginalized section of society, such as the elderly, infirm, or widowed, they rarely accrued either wealth or influence. However, large-scale changes to English society, combined with specific factors such as the growth of gin drinking, led to both pressures and opportunities. Rudimentary alehouses developed into more sophisticated, substantial, and overtly commercial concerns. Moreover, as they did so, the status of the individuals who ran them grew, a process hastened by increased legal scrutiny of the drinks trade during the eighteenth century. As alehouses became public houses, so their keepers became publicans.

This shift was neither simple nor rapid. Although the difference between some taverns and lesser inns and the newer public houses lessened, some unreformed alehouses continued to operate well into the eighteenth century. Moreover, the manner and speed at which the figure of the publican came to replace that of the alehouse keeper was subject to significant variation. Unlike their predecessors, who normally produced the drink they sold, publicans tended to focus solely on retailing. In London and the south, many traders had elected to purchase their beer from common brewers based in and around the capital from before 1700. Yet, in contrast, the old practice of brewing onsite continued in many northern areas well into the nineteenth century and was widespread in Birmingham as late as the 1870s.

Similarly, the nature of the relationship between publicans and their suppliers varied according to local factors. From the late eighteenth century on, it became increasingly common, yet never universal, for nonbrewing publicans to agree to purchase their beer from a single supplier. Often this agreement was based on the "tied-house" system. Publicans were able to rent brewery-owned licensed premises for considerably below the market value if they agreed to sell only beers produced by the owners. Alternatively, publicans could be bound by the "London" system, named after its popularity in the capital, whereby suppliers would provide them with a mortgage to purchase a house. Although this practice required publicans to be responsible for the upkeep of their premises, it allowed them the possibility of changing suppliers by securing a new loan from a different brewery. Although not as independent as publicans who operated a free house untied to a brewery, those bound by the London system enjoyed a greater level of commercial freedom than those in possession of a provincial tied house.

Despite such variations, publicans had become the main retailers of drink by the early nineteenth century. By the 1830s, they were faced with a number of commercial pressures. In addition to experiencing a decline in traditional sources of revenue, such as the coaching trade, they saw a substantial increase in competition brought about by passage of the 1830 Beer Act. Whereas publicans had to go before local magistrates to secure a license to supply alcohol to the public, an individual wishing to retail only beer could do so on payment of a fee to the customs and excise authorities. Consequently, publicans had to offer an increasingly varied range of services to retain their customers. Some of those offered, such as providing sponsorship and facilities for sporting events and rooms for societies, built on existing practices; others, such as providing entertainment at what came to be known as "singing saloons" and, later, "music halls," were new commercial developments. In addition, many publicans enlarged and remodeled their premises, creating, on occasion, spirits-focused "gin-palaces" to differentiate them from the newly opened beerhouses.

The expansion in both services and facilities was indicative of the ambiguous position of publicans in mid-nineteenth-century society. Ever more business minded, their long-term survival was dependent upon their ability to present themselves as something more than a simple retailer. Although often key figures in the neighborhoods they served, few publicans were either as prosperous or as powerful as they were believed to be.

The overwhelming majority of publicans trading during the nineteenth century were male. Although no general bar to the granting of licenses to women existed, the majority of magistrates refused to do so unless the female applicant was a direct relative of a deceased licensed holder. Nevertheless, many women were required to act as de facto publicans while their license-holding husbands or fathers undertook necessary additional paid work. This arrangement was common in smaller houses, particularly in rural areas. Moreover, outside of large-scale concerns in the city, where waged bar staff were becoming the norm, the bulk of publicans were dependent, at least in part, upon cheap familial labor to run their houses.

Just as changes to the law placed pressures on publicans during the 1830s, so they did during the late nineteenth and early twentieth centuries. The period witnessed a raft of new legislation, starting with the 1869 Wine and Beerhouse Act, which placed beerhouses under magisterial jurisdiction. Not only did the statute herald a period of increased legislative control of the licensed trade and growing support for the temperance cause, it paved the way for an expansion of the tied-house system. With substantially fewer beerhouses trading, and magistrates increasingly willing and able to refuse to grant new licenses, breweries moved to purchase public houses to secure their share of the retail market. Although a number of publicans' trade organizations campaigned, at times successfully, against calls for increased limitations on their freedom, few licensees were able to resist becoming tied to a brewery.

Support for the temperance cause fell significantly after the Edwardian period, but the fortunes of publicans continued to wane. A general decline in public drinking combined with shorter opening hours, initially instigated during World War I, had a significant impact on their trade. In addition, publicans found themselves competing for customers against newer communal spaces and commercial leisure activities such as the cinema, dance halls, greyhound racing, and tea houses. From the interwar period on, brewery-led attempts to improve the layout of premises and the skills of their tenants met with some success in encouraging new kinds of customers, such as women and the middle classes, who previously tended not to frequent public houses. The cost to publicans of such developments was allowing the breweries to dictate to an ever-increasing degree the nature of their businesses.

World War II brought about a brief upturn for publicans, with increased beer consumption and governmental acknowledgment of the importance of their work to maintaining civilian morale. Yet, with the end of the conflict, the fortunes of publicans as a whole declined once more. During the second half of the twentieth century, the number of public houses fell steadily. This trend came to a head when the 1989 Supply of Beer (Tied Estate) Order was fully implemented in 1992. The Beer Order restricted the number of public houses that could be tied to a brewer. This measure proved to have both negative and positive consequences for publicans. On the one hand, it resulted in established publicans leaving the trade. A sizable number of houses closed, and many of the new public-house chains utilized salaried managers to a greater degree than the brewery estates they replaced. By 2001, the number of pubs had declined 8 percent from 1990. On the other hand, by opening up the market, the order enabled publicans in possession of a free house and those tied to smaller breweries to gain a more secure and stable position.

Kieran Foley

See also: Beer Orders (United Kingdom); Music Halls; Pubs and Beerhouses (United Kingdom)

References

Clark, Peter. *The English Alehouse: A Social History.* London: Longman, 1983.

Gutzke, David W. *Protecting the Pub: Brewers and Publicans against Temperance.* Woodbridge, Suffolk: Boydell, 1989.

Harrison, Brian. *Drink and the Victorians: The Temperance Question in England, 1815–1872.* London: Faber & Faber, 1971; 2d ed., Keele: Keele University Press, 1994.

———. "Pubs" in *The Victorian City: Images and Realities,* edited by H. J. Dyos and Michael Wolff, 161–190. 2 vols. London: Routledge & Kegan Paul, 1973.

Winstanley, Michael. "The Rural Publican and His Business in East Kent before 1914" in *Oral History.* 4 (1976): 63–78.

Pubs and Beerhouses (United Kingdom)

Public houses, or pubs, and beerhouses were the two chief venues for consuming alcohol in England and Wales during the nineteenth century. They peaked at nearly 120,000 in 1869, the same year magistrates gained control over new beerhouses and began steadily eliminating them.

Typology of Drinking Premises

For centuries, inns, taverns, and alehouses—only the last unable to sell spirits—had formed the descending hierarchy of licensed drinking premises. In the early nineteenth century, the term "public houses" came into wide use for any on-premise establishment open to the public and licensed to sell beer, wine, and spirits. Over the next two centuries, a new hierarchy of popular licensed drinking outlets emerged.

"Dramshops" began operating in bigger towns and cities from the 1820s. Often adjoining pubs and selling spirits (and hence frequently called "spirit vaults"), they were retail shops that occupied strategic sites in the city center for serving principally working-class customers. Their focus on spirits and introduction of a sales counter distinguished them from predecessors. The counter transformed a residential dwelling virtually identical to adjacent buildings into a retail shop; goods were now dispensed to "patrons" rather than "visitors."

Beerhouses, licensed for beer and wine only, served no spirits. Created specifically by parliamentary legislation in 1830 and imbued with free-trade principles, they were a legislative experiment outside magisterial control. They were deliberately designed to appeal to a working-class clientele and required only an excise license, which was cheaper than a publican's liquor license. This expectation was not unfounded: Beerhouses in such big industrial cities as Blackburn, Manchester, Sheffield, and Leeds, as well as in various seaports (notably Portsmouth), exceeded the number of pubs. They also proved more receptive to giving an often hard-pressed clientele credit and allowing bartering for sales. Though beerhouse keepers were, as members of the upper working class, or even the lower middle class, thoroughly respectable, their premises (inseparable from adjoining houses) became the target of widespread criticism. To detractors they became synonymous with crime, drunkenness, violence, prostitution, and blood sports. To subsequent historians, their dismal reputation seems unwarranted, certainly exaggerated.

Beerhouses virtually disappeared by 1980, 150 years after Parliament created this separate category of license as a panacea for curbing gin drinking and the brewers' concerted efforts to corner the market in retail outlets. No other type of license opened to the public provoked so much controversy and received such public criticism as beerhouses.

A waitress takes an order at a table outside a pub in Chelsea, London, 1967. (Hulton/Archive)

Public houses, at the top of the status hierarchy, always outnumbered beerhouses in the country at large and stood atop the status hierarchy. Customarily called "pubs" only late in the century, they derived higher prestige from several legal factors: a license enabling sale of all types of alcoholic beverages, longer licensing hours, and costlier premises. As a status group, publicans outranked beerhouse keepers because pubs required a much bigger investment, which generally only the lower middle class could afford. Successful beerhouse keepers often aspired to, and sometimes achieved, social mobility, though primarily in the provinces, where less capital was necessary for assuming a pub tenancy than in London.

First appearing in the 1830s and soon immortalized by Charles Dickens in his *Sketches by Boz* (1836), gin palaces became an elite subgroup of pubs in the status hierarchy. They incorporated the latest technology, and their large exterior plate-glass windows, huge, multi-gas lights, etched mirrors and glass, mahogany fronts, and carvings established a distinctive identity in the public mind. So did huge painted signs advertising their presence at rooftop level. Gin palaces specialized in catering to drinkers wanting fast service and maximized space and time by eliminating seats, serving no food, and substituting counter service along a lengthy bar for waiters.

Interwar "improved pubs" displaced gin palaces in the status hierarchy and adopted a different architectural style and assumed wider functions. Brewers invested heavily in gentrification, giving newly built pubs restrained, dignified facades with modest inn signs and eliminating the numerous compartments subdividing interiors. Such pubs looked to the preindustrial age for inspiration, and reintroduced tables, chairs, popular games, and gardens. White tablecloths, shaded table lamps, food, tasteful prints, carpeted floors, upholstered chairs, and a special new room called the "lounge" (evoking upmarket hotels and cruise ships) suggest to some historians that brewers sought respectable upper working- and middle-class women as customers, though others dispute this purpose. Thousands of reformed pubs were built and tens of thousands substantially altered before World War II, almost exclusively in

England, especially in the capital, suburban London, Birmingham, and Oxford.

The term "roadhouse" became appropriated for a subgroup of reformed pubs with social pretensions in the mid-1930s. Imported from the United States, it initially applied to the new fashionable country clubs, resorts, hotels, and restaurants featuring dancing, dining, and recreation (sometimes including swimming) that were built along Britain's expanding trunk roads and dual carriageways as playgrounds for the rich. To entice this socially exclusive group, brewers purpose-built and lavishly equipped some new premises with huge car parks, cocktail lounges, chromium chairs and tables, West End chefs, and orchestras as dominant features. By the mid-1950s, both the category and concept had vanished.

Another term that acquired new meaning in the interwar years was "hotel." It had been used in the Victorian years to denote not only vast buildings with numerous letting rooms but also inns and pubs with social aspirations, even when offering just several rooms for accommodation. As part of gentrification, brewers and publicans began more systematically to elevate new or rebuilt premises by substituting "hotel" for "tavern," "inn," or "arms." Brewers further obscured terminology by building or buying genuine hotels of enormous size in which alcohol generated a small fraction of overall profits. This trend toward entering the hotelier business accelerated in the 1950s.

Two new types of pubs appeared in the postwar years. From the 1960s, brewers introduced theme pubs, where one motif dominated a pub's decor. The theme pub continued well into the 1990s; innumerable Scruffy Murphy Irish pubs attest to its enduring vitality. In this same decade came the superpub, as much as eight times bigger than conventional pubs with commensurate turnovers, serving food throughout the new unbroken licensing hours. Distinguishable by their wide entrances and transparent glass, their unobstructed layouts, and their irreproachable hygienic standards, superpubs occupy city-center sites and promote brand loyalty. In 2001, there were an estimated 49,500 pubs in the UK.

Patrons

The nineteenth-century pub became the refuge of a predominantly working-class male patronage, and to a lesser extent, of the lower middle class and working-class women. Women were never excluded entirely from drinking establishments, but a woman's respectability could be diminished if she entered a pub. After all, prostitutes regularly used the public houses and beerhouses to troll for men and to relax between clients. The association between prostitution and the lower establishments therefore was liable to taint the "respectable" working-class women who visited pubs with family and friends. For this reason, many women preferred to use the back doors and special rooms that afforded them some anonymity and separated them from the mostly male drinkers located in the taproom of the pub. These patterns of class and gender typified the general clientele of most pubs and beerhouses until World War I. However, during the interwar years, publicans made a conscious attempt, with some success, to draw in the younger and more respectable men and women of the middle classes.

Services and Amusements

Some pubs not only served drinks but also provided food, newspapers, tobacco, and cigars as well as space for meetings and leisure. Public houses were often the only buildings neighborhoods had with rooms big enough to accommodate a large number of people. Until the construction of large meeting halls and public buildings near the end of the nineteenth century, pubs provided meeting rooms for clubs, societies, unions, and public lectures. Some employers utilized the pub as a Saturday pay office, handing out the week's pay; this practice was ended during the nineteenth century at the urging of reformers, who worried that workers were wasting their earnings on drink. In a more official capacity, the pub even served as a site for the occasional coroner's inquest or political meeting.

The pub continued to provide a community space for cultural and social continuity in the rapidly changing industrial postwar world. Men and women often frequented the same neighborhood establishments, usually one close to work or home. Many benefited from the familiarity between publican and patrons. During tough times, publicans often helped those in need by extending credit and providing free food and a warm place to rest. Patrons could spend what little leisure time they might have in a multitude of ways, from casual socializing to more organized entertainments. Casual amusements included singing, talking—sometimes in lively debates—and games such as skittles, shove ha'penny, cards, bagatelle, billiards, bowling, cards, draughts, chess, darts, and dominoes. A few pubs sponsored more organized sports, such as cricket, boxing, and animal sports, until their ban mid-century (animal baiting was outlawed in 1835 and cockfighting in 1849) and were attended by a largely male clientele. Besides games, publicans sponsored theatrical or musical entertainments, dances, and a range of carnival-like acts appealing to both men and women. Many of these amusements disappeared from pubs by World War I, but a few remain in various forms, such as darts and skittles. Pubs provided a wide range of amusements that conformed to patrons' lifestyles, be they the more working-class drinkers of the nineteenth century or the varied social mix of the twentieth.

David W. Gutzke and Susan M. Kling

See also: Barmaids (England); Drinking Sites and Culture (Australia); Gothenburg Schemes/Disinterested Management; Saloons and Taverns (United States); Saloons, Taverns, Pubs, and Beerhouses (Canada)

References

Aslet, Clive. "Beer and Skittles in the Improved Public House" in *Thirties Society Journal*. 4 (1984): 2–9.

Bailey, Peter. "Will the Real Bill Banks Please Stand Up? Towards Role Analysis of Mid-Victorian Working-Class Respectability" in *Journal of Social History.* 12 (1979): 336–353.
Clark, Peter A. *English Alehouse: A Social History, 1200–1830.* London and New York: Longman, 1983.
Girouard, Mark. *Victorian Pubs.* London: Studio Vista, 1975.
Gourvish, Terence R. "The Business of Alcohol in the US and the UK: UK Regulation and Drinking Habits, 1914–39" in *Business and Economic History.* 26, no. 2 (1997): 609–616.
Gutzke, David. "Gender, Class, and Public Drinking in Britain during the First World War" in *Histoire sociale/Social History.* 27 (1994): 367–391.
———. *Progressivism in Britain: The Public House Reform Movement, 1890–1939.* Unpublished book manuscript.
Jennings, Paul. *The Public House in Bradford, 1770–1970.* Keele, Staffordshire: Keele University Press, 1995.
Jones, Peter. "Enter the Superpub" in *Town & Country Planning.* 65 (1996): 110–112.
Jones, Peter, and David Hillier. "Changes Brewing: Superpub Developments in the UK" in *Geography Review.* 10 (January 1997): 26–28.
Nugus, David. "Theme Pubs from the 1960s" in *Cheers! The Story of the Modern Pub,* edited by Angela Dimitrakaki. MA Gallery Studies Students at the University of Essex, 1992.
Walkowitz, Judith. *Prostitution and Victorian Society: Women, Class, and the State.* New York: Cambridge University Press, 1980.

Pure Beer Movement (United Kingdom)

The pure beer movement was an agricultural pressure group that campaigned for the introduction of legal restrictions on the allowed ingredients of British beer similar to those stated in the *Reinheitsgebot,* the Bavarian Purity Law of 1516. By banning the use of brewing sugar, in particular, proponents hoped to force brewers to increase their consumption of barley malt and thereby revitalize this section of British agriculture. The leaders of the movement, all prominent agricultural members of Parliament (M.P.s), argued that pure beer would benefit both beer drinkers and the rural economy. Pure beer legislation posed a serious threat to the profitability of the dominant sector of the British brewing industry, however, which during this period had restructured itself on the basis of the faster brewing techniques allowed by malt substitutes and chemical additives.

The pure beer movement was founded following the repeal of malt duty in 1880, which allowed brewers greater freedom in the ingredients they could use to produce beer, beyond the basics of malt, hops, and water. Advocates of pure beer argued that by using substitutes and additives, brewers were abusing this freedom and endangering British beer drinkers. They claimed that harmful adulteration was widespread and that the only way to protect beer drinkers was through an enforced return to traditional methods of brewing. The brewing industry argued that substitutes and additives improved the quality and consistency of their products, especially the light, sparkling beers favored by most drinkers during this period.

Several unsuccessful attempts were made to introduce pure beer legislation during the 1880s and 1890s. The 1896 Pure Beer Bill was withdrawn when a governmental committee found no evidence of harmful adulteration and advised against changes to existing law. The pure beer movement came closest to achieving its legislative goal following the Manchester arsenic-in-beer epidemic of 1900–1901. By presenting the epidemic as proof of brewing-industry malpractice, the movement successfully translated widespread public concern into a dramatic increase in support for its own objectives. Mass meetings were held throughout Britain in support of pure beer legislation, and in Parliament sixty-one M.P.s joined the Committee for Promoting the Purity of Beer.

Another beer bill was introduced in 1901, but in spite of winning substantial majority support, it was blocked by the government and eventually withdrawn. The government opposed the bill for three reasons: (1) because it would be less effective at protecting beer drinkers from arsenical poisoning than the analytical safeguards already put in place by brewers and the revenue authorities; (2) because it had appointed a royal commission to investigate the arsenic epidemic and did not want it forestalled by pure beer legislation; and (3) because the bill threatened to undermine the profitability of an industry responsible for fully one-third of the total United Kingdom tax revenue.

The bill's withdrawal marked the end of the pure beer movement. Popular support faded away even faster than it had accumulated. Further attempts were made to introduce pure beer legislation throughout the early years of the twentieth century, but, lacking significant political or public backing, they were all rejected by Parliament.

Matthew Copping

See also: Arsenic-in-Beer Scare

References

Copping, M. C. "Death in the Beer-glass: The Manchester Arsenic-in-Beer Epidemic" (M.A. thesis, University of Kent at Canterbury, 1999).
Phillips, Jim, and Michael French. "The Pure Beer Campaign and Arsenic Poisoning, 1896–1903" in *Rural History.* 9, no. 2 (1998): 195–209.

R

Rae, John Turner (1851–1929)

John Turner Rae (1851–1929) served as secretary of the British National Temperance League from 1899 until his death. For the ten previous years he was assistant secretary. His father, John Rae (1823–1900), had been the league's secretary from 1861 until 1899. After an accident in 1898 made him an invalid, his son took over the responsibilities of the office as acting secretary. According to William Gourlay, the league's historian, both father and son earned reputations for "tactfulness."

In addition to illustrating the widespread phenomenon of a family commitment to temperance reform, John Turner Rae represented the common practice of many temperance advocates serving different organizations concurrently. He was active in the Blue Ribbon gospel temperance movement, the total abstinence union for traveling showmen, the Native Races and Liquor Traffic United Committee, the Anglo-Indian Temperance Association, the Temperance Collegiate Association, the International Temperance Bureau, the Royal Army Temperance Association, and the National Council to Promote Social Centers. For many years he edited the *National Temperance Quarterly and Medical Review.* He also edited many books on medical and scientific aspects of temperance and encouraged the British government inquiry into the connection between alcohol and physical deterioration.

Rae was active in the international temperance movement as well. For instance, he helped organize the twelfth International Congress on Alcoholism, held in London in 1909, and edited its proceedings.

David M. Fahey

See also: National Temperance League (NTL)

References

Cherrington, Ernest Hurst, et al., eds. *Standard Encyclopedia of the Alcohol Problem.* 6 vols. Westerville, OH: American Issue Publishing, 1925–1930: 2247–2248.

Gourlay, William. *"National Temperance": A Jubilee Biography of the National Temperance League, Instituted 1856.* With an Introduction by John Turner Rae. London: Richard J. James, 1906.

Rand Report

The Rand Reports, which addressed the controversy of whether it is possible for individuals who have been diagnosed with a serious drinking problem to return to a nonproblematic drinking pattern, mark one of the most politicized periods in alcoholism treatment history. The first Rand Report was published in 1976 and the follow-up in 1981. These studies have been hailed by some social scientists as landmark studies attacking the hegemony of the disease concept of alcoholism championed by Alcoholics Anonymous. Alternatively, the studies have been attacked as reckless and dangerous by traditional alcoholism researchers.

This controversy began when the National Institute on Alcohol Abuse and Alcoholism (NIAAA) commissioned the Rand Corporation to examine the efficacy of its treatment centers. Rand followed a large sample of 2,000 alcoholic patients who received a variety of treatments at eight NIAAA treatment centers and systematically assessed patterns of alcohol consumption at six and eighteen months posttreatment. At the eighteen-month follow-up, 24 percent had been abstinent for at least six months, 21 percent reported being abstinent for one month, and the remaining 22 percent reported drinking in what was characterized as a "normal drinking pattern." This last assertion was the cause of much of the controversy. According to the traditional disease concept of alcoholism endorsed by Alcoholics Anonymous and pioneers such as E. M. Jellinek, once an individual is an alcoholic, it is impossible to return to "normal drinking patterns." Some scientists described this finding as empirical support for the rejection of the disease concept and concluded that the Rand study suggested that treatment centers should offer a harm-reduction model of treatment rather than one based solely on abstinence.

The first Rand Report was attacked as being methodologically weak. Specifically, the study was flawed because it suffered from: sampling bias (80 percent of the subjects dropped out); lack of validity of the drinking measures (it relied heavily on self-report of drinking behavior within the thirty days prior to the survey); and insufficiently stringent criteria for "normal drinking."

In response to these criticisms, the Rand Corporation did a follow-up study four years later and attempted to address some of these criticisms. In this second study, self-reports were augmented with interviews with family members and breath alcohol concentration tests. More stringent, empirically based definitions of normal drinking were also used. The second study stated that 14 percent of the sample continued to drink in a nonproblematic manner. Other researchers reanalyzing the same data arrived at a corrected estimate of 3–4 percent of the sample who were drinking in a nonproblematic manner. Critics pointed out that one of the primary weaknesses of the original study (assessing drinking behavior within the thirty days prior to the survey) persisted in the second study. Traditional disease-concept proponents argued that alcoholics can be expected to drink in a nonproblematic manner for a brief period. The anti-disease contingent of scientists complained that the NIAAA had removed the study from the Rand Corporation and analyzed the data on its own, coming to the biased conclusion, despite their own findings, that treatment centers should maintain an abstinence approach to the treatment of alcohol problems.

In the end, the Rand Reports stand out not as examples of sound research but of how politicized the disease concept had become. Scientists who continue to embrace the results are in the harm-reduction camp and believe that the abstinence approach is restrictive. This perspective is best exemplified by the "moderation management" model of alcohol treatment.

Paul E. Priester

See also: Alcoholics Anonymous (AA); Alcoholism; Jellinek, Elvin Morton; National Institute on Alcohol Abuse and Alcoholism (NIAAA)

References

Armor, D. J.; J. M. Polich, and H. B. Stambul. *Alcoholism and Treatment.* Santa Monica, CA: Rand Corporation, 1976.

Polich, J. M.; D. J. Amour, and H. B. Braiker. *The Course of Alcoholism: Four Years after Treatment.* Report prepared for the National Institute on Alcohol Abuse and Alcoholism, U.S. Department of Health, Education and Welfare. Santa Monica, CA: Rand Corporation, 1981.

Wallace, John. "Abstinence and Non-abstinence in Treatment: A Case Study in the Sociology of Knowledge" in *Controversies in the Addictions Field,* edited by Ruth Engs, 192–202. Dubuque, IA: Kendall/Hunt, 1990.

Raper, James Hayes (1820–1897)

James Hayes Raper (1820–1897), a Cumberland stonemason's posthumous son, was arguably the most effective agent ever to serve the prohibitionist United Kingdom Alliance. He became a teetotaler in 1837 in part in order to encourage temperance in the Sunday school where he taught. After being a local temperance activist, he served as the alliance's parliamentary agent from 1860 to 1878. In 1878, he resigned this paid position and joined the alliance's executive council. Unsalaried, he continued to do the work of parliamentary agent until 1886. He was a man of many talents who at a public meeting could deliver a rousing speech laced with humor, in private guide a committee meeting toward realistic decisions, and maintain cordial working relations with MPs who lacked his enthusiasm for prohibition. When explaining Raper's success in lobbying politicians, his biographer, J. Deane Hilton, emphasized his "frank, social audacity," his "extraordinary self-possession," and "presence of mind" (Hilton 1898: 86). Although a militant in his objectives, Raper could be practical about tactics. Some contemporaries claimed that the alliance's president, Sir Wilfrid Lawson, became less willing to compromise after Raper's death ended his moderating counsels.

Raper had radical views on many questions. He was a strong sabbatarian who on Sundays traveled by foot only. He supported parliamentary reform, the repeal of the Corn Laws, restrictions on factory hours, and the peace movement and opposed slavery and the Contagious Diseases Acts, which protected the customers of prostitutes while demeaning women. In 1878, he stood for Parliament as an advanced Liberal but finished third behind moderate Liberal and Conservative candidates.

He traveled overseas on behalf of temperance several times. In 1873, he visited Sweden to investigate the Gothenburg system in which quasi-governmental companies managed the retail drink trade. He also went to the United States and Canada in the mid-1870s to study the operation of prohibition and attend temperance meetings. He was a delegate to an international temperance congress in Paris in 1877.

Raper enjoyed great popularity in the temperance world. In 1880, a testimonial presented him with £1,600 and a silver tea and coffee service. Sir Wilfrid Lawson contributed £200 to the fund. Another close friend, W. S. Caine, contributed £100 pounds. A lecture series jointly honored Raper and F. R. Lees after their deaths.

Raper was a North of England Puritan who lived austerely. J. Deane Hilton wrote that Raper often made his midday meal out of "a roll of brown bread and a glass of cold water" (Hilton 1898: 25). Although not literally a vegetarian, Raper did subscribe to the Vegetarian Society. He probably never entered a theater. Although an affable conversationalist, he was too reserved in personality ever to reveal much about himself. He was reared as a Wesleyan Methodist.

David M. Fahey

See also: Caine, William Sproston; Gothenburg Schemes/Disinterested Management; Lawson, Sir Wilfrid; Lees, Frederic Richard; Liberal Party (United Kingdom); Lucas, Margaret Bright; United Kingdom Alliance (UKA); Vegetarianism and the Anti-Tobacco Movement (United Kingdom)

References

Harrison, Brian. *Dictionary of British Temperance Biography.* Aids to Research, no. 1, *Bulletin Supplement.* Coventry, England: Society for the Study of Labour History, 1973: 101–102.

Hilton, J. Deane. *A Brief Memoir of James Hayes Raper, Temperance Reformer, 1820–1897.* London: Ideal Publishing Union, 1898.

Raskob, John Jakob (1879–1950)

John Jakob Raskob (1879–1950) was a prominent member of the Association Against the Prohibition Amendment (AAPA) who played an important role in aligning the Democratic Party with the movement to repeal the Eighteenth Amendment to the

U.S. Constitution. He was chairman of the Democratic National Committee (DNC) from 1928 to 1932 and had a close association with New York Governor Alfred E. Smith, who lost to Herbert Hoover in a landslide election for U.S. president in 1928.

Like Al Smith, Raskob was the child of poor immigrant Roman Catholic parents. Forced by his father's death into the role of provider for his mother and five younger siblings while still in his teens, Raskob mastered stenography and bookkeeping. At age twenty-one, he became secretary to Pierre du Pont and assisted him in gaining control of, reorganizing, and refinancing the Du Pont Company of Wilmington, Delaware. Becoming Du Pont corporate treasurer in 1914, Raskob invested heavily in the struggling Detroit automobile manufacturer General Motors (GM) and then encouraged Pierre du Pont to do likewise. Du Pont emerged from World War I with control of the auto maker. As GM treasurer and as architect of the General Motors Acceptance Corporation, Raskob acquired a reputation as a capable and progressive business leader.

Viewing National Prohibition as hostile to immigrant cultures such as his own, Raskob in June 1922 accepted the invitation of William H. Stayton, whom he had met during World War I, to join the AAPA. Perceiving Prohibition as an anti-Catholic policy, Raskob regarded the law as a curtailment of personal liberty. Repealing the Eighteenth Amendment was to Raskob a means of reestablishing social and religious tolerance. By the mid-1920s he was making substantial contributions to the AAPA and committing considerable energy to the campaign against Prohibition.

Raskob met Al Smith in 1927 and discovered the similarity of their views on alcohol. He made large contributions to Smith's presidential campaign. After Smith won the Democratic nomination, Raskob became his campaign manager and chairman of the DNC. Stung by Smith's loss to Hoover, Raskob determined to make the DNC an ongoing organization rather than one that merely functioned in the few months prior to a presidential election. Personally funding a full-time national committee staff, Raskob strengthened the Democratic Party significantly for the 1932 election. He also made clear on various occasions that he thought the party should embrace Prohibition repeal.

Raskob's efforts were opposed by Democratic presidential aspirant Franklin D. Roosevelt, who replaced Raskob as party chairman as soon as he gained the 1932 nomination. By that time, however, the party's convention had voted by more than four to one to adopt a platform plank embracing repeal. The Democratic endorsement of repeal, in sharp contrast to the Republican commitment to continued "dry law" enforcement, was widely perceived as a major reason for the Democratic victory in 1932 and propelled the adoption of a repeal amendment in 1933. Raskob remained an outspoken critic of Roosevelt thereafter but never again achieved the political influence that he had demonstrated in the political battle for Prohibition repeal.

David E. Kyvig

John Jacob Raskob, Chairman of the Democratic National Committee, testifies before the Caraway Lobby Investigation Committee at the Capitol in Washington about his contributions to the Association against the Prohibition Amendment, 1930. (Library of Congress)

See also: Association Against the Prohibition Amendment (AAPA); Democratic Party (United States); Du Pont, Pierre Samuel; National Prohibition (United States); Smith, Alfred Emanuel; Stayton, William H.

References

Kyvig, David E. *Repealing National Prohibition.* 2d ed. Kent, OH: Kent State University Press, 2000.

Slayton, Robert A. *Empire Statesman: The Rise and Redemption of Al Smith.* New York: Free Press, 2001.

Rechabite Friendly Society

A number of temperance organizations have been named after the original Rechabites, a biblical tribe who dwelt in tents and were required never to plant vines or drink wine (see, for example, 1 Chron. 2:55; Jer. 35:6–8, 12–17). The oldest and largest of these temperance societies is the Independent Order of Rechabites (Salford Unity).

The Independent Order of Rechabites (IOR) was the first teetotal friendly society with numerous local affiliates. It was founded in 1835 at Salford, England, and spread rapidly throughout Britain and the Empire. It was most successful in England, Scotland, Wales, and Australia; but there were also branches in Ireland, South Africa, India, Canada, New Zealand, and elsewhere. A rival society with a similar name was founded in the United States in 1842, and the British society itself later established U.S. branches.

A member of the Rechabite fraternal temperance society wearing lodge regalia (Library of Congress)

Members of the Salford Temperance Burial Society founded the IOR. As a friendly society, its original purpose was to provide funeral benefits and to pay sickness benefits to members unable to work because of illness or injury. It combined this role with that of a temperance society. From about 1841, medical benefits were added when some of its branches began employing doctors to treat members who were sick. In common with friendly societies generally, the IOR was a "secret" society with a formal initiation process, colorful emblems and regalia, and semireligious rituals. Friendly societies also provided a social forum, and good fellowship was an important element in their success. They traditionally met in public houses—an arrangement unsuitable for abstainers. The IOR therefore sought to provide a teetotal alternative, holding its first meetings in a Salford coffeehouse.

The IOR drew its members mainly from the skilled working classes. This was also true of temperance and friendly societies in general. They were grassroots associations, run on democratic principles, with working-class social and political aspirations. The IOR was formed during a vital period of working-class history notable for the emergence of trade unions, a radical press, and Chartism (a popular movement for parliamentary reform). However, it followed the mainstream temperance and friendly societies in remaining politically moderate, promoting social change through moral reform and cooperation for mutual benefit.

The branches of the IOR were called "tents" and were given inspiring names from the Bible and from popular morality. Within a few months of its foundation, the society had three tents (Ebenezer and Good Samaritan in Salford and Felicity in Manchester). Five years later, it was firmly established in Scotland and Wales, and there were more than 500 tents. By then it was publishing its own journal, the *Rechabite Magazine.* In 1842, several colonial branches were established. By 1843, there were more than 1,000 tents and nearly 30,000 members.

The IOR was unique among friendly societies in admitting women. From 1836 to 1838, the female tents were reconstituted as a separate organization known as the United Order of Female Rechabites (UOFR). By 1842, the UOFR had 140 tents in twenty-two districts, including about 6,500 members. The females rejoined the IOR in 1856. Juvenile tents were introduced in 1840. At the same time, a widow and orphans fund was established to complement the existing sickness benefit and funeral funds.

The IOR suffered two severe setbacks in 1843. First, an ill-advised political campaign intended to reform the law governing the registration of affiliated societies proved to be a costly failure resulting in considerable adverse publicity and a general loss of confidence in the society. Second, an attempt to increase central control over the branches led to widespread discontent and many defections. By 1854, membership was reduced to 6,000—less than a quarter of the total ten years earlier.

The IOR recovered, however, and grew steadily in the following decades. Australia had become a stronghold of the society by 1870—the colony of Victoria alone had about a quarter of the total IOR membership, and the strength of the colonial districts had become a key factor in sustaining the organization. In 1875, membership surpassed 30,000 for the second time. By 1900, total membership was about 250,000. The society was then very strong in the industrial areas of Scotland, where Glasgow had about 30,000 members. By 1910, the IOR had about 460,000 members, three times as many as its nearest rival, the British division of the Sons of Temperance Friendly Society.

Friendly societies got a boost with passage of the National Insurance Act of 1911, which introduced compulsory health insurance in Britain. Although individuals could purchase the required insurance through commercial insurance companies, they could also obtain it through membership in a friendly society. The legislation increased competition between commercial insurance companies and increased society membership. It also greatly increased the remuneration and power of the doctors, many of whom were directly employed by the societies. By 1914, the IOR had enlisted about 350,000 "state members."

Over the next three decades, changing social values, along with new economic and political factors, began to undermine the rationale of friendly societies in general, and temperance friendly societies in particular. When the welfare state became a central feature of Western economies after World War II, it radically altered the environment in which friendly societies operated. In Britain, within a few decades the number of friendly societies declined from 18,000 to 300. But there, and in Australia and New Zealand, the IOR has carried on into the twenty-first century. Greatly reduced in scale, but still adhering to temperance principles and the friendly society philosophy, it has once more adapted itself to a changing environment. The British society has added life endowment and mortgage protection assurance to its products and still serves about 20,000 members, all of whom must be total abstainers. In 1995, after 160 years of continuous existence, it changed its name to the Rechabite Friendly Society. The society maintains its archives and a museum at its head office in Manchester, England.

Stephen C. Due

See also: Chartism; Good Templars (IOGT)

References

Campbell, Richardson. *Rechabite History.* Manchester: IOR, 1911; 2d ed. by Robert Highet, 1935.

Checkland, Olive. *Sobriety and Thrift: John Philipson and Family.* Newcastle upon Tyne: Philipson, 1989.

Denny, Norma. "Self-Help, Abstinence and the Voluntary Principle: The Independent Order of Rechabites, 1835–1912" in *Journal of the Scottish Labour History Society.* 24 (1989): 24–46.

Gorsky, Martin. "Self Help and Mutual Aid: Friendly Societies in Nineteenth Century Britain" in *Recent Findings of Research in Economic and Social History.* 28 (Spring 1999): 1–4.

Gosden, P. H. *Friendly Societies in England, 1815–1875.* Manchester: Manchester University Press, 1961.

Green, David G. *Welfare State, for Rich or Poor?* London: Institute of Economic Affairs, 1982.

"Independent Order of Rechabites" in *Standard Encyclopedia of the Alcohol Problem,* edited by Ernest Hurst Cherrington. Westerville, OH: American Issue Publishing, 1925–1930: 1292–1294.

Russell, Rex C. *The Water Drinkers in Lindsey, the Earlier Temperance Movement, 1837–1860: Rechabites, Teetotallers, Moral Revolutionists, Rational Recreation.* Barton-upon-Humber: Barton Branch, Workers' Educational Association, 1987.

Reformed Men's Clubs

During the 1870s several religiously oriented groups were established to "reform the inebriate" and encourage total abstinence. These societies were primarily aimed at men who drank heavily. Their formation instituted a gospel reform club movement that swept North America for the next generation. The three most prominent groups originated in Maine. These were the Gardiner Reform Club, founded in 1872 by J. K. Osgood; the Blue Ribbon Movement, launched by Francis Murphy in 1873; and the Red Ribbon Reform Club, inaugurated around 1874 by Dr. Henry A. Reynolds. The structure and activities of these groups were similar to the modern Alcoholics Anonymous, except that members were required to sign an abstinence pledge. In addition, members of the Red Ribbon and Blue Ribbon clubs often wore ribbons to signify their abstinence. Their emergence testifies to the persistence of a self-help approach among drinkers during a period when most temperance reformers embraced legal solutions to drinking problems.

Joshua Knox Osgood (1816–1895), who had lost his business and fortune from problem drinking, became interested in the temperance movement after a religious conversion experience, when he and a former drinking companion began mission work to reform "occasional drinkers, constant drinkers, hard drinkers, and young men who are tempted to drink" (Cherrington 1925–1930, 1: 359). On January 19, 1872, they called a meeting, and eight drinking companions signed a pledge not to drink. The Gardiner Reform Club had several hundred members within a few months. Osgood then organized numerous clubs in Massachusetts and New Hampshire. Subsequently he moved his base of operations to Massachusetts under the sponsorship of the Massachusetts Temperance Alliance. Men who signed the pledge began to wear blue ribbons in their lapels, and the crusade became known as the "temperance reform club and blue ribbon movement." Osgood died, however, before the clubs reached their peak of popularity.

The expansion of the movement is credited to the efforts of Francis Murphy (1836–1907). Murphy, an Irish American and well-known Portland, Maine, saloon keeper, was sent to jail for illegally selling liquor in 1870. He experienced a religious conversion and took an abstinence pledge. After being released from jail, he resolved to help others and gave a temperance speech in Portland in April 1871 to friends and neighbors. The result of this meeting was the formation of a reform club with old drinking companions who signed an abstinence pledge. Having a natural gift for oratory, he was invited to speak at other communities around New England. In 1874 Frances Willard, then a local temperance activist in Chicago, asked him to carry the movement to her city, where he spoke at numerous churches. Murphy's crusade adopted the same symbol, a blue ribbon, as Osgood's club had. The "Murphy Movement" spread nationally and internationally. During the late 1870s and 1880s, hundreds of active local societies existed in twenty-eight states. Many of his reform clubs created reading rooms and arrangements for entertainment as a substitute for the saloon. One of his most successful campaigns took place in the fall and winter of 1876–1877 in the Pittsburgh area. By the end of this campaign, an estimated 40,000 people had taken the temperance pledge with the motto, "With malice toward none, and charity for all." Murphy made his headquarters in Pittsburgh. For the next decade, with the help of his two sons and second wife, he evangelized in many states as well as in Canada, the United Kingdom, and Australia. In 1901, Murphy moved from Pittsburgh to Los Angeles for his declining years.

Unlike many temperance reformers of the day, he did not join the drive for legislative prohibition.

The Red Ribbon Reform Club was instigated by Henry A. Reynolds (1839–1922), a Harvard-educated physician and surgeon in the Civil War who had lost his medical practice on account of drinking. He began to recover upon signing an abstinence pledge at a temperance meeting held by women activists around 1874. Grateful for his recovery, he began speaking about his experiences. He formed a reform club in Bangor, Maine, with eleven men based upon Christian principles around 1874 or 1875 (sources differ concerning the date). The club was originally limited to alcoholic men who would support each other in their efforts to stop drinking, and their gatherings served as a substitute for the saloon. They adopted the motto, "Dare to do right." Reynolds also formed an early club in New Brunswick. At a meeting of delegates from out of town he suggested that members of his clubs wear red ribbons for the ease of identifying each other. Invited by the Woman's Christian Temperance Union (WCTU) to form clubs in Massachusetts, he made the red ribbon a badge of membership in his campaign beginning in 1876. Within a year, more than seventy clubs had been established in that state. Regular meetings of the club, which were similar to nondenominational Protestant prayer assemblies, were held on a weeknight, and public meetings with the WCTU on weekends. The WCTU asked Reynolds to expand his campaign, and the movement spread to several other states. In Michigan, forty-five clubs were formed between 1876 and 1877. Reynolds was aided in his campaign by Methodist, Congregationalist, and Baptist churches. The movement was also given support by the Young Men's Christian Association (YMCA). Reynolds did not join prohibition forces but instead attempted to convince listeners that abstinence was good for the individual and the family. In the closing years of the 1870s, however, interest in the club declined and its influence began to wane. In some communities the clubs no longer had much of a following by the late 1880s.

The commonalities of all the reform clubs were total abstinence, reliance upon God's help in all things, and missionary work to induce others to sign the pledge. However, by the turn of the twentieth century the work of temperance had been taken over by the Anti-Saloon League of America and other organizations that crusaded for local, state, and national prohibition legislation.

Ruth Clifford Engs

See also: Blue Ribbon Movement; Murphy, Francis; Woman's Christian Temperance Union (WCTU) (United States)

References

Cherrington, Ernest Hurst, et al. eds. *Standard Encyclopedia of the Alcohol Problem.* 6 vols. Westerville, OH: American Issue Publishing, 1925–1930, vol. 1: 359–360; vol. 4: 1839–1840; vol. 5: 2258–2259, 2268–2269.

Daniels, W. H. *The Temperance Reform and Its Great Reformers.* Cincinnati: Nelson and Phillips, 1878.

Fehlandt, August F. *A Century of Drink Reform in the United States.* Cincinnati: Jennings and Graham, 1904.

Lender, Mark Edward. *Dictionary of American Temperance Biography.* Westport, CT: Greenwood, 1984.

Malone, Dumas. "Murphy, Francis" in *Dictionary of American Biography,* edited by Allen Johnson and Dumas Malone. 20 vols. New York: Charles Scribner's Sons, 1928–1937, vol. 7: 349–350.

Murphy, Rebecca Fisher. *Memoirs of Francis Murphy, the Great Temperance Apostle, by His Wife.* Long Beach, CA: Graves and Hersey, 1908.

Van Woerkom, Gerald. "They 'Dared to do Right': Prohibition in Muskegon" in *Michigan History.* 55 (1971): 41–60.

Republican Party (United States)

Founded in 1854, the Republican Party (Grand Old Party, or GOP) endured an almost constant tension because of the many temperance supporters in its ranks and the party's refusal to take firm positions on the prohibition of alcoholic beverages during the three-quarters of a century between the party's founding and the last years of National Prohibition. Struggle between knots of prohibitionist Republicans and those party supporters opposed to strict liquor regulation—most notably Protestant Germans—led party leaders to deemphasize the issue in the name of party unity throughout the late nineteenth century. Nevertheless, the party came to be associated with prohibition as bands of "dry" Republicans seized party organizations and enacted state prohibition in the 1880s, and GOP managers actively campaigned to win the loyalty of temperance voters. The drive toward National Prohibition after 1913 cemented ties between Republican officeholders and the lobbyists of the Anti-Saloon League of America (ASLA). Not until 1928, however, did the Republican Party make a brief and unsuccessful commitment to vigorously enforce National Prohibition.

Origins to 1900

Prohibition was a divisive issue among Republicans from the party's beginnings. Dissension over the Maine Laws, state prohibition statutes that spread through the Northeast and Midwest in the early 1850s, was one of the factors in the destruction of the Second-Party system that, in turn, gave rise to the Republican Party. Proponents of the Maine Laws, many of them refugees from the collapsed Whig Party, carried their reform enthusiasm into Republican ranks. They met resistance from others drawn into the new party by its positions on slavery expansion, economic development, and other reforms. Sectional issues, and, eventually, Civil War, overrode temperance reforms in the late 1850s and 1860s.

Renewed temperance legislation, often in the form of more stringent Sunday closing laws, attracted the support of dry Republicans after the Civil War. But the alienation of German Republicans, who enjoyed beer in the relaxed tradition of the continental Sunday, and a string of state-level electoral defeats persuaded party leaders to neutralize the volatile liquor question. The 1872 national party platform explicitly retreated from advocacy of temperance legislation. Republican officials encouraged local governments to handle liquor regulation through licensing decisions, thus removing the matter from party politics. Some disgruntled dry Republicans withdrew from the GOP and helped found the Prohibition Party in 1869.

Nevertheless, links between temperance reform and Republicanism remained persistent and fractious. Republicans saw their party as a force for moral improvement and were more willing than Democrats to use the state as an instrument of reform. Between 1879 and 1882, dry Republican activists in Kansas and Iowa revolted against cautious party leaders and orchestrated the adoption of state constitutional prohibition. For these dry reformers, "Republicanism . . . meant a schoolhouse on every hill, and no saloon in the valley" (Kleppner 1979: 317). By 1890, often stirred by dissident Republican factions, sixteen other states had voted on prohibition amendments. The prohibition issue influenced the presidential candidacy of Republican James G. Blaine in 1884. Dry Republicans in New York deserted Blaine, who was evasive on the liquor question, and voted for the Prohibition Party candidate in sufficient numbers to help deny the Republican nominee victory in the tight election. In the aftermath of that defeat, Republican strategists appealed to dry voters. The party employed J. Ellen Foster of the Woman's Christian Temperance Union (WCTU) to reinforce Republican sympathies among dry women and their families. Republican operatives obtained the mailing list for Prohibition Party publications and sent literature imploring drys to return to the Republican fold. The 1888 Republican Party platform expressed sympathy for temperance without committing itself to specific legislation. Yet the pendulum also swung the other way. Incautious statements against liquor and Catholics on the part of dry Republicans had hurt Blaine with some voters in 1884. "Wet" and German Republicans retaliated against prohibition laws and nativist sentiments in the Midwest by bolting the GOP, resulting in a series of dispiriting Republican defeats in Kansas, Iowa, Ohio, Illinois, and Wisconsin between 1882 and 1890. During the depression decade of the 1890s, the national Republican Party emphasized economic issues and downplayed temperance.

The Republican Party and National Prohibition

In the early twentieth century, the pressure for prohibition moved outside the parties to nonpartisan lobbying on the part of the ASLA and the WCTU. Rather than pressing the Republican party to endorse prohibition, dry lobbyists concentrated their attention on individual officeholders. In 1905, the Ohio ASLA unseated Republican governor Myron T. Herrick and thereafter enjoyed the courteous attention of the state's ambitious Republican politicians. Republican state officials elsewhere felt similar pressure. Wet national Republican figures such as Speaker of the House of Representatives Joseph Cannon withstood ASLA attacks, but enough congressmen responded to dry lobbying that the ASLA launched a campaign for national constitutional prohibition in 1913.

Since the new dry strategy did not require political parties to issue platform statements in support of prohibition until 1916, prohibition did not dominate and divide GOP conventions. Once the ASLA and the WCTU did request party statements, the accumulation of state prohibition victories and dry votes in Congress compelled Republicans (as well as Democrats) to remove the prohibition question from national party debate by submitting it instead to the states. Thus, in 1917 Congress passed the Eighteenth Amendment establishing National Prohibition.After ratification and until his death in 1927, ASLA general counsel Wayne Wheeler attempted to use his personal relationships with presidents Warren Harding and Calvin Coolidge, as well as his influence over the federal enforcement agency for Prohibition, the Prohibition Bureau, to construct closer ties between Republicans and the ASLA. Still, Harding, Coolidge, and the GOP took no extraordinary steps to strengthen Prohibition enforcement or make Prohibition a party issue. In 1928, with presidential candidate Herbert Hoover as a more willing enforcer of the law, dry Republicans committed their party to stricter enforcement of National Prohibition. The candidacy of wet Democrat Al Smith as Hoover's rival in the 1928 election furthered the impression that the Republicans had become the party of Prohibition. Depression and the dismal record of Prohibition enforcement led several state Republican organizations to endorse Prohibition repeal by 1932, but Hoover kept a repeal plank out of the GOP platform. Repeal came in 1933, with Republicans once more divided over prohibition.

Thomas R. Pegram

See also: Anti-Saloon League of America (ASLA); Foster, Judith Ellen Horton; Hoover, Herbert Clark; Maine Law; National Prohibition (United States); Prohibition Bureau; Prohibition Party (United States); Smith, Alfred Emanuel; State Prohibition (United States); Wheeler, Wayne Bidwell; Woman's Christian Temperance Union (WCTU) (United States)

References

Gienapp, William E. *The Origins of the Republican Party, 1852–1856.* New York: Oxford University Press, 1987.

Kerr, K. Austin. *Organized for Prohibition: A New History of the Anti-Saloon League.* New Haven, CT: Yale University Press, 1985.

Kleppner, Paul. *The Third Electoral System, 1853–1892: Parties, Voters, and Political Cultures.* Chapel Hill: University of North Carolina Press, 1979.

Mayer, George H. *The Republican Party, 1854–1966.* New York: Oxford University Press, 1967.

Pegram, Thomas R. *Battling Demon Rum: The Struggle for a Dry America, 1800–1933.* Chicago: Ivan R. Dee, 1998.

Research Council on Problems of Alcohol (RCPA)

The Research Council on Problems of Alcohol (RCPA) was formally organized in the United States in 1938. Its participants included scientists, educators, and prominent public-policy and business figures. Early RCPA members founded the organization in an effort to secure philanthropic grants for scientific research into alcohol problems and to put such research in the service of public education. During the late 1930s and early 1940s, the RCPA helped formulate important new perspectives on the problems associated with alcohol consumption.

The RCPA emerged five years after the repeal of National Prohibition in the United States. At that time, controversial debates between "dry" prohibition advocates and "wet" prohibition opponents continued to dominate discussions of alcohol issues. The Research Council came onto the scene hoping to chart a neutral course amid the morally and politically charged debates between drys and wets, which more and more Americans either ignored or viewed as a tiresome stalemate. In an effort to reformulate the terms of discussion, the RCPA sought to develop more scientifically objective approaches to understanding and addressing alcohol problems. By 1939, the Research Council had come to define the most pressing of the alcohol problems in essentially medical terms, promoting and sponsoring scientific investigation of the emerging modern disease concept of alcoholism. Along with Alcoholics Anonymous, the Yale Center of Alcohol Studies, and the National Committee for Education on Alcoholism (NCEA), the RCPA helped shift American attention away from dry-wet debates and refocus it upon alcoholism during the late 1930s and the 1940s. If the 1935 founding of Alcoholics Anonymous marks the popular source of what scholars have described as the alcoholism movement, the formation of the Research Council represents the founding moment of the scientific and public-health branches of the alcoholism movement. Although the RCPA played a central role in laying the foundations of alcoholism science, the organization found itself drifting to the margins of activity by the mid-1940s.

Early Years

Although scholars have produced conflicting accounts of the RCPA's origins, sociologist Ron Roizen (1991) has offered the most well-documented study of the organization's early years. The RCPA was officially born in January 1938, when the education-oriented Sponsoring Committee of the National Conference on Alcohol—a short-lived organization led primarily by the moderate but dry-leaning reformer Harry H. Moore—became an associated society with the American Association for the Advancement of Science (AAAS). Early on, the Research Council failed to secure a grant from the Rockefeller Foundation. As it turned out, the Rockefeller Foundation's wariness of the controversy surrounding alcohol outweighed its interest in supporting scientific research on alcohol problems. The RCPA remained financially unstable throughout its first year. The council relocated its headquarters from Washington to New York when New York Medical College psychiatrist Karl Bowman joined the fledgling group and offered it office space free of charge at the New York Academy of Medicine. Thereafter, Bowman began a rapid rise to power in the RCPA. By September 1938, Bowman had become chairman of the RCPA Executive Committee and established a thirty-two-member Scientific Committee. At that point, both Howard H. Haggard, M.D., director of the Laboratory of Physiology at Yale University, and Dr. Norman Jolliffe, a researcher at New York's Bellevue Hospital, joined the group. Jolliffe became an RCPA member soon after completing a review of European approaches to alcohol problems, a project that received funding from the Rockefeller Foundation before the RCPA's formation. During its first year in existence, the RCPA received valuable publicity when *Science* and the *New York Times* reported on the organization's new scientific approach to alcohol problems.

Successes and Frustrations

Under Bowman's leadership, the RCPA scored an early success when the Scientific Committee won a grant of $25,000 from the Carnegie Corporation to support Jolliffe's research into the relationship between vitamin B1 and physical recovery from alcohol addiction. The first stage of the project entailed a thorough review of previous scientific studies on alcohol and its effects on the human body. To conduct this review, Jolliffe recruited E. M. Jellinek from Worcester State Hospital in Massachusetts, where he had been studying the neurology of schizophrenia. Jellinek would subsequently become perhaps the leading mid-twentieth-century researcher of alcohol addiction in the United States. At Jellinek's suggestion, Howard Haggard secured Yale University funds to launch the *Quarterly Journal of Studies on Alcohol,* which became the official journal of the RCPA in 1940. In fact, 1940 marked a high point for the Research Council. That year, it organized several meetings and symposia that helped raise public awareness of alcoholism, which had become the group's central concern.

In time, however, internal conflict compounded problems stemming from the RCPA's financial insecurity. After the initial grant for Jolliffe's project, the RCPA was unable to win additional funds from the Carnegie Corporation, and it was equally unsuccessful in its attempts to solicit support from wealthy individuals. Coinciding with the RCPA's 1939 decision to focus on alcoholism in the near future, the organization announced that it would accept funding from the liquor industry. During the 1940s, the RCPA became increasingly dependent on relatively small donations from the liquor industry. Eager to separate alcohol from addictive drinking, liquor-industry donors soon began pressuring the RCPA to abandon the term "alcoholism" in favor of "problem drinking." Angered by these developments, Haggard and Jellinek broke with the organization in 1943, and Haggard struck the RCPA's name from the editorial page of the

Quarterly Journal of Studies on Alcohol. Indeed, by that time, Haggard's university-funded alcohol section of Yale's applied physiology lab had become the Yale Center of Alcohol Studies, which Jellinek headed. While the Yale Center achieved preeminence in the scientific study and treatment of alcohol addiction, the RCPA languished. Both the National Committee for Education on Alcoholism, founded by reformed alcoholic Marty Mann in 1944, and Alcoholics Anonymous proved more successful than the Research Council at raising public awareness of alcoholism. As it turned out, by focusing its energies on researching and publicizing alcoholism, the RCPA committed itself to work that other institutions and organizations performed more effectively. The RCPA survived into the late 1940s. In January 1949, the Research Council's president, Dr. Anton J. Carlson, announced its termination.

Timothy Yates

See also: Alcoholics Anonymous (AA); Alcoholism; Haggard, Howard Wilcox; Jellinek, Elvin Morton; *Journal of Studies on Alcohol;* Mann, Marty; National Council on Alcoholism and Drug Dependence (NCADD); Yale Center of Alcohol Studies

References

Johnson, Bruce N. "The Alcoholism Movement in America: A Study in Cultural Innovation" (Ph.D. dissertation, University of Illinois at Urbana-Champaign, 1973).

Keller, Mark. "Problems with Alcohol: An Historical Perspective" in *Alcohol and Alcohol Problems: New Thinking and New Directions,* edited by William J. Filstead, Mark Keller, and Jean J. Rossi, 5–28. Cambridge, MA: Ballinger, 1976.

Roizen, Ron. "The American Discovery of Alcoholism, 1933–1939" (Ph.D. dissertation, University of California at Berkeley, 1991).

———. "Norman Jolliffe, the Rockefeller Foundation, and the Origins of the Modern Alcoholism Movement" in *Journal of Studies on Alcohol.* 55, no. 4 (July1994): 391–400.

Resorts

Resorts in the United States do not have a reputation for sobriety. From the inception of vacationing in America, alcohol served as the great lubricant of social discourse. Drinking also helped relax inhibitions in resort settings where strangers met partially outside the realm (and rules) of mundane existence.

The earliest American "watering places" were generally located in the vicinity of mineral water springs and intrinsically related to the act of drinking. At the springs in western Virginia, the southern gentry took the waters, gambled on cards, drank corn whiskey, bet on horses, and flirted to an extent that astonished more sedate visitors. Visitors who found mineralized spring water an acquired taste thought the libation improved when mixed with sherry or whiskey. Even the smallest resort hotel furnished a place to drink alcohol, a logical development given the hotel's evolution from the tavern in the early 1800s.

Often resorts became a temperance battleground when urbanized visitors looking for a good time away from home confronted more conservative local residents. At Saratoga Springs, New York, the entrepreneurial desire for profit conflicted with the fact that New York's first temperance society was located in the county. Despite numerous complaints, Saratoga's town government in the 1820s granted tavern licenses and allowed grocery stores to sell intoxicating liquors by the gallon jug, then waived several fines imposed on tavern owners for peddling liquor in violation of the rules that had been set for them. Enticed by the opportunity to drink and gamble, tipplers helped make Saratoga Springs the most popular nineteenth-century resort in the United States. The charismatic Christianity of the Second Great Awakening completely passed over most northeastern resorts, although revivals shook neighboring communities.

The passage of "Maine Laws" prohibiting the sale of alcohol in many states in the 1850s threatened the antebellum resort community. The *Newport Mercury* pleaded with locals to uphold the Rhode Island prohibition law of 1852, but most Newport residents, whatever their personal attitudes, viewed sumptuary legislation as a threat to the city's prosperity. Newport quickly acquired a reputation for its civil disobedience; stubborn summer visitors, with the forbearance of hotel keepers, packed sherry and port in with their luggage and helped subvert the state legislation.

After the Civil War, seaside resorts such as Revere Beach, Massachusetts, and Coney Island, New York, typically allowed Sunday drinking and dancing; the easy sale of alcohol popularized a commercialized version of German festival culture with its bands and beer gardens (often sponsored by breweries). In 1888, the Mercantile Agency Reference Book listed forty-two hotels and saloons in Coney Island and only eleven other businesses. Philadelphia had 1.15 saloons for every 1,000 people in 1893, but nearby Atlantic City, New Jersey, had 14.55 per 1,000. Periodic outcries from reformers generated occasional farces of enforcement, but almost no local jury in a resort town would indict or convict on an alcohol charge, regardless of the evidence.

Some post–Civil War "Christian" resorts banned alcohol. Old Orchard Beach, Maine; Ocean Grove, New Jersey; Round Lake, New York; and Wesleyan Grove, Massachusetts, all represented conservative entrepreneurial adaptations of the older revival camp-meeting and forbade intoxicating spirits as a sin. Chautauqua (New York) and its imitators lacked evangelistic services but sponsored an interest in popular education and wholesome recreation in a resort atmosphere that supposedly required the absence of alcohol. These respectable resorts introduced sober and industrious middle-class Protestants to the leisure revolution in the late 1800s. But although covenant deeds might ban alcohol and require rigorous Sabbath observance, eventually most of these religious resorts compromised on dancing, drinking, and the construction of amusements.

After 1900, amusement-park entrepreneurs generally banned the sale of alcohol as a basic step in rationalizing leisure by regulating crowd behavior and attracting a middle-class clientele. Kennywood, an amusement park outside of Pittsburgh, advertised, "No fakes, no liquor, no gambling, no disorder." Nonetheless, beer gardens and "Bowerys" flooded the areas adjacent to these enclosed parks, allowing visitors a place to imbibe. Amusement areas often clashed with their more staid townships over the issuance of liquor licenses. One example is Venice, California, almost the only place in the vicinity of Los Angeles where someone could buy a legal drink in the 1910s, which encountered opposition from the town of Ocean Park in 1907.

The Eighteenth Amendment drove alcohol underground at hotel-oriented resorts but particularly hurt the amusement-park industry. At some beach resorts, such as Cape May, New Jersey, the coastline became an ideal landing area for rum runners. After the demise of National Prohibition, liquor and beer were widely available and their use was socially acceptable throughout most of the United States. Widespread advertising in the late twentieth century, especially on television, linked alcohol consumption with the pleasure and self-discovery of leisure time and allied drinking and vacations. Brewers updated their tradition of sponsoring beer gardens, and the Anheuser-Busch Corporation opened large theme parks in Williamsburg, Virginia, and Tampa Bay, Florida.

In popular late-twentieth-century resorts based around gambling and casinos, such as Atlantic City and Las Vegas, Nevada, alcohol was typically served at no charge, both as an inducement to play at the tables and as the house's way to erode good judgment. Many new Indian casinos, however, mindful of the damage alcohol has historically done to Native American culture, prohibit alcohol completely; the Jackson Rancheria, one of the biggest and most successful casinos in Northern California, does not serve alcohol.

Modern resorts vary greatly regarding their policy on the role of alcoholic beverages. College students celebrating spring break in beachfront resorts such as Fort Lauderdale, Florida, associate excessive alcohol consumption with freedom. Club Med, which runs supposedly one-fee "all-inclusive" vacation villages around the world, offers unlimited beer and wine at lunch and dinner but also allows guests the option of purchasing an "open-bar" or "non-alcoholic open bar" package. In general, the restriction of alcohol at a resort, such as at Disneyland (California) or Disney World (Florida), now symbolizes a "family" atmosphere free of sexual frisson.

Jon Sterngass

See also: Anheuser-Busch; Maine Law

References

Paulsson, Martin. *The Social Anxieties of Progressive Reform: Atlantic City, 1854–1920.* New York: New York University Press, 1994.

Sterngass, Jon. *First Resorts: Pursuing Pleasure in Saratoga Springs, Newport, and Coney Island.* Baltimore: Johns Hopkins University Press, 2001.

Weiss, Ellen. *City in the Woods: The Life and Design of an American Camp Meeting on Martha's Vineyard.* New York: Oxford University Press, 1987.

Roadhouses

Although the term "roadhouse" was first coined in the mid-nineteenth century to denote a type of rural inn or tavern, what is today commonly thought of as a roadhouse came into being in the early twentieth century, arising in tandem with the urban nightclub. The roadhouse per se was a drinking establishment offering dining and dancing to a diverse and fluid patronage that flourished with the spread of the automobile. It peaked in popularity during the years of U.S. National Prohibition, often selling alcohol by the drink illegally, and remained popular during much of the Great Depression. As distinct from the bar, honky-tonk, or juke joint, the roadhouse was larger and less socially intimate. It was a place of social drinking and live musical entertainment located outside of the city limits along major motor routes and generally attracted a clientele of city or town dwellers on an evening out.

Roadhouses appeared all across North America and had analogues in Central and South America. Before National Prohibition, their location outside incorporated towns allowed them to avoid blue laws regulating liquor sales and opening hours on Sundays; they were often located near the borders of adjacent "dry" counties, particularly in, but not limited to, the southern United States. This allowed easy access to alcohol for patrons residing in temperance towns and counties.

The clientele of the roadhouse, mostly couples, helped usher in a change in North American drinking patterns. Like their urban, nightclub counterparts, roadhouses quickly became acceptable public drinking places for women. Unlike pubs, bars, and honky-tonks, which were almost exclusively male in patronage, the roadhouse's inclusion of dining and entertainment actively encouraged the attendance of women and lent them a veneer of respectability in an age of changing social mores. During National Prohibition in the United States, women actively operated many roadhouses.

National Prohibition actually created further business for roadhouses. Ideally located just out of the view and jurisdiction of many law-enforcement agencies, they became the country cousin of the speakeasy. Rural electrification and the spread of the automobile worked in tandem to increase their numbers and usage. Their legitimate role as dining establishments provided a law-abiding front.

The repeal of National Prohibition and the deepening of the Great Depression signaled the beginning of the end of the heyday of the roadhouse. With alcohol again available in plain view and money in short supply, fewer people engaged in a "night out." Live entertainment became too expensive for most roadhouses to provide, and even the smallest establishments could install a jukebox for dancing. By mid-century, the roadhouse

A 1938 photograph of teens cavorting at a roadhouse in Louisiana (Library of Congress)

was viewed as either seedy or quaint or both. The end of the twentieth century saw the name and image of the roadhouse appropriated and sanitized as marketing tools when a national chain of U.S. family-oriented theme restaurants, located in primarily suburban areas and designed and decorated to look rural, took "Roadhouse" as part of their name.

John P. Hundley

See also: Juke Joints; Saloons and Taverns (United States)

References

Clinard, Marshall B. "The Public Drinking House and Society," in *Society, Culture, and Drinking Patterns,* edited by David J. Pittman and Charles R. Snyder, 270–292. Carbondale: Southern Illinois University Press, 1962.

Murphy, Mary. "Bootlegging Mothers and Drinking Daughters: Gender and Prohibition in Butte, Montana" in *American Quarterly.* 46, no. 2 (June 1994): 174–194.

Robinson, Sarah (1834–1921)

Sarah Robinson (1834–1921) was an English moral reformer who earned the title "Soldier's Friend" for her Christian and temperance work on behalf of British soldiers. Robinson's efforts reflected the activities of a wider social purity movement of the late nineteenth century that married moral reforms with Christian conversion. In 1874, she founded the Soldier's Institute at Portsmouth, which catered to the physical and spiritual welfare of soldiers and their families. Robinson also lectured to soldiers about moral hygiene and received a mention in the 1870 Parliamentary Blue Book on military education. In the early 1880s, she helped to found the Soldier's Institute in Alexandria, Egypt. By then, Robinson had broadened her temperance and religious work to include the working classes of Portsmouth. Due to ill health, she retired from active missionary work in 1892.

Robinson was born in 1834 in Blackheath, England, to a wealthy family as the fourth of six children. After her father moved the family to an estate near Lewes, Robinson briefly attended a girls' boarding school. She withdrew because of her mother's death and her own illness. Raised a Calvinist, Robinson attested that she had undergone a Christian conversion experience at the age of seventeen. In 1862, her family moved to Guildford, where she taught singing and Bible classes in Sunday school. In addition, she engaged in Christian mission work, visiting the homes of the sick and impoverished. In 1865, she embarked upon mission work to soldiers who were stationed in nearby Aldershot. With permission from military authorities, Robinson held Christian and temperance meetings with soldiers in their barracks. Her work with soldiers convinced her that true Christian conversion was impossible without total abstinence. In addition to her work with soldiers, she concurrently visited brothels in her attempts to improve the physical and spiritual condition of both prostitutes and their customers.

In 1873, with the backing of the National Temperance League and the permission of army officials, Robinson set up a temperance canteen for soldiers at Dartmoor during army maneuvers. After the success of this venture, Robinson extended the scope of her mission by establishing a permanent temperance canteen and home in Portsmouth designed to cater to the multitudes of soldiers leaving for and returning from campaigns abroad. The Soldiers' Institute, opened in 1874, was an establishment that provided accommodation for soldiers, sailors, and their families. Robinson's strategies were similar to those of other social purity activists whose work combined practical reform with earnest Christianity. In 1876 she wrote *Christianity and Teetotalism: A Voice from the Army.* In addition to her efforts on behalf of soldiers and their families, Robinson worked to spread temperance and Christianity among the working classes of Portsmouth. In the 1880s, she founded "A Helping Hand," a Portsmouth temperance organization. She opened a coffeehouse, the Blue Ribbon Coffee Tavern, in a building that formerly had housed a pub. In 1892 Robinson retired from the Soldiers' Institute at Portsmouth. During her retirement, whe wrote about her work on behalf of soldiers and temperance. Her memoirs included *Yarns; Being Sundry Reminiscences* (1892), *A Life Record* (1889), *The Soldier's Friend: A Pioneer's Record* (1913), and *"My Book": A Personal Narrative* (1914). Other than those writings, Robinson lived quietly after she retired from public work in 1892.

Mary A. Conley

See also: Army Temperance Association; National Temperance League (NTL); Weston, Agnes

References

Hopkins, J. E. *Active Service, or Work among Our Soldiers.* London, 1872.

Robinson, Sarah. *Christianity and Teetotalism: A Voice from the Army.* Comp. by John Guthrie. London: Published for the National Temperance League by W. Tweedie, 1876.

———. *A Life Record.* London: J. Nisbet, 1898.

———. *"My Book": A Personal Narrative.* London: S. W. Partridge, 1914.

———. *The Soldier's Friend: A Pioneer's Record.* London: R. Fisher Unwin, 1913.

———. *Yarns; Being Sundry Reminiscences.* London: Grafton Printing Works, 1892.

Shiman, Lilian Lewis. *Crusade against Drink in Victorian England.* New York: St. Martin's, 1988.

Ross, Sir Henry (James) (1893–1973)

Sir Henry James Ross (1893–1973) was chairman of the Distillers Company Limited (DCL) from 1948 to 1958. He served DCL in various capacities for more than fifty years.

He was born in Edinburgh, the son of William Henry Ross, the chairman of DCL from 1925 to 1935. Henry Ross was educated at George Watson's College, Edinburgh, the Leys School, Cambridge, and the Institut Tully, Berlin. He joined DCL as an apprentice in 1910, starting at DCL's Glenochil distillery as a maltman. In 1911, he took the brewing and distilling course at Heriot-Watt College, Edinburgh, before continuing his apprenticeship at DCL's Carsebridge and Caledonian distilleries. He moved to the head office and spent two years in the export department; because of the travel involved in this position, before he was twenty-one he had visited Europe, Egypt, Algeria, India, Ceylon, and South Africa. During World War I, he served in the Royal Naval Division, was invalided out in 1916, and rejoined DCL as assistant to his father, who was general manager. He became a deputy general manager in 1924 and was appointed to the board in 1925, becoming, at thirty-two, its youngest member.

Ross's responsibilities spanned both the potable and "industrial" (that is, industrial alcohol and chemicals) sides of DCL. On the potable side, his main interest lay in exports. In the 1930s, he assisted Thomas Herd, DCL's managing director (1932–1935) and chairman (1935–1937), in reestablishing trade with the United States after the repeal of Prohibition and in circumventing protectionist legislation in Canada. On the industrial side, he was a member of the Technical and Research Committee, which initiated diversification into chemicals and laid the foundation for a much greater development after 1945.

Appointed to the influential Management Committee in 1939, he became chairman of its Production Committee in February 1941, becoming responsible for potable spirit and whiskey stocks. With the wartime cereal shortage and restrictions on distilling, his task was one of rationing and attempting to conserve mature stocks in the face of an unprecedentedly high demand. In 1939, DCL's stocks were 94 million proof gallons (mpg); by 1945, they had fallen to 53 mpg. Despite acute shortages, there were few complaints about DCL's rationing system, and Ross was credited with operating it fairly. In March 1948, Ross succeeded Lord Forteviot as DCL's chairman.

Opinion about the Scotch whiskey industry's prospects was decidedly pessimistic. Ross played an important role in negotiations with the government that led to an agreement under which cereals were allocated to the industry on condition that it increased exports to hard-currency markets. In 1947, the allocation was equal to 55 percent of prewar production, and for every 3 imperial gallons that were exported, 1 could be released to the home market. DCL led the industry in meeting its export targets, though in public Ross protested about the "strangulation of our home trade interests" (DCL 1949). Although by 1954 stocks had been rebuilt to almost the same level as in 1939, a large proportion was not yet adequately matured and ready for sale. Ross insisted that the maintenance of quality was crucial and refused to compromise the standard of the brands; supplies were, therefore, restricted within the limits of adequately matured stock. In 1953 he argued that: "The essence of success in our Whisky business lies in quality, and there can be no departure from the standards which we have set ourselves in this respect, and have so firmly maintained throughout the difficult postwar years. This standard rests on a basis of unchallenged excellence, a true and unvarying hall-mark which is universally recognised and respected" (DCL 1953).

From 1 January 1954, the government left firms free to adopt their own pattern of trade but on the understanding that every effort would be made to increase exports to the dollar area. By 1953, the industry's export sales amounted to 13.2 mpg, the highest ever, compared to a prewar figure of 8 mpg. Home trade releases amounted to 4.3 mpg, the highest since 1947 but markedly less than the prewar home trade of 7 mpg. The prolonged effect of wartime restrictions was still evident when Ross relinquished the chairmanship in 1958; only in 1959 did the home trade approach free supply. Gin also enjoyed buoyant demand, though exports faced competition from local production, a development that led DCL to set up gin distilleries in South Africa in 1952–1953 and in Brazil in 1957–1958, and from a new consumer preference for vodka.

DCL undertook an ambitious postwar expansion program in chemicals, plastics, and pharmaceuticals, but this owed more to Graham Hayman, who succeeded Ross as chairman, than to Ross himself. Ross's contribution was to prevent any internecine conflict between the potable and industrial sides. The high level of capital commitments in the mid-1950s, with the potable side rebuilding stocks and modernizing distilleries and the industrial side constructing large-scale chemical plants, could easily have provoked disagreements. Ross argued that "there must not be any question of competition between groups within the organisation" and set a system of priority for the company as a whole (DCL 1955b).

Ross was appointed "Life President" of DCL in December 1957. He attended his last meeting of the board on 4 February 1963; his service with DCL had extended over fifty-three years, thirty-eight of these as a board member.

Ross was prominent in the wider affairs of the Scotch whiskey industry as a member of the Council of the Scotch Whisky Association (1942–1958), chairman of the Scotch Whisky Association (1945–1958), and its president (1959–1968). In 1950, he tried, unsuccessfully, to secure the amalgamation of the Pot Still Malt Distillers' Association with the Scotch Whisky Association. He supported "the gentleman's agreement" not to advertise spirits on commercial television and was against exporting in bulk rather than by bottle. He campaigned against the export of immature spirits (1952) and for the free entry of Canadian and U.S. whiskeys into the United Kingdom, a measure achieved in November 1956. He believed strongly that ownership of the Scotch whiskey industry should remain in native hands as "Scotch whisky is the best ambassador this country possesses" (DCL 1955a).

Ronald Weir

See also: Distillers Company Limited (DCL); Whiskies

References

Distillers Company Limited (DCL). Annual Report, 1949.

———. Annual Report, 1953.

———. Annual Report, 1955a.

———. Board Minute Book 26. 9 November 1955b.

Weir, Ronald B. *The History of the Distillers Company, 1877–1939: Diversification and Growth in Whisky and Chemicals.* Oxford: Clarendon Press, 1995.

Rowntree, Joseph (1836–1925)

Joseph Rowntree (1836–1925), British cocoa manufacturer and social reformer, consistently devoted his wealth, his energies, and his business acumen to promoting schemes of practical philanthropy. He was an important figure in the temperance movement in Britain in the early years of the twentieth century by virtue of his research into the drink question and his advocacy of the Gothenburg system of disinterested management as an alternative to prohibition or the local veto.

Rowntree was the second son of Joseph Rowntree, a prominent Yorkshire Quaker. In 1869, he entered into the Rowntree cocoa company, becoming sole owner in 1883 and remaining chairman until 1923. As an enthusiast for adult education, Rowntree quickly became concerned about how the temptation to drink blighted the lives and aspirations of members of the working class. He consistently maintained that the drink problem had to be viewed in the broader context of poverty and adverse social conditions. In about 1892, he examined the alleged triumphs of prohibition in areas of New England but came to the conclusion that such a solution only worked in remote rural areas. He published a paper on this, "A Neglected Aspect of the Temperance Question," in 1896.

Rowntree then commissioned Arthur Sherwell, who came to York to help with the adult education school, to undertake extensive research into the operation of the local veto in the United States and other aspects of the drink issue. Five jointly authored volumes resulted from their investigations. The most influential was *The Temperance Problem and Social Reform* (1899), in which they emphasized the deleterious effect of intemperance upon Britain but maintained local veto on its own to be an inadequate solution. More attention needed to be paid to "constructive" as opposed to "restrictive" measures. They also stressed the "social and political menace" that the drink trade posed to public and political life in Britain. They advocated that municipalities should be able to run public houses along lines of disinterested management whereby the publican had no vested interest in sales. Profits should not go directly to the municipalities, as happened in Scandinavia, however, but to a national state authority, whence they would be distributed to provide for counterattractions to drink.

Rowntree and Sherwell's ideas attracted considerable attention and swiftly led to a bitter, major, and long-lasting split in the British temperance movement between the advocates of disinterested management and the orthodox prohibitionists. The Carlisle scheme and the work of the Central Control Board (Liquor Traffic) during World War I reflected the influence of Rowntree and were enthusiastically supported by him.

Rowntree established a model housing project for lower paid workers at New Earswick, near York, in 1904, where alcohol was prohibited, and this experiment influenced the later Garden City movement. Rowntree's various trusts financed many charitable and educational projects throughout the twentieth century. Apart from temperance, the causes closest to Rowntree's heart were adult education, pacifism, and the League of Nations. Unpublished temperance reminiscences for his grandchildren are to be found among Rowntree's papers at the Joseph Rowntree Foundation library in York.

John Greenaway

See also: Carlisle System; D'Abernon, Lord (Sir Edgar Vincent); Gothenburg Schemes/Disinterested Management

References

Greenaway, John. *Drink and British Politics since 1830: A Study in Policy-Making.* Basingstoke, UK, and New York: Palgrave Macmillan, 2003.

Rowntree, Joseph, and Arthur Sherwell. *The Temperance Problem and Social Reform.* 7th ed. London: Hodder & Stoughton, 1900.

———. *Public Control of the Liquor Traffic.* London: Grant Richards, 1903.

———. *The Taxation of the Liquor Trade.* 2d ed. London: Macmillan, 1908.

Vernon, Anne. *A Quaker Business Man: The Life of Joseph Rowntree, 1836–1925.* London: Allen & Unwin, 1958.

Royal Commission on the Liquor Traffic (Canada)

The Royal Commission on the Liquor Traffic investigated the economic, political, and social aspects of the liquor traffic in Canada and considered what effect a prohibitory liquor law would have on the country. The commission was convened in 1892 and consisted of five commissioners, Sir Joseph Hickson (Montreal), Herbert S. McDonald (Leeds and Grenville County, Ontario), Edward F. Clarke (Toronto), George Auguste Gigault (St. Cesaire, Quebec), and the Rev. Dr. Joseph McLeod (Fredericton, New Brunswick). Its mandate was to collect data regarding the effect of the liquor traffic on Canada, measures that had been adopted to lessen, regulate, or prohibit the traffic, and the results of these measures; the effects that a prohibitory law would have on commerce, society, and revenue; and "all other information bearing upon the question of Prohibition."

Between 1892 and 1894, the commission sent questionnaires to various professionals and officials, invited notables (including non-Canadians such as Neal Dow of Maine) to be interviewed by the commissioners, and most especially, held public hearings across the country, beginning in Halifax, Nova Scotia, from 1893 to 1894. The brewers and distillers engaged a Mr. Louis P. Kribs to look after their interests during the inquiry. The temperance forces, led by the Dominion Alliance for the Total Suppression of the Liquor Trade, engaged Francis Spence, the alliance's secretary, to participate on behalf of the temperance interests. Spence, Kribs, and other incidental observers were not permitted to question witnesses, address the commission, or call their own witnesses, although they tried to defend their interests when they found it necessary.

The commission was beset by problems. Not all commissioners attended the hearings, many questionnaires were never returned, the long intervals that elapsed between the sittings of the commission were criticized in Parliament, and temperance supporters complained that only one commissioner, McLeod, was not hostile to prohibition before the commission began. The final report, submitted in March 1895, was fully supported by only three commissioners (Hickson, McDonald and Clarke), with Gigault signing it but dissenting from a few items in the report. McLeod wrote his own minority report in April. Kribs and Spence also wrote their own reports, which were published by their supporting organizations.

The results of the commission's work are difficult to assess. The majority report argued against prohibition, but it did not stop the temperance advocates from continuing their campaign. In 1896, the Liberal Party was elected to federal Parliament, having promised a plebiscite on the issue. Two years later, in 1898, a general plebiscite on prohibition was held throughout Canada. The results of this plebiscite found a 52.5 percent majority of voters in favor of prohibition, but only 44 percent of the electorate—well below the average for general elections—had voted; prohibition had won by fewer than 13, 687 votes. Prime Minister Sir Wilfrid Laurier argued that this margin was too slim to act upon and that the plebiscite was not representative of the general will. Laurier and his Liberal Party had strong support in the province of Quebec, the only province in which the vast majority had voted against prohibition.

Daniel J. Malleck

See also: Dominion Alliance for the Total Suppression of the Liquor Traffic; Dow, Neal; Laurier, Sir Wilfrid; Spence, Francis Stephens

References

Kribs, Louis. *Report of Louis P. Kribs in Connection with the Investigation Held by the Canadian Royal Commission on the Liquor Traffic.* Toronto: Murray Print Co., 1894.

Report of the Royal Commission on the Liquor Traffic in Canada. Ottawa, 1895.

Smart, Reginald, and Alan Ogborne. *Northern Spirits: A Social History of Alcohol in Canada.* 2d ed. Toronto: Addiction Research Foundation, 1996.

Spence, Francis. *The Facts of the Case: A Summary of the Most Important Evidence and Argument Presented in the Report of the Royal Commission on the Liquor Traffic Compiled under the Direction of the Dominion Alliance for the Total Suppression of the Liquor Traffic.* Toronto: Newton & Treloar, 1896; reprint, Toronto: Cole's, 1973.

Traynor, Brian P. "Towards a Genealogy of Temperance: Identity, Belief and Drink in Late Victorian Ontario" (M.A. thesis, Queen's University, Kingston, 1993).

Rum

Rum is the distilled alcoholic beverage made from sugarcane juice and the waste products of sugar making. It is made in sugarcane-growing regions of the world, but the name has also been applied to alcoholic beverages made from sugarcane byproducts in other areas. For example, New Englanders, who distilled molasses purchased from Caribbean sugar islands, also called their spirits rum. To a lesser extent, European distillers, who distilled syrup, the waste of metropolitan sugar refineries, also used the term for their product.

Origins

The high sucrose content of sugarcane *(Saccharum officinarum)* makes it an ideal source for the production of alcoholic beverages. Ancient Sanskrit texts refer to fermented varieties of sugarcane wine in India. Sugarcane wine was probably also produced in parts of the Arab world during the Mediterranean phase of sugar production. Distilled rum, however, was first produced in Brazil, Mexico, and the Caribbean in the late sixteenth and early seventeenth centuries.

Numerous terms have been used for alcoholic beverages made from sugarcane. *Cachaça* is the most common name for distilled alcohol made from sugarcane in Brazil. In the French Caribbean, *tafia, eau de vie de canne,* and *clarin* all refer to alcoholic beverages made from sugarcane. In the Spanish Americas, *aguardiente de caña* and *chingurito* have been used. *Kill devil* referred to distilled sugarcane-based alcoholic beverages in the early British Caribbean, and this term transferred to the French as *guildive.*

"Rum" eventually became the most common term for a distilled sugarcane-based alcoholic beverage outside of Brazil. It originated in the British Caribbean in the seventeenth century and derived from the English word "rumbullion." In 1651, Giles Silvester, a resident in the British colony of Barbados, made the earliest, and possibly only, reference linking rum and rumbullion. Silvester wrote, "The chiefe fudling they make in the Iland is Rumbullion, als Kill-Divill, and this is made of suggar cones distilled a hott hellish and terrible liquor" (Harlow 1925: 46). "Rumbullion" was a word commonly used in Devonshire, England, to mean "a great tumult" and its use in the British colony probably reflects the large number of West Country English who settled Barbados in the early seventeenth century (Davis 1885: 77). By the mid-1650s, "rumbullion" had been shortened to "rum." Sugar planters in the French and Spanish Caribbean adopted the term for a distilled sugarcane-based alcoholic beverage, translating it to *rhum* and *ron,* respectively.

Early Caribbean Expansion

In the seventeenth century, Barbados and the French colony of Martinique developed the most sophisticated rum industries. In the early years of rum making, distillers in these islands relied heavily on *scum,* the material skimmed off during the sugar boiling process. By the late seventeenth century, *molasses,* the material that dripped from sugar molds during the sugar curing process, became a primary ingredient in rum making. Alcohol distillation was still a relatively new art, and the capacity of early Caribbean stills was small. However, by the late seventeenth century, Barbados was producing about 1 million U.S. gallons (3.785 million liters) annually and Martinique was probably producing about half that amount.

Three of the rums produced by the Saint James distillery of Martinique, ranging from white to dark (Philip Gould/Corbis)

Rum making emerged to meet the alcohol needs of Caribbean colonists. The high cost and limited availability of imported European alcoholic beverages led colonists to the search for local alternatives. Early settlers in Barbados and Martinique produced a wide variety of alcoholic beverages from local resources, especially cassava, and turned to rum making at the start of the sugar revolution in the 1640s. Sugar production provided an enormous amount of waste material, which colonists distilled into rum. The growth of the Atlantic trade also fueled the rise of rum making in the Caribbean. In early modern Europe, seamen considered alcohol a necessary provision on trading ventures. Although wine and brandy filled the hulls of ships departing Europe, Caribbean sugar planters exploited the maritime demand for alcohol on the other side of the Atlantic and sold rum to traders for the return voyage.

The early rum trade catered to markets at the margins of the emerging Atlantic world. British and French Caribbean planters exported rum to non-sugarcane-growing regions of the Spanish Americas and sold it to Carib Indians in the Lesser Antilles. Caribbean merchants exported rum to colonists in British and

French North America, where it was exchanged for plantation supplies, especially grain and lumber. Ireland also received its share of Caribbean rum. And, by the end of the seventeenth century, sloops from Barbados and Brazil carried rum to the coasts of West and Central Africa. African chiefs and traders valued the novelty of the beverage, and it became a central item in gift-giving ceremonies and a secondary item of trade.

Eighteenth-Century Mercantilism

In the eighteenth century, mercantilism shaped the growth of New World rum industries. In 1713, Louis XIV of France issued a royal decree that prohibited, except for ports in Normandy, the import of French Caribbean rum into France. The declaration specifically argued that rum was pernicious to health and threatened to compete with French wine and brandy. French Caribbean sugar planters continued to export rum to markets at the margins of the Atlantic world, especially French Canada, where its exchange helped feed the fur trade. The small rum-making operations in the Spanish Americas also faced opposition. Spanish colonial officials instituted local ordinances that sought to curb the production and use of rum. They argued that rum was a threat to colonial import revenues from Spanish wine and brandy, and many perceived excessive rum consumption, especially by African slaves and Indians, as the cause of social disorder. The Real Cédula (Royal Edict) of 8 June 1693 prohibited rum making in all the Spanish colonies, but the constant reiteration of the ban in the following century suggests that officials were unable to control illicit distilling.

In contrast, British officials embraced British Caribbean rum as an ally in their war against foreign spirits that had drained England of capital for centuries. They opened the home market to British Caribbean rum and offered incentives to rum makers, including low import duties. In 1719, British imports of rum surpassed those of brandy for the first time. After 1741, rum imports regularly exceeded those of brandy for the rest of the eighteenth century (Schumpeter 1960). Rum contracts with the British Royal Navy also stimulated the growth of British Caribbean rum industries. Jamaica emerged as the leading rum-making colony in the Caribbean, exporting more than 2 million U.S. gallons per year in the early 1770s (Long 1774, 2: 496–497). Jamaican rum found a strong market in Britain, while rum makers in the smaller British Caribbean colonies, such as Barbados and Antigua, primarily fed the huge Irish and North American demand.

British North American traders also purchased rum and molasses from the French Caribbean. As with their British Caribbean counterparts, French Caribbean planters relied on rum and molasses to help cover the cost of plantation supplies. French Caribbean sugar planters, who had no home market for rum, had plenty of rum and molasses for North American traders. However, the trade was problematic for British Caribbean sugar planters. If North Americans bought cheap rum, or produced their own rum from cheap foreign molasses, then the value of British Caribbean rum decreased. In 1733, British Parliament, backed by British Caribbean interests, passed the Molasses Act, which imposed a six-pence per gallon tax on foreign molasses entering North American ports. The tax was especially aimed at curbing the importation of French Caribbean molasses, which New Englanders used to produce rum. The Molasses Act also barred French Caribbean rum from ports in Ireland. However, the Molasses Act was rarely obeyed or enforced and traders found numerous ways of circumventing the tax through bribes and smuggling.

In 1763, at the end of the Seven Years' War, Britain's Parliament took an interventionist role in developing the economy of the continental colonies. The Sugar, or American Revenues, Act of 1764 established greater metropolitan judicial control that strengthened enforcement of the Molasses Act. The act reduced the duty on foreign molasses to three pence per gallon in an attempt to discourage bribery and smuggling, but it also levied an equal tax on British Caribbean molasses. The continental colonists' reaction was to boycott all British goods. Within two years, the Sugar Act was repealed and a reduced duty of a penny per gallon levied on all imported molasses. The clash over the Molasses Act and the Sugar Act helped fuel the revolutionary fervor of North Americans, which led to the eventual separation of the thirteen continental colonies from Britain.

Impact of American Revolution

The American Revolution and the trade restrictions that followed the conflict devastated rum makers in the smaller British Caribbean islands, which had traditionally supplied the North American market. Rum also faced competition from North American whiskey. According to American historian William Rorabaugh rum was associated with colonial dependence and therefore "suffered from rising nationalism" (Rorabaugh 1979: 67). In the late eighteenth and early nineteenth centuries, whiskey, made from American-grown corn, replaced rum as the national drink. The repeal of the whiskey excise tax in 1802 hastened the demise of the North American rum market.

Rum followed the spread of sugar production. In the late eighteenth and early nineteenth centuries, colonists in British, French, and Portuguese settlements in India and Southeast Asia began distilling rum. Rum making emerged in Mozambique and in the Indian Ocean colonies of Madagascar, the Seychelles, and the Mascarene islands, especially Mauritius and Réunion. Distillers often referred to their product as *arrack,* a traditional term for alcohol in the Indian Ocean region. Rum making also emerged in the South Pacific, Australia, and in the new sugar-cane-growing regions of the Caribbean, including Demerara and Trinidad and Tobago.

Nineteenth-Century Changes and Modern Developments

In the nineteenth century, the rise of government-subsidized beet-sugar industries in Europe glutted world sugar markets.

The profitability of Caribbean sugar production declined and planters turned increasingly to rum production to help keep their sugar plantations solvent. In the mid-nineteenth century, a species of fungus known as *Oïdium tuckerii* and an aphid known as *Phylloxera vastatrix* attacked grape vineyards in Europe and greatly reduced wine and brandy production. France was hit especially hard. In 1854, Napoleon III suspended the duty on French Caribbean rum imports in order to replenish alcohol supplies. The move helped introduce rum to the French public on a wider scale. In the late nineteenth century, the French Caribbean exported millions of gallons of rum to France annually and Martinique emerged as one of the world's leading rum producers.

Rum making also followed the expansion of sugar production in Cuba in the nineteenth century. In 1830, Facundo Bacardi, a Catalonian, immigrated to Cuba and settled in Santiago de Cuba. With financial backing from his brother José, he purchased a distillery in 1862 and started what was to become a rum empire. In 1876, at the International Centennial Exposition in Philadelphia, Bacardi rum won its first international award, beating out Jamaican contenders who at the time were considered the world's best rum producers.

Cuban rum making expanded in the early twentieth century and found a strong market in the United States. In 1898, Cuba won independence from Spain. In order to stimulate the Cuban economy after the Spanish-American War, Cuban goods, including rum, received favored trade status in the United States. Between 1898 and 1902, U.S. forces stationed in Cuba were introduced to Cuban rum and the specialty rum-based drink known as the *daiquiri,* named after the southern port town of the same name. During Prohibition (1920–1933), thousands of American tourists flocked to Cuba to indulge their alcoholic fantasies. As with U.S. troops two decades earlier, they brought back a taste for Cuban rum. Smuggled Cuban rum was one of the few spirits available during Prohibition. Although whiskey remained the staple American drink, concerns about grain shortages during World War II led to tight restrictions on whiskey making, which further opened the door for Cuban rum.

In 1898, Puerto Rico was ceded to the United States and American corporations quickly financed the expansion of sugarcane cultivation in the island. The new interest in sugar production fueled the growth of Puerto Rican rum making, which challenged Cuba for control of the North American rum market. The Cuban Revolution (1959) closed trade between the United States and Cuba and further spurred the growth of Puerto Rican rum making. Moreover, the Bacardi family, which left Cuba after the revolution, reestablished its operations in Puerto Rico. In the 1980s, the U.S. government began public hearings on the Caribbean Basin Economic Recovery Act, part of Ronald Reagan's Caribbean Basin Initiative (CBI). CBI sought to bolster the economies of the favored Caribbean countries. The act offered discretionary tariff reductions and tax incentives to Caribbean rum makers and transferred all revenues from rum excise taxes to Puerto Rico and the Virgin Islands.

At the beginning of the twenty-first century, rum continues to foster economic growth in a number of sugarcane-growing regions. Rum controls about 11 percent of the world's spirits market. It represents about 10 percent of all spirits consumed in the United States, putting it in third place behind whiskey and vodka. In addition, sugarcane juice and the byproducts of sugar making are distilled to produce many industrial forms of alcohol, including ethanol for motor fuel.

Frederick H. Smith

See also: Australia; Caribbean; India; National Prohibition (United States); Processed Sugar Ferments

References

Davis, N. Darnell. "The Etymology of the Word Rum" in *The Journal of the Royal Agricultural and Commercial Society of British Guiana.* 4 (1885): 76–81.

Harlow, V. T. *Colonising Expeditions to the West Indies and Guiana, 1623–1667.* Reproduced in 1967 by permission of the Hakluyt Society from the edition originally published by the society, Kraus Reprint, 1925.

Long, Edward. *History of Jamaica, or The General Survey of the Ancient and Modern State of That Island.* 3 vols. London, 1774.

Rorabaugh, W. J. *The Alcoholic Republic: An American Tradition.* New York: Oxford University Press, 1979.

Schumpeter, Elizabeth. *English Overseas Trade Statistics.* Oxford: Clarendon Press, 1960.

Rush, Benjamin (1746–1813)

A signer of the Declaration of Independence, Benjamin Rush (1746–1813) was surgeon general during the American Revolution and one of the first prominent Americans to speak out against excessive drinking. Born on 24 December 1746 in Byberry, Philadelphia, he was educated at Samuel Finley's academy at Nottingham, Maryland. In 1759, he entered Princeton, and he obtained his bachelor of arts degree a year later. After studying medicine in Philadelphia, he went to Edinburgh for further study, graduating with an M.D. in 1768. At the age of twenty-three, he became the first professor of chemistry in America at the College of Philadelphia and opened his private medical practice. Involved in efforts to stop the slave trade, he began to associate with Thomas Paine, John Adams, Thomas Jefferson, and Benjamin Franklin. After serving as surgeon general of the Middle Department of the Army, he returned to his position and founded the Philadelphia Dispensary to aid the poor. He founded Dickinson College and worked heroically during the yellow fever epidemic that hit the city in 1793. His greatest contribution to the medical field was the reforms he instituted in the care of the mentally ill at the Pennsylvania Hospital, for which he is known as the "Father of American Psychiatry."

Rush's thought on drinking was influenced by the Quaker intellectual Anthony Benezet's pamphlet "The Mighty Destroyer

Benjamin Rush (1746–1813) (Library of Congress)

Displayed" (1774), which exposed the detrimental effects of alcohol. In 1784, Rush published his "moral thermometer" in *An Inquiry into the Effects of Ardent Spirits Upon the Human Body and Mind.* This visual depiction of the horrors that awaited drunks energized the U.S. temperance movement and sold more than 170,000 copies by 1850. It showed the reader that as dependence on alcohol increased, the drinker became vulnerable to more and more threatening diseases, vices, and punishments. Those who only drank nonalcoholic drinks such as water were shown to be healthy, wealthy, in possession of a serenity of mind, and likely to live a long and happy life. Moderate drinkers of cider, wine, port, or strong beer taken only with meals were strengthened and gained nourishment while remaining cheerful. Yet as drinkers increased intake of punch, toddy, grog, bitter sprits, morning drams, and rum, they left themselves open to vices such as quarreling, lying, fraud, hatred, anarchy, and murder, all of which could end in suicide. They could expect to suffer from gout, tremors, sore legs, jaundice, dropsy, epilepsy, palsy, and death. Drunks could count on facing debt, rage, and hunger, winding up in the almshouse, the poorhouse, or jail, and finishing at the gallows. Rush believed that strong drink corrupted republican virtue and thereby threatened American freedom. He carried on his "one-man crusade" against alcohol and tobacco until the end of his life.

Rush died on 19 April 1813 in Philadelphia, still believing that by the twentieth century the use of spirits would be "as uncommon in families as a drink made of solution of arsenic or decoction of hemlock" (Hawke 1971: 370–371).

T. Jason Soderstrum

References

Binger, Carl Alfred Lanning. *Revolutionary Doctor: Benjamin Rush, 1746–1813.* New York: Norton, 1966.

D'Elia, Donald J. *Benjamin Rush, Philosopher of the American Revolution.* Philadelphia: American Philosophical Society, 1974.

Hawke, David Freeman. *Benjamin Rush: Revolutionary Gadfly.* Indianapolis: Bobbs-Merrill, 1971.

Rush, Benjamin, "An Inquiry into the Efffects of Ardent Spirits upon the human Body and Mind, with an Account of the Means of Preventing, and the Remedies for Curing Them" in *medical Inquiries and Observations,* 4 vols. Philadelphia: Anthony Finley, 1819, 2: 151–176.

Russell, Howard Hyde (1855–1946)

Howard Hyde Russell (1855–1946) founded the Anti-Saloon League of America (ASLA) in 1893 and served its cause for the rest of his life. Russell had several careers. He began adulthood as a teacher, worked for a short while as a newspaper editor, and then became an attorney, building a successful practice in Corning, Iowa. He married Lillian Davis and under her influence gave up drinking and vowed to become a clergyman. In 1888, Russell graduated from the seminary at Oberlin College, where he helped lead a campaign for local prohibition laws in Ohio. Infused with the temperance spirit, Russell embarked on a career in the ministry of the Congregational Church, working first in the stockyard district of Kansas City before accepting a call to the Armour Mission in Chicago, again to serve stockyard workers and their families. Russell left the mission in 1893, met with friends in Oberlin, and founded the Anti-Saloon League of Ohio in 1893.

An effective public speaker and an inspiring, visionary leader, Russell's special goal was to bring professional and bureaucratic efficiencies to prohibition work. He and his friends at Oberlin had marveled at the effectiveness of the nation's new large businesses, and they dreamed of building a reform organization on the model of a corporation. After graduation, Russell observed large meat-packers, then among the most innovative and effective organizations. In 1893, he set out to build the Anti-Saloon League as an ongoing organization capable of dedicating professional expertise to enacting prohibition laws. Although personally a committed Republican, Russell believed that a nonpartisan approach, such as advocated by the Non-Partisan Woman's Christian Temperance Union (WCTU), was most likely to win victories. He set out to change Ohio and soon, in 1895, inspired other reformers to construct a national organization.

Russell's leadership was most important to the league in the first decade of its life. The ASLA enjoyed some success in changing Ohio's liquor regulations, making it easier for citizens to outlaw the liquor trades in local areas. Nevertheless, survival was a constant struggle. Russell was effective in persuading congregations to donate funds, and he befriended John D. Rockefeller, whose donations came at critical times in the early years.

Russell recruited able men to temperance work, including Purley Baker and Wayne Wheeler. He was himself less suited to the day-to-day work of running an organization, however, and he left the national organization in Baker's charge in 1903. Thereafter, Russell served the ASLA in various capacities. He took over the troubled organization in New York in 1901, then the largest and politically most powerful state; the league was never successful in controlling New York politics, however. For the rest of his active years, Russell traveled widely to speak for the national organization and helped its leaders raise funds. Russell founded the Lincoln-Lee Legion in 1903 as a league affiliate to encourage people to pledge to abstain from consuming alcoholic beverages. In 1909, he was named associate superintendent; thereafter he was closely associated with the national officers, although never again commanding the organization.

K. Austin Kerr

See also: Anti-Saloon League of America (ASLA); Baker, Purley Albert; Non-Partisan Woman's Christian Temperance Union; Wheeler, Wayne Bidwell

References

Blocker, Jack S., Jr. *Retreat from Reform: The Prohibition Movement in the United States, 1890–1913.* Westport, CT: Greenwood, 1976.

Chalfant, Harry Malcolm. *These Agitators and Their Idea.* Nashville, TN: Cokesbury Press, 1931.

Kerr, K. Austin. *Organized for Prohibition: A New History of the Anti-Saloon League.* New Haven, CT: Yale University Press, 1985.

Russia, Alcohol Consumption in

See Alcohol, Consumption of (Russia); Soviet Union and Russia since 1917, Alcohol and Temperance in

Russia, Imperial, Temperance in

Arising initially in Western Europe and the United States in the late eighteenth and early part of the nineteenth century, the issue of temperance (that is, the partial or total abstinence from the consumption of alcoholic beverages) did not reach Russia and did not become a matter of civic concern until the last quarter of the century. Although Russian clerics had inveighed against drunkenness (*p'ianstvo*) since the tenth century, such abuse of alcohol was seen primarily as a vice (*porok*) and not as an illness. The earliest manifestation of temperance concern arose in Russia's Baltic Provinces, especially those where Lutheranism was most widespread. The ethnically Russian provinces did not see any manifestation of an appreciable temperance concern until the late 1850s and early 1860s; temperance development there was slowed to a considerable extent by the Russian state's heavy financial profit deriving from the sale of alcoholic beverages in general and vodka in particular. These temperance efforts, however, tended to be confined to individual landlords who, during the period of serfdom, were intent on rooting out drunkenness among their serfs. In addition, the often violent antivodka peasant boycotts in the latter decade were not so much inspired by a real temperance sentiment as by resentment over the relatively high price of the drink.

Origins

Real temperance concern began to emerge in the 1870s through the writings of Russian public-health physicians associated with the *zemstvo* (*zemstva* were local public health councils) reforms of Alexander II. Concerned with the association between drunkenness (p'ianstvo) and ill health and deplorable housing conditions, this sentiment became amplified in the 1880s and the 1890s among some members of the Russian medical community and press. In the 1890s, Russian psychiatrists, lawyers, public-health physicians, and other members of Russian civil society became increasingly involved in the issue—especially after the creation in 1894 of the state monopoly on vodka sales, which prohibited the sale of vodka except from specially designated state shops and in bottles of a state-specified volume and affixed with a state-issued stamp.

Alcohol Commission of the Committee for Public Health

An additional step in the temperance direction was the creation in 1898 of the Alcohol Commission of the Committee for Public Health. A quasi-official but largely civil grouping of individuals in the professions noted above—including teachers at all levels in the Russian secular school system, journalists, and temperance-minded priests of the Russian Orthodox Church—the commission continued through the outbreak of World War I in 1914. The commission's debates and sessions during this decade and a half touched on a variety of subjects intimately connected in the participants' minds with questions of alcohol abuse and concomitant alcoholism. Among these subjects were the living conditions of workers as a factor contributing to alcoholism, the role of the workplace in the development of alcoholism, and the specific relationship of alcohol abuse to a wide variety of medical ills and social dysfunctions (including, among other things, epilepsy, various psychoses, neurasthenia, tuberculosis, racial degeneration, violence, hooliganism, and crime). The first two areas of concern were directly influenced by Friedrich Engels's writings, chief among them *The Condition of the English Working Class in 1848,* while the latter concern was heavily influenced by the writings of the Italian Cesare Lombroso and the Frenchman Benedict Morel. Although the Commission did not enact any legislation or play a significant role in the formulation of policy, it was nonetheless important for its role as a forum in which various professionals from Russian civil society could bring to the attention of an educated public the consequences of Russia's "problem" in a more pubic manner and to identify alcoholism (*alkogolizm*) as a distinct illness separate and apart from the traditional concern for mere "drunkenness" (p'ianstvo).

Temperance Societies

Also in the early years of the nineteenth century, a number of temperance societies were springing up in various parts of the Russian Empire. These ran the gamut from church-sponsored groups that swore religious oaths of abstention from alcohol, to school temperance groups. Private (that is to say nonsectarian and noninstitutional) temperance organizations were few except in Russia's heavily German Baltic Provinces and in the Grand Duchy of Finland. These groups were identified in the 1890s by Dr. N. I. Grigor'ev in a pamphlet entitled *Russian Temperance Societies. Their Organization and Activity in the Years 1892–1893* (*Russkie obshchestva trezvosti. Ikh organizatsiia i deiatel'nost v 1892–1893*). The few that were important in terms of sheer numbers of members were either church-connected, for example the Kazan Temperance Society and Saint Petersburg's Alexander Nevskii Temperance Society, or were closely associated with the Russian state's temperance efforts, launched after the introduction of the vodka monopoly in 1894 under the title of Guardianships of Popular Sobriety (Popechitel'stva o Narodnoi Trezvosti). Whatever their association, however, each and every temperance group had perforce to receive permission to function from the Ministry of the Interior, which demanded of them as a requirement for certification a statement of their aims, the "regulations" that guided their activities, and their various levels of membership. Thus, there were a number of the activities of both church- and state-based societies that were similar in inspiration and execution. Since one of the assigned missions of the guardianships at the imperial, provincial, and local levels was to render financial and other support to other temperance groups, church and state societies can also be seen as mutually reinforcing.

Activities of Temperance Societies

Among the many activities sponsored by both church- and state-based societies were the organization of tea and reading rooms as alternative gathering sites for workers and their families instead of taverns or, in the case of the men, on street corners near the state-run vodka shops. In addition, frequent offerings included lectures, public readings, the organization of choral groups, and the presentation of choral performances of religious and folk songs, as well as theatrical presentations and "living tableaux" of historical and religious events. The latter in particular were sources of conflict between members of the clergy and the secular intelligentsia involved in such temperance activities. Some offered free legal advice, others maintained something resembling labor exchanges, where peasants newly arrived in cities and towns, or unemployed workers, could find referrals to available jobs. All sought to offer both peasants and workers advice on personal hygiene, thrift, and budgetary practices. Others organized special activities for women—sewing cooperatives or job referral services to young peasant women newly arrived in urban centers to prevent them from falling into prostitution. Ironically, however, aside from a mere handful of activists—the most notable of whom was Elizaveta Chebyshëva-Dmitrievna—women did not play a significant role in these institutions, exerting, certainly, nowhere near the leadership and influence of their American and British sisters. The Kazan and the Alexander Nevskii temperance societies offered religious outreach, whereas the guardianships did not involve themselves directly in religious instruction, although many church officials at various government levels were on their administrative boards. This avoidance of religious instruction would, by 1914, come to be an additional source of friction between the clergy and the secular elements of the temperance movement.

Just as important as the activities and assistance offered to abstainers were the efforts by the guardianships and other temperance groups organized in the empire's major cities to provide medical assistance to men and women who regularly abused alcohol. Ambulatory clinics were thus established by the Saint Petersburg Guardianship of Popular Sobriety and the Kazan Temperance Society. The former in particular witnessed an exponential growth in these institutions between 1894 and 1914, when there were seven clinics scattered throughout the capital. In addition to providing diagnoses, constructing detailed questionnaires about their patients that would provide the first real picture of the landscape of Russian alcoholism, and treating patients with a variety of medicinal and dietary methods, the Saint Petersburg Guardianship clinics, under the direction of Dr. A. Iakov Mendel'ssohn, used hypnosis or hypnotic suggestion as a method of building the individual's resistance to the lure of alcohol. A similar approach was used by one of Russia's leading psychiatrist-neurologists, Dr. V. M. Bekhterev, at his outpatient clinic located in Saint Petersburg's prestigious Military-Medical Academy. Indeed, he published the results of his works not only in Russian, but in German as well. The problem of alcohol abuse and the need for temperance in the Russian army and navy was equally a subject of concern—especially after the military debacle of Russia's armed forces in the Russo-Japanese War of 1904 through 1906. A particular target of temperance advocates both in the Alcoholism Commission and other groups was the vodka ration issued periodically to Russian troops in order to commemorate regimental holidays, and royal birthdays and name days. A similar ration was issued to members of the fleet on "hazardous duty," cold weather, and long voyages. Finally, in an effort to address concretely the problems that temperance advocates considered to be conducive to the development of alcohol abuse, the Commission on Alcoholism formed an entity headed by the jurist D. A. Dril' to construct "healthful" and clean workers' residences in Saint Petersburg's Vasiliiostrovsk region in 1911 and 1912, an effort that led to the successful construction of five such units.

The Orthodox Church

Church-related temperance groups largely endeavored to provide distractions for workers in Russia's major urban industrial centers by conducting pilgrimages and outings to monasteries, cathedrals, and other holy shrines in order to promote solidarity among nondrinking workers and to keep them in contact with the roots of their faith. The church also jealously guarded its temperance role, and this territoriality sparked a minor scandal in the years just before the outbreak of World War I, when the church hierarchy conducted an ultimately successful campaign against several lay brothers who attempted to invest their efforts at rescuing alcoholic Russians in both Saint Petersburg and Moscow with religious trappings. The church's condemnation and punishment of the "brothers" Ivan Churikov in Saint Petersburg and Dmitrii Gregor'ev and Ivan Koloskov in Moscow, in fact, aroused the ire of Russia's Duma liberals and moderates alike. The combined incidents were exemplary of the growing tension between secular temperance advocates and those in the church—a phenomenon unlike temperance movements in Great Britain and the United States. Indeed, after the 1909–1910 All Russian Temperance Congress held in Saint Petersburg, the Church aggressively pursued its own independent temperance line, a direction that culminated in the 1911 Congress of Anti-Alcohol Activists in Moscow and conducted exclusively under the aegis of the church.

The Position before World War I

Ironically, both of these congresses, long viewed by Russian temperance advocates as *desiderata* in bringing the questions of alcoholism and temperance into the Russian public forum, addressed many of the issues noted above as representative of the efforts of the movement in general, but they were so divided in terms of their respective approaches to achieving these goals as to almost constitute two separate phenomena. Nonetheless, the combined secular and ecclesiastical efforts produced a rich stream of temperance pamphlets, broadsides, newspapers, and journals in the years between 1894 and 1914. Whatever their mutual differences between secular and ecclesiastical approaches on the one hand and the division between proponents of "moderate consumption" (*umerennoe potreblenie*) and "teetotalers" (*polnoe vozderzhani*) on the other, however, most temperance advocates agreed on the deleterious effect that the bureaucratic composition of the Guardianships of Popular Sobriety had on wide-ranging temperance initiatives—especially after 1907, when fear of the revolutionary fervor of the preceding years led the imperial government sharply to curtail the revenues that it channeled to these and other temperance organizations. This situation was further complicated by the fact that, until 1914, at least a third of all state revenues came from its sale of vodka.

Impact of World War I

The issue of total abstinence, and by association the temperance issue, was resolved temporarily with the outbreak of war in 1914 and the declaration of Prohibition by the Russian government—anticipating the American effort by a full six years. Justifying the action on the moral grounds so frequently initially advanced by all participants in this conflict, the Russian government had the more basic reason of avoiding the alcohol abuse that had characterized the conduct of mobilized soldiers in the war with Japan just ten years previously. Thus, initial Prohibition for the duration of mobilization was followed by Prohibition for the duration of the war and, finally, total Prohibition to extend to the postwar period. Whether such an effort would have succeeded is problematic; but the Bolshevik seizure of power in November 1917 rendered the point moot. There is persuasive evidence, however, that the level of bootlegging, illegal distillation, and the abuse of surrogates of ethyl alcohol had grown with each passing year of the war. Thus, efforts in Russia toward prohibitory measures yielded the same mixed achievements and ultimate failure as the longer American effort from 1920 to 1933.

George Snow

See also: Alcohol, Consumption of (Russia); Soviet Union and Russia since 1917, Alcohol and Temperance in; Vodka

References

Christian, David. *Living Water: Vodka and Russian Society on the Eve of Emancipation.* New York: Oxford University Press, 1990.

Herlihy, Patricia. *The Alcoholic Empire: Vodka and Politics in Later Imperial Russia.* New York: Oxford University Press, 2002.

Hutchinson, John. "Medicine, Morality and Social Policy in Imperial Russia: The Early Years of the Alcohol Commission" in *Social History.* 7 (November 1974): 201–225.

———. "Science, Politics and the Alcohol Problem in Post-1905 Russia" in *Slavonic and East European Review.* 58 (April 1990): 232–254.

Snow, George E. "Alcoholism in the Russian Military: The Public Sphere and the Temperance Cause" in *Jahrbücher für Geschichte Osteuropas.* 45, no. 3 (1997): 417–433.

———. "Drinking and Drunkenness in Russia and the Soviet Union. A Review Essay" in *Social History of Alcohol Review.* 17 (Spring 1988): 7–15.

———. "Perceptions of the Link between Alcoholism and Crime in Pre-Revolutionary Russia" in *Criminal Justice History.* 8 (1988): 37–51.

———. "Socialism, Alcoholism and the Working Classes before 1917" in *Drinking. Behavior and Belief in Modern History,*

edited by Susanna Barrows and Robin Room, 243–264. Berkeley, CA: University of California Press, 1991.
———. "Temperance Materials in Russian Archives" in *Social History of Alcohol Review.* 30/31 (Spring/Summer 1995): 25–27.

Rutgers Center of Alcohol Studies

The Yale Center of Alcohol Studies relocated to Rutgers University in 1962 after a ten-year struggle with a Yale administration that sought to divest itself of applied research projects. The National Institute of Mental Health (NIMH) provided support for the transition, and philanthropist R. Brinkley Smithers helped fund construction of a new facility on the university's science campus in Piscataway, New Jersey. The *Journal of Studies on Alcohol* relocated with the center, and the center's researchers became Rutgers University faculty. The center continued its basic and applied research on the nature and consequences of alcohol use, gradually broadening its scope to encompass other psychoactive substances. In addition, the center created a new division for clinical research and services. It also expanded efforts to disseminate research and clinical information through a variety of professional education programs and information services.

Center research has included a wide range of projects in the biological sciences (especially biochemistry and neuropharmacology), clinical and experimental psychology, and sociology. In 1978, the center became one of the first institutes to receive an Alcohol Research Center award from the National Institute on Alcohol Abuse and Alcoholism (NIAAA) to conduct a longitudinal study of adolescent drinking patterns. The study has continued as the Health and Human Development Project with additional grants from NIAAA and the National Institute on Drug Abuse. In 1983, the center established a new division for clinical research, outpatient services, and clinical training of graduate and professional students. A second Alcohol Research Center award initiated a major study of alcoholism treatment outcomes in 1991.

Another substantial Smithers gift to the Center in 1987 established a new division for prevention research and expanded the center's physical facilities. As the research programs grew, so did documentation and dissemination efforts. The *Journal of Studies on Alcohol* expanded from a quarterly to a monthly publication in 1975, and the center's library developed one of the world's foremost collections of alcohol research material. The center became the first to offer a commercially available on-line alcohol research database in 1978; however, the venture proved economically unviable and was short-lived. For many years, federal grants from NIMH supported most of the center's publications program, including its journal and abstracts, as well as the library. Support shifted to NIAAA when that agency was created in 1971. In 1982, however, NIAAA support for the documentation activities ended. The *Journal of Studies on Alcohol* discontinued its abstracts of published literature and became a bimonthly publication with original research articles only. Rutgers pledged support for the center's library, and in 2000 the library offered its alcohol research database on-line via the Internet.

The Summer School of Alcohol Studies, which began at Yale in 1943, also moved to Rutgers with the center and continued its programs to train professionals on substance abuse issues. The education and training division also offered refresher courses for alumni and developed special programs for physicians, medical students, and school nurses. In 1976, state funding helped create the New Jersey Summer School of Alcohol and Drug Abuse Studies to provide introductory training for new substance abuse personnel. A series of one-day continuing education workshops followed, and in the late 1990s the center inaugurated a periodic summer training program outside the United States. By that time, however, increased competition for students and funding had forced reductions in the center's training programs, and the annual summer school had been reduced from three weeks to one.

Sociologist Selden Bacon, who negotiated the center's move from Yale to Rutgers, remained as director until 1975, when he was succeeded by psychologist John A. Carpenter, who had also begun his research career at Yale. They were the last two alumni of the Yale program to head the center. Psychologist Peter E. Nathan of Rutgers became director in 1983, followed by psychologist Robert J. Pandina in 1993.

Penny B. Page

See also: Bacon, Selden Daskam; Christopher D. Smithers Foundation; *Journal of Studies on Alcohol;* Keller, Mark; National Institute on Alcohol Abuse and Alcoholism (NIAAA); Yale Center of Alcohol Studies

References

Milgram, Gail Gleason. "The Summer School of Alcohol Studies: An Historical and Interpretive Review" in *Alcohol Interventions: Historical and Sociocultural Approaches,* edited by David L. Strug et al., 59–74. New York: Haworth Press, 1986.
Nathan, Peter E. "Reports from the Research Centres. B1. Rutgers: The Center of Alcohol Studies" in *British Journal of Addiction.* 82, no. 8 (1987): 833–840.

Rutherford, Derek (1939–)

Derek Rutherford (1939–) is an English temperance reformer who during the late twentieth and early twenty-first centuries served various antialcohol organizations. Recognizing the decline of the temperance movement in Europe and North America, he sought allies in the public health movement and found hope for reform in the agreement among abstainers and moderate drinkers on the need to attack alcohol-related problems. Abandoning the traditional temperance program of prohibition, he argued for policies based on scientific research that were politically practicable and culturally sensitive.

Although not from an old temperance family, Rutherford at the age of ten joined a children's auxiliary of the International Order of Good Templars (IOGT) in a mining village in northeastern England. After receiving an honors degree in theology from the University of Leeds and postgraduate certification in education from the University of London, he taught in secondary schools for several years.

Rutherford left teaching for sobriety work in 1969. He became education officer for the United Kingdom Temperance Alliance and soon was also founding director of the Teachers Advisory Council on Drug Education (now known as TACADE). In 1973, he was appointed director of the National Council on Alcoholism. With the help of public funds from the Department of Health, he expanded the number of alcohol information and advice centers in England and Wales from six in 1969 to forty-four in 1982, when he left the National Council. He also established a national voluntary counseling training scheme. In 1977, he wrote the first report on alcohol and work, and from 1975 to 1979 he was a member of a governmental advisory committee on alcoholism.

Rutherford returned to the United Kingdom Temperance Alliance as its chief executive in 1982. In collaboration with Andrew McNeill, formerly industrial officer at the National Council on Alcoholism, he organized the educational Institute of Alcohol Studies in 1983. Rutherford started to organize Eurocare, an alliance of voluntary, nongovernmental organizations representing a diversity of views and cultural attitudes, in 1989. Eurocare was concerned with the impact of the new European Union on alcohol policy in its member states. The European Commission helped finance the Eurocare report on *Alcohol Problems in the Family: A Report to the European Union* (1998). A number of Eurocare reports (for instance, *Marketing Alcohol to Young People,* published in 2001) have influenced European Union policy.

In 1990, Rutherford became honorary secretary of the IOGT International, the world headquarters of the Good Templar organization, and he became its executive director in 1998. He retired from IOGT International at the end of 2002.

With the help of public health advocates in the United States, Rutherford helped organize the Global Alcohol Policy Alliance. In 2000, the alliance held a preliminary conference at Syracuse, New York, the birthplace of the IOGT, and soon afterward it set up offices in London and Washington. With focus on the developing world, the alliance hopes to respond to the marketing strategies of the international alcoholic beverages industry. A magazine produced by the Institute of Alcohol Studies for the IOGT International, called the *Globe,* became the organ of the Global Alcohol Policy Alliance. Rutherford has served as a magistrate and has been active in the Labour Party.

David M. Fahey

See also: Good Templars (IOGT); Institute of Alcohol Studies (IAS); IOGT International; United Kingdom Temperance Alliance (UKTA)

Reference

Rutherford, Derek. "The Drinks Cabinet: UK Alcohol Policy" in *Contemporary Record.* 5, no. 3 (1991): 450–467.

Rye

See Whiskies

S

Sabin, Pauline Morton (1887–1955)

Pauline Morton Sabin (1887–1955) was founder and president of the Women's Organization for National Prohibition Reform (WONPR). Her efforts on behalf of Prohibition repeal represented the most notable stage of a long public career.

Sabin spent her life in an environment of affluence and politics. Her grandfather, J. Sterling Morton, founded Morton Salt and served as governor of Nebraska, U.S. senator, and secretary of agriculture. Her father, Paul Morton, a railroad and insurance executive, became secretary of the navy. She grew up in Washington, D.C., with Theodore Roosevelt's daughter Alice as a friend. After a first marriage ended in divorce, she wed Charles H. Sabin, president of Guaranty Trust Company of New York. Independent and energetic, she chose involvement in the Republican Party of her father and grandfather rather than the Democratic Party favored by her husband. She was a delegate to three Republican National Conventions, founded the Women's National Republican Club in 1921, served as New York State's first female member of the Republican National Committee, managed New York political campaigns for Calvin Coolidge, James Wadsworth, Herbert Hoover, and Fiorello La Guardia, and earned a reputation as an extraordinary political fund-raiser. Later, she would join the executive committee of the American Liberty League, direct volunteer activities for the American Red Cross, and consult on Harry S. Truman's redecoration of the White House. None of these activities, however, brought her the attention she gained in her campaign for Prohibition repeal.

Pauline Sabin began criticizing Prohibition in 1926 while campaigning for the reelection of New York Senator James Wadsworth, an opponent of the Eighteenth Amendment. After Wadsworth's loss, she began complaining that the liquor issue was diverting and corrupting public officials as well as putting excessive power into government hands. Also, she charged, rather than protecting children from temptation, the widely violated alcohol ban was causing them to grow up without respect for law and the Constitution. In an April 3, 1929, speech at a Women's National Republican Club luncheon in her honor, Sabin denounced the new Hoover administration's support for Prohibition, announced her resignation from the Republican National Committee, and declared her intention to work against Prohibition. Rather than join the established Association Against the Prohibition Amendment, in which her husband was active, she set out to create an independent women's anti-Prohibition organization.

Employing her extensive social and political contacts, Pauline Sabin quickly brought together a group of women of similar backgrounds and formed the WONPR. Stylish, photogenic, well-spoken, and experienced in dealing with the press, Sabin proved effective at generating publicity about women's opposition to Prohibition. Magazines such as *Vogue* and *The New Yorker* carried feature stories about her crusade against the "dry law," drawing more women to the rapidly growing organization. When *Time* put Sabin on its cover on the eve of the 1932 national political conventions, any lingering image of unified female support for Prohibition disappeared. Sabin's importance in the successful repeal crusade was acknowledged by her election to the New York State convention to ratify the repeal amendment and her selection to present the resolution by which New York voted to end National Prohibition.

David E. Kyvig

See also: Association Against the Prohibition Amendment (AAPA); National Prohibition (United States); Women's Organization for National Prohibition Reform (WONPR)

References

Kyvig, David E. "Pauline Morton Sabin" in *Notable American Women: The Modern Period,* edited by Barbara Sicherman and Carol Hurd Green, 617–618. Cambridge: Harvard University Press, 1980.

Murdock, Catherine G. *Domesticating Drink: Women, Men, and Alcohol, 1870–1940.* Baltimore: Johns Hopkins University Press, 1998.

Rose, Kenneth D. *American Women and the Repeal of Prohibition.* New York: New York University Press, 1996.

Sake

Sake (pronounced "sah-keh") is internationally identified as *the* Japanese alcoholic beverage, but it constitutes only about 12 percent of the volume of alcoholic beverages sold in Japan (Kokuzeicho 1997: 16). The common translation of sake as "rice wine" probably derives from its appearance, which is rather like white wine: clear, colorless or pale amber, and noneffervescent. The standard alcohol content of sake, at about 16 percent, is also closer to the 12 percent of wine than it is to the 5 percent of beer.

In terms of production, however, sake is closer to being "rice beer." Alcoholic beverages are produced through fermentation. Fermentation is simply the process in which yeasts consume sugar and give off alcohol and carbon dioxide. Wines are made from fruits, which are naturally high in sugar. Unlike wine, beer and sake are made from grains, and grains have little sugar. Instead, grains are rich in starches, more complex molecules that must be broken down into sugars—a process called saccharification—for fermentation to take place. In the case of beer, barley is germinated, cracked, and mixed with water, then naturally occurring enzymes transform starches into malt sugar. After saccharification, yeast is introduced to begin fermentation. In the earliest Japanese brewing, saccharification of rice was brought about by chewing and spitting out the rice so that diastatic enzymes in saliva would break down the starches. The modern technique for carrying out saccharification, however, is the inoculation of the rice with a *koji* mold *(Aspergillum oryzae)* along with the yeast. Unlike the sequential saccharification-fermentation of beer, the sake processes run in parallel, the yeast creating alcohol as the koji mold creates the sugars.

Modern sake begins when short-grain brown rice *(Oryza japonica)* is polished to white rice, removing the husk and surface fats, amino acids, and minerals, which would disturb the clarity and the flavor of the final product, leaving a core that approaches pure starch. Most sake rice is reduced about 30 percent, but more polishing is associated with superior quality and premium brewers reduce their rice by as much as 65 percent. After polishing, the rice is washed, soaked, steamed, and cooled. Koji mold is then mixed with about 20 percent of the steamed rice (Kondo 1984: 45). After about two days of allowing the mold to develop, the producer creates a starter mash by mixing some of the mold rice with water, more steamed rice, and yeast. The yeast in the starter reaches its maximum strength after about two weeks. At that time, additional large quantities of steamed rice, koji rice, and water are added to produce the final mash. The final mash is kept cool to prevent the growth of undesirable bacteria, as well as to slow the action of the mold, thus allowing optimum growth of the yeast. After two to three weeks, the alcohol content of the mash has reached such a level that yeast growth is inhibited and fermentation is deemed complete. The lees are strained from the mash by pouring the mash into cloth bags and then pressing the bags. The strained sake is chilled and allowed to sit for another ten days while enzymatic changes continue to occur. The sake may be filtered at this point by mixing in and then removing powdered, activated charcoal. Next, most sake is pasteurized by briefly raising the temperature to 140°F (60°C), which kills undesirable bacteria and wild yeasts and denatures enzymes that destroy desired qualities. The sake is then stored in large vats for six to twelve months while it further mellows. As with beer, and unlike wine, sake has a short aging period; sake is considered best in the year following brewing. (The short aging of Japanese sake distinguishes it from its Chinese counterparts, which are generally considered to be better the longer they have been aged and may, therefore, be aged for decades.) Finally, before bottling, different batches of sake may be blended together to try to achieve a consistent product. Because top-fermenting ale yeast *(Saccharomyces cerevisiae)* is used, and because of the slow release of sugars by the mold, an alcohol content of about 20 percent is reached, so water may be added to adjust the alcohol level to about 16 percent. Some brewmasters, however, choose to add alcohol to lighten the taste or brighten the aroma of their sake.

There are a number of standard categories of sake. *Jummaishu,* or "pure rice sake," is the legal designation for sake made with no other ingredients than rice, water, mold rice, and yeast. The taste tends to be heavy. When brewer's alcohol is added to the mash to get a milder flavor, the sake is known as *honjozukuri.* Sake that is not diluted with water to bring it to 16 percent alcohol is called *genshu.* The premium designation of *ginjozukuri* is reserved for only that sake made of rice that has had at least 40 percent polished away (Gauntner 1997: 49). If the sake is not strained, or is only partially filtered, it is classed as *nigorizake,* or "cloudy sake." Unpasteurized sake—which must be drunk soon after uncorking—is known as "raw sake" *(namazake).* And when sake is aged more than a year, it becomes what is known as "old sake" *(koshu).*

There are more than 2,000 breweries in Japan, and brewmasters have a great number of options to choose from in materials, timing, and temperature. Thus, not only are there major sake brands with national and international distribution, there is also a great array of small local brews *(jizake).* Sake, like wine, is understood to have many different possible characteristics. The familiar qualities of sweet and dry are found in sake, as well as bitterness, astringency, and acidity. Sake connoisseurs look for hints of grain, yeast, and mold while trying to poetically articulate ineffable tastes analogous to woods and metals, fruits and flowers. Likewise, those in the know can speak extensively on the visual appeal and the mouth qualities of sake.

Sake is famous in the West for being served hot in 0.18 liter (6 fluid ounces) flasks and drunk from thimble sized cups *(sakazuki).* This, however, is certainly not the only way that sake is consumed. Hot sake is also drunk from larger, teacup-sized cups by drinkers who wish to forgo the usual etiquette that surrounds the pouring of drinks. Sake is also drunk at room temperature and chilled. When not served hot, drinking from a small tumbler or a lowball glass is the common practice. Another special kind of sake is known as *taruzake,* or "barrel sake," and is stored in cyprus wood casks. The wood imparts a pleasant aroma to the beverage, which is reinforced by drinking from a wooden box. When drinking from a box one places a pinch of salt on the edge of the box or the back of the hand, and little tastes of the contrasting salt are taken as the sake is sipped. An unusual variant of hot sake is *hirezake* (literally "fin sake") in which the fin of a blowfish or a red snapper is grilled and dropped into a glass of sake to steep.

Stephen Smith

See also: Japan

References

Gauntner, John. *The Saké Handbook.* North Clarendon, VT, and Tokyo: Yenbooks, Charles E. Tuttle, 1997.

Kokuzeicho. *Dai 123 kai Kokuzeicho Tōkei Nenpōsho* (The 123rd National Tax Office statistical report). Tokyo: Kokuzeicho Kanzeibu Shuzeika (National Tax Office, Indirect Tax Bureau, Alcohol Tax Section), 1997.

Kondo, Hiroshi. *Sake: A Drinker's Guide.* Tokyo, New York, and San Francisco: Kodansha International, 1984.

Saloons and Taverns (United States)

Alcohol has been a part of American culture since the arrival of the first colonists. Taverns were among the first permanent structures built and among the most important buildings in the colonies; in fact, often the tavern was a colony's only public building in the early years. It continued to be an important gathering place in communities across America up until the Industrial Revolution, when it evolved into the urban saloon, an institution that generated intense debates during the late nineteenth and early twentieth centuries, culminating in the "noble experiment" of National Prohibition. Today, taverns are redefining their roles and attempting to find a place in modern society.

Over the years, there have been many types of taverns, generally defined by their customers and serving several distinct functions. Changing roles are evident in the three eras of the American tavern. The colonial era was marked by local taverns and inns serving mainly as community social and business centers as well as hotels. It lasted until the middle of the nineteenth century. The "Saloon Era," marked by increasing class and gender separation, found the urban tavern among the most controversial institutions in America. It lasted from about 1870 until the coming of National Prohibition. The modern era started after the repeal of National Prohibition and has been a time of constant redefinition of the tavern's functions and forms.

Although drinking has always been the most important element in making America's taverns and saloons profitable, in all eras taverns have been more than just drinking establishments. They have been community centers, a place for Americans to gather, socialize, and build the ties that bind communities. At times, as community centers they have defined particular eras and segments of society itself. For example, the western saloon is the most common image of the cattle towns of the late nineteenth century. Likewise, discos defined an era.

The Colonial Era

Colonial villages clustered around taverns, especially in the New England and Middle Atlantic colonies. In the southern colonies, where urban places were less common, taverns were generally erected only when a sizable community developed. In New England and the Middle Atlantic colonies, taverns served a public function as well as a private one. In addition to providing a place for colonists to gather, gossip, socialize, and argue about politics, they often offered overnight rooms for visitors. This gave colonial taverns—often called inns—a key role in developing early business and commerce. They also provided a public space for meetings, trials, and even some religious services before specific structures for those functions could be built. The size, location, and functions of village taverns differed according to the type of colonist, with ethnicity and religious preference affecting tavern development. Colonial taverns were visited throughout the day, all week long, except for the Sabbath. This was an era of so-called "dram drinking," with colonists drinking small amounts of alcoholic beverages several times a day. Tavern-keepers were respected within the communities and often were among the most financially successful citizens.

Taverns retained their importance as the colonies merged into states and the new United States became more financially stable, offering a reliable currency and a citizenry with more money to spend. American drinking tastes changed. German immigrants introduced lager style beers, which were more difficult to brew than the ales and ciders that virtually all colonists made at home. Visits to the local tavern were still somewhat of a luxury during this period, however. After the American Revolution, the newly prosperous citizens of the new republic began to drink their beers, rums, and ciders in public, increasing the tavern's role as defining culture. Americans building a new society found the public space important for discussion of the issues of the day. Political matters were favorite subjects of long, loud, and frequent discussions in taverns in the first half of the nineteenth century.

The Saloon Era

The Industrial Revolution, under way by the mid-nineteenth century, made taverns both more important and more controversial. With industrialization, the style of work changed from one based upon individual efforts and apprenticeships in trades, often located in workers' homes and with flexible hours, to one in which work was performed at specific times at locations away from homes. This separation of work from the family and home resulted in new distinctions among three kinds of spaces: the home, where a worker would spend time with his family; the workplace, where he or she was employed; and a new, third space for socializing. With this change came a new drinking style as well. Whereas before, workers drank small amounts frequently throughout the day, now, with employers discouraging drinking on the job, workers began to drink more after working hours and needed a place to socialize. This was the start of what has come to be known as the Saloon Era.

With continuing urbanization and industrialization, that socializing space became even more important. The term "saloon" developed from the French word *salon,* which was a space for socializing and discussion. As they became more popular, several types of saloons competed for Americans' attention. Just as the workforce was increasingly divided into owners and managers, on one hand, and workers, on the other, saloons evolved into what were called "businessman's clubs" and "workingman's saloons (or clubs)," with the latter far more numerous. The two prevalent types of workingman's saloons were the Yankee drinking saloon and the German *Bier Garten.* These taverns became the models for a wave of saloons that both grew in popularity and sparked the reforms that ended in National Prohibition. The differences between the two types were marked, and the eventual success of the Yankee drinking saloon had profound effects upon American tavern life.

Bier Gartens were an attempt by German immigrants to recreate their village life in the New World. Attempting to emulate German parks, they were generally located at the edges of cities and were usually large, park-like areas filled with entertainment, often including such things as carousels, swimming areas, and even zoos. Today their influences are found in the growing number of theme parks. Family-friendly, these were areas for leisurely drinking and socializing, with entire families gathering at long tables to eat and drink (mostly) beer. Most of the drinking was done at these tables. German Bier Gartens were inclusive, welcoming all ethnicities, ages, classes, and both married and single alike. Indeed, that welcoming attitude of the German Bier Garten was one of the factors in its decline. The growing temperance movement of the late nineteenth century saw the Bier Gartens as a threat because they supposedly induced women and children to drink. The Bier Garten also drew the ire of various church leaders because German immigrants liked to socialize in them on Sunday afternoons, a time church leaders wanted devoted to church activities.

Yankee drinking saloons, in contrast, existed mainly for the purpose of drinking and camaraderie. They almost always featured a long bar where men stood and drank. Some had stools, and a few featured small tables, but almost all the drinking in Yankee saloons was by men standing at the long bars. In contrast to the German Bier Gartens, Yankee saloons quickly evolved into male-only space that was distinctly unfriendly to women. Not only did male customers oppose most women (women entering saloons often would be greeted with catcalls and insults), but saloon decor was distinctly unwelcoming, featuring large painted pictures of sports figures, such as the Irish heavyweight boxing champion John L. Sullivan or baseball star Babe Ruth, and female nudes. To soak up spilled drink, many floors were covered with sawdust, which soiled the long skirts of women. In addition, these saloons often featured gambling and rough athletic contests. They also became tarred because of the frequent presence of prostitutes, and they were known as centers of urban political machines (especially the Irish machines ruling Boston, New York, and Chicago). In short, these saloons frequently were fearsome, unpleasant places.

Nevertheless, the Yankee drinking saloon became the more successful model of the Saloon-Era tavern for several reasons. It was less expensive to build and operate than the more elaborate German Bier Garten. It was more profitable, because of the emphasis on drinking quickly. Bar customs—especially treating, when customers were expected to purchase rounds for others—led to more alcohol consumption than in German Bier Gartens, where drinking was in moderation, and treating, with its obligation to drink at someone else's instigation and the implied obligation to treat in return, stood in direct opposition to the leisurely style of moderate drinking favored by German immigrants. Perhaps most important, though, was the development of a brewing industry that financed "tied-houses," bars with financing contingent upon selling only a particular brewery's products.

The U.S. brewing industry evolved from a system of mostly local breweries and taverns at the end of the Civil War to national "shipping" brewers selling their beer in the new and fast-growing towns along railroad tracks as well as in urban American cities. The brewing industry fed the rapid growth of saloons, with brewery financing supporting the Yankee drinking saloon as the standard type of bar because of the quantities of beer that saloon customers could drink. Thus, by the latter part of the nineteenth century, urban saloons became extremely numerous, with large areas in some cities featuring dozens of taverns within a block of each other. The saloon culture set the stage for the most volatile political issue of the next half century—temperance and prohibition.

The urban workingman's saloon grew out of a need by men to create a space of their own. With industrialization, workingmen found themselves in an especially unstable condition. Although the American ideal was that of independence, the workingman found himself without independence in the two places where he spent the bulk of his time: in the industrial workplace, where supervisors were in charge, and at home, the province of their wives. Therefore they gravitated toward saloons as a location where they could be solely responsible for themselves. The urban saloon developed a male culture where more and more men began to spend more and more time. Women did their drinking in private, often with other women, but in their homes. Beer, wine, and liquor could be purchased at the saloon (most had separate rooms and entrances for women to make those purchases where they could sit and drink themselves).

However, as the saloons became more male-oriented and more popular, they began to attract serious opposition. Women came to view the saloon as competition to their families, a rival for family funds, and a threat to their marriages. Employers, too, began to voice opposition to the drinking customs, which made many men unable to work after a drinking weekend. From the earliest days on, Americans often had drunk throughout the day. This custom was continued into industrialization, with many factories, printing plants, and shops sending "bucket boys" to taverns for buckets of beer. As industrialists grew in strength, they became more opposed to these practices, vocally attacking the saloon culture. Although many localities had passed Sunday closing laws by the turn of the twentieth century, these laws were often openly ignored by saloons. This drew opposition from church leaders. These three elements—wives, businessmen, and church leaders—became united in opposition to the saloon culture, which they linked to crime, poverty,

A saloon in Beowawe, Nevada (Library of Congress)

prostitution, and political graft. Saloons became the symbol of all that was wrong with urban culture.

Prohibition ended the saloon culture in 1920. Although some private clubs remained open, serving liquor legally purchased before implementation of Prohibition, and illegal "speakeasies" offered liquor during Prohibition, the "noble experiment" ended the saloon culture. Following Prohibition, Americans changed their drinking patterns, and taverns changed as well. The term "saloon" had come to be so widely detested that few of the reopened or new taverns used that term. Instead, they hearkened to their neighborhood community function and reverted to the earlier term of "tavern" or called themselves "bars." A new term also emerged: Americans had become more accustomed to drinking cocktails during Prohibition, and thus the respectable and expensive "cocktail lounge" was born.

The Modern Era

The years since National Prohibition have seen American taverns searching for an identity. The neighborhood bar came under pressure from Americans' changing lifestyles, especially the growth of suburbs, which frequently had no place for taverns. As women and younger people have joined older males as public drinkers, taverns sought a new image. No longer dark spaces with pictures of nude women on the walls and a line of men standing at a bar drinking, the tavern of the past half-century has become brighter and more likely to feature popular music and themes intended to draw younger drinkers. Ladies nights, sports bars, nostalgia bars, taverns featuring many imported beers, martini bars, cigar bars: Such innovations are intended to offset another change in American society that directly impacts taverns—a change in drinking habits away from drinking in public. Today, most American drinking is done in private residences, not in taverns. Less than 25 percent of the total of alcoholic beverages drunk in the United States is now consumed in taverns, restaurants, or other public locations. This trend toward drinking in private began with National Prohibition and shows no signs of abating. The number of taverns has declined 40 percent since World War II.

Stephen R. Byers

See also: Juke Joints; Pubs and Beerhouses (United Kingdom); Roadhouses; Saloons, Taverns, Pubs, and Beerhouses (Canada); Shebeens; Shebeens (Southern Africa)

References

Anbinder, Tyler. *Five Points: The 19th-Century New York City Neighborhood That Invented Tap Dance, Stole Elections, and Became the World's Most Notorious Slum.* New York: Free Press, 2001.

Barr, Andrew. *Drink: A Social History of America.* New York: Carroll & Graf, 1999.

Chudacoff, Howard P. *The Age of the Bachelor: Creating an American Subculture.* Princeton, NJ: Princeton University Press, 1999.

Duis, Perry R. *The Saloon: Public Drinking in Chicago and Boston, 1880–1920.* Urbana: University of Illinois Press, 1983.

Erdoes, Richard. *Saloons of the Old West.* New York: Alfred A. Knopf, 1979.

Jersild, Devon. *Happy Hours: Alcohol in a Woman's Life.* New York: Cliff Street Books, 2001.

Lender, Mark Edward, and James Kirby Martin. *Drinking in America: A History.* New York: Free Press, 1987.

Oldenburg, Ray. *The Great Good Place.* New York: Paragon House, 1989.

Powers, Madelon. *Faces along the Bar: Lore and Order in the Workingman's Saloon, 1870–1920.* Chicago: University of Chicago Press, 1998.

Smith, Gregg. *Beer in America: The Early Years, 1587–1840.* New York: Siris Books, 1998.

Saloons, Taverns, Pubs, and Beerhouses (Canada)

Taverns, saloons, pubs, and beerhouses have occupied a prominent place in Canadian society. Much of the social and cultural life of early settlers centered upon taverns, and they continue to perform a valuable function in society to this day. Taverns served a number of functions in frontier society and were often the first structures built in any community. They were used for a myriad of purposes beyond just drinking, including polling stations, courthouses, government balls, militia parades, weddings, banquets, inns, meeting places, and even schools and gathering places for religious functions until such time as proper schools and churches could be built.

Early Years

In the 1830s, Anna Jameson, a pioneer settler from Upper Canada, observed that "taverns and low drinking places" were often the only place of assembly or amusement and "taverns abounded everywhere." The tavern barroom provided a warm fire on a cold night and frequently was the only place for discussion and debate about events of the day. It was travelers who brought news of the outside world to the local inhabitants, since few residents ever saw a newspaper. Taverns, then, served as the place for discussion, debate, and dissent. And, since most meetings or festivities were located in the tavern, it should come as no surprise that alcohol was the beverage of choice.

Most of the alcohol consumed in those days was whiskey, brandy, or rum. There were few imported wines, and beer was not consumed much until the 1840s. Data about consumption are fragmentary and incomplete, but settlers' accounts suggest that heavy drinking seemed to be the rule in pioneer taverns. One observer noted that "immigrants will find every inn and tavern and beer shop filled at all hours with drunken, brawling fellows and the quantity of ardent spirits consumed will truly astonish you." Several factors might explain the prevalence of drinking in the society. First, alcohol was considered safer than water—water had a bad reputation in Europe, where much of it was polluted, and many settlers expected to find the same conditions in North America. Second, alcohol was considered a restorative with medicinal powers. Many believed that liquor eased childbirth, fevers, indigestion, and fatigue, to name but a few ailments. Third, alcohol was seen by many as a source of energy and was therefore particularly appealing to laborers, farmers, and those in the military. Finally, alcohol—especially whiskey—was cheap, and taverns dotted the landscape. And, prior to the advent of railroads virtually all land travel was by horse or by horse and carriage. On a good day a horse could travel twenty-five miles, and roadside inns and taverns were needed every few miles to provide food, drink, rest, and lodging. In the 1840s, a traveler from Toronto to Barrie, a distance of fifty miles, would pass some sixty-six licensed taverns and a number of unlicensed establishments, more than one bar per mile. Perhaps the best evidence of the importance of alcohol in the early history of Canada was found in Toronto. In 1846, the city had 500 beer shops, 200 liquor stores, and 150 unlicensed taverns for a population of 20,000 (850 establishments in all). Or, viewed another way, Toronto contained one establishment for every twenty-three people. This is not to say, however, that drunkenness characterized pioneer society, for not everyone drank every day.

During those early days, taverns appealed to the poor and well-to-do alike. Liquor, and specifically taverns, served to bring the classes together in one place. For the poor, taverns provided many comforts not found in their own homes, such as heat, light, furniture, and newspapers. Men of prestige were also heavy imbibers. Magistrates, who often held court sessions in taverns, frequently drank liquor during the proceedings. Even clergy frequented the taverns, since no religious denomination required abstinence before 1830. As Joseph Gusfield (1987) observed, in North American culture alcohol acted as a symbol of leisure and served as a disinhibiter breaking down notions of hierarchy and structure. Equally important was the tradition of "treating," a ritual that required patrons—usually males—to buy rounds for their friends, regardless of the size of the group or their socioeconomic standing. The buying of rounds was an expression of hospitality and "manliness," and this ritual is firmly entrenched to this very day. This practice increased consumption. Most early regulations dealt with license fees and their collection, not with drunkenness or other controls on consumption. Taverns were usually open seven days a week and, in some cases, twenty-four hours a day. By the 1830s, the behavior of tavern patrons drew the ire of the temperance movement across Canada. The movement, at least at the outset, became more "antitavern" than "antidrinking."

By the 1850s, the temperance movement had redirected its efforts to advocate complete abstinence and pressured local and provincial governments to regulate the taverns. Legislation included doubling the license fees, so as, reformers hoped, to cut down on the number of taverns, rejection of tavern license applications, local option, establishing quotas on the number of taverns in a given district, the closing of the more disreputable establishments, holding tavern owners responsible for the actions of their patrons, regulation of tavern hours, imposition of fines for drunkenness, and prohibition. Not surprisingly, tavern owners opposed stricter controls. Others who opposed the temperance movement included distillers and brewers and their employees, farmers, politicians, and members of the upper classes who did not frequent the rowdier taverns, which were the real focus of the teetotalers' campaigns. Collectively, these measures did bring about a drop in the consumption of alcohol, but these policies failed to bring about total abstinence and the elimination of taverns altogether.

Taverns still served a useful function in Canadian society, especially for the working class, including agricultural workers and those employed in the extractive industries such as logging and mining. For the working class, the tavern served as the neighborhood bank, since the tavern was the only place where one could cash one's paycheck, and labor unions were loath to oppose any establishment that provided a service for their members. Equally important, labor unions saw policies directed against drinking establishments as class legislation. In the twentieth century, taverns continued to offer refuge from the abuses of the factory system and amenities that the working classes could not find in their own homes, such as telephones and washrooms. Equally important, taverns, saloons, and beerhouses served as information center, employment office, and union hall. And, they gave men the opportunity to escape from the home environment and bond with other males. Going to the tavern provided welcome leisure for many and a "time-out" from responsibility.

Twentieth Century

By the turn of the century, public drinking was unacceptable for the respectable woman and the taverns and saloons had become the preserve of men. Women who frequented such establishments were assumed to be either prostitutes or "gin-shop derelicts." In the late nineteenth century, the term "saloon," coined from the French *salon,* was transplanted from the United States to Canada to enhance the image of the tavern. Similarly, beerhouses first appeared in Canada in the late 1800s. For women in the temperance movement, regardless of the term used to describe the local drinking establishment saloons represented all that was evil in society. Moreover, they were frequently associated with the destitute immigrants arriving by the thousands in Canada every year.

Saloons were supported by the breweries and flourished across the country. Normally associated with hotels, they sold beer by the glass or bottle. Through their decor and configuration, they encouraged heavy drinking. Most featured a long bar where the patrons stood to drink. Some had stools and perhaps a few tables, but the bulk of the drinking was done leaning against the bar. The saloons were not particularly well-appointed, and most featured sports and hunting memorabilia hanging on the walls and behind the bar, sawdust-covered floors, and entertainment that was typically male oriented, usually gambling or athletic contests. Some of the more disreputable establishments, such as Joe Beef's Canteen in Montreal, kept bears on the premise to entertain patrons by playing pool with the proprietor, fighting with the owner's dogs, or drinking. The bears, like the patrons themselves, were particularly heavy drinkers; Tom, one of the bears, frequently consumed twenty pints of beer a day.

Not surprisingly, saloons attracted the attention and the ire of temperance societies. Prohibition did not ban saloons, however, and during Prohibition many of them were turned into beer parlors. As Robert Campbell's study of beer parlors in Vancouver has shown, "Beer was [now] served in glasses and only to seated patrons. No entertainment was allowed, not even singing, and eventually there were separate entrances and seating for women" (Campbell 2001). The beer parlors catered to the working class, and "class, gender and sexuality, race, age, and decency were regulated as well as alcohol" (ibid.).

In the 1940s, cocktail lounges began opening across the country, broadening the choices available to Canadians not only in terms of types of establishments available but also in terms of types of drinks offered. Cocktail lounges were upscale compared to the traditional beer parlors and catered to a different clientele. More recently, taverns have changed their image as well to counteract a change in the drinking habits of Canadians away from drinking in public. Bars are no longer the dark and dingy establishments they once were. Neighborhood pubs have sprung up across the country and are usually warm, inviting places that offer a wide variety of alcoholic beverages, good food, and often musical entertainment. Other bars—sports bars, for example—cater to specific clienteles and provide big-screen televisions in order to draw in customers.

The bars today hearken back to an earlier time in Canadian history when taverns were inclusive rather than exclusive places. Today, neighborhood pubs appeal to both genders, all classes, and those of all ages. Equally important, bars continue to serve an important role in today's society, albeit on a reduced scale: They provide a respite from work and a place to socialize.

Shawn Cafferky

See also: Alcohol, Consumption of, per Capita (Canada); Licensing (Canada); Provincial Liquor Boards (Canada); Provincial Stores (Canada)

References

Campbell, Robert A. *Sit Down and Drink Your Beer: Regulating Vancouver's Beer Parlours, 1925–1954.* Toronto: University of Toronto Press, 2001.

DeLottinville, Peter. "Joe Beef of Montreal: Working-Class Culture and the Tavern, 1869–1889" in *Labour/Le Travailleur.* 8/9 (1981–1982): 9–40.

Guillet, Edwin Clarence. *Pioneer Inns and Taverns.* Toronto: Ontario Publishing, 1964.

Gusfield, Joseph. "Passages to Play: Rituals of Drinking Time in American Society" in *Constructive Drinking: Perspectives on Drink from Anthropology,* edited by Mary Douglas. Cambridge: Cambridge University Press, 1987.

Smart, Reginald G., and Alan C. Ogborne. *Northern Spirits: A Social History of Alcohol in Canada.* Toronto: Addiction Research Foundation, 1996.

Warsh, Cheryl Krasnick. *Drink in Canada: Historical Essays.* Montreal and Kingston: McGill-Queen's University Press, 1993.

Salvation Army

The Salvation Army is an international Christian organization whose activities are rooted in the principles of social Christianity. Since its formation in nineteenth-century England by William and Catherine Booth, a central tenet of the Salvation Army always has been a commitment to total abstinence on the part of its officers and members, combined with programs of youth education and rehabilitation of alcoholics. The Salvation Army's social services program is now active in over 100 countries with more than 130 homes and centers for alcoholics and a network of drug rehabilitation programs operating worldwide.

William Booth (1829–1912), founder of the Salvation Army (Library of Congress)

The Salvation Army's ongoing commitment to temperance is a legacy of the personal experiences, beliefs, and life work of its founders. For Catherine Booth (1829–1890), temperance implied total personal abstinence and sympathetic but aggressive rescue of untouchable drunkards. William Booth (1829–1912) believed that his mission to save the masses from their "moral degradation" and "spiritual destitution" was doomed without first improving their economic and physical conditions. Dealing head-on with the drink difficulty was therefore essential.

Catherine Booth (née Mumford) was born in 1829 at Ashbourne in Derbyshire, England. In 1834, her father joined a local temperance society in Lincolnshire, sparking off his daughter's lifelong interest in the temperance movement. However, from 1846 he went through a series of religious, emotional, and financial crises that often resulted in him being penniless and prone to heavy bouts of drinking. The combination of her father's drinking and his sporadic involvement with the temperance movement had a lasting and fundamental influence, particularly with regard to Catherine's opinions on drunkards and temperance campaigning. She promised herself that she would only marry a man who was as committed to total abstinence as she was.

William Booth was also born in 1829, in Nottingham. His father died when William was still a child, leaving the Booth family impoverished. To help support his mother and sisters, William was apprenticed at the age of thirteen with a local pawnbroker. Here he observed firsthand the cycle of poverty in which many people lived and how they suffered humiliation and degradation because of it. In spite of having been a local Wesleyan Methodist preacher since the age of seventeen, pawnbroking was still William's livelihood when he moved to London seven years later.

By the time he met Catherine in 1851, he was a full-time paid preacher on the Methodist Reform circuit. Catherine had been a confirmed teetotaler for a number of years, but William's commitment to temperance at this point was more ambiguous. He had signed the pledge as a child, and apparently remained teetotal until a teenager, but letters between Catherine and William early in their relationship show that William took alcohol for medicinal reasons. It was not long, however, before Catherine had convinced him to "be a teetotaller in principle and practice" (Mumford to Booth 1852), and from this point on their mutual commitment to total abstinence was absolute.

In August 1853, Catherine heard the American temperance crusader and reformed drunkard John Bartholomew Gough (1817–1886) speak at Exeter Hall in London. Gough's life story convinced her of the need to reclaim drunkards by encouraging them to attend temperance meetings. Her initial attempts would turn into the more direct public action she later referred to as "house-to-house visitation," a practice that would become a central component of Salvation Army rescue work (Booth 1879: 92).

After Catherine and William Booth were married on June 16, 1855, Catherine began to work with William as he traveled around the country preaching. Her work in public began when she gave a temperance lecture to a Juvenile Band of Hope in 1857. The success of this first lecture encouraged Catherine to continue, and it is clear from her letters to her parents that she was motivated as much by a desire to spread the temperance message as by a need to contribute to the family income. By the end of the 1850s, William was holding open-air revival meetings for the Methodist New Connexion. His sermons invariably contained damning accounts of the effects of the demon drink. Vocal and often violent demonstrations from brewers, publicans, and gangs of their customers marked the beginning of this kind of resistance to the Booths' salvationist mission, which now included more aggressive attempts on Catherine's part to reclaim drunkards.

By the time the Booths had established the East End Mission in London in 1865, teetotalism and a commitment to alleviating the human cost of excessive drinking lay at the heart of much of the Booths' work. In 1870, the constitution presented at the first conference of what at this time was called the Christian Mission stated that total abstinence from alcoholic beverages was a condition of holding office. Insisting that the entire organization be teetotal was more of a problem. In 1878, when the Christian Mission was finally transformed into the Salvation Army, all members (now called "soldiers") were invited to sign the pledge. In part because of the high cost of tea and coffee in comparison with beer, and in part because of a fear of losing potential recruits, the pledge was not obligatory. Total abstinence only became mandatory when the "articles of war" were introduced in 1882. Although as early as 1869, local circuits of the mission established the "Drunkards' Rescue Society," it was in 1874 that the "Drunkards' Rescue Brigade" was established in order to formalize the mission's ad hoc temperance work along the lines Catherine had practiced since the late 1850s.

The strategies of "aggressive Christianity," outlined in Catherine's articles—"Compel Them to Come In" (1869), "To the Rescue; or, What Shall We Do In 1874?" and "Strong Drink versus Christianity" (the latter two published in 1874)—demonstrate her central role in the creation of Salvation Army policy, method, and ideology. The formulation of her temperance ideology was not carried out in isolation from the mainstream nineteenth-century temperance movement in Britain. Her speech to the Mission Conference in 1874 concluded with her begging the conference to "Haste to the Rescue, for Jesus' sake," a clear reference to Mrs. Julia Wightman's *Haste to the Rescue; or Work While It Is Day.* First published in 1858, Wightman's book has been credited with being a major influence in the formation of the Church of England Total Abstinence Society in 1862. Similarly, nineteenth-century temperance historians signaled the importance of Catherine's work by including "Strong Drink versus Christianity" in a list of the key temperance publications for 1874. Catherine was, however, often overtly critical of what she perceived to be the failings of the temperance movement: its emphasis on prevention rather than salvation and its unwillingness to "go down" and save the masses. Catherine Booth's temperance ideology became enshrined within the Salvation Army from its inception and remains so to this day. She died of cancer on 4 October 1890, and more than 30,000 people attended her funeral on 13 October.

Published in the year of his wife's death, *In Darkest England and the Way Out* (1890) outlined William's plan to save the masses from poverty and unemployment, but throughout it also emphasized the Booths' fundamental conviction that many of society's "evils would dwindle away and die if they were not constantly watered with strong drink" (Booth 1890: 47–48). The Booths' mission, and subsequently the Salvation Army's, was not limited to Britain. From their first visit to the United States in 1880, the Salvation Army began to establish branches and outposts around the world. The Salvation Army's International Congress of 1914 attracted 2,000 delegates from more than thirty-five countries. Between 1890 and 1907, William traveled throughout Europe and to the Middle East, Japan, and Australia, as well as making several visits to the United States. He spoke to large crowds wherever he went. He was still regularly speaking in public only four months before his death on 20 August 1912.

Kristin G. Doern

See also: Gough, John Bartholomew; Nonconformists (United Kingdom); Wightman, Julia Bainbrigge

References

Bailey, Victor. "Salvation Army Riots, the 'Skeleton Army' and Legal Authority in the Provincial Town" in *Social Control in Nineteenth Century Britain,* edited by A. P. Donajgrodzki, 231–253. London: Croom Helm, 1977.

Begbie, H. *Life of William Booth, the Founder of the Salvation Army.* London: Macmillan, 1920.

Booth, Catherine. "Compel Them to Come In" in *Papers on Practical Religion.* London: S. W. Partridge, 1879.

———. *Strong Drink versus Christianity.* London: S. W. Partridge, 1874.

Booth, William. In *Darkest England and the Way Out.* London: McCorquodale, 1890.

Booth-Tucker, F. de L. *The Life of Catherine Booth: Mother of the Salvation Army.* London: Salvationist Publishing, 1893.

Doern, Kristin G. "Temperance and Feminism in England, c.1790–1890: Women's Weapons—Prayer, Pen & Platform" (Ph.D. dissertation, University of Sussex, 2001).

Green, Roger J. *Catherine Booth: A Biography of the Co-Founder of the Salvation Army.* London: Salvation Army and Monarch Publications, 1997.

Hattersley, Roy. *Blood & Fire: William and Catherine Booth and Their Salvation Army.* London: Little, Brown, 1999.

Mumford, Catherine, to William Booth, 27 to 30 December 1852. British Library, Add. MS3, 64799.

Salvation Army International Heritage Centre website. http://www1.salvationarmy.org/heritage.nsf/All?openView.

Sandall, R., et al. *The History of the Salvation Army.* Vols. 1–7. London: Thomas Nelson, 1947–1986.

Schlitz, Joseph, Brewing Company

In 1849, German immigrant Georg August Krug began operations as the owner of a saloon and brewery in Milwaukee, Wisconsin. After receiving an $800 gift from his father, Krug sold the saloon and brewing business and began construction of a full-time brewing operation. In 1850, Krug took on Joseph Schlitz as his bookkeeper. About the same time, brewer Krug and his wife Anna Maria Krug assumed responsibility for the care of their nephew, August Uihlein, who began work as a laborer in the Krug brewery.

In December 1856, G. August Krug died and Joseph Schlitz took on the management of Krug's growing business. Two years later, the enterprising Schlitz married the widow Krug, renamed the business the Joseph Schlitz Brewing Company, and began expansion of the lager beer brewery. In 1858, August Uihlein assumed the title of bookkeeper at the brewery and was later joined in the business by his brothers Henry, Alfred, and Edward.

The company's greatest opportunity came with its expansion to Chicago after the Great Fire of October 1871, which had partially destroyed the city along with nineteen of its breweries. Named as the vice president of export operations for the Schlitz brewery, Edward Uihlein would later write in his personal journal that the business in Chicago was almost overwhelming, as Chicagoans consumed record levels of Schlitz beer.

In 1873, Joseph Schlitz made the business a stock company with himself as president, August Uihlein as secretary, and Henry Uihlein as superintendent. Using the vast railroad hub in Chicago as a springboard for further growth, the brewery began expansion into distant markets.

In May 1875, brewer Joseph Schlitz drowned after the SS *Schiller* sank off the coast of Land's End, England. Control of the brewery was turned over to the Uihleins, with August as chairman of the board, Henry as president, Alfred as brewmaster and superintendent, and Edward as vice president. Under the guidance of the Uihlein brothers, especially Edward in Chicago, prime corner locations were purchased and retail saloon outlets established. These real-estate holdings would later prove a worthwhile investment when the brewery ceased brewing operations during National Prohibition. Many of the choicest spots were sold to oil companies for gas stations. Until Prohibition, beer exportation continued with the utilization of bottling and pasteurization of the product.

In December 1902, the Joseph Schlitz Brewing Company laid claim to a ranking as the world's largest brewer with an annual output of more than 1 million barrels. Use of brown bottles and cork-lined crown caps to preserve freshness added to the perception of Schlitz beer as a quality product.

With a growing wave of temperance and prohibition movements throughout the United States, the succeeding second generation of Uihleins moved to further diversify their holdings into nonbrewing-related businesses, such as timber and the manufacturing of carbon electrodes for the steel industry. Because of their established positions in railroads, banking, and real estate, the family managed to hold on to most of its fortune during the "dry" years of National Prohibition. Production at the brewery during the dry years was limited to malt extract and cereal beverages.

1894 advertisement for Schlitz (Library of Congress)

With the return of legalized beer on 7 April 1933, the Joseph Schlitz Brewing Company embarked on an ambitious plan of expansion and modernization. By 1947, the brewery had reestablished its claim as the world's largest producer of beer. In 1949, the Schlitz brewery expanded its operations by purchasing the George Ehert brewery in Brooklyn, New York. The following year, Schlitz sold more than 5 million barrels of beer. A strike in 1953 by employees at Schlitz's Milwaukee plant, however, and a subsequent falloff in production, allowed the Anheuser-Busch Brewing Company to move into position as the national sales leader.

Schlitz continued its geographic expansion in the 1960s, purchasing the Hawaiian Brewing Company, S.A., Brasserie de Ghlin in Belgium, and three Spanish breweries, the La Cruz del Campo, S.A., of Seville, Henniger Espanola, S.A., at Madrid, and the Cereceras Asociadas, S.A., in Barcelona.

During the early 1970s, now led by fourth-generation Robert Uihlein Jr., the company implemented the use of a brewing process known as accelerated batch fermentation, sometimes known as accurate balanced fermentation (ABF). This process was felt to be capable not only of doubling brewery output, but with the substitution of cheaper corn grits or corn syrup, of lowering production costs by $1 per barrel. The conversion proved disastrous as customers started to complain that the beer poured flat. A weak attempt to correct this problem led to further complaints by customers of a snowflake-like substance in the finished product. As sales fell, Schlitz did little to correct the problem, resorting instead to a poorly received new advertising campaign and fighting allegations by a Milwaukee federal grand jury that Schlitz had made cash payments to a number of its largest accounts for exclusive sales rights. Charged with three counts of felony tax fraud, one count of conspiracy to violate the Federal Alcohol Administration Act, and more than 700 counts of actually violating the act, the brewery eventually settled with the federal government, pleading nolo contendere, and paid a fine of $750,000.

Weakened by a further decline in sales, the Joseph Schlitz Brewing Company accepted a buyout offer in 1981 from the G. Heileman Brewing Company of La Crosse, Wisconsin. The federal government stopped the proceedings, charging that a merger of Schlitz and Heileman would violate antitrust laws. A subsequent offer by the Pabst Brewing Company of Milwaukee was rebuffed by the Schlitz board of directors.

The brewery eventually fell into the hands of the Stroh Brewing Company of Detroit, Michigan, for a reported purchase price of $495 million. With the exiting of Stroh in 1999, the Schlitz label is now owned by Pabst but contract-brewed by the Milwaukee-based Miller Brewing Company

Bob Skilnik

See also: Anheuser-Busch; Beer; Heileman, G., Brewing Company; Miller Brewing Company; Pabst Brewing Company; Stroh Brewing Company

References

A Memoir of Edward G. Uihlein. Translated by Rosina L. Lippi and Jill D. Carlisle. Chicago: Chicago Historical Society, 1917.

Skilnik, Bob. *The History of Beer and Brewing in Chicago,* vol. 2. Haverford, PA: Infinity, 2002.

Smith, Alice E. *The History of Wisconsin,* vol. 1, *From Exploration to Statehood.* Madison: State Historical Society of Wisconsin, 1997.

Scientific Temperance Federation

Formed to collect and disseminate the "scientific facts about the alcohol question," the Scientific Temperance Federation (STF) published research studies that detailed the deleterious effects of alcohol on humans. This prohibition education organization was reorganized out of the Department of Scientific Temperance Instruction of the Woman's Christian Temperance Union (WCTU) by several individuals, including Cora Frances Stoddard, who became its lifetime executive secretary. Throughout its long career, the STF worked assiduously to cover the advocacy efforts of temperance and prohibition reformers with the mantle of science.

The WCTU, through Mary Hunt, head of the Department of Scientific Temperance, had established around 1880 a campaign to institute the compulsory teaching of the negative effects of alcohol in the public schools. By 1901, every state required "scientific temperance instruction." Hunt in the meantime formed the Scientific Temperance Association to study and endorse temperance education texts. From her evaluation of materials, she received a portion of the royalties. Hunt died in 1906, and disputes over property rights led the WCTU to disavow its association with the group. Hunt's secretary, Cora Frances Stoddard, and others reincorporated the association as an independent organization headquartered in Boston on 18 December 1906. Charter members in addition to Stoddard included T. D. Crothers, M.D., the Rev. Perley B. Davis, Mrs. Susan S. Fessenden, Mrs. A. J. Gordon, John Herbert, William C. Lilley, Robert H. Magwood, Elvert O. Taylor, D.D., and Mrs. Emma L. Transeau. The federation compiled statistical studies concerning the effects of alcohol use upon the individual and society. It encouraged research on alcohol and disseminated information in its quarterly *Scientific Temperance Journal,* which had developed from a newsletter launched by Hunt in 1892. Stoddard served as editor of the journal over her lifetime.

The STF became well known for its statistics and emerged as a center for information. Temperance and other organizations desiring reliable material frequently consulted with the federation, which had an extensive library and kept abreast of the latest findings in North America and internationally. It also published material in popular books, posters, and pamphlets and presented exhibitions at health conferences. In 1912, the federation organized a traveling exhibit, first presented at the International Hygiene Congress held at Washington, D.C., that used a set of "stereoptic" slides on the effects of alcohol. Another traveling educational program used a "talking machine"—records—in several languages along with a poster exhibition. Small store-window exhibits and posters on the effects of alcohol on health were also developed. These educational programs were widely disseminated throughout the nation until around 1920. This concerted effort of alcohol education likely influenced public sentiments in favor of National Prohibition. However, the federation struggled financially, and in 1913 Stoddard arranged with Ernest Cherrington, head of the Anti-Saloon League of America's American Issue Publishing Company, to assume formal control of the group and publish the *Scientific Temperance Journal.* Stoddard continued to operate independently and served for her lifetime as executive secretary.

Publication of statistical studies that discussed the dangers of alcohol achieved wide circulation and helped to persuade scientists, physicians, social workers, and industrialists to support Prohibition. One of the most notable works of the federation was a series of five reports "on the operation of Prohibition in

Massachusetts" from 1920 through about 1930. The final publication of this series was *Fifteen Years of the Drink Question in Massachusetts* (1929), coauthored by Stoddard and Amy Woods, the federation's social-welfare secretary. This publication compared seven consecutive years before and after National Prohibition went into effect. The report suggested that "drunkenness arrests dropped 68 percent" in the first year of Prohibition. Over the fifteen-year span of the study, the STF reported that many social problems associated with drinking, including arrests for drunkenness, neglect of children, youthful delinquency, alcoholism, cirrhosis of the liver, homicides, and suicides had declined, in particular during the first three or four years. However, in the late 1920s an upward trend of alcoholism deaths began. Some scholars suggested this was due to a tendency to heavy drinking when alcohol became available. The report concluded that in Massachusetts National Prohibition had benefited public order, health, and the home environment and had reduced other problems related to alcohol.

During World War I, the federation produced material for the U.S. military, and in 1919, subsequent to the creation of the World League Against Alcoholism, the STF served as its scientific advisory board. Honorary members included Irving Fisher, the Yale economist; Adolf Meyer, a public health reformer; and Harvey Wiley, father of the pure food and drug laws. Fisher became honorary president of the federation in 1920 and served in this capacity into the 1930s.

After passage of the Eighteenth Amendment mandating National Prohibition, the WCTU shifted its emphasis from working with public schools toward sponsoring essay contests among youth. Concerned about this decline in school programs, the STF embarked in 1923 on a long-range strategy of committing teachers and school administrations to a program of instruction. New curricular materials were prepared, and in 1931 they were approved by the National Education Association. However, in 1933, due to the decline in interest in prohibition and the beginning of the Great Depression, the STF experienced deep financial trouble. The following year, it merged with the newly formed Temperance Education Foundation, headed by Ernest Cherrington; thereafter, it functioned as the research department of the foundation. The federation continued to publish the *Scientific Temperance Journal* until 1952. The journal was then issued by the Temperance Education Foundation until December 1970, when the organization folded.

Ruth Clifford Engs

See also: Anti-Saloon League of America (ASLA); Cherrington, Ernest Hurst; Crothers, Thomas Davison; Fisher, Irving; Hunt, Mary Hannah Hanchett; Stoddard, Cora Frances; Woman's Christian Temperance Union (WCTU) (United States); World League Against Alcoholism (WLAA)

References

Cherrington, Ernest Hurst, et al., eds. *Standard Encyclopedia of the Alcohol Problem.* 6 vols. Westerville, OH: American Issue Publishing, 1925–1930, vol. 5: 2379–2380; vol. 6: 2535–2536.

Kerr, K. Austin. *Organized for Prohibition: A New History of the Anti-Saloon League.* New Haven, CT: Yale University Press, 1985.

Timberlake, James H. *Prohibition and the Progressive Movement, 1900–1920.* New York: Atheneum, 1970.

Zimmerman, Jonathan. *Distilling Democracy: Alcohol Education in America's Public Schools, 1880–1925.* Lawrence: University Press of Kansas, 1999.

Scotch

See Whiskies

Scotland

The conjunction of alcohol as a social problem and temperance as a practical solution began to be forcefully articulated in Scotland during the 1820s. Temperance campaigners saw drunkenness as the visible manifestation of social dislocation, especially in urban and industrial districts. The early movement derived its missionary determination to regenerate society from religious evangelicalism, but from the 1830s the temperance campaign gathered a mass following. Although the profile and objectives of the movement altered continually during the course of the nineteenth century, temperance influence on Scottish politics and popular culture was profound. The success of the crusade to eradicate alcohol and promote a sober lifestyle was reflected in the unusually restrictive Scottish licensing law, which prevailed until the late twentieth century.

The strength of the Scottish temperance movement was paradoxical, given the robust image of conviviality that had been fostered during the eighteenth century and beyond. The writings of Robert Fergusson, Robert Burns, and countless minor poets buoyantly celebrated the earthier qualities of tavern life. Social drinking was closely bound up with the calendar of traditional Scottish festivals, such as New Year, May Day, Halloween, and fair time. An elaborate dining culture emerged among urban elites as a symbol of conspicuous consumption, with alcohol an essential accompaniment. When King George IV came to Edinburgh in 1822, paying the first visit of a British monarch to Scotland since 1650, his taste for Glenlivet malt whiskey set a trend for the fashionable to follow.

Scottish licensing law, which was and remains separate from that of England and Wales, did little to impose constraints until the nineteenth century. From 1756, licensing powers had been vested in the justices of rural districts and the magistrates of towns and cities. Local regulations, such as Edinburgh's 10 P.M. curfew, were aimed at curbing excessive drinking. However, the means of enforcing such strictures, through police control, was undeveloped and notoriously inefficient. The quality and supply of drink was also barely regulated. In parts of Scotland, illegal spirits, whether contraband gin and brandy from Europe or illicit Highland whiskey, were commonly sold in taverns. The licensing law was tightened in 1828, but any benefit for sobriety was offset by government fiscal policy during the 1820s, which drastically reduced the duty on spirits.

Given the ubiquity of drink in Scotland, the virtues of temperance were disseminated long before the organized movement appeared. For instance, *Whiskiana, or, The Drunkard's Progress* was an anonymous vernacular poem, published in 1812, that used humor to convey a serious moral message about alcohol abuse. Attempts to start a Scottish movement began in earnest during the 1820s, at the time of the licensing reforms. According to John Dunlop, one of the pioneers, it was not sufficient simply to change the law; education was necessary to reverse the corrosive effects of drink on society. Transatlantic role models of temperance organization featured prominently in his vision of reform. France also inspired Dunlop during a visit in 1828, when he was impressed by the population's abstemious habits compared with spirit-drinking Scots.

Early Temperance Organization

Dunlop's organizational ability owed much to his training as a lawyer, but he was also driven by his evangelical commitment. The missionary impulse had been crucial for the success of the North American temperance campaign, and during the 1820s there was a similar mood of moral revivalism in Scotland that derived impetus from the ideas of the Presbyterian theologian Thomas Chalmers. Dunlop consciously incorporated temperance into the evangelical agenda, taking direct action to further the cause during 1829. In October that year, Scotland's first organized society, exclusively for women, was formed in Gairbraid, near Glasgow, under the direction of Dunlop's philanthropist aunt, Lillias Graham. Shortly afterward, a society was established in Dunlop's home base of Greenock, a populous port town on the Firth of Clyde.

Dunlop's main ally from 1829 was William Collins, a self-made Glasgow publisher and close personal associate of Chalmers's. The two men took their campaign a significant step forward in 1830, when they inaugurated the Glasgow and West of Scotland Temperance Society. This group aimed to provide a coordinating body for the network of other local societies. Wealthy Tory evangelicals were prominent patrons of the new society, and the presence of iron masters and textile manufacturers among the subscribers indicated their unease about the debilitating effects of alcohol on the industrial workforce. The temperance pledge was qualified, relating only to abstention from "ardent spirits," except for medicinal purposes. This anti-spirits focus characterized the early British movement, but the national predilection for whiskey gave it particular resonance in Scotland.

In 1831, the society extended its network and was renamed the Scottish Temperance Society. With a rapidly rising membership, predominantly in Glasgow and Edinburgh, the leadership was encouraged to look toward England as a sphere of influence. That same year, Collins played a key role in founding the London-based British and Foreign Temperance Society. However, activity continued in Scotland, especially in promoting propaganda work. Collins edited and published a newspaper, *The Temperance Society Record,* and in 1833 the first edition of Dunlop's *Artificial Drinking Usages in North Britain* appeared. Now recognized as authorities on drinking practices, they made substantial contributions to the 1834 House of Commons Select Committee on Drunkenness, a parliamentary inquiry into the extent of the problem, especially in industrial districts.

Despite this evidence of unity and purpose, personal differences were emerging between Collins and Dunlop. Their estrangement was exacerbated by the debate about total abstinence. The teetotal movement was not new in Scotland, its organizational presence dating arguably from 1830, when a society was formed in Dunfermline. More important, Dunlop, Robert Kettle, and others in the Scottish Temperance Society became convinced by the argument that beer and wine were as pernicious as spirits. Politics was also a contributing factor to the tensions. The opening out of the Scottish franchise in 1832 and 1833 had buoyed the reform cause, and total abstinence was seen as part of the political cleansing process. The radicalizing impact of teetotalism was a major reason why the Scottish Temperance Society, with its paternalistic origins, folded in 1835.

Total Abstinence and the Legislative Solution

From the mid-1830s, teetotal missionaries arrived in Scotland from the north of England, where the strength of the total abstinence movement was concentrated. Particularly influential was John Finch from Liverpool, whose evocation of the "demon drink" had a vivid, folk-drama quality to it intended both to entertain and unsettle his audiences. Finch's first Scottish success was the formation of a total abstinence society in Annan, Dumfriesshire, in 1836. Societies in Glasgow and Edinburgh soon followed, and within a year the movement had reached Aberdeen and the northeast. Although the teetotal organizations symbolized new beginnings, they were not necessarily a complete break with the past. The influential Western Scottish Temperance Union, constituted in 1839, was structurally similar to the Scottish Temperance Society, and both Kettle and Dunlop were among its office bearers.

By the early 1840s, there was an almost millenarian fervor about temperance activity in Scotland. The period was economically depressed, particularly in the industrial districts, and sobriety seemed to provide reassurance in the midst of uncertainty. The Irish dimension was also crucially important, given the nation's close proximity to Scotland and the high incidence of immigration. Father Theobald Mathew's crusade against drink spread to Irish communities throughout mainland Britain, and he was rapturously welcomed to Glasgow in 1842. Teetotalism was also espoused by followers of Chartism, a radical political movement that campaigned for universal male suffrage. The Chartist journalist John Fraser's *Edinburgh Monthly and Total Abstinence Advocate,* for example, promoted the dual ideals of democracy and temperance.

The radical appeal of temperance reached a climax around 1842, but thereafter the movement took off in a different direction. In an effort to overhaul the Scottish licensing law of 1828, campaigners began to press for reform. They claimed that publicans could too easily set up in business because magistrates were not sufficiently rigorous in their approach. The proliferation of urban dramshops was especially disturbing, and an 1846 House of Commons Select Committee, under William Forbes Mackenzie, recommended a drastic reduction. The mechanism was thus set in motion for stricter regulation, which was achieved, after a fierce public debate, when the Forbes Mackenzie Act was passed in 1853. The legislation, which fixed opening hours and introduced Sunday closing, served as the basis of Scotland's licensing law until 1976.

The Forbes Mackenzie campaign generated the growth of new societies, usually with a more middle-class and overtly religious profile than hitherto. The Scottish Temperance League, established in 1844, exemplified the stress on respectability. The United Presbyterian Church Abstainers' Society was formed in 1847, and two years later William Collins's son, also named William, founded the Free Church of Scotland Abstainers' Society. John Hope, an Edinburgh lawyer and Church of Scotland evangelical, formed the British League of Juvenile Abstainers in 1847. By the 1850s, temperance embraced a variety of opinions, from political radicals to staunch Conservatives, such as Hope. For many supporters, however, the success of Forbes Mackenzie legitimized the legislative solution and created a base to agitate for measures that would forbid the manufacture and sale of alcohol.

Prohibition, Politics, and Temperance Culture

Prohibition inevitably provoked divisions in the temperance movement. The leaders of the Scottish Temperance League eschewed compulsion, believing that individual example and collective association along "moral suasionist" lines was the most constructive way forward. In response, a succession of visiting prohibitionists from the United States presented their cause as liberating and gave practical examples of its operation in pioneering "dry" states, notably Maine. The Scottish prohibitionists eventually came together in 1858, when the Scottish Permissive Bill and Temperance Association was launched in Glasgow. The notion of permissive legislation, through local "veto polls" of electors—not full-blown, Scottish-wide prohibition—established a more realistic objective that could serve as a springboard for the bigger campaign.

The Scottish drink trade recognized the dangers of the Permissive Bill but was initially ineffectual in fighting back. It did not have the clout of its English counterpart, which was financed and organized by large brewers with vested interests in the control of public houses. The more typical representative of the business in Scotland was the independent proprietor of a small dramshop who could not afford to pour money into trade defense. It was also difficult for trade activists to counter the temperance propaganda onslaught. Charismatic platform speakers, such as John B. Gough from the United States, attracted enormous audiences. As well as a prolific temperance literature, there were a number of supportive newspapers; the Glasgow-based *North British Daily Mail,* for example, was particularly influential from 1864.

The Permissive Bill campaign intensified the links between temperance and politics in Scotland. This was most apparent in local government, where temperance sympathizers had already been growing in strength from the 1830s. By mid-century, the industrial city of Dundee had elected George Rough as Scotland's first teetotal Lord Provost (mayor), while Duncan Maclaren, his Edinburgh equivalent, had played a leading part in the Forbes Mackenzie campaign. Prohibitionists quickly made municipal inroads in Glasgow, but their presence was most visible after the urban male vote was extended in 1868. Led by William Collins, Jr., their argument was straightforward; in solving the drink problem, all other social problems would be solved. By 1877, Collins had become Lord Provost and was one of the most influential figures in the Scottish Liberal Party.

Scottish women won the municipal vote in 1882, and this victory was seen as beneficial for temperance, given the movement's entrenched female following. As the Gairbraid example from 1829 has shown, women were an integral part of temperance organization from the outset. John Dunlop's emphasis on sobriety as a way of life necessarily involved women in their nurturing role, and the domestic mission was a central theme of Scottish temperance culture. Recreational activities, in particular, were promoted as nonexclusive and family oriented. Organizations such as the Scottish Association for the Suppression of Drunkenness and the Glasgow Abstainers' Union prioritized drink-free forms of leisure ranging from coffeehouses to concerts and excursions. For over a century from the 1850s, the Band of Hope was Scotland's main temperance forum for children.

The Twentieth Century

By the beginning of the twentieth century, the temperance movement was one of the most vigorous and diverse components of Scottish associational culture and enjoyed a powerful political influence. It was closely identified with radical liberalism, and temperance adherents were often proponents of Scottish home rule. Their rationale was that the London-based Parliament was reluctant to legislate for reforms such as the Permissive Bill, even though there was an overwhelming measure of support among Scottish representatives. The pioneering Labour politicians, notably James Keir Hardie, absorbed elements of this radicalism. In 1888, the founding objectives of the Hardie-led Scottish Labour Party, a forerunner of today's Labour Party, included prohibition and home rule.

Scotsman drinking whiskey, ca. 1901 (Library of Congress)

At the same time the drink trade was more combative against temperance pressure. The Scottish Licensed Trade Defence Association had recruited the vast majority of the nation's licensees, around 10,000 retailers and publicans. The industry had expanded to meet consumer demand, especially for beer and blended whiskeys, and was thus more generous in support of trade interests. Yet as the century progressed, temperance influence was demonstrated by the steady tightening of licensing arrangements. An enactment of 1903 aimed to reduce drunkenness by allowing for general 10 P.M. closing in public houses. Then, in 1906, the landslide return of the Liberal government provided the first meaningful opportunity to legislate for the veto poll in Scotland. This controversial measure was approved by Parliament in 1913 but did not come into effect until 1920.

In the interim, the outbreak of war in 1914 provided the opportunity for a series of extraordinary measures to restrict the availability of intoxicating liquor. Scotland had two of Europe's largest munitions centers, and the government was anxious to ensure that nothing would slow down the momentum of production. The wartime combination of temperance and patriotism gave the movement confidence to raise its profile after the armistice of 1918. The introduction of Prohibition in the United States further galvanized Scottish activists, and the implementation of the Temperance (Scotland) Act in 1920 afforded a platform to persuade electors to opt for "no license" in their localities. An extraordinary victory for the movement took place in 1922, when Edwin Scrymgeour, campaigning as a prohibitionist, was elected to Parliament for the Dundee constituency.

Hopes of the temperance reformation were premature. Scrymgeour's success was isolated, and he was defeated in the 1931 general election. Voters were also reluctant to endorse change through the local veto, especially in working-class districts. The failure of U.S. National Prohibition, and its repeal in 1933, was a major setback for the Scottish campaign. By the time of World War II, the temperance movement was seen as anachronistic and irrelevant, and there was no restriction of drinking activities as in the previous conflict. However, the temperance ethos remained pervasive. It was not until the 1960s that a relaxation of Scotland's licensing laws could be contemplated, leading in 1976 to the wholesale dismantling of the structure originally set in motion by Forbes Mackenzie.

Drinking

In the nineteenth and early twentieth centuries, Scotland had a powerful temperance movement but continued to live up to its reputation for heavy drinking, particularly in industrial Glasgow. Licensed premises were abundant: In the 1840s, Glasgow offered one liquor outlet of some sort for every 150 men, women, and children; these were supplemented by illegal shebeens, the worst of which sold a mixture of whiskey and methylated spirits. In the period 1831–1851, according to a report to Parliament on arrests for drunkenness, "Glasgow [was] three times more drunken than Edinburgh and five times more drunken than London" (quoted in Devine 1999: 353). Like the Irish and unlike the English and the Welsh, ordinary people in Scotland drank a great deal of whiskey. Originally the drink of the Highlands, whiskey became popular in the rest of Scotland starting in the second half of the eighteenth century, mostly in the form of blended whiskey. In part this was the result of Highlanders and Irish migrating to the industrial towns in the Lowlands. By the twenty-first century, Scottish per capita consumption of whiskey had shrunk to only a seventh of that consumed in the first half of the nineteenth century. Scottish whiskey (or Scotch) continues to be a valued export, with single malts especially prized. Scotland also is a major brewing and beer-drinking country. Typically, Scotch ales have been maltier and less bitter than English beers, and it became common practice to "chase" a shot of whiskey with light ale (what in America is called a "boilermaker" or in Scotland "a schooner and a stick").

Scotland has a research center for brewing history. Founded in 1982, the Scottish Brewing Archive was originally housed at Heriot-Watt University and later at Glasgow University.

Irene Maver

See also: Band of Hope; Blaikie, Margaret Catherine; Brewers Association of Scotland (BAS); Chartism; Courage Brewery; Cumming, Sir Ronald Stuart; Distillers Company Limited (DCL); Dunlop, John; Erroll and Clayson Committees (United Kingdom); Good Templars (IOGT); Gough, John Bartholomew; Hayman, Sir (Cecil George) Graham; Honeyman, Tom; Liberal Party (United

Kingdom); Malt Distillers' Association of Scotland (MDA); Mathew, Father Theobald; McEwan, William; Ross, Sir Henry (James); Scottish & Newcastle Breweries Limited; Scottish Christian Union (SCU); Scottish Temperance League (STL); Scrymgeour, Edwin; Tennent, Hugh; White, Mary

References

Aspinwall, Bernard. *Portable Utopia: Glasgow and the United States, 1820–1920.* Aberdeen: Aberdeen University Press, 1984.

Devine, T. M. *The Scottish Nation: A History, 1700–2000.* New York: Viking, 1999.

Donnachie, Ian. "Drink and Society, 1750–1850: Some Aspects of the Scottish Experience" in *Scottish Labour History Society Journal.* 13 (1979): 5–22.

———. *A History of the Brewing Industry in Scotland.* Edinburgh: John Donald, 1970; 2d ed., 1998.

Kenna, Rudolph, and Anthony Mooney. *People's Palaces: Victorian and Edwardian Pubs of Scotland.* Edinburgh: Paul Harris, 1983.

King, Elspeth. *Scotland Sober and Free: The Temperance Movement, 1829–1979.* Glasgow: Glasgow Museums and Art Galleries, 1979.

Moss, Michael S., and John R. Hume. *The Making of Scotch Whisky: A History of the Scotch Whisky Distilling Industry.* Edinburgh: Collins, 1981.

Smitley, Megan. "'Woman's Mission': The Temperance and Women's Suffrage Movements in Scotland, c. 1870–1914" (Ph.D. dissertation, Glasgow University, 2002).

Sweeney (later Mauer), Irene. "Local Party Politics and the Temperance Crusade: Glasgow, 1890–1902" in *Scottish Labour History Society Journal.* 27 (1992): 44–63.

Weir, Ronald B. *The History of the Distillers Company, 1877–1939: Diversification and Growth in Whisky and Chemicals.* New York: Oxford University Press, 1996.

Scottish & Newcastle Breweries Limited

By the 1980s, Scottish & Newcastle Breweries Limited had become one of the biggest players in British brewing. Formed in 1960, it brought together Scottish Breweries Limited—itself a 1931 linkup between William Younger and Company Limited and William McEwan and Company Limited—and Newcastle Breweries Limited. The amalgamation was the culmination of a close relationship that had evolved over the previous century between the brewing industries of Scotland and northeastern England. The origins of Scottish and Newcastle's constituent firms lay in the enterprising Edinburgh brewers William McEwan and the Younger family and the prominent Tyneside dynasty of the Barras/Reeds.

The first William Younger began brewing in Leith, Lothian, in 1749. His sons, Archibald, William II, and Richard, all brewed on their own accounts. By 1821, with the deaths of his mother, brothers, and business partners, William Younger II was able to consolidate the family's many brewing interests into William Younger and Company. His eldest son, William III, was admitted as a partner in 1836. William Younger IV joined in 1849, followed by his brother Henry Johnston Younger in 1852. In 1858, the adjacent premises of Alexander Berwicke were bought and the new Holyrood Brewery, Edinburgh, established. William McEwan began brewing at the Fountain Brewery, Edinburgh, in 1856.

Edinburgh brewers used coastal shipping routes to exploit the demand for their superior beers in both England and Scotland, but Younger's and McEwan's also developed a growing export trade in continental Europe and the colonies. Both firms were among the first to incorporate: William Younger and Company Limited was registered in 1887 and became a public company in 1889, while William McEwan and Company became a limited company in 1889 when its owner, returned as member of Parliament for Edinburgh Central in the 1886 general election, was concentrating on politics. In 1907, McEwan's took over the Melvin and Company's Borough Loch Brewery in Edinburgh.

The two companies continued to flourish, and in 1931 they were formally joined as Scottish Brewers Limited. The new organization purchased the Collin Croft Brewery Company Limited, Kendal, Cumbria, in 1933. In the 1950s, Scottish Brewers took over whiskey distillers John E. McPherson and Sons, the Tower Lager Brewery Limited, Manchester, and two Tyneside businesses, those of Alexander Deuchar and W. B. Reid and Company Limited of Newcastle (in which Younger's had held a minority interest since its incorporation in 1891). In 1960, the Edinburgh brewing firms of T&J Bernard and Company Limited of the Edinburgh Brewery, J&J Morison Limited of the Commercial Brewery, and Robert Younger Limited of the St. Anne's Brewery were acquired.

Until its latter years, Scottish Brewers and its antecedents evolved into large-scale operators through organic growth. In northeastern England, however, the tied-house system required brewers to control licensed premises in order to protect and expand markets. Newcastle Breweries Limited grew in importance through a process of merger and acquisition. Its origins went back to 1775, when John Barras established a brewing business in Gateshead, County Durham. On his father's death in 1811, John Barras Jr. took over the reins. None of his sons were able to take on the management of the family business, so in 1848 John Barras Jr. took his nephew, Charles Reed, into the firm, and after John Jr.'s death in 1861 Reed leased the brewery from the Barras trust. In 1882, the brewery and licensed houses went onto the market; Reed acquired nineteen licensed houses but moved to Newcastle to buy and then refit the vacant Tyne Brewery. John Barras and Company Limited became a public company in 1889 and a year later became the prime mover in the creation of the Newcastle Breweries Limited, a grouping of Barras and Company, Carr, Bros. and Carr, J. J. and W. H. Allison, W. H. Allison and Company, and Swinburne and Company.

The new company controlled 215 licensed properties, and brewing operations were gradually centralized in Newcastle. In the mid-1890s, Newcastle Breweries took over near neighbors W. A. Falconar and Company of the Howdon Brewery, Sanderson and Son of the Haymarket Brewery, and the licensed houses of wine and spirits merchant James Routledge. Forsters' Bishop Middleham Brewery Company Limited of County Durham was added in 1910. In the early 1920s, the company absorbed brewers Matthew Wood and Sons of South Shields, County Durham, and took over Newcastle wine merchants Turnbull and Wood. Two brewing companies that had originally operated in northeastern England but moved operations to Scotland were acquired during the 1950s: Robt Deuchar Limited of the Duddingston Brewery, Lothian (with more than 300 tied houses), and James Deuchar Limited of the Lochside Brewery, Montrose (with 125 public houses and hotels). John Rowell and Son Limited of the New Brewery, Gateshead, was bought in 1959.

In 1960, Newcastle Breweries Limited merged with Scottish Brewers Limited to form Scottish & Newcastle Breweries Limited. Early acquisitions by the new group were distillers or wine merchants, but in the late 1980s the Home Brewery plc of Nottingham and Matthew Brown plc of the Lion Brewery, Blackburn, became part of Scottish & Newcastle. The group accounted for 10 percent of the United Kingdom beer market and owned 2,300 public houses in Scotland, northern England, and the East Midlands. In the early 1990s, Scottish & Newcastle took over the Chef and Brewer chain of public houses and the word "Breweries" was removed from the company name to reflect its wider interests across the leisure industries. With the £425 million takeover of Courage in 1995, Scottish and Newcastle became Scottish Courage, Britain's biggest brewer.

Brian Bennison

See also: Beer; Beer Orders (United Kingdom); Courage Brewery; McEwen, William; Scotland

References

Donnachie, Ian. *A History of the Brewing Industry in Scotland.* Edinburgh: John Donald, 1979; 2d ed., 1998.

Richmond, Lesley, and Alison Turton. *The Brewing Industry. A Guide to Historical Records.* Manchester: Manchester University Press, 1990.

Ritchie, Berry. *Good Company: The Story of Scottish & Newcastle.* London: James & James, 1999.

Scottish Beer and Pub Association

See Brewers' Association of Scotland (BAS)

Scottish Christian Union (SCU)

The Scottish Christian Union (SCU), the largest single-sex temperance organization in Scotland, was part of the late nineteenth-century florescence of women's temperance organizations first embodied by the Woman's Christian Temperance Union (WCTU) in the United States. Although the SCU maintained its place in temperance reform until it disbanded in the early 1990s, the height of its work was in the late nineteenth and early twentieth centuries.

The Scottish Christian Union was established in 1876, the same year that the British Women's Temperance Association (BWTA, later the National BWTA) was formed in England. Although the SCU was an independent organization and ran its own program of temperance work, it was affiliated with the BWTA during the nineteenth century. In 1900, the SCU hosted the biennial conference of the World's Woman's Christian Temperance Union (WWCTU) in Edinburgh. Because the SCU paid its WWCTU affiliation fee through the BWTA, it could not be represented as an independent organization. The SCU's position at the WWCTU conference precipitated the SCU's decision to sever all formal links with the BWTA. In 1904, the SCU ceased to be affiliated with the BWTA and paid its £40 affiliation fee directly to the World's Woman's Christian Temperance Union.

The Scottish Christian Union was organized around the Edinburgh Central Branch (ECB), which acted as its headquarters. In 1878, the ECB suggested that all ladies' temperance prayer unions in Scotland join under its administration. By 1879, twenty-one prayer unions had affiliated. The number of branches increased to 332 by 1908 and included a membership of 80,000. In 1933, the SCU had 450 branches, but the union went into decline in the 1950s. The ECB published the annual reports of all SCU branches, and, from 1896, the Scottish Christian Union published its own official journal, the *Scottish Women's Temperance News,* under the editorship of Miss Christina E. Robertson. In 1902, a headquarters office was set up in Edinburgh, and in 1906 a suboffice was opened in Glasgow.

In order to lessen the headquarter's responsibilities, district unions were established in 1902. Eleven district unions were formed: Aberdeen, Ayrshire, Borders, Dumfries and Galloway, Dundee, Edinburgh and Lothians, Fifeshire, Glasgow, Moray and Ross, Perth, and Stirling. Each district union had proportional representation in the SCU and voted by ballot at district union conferences. The Glasgow District Union (GDU) was second in importance only to the ECB. The GDU was formed out of the Glasgow Prayer Union, which had been established in 1874, and was largely responsible for overseeing temperance work in the west of Scotland.

As a result of its early affiliation with the BWTA, the members of the Scottish Christian Union were called the "British Women." Drawn from the middle and upper middle classes, many of them were the wives of ministers, manufacturers, businessmen, politicians, and professionals. For example, Mrs. George Stewart, one of the founders of the GDU, was the wife of the Ewing Place Church minister. Another prominent member, Mrs. Wilhemina Woyka, was married to John Woyka, a timber merchant. The prestigious reforming family of Miss Eliza Wigham, a member of the ECB, was involved in cotton manufacture. The SCU membership also included the wives of members of Parliament such as Mrs. Hunter Craig, Mrs. Alice Mary

Cameron Corbett, and Mrs. Priscilla McLaren. The SCU included representatives from across the Presbyterian denominations but few Roman Catholics.

The Scottish Christian Union viewed itself as strongly influenced by the WCTU in the United States. Many prominent members recorded their admiration for the American women involved in the Women's Temperance Crusade in the winter of 1873–1874. Eliza "Mother" Stewart was particularly loved by the SCU. Stewart was invited to tour Scotland by Mrs. Margaret Parker of Dundee. In 1876, Stewart spent six weeks touring Scotland and attended the meeting at Newcastle-upon-Tyne where the BWTA and the SCU were formed.

The SCU was less conservative than the BWTA. Without the controversy that afflicted its English counterpart, the Scottish Christian Union embraced Frances Willard's "Do Everything" policy in the 1890s. This resulted in the systematization of the SCU's temperance work by the creation of separate departments. The SCU had thirteen departments of work in 1893, and this number had expanded to twenty by 1908. In line with Willard's emphasis on "women's issues," the Scottish Christian Union was overwhelmingly in favor of women's suffrage and included a department devoted to women's right to vote.

The Scottish Christian Union was funded mainly through affiliation fees and contributions to the extension fund. Two special fund-raising schemes took place in the early part of the SCU's history. In 1898, the Victoria extension fund raised £9,993 for extension work, or for the development and maintenance of new branches. In 1905, the GDU organized a bazaar at St. Andrew's Halls, Glasgow. Through this sale of women's crafts or "fancy work," the bazaar raised £7,700.

The SCU was particularly concerned with temperance work among poorer women. Its most important response to female drunkenness was its use of women's "inebriate homes." Both the ECB and the GDU managed homes for female drunkards: the Brownsland Temperance Home for Women (also known as the Navitie Home) and the Whitevale Mission Shelter (also known as the Glasgow Mission Shelter and originally as the Prison Gate Mission), respectively. The "inmates" of these homes were drawn from the work of the SCU's prison-visiting department. The British women visited the prisons with the aim of bringing women convicted of drunkenness to temperance homes.

Megan Smitley

See also: Blaikie, Margaret Catherine; British Women's Temperance Association (BWTA); National British Women's Temperance Association (NBWTA); Parker, Margaret Eleanor; Willard, Frances Elizabeth Caroline; Woman's Christian Temperance Union (United States); Women's Temperance Crusade; World's Woman's Christian Temperance Union (WWCTU)

References

Barrow, Margaret. "Teetotal Feminists: Temperance Leadership and the Campaign for Women's Suffrage" in *A Suffrage Reader: Charting Directions in British Suffrage History,* edited by Claire Eustance, Joan Ryan, and Laura Ugolini, 68–89. London: Leicester University Press, 2000.

King, Elspeth. *Scotland Sober and Free: The Temperance Movement, 1829–1979.* Glasgow: Glasgow Museums and Art Galleries, 1979.

Logan (later Denny), Norma Davies. "Drink and Society: Scotland, 1870–1914" (Ph.D. dissertation, University of Glasgow, 1983).

Robertson, Christina E. *BWTASCU: Its Origins and Progress.* Ayr, Scotland: Scottish Christian Union, 1908.

Smitley, Megan. "'Inebriates,' 'Heathens,' Templars and Suffragists: Scotland and Imperial Feminism, 1870–1914." *Women's History Review.* 11, no. 3 (2002): 455–480.

———. "'Woman's Mission': The Temperance and Women's Suffrage Movements in Scotland, c. 1870–1914" (Ph.D. dissertation, University of Glasgow, 2002).

Scottish Temperance League (STL)

The Scottish Temperance League (STL) was established at Falkirk in east central Scotland on 5 November 1844 in a meeting that included James A. Johnston, Dr. William Menzies, J. Ballantyne, Archibald D. Campbell, and George McWhirter from Edinburgh and Robert Reid, W. T. Templeton, A. H. MacLean, and William Logan from Glasgow. The STL was the first nondenominational total abstinence society in Scotland. Although the league operated throughout Scotland, its work and leadership was firmly based in Glasgow. The STL was known for its broad temperance reform program that emphasized moral suasion but accommodated demands for legislative prohibition.

The membership and leadership of the Scottish Temperance League was overwhelmingly middle class. It had a hierarchical structure of honorary directors and vice presidents as well as an executive council composed of directors. A third of the STL leadership was made up of prominent businessmen. Rather than relying on an honorary board of aristocrats for financing—as did the Scottish Society for the Suppression of Drunkenness—the leaders of the STL were evangelical businessmen who used their own time and money to fund the league. The STL leadership also included many Presbyterian ministers who strongly influenced STL policy and expansion. Although the directors were traditionally a mixture of businessmen and ministers, by the 1890s women, medical doctors, and academics had become league officials.

The creation of the Scottish Temperance League was the result of a schism between total abstainers and moderationists. In contrast to total abstainers, who rejected the use of all alcoholic drinks, moderationists advocated abstinence from "spirituous liquors" and an avoidance of drunkenness but allowed for the moderate use of wine and beer. The STL wanted total abstinence.

The league's strong ties to the Christian Institute brought the STL under the indirect influence of the famous evangelical Dr. Thomas Chalmers. The STL's connections with Chalmers and the religious establishment more generally influenced its focus on moral suasionist tactics. In other words, the STL was more interested in bringing about a change in attitudes toward alcohol and drinking than in passing prohibition legislation.

The Scottish Temperance League's main goal was to bring piety, in the form of sobriety, to the lapsed urban masses. STL presidents enthusiastically pursued this goal; their interest in temperance work often spilled over to their families or to other areas of their lives. For instance, Robert Smith, J.P., president from 1852 to 1873, was a City Line shipping magnate whose descendants included the notable temperance family the Allans of the Allan and State Line. The Allans would not hire drinkers, supported the Glasgow United Evangelical Association, United Presbyterian Missions, socialism, and women's suffrage. Another president, (Sir) William Collins II, insisted on "dry" public functions during his time as Lord Provost of Glasgow. Collins was associated with the Good Templars, the Rechabites, the Bands of Hope, and the United Kingdom Alliance. Alexander Forrester Paton, an Alloa mill owner who worked with the National Commerce League, was a renowned philanthropist. The temperance beliefs of these evangelical businessmen reflected a combination of moral suasion and prohibitionist tendencies.

The Scottish Temperance League pursued its aim of challenging popular attitudes to alcohol and drinking through a concerted propaganda campaign. The STL argued that legislative prohibition, to be effective, had to be complemented by a shift in drinking attitudes. From its publishing center in Hope Street, Glasgow, the STL produced masses of literature such as temperance novels, songbooks, tracts, medical publications, and periodicals. The extent of the propaganda onslaught can be seen in the league's claim to have undertaken 2,300 lectures and to have sold approximately 16,000 temperance "volumes" and 430,000 tracts in the course of 1898.

Journals included the *Scottish Temperance Review,* later the *League Journal,* a weekly publication that provided detailed accounts of the Scottish, British, and international temperance movement. The league also published *The Adviser,* a monthly periodical aimed at children, and an annual *Scottish Temperance Register and Abstainers' Almanack.* Some of the major themes for articles in these publications were the pernicious nature of "custom" or social practices that included alcoholic drink, pseudoscientific discussions of the physiological harm caused by alcoholic drink, and the importance of temperance education in schools.

Although the Scottish Temperance League was keen to promote the total abstinence life without recourse to coercive legal methods, the league did also undertake work for prohibition. Many of the STL's leaders were active as temperance politicians on town and parish councils and on school boards. The proportion of the league's directors who were members of local government bodies increased in the period from 1896 to 1909 owing to the STL's involvement in campaigns for licensing restrictions and local-veto legislation. For example, in the late nineteenth century, the league was involved in campaigning successfully for the Public Houses (Amendment) (Scotland) Act, the Publicans' Certificates (Scotland) Act, and the Passenger Vessels Licensing (Scotland) Act. In the 1920s, the league joined other temperance groups within the National Citizens' Council in order to encourage the implementation of local option under the Temperance (Scotland) Act, 1913.

The broad reform program of the Scottish Temperance League was one reason for its preeminence among Scottish temperance societies in the period 1844–1924. During this time, the STL was the most successful organization in terms of membership numbers and finances. The league was adversely affected by the 1878 Glasgow Bank crash and commercial depression, however, and 1880–1914 saw its fluctuating fortunes exacerbated by new competition from medical charities.

Megan Smitley

See also: Band of Hope; Good Templars (IOGT); Rechabite Friendly Society; Scotland; United Kingdom Alliance (UKA)

References

Honeyman, Tom. *No Licence! The New Campaign: The Handbook of the National Citizens' Council Temperance (Scotland) Act.* Glasgow: National Citizens' Council, 1921.

Johnston, William. "The Scottish Temperance League" in *Scottish Temperance Annual.* (1899): 60–63.

King, Elspeth. *Scotland Sober and Free: The Temperance Movement, 1829–1979.* Glasgow: Glasgow Museums and Art Galleries, 1979.

Logan (later Denny), Norma Davies. "Drink and Society: Scotland, 1870–1914" (Ph.D. dissertation, University of Glasgow, 1983).

Scrymgeour, Edwin (1866–1947)

Edwin Scrymgeour (1866–1947) is best known for his establishment of the Scottish Prohibition Party (SPP) in 1901 and his successful campaign in 1922 against Winston Churchill for election as member of Parliament (M.P.) for Dundee. Scrymgeour was a local politician of some note prior to his election to Parliament. His politics were marked by an emphasis on prohibition and social reform for the working classes. Scrymgeour's lifelong pursuit of prohibition was fueled by a strong Christian faith marked by millennialist ideology.

On 28 July 1866, Scrymgeour was born in Dundee, the nineteenth-century center of jute manufacture. Edwin's father, James, was particularly influential on Edwin's political career. James himself was a prominent figure in Dundee's nineteenth-century temperance movement who served as superintendent of the Dundee Band of Hope and as worthy councilor in a lodge of the International Order of Good Templars (IOGT). By the age of fourteen, Edwin was a member of the IOGT, and he later rose to the rank of worthy chief, heading a local lodge. He was educated at the West End Academy until age fourteen, when he took a job as a clerk. In 1892, he married Margaret Croston; they had no children.

In 1895, Scrymgeour embarked on his political career as a Labour member of the Dundee Parish Council. When reelected to the parish council in 1898, he left full-time work and concentrated on politics. In 1905, he was elected to the Dundee Town Council, a position he held almost continually until 1919. In 1898, he became embroiled in a dispute within the IOGT, and in 1901 he led a breakaway group to form the SPP. He edited the SPP's *Scottish Prohibitionist,* which had a readership of 10,000.

From 1908, Scrymgeour stood for Parliament as "Prohibitionist Labour." He defeated Churchill in 1922 and served as M.P. for Dundee until 1931. Scrymgeour's was the only parliamentary success for the SPP. His win has been attributed to a variety of factors, including his personal tenacity, his ability to gain cross-party support by not standing as a Labour candidate, support from Irish voters based on his home rule stance, additional support in the 1929 election because of the 1928 female suffrage legislation, and Dundee's intense pride over its local celebrities.

Edwin Scrymgeour, and through him the SPP, held strong millennialist views. Millenarianism refers to the belief, rooted in the Book of Revelation, of 1,000 years of earthly peace and prosperity before the Last Judgment. He exhorted his supporters to "Vote as You Pray." SPP members wore "purity badges," thereby demonstrating their preparedness for peace on earth. Scrymgeour's reputation as an extremist in local and parliamentary politics can be traced to his millenarianism. Indeed, he believed prohibition was key to alleviating social ills and styled himself the "divinely commissioned avenger" against human wickedness.

Megan Smitley

See also: Churchill, Winston; Good Templars (IOGT)

References

Knox, William, and John Saville. "Edwin Scyrmgeour" in *Dictionary of Labour Biography,* vol. 7, 214–218. London: Macmillan, 1972–1993.

Walker, William M. "Dundee's Disenchantment with Churchill: A Comment upon the Downfall of the Liberal Party" in *Scottish Historical Review.* 49, nos. 147–148 (1970): 85–105.

———. "The Scottish Prohibition Party and the Millennium" in *International Review of Social History.* 18 (1973): 353–379.

Seagram

Initially a distilling company founded in 1883 by Joseph Emm Seagram (born on 15 April 1841 at Fisher Mills, Ontario; died 18 August 1919) of Waterloo, Ontario, Canada, Seagram was purchased in 1928 by the Bronfman family, who developed the business into a global liquor and entertainment empire before selling most of it off in 2001.

The original distillery was built on the Grand River in Waterloo, Ontario, in 1857 by William Hespeler and George Randall, who operated the distillery to deal with surplus grain from their Granite Mills milling operation. Seagram, the son of an English immigrant farmer and tavern owner, joined the distillery in 1864 and purchased Hespeler's share soon afterward (and married Hespeler's niece, Stephanie Urbs, in 1869). In 1873, he bought Randall's share in the company, and in 1883 he bought out the remaining partner, William Roos, becoming sole owner of the renamed Joseph Seagram Flour Mill and Distillery Company.

Seagram developed the distilling operation, marketing Seagram's Old Rye in Ontario and Seagram's White Feather in Quebec. He also introduced blended whiskeys to Canada. In 1883, he distilled a special blend to commemorate his acquisition of the distillery, and Seagram's '83 was first sold in 1887. It became one of Canada's most popular whiskeys and Seagram's best-seller during his lifetime. Seagram pursued business in European and U.S. markets after acquiring the distillery, and by the 1890s No. 83 was one of the most popular Canadian whiskeys in the United States. In 1911, he incorporated as Joseph E. Seagram and Sons Limited, and his sons, Edward and Thomas, became more active in the operation. Around 1913, Seagram produced Seagram's VO (standing for "Very Old" or "Very Own"), which grew to become one of the best-selling Canadian whiskeys in the world.

Seagram was also involved in horse breeding and politics. His horses won the Queen's Plate a total of ten times and five King's Plate victories before Seagram's death. He was first elected to Waterloo town council in 1881, and in 1896 he became Conservative member of Parliament for Waterloo North, a post he held until 1908. Seagram died in 1919, but the company remained a family operation until 1926, when it went public.

In 1928, the Bronfman family's Distiller's Corporation Limited purchased the distillery, which was foundering under U.S. National Prohibition. The Bronfmans incorporated their business as Distiller's Corporation–Seagram's Limited and operated under this name until 1975, when it was renamed The Seagram Company Limited. From 1928 until the end of the century, Seagram's history is inextricably linked with the Bronfman family.

In the period immediately following the repeal of National Prohibition in the United States, Seagram fell under suspicion for bootlegging activities purportedly undertaken by various members of the Bronfman family. In 1934, charges brought by the Royal Canadian Mounted Police were dismissed by Canadian courts. In the next year, U.S. government prosecutors settled their claims for more than $25 million in retroactive customs duties with main Canadian distilleries for $6 million, of which Seagram paid half.

In 1934, a year after the repeal of Prohibition, Seagram introduced its medium-priced Five Crown and Seven Crown whiskeys, which quickly became the leaders in volume of all brands on the U.S. market. Sales reached 1 million cases, an unprecedented record, in 1935. The Crown brands remained leaders for the next six years. In 1939, to commemorate the visit of the king and queen of England to Canada, Seagram introduced Crown Royal.

Seagram's marketing strategies were also innovative. Sam Bronfman, who reportedly was obsessive about the quality of his products, hoped to elevate the image of whiskey to portray it as a beverage of choice for the sophisticated class. He wanted to project the image of respectability, responsibility, and success within a distinctly male social culture. In 1938, Seagram's U.S. subsidiary was presented with the *National Ad-Views* "Socrates' High Award" for the quality and character of its advertising campaigns.

During World War II, Seagram acted patriotically and strategically. The Canadian government ordered distillers to produce alcohol used in the manufacture of military supplies such as smokeless powder. Seagram scientists also pioneered techniques for producing high-proof alcohol for the manufacture of synthetic rubber. Also during the war, Seagram expanded its liquor and spirits manufacturing holdings, moving into West Indian rum manufacturing, California wineries, and buying a number of Canadian and U.S. distilleries.

The postwar period saw continued expansion in diverse liquor interests. Seagram continued to expand rum production in the Caribbean, establishing Captain Morgan and acquiring controlling interest in the Puerto Rico Distiller's Group, Myer's Rum, and the Puerto Rico Rum Company in the 1950s. The company purchased Chivas Brothers in 1950, a Scottish grocer and liquor merchant with a distinguished name, which led to the introduction of Chivas Regal two years later. In 1956, Seagram established the Seagram Overseas Corporation, and a decade later the Seagram Overseas Sales Company (1965). Seagram was becoming a global organization.

Prosperity and growth in the industry in the 1970s gave way to decline in the 1980s, when rapid reductions in spirit sales led to the closure of distilleries across the country. The Seagram plant in Waterloo closed in 1992. The Seagram Museum, which had opened in 1984 on the site of the Seagram distillery in Waterloo and was reputedly the only integrated museum devoted to the history and technology of wine and spirits, closed in 1995. Its records were divided between the libraries at the University of Waterloo, Wilfrid Laurier University (Waterloo), Brock University (St. Catharines, Ontario), and the City of Waterloo Library. The archives of the Seagram Company after 1928 are housed at the Hagley Museum, Wilmington, Delaware.

In 2000, the Bronfmans announced the sale of the entire Seagram empire to Vivendi, which in turn divided the spirits and wine interests and sold them to Diageo and Pernod Ricard. Each company took some of the products. For example, Diageo acquisitions included Crown Royal, Seagram's VO, Captain Morgan Rums, 7 Crown, and Seagram's wine business. Pernod Ricard acquired Chivas Regal, The Glenlivet Scotch Whisky, Martell Cognac, and others. Except as a brand name, Seagram effectively ceased to exist on 21 December 2001.

Daniel J. Malleck

See also: Bronfman Family; Whiskies

References

Bianco, Anthony. "Deal Time at Seagram." *Business Week International Edition* website. http://www.businessweek.com/2000/00_26/b3687039.htm.

Marrus, Michael R. *Mr. Sam: The Life and Times of Samuel Bronfman.* Toronto: Penguin, 1991.

Newman, Peter C. *The Bronfman Dynasty: The Rothschilds of the New World.* Toronto: McClelland and Stewart, 1978.

Rannie, William F. *Canadian Whisky: The Product and the Industry.* Lincoln, Ontario: Rannie, 1976.

The Seagram Plant in Waterloo. Waterloo, Ontario: Seagram Museum, 1996.

Stone, S. I. "Throwing Stones: The Elements That for Approximately 75 Years Defined the Beverage Alcohol Industry in the United States Are, Obviously, Disappearing." *Massachusetts Beverage Business* website. http://www.beveragebusiness.com/art-arch/mmthrowingstone08a.html.

Thomson, Andrew McCauley. "Joseph Emm Seagram." Draft of entry completed for the *Dictionary of National Biography,* in Porter Rare Book Room, University of Waterloo, #G11830.

Server Responsibility Laws (United States)

Server responsibility laws (commonly referred to as "dramshop laws") attempt to reduce negative consequences of overconsumption of alcohol by making the party who provided the alcohol to the consumer partially responsible for the actions of the intoxicated individual. There are two primary goals of dramshop legislation: stopping the sale of alcohol to minors or intoxicated individuals and the compensation of innocent third parties who are injured as a result of the intoxicant's actions.

Prior to the temperance movement of the mid-nineteenth century, civil courts held individuals personally responsible for acts that were committed while inebriated. As the power of the temperance movement grew, legislators began to pass state laws to counteract the destructive influence of alcohol on the family. The focus of this early legislative action was to provide economic relief to children and family members of individuals who were injured or killed owing to the actions of an intoxicated individual. The vendors of the alcohol were now fiscally liable for the actions of their patrons. At the end of National Prohibition, these laws were repealed. In the 1980s, groups such as Mothers Against Drunk Driving (MADD) began pressuring legislators to reintroduce dramshop legislation in the hope that it would decrease the incidence of drunk driving. Currently, forty-four out of the fifty states in the United States have active dramshop laws in effect.

These states have similar dramshop legislation, with some minor variation. In order for an alcohol-procuring establishment to be liable under these laws, one of four conditions needs to be met. First, the sale falls under dramshop legislation if it is illegal under other standing alcohol-related laws. Even selling alcohol on a Sunday in violation of local ordinance can create liability if the intoxication leads to injury. Second, the sale of alcohol is subject to dramshop laws if it is made to a visibly intoxicated individual. Even though the initial intoxication may be the responsibility of the alcohol-consuming individual, adding to the danger of the situation by increasing the drunkenness makes the seller subsequently liable. Third, the sale of alcohol to a minor creates liability under the legislation. Minors are considered to lack the emotional and physical maturity needed to manage the use of alcohol, and any consumption is considered inherently dangerous in the eyes of the court. The fourth condition is one in which the purveyor of alcohol was misleading the consumer, generally in either misrepresenting the strength of the alcohol or not disclosing that the product contained alcohol.

Dramshop laws clearly limit who is qualified to seek remedy from alcohol-related injury. Inebriates as well as coadventurers (that is, drinking buddies) are typically exempt. Some states have allowed family members of intoxicated individuals to use these laws, but this is an exception. Primarily, dramshop laws serve as a remedy for innocent individuals who are injured as a result of the actions of the intoxicated individual.

Research has suggested that dramshop laws have been effective. In states with dramshop laws, there has been a 10 percent decrease in alcohol-related fatalities. Other studies demonstrate an increase in commercial establishments participating in beverage server training programs as well as alcohol vendors taking greater precautions against serving minors or intoxicated individuals.

Paul E. Priester

See also: Civil Damage Laws

References

Holder, Harold, Kathleen Janes, James Mosher, and Robert Saltz. "Alcoholic Beverage Server Liability and the Reduction of Alcohol-Involved Problems" in *Journal of Studies on Alcohol.* 54 (1994): 23–36.

Hunnicutt, D., M. Mann, K. Wagner, and J. Leutzinger. "Using Environmental Strategies to Reduce Drinking and Driving among College Students" in *NASPA Journal.* 33 (1996): 179–191.

Parsons, Elaine Frantz. "Slaves to the Bottle: Smith's Civil Damage Law" in *Annals of Iowa.* 59, no. 4 (2000): 347–373.

Sloan, Frank, Bridget Reilly, and Christopher Schenzler. "Effects of Prices, Civil and Criminal Sanctions, and Law Enforcement on Alcohol-Related Mortality" in *Journal of Studies on Alcohol.* 55 (1993): 454–465.

Smith, Richard. "A Comparative Analysis of Dramshop Liability and a Proposal for Uniform Legislation" in *Journal of Corporation Law.* 25 (2000): 553–601.

Shea's Winnipeg Brewery Limited

Shea's Winnipeg Brewery Limited was one of western Canada's largest independent brewers during the late nineteenth and early twentieth centuries. Shea's should be rightly recognized as having established the famous stable of horses that are now prominently associated with Anheuser-Busch, brewers of Budweiser. Shea's Winnipeg Brewery traces its origins to the brewery established by French émigré Celestin Thomas in the Middlechurch area of the Red River Colony, Manitoba, around 1860–1862. In 1873, Thomas, seeking better prospects, relocated to Colony Street in the rising town of Winnipeg. The business was rechristened the Winnipeg Brewery. For the next decade and a half, Thomas either ran the brewery himself or rented it out to various tenant-brewers. However, the last of Thomas's tenants, Torontonian John Cosgrove, short on capital but long on ambition, overextended himself through plant renovation, went insolvent, and surrendered the business to his chief creditor, the Commercial Bank of Winnipeg. The bank later bought the brewery land from Celestin Thomas and in September 1887 sold the Winnipeg Brewery and the brewery lot to Irish immigrants John McDonagh and Patrick Shea for $16,000. To finance the purchase, McDonagh and Shea sold the Waverley Hotel, a business they had owned and operated jointly since 1884.

In a partnership styled McDonagh and Shea, the former hotelkeepers used their working knowledge of the local drink scene to put their new acquisition on a sound financial footing. Sales benefited from Winnipeg's booming resident population; the city numbered around 2,000 inhabitants prior to city incorporation in 1874 and well over 30,000 some dozen years later. Winnipeg's position as the gateway city to Canada's expanding prairie west also served the brewers well, since thousands annually passed through the city on their way to new homes further west. Moreover, Winnipeg, as Canada's western railway hub, gave McDonagh and Shea convenient shipping access to new and growing prairie markets. Their brewery's fortunes, although substantial, must be seen in relative terms, however; indeed, the Winnipeg Brewery far from rivaled the size of much larger breweries in central and eastern Canada. Nevertheless, McDonagh and Shea were major regional players in the province of Manitoba.

In 1894, John McDonagh died, leaving his share of the business to Patrick Shea, who continued to brew under the McDonagh and Shea name. Over the next two decades, Shea, faced with stiffening competition from local, eastern Canadian, and even U.S. breweries, increasingly cultivated sales agencies throughout Manitoba and Saskatchewan and in parts of southeastern Alberta and northwestern Ontario. Shea's operative assumption was that securing markets in newly established centers before his competitors did was his best means of survival. When and where he could, the Winnipegger employed various means of leverage over hotels—ranging from discount incentives to holding hotel mortgages—to arrive at exclusive retail arrangements in order to boost his sales. To service his growing agency trade, Shea incorporated a distribution subsidiary, Brewery Products Limited, in 1906. Underpinning Shea's agency development was his turn to a new promotional strategy that incorporated customized embossed bottles, full-color labels, brand-driven newspaper advertisements, bar posters, and a variety of advertising giveaways. Shea also routinely funneled a portion of his earnings into plant improvement, such as physical expansion and electric automation. Shea's overall marketing plan worked. On the eve of province-wide prohibition in Manitoba in 1916, the McDonagh and Shea brewery reigned as the largest on the Canadian prairies.

Provincial prohibition forced Shea to rely on distant sales. Accordingly, he began to ship his beer overland by rail to Port Arthur/Fort William (now Thunder Bay, Ontario) and thence by lake steamer to reach the heart of Ontario. When this venture proved more difficult than anticipated—Ontario's brewers tightly controlled their home market—Shea looked south, a move that became all the more promising when the Volstead Act went into effect in the United States on 1 January 1920. To overcome his brewery's lack of recognition outside western Canada, Shea created a clever promotional device to place his name before a continental public. In 1921, despite having earlier adopted gasoline-powered delivery trucks, the brewer purchased several purebred Clydesdales from a rancher in Arcola, Saskatchewan. To these horses he coupled a specially commissioned show wagon, which, of course, prominently sported his beer's name. As the Shea horse-team claimed prize after prize at major fairs and exhibitions from Ottawa to Chicago, recognition of the Shea name spread across central North America. In addition, press coverage of Shea's contest winnings came at no cost, and it circumvented the ban on alcohol advertising in the myriad North American jurisdictions that Shea had identified as his beer markets. The Clydesdales won the brewer renown, and the rather porous land border between western Canada and the western United States facilitated his "export" sales. Meanwhile, Shea incorporated as Shea's Winnipeg Brewery Limited in 1926. Seven years later, eight horses from the Shea stable were sold to Anheuser-Busch of St. Louis, Missouri. Today, descendants of Shea's original Clydesdales continue in the promotional service of the U.S. brewer and are a worldwide marketing phenomenon.

In 1933, Patrick Shea died. His son, Frank, who assumed the brewery's presidency, also died that year, leaving a non–family member, but long-term brewery employee, John T. Boyd, in charge. Boyd, a capable manager in his own right, steered the brewery through the Great Depression and World War II. Most notably, over the 1930s, Shea's added to its subsidiary hotel chain, an inheritance from earlier opportunistic acquisition during provincial prohibition. The majority of stock in Shea's Winnipeg Brewery Limited remained in the Shea family until 1953, when John Labatt Limited of London, Ontario, secured controlling interest in the company as part of its postwar national expansion program. Five years later, Shea's was reincorporated as Labatt's Manitoba Brewery Limited. In 1974, a new Labatt plant, at 1600 Notre Dame Street in Winnipeg, replaced the old Shea brewery on Colony Street. Amidst a plant reduction program in the mid-1990s, Labatt closed its Winnipeg facility, and it no longer bottles under the Shea brand name.

Glen C. Phillips

See also: Anheuser-Busch; Labatt Brewing Company Limited

References

The Brewing Industry of Canada, various volumes. Ottawa: Dominion Bureau of Statistics (later Statistics Canada), 1925–1975.

Douglas, William. *The House of Shea: The Story of a Pioneer Industry.* Winnipeg, Manitoba: Shea's Winnipeg Brewery, 1947.

Winnipeg Free Press, 1874–1974.

Shebeens

The exact origins of the word *shebeen* are unknown, but it most likely came into popular use in Ireland from the Irish-Gaelic language before spreading to the English-speaking regions of Ireland. The Irish Catholic diaspora in the nineteenth century appears to have expanded its use across Britain, the northeastern United States, Australia, and perhaps as widely as to all locations where large numbers of Irish Catholics migrated.

Defined as a low beerhouse or unlicensed house that sold beer or spirits, the shebeen was common throughout Ireland in the nineteenth and early twentieth centuries. Though the term usually referred to a fixed location (house, tenement room, garret, stable, or even cellar), a vendor of beer or spirits on foot or horseback or using a horse-drawn carriage was also at times known as a "traveling shebeen." This was especially true in the latter decades of the nineteenth century, when efforts were made to stamp out the unofficial sale of intoxicating beverages in Ireland, and when "shebeen" became an increasingly derisive word suggesting an illegal house of a low or criminal character. Because of the clandestine nature of shebeens, it is impossible to know with any certainty the number that existed at any time. Their numbers were considerable, especially in the poorer quarters of Dublin and Belfast, where they posed a serious threat to the public-house trade in the middle decades of the nineteenth century.

Local magistrates, the police, and publicans' trade associations worked together to root out shebeens and their proprietors. The state viewed shebeens as a threat to public order and an incubator of seditious activities. Licensed vintners saw the shebeen not only as an economic threat to the legitimate liquor retail trade but also as the source of a stigma tarnishing the image of the respectable licensed house. Condemned as sanctuaries for thieves, prostitutes, and the purveyors of *pocheen* (contraband Irish whiskey), shebeens became synonymous with the evils potentially arising from unregulated public drinking spaces. Shebeen proprietors were mostly lower-class women of questionable morality who allegedly lured unsuspecting patrons into their establishments with relatively inexpensive drink that was adulterated or of a poor quality.

Irish publicans criticized almost all forms of temperance legislation as a boon to the shebeen trade. Curtailing or prohibiting the respectable licensed trade, they contended, would serve only to force drinking into the back alleys and tenements of the lowest urban neighborhoods and far-off rural environs where shebeens could operate most freely.

With advancements in communication, transportation, and police supervision, the presence of shebeens began to wane in Ireland by World War I. It was not until the 1950s, however, that local pub and home consumption completely supplanted this underground Irish working-class institution.

In the United States, the pattern was much the same. Emerging in the 1830s with the influx of Irish immigrants in the northeast, shebeens, often run by nearly destitute Irish widows, emerged as competitors to the officially licensed liquor trade. Their demise in America appears to have predated their decline in Ireland. They became increasingly scarce in the 1890s because of government and saloon trade efforts.

Bradley Kadel

See also: Ireland; Shebeens (Southern Africa)

References

Kearns, Kevin C. *Dublin Pub Life and Lore: An Oral History.* Dublin: Gill and Macmillan, 1996.

Malcolm, Elizabeth. *"Ireland Sober, Ireland Free": Drink and Temperance in Nineteenth Century Ireland.* Dublin: Gill and Macmillan, 1986.

Powers, Madelon. *Faces along the Bar: Lore and Order in the Workingman's Saloon, 1870–1920.* Chicago: University of Chicago Press, 1998.

Shebeens (Southern Africa)

Shebeens emerged as a product of the rapid industrialization and urbanization of early-twentieth-century South Africa. This term, borrowed from the Irish, came to describe a highly characteristic institution that became a haven for African drinkers across southern Africa. With the development of the mining industry in the Witwatersrand from the 1880s, distilleries flooded local markets and created what many regarded as a serious alcohol problem among Africans. Concerned about labor discipline and productivity, public officials and employers imposed prohibition on Africans in 1897. An illicit liquor market soon developed, and drinking establishments proliferated. By the 1920s, there were hundreds of shebeens in the black townships around Johannesburg.

In Durban on the Indian Ocean coast, the city fathers reacted to the emergence of beer bars operated by local household brewers by imposing a state monopoly on beer in 1908. The municipality erected a series of beer halls that provided the only legal access to alcohol for Africans and directed the large profits to the maintenance of racial segregation. In Durban, as on the Rand, individual entrepreneurs, often female, opened shebeens to provide for their families. The owners operated in the shadows and had warning networks to guard against ubiquitous police raids; women clubbed together to bribe officials and pay fines for those unlucky enough to get arrested. Liquor had to be produced quickly and in secret and kept buried in drums waiting for the customers who sought out these establishments in preference to the sterile mass beer halls (which were also introduced on the Rand in the 1930s). In this atmosphere, traditional brewing was discarded in place of rapid infusions of cheap rotgut obtained from white liquor sellers.

Shebeens offered both a harbor for men cut off from family and rural life and a rapid tutorial in the styles and mores of South African cities. However humble, no shebeen could survive without music, dancing, and sharply dressed young men; amid this Marabi culture the distinctive South African jazz style was born. The image of drinkers enjoying music and drinks such as "skokiaan" (homemade sorghum beer) or other home brews such as "barberton," often called "kill me quick" because of its dangerous methanol content, obscured a reality that included fear of police, violence, alcohol abuse, and desperate poverty. After 1945, low-class shebeens nurtured drunkenness and provided bases to bands of *tsotsi* gangs that terrorized the black townships. At the same time, in the pages of *Drum* magazine writers such as Can Themba celebrated life in the posh shebeens of Joburg, where successful hoodlums adopted the names of American film stars and a distinctive subculture thrived among a thoroughly urbanized population.

Yet the individual "Shebeen Queens" that dominated the commerce still faced the constant threat of arrest. By the mid-1950s, more than 250,000 South Africans were being convicted annually of liquor offenses. Brewers aggressively opposed the state monopoly, and their protests climaxed in mass demonstrations in 1959 at Cato Manor outside Durban. But this movement was soon crushed, as almost all black political activity was suppressed. Ironically, in 1961, as the government pressed its violent policies of political repression and mass black population removals, prohibition was repealed, making it legal for Africans to purchase European-type beer, wine, and spirits. This action, in conjunction with the imposition of strict urban apartheid, marked the end of the era of the shebeen and the subculture it had spawned.

Charles Ambler

References

Bailey, Jim, and Adam Seftel, eds. *Shebeens Take a Bow!! A Celebration of South Africa's Shebeen Lifestyle.* Johannesburg: Bailey's African History Archives, 1994.

Crush, Jonathan, and Charles Ambler, eds. *Liquor and Labor in Southern Africa.* Athens: Ohio University Press, 1992.

La Hausse, Paul. *Brewers, Beerhalls and Boycotts: A History of Liquor in South Africa.* History Workshop Booklet. Johannesburg: Ravan, 1988.

Sheppard, Catherine ("Kate") Wilson (1847–1934)

Kate Sheppard (1847–1934) led the campaign for women's suffrage in New Zealand as national superintendent of the franchise departments of the Woman's Christian Temperance Union (WCTU; known in New Zealand as the Women's Christian Temperance Union). A persuasive writer and speaker and a skilled strategist and organizer, Sheppard led a successful movement. Women won the vote in 1893, making New Zealand the first nation-state with the female franchise (available at the same age as the male and to indigenous Maori as well as the white population). Kate Sheppard continued to be a leading advocate for women's rights, temperance, and a broad range of social reform in New Zealand and internationally.

Born Catherine Wilson Malcolm in Liverpool in 1847 to Scots parents, Kate (as she preferred to be known) traveled to New Zealand in 1868 with her widowed mother and two brothers (her father had died in 1862). She was well educated, and her lifelong adherence to religious principles and interest in Christian socialism is attributed to the influence of an uncle who was a minister in the Free Church of Scotland in Nairn. At age twenty-four, Kate married Walter Sheppard, a grocer and general merchant; their only child, Douglas, was born in 1880. In these years, she was active in the Trinity Congregational Church, where her capacity for leadership was recognized early. Sheppard became secretary of the Ladies Association as well as a participant in Bible classes and fund-raising.

The WCTU was founded in New Zealand in the wake of Mary Leavitt's 1885 mission. Two years later, Kate Sheppard was appointed national superintendent for the newly established franchise and legislation departments. She argued strongly that women would have little influence over liquor legislation while excluded from direct political power. Her arguments for the vote were advanced with a keen intelligence—she saw no reason why women should be excluded from the political and social world and felt their position within the family was devalued by their lack of political and social rights. Speaking at public meetings, corresponding extensively both publicly and privately, and circulating petitions, Sheppard built an effective nationwide campaign. Her acumen was vital in building a broad coalition incorporating temperance and non-temperance supporters of the women's vote and in linking a broad activism with tactical moves by key allies in the House of Representatives. Ten weeks after the Electoral Act was passed in September 1893, 65 percent of eligible women voted in the general election.

In 1894, Kate Sheppard traveled to England, where she was in demand as a speaker in debates on women's franchise. She also attended the World's Woman's Christian Temperance Union (WWCTU) convention in London in 1895, meeting with the union president Frances Willard. Sheppard was also a prolific writer, becoming founding editor of the WCTU's paper, the *White Ribbon,* in 1895, a position she held until 1903. In 1896, she was elected inaugural president of the National Council of Women (NCW) of New Zealand, a federation of women's political organizations dedicated to pressing the women's rights agenda forward in the postfranchise era.

Suffering ill health, Kate Sheppard withdrew from much public life in her mid-fifties, but she traveled again to England via North America in 1903 to meet with Carrie Chapman Catt and other suffrage leaders. She continued to contribute to women's activism in New Zealand and internationally through her writing and was instrumental in reviving the NCW after a period of recess following World War I.

Both her son and husband had died by 1915; in 1925, she married William Lovell Smith, author of *Outlines of the Women's Suffrage Movement in New Zealand* (1905). Sheppard died in Christchurch in July 1934. Her memory is kept alive at a popular level still today, as her face features on the current New Zealand $10 banknote.

Charlotte Macdonald

See also: Leavitt, Mary Greenleaf Clement; New Zealand; World's Woman's Christian Temperance Union (WWCTU)

References

Devaliant, Judith. *Kate Sheppard: A Biography.* Auckland: Penguin, 1992.

Grimshaw, Patricia. *Women's Suffrage in New Zealand.* Rev. ed. Auckland: Auckland University Press/Oxford University Press, 1987.

Macdonald, Charlotte. "Kate Sheppard" in *The Book of New Zealand Women,* edited by Charlotte Macdonald, Merimeri Penfold, and Bridget Williams, 604–607. Wellington: Bridget Williams, 1991.

Malcolm, Tessa K. "Katherine Wilson Sheppard, 1847–1934," *Dictionary of New Zealand Biography,* vol. 2, 459–462. Wellington: Bridget Williams Books/Department of Internal Affairs, 1993. Also at http://www.dnzb.govt.nz.

Sheppard, John Morris (1875–1941)

John Morris Sheppard (1875–1941), Texas prohibitionist and Senate sponsor of the Eighteenth Amendment, was born in Wheatville, Texas. He served in the U.S. House of Representatives from 1902 to 1913, and in the Senate from 1913 until his death in 1941.

Sheppard was a graduate of the University of Texas and the University of Texas Law School. He received a master's degree in law from Yale in 1898. From early in his career, he was an ardent prohibitionist, abstaining from alcohol and tobacco while at New Haven. Upon his return to Texas, he worked in the family law firm, was active in the Woodmen of the World Life Insurance Association, and aligned himself with the progressive wing of the Texas Democratic Party. He was elected to the U.S. House of Representatives in 1902, filling the seat vacated by the death of his father, Congressman John Levi Sheppard. In the House he was an early sponsor of the Webb-Kenyon Act, a significant piece of legislation that set the stage for National Prohibition by banning the commercial transportation of liquor from "wet" to "dry" states.

In 1913, Sheppard was elected by the Texas legislature to fill the U.S. Senate seat vacated by the resignation of the controversial

Joseph Bailey. He won that vote despite the opposition of many prominent Democrats in the state who disliked his progressive politics. In his first speech after election to the Senate, he promised to support the policies of President-elect Woodrow Wilson and to champion prohibition. During his tenure in the House, he also supported child-labor protection, antitrust legislation, and women's suffrage.

In 1913, when the Anti-Saloon League of America (ASLA) presented its draft of a proposed constitutional amendment to prohibit the manufacture or sale of alcohol, Sheppard was on hand to receive the ASLA delegation and to present the proposal to the Senate. He lobbied President Wilson to support the measure, writing to him after U.S. entry into World War I that prohibition would increase recruitment numbers and help to conserve resources for the war effort. Sheppard introduced the Eighteenth Amendment on 4 April 1917 and led the prohibition forces during the debate, echoing William Lloyd Garrison in decrying the Constitution without a prohibition amendment as "still a covenant with death and an agreement with Hell." After the amendment passed, Sheppard assisted in the drafting of the Volstead Act and worked to keep intact its strict enforcement requirements.

During the 1920s, with his party out of power, Sheppard continued to work on reform issues in the Senate, rose through the Democratic ranks, and was selected to be Democratic whip in 1929. With the return to power of the Democrats in 1932, Sheppard became an important supporter of Roosevelt's New Deal legislation, opposing the president only on the issue of repeal. After the ratification of the Twenty-First Amendment repealing Prohibition, Sheppard did not abandon the cause. He continued to oppose the sale of alcohol, even endorsing a second prohibition amendment. He delivered a prohibition speech each year on the anniversary of the ratification of the Eighteenth Amendment. He died while still serving in the Senate in 1941.

James D. Ivy

See also: Democratic Party (United States); Federal Regulation of Alcohol before 1920 (United States); National Prohibition (United States)

References

Gould, Lewis L. *Progressives and Prohibitionists: Texas Democrats in the Wilson Era.* Austin: University of Texas Press, 1973.

Hamm, Richard F. *Shaping the Eighteenth Amendment: Temperance Reform, Legal Culture, and the Polity, 1880–1920.* Chapel Hill: University of North Carolina Press, 1995.

"Sheppard, John Morris," The Handbook of Texas website. http://www.tsha.utexas.edu/handbook/online/articles/view/SS/fsh24.html.

Simonds Family

The Simonds family of Reading, Berkshire, was involved in brewing both in Britain and in the British Empire for more than 200 years. The Simonds had lived in Berkshire since Saxon times and had prospered in a variety of trades and professions as maltsters, millers, farmers, landowners, lawyers, and bankers. William Simonds, a maltster, began brewing as early as 1768, and his son, William Blackall Simonds, inherited the business in 1782. William Blackall Simonds then established a new brewery in Broad Street in 1785. He also played a large part in the social and economic life of Reading, as town treasurer in 1793, 1802, and 1817 and as receiver general of taxes for Berkshire. From 1790, William Blackall Simonds was also a partner in the family bank, J&C Simonds Bank of Reading. In about 1800 he brought his three sons—Blackall Simonds (1784–1875), Henry Simonds (1785–1874), and George Simonds (1794–1852)—into the business, and they continued running the family company after his retirement in 1816. By the 1850s, the firm was being led by William Blackall Simonds's grandsons: Henry John Simonds, Henry Adolphus Simonds, George Blackall Simonds, Louis de Luze Simonds, and another Blackall Simonds. Output was 115,000 barrels in 1885, the same year the family firm became a limited liability company called H&G Simonds Limited.

Louis's eldest son, Frederick Adolphus "Eric" Simonds, entered the business in 1902 and became the driving force behind the company until his retirement in 1952. The success of Simonds's business had been built on supplying beer to the rapidly expanding network of railway refreshment rooms and on sales of beer to the army garrisons at nearby Sandhurst. As the British Empire expanded, so did the firm's trading area; a lucrative trade with the armed forces overseas was established, with branches at Malta, Gibraltar, Egypt, and South Africa. Then, in 1929, H&G Simonds became part owners of Simonds Farsons Limited on Malta and sold beer to the constantly changing garrisons on the island as well as along the coast of North Africa. Under the leadership of Eric Simonds, the firm also expanded its interests within Britain and acquired a string of smaller brewers: the Tamar Brewery, Plymouth, in 1919; South Berks Brewery Company of Newbury, Berkshire, in 1929; Ashby's Brewery at Staines, Middlesex, in 1930; W. J. Rogers of Bristol in 1935; the Cirencester Brewery, Gloucestershire, in 1938; Lakeman's Brewery of Brixham, Devon, in 1938; R. H. Stiles of Bridgend, South Wales, also in 1938; and the Dorsetshire brewer of J. L. Marsh in 1939.

After World War II, Eric Simonds's son, E. D. (Duncan) Simonds (1920–2002), joined the board of directors. H&G Simonds continued to expand at home and abroad, acquiring Bowley's Brewery of Swindon, Wiltshire, in 1945; a substantial interest in East African Breweries Limited of Nairobi, Kenya, in 1947; John May of Basingstoke, Hampshire, in 1948; Phillips Brewery of Newport, South Wales, in 1949; and the Octagon Brewery of Plymouth, Devon, in 1954. When Eric Simonds retired as chairman in 1952, he was succeeded by General Sir Miles Dempsey, this being the first time control of the firm had passed out of the hands of a family member. H&G Simonds found itself on the defensive in the "merger mania" that swept the British brewing industry of the 1950s, and in 1960 a merger was announced with London brewers Courage and Barclay. The new firm of Courage Barclay and Simonds Limited became one of the so-called "Big Six" brewers of 1960s Britain, and E. Duncan Simonds, son of Eric Simonds and a direct descendant of the founder William Blackall Simonds, represented the family on the board throughout this period.

Kenneth Thomas

See also: Beer; Courage Brewery

References

Corley, T. A. B. *The Road to Worton Grange.* Reading: Courage, 1980.

Pudney, John. *A Draught of Contentment: The Story of the Courage Group.* London: New English Library, 1971.

Thomas, Kenneth. "The Adventures of H & G Simonds in Malta and East Africa" in *Business Archives.* November 1992: 40–54.

Sinclair, Upton (1878–1962)

Upton Sinclair (1878–1962), novelist, reformer, and political activist, was born in Baltimore, Maryland. Like many American reformers, and unlike many writers of his generation, Sinclair was a lifelong teetotaler. His father worked for a time as a wholesale liquor dealer, and when he gave up that line of work he was unable to give up the drink. Sinclair reported childhood memories of his father's unsuccessful efforts to reform. His father's brother died an alcoholic, and his mother's brother, also a heavy drinker, committed suicide when Sinclair was a child. For the rest of his life he condemned social drinking and warned of the dangers of alcohol. The crusade against alcohol was a central feature of his literary career.

In his autobiography, Sinclair listed "the gifted people known to me who fell into the grip of John Barleycorn," including Jack London, George Sterling, Eugene O'Neill, Scott Fitzgerald, O. Henry, Stephen Crane, Finley Peter Dunne, Isadora Duncan, Edna St. Vincent Millay, Dylan Thomas, and Sherwood Anderson (Sinclair 1962). In his fiction, Sinclair continued his critique of alcohol. When the laborers of *The Jungle* (1906) and the miners of *King Coal* (1917) succumb to the temptation of drink, it unfailingly contributes to their miseries. Even the temperate and hard-working Jurgis Rudkis, protagonist of *The Jungle,* eventually turns to drink at his peril.

In 1931, Sinclair published *The Wet Parade,* a prohibition novel reportedly inspired by a *Literary Digest* poll indicating that most Americans supported repeal of the Eighteenth Amendment. The protagonists are Maggie May Chilcote and her husband Kip Tarleton. Chilcote's family background is clearly drawn from Sinclair's personal experiences. She is southern by birth, living in New York. Her father, an alcoholic, committed suicide, and her brother is an alcoholic poet, not unlike Sinclair's close friend George Sterling. After she and Tarleton wed, she becomes a temperance lecturer and Tarleton a Prohibition agent. Before the novel ends, he is killed in the line of duty. The novel was not a critical success, but it was less maudlin than its nineteenth-century predecessors and sold well. It became Sinclair's only novel produced as a major motion picture. In 1932, the MGM film, starring Robert Montgomery and Myrna Loy, premiered at Grauman's Chinese Theater in Hollywood.

Sinclair's political activism was not limited to the fight against repeal. In 1933, he broke with the Socialist Party and began a campaign to become the Democratic nominee in California's 1934 gubernatorial race. He published a pamphlet entitled *I, Governor of California and How I Ended Poverty: A True Story of the Future* and founded the organization End Poverty in California (EPIC) to organize the campaign. He outlined a plan to confiscate and redistribute wealth and may have had some influence on the liberals among the New Deal architects. To the surprise of many, he won the nomination, but he lost the race for governor to Republican Frank Merriam. Had it not been for a third-party progressive candidate, he may well have been victorious, however.

In 1956, long after the fight against repeal had been lost, Sinclair published *The Cup of Fury,* a cautionary catalog of the disastrous impact of drink on cultural giants of the twentieth century and a paean to wide-eyed sobriety. In it he mourns the tragic deaths of great writers and castigates his lifelong nemesis and friend H. L. Mencken for his advocacy of drink. Upton Sinclair died in 1962 at the age of ninety.

James D. Ivy

See also: Literature, Representations of Drinking in

References

Bloodworth, William A., Jr. *Upton Sinclair.* Boston: Twayne, 1977.

Harris, Leon. *Upton Sinclair: American Rebel.* New York: Thomas Y. Crowell Co., 1975.

Sinclair, Upton. *The Autobiography of Upton Sinclair.* New York: Harcourt Brace, 1962.

———. *The Cup of Fury.* Great Neck, NY: Channel Press, 1956.

———. *The Wet Parade.* New York: Farrar & Rinehart, 1931.

Slack, Agnes Elizabeth (1857–1946)

Agnes Elizabeth Slack (1857–1946) was an international temperance organizer and an indefatigable worker for the British Women's Temperance Association (BWTA), the National British Women's Temperance Association (NBWTA), and the World's Woman's Christian Temperance Union (WWCTU). She came from a Nonconformist and Liberal family. Her early temperance career included working with women at the Ripley Wesleyan Church, where she was the organist and led the choir. She organized weekly teas for women, a savings bank, Bible readings, and sewing circles. In 1893, when she was elected to the Board of Guardians, she was the only female guardian in Derbyshire. Her experience in local politics was to stand her in good stead in her later career. A political woman, as a member of the executive of the Women's Liberal Federation she was responsible for the women's Liberal associations in Derbyshire and adjacent counties. She also was a member of the Central Society for Women's Suffrage and on the executive council of the National Free Church Council. She supported W. T. Stead and his campaigns to increase the age of sexual consent. After World War I, she attended meetings of the League of Nations.

On the recommendation of Lady Henry Somerset, Slack became honorary secretary for both the NBWTA and the WWCTU in 1895. She served the latter organization for nearly fifty years. These posts required Slack to travel widely in the British Isles and abroad. One of her earliest journeys was to Ireland in 1895 at Frances Willard's request. During her visit she founded the first white ribbon organization in Dublin. She rallied Methodists to her cause but had less success in recruiting Catholics. Her Irish tour brought her to schools, coffeehouses, halls, and the barracks of the 15th Hussars, where she preached on temperance. She returned to Ireland in 1901, visiting County Mayo, where she was shocked by the number of licenses granted for the sale of drink and the amount of alcohol drunk. Invited by the American WCTU, she visited the United States in 1896. She spoke at state conventions in place of Frances Willard, who was in England and ill. Slack took the opportunity to learn about social issues in the United States as well as studying American methods of reform to prevent the abuse of alcohol. In the following year, she became editor of the WWCTU's *White Ribbon Bulletin,* a four-page monthly. She also contributed frequently to other women's temperance newspapers.

Slack worked with Frances Willard, Lady Henry Somerset, and Rosalind Howard, Countess of Carlisle. Carlisle succeeded Somerset as NBWTA president in 1903 and as WWCTU president in 1906. The relationship between Carlisle and Slack, the WWCTU secretary, was strained. Both women were confident and determined, and they clashed on matters of methodology and policy.

Slack was a woman of her class and time. She published books on her travels that seem in the twenty-first century at the very least patronizing. One of her books, *My Travels in India* (1908), dealt with her extensive travels in the subcontinent, where she visited Christian missions, observed the temperance movement, and studied alcohol abuse and Indian life in general as well as Britain's role. Her later travels took her to Scandinavia several times to study and evaluate the effects of drink experiments in the Nordic countries and especially the Gothenburg scheme of disinterested management. She rejected the Gothenburg experiment as a solution for the United Kingdom, however.

Slack was the last president of the NBWTA in 1925–1926 and the first president of the National British Women's Total Abstinence Union in 1926–1928, after the reunion of the NBWTA and the Women's Total Abstinence Union. She remained active in the "British Women" until shortly before her death in 1946.

Many of Slack's letters can be found in the papers of Rosalind Howard, Countess of Carlisle, at the Castle Howard archives in Yorkshire.

Margaret Barrow

See also: British Women's Temperance Association (BWTA); Carlisle, Countess of (Rosalind Howard); Gothenburg Schemes/Disinterested Management; National British Women's Temperance Association (NBWTA); Somerset, Isabella Caroline (Lady Henry Somerset); Willard, Frances Elizabeth Caroline; Woman's Christian Temperance Union (United States); World's Woman's Christian Temperance Union (WWCTU)

References

Slack, Agnes E. *My Travels in India.* London: NBWTA, 1908.

———. *People I Have Met and Places That I Have Seen: Some Memories of Agnes E. Slack.* Compiled by Ernest Winterton. Bedford: privately printed, 1942.

Tillyard, Aelfrida. *Agnes E. Slack: Two Hundred Thousand Miles: Travels for Temperance in Four Continents.* Cambridge: W. Heffer, 1926.

Tyrrell, Ian. *Woman's World, Woman's Empire: The Woman's Christian Temperance Union in International Perspective, 1880–1930.* Chapel Hill: University of North Carolina Press, 1991.

Smith, Alfred Emanuel (1873–1944)

Alfred E. Smith (1873–1944), four-term governor of New York and Democratic presidential nominee in 1928, was the most prominent U.S. politician opposed to National Prohibition and, because of his Catholicism and identification with the interests of immigrants, a symbol of the ethnic and religious divisions that accompanied temperance reform. Born to Irish immigrant parents and raised in the Bowery in New York City, Smith served his political apprenticeship in the city's Democratic Tammany Hall organization and entered the New York Assembly in 1904. He served there until 1915, compiling a record distinguished by concern for the immigrant, industrial reform, and administrative efficiency.

Although unfriendly to temperance legislation and denounced by the New York Anti-Saloon League, Smith did not become a "wet" celebrity until he was elected governor in 1918. In 1919, he opposed the ratification of the Eighteenth Amendment by the state legislature, calling instead for a popular referendum on National Prohibition. In 1920, he signed a bill allowing the manufacture and sale in New York of beer with an alcohol content of 2.75 percent, a position that disregarded the national standard of 0.5 percent established by the Volstead Act. The U.S. Supreme Court soon ruled against Smith's viewpoint on both issues, and Smith was defeated for reelection. In 1921, New York adopted a state prohibition enforcement law, the Mullan-Gage Act, in keeping with the concurrent state and federal enforcement strategy of the Eighteenth Amendment.

Returning as governor in 1923, Smith again endorsed the modification of Prohibition to allow for 2.75 percent beer and, after some hesitation, signed the repeal of the Mullan-Gage Act, thus ending state enforcement of Prohibition. That action, along with press reports of Smith's nostalgic comments about saloons, made the New York governor the symbol of opposition to

National Prohibition. Smith contended for the Democratic presidential nomination in 1924, but the raucous, bitterly divided New York convention conveyed the impression that the party was split between a rural, "dry" wing unwilling to condemn the Ku Klux Klan and an urban, ethnic, wet faction devoted to Al Smith, and Smith lost.

In 1928, Smith received the Democratic presidential nomination. The candidate surprised party officials, who had issued a noncommittal platform statement on Prohibition, by immediately announcing his enthusiasm for fundamental changes in National Prohibition. Some dry Democrats, such as the Anti-Saloon League's Bishop James Cannon, denounced Smith's wet convictions and his Catholic faith, and the New York governor was soundly defeated by Herbert Hoover, the Republican candidate, in the national election.

Smith's personal style reinforced his image among drys as an apologist for urban, ethnic defiance of Prohibition. A talented administrator and master of legislative detail, Smith's colloquial speech, sharp Bowery accent, and gravelly voice led his foes to see him as a Tammany chieftain. Smith also drank moderately. He echoed arguments of the Association Against the Prohibition Amendment (AAPA) and named the AAPA's John J. Raskob, a former Republican, as his 1928 campaign manager. Wary of growing federal power, Smith later joined former AAPA leaders in the anti–New Deal American Liberty League.

Thomas R. Pegram

See also: Association Against the Prohibition Amendment (AAPA); Cannon, James, Jr.; Democratic Party (United States); National Prohibition (United States); Raskob, John Jakob

References

Kyvig, David E. *Repealing National Prohibition.* 2d ed. Kent, OH: Kent State University Press, 2000.

Slayton, Robert A. *Empire Statesman: The Rise and Redemption of Al Smith.* New York: Free Press, 2001.

Smith, Robert Holbrook (1879–1950)

Robert Holbrook Smith (1879–1950), who would be known as "Dr. Bob, co-founder of Alcoholics Anonymous," was born on 8 August 1879 in St. Johnsbury, Vermont, the son of Judge and Mrs. Walter Perrin Smith. Bob made the acquaintance of "Demon Rum" at St. Johnsbury Academy, from which he graduated in 1898. Partly because of his drinking while at Dartmouth College, Smith took a three-year hiatus after graduation before entering the University of Michigan as a premed student in 1905. Smith's drinking increased at Ann Arbor, however, and he took further time off from his studies before receiving his medical degree from Rush Memorial College in Chicago in 1910. Smith interned at City Hospital in Akron, Ohio, and practiced in that city until his retirement in 1948.

After marrying Anne Ripley in 1915, Smith stayed dry for a time, but his drinking increased, facilitated by National Prohibition's loophole for physicians obtaining "medicinal alcohol." In 1933, his efforts to moderate his drinking led him to join the Oxford Group, a nondenominational Christian fellowship with a prominent presence in Akron, largely through the efforts of the city's rubber barons. Smith took to the group with enthusiasm, reading its literature, following its practices, and attending its meetings. His binges nevertheless continued.

On 11 May 1935, Anne Smith received a telephone call from her Oxford Group friend Henrietta Seiberling inviting her and her husband to meet a visiting New York Grouper, William Griffith Wilson. The meeting took place the next day, Mother's Day. The men spoke for over five hours, and Smith identified with Wilson's story of his drinking. Dr. Smith went "on one last toot" at a medical convention in Atlantic City the next month, but he took his last drink on 17 June 1935. Due to a confusion over the date, most members of Alcoholics Anonymous celebrate Dr. Bob's "last drink" as taking place on 10 June 1935, which is the accustomed date for the "birthday" of Alcoholics Anonymous. Whatever the exact date, the meeting between Smith and Wilson was truly historic.

Bill Wilson, Wall Street habitué and born promoter, was sixteen years younger than Dr. Smith and outlived him by twenty-one years. Although others might have claimed to be "co-founders" of Alcoholics Anonymous, AA's memory cherishes Wilson and Smith as cofounders, perhaps because the two men were so very different.

Smith carried into his work with alcoholics his surgical habit of working on one person at time. His constant theme was "Keep It Simple," and it seems that Bill Wilson realized he needed this quiet check on his own promotional enthusiasm. For his part, Smith defended Wilson against attackers in both Cleveland and Akron, as well as from New York. In their common alcoholism, these two men—so very different despite being fellow Vermonters—found a combination of individuality and universality that made possible and gave birth to and nourished an organization unique in human history for precisely that combination of qualities—Alcoholics Anonymous.

Riven by cancer, Dr. Robert Holbrook Smith died on 16 November 1950. He was buried at the family gravesite under a stone marked only with his name and the dates of his birth and death.

Ernest Kurtz

See also: Alcoholics Anonymous (AA); National Prohibition (United States); Wilson, William Griffith

References

Anonymous. *Alcoholics Anonymous.* Various editions. New York: AAWS, 1939, 1955, 1976, 2002.

Anonymous. *Dr. Bob and the Good Oldtimers.* New York: AAWS, 1980.

Smithers Foundation

See Christopher D. Smithers Foundation

Society for the Study of Addiction (SSA)

The Society for the Study of Addiction (SSA) is a British organization founded by physicians and moral reformers in 1884 to study addiction and influence legislation. Originally the Society for the Study and Cure of Inebriety ("and Cure" was removed in 1887), the society was spearheaded by Dr. Norman Kerr, a highly regarded temperance advocate who was president of the organization until his death in 1899. The society was preceded by the Society for Promoting Legislation for the Control and Cure of Habitual Drunkards, which had been formed in 1876.

Although an alliance of reformers, clergy, and physicians, the society was decidedly medical. Only medical practitioners could have full membership; all others were "associates" with no voting privileges or membership in the society's council. This medical focus was reflected in the society's adherence to the disease concept of inebriety with its concomitant implication that inebriates should be treated in a hospital or asylum. Early activities of the society lay in presenting and discussing papers on inebriety at meetings, publishing its *Proceedings,* and supporting efforts to create or extend British inebriate legislation. The society was international in scope, including American and European members, and held an International Congress on Inebriety in 1887.

At the turn of the century, the society expanded its activities. In 1899, it appointed a committee to investigate the relationship between alcohol and heredity, which reported in 1901. Society members also gave evidence to the Inter-Departmental Committee on Physical Deterioration of 1904. In 1903, it replaced its *Proceedings* with the *British Journal of Inebriety.* The society established a Norman Kerr Lectureship in 1905, which continued every second year until 1943.

These activities of the society created a greater public presence for its work and in turn drew members into its ranks. At the beginning of the twentieth century, the society had only 52 medical members and 37 associates. Within six years it had increased its membership base by about 700 percent, to 262 medical and 395 associate members. New members of the society saw themselves as more "scientific" than their predecessors and were interested in looking for the scientific causes and effects of inebriety rather than agitating for abstinence. The key issues in the decade before World War I included heredity, motherhood, and inebriety asylums, and some interest developed in psychology and psychotherapy.

Membership expanded again during World War I, partly owing to the presidency of Sir William Collins, a physician who challenged physical causation models of addiction. Believing that inebriety was a disease of the will, Collins emphasized legislative over medical intervention. This approach attracted nonmedical, associate members. Collins argued that the society should remodel its constitution to include nonmedical people as full members, but constitutional changes made in the early 1920s did not include this expansion of membership rights.

In the interwar years, the society was affected by the decline of alcohol as a public issue. By this time, the society was calling itself the Society for the Study of Inebriety (Alcoholism and Drug Addiction) and had considered renaming itself and the journal to replace the outdated word "Inebriety" with the modern "Addiction." No formal changes were made, however.

The society faced several challenges during World War II. First, the long-standing editor of the *British Journal of Inebriety,* Theophilus Kelynack, retired in 1941. He was replaced by Dr. John Yerbury Dent. Second, partly as a result of Dent's editorship and partly owing to wartime paper restrictions, the publication schedule of the journal became erratic. Third, membership in the society plummeted. As a result, the status of associate members began to change. In 1942, four associate members were added to the society's council. The distinction between medical and lay membership was fully eliminated in 1959.

In the postwar period, the society underwent significant changes. It secured controversial funding from the brewers Monthly Bulletin Fund (after the brewers periodical) to finance addiction research in 1945. The society changed its name to the Society for the Study of Addiction (referring to alcohol and other drugs) and the journal became the *British Journal of Addiction* in 1947. A new lecture series, the Kelynack lectures, began in 1947, three years after Kelynack's death.

The SSA struggled through the 1950s. Membership remained low. An attempt to gain "Royal" status was rejected in 1955. A proposed collaboration between the society, the National Association of Mental Health, and the Church of England Temperance Society to create a clinic for alcoholics caused discord within the society. An attempt to create a research prize, the Hubert Norman Prize in 1953, received only one entry, and that based on previously published research. The Kelynack Fund was nearly exhausted, and the Monthly Bulletin Fund for research was not going to be renewed.

A resurgence in public interest in addiction during the 1960s presented the SSA with new challenges. It was involved with the "Brain Reports" on drug addiction in 1961 and 1965. It also influenced the establishment of treatment clinics in the late 1960s, mostly through its work with the London Committee for the Study of Drug Addiction. In 1961, Max Glatt became editor of the *British Journal of Addiction* and altered its scope. Reflecting the direction of alcohol and drug research, the journal included more research-based social science and epidemiological papers. With renewed interest in addiction, however, came competing organizations, such as the National Council on Alcoholism (1962), the Medical Council on Alcoholism (1967), and the Institute for the Study of Drug Dependence (1968).

At the end of the 1970s, the society began to reshape itself into a learned scientific society. In 1976, D. L. Davies became president and appointed individuals from prestigious research units as officers. Notably, Griffith Edwards became editor of the journal in 1978. Edwards expanded the journal's scope, increasing participation by a wider range of researchers in its publication. From being a financial liability, the journal became an asset. With financial stability and expanded visibility, the SSA began to sponsor more research work. In 1985, it began to offer funds to groups to organize meetings on alcohol or addiction, and in 1988 it began to fund a series of high-level workshops on addiction.

The 1990s continued to see expansion and change. In 1993, the journal was renamed *Addiction* and began to be published jointly with the National Addiction Centre. In 1994, the SSA began to publish *Addiction Abstracts,* and in 1996 it created a sister publication to *Addiction. Addiction Biology* has a scientific focus that enables *Addiction* to emphasize social-science research. The society established the International Research Monographs in the Addictions with Cambridge University Press, and the first of the series appeared in 1998.

Daniel J. Malleck

See also: Kerr, Norman Shanks

References

Berridge, Virginia. "Editorial: The Centenary Issue" in *British Journal of Addiction.* 79 (1984): 1–6.

———. "The Society for the Study of Addiction, 1884–1988" in *British Journal of Addiction.* 85 (1990): 987–1077.

Edwards, Griffith. "Addiction Biology—A Significant Addition to the Range of *Addiction* Journals" in *Addiction.* 91 (1996): 15–16.

Rolleston, Sir Humphry. "The Jubilee of the Society and the Centenary of the Birth of Its Founder" in *The British Journal of Inebriety (Alcoholism and Drug Addiction).* 32, no. 1 (July 1934): 1–13.

Sociological Theories of Drinking and Temperance

Alcohol use (or ethanol ingestion), a widespread, costly, and controversial activity that both affects and is affected by the human world in which the individual must live, has yet to arrive fully as an object of sociological research and theorizing. Although sociologists have made some important contributions to knowledge about drinking and temperance, those contributions have emerged from a discipline that has sometimes overlooked, ghettoized, and merely toyed with the deep sociomaterial impacts of alcohol. Alcohol scholars and counselors certainly should know what sociologists have learned about drinking, and sociologists ought to strive to deepen their contributions on this vital topic.

Alcohol and Society

Sociologists attempt to explain how supraindividual relationships, experiences, places, and things shape human lives. Given its ubiquity and its intoxicating and potentially addictive effects on the human individual, alcohol use is a natural and logically mandatory subject for sociological research. How does alcohol alter the ways in which some people act, and through their altered behavior, affect the social settings in which others must live? What social environments, ideologies, and processes have induced or restricted alcohol drinking, and for what reasons? How and why have the costs and benefits of alcohol use been politicized, negotiated, and battled over in groups of varying sizes? To what extent do particular social and environmental factors facilitate or hinder alcohol abuse and alcohol addiction? Whose interests are served by widespread alcohol use and abuse, and how do those interests affect the way people think and talk about the topic? Developing robust debate and powerful individual and democratic answers to these and other questions hinges in part on the quality and quantity of the work of sociologists.

The Age of Suggestive Fragments: Alcohol in Early Sociology

The main impetus behind the consolidation of modern social science was the European Enlightenment of the seventeenth and eighteenth centuries. In that era, leading intellectuals began systematically questioning inherited metaphysical theories of natural and social processes. As medieval dogmas dissolved, more and more scholars began to investigate the ways in which human events could be better understood as products of material forces, environmental experiences, and socially embedded human decisions.

Against this background, sociology as a distinct discipline crystallized amid nineteenth-century European debates over "the social question," which asked why poverty, community dysfunction, and political disorder all seemed to be advancing in step with industrialism. Before the early 1800s, European elites had usually interpreted poverty and its attendant forms of human degradation as an important exception to, rather than a logical by-product of, their basic institutions. By the 1830s, however, new realities, including new forms of popular upheaval and resistance, were forcing Europeans of all ranks to admit that deprivation and dislocation were somehow intrinsic to industrialization itself. The new pattern of troubles was tremendously counterintuitive to established elite worldviews. Among the many intellectual and political novelties the ensuing foment produced was "sociology," or the formal effort to subject society and its component processes to scientific methods of inquiry, analysis, and explanation.

What society now calls "alcoholism" or "alcohol abuse" was very much part and parcel of the new complex of industrialization, urbanization, and poverty at the heart of "the social question." Public discussion of the problem was certainly rife. William Hogarth's *Gin Alley* and *Beer Lane* sketches had long been famous. Charles Dickens depicted alcohol-troubled characters such as Jenny Wren (*Our Mutual Friend,* 1865), whose woes stemmed from being "surrounded by drunken people from her cradle—if she ever had one." Public health investigators reported woeful findings to parliamentary and royal commissions. Early temperance activists were beginning to mobilize.

Unfortunately, probably due to their immersion in the long-established European tendency to view alcohol-drinking as a natural part of life, early sociologists met the growing public problem and concern with only the occasional valuable comment on drinking and temperance. Systematic research was not yet on the sociological agenda. One might, for example, have expected Auguste Comte (1798–1857), who coined the word "sociology," to pay some attention to drinking and temperance. Comte was convinced that an improved knowledge of (and thus a better life for) humanity could only come through deeper understanding of how "External Order" impinged upon individuals. People, Comte wrote, "have been for a long time ignorant of this Order. Nevertheless we have been always subject to it; and its influence has always tended, though without our knowledge, to control our whole being; our actions, . . . thoughts, and even our affections. As we have advanced in our knowledge of it, our thoughts have become less vague, our desires less capricious [and] our conduct less arbitrary" (Comte 1957 [1842]: 58). Despite this seeming call for the careful study of external behavioral influences, and despite Comte's life amid the tumults of postrevolutionary, industrializing France, Comte said nothing substantial about the sociology of alcohol, preferring instead to philosophize about universal knowledge.

Other early sociologists were somewhat less insensitive to the social context of questions of drinking and temperance. In his 1845 report of his ethnographic wanderings through industrial Manchester, Friedrich Engels (1820–1895) mentioned, though never genuinely explored, the place of alcohol in modern poverty. Unlike Comte, Engels was intensely interested in describing and understanding the daily lives of his plebeian contemporaries. As a result, his famous observations were at least empirically rich enough to provide the beginnings of theoretical models and research agendas on the social roots of drinking and temperance. "The filth and stagnant pools of the working-people's quarters in the great cities have . . . the worst effect upon the public health," Engels wrote. He described society's treatment of the "great multitude of the poor" as "revolting":

Friedrich Engels (1820–1895) wrote at length on the social roots of drinking and temperance. (Library of Congress)

> They are drawn into the large cities where they breathe a poorer atmosphere than in the country; they are relegated to districts which, by reason of the method of construction, are worse ventilated than any others; they are deprived of all means of cleanliness, of water itself, since pipes are laid only when paid for, and the rivers so polluted that they are useless for such purposes; they are obliged to throw all offal and garbage, all dirty water, often all disgusting drainage and excrement into the streets, being without other means of disposing of them; they are thus compelled to infect the region of their own dwellings. . . . They are *deprived of all enjoyments except that of sexual indulgence and drunkenness,* are worked every day to the point of complete exhaustion of their mental and physical energies, and are thus *constantly spurred on to the maddest excess in the only two enjoyments at their command.* (Engels 1984 [1845]: 128–129, emphasis added)

Engels retained a lifelong conviction that there were powerful connections between class dynamics, individuals' life circumstances, and drinking and temperance issues. "Under existing circumstances," Engels wrote in 1872, "drunkenness among the workers is an inevitable product of their living conditions, just as inevitable as typhus, crime, vermin, the bailiff and other social ills, so inevitable in fact that the average figures of those who succumb to chronic drunkenness can be calculated in advance" (Engels 1970 [1872]: 44). Precisely how and why this was so, however, Engels left to future thinkers to explain.

Like his friend Engels, Karl Marx (1818–1883) was keenly interested in commoners' daily experiences and problems yet also left only a few suggestive fragments on the place of alcohol in them. Perhaps the best-known and most important of these fragments was Marx's suggestion that capitalism's authoritarian, stultifying control of paid labor processes impelled workers to fixate upon alcohol. "The fact," Marx argued, is that, under capitalism, "labour is external to the worker. . . . Hence, the worker feels himself only when he is not working. . . . He is at home when he is not working, and not at home when he is working. . . . The result is that the individual feels that he is acting freely only in his animal functions—eating, drinking, and procreating" (1988 [1844]: 71).

Another important early sociologist who left an interesting crumb on drinking and temperance was Emile Durkheim (1858–1917). *Suicide* (1951 [1897]) was Durkheim's attempt to demonstrate that socially determined levels of group "solidarity" are the main factor driving societies' varying suicide rates. In building this argument, Durkheim defined alcoholism as an "extra-social" cause of suicide and argued that it was but a minor part of the suicide phenomenon. Of course, from a modern perspective, virtually every aspect of Durkheim's consideration of alcohol in *Suicide* was unfortunate, at best. He had based his conclusions upon the statistics available to him, which showed no correlation between rates of alcohol consumption and suicide. Nevertheless, recalling the erroneous and somewhat dismissive attitude taken to drinking and temperance issues by this "founding father" of modern sociology helps sociologists understand why there is still so much work to be done on the topic today. In Durkheim's hands, the real world of alcohol drinking became a mere footnote to a Comtean search for grand theories of timeless knowledge.

Beyond Fragments: Alcohol in Modern Sociology

Given the newness and incompleteness of scientists' understanding of the neurochemistry of the human brain, it should come as no surprise that sociologists have not yet produced a truly mature, integrated body of thought on drinking and temperance. The extent to which advances in knowledge of brain functions will require new sociological theories of alcohol usage remains unknown.

Since roughly the end of World War I, sociologists have, however, turned more serious and direct attention to alcohol use. The case for developing modern sociological inquiry into drinking and temperance has been aptly stated by sociologist Erich Goode: "No drug," he said, even one "with physically addicting properties, can [itself] *dictate* that human beings become addicted to it" (Goode 1984: 22). As Goode noted, addictive substances do not just leap into someone's bloodstream. They get there only via the particular social processes that discover, produce, promote, distribute, and ethically inflect them. Hence, understanding a drug as a substance is never enough to understand how and why people use or abstain from it. Good knowledge of drinking and temperance (which are drug-use issues) thus requires a major element of good sociology.

Social environments also heavily constrain drinking practices and intoxicated behavior. Sociologists, along with anthropologists and social historians, have found that the sociocultural conditions in which ethanol ingestion occurs have a very significant impact on how alcohol imbibers behave. Just as alcohol and other drugs do not of themselves leap into people's bodies, neither does alcohol intoxication in itself determine how the intoxicated person will act, or how intoxication will be interpreted by the drinker's peers. To comprehend and predict these things, sociologists argue, one must understand the specific cultural elements, social institutions, and power relations that give alcohol intoxication its differing costs, benefits, and meanings in different times and places.

Sociological proofs of the importance of the social context of alcohol-related behavior among drinkers have taken two main forms. First, sociologists and anthropologists have cataloged the historical and crosscultural variety of human alcohol regimes (see, for example, Marshall 1979). Second, there have been controlled experiments very strongly suggesting that socially learned expectations are, at least at low to moderate levels of intoxication, powerful enough to get alcohol imbibers to act out alcohol routines not in accordance with the facts of their bloodstreams. As Dwight B. Heath reported: "Numerous experiments conducted under strictly controlled conditions (double-blind, with placebos) on a wide range of subjects and in different cultures have demonstrated that both mood and actions are affected far more by what people think they have drunk than by what they have actually drunk. That is, when people consume a non-alcoholic beverage that they think contains alcohol, then they tend to become 'intoxicated.' But when they consume an alcoholic beverage that they think is non-alcoholic, they tend to act 'sober'" (Heath 2003).

Judged in terms of its subsequent influence, the single most important event in modern sociologists' treatment of drinking and temperance was the publication of an obscure journal article not directly on the topic. In 1938, Robert K. Merton published his essay "Anomie and Social Structure" in the *American Sociological Review.* In that piece, Merton attempted to counter the prevailing assumption that "deviate behavior" was exclusively "biologically derived." On the contrary, Merton asserted, "antisocial behavior is in a sense 'called forth' by certain conventional values of the culture *and* by the class structure involving differential access to the approved opportunities for legitimate, prestige-bearing pursuit of . . . culture goals" (Merton 1938: 679). In other words, some people break rules because their nonegalitarian society fails to provide them with easy avenues to success.

Merton's essay eventually spawned a "deviance" industry within sociology. In recent decades, the popularity of the Mertonian investigation of why people fail to conform to norms and rules has been immense. As more and more research and teaching have been undertaken on this conceptual basis, the large majority of post–World War II sociologists' examinations of drinking and temperance have been subsumed under "deviance" studies. The effect has probably not been entirely positive. Treating alcohol drinking as an issue of deviance sometimes tempts sociological researchers, teachers, and students to pay attention to extreme drinking-related cases and issues at the expense of more "normal" ones. It also tends to downplay the role of institutions and organizations in the explanation of the problems at hand. Likewise, despite sociologists' nearly universal assent to the proposition that good sociology must be rooted in solid explanations of relevant historical contexts (see Mills 1959), one temptation and tendency of sociological deviance studies is to fix too much attention on abstract norms and rules, and too little on actual history.

This is not to say that sociologists have entirely ignored the history of drinking and temperance. On the contrary, one of the most important sociological works on the topic is Joseph R. Gusfield's *Symbolic Crusade: Status Politics and the American Temperance Movement* (1963). In this widely cited book, Gusfield reviewed the political struggles between "dry" and "wet" status-group factions in the post–Civil War United States up to the time of Prohibition and repeal. Seriously dated and perhaps distorted by his early Cold War assumption that social class schisms were less important in the United States than elsewhere, Gusfield's book nonetheless provided a model for the historical-sociological analysis of the cultural and political governance of alcohol distribution and use.

Gusfield's narrow social history of drinking and temperance in the late-nineteenth- and early-twentieth-century United States is counterbalanced by the recent panoramic work of David T. Courtwright. Contrary to Gusfield, who contended that drinking has been relatively isolated from social class dynamics, Courtwright treated alcohol as part and parcel of a "psychoactive revolution" both historically and logically promoted by capitalism. The rapid global spread of illicit drug use, Courtwright said, was "one of the signal events in world history" and most definitely "had its roots in the transoceanic commerce and empire-building of the early modern period" (Courtwright 2001: 2). Moreover, he suggested, despite the intervening rise of public health knowledge and advocacy, from both the supply side and the demand side, contemporary corporate capitalism continues to be heavily dependent upon the masses' recreational chemical use, abuse, and dependencies.

Another major sociological contribution to the understanding of drinking and temperance issues has been the rise of close-up recording and analysis of the social influences on and meanings of alcohol drinkers' experiences. In 1987, Norman K. Denzin, for example, reported his findings from a series of participant observations of those who defined themselves as "alcoholics." Setting himself the task of making empathetic yet critical sense of "the self-definitions and self-feelings of the alcoholic," Denzin discovered that adopting the label "alcoholic" caused individuals to experience a "divided sense of self" (Denzin 1987: 11). As some alcoholics managed to gain mastery over their feelings of internal division and to "recover" from their alcoholism, he wrote, they often found themselves becoming critics of their larger social worlds in order to try to understand the human forces driving them to drink.

Denzin's finding suggests that, as the twenty-first century proceeds, there may very well be an increasing number of nonprofessionals seeking solid sociological theories of drinking and temperance. Certainly, the rise and rapid growth of popular interest in new, deeper, supramoralistic questioning of issues of drug and alcohol use was one of the main social trends of the twentieth century. As bioscientists continue to help us discover which terminologies are and are not valid for describing the effects of alcohol on the human mind and body, sociologists will certainly continue to work at explaining the social dimensions in which drinking and temperance get acted out.

Michael Dawson

See also: Anthropological Theories of Drinking and Temperance; Beer; Gin; Gusfield, Joseph R.

References

Comte, Auguste. *A General View of Positivism/Discours sur l'ensemble du positivisme.* Translated by J. H. Bridges. New York: R. Speller, 1957 [1842].

Courtwright, David T. *Forces of Habit: Drugs and the Making of the Modern World.* Cambridge: Harvard University Press, 2001.

Denzin, Norman K. *The Alcoholic Self.* Newbury Park, CA: Sage, 1987.

Durkheim, Emile. *Suicide: A Study in Sociology.* New York: Free Press, 1951 [1897].

Engels, Friedrich. *The Condition of the Working Class in England.* Chicago: Academy Chicago Publishers, 1984 [1845].

———. *The Housing Question.* Moscow: Progress, 1970 [1872].

Goode, Erich. *Drugs in American Society,* 2d ed. New York: Alfred A. Knopf, 1984.

Gusfield, Joseph R. *Symbolic Crusade: Status Politics and the American Temperance Movement.* Urbana: University of Illinois Press, 1963.

Heath, Dwight B. "What We Can Learn from Others: An Interview with Dwight B. Heath." http://www2.potsdam.edu/alcohol-info/InTheirOwnWords/HeathInterview.

Marshall, Mac, ed. *Beliefs, Behaviors, and Alcoholic Beverages: A Cross-Cultural Survey.* Ann Arbor: University of Michigan Press, 1979.

Marx, Karl. *The Economic and Philosophic Manuscripts of 1844.* Amherst, NY: Prometheus, 1988 [1844].

Merton, Robert K. "Anomie and Social Structure" in *American Sociological Review.* 3 (1938): 672–682.

Mills, C. Wright. *The Sociological Imagination.* New York: Oxford University Press, 1959.

Soft Drinks

Soft drinks, by definition, came into being when people explicitly sought liquid refreshment without the stimulus of alcohol. Coca-Cola, "The Real Thing" by which all other soft drinks are measured, claimed to be "The Great National Temperance Beverage" when it first found its market in turn-of-the-century American soda fountains. However, the genealogy of such familiar commercial soft drinks extends back to the "small beers," fruit juice cordials, and sparkling spa waters long consumed in premodern Europe. Although temperance sentiments facilitated popular acceptance of modern soft drinks, the genesis of today's nonalcoholic beverage industry came when the Enlightenment's quest to better Nature's handiwork was complemented with the marketing tricks of nineteenth-century patent medicine entrepreneurs. During the twentieth century, advancements in packaging technologies and the development of franchised business organizations enabled U.S. trademarked soft drinks to expand their market internationally. With such "Coca-Colonization," soft drinks became the most ubiquitous symbol of contemporary globalized consumer capitalism.

Soft drinks, following the American usage of the term as coined in the late 1800s, are water-based beverages, usually carbonated, sweetened, and flavored with edible acids and essences derived from fruits, nuts, berries, roots, and herbs. Coffee, tea, cocoa, milk, and undiluted fruit and vegetable juices, although nonalcoholic beverages, are not typically considered soft drinks. However, this modern designation for refreshing beverages does make sense when applied to the countless varieties of slightly alcoholic hard ciders and "small beers" (such as spruce, dandelion, and ginger beers) that once regularly quenched the thirst of Europeans and North Americans. Water itself was long considered a suspect beverage unless it came from a proven uncontaminated source. Instead, casual liquid refreshment was found in such homemade fruit and herbal concoctions that were allowed to ferment partially as a preservative. (Similar drinks, ranging from the "vinegars" of biblical times to the kava still consumed in Polynesia, also might be deemed soft drinks but fall outside the purview of this entry.)

Early truly "soft" alternatives, whether natural spring waters, lemonades, or herbal cordials, usually were understood as both "treatment" and "treat," their refreshing qualities subsumed under medicinal expectations. Already in ancient Greece, Hippocrates, the father of medicine, praised the therapeutic properties of the mineral-rich waters found in natural springs. Although such spa waters were bottled and shipped to those who could afford it, the first commercial soft drink was probably a kind of lemonade sold in Renaissance Italy. Nonetheless, by the eighteenth century the pleasant taste of such beverages was complemented by knowledge that lemon juice helped prevent scurvy. More significantly, it was the supposed health benefits of spa mineral waters and herbal concoctions that sparked the creation of modern soft drinks.

Soda Water and Soda Fountains

In 1767, capping centuries of inquiry into why sparkling spring waters sparkled, the British *philosophe* Joseph Priestley made the first glass of potable artificially carbonated water. Others, such as the Swedish chemist Torbern Bergman, expanded upon Priestley's discovery, and soon pharmacists throughout Europe were manufacturing sparkling mineral-fortified waters that they expected to deliver the healthful benefits long associated with famous spa waters. Among the most successful at bringing such beverages to the mass public was a German entrepreneur from Geneva, Jacob Schweppe, who during the 1790s established in London what would become the first enduring soft-drink enterprise, Schweppe and Company. In addition to copies of spa waters like those of Seltzer and Pyrmont, Schweppe marketed generic "Acidulous Soda Waters" that he advertised as "not ungrateful to the palate" and good for "invigorating the system and exalting the spirits." In 1806, Benjamin Silliman, the founder of Yale University's chemistry department, and Philip Physick, the University of Pennsylvania's first professor of surgery, were among those who began to market similar artificially carbonated waters in the United States.

Although such soda waters were still considered medicinal, innovations by two French expatriates in Philadelphia set the stage for modern soft-drink refreshment. In 1825, Elias Durand outfitted his pharmacy with mirrors, a marble counter, fancy glassware, and a dispenser for carbonated mineral waters. At the time, American inventors, including John Mathews, Gustavus Dows, and James Tufts, were beginning to transform primitive keg-like carbonating devices into elegant countertop soda fountains. Meanwhile, around 1838, Eugene Roussel began to market bottled carbonated water that was flavored with lemon syrup. However, spoilage, unwanted fermentation, and primitive closures ensured that the "pop" of sweetened bottled drinks was often violent and unexpected. Although such bottling enterprises remained a risky venture until the end of the century, American pharmacists quickly adapted the two innovations into popular drugstore soda fountains where, in addition to medications, customers could find sociability and a variety of refreshing sundries. Consequently, American drugstores became increasingly ostentatious as proprietors competed for customers with ever more elaborate soda fountains. Moreover, the initial handful of soda-water flavorings (such as lemon, sarsaparilla, and strawberry) expanded exponentially with the creation of new flavoring syrups and the addition, in 1874, of ice cream as a popular soda-water complement.

By the 1890s, working as a "soda jerk" was a skilled vocation as customers demanded an endless variety of creative flavor combinations. Meanwhile, some entrepreneurs hoped to strike it rich by creating a popular "specialty" flavor that only they could provide. Following the trail blazed by similar patent medicines, these proprietary flavors were promoted directly to consumers with extensive and trend-setting advertising. Although by this time soda water itself had lost its medicinal pretensions (as well as any residual sodium bicarbonate), the claims made for such patent flavorings often reflected the drink's origins in a pharmacist's mortar. For example, Moxie Nerve Food, a beverage from New England created in 1876, promised to recover loss of manhood, imbecility, locomotor ataxia, and softening of the brain. Dr. Pepper, created in 1885 by two pharmacists in Waco, Texas, was sold as a "tonic, brain food, and exhilarant." The following year, John Pemberton, an Atlanta druggist, introduced a new soda-fountain flavor named after two ingredients then celebrated in pharmacy journals, Coca-Cola. The label on its syrup bottles explained, "This Intellectual Beverage and Temperance Drink contains the valuable Tonic and Nerve Stimulant properties of the Coca plant and the Cola (or Kola) nuts, and makes not only a delicious and refreshing and invigorating beverage . . . but a valuable Brain Tonic and cure all for nervous affections—Sick Head-Ache, Neuralgia, Hysteria, Melancholy, etc." Such overblown claims were standard advertising fare until the Pure Food and Drug Act of 1906.

The growing temperance movement encouraged the consumption of soft drinks and helped the soda fountain eclipse the saloon as the parlor of U.S. society. Meanwhile, innovations in bottling, such as the crown cap in 1892, made "soda pop" a refreshment that could be taken anywhere. Although the need for a sober workforce caused a variety of early-twentieth-century European industrialists to manufacture soft drinks in-house, Coca-Cola pioneered a system of franchised independent local bottling operations that supplied its vast North American market. Nonetheless, in 1920 the advent of Prohibition in the United States was not the boon to soft drinks one would expect. The ongoing rise in Coca-Cola's sales figures was driven more by its advertising than by changing patterns of beverage consumption. Moreover, the soft drink that did see its popularity skyrocket unexpectedly in the 1920s sold well because it was deemed the ideal mixer: "Canada Dry Ginger Ale," quipped *Fortune* magazine in August 1931, "had class and cheap gin needed class badly."

A 1953 Coca-Cola sign (Library of Congress)

Coca-Colonization

If the costs of transporting heavy bottles can be avoided, the universality of human thirst makes refreshing soft drinks a readily traded commodity. Before World War I, Sinalco, a brand-name soft drink from Detmold, Germany, was able to achieve international distribution by exploiting the network of German-trained *Braumeistern* who brewed beer from Australia to Buenos Aires. Coca-Cola began to push its foreign sales in the 1920s but only established its worldwide dominance after World War II. Capitalizing on a global beachhead that was established while supplying the far-flung U.S. armed forces in wartime, Coca-Cola transplanted its franchise structure abroad where it complemented Coke's marketing power with local capital and entrepreneurial enthusiasm. When in the 1950s local interest groups in France and elsewhere tried to use the rising tide of Coca-Cola consumption to sound alarm about the dangers of so-called Americanization, Coca-Cola was able to undermine the opposition by offering bottling franchises to powerful domestic producers of (usually alcoholic) beverages.

The "brand power" generated by the soft-drink industry's success superimposed the "Cola Wars" onto the Cold War. In 1972, as Communist authorities tried to curb their problem with rampant alcoholism, Pepsi-Cola, Coke's most serious domestic rival, was introduced into the Soviet Union. In a modified barter agreement, Pepsi exported Stolichnaya vodka in exchange. Following the fall of communism in 1990, this sweet deal left a sour aftertaste when Coca-Cola was able to capitalize on its untainted Western image to quickly gain a significant market share in Eastern Europe. Nonetheless, after the turn of the millennium both Coca-Cola and Pepsi suffered when activists in the Middle East promoted a boycott of goods identified as American. However, the inroads made by such soft drinks into what Coca-Cola executives call the "stomach share" was demonstrated when consumers turned not to traditional local beverages but to Zamzam, an "Islamic cola" from Iran.

Internationally, the popularity of soft drinks continues to climb. Although Americans, who in 2000 drank an average 210 liters (75 U.S. gallons) of soft drinks per capita, still consume about a third of the world's carbonated soft drinks, the global rate of annual per capita consumption has reached 30 liters (11 U.S. gals.). The Coca-Cola Company commands half of the world's soft-drink market, followed by PepsiCo with about 20 percent and Cadbury Schweppes with almost 10 percent. However, the most dynamic area of growth in nonalcoholic beverages is not with major brand-name products such as Coke, Pepsi, and Canada Dry. Instead, in an echo of the industry's origins, bottled water and "new age" nonalcoholic beverages, often with medicinal pretensions, portend the future.

Jeff R. Schutts

See also: Kava; National Prohibition (United States)

References
Allen, Federick. *Secret Formula: How Brilliant Marketing and Relentless Salesmanship Made Coca-Cola the Best-Known Product in the World.* New York: HarperBusiness, 1994.
Dietz, Lawrence. *Soda Pop: The History, Advertising, Art and Memorabilia of Soft Drinks in America.* New York: Simon and Schuster, 1973.
Louis, J. C., and Harvey Yazijian. *The Cola Wars.* New York: Everest House, 1980.
Martin, Milward. *Twelve Full Ounces.* 2d ed. New York: Holt, Rinehart and Winston, 1969.
Miller, Daniel. *Capitalism: An Ethnographic Approach.* New York: Berg, 1997.
Pendergrast, Mark. *For God, Country and Coca-Cola: The Definitive History of the Great American Soft Drink and the Company That Makes It.* Rev. ed. New York: Basic, 2000.
Riley, John. *A History of the American Soft Drink Industry.* Washington, DC: American Bottlers of Carbonated Beverages, 1958.
Simmons, Douglas. *Schweppes: The First 200 Years.* Washington, DC: Acropolis, 1983.
Woodroof, Jasper Guy, and G. Frank Phillips. *Beverages: Carbonated and Noncarbonated.* Rev. ed. Westport, CT: Avi, 1981.

Isabella Caroline Somerset (1851–1921) (Library of Congress)

Somerset, Isabella Caroline (Lady Henry Somerset) (1851–1921)

Isabella Caroline Somerset (1851–1921), born Isabella Caroline Cocks and generally known as Lady Henry Somerset, became president of the British Women's Temperance Association (BWTA) following the death of Margaret Bright Lucas in 1890. When as president she changed the culture of the society, the BWTA split into two organizations: the National British Women's Temperance Association (NBWTA) and the Women's Total Abstinence Union (WTAU). It was to be thirty years before the two organizations reunited.

Lady Henry Somerset was born into a wealthy family and married Henry Somerset, son of the Duke of Beaufort. Her marriage was not successful. She left her husband after having discovered his homosexuality. She then retired to her estates in Ledbury to begin a life of religion and philanthropy.

Following the death of Margaret Bright Lucas, the executive committee of the BWTA decided that the prestige of the organization would be enhanced with a titled lady as president. Lady Henry Somerset was elected despite her lack of experience in the politics of temperance. At her election there was concern that she might become more than the figurehead the membership expected. When Lady Henry Somerset sailed to the United States in 1891, there was alarm among the executive committee who feared American influence. Somerset's return was closely followed by the arrival in Britain of Frances Willard, national president of the Woman's Christian Temperance Union (WCTU) in the United States.

These events coincided with the British temperance movement's attempts to form a common agenda during the run up to the 1892 parliamentary elections. In the furtherance of this aim, Somerset began to form alliances with other temperance leaders, sending out a call to temperance societies and their workers to demand a Sunday closing bill. The campaign combined the fervor of evangelical meetings with political rhetoric. Despite their efforts, no bill was immediately forthcoming, and the BWTA, like other societies and organizations, were disappointed. There was a long delay before Sir William Harcourt eventually drafted the Liberal government's bill in 1893.

When Somerset suggested to the executive committee that the name of the BWTA should be changed to the National Women's Temperance Association and that the World's Woman's Christian Temperance Union (WWCTU) should be renamed the WWCTU of the British Empire, the BWTA was thrown into turmoil. Many members believed that Somerset was under Willard's influence. Somerset's closeness to Willard caused the executive committee to have doubts about Somerset's ability to ensure that the BWTA would remain a British organization. It became obvious that Somerset was not the president that some members had expected.

Somerset believed that the BWTA needed to widen its program and include women's suffrage and social purity, which would mirror the WCTU's "Do Everything" policy. The relationship between Somerset and her executive committee deteriorated so that by 1893 neither side trusted the other. Things came to a head at the 1893 annual meeting, when, after a long debate, the majority of the executive committee left the room and immediately formed the Women's Total Abstinence Union, a single-issue society supported by a minority of the old BWTA. The BWTA renamed itself the National British Women's Temperance Association. Somerset's attempt to remodel the NBWTA on Willard's ideas was only partly successful. She attempted to make the membership more politically aware, but her ambition to turn the county associations and branches into political units failed.

The political debate over temperance and prohibition continued. The Royal Commission on Liquor Licensing Laws, which sat from 1896 to 1899, took evidence from both the liquor trade and the temperance movement. Somerset answered more than 640 questions covering a wide range of issues, including inebriated women, distribution of public houses, increased female drunkenness, grocers' licenses, and employment of women in bars. This session should have provided her with an opportunity to advance her political agenda for temperance reform, but unfortunately some of her answers were controversial, particularly her surprising comment that drink in itself was not wrong. This failure to clearly state the temperance movement's views provided further evidence that her political judgment was not to be trusted.

This was not the first time that she had showed a lack of political acumen. Both Somerset and Willard were censored for not condemning lynching in the southern United States. More important, arguably, were her imprudent remarks favoring regulated prostitution in the *Times* (21 April 1897) on the introduction of Contagious Diseases Acts in India. The fight against the Contagious Diseases Acts was a longstanding issue, and temperance women were aware of the difficulties. Josephine Butler, a seasoned campaigner for the repeal of the acts and a supporter of the women's temperance movement, condemned Somerset's letter. Eventually Somerset was forced to withdraw her remarks, but she had lost the confidence of many former supporters.

Willard's death in 1898 left Somerset with personal grief and no political mentor. She began the Duxhurst Colony project, a home for inebriate women located between Reigate and Horley. The women were divided into three categories. The poorest women lived six to a cottage, women who could pay for their treatment were housed in a manor house, and those who could pay only part of their costs were housed in Hope House about a mile away. Funding came in the main from the NBWTA. As with many other similar institutions, success was limited, and the colony was finally put up for auction in 1936.

Somerset had made too many mistakes. When she suggested that she did not support prohibition and instead favored disinterested management (the Gothenburg scheme), this was too much of a change in policy for the majority of the NBWTA membership. She resigned as NBWTA president in 1903 but remained as president of the WWCTU until 1906. She then returned from Methodism to the Church of England and spent the rest of her life at Duxhurst.

Two hundred of her letters, dated 1894 to 1914, are in the papers of Rosalind Howard, Countess of Carlisle, her successor as NBWTA and WWCTU president, at the Castle Howard archives.

Margaret Barrow

See also: British Women's Temperance Association (BWTA); Carlisle, Countess of (Rosalind Howard); Lucas, Margaret Bright; National British Women's Temperance Association (NBWTA); Slack, Agnes Elizabeth; Willard, Frances Elizabeth Caroline; Women's Total Abstinence Union (WTAU); World's Woman's Christian Temperance Union (WWCTU)

References

Bordin, Ruth. *Frances Willard.* Chapel Hill: University of North Carolina Press, 1987.

Chappell, Jennie. *Noble Workers.* London: S. W. Partridge, 1910.

Fitzpatrick, Kathleen. *Lady Henry Somerset.* London: Jonathan Cape, 1923; Boston: Little, Brown, 1923.

Gordon, Anna A. *The Beautiful Life of Frances Willard.* Evanston, IL: Woman's Temperance Publishing Association, 1898.

Tyrrell, Ian. *Woman's World, Woman's Empire: The Woman's Christian Temperance Union in International Perspective, 1880–1930.* Chapel Hill: University of North Carolina Press, 1991.

Songs, Drinking

See Drinking Songs (United States); English Songs, Representations of Drinking in (1600–1900); Music, Popular, Representations of Drinking in

Sons of Temperance

The Sons of Temperance was established on 29 September 1842 by members of the Washingtonian movement in New York. Eighteen men, meeting at Teetotaller's Hall at 71 Division Street in New York, pledged to strictly abstain from alcoholic beverages, to live moral and sober lives, and to provide mutual aid. From these modest beginnings, the Sons of Temperance claimed a dues-paying membership of nearly a quarter of a million by the end of the decade, ranking it among the largest temperance organizations in antebellum America.

In the early nineteenth century, the temperance movement was fractured along socioeconomic lines. The American Temperance Society, founded in 1826 by leading evangelical clergymen, physicians, and academics, attempted to "improve" and reform drunkards by instruction, and eventually by legal coercion. In contrast to this hierarchical vision of reform, the Washingtonians embodied an egalitarian model, drawing membership from workers and artisans, and often from the ranks of reformed drunkards, who pledged to refrain from beverage alcohol. The Sons of Temperance, founded by New York printers John and Isaac Oliver, bridged the gap between these two wings of U.S. temperance reform. The Washingtonian meetings were noisy affairs, spontaneous and rowdy, if technically sober. The Olivers organized meetings with an eye toward respectability. There were formal speeches and unison singing with song lyrics printed on cards (courtesy of the Olivers' printing office) for audience benefit. Distancing themselves from the Washingtonians, who were criticized for accepting as members inn and saloon keepers who pledged personal abstinence, the Sons of Temperance required members to promise "neither [to] make, buy, sell, nor use as a beverage, any spiritous or malt liquors, or wine or cider."

The format of meetings was not the only change made by the Olivers. Other innovations reflected elements typical of many nineteenth-century fraternal organizations. The Sons of Temperance adopted secret passwords, pledges, signs, hierarchical orders, and regalia to appeal to prospective members and retain membership. This veil of secrecy also served to protect the fledgling organization from the discredit of backsliding members, an issue that had plagued the more transparent Washingtonians. The mutual aid provided by the Sons of Temperance extended beyond support in avoiding the temptation of drink. Again, following a practice common to other fraternal associations of the nineteenth century, the Sons of Temperance set aside a portion of collected dues to provide death and disability benefits for members and their wives.

These changes met with resistance in some quarters. Some older Washingtonians chafed at what they perceived as pretensions to respectability by the Sons of Temperance and questioned the motives of the Oliver brothers, who parlayed their more civilized meetings and temperance printing contracts into positions of social standing. In some communities, working-class splinter groups formed to recapture the camaraderie and spirit of the older meetings. Among more conservative reformers, the secrecy and ritual of Sons of Temperance meetings smacked of Masonry and seemed to threaten the status of Christian religious institutions in the lives of members. In response to these latter critics, the Sons of Temperance soon dropped the more elaborate rituals and abandoned the signs and pledges to secrecy. Dissidents within the organization who wanted to keep the more formal rituals left and formed the Temple of Honor. Despite these controversies, the organization grew quickly. Less than one month from the founding of the organization, local organizers in New Jersey applied for, and were duly granted, a charter to form the Newark Division No. 1. Within two years, more than thirty new divisions had been chartered, and the organization began to establish itself throughout the country, although its greatest strength remained in the cities and towns of the northeastern states.

Expansion was not limited to the United States. In 1847, the Acadia Division No. 1 was chartered in Nova Scotia by the Reverend William Ashley, who also served as the deputy most worthy patriarch for the National Division. In 1849, a division was chartered in Liverpool, England, with an affiliated "Cadets of Temperance" for young boys organized the following year. The National Division of Great Britain and Ireland was established in 1855. The Sons of Temperance Friendly Society still maintains headquarters in London.

As important as the changes in structure and membership that the Sons of Temperance imposed on the Washingtonian movement was its adoption of the principle of legal prohibition. Earlier, Washingtonians had clashed with the American Temperance Society over the issue of compulsory temperance, stressing moral suasion and mutual support rather than coercion. The Sons of Temperance, in contrast, endorsed "no-license" and statutory prohibition campaigns in a number of states. Often the process of chartering a new division would coincide with a local or statewide campaign. The Sons of Temperance, through local and state divisions, petitioned legislatures, printed campaign materials, and provided venues for speakers endorsing prohibition.

Membership numbers in the Sons of Temperance peaked ten years before the Civil War. Internal conflicts over membership policy and ongoing disputes regarding the relative importance of moral suasion and prohibition took their toll. The deepening sectional conflict hurt the cause of legal prohibition, as politicians increasingly viewed the reform as a threat to the delicate political balance. The organization emerged in the decades after the Civil War much smaller and with little national structure. As sympathy for prohibition shifted from the northeast to the southern and western states, the Sons of Temperance remained an important force, and its membership an important constituency, in local prohibition efforts. But often local and statewide prohibition efforts were directed by other organizations that did not have an association with antebellum northern reform. By the end of the nineteenth century, the Sons of Temperance, like many other temperance organizations from the early decades of the republic, was eclipsed first by the remarkable development of the ascendance of women in temperance reform, and later by the politically sophisticated Anti-Saloon League.

James D. Ivy

See also: American Temperance Society (ATS); Anti-Saloon League of America (ASLA); Daughters of Temperance; Washingtonians

References

Beattie, Donald W. "Sons of Temperance: Pioneers in Total Abstinence and 'Constitutional' Prohibition" (Ph.D. dissertation, Boston University, 1966).

Pearson, C. C., and Edwin J. Hendricks. *Liquor and Anti-Liquor in Virginia, 1619–1919.* Durham, NC: Duke University Press, 1967.

Sons of Temperance Friendly Society website. www.sonsoftemperance.co.uk/.

Tyrell, Ian R. *Sobering Up: From Temperance to Prohibition in Antebellum America, 1800–1860.* Westport, CT: Greenwood, 1979.

South African Breweries (SAB)

The South African Breweries (SAB) was founded in 1895, a by-product of the gold-mining revolution that created a major urban and industrial center in the area surrounding Johannesburg. The company offered the first shares traded on the fledgling Johannesburg Stock Exchange in 1897, and from this very early stage the SAB played a critical role in South African industrialization. As white immigrants flooded into the Witwatersrand mining region, the breweries expanded production to meet rapidly growing demand. A web of regulations effectively prohibited almost all Africans from buying and consuming European-type spirits, beer, and wine, and as a result the SAB focused on serving the white market. Pushing its Lion and Castle brands, the company spread its reach across South Africa and into the neighboring British dependencies of southern and central Africa. In the highly racialized environment of South Africa, consumer products were often highly defined in race terms, and many whites fought bitterly to resist suggestions that blacks be allowed access to products and brands defined in the popular consciousness as "white."

The steady expansion of the SAB accelerated in the aftermath of World War II with the aggressive acquisition of competing breweries in the 1950s. From that point, the SAB moved to consolidate its monopolistic position in the South African beer market. By 2000, beer produced by the SAB had captured more than 98 percent of the South African market for bottled beer. The scale of production and profits rose rapidly during the 1960s as the South African state repealed prohibition and gave Africans the legal right to purchase beer (and other liquor) at package stores. In the highly politicized and repressive atmosphere of the apartheid era, many blacks associated the old municipal sorghum beer monopolies that supplied Africans with industrially produced "traditional" beer with the policies of systematic race segregation advanced by the white regime. In 1976, when residents of the vast black residential ghetto of Soweto rose in rebellion, the first targets of their attacks were the government-owned liquor outlets. In this environment, the lager beers produced by the SAB represented a "modern" alternative to tainted pseudotraditional drinks, and the SAB nurtured and capitalized on those sentiments as it pitched its products to an expanding African consumer base. During the 1980s, the steady pressure to exclude South Africa from global political and economic spheres forced the company into diversification, but as the structures of apartheid began to crumble in the late 1980s, the company pulled back from that approach to focus on its core businesses.

The SAB was well positioned to take advantage of the end of the apartheid regime in the early 1990s. The company had been in fact a leader within the "realist" camp in South Africa's corporate community, arguing that securing a profitable future required a gradualist policy of increasing black economic opportunities. As then President Nelson Mandela noted at SAB centenary ceremonies in 1995, "SAB was amongst those leaders of business who embraced the future, even when it was less fashionable to do so." As early as 1971, the company began an affirmative-action or equity initiative. At that time, only 1 percent of all salaried staff were black (including Asians, "coloured" people, and Africans). By 1978, that number had risen to 13 percent. In 2003, 55 percent of total salaried staff and 22 percent of executives were black.

With the end of apartheid and the elimination of restrictions on trade with South Africa and South African investment, the SAB moved quickly to gain dominance over the southern African beer market and to establish itself as a major player in beer production and distribution across the continent. The company also repositioned itself as a global brewing giant. With the recent acquisitions of several European and Latin American breweries and the U.S.-based Miller Brewing Company, SAB has moved its headquarters to London and is now among the four largest international brewing companies.

Charles Ambler

References

Crush, Jonathan, and Charles Ambler, eds. *Liquor and Labor in Southern Africa.* Athens: Ohio University Press, 1992.

International Labour Organization, High Performance Working Research Project. "The South African Breweries, Ltd.," 30 March 2002. http://www.ilo.org/public/english/employment/skills/training/casest/brewery.htm.

Mandela, Nelson. Speech at the Opening of the South African Breweries Centenary Center, 15 May 1995. http://www.anc.orgza/ancdocs/history/mandela.

South America

South America is a vast and varied continent with ethnically and culturally diverse populations in almost every country. The appreciable variation in ecology, economic systems, and political and social organization is overlaid by some cultural similarities based on centuries of Spanish and Portuguese colonial rule. Beverage alcohol plays many different roles in these cultures. Alcohol and temperance in the modern history of South America have not been subjects of much systematic research, but in the work that does exist, some distinctive patterns emerge highlighting both theoretical and practical similarities and differences between this continent and other major world areas. The best approach to understanding relations between alcohol, temperance, and history in these many countries is to outline the diversity with some illustrative examples.

Beverages

The species of grapes that are indigenous to South America generally do not contain enough sugar to ferment readily. Viticulture was introduced early in the Spanish and Portuguese colonial period by Roman Catholic missionaries to supply the Mass and rapidly expanded in a few favorable settings (parts of Argentina, Chile, and Peru). Wine became an important daily beverage there, but not in most other areas. Some of the better quality wines are now exported to North America and Europe.

Throughout much of South America, indigenous peoples have traditionally produced a wide variety of beers (or *chichas*) from various fruits, vegetables, grains, and honey. Recipes and flavors are highly variable, but they are generally low in alcohol (5–8 percent by volume) and short-lived. Many are rich in vitamins and minerals that are scarce in indigenous diets, providing an important supplement when consumed, often as the primary beverage. A few communities are famous for their chicha, and some chichas are available only on specific occasions; others are laced with medical or hallucinogenic additives, often playing important roles in rituals, social exchange, and celebration. Industrially produced lager and pilsen beers gained popularity as refrigeration and transportation improved in the late nineteenth century, and they are the most popular drinks in many areas today.

For reasons that are still unknown, distillation did not appear in the Americas until the coming of Europeans (beginning in the sixteenth century, but later in a few isolated frontier areas). With distillation came a variety of spirits drinks: alcohol, or *aguardiente,* from sugarcane, which becomes rum, or *cachasa,* if aged; tequila from mezcal sap; *pisco,* or brandy, from grapes, among others. As elsewhere, liquor was often used as a tool of colonization. Small portions were valuable both as trade goods and as partial payment for work. In a few societies, rum became domesticated in strictly ritualized toasting, whereby members of a loosely organized society found comfort and satisfaction that could not be found otherwise among kin or neighbors in a drinking bout. In recent decades, imports such as Scotch whisky or U.S. whiskey, vodkas, and gins are gaining popularity among those who can afford them.

Drinking Patterns

In general, there is a dearth of detailed information about alcohol use and its outcomes in every South American country. Occasional articles that purport to provide statistical information about per capita consumption, psychiatric effects, traffic fatalities, or other alcohol-related issues tend to be narrowly local in scope, using samples and methods that instill little confidence among an international readership, and they are often frank and explicit about their limitations. Ethnographic or anthropological studies, where they occur, are often richly detailed but unrepresentative of the nation. No South American country has a homogeneous population, and in some a majority speak indigenous languages. Within each country, different segments of the population have different cultures, and some cultures spread across national boundaries. Striking diversity occurs in those many cultures, but a few generalizations can be made about drinking patterns.

As in virtually every population that has been studied, males drink more—and more often—than do females throughout South America. In most countries, the percentage of men who drink is at least double that of women, and those men generally drink at least twice as much and twice as often as the women who drink. Also nearly universal is that the difference is slowly narrowing in recent years, although a double standard persists by which male drinking is generally accepted, whereas female drinking is often construed as symptomatic of immorality. Many women drink only on rare and special occasions, although they tend not to be outspokenly opposed to drinking on the part of others.

Traditionally, rural populations drink more than urban ones, but that trend appears to be changing in recent years with increasing rural-to-urban migration. There is often marked stratification of drinking patterns according to socioeconomic class. It is commonplace for members of the lower class to drink more, and to drink more often, than do members of the middle and upper classes. Their choice of beverage often has to do more with cost than with taste; thus beer is the predominant beverage where it is cheaper, but rum prevails where it costs less than beer. Episodic heavy drinking is often recognized as a stereotypical characteristic of laborers, some of whom boast about drinking to the point of drunkenness every weekend.

With few exceptions, indigenous brews tend to be ignored by middle- or upper-class urbanites, and other drinks are popular approximately in proportion to their cost. If moderate drinking occurs, it tends to be among these groups. On special celebratory occasions, however, beverage alcohol can be an important item in conspicuous consumption, with middle-class drinkers in a public place ostentatiously displaying as many bottles as possible on the table. In much the same way, an imported beer or rum can serve as a relatively inexpensive occasional luxury for a laborer.

Adolescence and young adulthood tend to be the periods of maximum drinking, with most individuals tapering off rapidly once they marry or reach the age of thirty. Those few who become dependent on alcohol are often pitied. Alcoholics Anonymous is spreading, but few other modes of treatment are available in most areas.

Public sales outlets include not only bars and restaurants but also, in many instances, supermarkets, sidewalk kiosks, general stores, and gasoline stations. Males tend most often to drink with male friends away from home, frequently in bars or cantinas. On those occasions when women do drink, it tends to be in the company of other women or of family, and in the home. In general, there is a high correlation between level of schooling, prestige of occupation, and income as joint markers of socioeconomic class standing, and all such measures tend to be inversely correlated with levels of consumption of beverage alcohol.

Drinking often accompanies dancing, the watching of sports, or participation in other leisure activities (such as singing or playing cards, dominoes, or dice-poker). For many South Americans, it is difficult to imagine a party without drinking (and often drunkenness), and the act of drinking itself is often construed as festive or lighthearted. By contrast, in some South American cultures drinking is a serious adjunct to divination, curing, prayer, and so forth. Variation beyond a basic overall pattern is striking, so that understanding the context and meanings of the act is imperative.

The offer of a drink in South America is often an important symbol of friendship or hospitality, to the point where one may cause offense by declining such an offer. Toasting is a common pattern. Sometimes it involves elaborate word-play or teasing in informal settings, or elaborate speechmaking in more formal contexts. Games and competition are often closely associated with recreational drinking, and in these contexts few drinkers pay much attention to flavor, color, or other aesthetic aspects of beverages, preferring whatever delivers the most alcohol for the price. Drunkenness tends to be viewed as exhilarating and confirming of one's masculinity, especially among those who have little wealth, power, or social status. For those reasons, many men drink primarily for the psychoactive effect; this type of drinking is probably also connected to the fact that little attention is paid to low levels of inebriation or blood-alcohol concentration.

All such sociocultural factors have generally been ignored, except by a few historians and anthropologists. Paying additional attention to qualitative data, such as context and attitudes, may suggest meanings that have been lacking in studies of alcohol consumption in South America, given the overwhelming attention that has been paid to exclusively quantitative data. Recent interest in drinking patterns going beyond sheer measures of quantity and frequency of consumption and various types of alcohol-related harm hold some promise of expanding researchers' understandings of alcohol use and its outcomes.

Attitudes and Images of Drinking and Temperance.

During the early period of colonization, Europeans were often alarmed and disgusted by the frequency with which they encountered a pattern of episodic heavy drinking, often resulting in drunken stupor, among indigenous populations. Such behavior was construed as primitive or barbaric, and its association with indigenous deities and spirits was denounced as pagan. A stereotypical characterization of "the drunken Indian" emerged that was strikingly similar to that in North America. That view attributed to the "Indians" a congenital susceptibility to alcohol whereby they became intoxicated faster than "whites," stayed drunk longer, and became dangerously aggressive while under its influence. Ethnographic and physiological evidence have established that such a myth has little scientific basis, but the stereotype remains popular as an ethnic stigma. At various times during the colonial period, various administrative and ecclesiastical jurisdictions enacted prohibitions against alcohol, usually pertaining only to indigenous peoples. Such prohibitions were mostly ineffective and short-lived.

After the 1820s, the several newly independent republics often saw the use of beverage alcohol as a medium of exchange that allowed those who had easy access to distillates to trade cheaply with peasants for control of land, which became concentrated in the hands of a few. In addition, many indigenous populations used alcohol as an offering to deities and spirits, as a tool in divination and curing, and as a medium for celebration of various social events.

The drinking of beers in many indigenous cultures had important symbolic connotations of sharing with others, communicating with the ancestors, pleasing gods and spirits, and transcending workaday reality to expand one's consciousness. Many of those ideas persist and can even be found among lower-class drinkers of European descent.

In the tropical lowlands, homebrewed beers continue to be used both as food and occasionally as intoxicants. Although drunkenness is not uncommon, it usually takes place in a setting where it poses almost no problems to individuals or to the community. The Camba of eastern Bolivia exemplify this well, with probably the highest rate of drunkenness in the world and virtually no alcohol-related problems (whether social, psychological, economic, health, or in any other terms).

There is often an attitude among South Americans that hardworking men "have earned" or "deserve" to drink for relaxation, camaraderie, and enjoyment, even to intoxication. Depending on how they are paid, this often results in episodic heavy drinking on weekends, monthly, or at the end of a season's work, with little drinking between times. Holidays, both religious and secular, are usually marked by heavy drinking, as are weddings, birthdays, funerals, and various other special occasions.

The reasons or justifications that people give for drinking include pleasure, celebration, social interaction, relaxation, and change of mood, all positive from their perspective. Apart from isolated instances in which excessive drinking results in major problems in terms of health or economic or social relationships, most view drinking as a pleasant and rewarding activity, finding it difficult to comprehend the negative moral evaluation that Protestants give to it or the occasional alarm with which it is portrayed by public health authorities.

In recent decades, Protestantism (especially more fundamentalist or ascetic versions) is gaining ground in what had traditionally been overwhelmingly Roman Catholic populations. When abstention from alcohol is an article of faith, drinking (or not) can be an important boundary-marker between social groups. In small communities where the performance of syncretic folk-Catholic rituals, often accompanied by heavy drink-

Porters celebrate Cruz Velacuy in the Machu Picchu and Cuzco region. (John Van Hasselt/Corbis SYGMA)

ing, is important, Protestants may be excluded from positions of power and prestige, but they may also be able to accumulate more wealth by not being pressed to spend on the ceremonial cycle.

In many South American countries, serious poverty is the lot of significant numbers of people; malnutrition, undernutrition, and several diseases are endemic. People everywhere are subject to manifold pressures, including at various times and places unemployment and underemployment, political repression, runaway inflation, deforestation and pollution, dislocation by war and natural catastrophes, rapid and unplanned growth of cities, generally poor facilities and infrastructures, and a host of other problems. In such a context, it is little wonder that not much attention is paid to alcohol as a source of pathology but that it is widely appreciated for the enjoyment and social benefits that it provides.

This is not to imply that drinking is invariably free of harmful outcomes. Spouse abuse and other forms of verbal and physical aggression sometimes occur in association with extreme intoxication, and, as motor vehicles become more numerous, traffic accidents are increasing in many countries.

Patterns of Regulation

Temperance sentiment has rarely gained much popularity in South America. A few countries experimented briefly with state monopolies on the production of beverage alcohol, and few ever monopolized sales. Most levy high import duties on luxuries, so that counterfeiting of labels of expensive brands is commonplace. Excise taxes and licensing are not much used to curb drinking, except that in some jurisdictions heavy taxation has resulted in widespread production of illicit alcohol. This can pose a health problem when toxic agents are used during fermentation or distillation.

For the past three decades, a small but highly vocal group of self-styled public health advocates have promoted a variety of policies that they claim would reduce the availability of alcoholic beverages, thereby diminishing average per capita consumption, and consequently reduce or prevent dependence and various kinds of alcohol-related damage to drinkers and others. Although both conceptual and empirical bases for those assumptions remain questionable, the World Health Organization has vigorously promoted such policies and they have been widely adopted, even in areas where there was little previous tradition of control. It remains to be seen whether the more recent General Agreement on Trade in Services will render all or most of such restrictions moot. Among the controversial efforts at control, some refer to sales, others to advertising, and others to labeling as illustrated below.

Many who would restrict alcohol consumption presume that demand for alcoholic beverages is a function of advertising, and so a number of different types and degrees of regulation have been applied. The long panorama of history, including Soviet experience through most of the twentieth century, reinforces findings that advertising has little effect on alcohol consumption. There has been no such research in South America, however, and several countries have adopted a patchwork quilt of policies restrictive of advertising.

The categories of advertising that are considered include television, radio, cinema, print, outdoor signage, and sponsorship of events. The countries of South America are highly variable in their efforts to regulate advertising. Argentina allows all the named types of advertising, with some restrictions on content. Bolivia's only limitation is that televised liquor ads not appear before 9 P.M. Brazil lets only beverages below 26 proof be advertised on television and radio, and disallows that between 6 and 9 P.M. and during children's programming. Ads may appear at the cinema after 8 P.M. for audiences over age eighteen. Print ads are allowed except in children's magazines. Outdoor signs must be at least 20 meters from a school and should not depict Olympic sports.

In Colombia, beverage alcohol may be advertised on television only after 11 P.M., and for no more than 60 seconds in each 30 minutes, with a health warning required. Advertising is permitted in other media with some specific restrictions on content. Ecuador limits such ads to between 10 P.M. and 6:30 A.M. on television, and after 7 P.M. at cinemas; a few municipalities restrict outdoor advertising. Guyana has no restriction on alcohol ads.

Panama disallows alcohol advertising on TV before 6 P.M. and specifically prohibits depiction of the act of drinking. Paraguay allows television ads only after 8 P.M.; like alcohol ads on radio and at cinemas, they must be cleared through the Ministry of Health. The location and content of outdoor advertisements are also restricted. Peru requires that television advertising of alcohol be after 10 P.M. Uruguay allows all kinds of advertising, subject to an industry-written code of ethics; Venezuela allows only the showing of a brand name or the product itself.

Those who wish to promote temperance or prohibition in urban, industrial, and developed settings often point to traffic accidents, injuries, and deaths as vivid examples of the dangers of drinking. Alcohol is readily absorbed into the blood and its physiological impact tends to be dose-related. Although there is considerable individual variation in that respect, it is easy to rely on blood-alcohol level (BAL; or blood-alcohol concentration, BAC) as a supposedly objective measure of impairment. Law-enforcement agencies have often adopted tools for unobtrusive measurement of BAL, usually calibrated in grams (of ethanol) per liter (of blood), equivalent to fractions of 1 percent. Among South American countries, only Argentina has set a level (0.5 percent) as prima facie evidence of impairment for driving. Few countries in South America systematically collect data on traffic accidents, but those that do have found that most traffic accidents involving alcohol are due to a small group of heavy drinkers who are repeat offenders.

Some nations have set minimum ages for purchase and/or consumption of beverage alcohol in the hope of diminishing alcohol-related problems. The world's highest minimum age for drinking is twenty-one years, as in Chile (although it is not much enforced). Uruguay has a minimum age of eighteen for purchase, but no minimum for consumption. By contrast, there is no minimum age for purchase in Brazil, Colombia, and Peru, although they all set eighteen as the minimum age for drinking. In Argentina, eighteen is the minimum age for both; other countries have no legislation on the subject.

During the closing decades of the twentieth century, several South American countries joined a growing international trend by requiring that all beverage alcohol packages carry health warning labels. Specifically, they often indicate alcohol content by volume, and words to the effect that one should "Drink in moderation," or "Excessive consumption is dangerous to the health." Sometimes also it is indicated that "Sale is prohibited to those under *n* years of age." Argentina, Brazil, Colombia, and Venezuela include all of the above. In addition, Colombia warns against drinking by minors and by pregnant women. Ecuador's warning extends to mentioning damage to the family. Paraguay does not require health warnings, but some beers and whiskeys carry messages from the producers encouraging "moderation" or "responsibility." A few nations also have mandated that advertisements include health warnings. Research has yielded little evidence that health warnings are effective in reducing alcohol consumption in any population.

Research on Alcohol

Alcohol has rarely been viewed as problematic in South America, except inasmuch as a small minority of individuals, usually long-term heavy drinkers, create problems for themselves and others. Few jurisdictions keep the kinds of records that are used in other parts of the world as indices of drinking or of any of the various kinds of problems that are said to be associated with it. As in most developing areas, South America is hampered by both limited resources and a host of other pressing problems: shortage of skilled investigators, severe lack of funds and of a tradition in support of research, and the many other major concerns in the field of public health and social welfare that tend to eclipse any local interest in alcohol research.

In the 1960s, Chileans did some pioneering work on the epidemiology of alcohol use, but it was not long sustained. Since the 1990s, a new theme in Chilean research is the benefits of moderate wine consumption. Individuals have made a few ethnographic studies of drinking and its meanings among various local populations in different countries, and Brazil is beginning to produce some interesting physiological and psychological data. The recently organized Latin American Society for Biomedical Research on Alcoholism is an informal network of colleagues at various institutions, most of whom work independently.

Producers and Distributors

The beverage-alcohol trade is one in which many companies deal with more than a single class of beverage, for example, producing wine or beer and simultaneously distributing distilled spirits that were produced elsewhere. It is also a business in which actors come and go with some frequency, although this tends to be less the case in South America than in Europe or North America. Because of licensing agreements, mergers, and other arrangements, it is not always clear whether a company is indigenous to the country or acting as an agent for a larger international organization. That said, some of the major producers and distributors in the region at the beginning of 2002 were the following (listed alphabetically by country, and within that, grouped in the four beverage categories: wine, beer, spirits, and various).

In Argentina, wine is produced and distributed largely by Valentín Bianchi, SACIF, Bodegas Chandon, S.A., and Bodegas Trapiche, SAICA; spirits by Erven Lucas Bols, S.A.; and a variety of beverages by Compania International de Bebidas y Alimentos, S.A., Finca Flinchman, S.A., Hiram Walker, S.A., Peñaflor, S.A., and Seagram de Argentina, S.A. In Brazil, spirits are produced and distributed by Bols do Brasil Ltda, Hiram Walker Brasil Commerçio e Industria Ltda, and Industria de Bebidas Joaquim Thomaz de Aquino Filho, S.A.; a variety of beverages by Bacardi Martini do Brasil, Brasif Comercial Exportaçao e Importaçao Ltda, Franco-Suissa Importaçao, Exportaçao e Representaçoes Ltda, Heublein do Brasil, Indústrias Müller de Bebidas Ltda, Rémy Lacave do Brasil, and Seagram do Brasil Industria e Comercio Ltda.

In Chile, wine is a major agricultural product and export is handled largely by José Canepa y Cia Ltda, Viña Carta Vieja, S.A., Viña Cousiño-Macul, Cooperativa Agrícola Vitivinícola Curico Ltda, Discover Wine, S.A., Agrícola Viña Los Vascos, S.A., Viña Manquehue Ltda, Domaine Oriental, Viña Santa Inés, and Sociedad Anónima Viña Santa Rita. Companies that handle various beverages include Casa Lapostolle, S.A., Viña Concha y Toro, Viña Erráxuriz, S.A., Martini and Rossi, SAIC, Viña San Pedro, S.A., and Seagram de Chile.

In Colombia, the company for beers is Cervecería Unión, S.A.; that for spirits, Constain Ltda; others dealing in various beverages include Pedro Domeq Colombia, S.A., International Distillers Colombia, S.A., and Puyana y Cia de Bogotá, S.A. In Panama, spirits are handled largely by Bacardi Centroamérica, S.A. In Peru, most spirits are handled by United Distillers G&R del Perú. In Uruguay, Gilbey, S.A. dominates the spirits market. In Venezuela, beers are handled largely by Brahma Venezuela and Cervecería Polar, C.A.; spirits by Ron Santa Teresa, C.A.; and various beverages by Seagram de Venezuela, C.A., and by United Distillers de Venezuela, C.A.

Dwight B. Heath

See also: Anthropological Theories of Drinking and Temperance; Colonization, European, and Drinking Behavior among Indigenous Peoples (Portuguese America); Native Americans: Drinking Patterns and Temperance Reform; Rum; World Health Organization (WHO)

References

Acta Psiquiátrica y Psicológica de America latina. Special issue on alcohol in Latin America. 20, no. 2 (1974).

Heath, Dwight B. *Drinking Occasions: Comparative Perspectives on Alcohol and Culture.* Philadelphia: Brunner/Mazell, 2000.

———, ed. *International Handbook on Alcohol and Culture.* Westport, CT: Greenwood Press, 1995.

Horwitz, José, Juan Marconi, and Gonzalo Adis Castro. *Bases para una epidemiología del alcoholismo en América latina.* Monografía de Acta Psiquiátrica y Psicológica de América latina 1. Buenos Aires: Fondo para la Salud Mental, 1967.

Institute of Medicine. *Legislative Approaches to Prevention of Alcohol-Related Problems: An Inter-American Workshop.* Washington, DC: National Academy Press, 1982.

Marshall, Mac, ed. *Beliefs, Behaviors, and Alcoholic Beverages: A Cross-Cultural Survey.* Ann Arbor: University of Michigan Press, 1979.

Soviet Union and Russia since 1917, Alcohol and Temperance in

The control or prohibition of alcohol consumption has been a central feature of social reform in the Soviet Union and Russia throughout the twentieth century. The government formally instituted prohibition from 1914 to 1925, yet the population found ways to distill illicit alcohol. During World War I (1914–1917), the October Revolution (1917), and the Russian Civil War (1918–1921), drunkenness among the lower classes reached epidemic proportions. In the 1920s, the state repealed prohibition, revived a state monopoly on the production and sale of alcohol, and sponsored a nationwide temperance campaign that lasted into the 1930s. Every Soviet leader since Joseph Stalin has, in some fashion, tried to maximize liquor revenues while simultaneously reducing the incidence of alcohol abuse among the population.

Historically, Russian socialist parties had a strong temperance tradition, especially since the prerevolutionary drink trade was in the hands of the tsarist government itself. When the Bolsheviks came to power as a result of the October Revolution of 1917, they extended the prohibition measures that had been introduced as part of the mobilization in 1914. On 28 November 1917, the new government closed all remaining wine and spirits factories and prohibited the production or sale of alcoholic beverages. Within a month, the state established a Commissariat for the Struggle Against Alcoholism and Gambling, with representatives in major cities and the provinces, followed by the nationalization of all existing stocks of alcohol in late 1918.

These measures triggered violent battles between drinkers and antialcohol brigades and led to the development of a thriving trade in illegal spirits. According to some estimates, at least one-third of all rural households were engaged in distilling grain alcohol for sale on the black market in the early 1920s. In 1922 alone, there were more than 500,000 prosecutions in the Russian Republic for brewing or selling moonshine. As the courts began to collapse under the strain and the jails filled with poor peasants, the problems associated with prohibition intensified: declining state revenues, dwindling grain supplies, and no control over the trade in spirits. Because prohibition and the war had sharply curtailed alcohol revenues, many Soviet authorities saw an end to prohibition as a means for the fledgling state to attain financial solvency and gain a measure of control over illegal brewing. (In 1913, the population consumed nearly 340 million U.S. gallons of state-produced vodka yielding just over 953 million rubles, or 26 percent of total state revenues.) The idea of the new socialist state producing and selling alcohol for any reason, however, sparked heated debate among Communist Party leaders. In the end, financial considerations prevailed. In August 1921 the government legalized the sale of all wines, and six months later it legalized the sale of beer. In January 1923 the state legalized the production and sale of 40-proof liquor, and by 1 October 1925, it began making and selling 80-proof vodka, marking the end of prohibition and the reintroduction of the state liquor monopoly.

The Society for the Struggle against Alcoholism

The end of prohibition was met by a wave of drunkenness that engulfed the entire country. In an attempt to curb public drunkenness, state officials launched a nationwide temperance campaign. Led by Yuri Larin, the country's most renowned economist, on 16 February 1928 a group of concerned politicians and medical professionals met in Moscow to organize the first Soviet temperance society, the Society for the Struggle Against Alcoholism (Obshchestvo po bor'be s alkogolizmom [OBSA]). At the first meeting, Larin, by way of justifying state sale of vodka, explained that alcoholism was caused by social conditions and not by the availability of vodka. He urged careful examination of society as a whole in order to find ways to improve the country's material and cultural growth.

The first official task on OBSA's agenda was to enlist the help of various state agencies to adopt legislation limiting the production and sale of alcohol, controlling its consumption, and enforcing new legislation. OBSA's second task was to initiate a massive propaganda campaign aimed at swaying public opinion against drinking. Within the first five months of its existence, OBSA organized 2,000 mass meetings, lectures, and exhibitions attended by nearly 150,000 people. The society began to publish a monthly journal, *Trezvost' i kul'tura* (Sobriety and Culture), from July 1928 in which the bulk of its propaganda appeared. The most obvious feature of OBSA's propaganda was the extremism in definitions of alcoholism and calls for sobriety. According to OBSA literature, drinking necessarily led to alcoholism and alcoholism necessarily led to criminality, insanity, death, murder, mayhem, and disease. Displaying pictures of livers that resembled Jerusalem artichokes and swollen, blackened hearts, many of OBSA's journals and books detailed the physical horrors that awaited one in taking a few drinks.

OBSA membership consisted primarily of male workers who were required to practice total abstinence. A worker who still drank could be a "candidate" member until he or she gave up drink altogether. The first OBSA cell was organized in late 1928 at the Moscow factory Serp and Molot and immediately began a campaign under the slogan "The Unified Front in the Struggle against Alcohol and Tobacco." In early 1929, OBSA boasted over 250,000 members in more than 200 factory cells, thereby declaring itself a genuinely proletarian movement. In May that same year, OBSA's founding board organized the All-Union Council of Anti-Alcohol Societies (Vsesoiuznyi sovet protivalkogol'nykh obshchestv SSSR [VSPO]) to coordinate the various branches' activities. More than 100 delegates from factories as far away as Ukraine, Belorussia, Azerbaijan, and Turkmenistan attended the first session, held in Moscow. Within the year, VSPO organized more than 100 antialcohol demonstrations in different towns and some 50 local workers' conferences on how to struggle against alcoholism.

At the beginning of the 1930s, official policies changed and OBSA was given instructions to shift its focus from narrow antialcohol work to an all-out struggle for improvements in the conditions of everyday life. The temperance journal *Trezvost' i kul'tura* explained to its readers that the temperance movement had been misdirected—socialism and the socialist way of life would cure drunkenness. In keeping with this shift, the following month the journal changed its name to *Kul'tura i byt* (Culture and Everyday Life). In March 1931, OBSA officially changed its name to the Society for Healthy Living and ceased all antialcohol agitation and propaganda.

Alcohol and the Stalinist State

OBSA's close ties with state and party administrations had made it vulnerable to changes and shifts in politics and policy. By 1930, the Stalinist state had abandoned the quest for a sober society in the interests of financial expediency. Convinced that the Soviet Union needed liquor revenues to survive, Stalin directed factories to increase alcohol production and aim for maximum output. In 1933, he reversed the state's previous calls for the creation of a sober working class and encouraged workers to reward themselves for a job well done with a "little glass of champagne." Within a few months, discussion of workers' drunkenness in the press, which had occupied a central place in public and official discourse on creating a new socialist society, ceased. A large and vital temperance movement fell silent, all factory cells of the successor to OBSA were dissolved, and their leaders purged. In the ensuing years, the central government built more factories without taking measures to control the drinking of the increasing numbers of factory workers, and by 1940 the Soviet Union had more liquor stores than shops selling meat, fruits, and vegetables combined.

Because of the culture of secrecy that permeated the Soviet state as a legacy of the Stalinist era, it is nearly impossible to discern with any accuracy the amount of alcohol produced or consumed in the Soviet Union from the 1940s into the 1980s. Beginning in 1963, the state stopped publishing figures on alcohol output and obscured or altered statistics concerning sales of alcoholic beverages. Similarly, from the 1930s no serious studies of the social, economic, or legal issues related to abusive drinking were conducted and the subject of alcohol completely vanished from the press. Despite the paucity of information, one Western study found that alcohol consumption rose steadily from the 1950s: Annual per capita consumption of hard liquor (excluding beer and wines) was estimated at 7.3 liters (1.9 U.S. gallons) of liquor in 1955, rising to 15.2 liters (4 U.S. gallons) in 1979. Actual consumption levels were probably much higher, however, since the study did not adjust for differences owing to gender, age, ethnicity, or location. Alcohol revenues in the 1960s and 1970s accounted for more than one-third of all state revenues and one-ninth of the entire state budget. By the early 1970s, taxes collected from alcohol sales exceeded the officially declared defense budget.

Perhaps in response to high levels of alcohol consumption, each successive leader after Stalin introduced some type of temperance campaign. In December 1958, under Nikita Khrushchev, leader of the Communist Party from 1953 to 1964, top party and state organs called for a determined struggle against alcoholism and other survivals of the capitalist past. This campaign mainly took the form of propaganda and education, but the state placed some restrictions on the sale of alcohol in shops and restaurants and strengthened laws against home brewing. In 1960, the criminal code provided for compulsory therapy along with harsh sentences for those arrested for drunkenness more than once. Leonid Brezhnev, general secretary from 1964 to 1982, himself a notorious drinker, furthered

Customers at the counter in a liquor store in Moscow, Russia, ca. 1950 (Hulton/Archive)

efforts to control alcohol abuse. In 1966, a series of fines for public intoxication was introduced, followed the next year by the establishment of a network of labor rehabilitation centers (*lechebno-trudovye profilaktorii,* LTPs) that provided compulsory treatment and labor reeducation of problem drinkers. Amounting to little more than squalid jails for drunkards, the LTP system was extended in the 1970s. Despite repeated calls from the Brezhnev administration for a more determined struggle against alcoholism, state alcohol output increased, per capita consumption rose, and most officials deemed the alcohol problem insurmountable.

The brief period between Brezhnev's death in 1982 and the accession of Mikhail Gorbachev in 1985 was marked by a determined effort to strengthen public order and morality under Yuri Andropov (1982–1984) and Konstantin Chernenko (1984–1985). An attempt to reduce alcohol abuse was a central part of this campaign. In 1983, Andropov introduced new penalties for drinking at the workplace. His successor, Chernenko, expressed concern over the failure of earlier campaigns to effect changes in the population's drinking habits and began to make plans for a national antialcohol movement. There was even discussion of total prohibition and the creation of a national temperance society. He died, however, before implementing any of these plans.

Poster from Gorbachev's 1985 alcohol suppression program. Official toasts were to be made with juice, but the program failed because drinkers consumed homemade alcohol. (Shepard Sherbell/Corbis SABA)

The Anti-Alcohol Campaign of the Gorbachev Era

Upon succession as leader of the Soviet Union in 1985, Mikhail Gorbachev took up the issue of alcoholism as the first public priority of his administration. To that end, the state and party issued a series of decrees restricting the amount of alcohol that could be bought, the hours it could be sold, and the places it could be drunk. State production of vodka and wines was cut by more than 50 percent, from 29.5 million liters (7.79 million U.S. gallons) of vodka in 1980 to 14.2 million liters (3.75 million U.S. gallons) in 1988, and from 32.3 million liters (8.53 million U.S. gallons) of wine in 1980 to 17.9 million in 1988. In September 1985, officials established an All-Union Voluntary Society for the Struggle for Temperance, modeled after the earlier temperance society, to promote sobriety throughout the USSR. By May 1986, the society claimed 350,000 branches and more than 11 million individual members. That same year, the society revived publication of the journal *Trezvost' i kul'tura,* with more than 600,000 subscribers.

The results of the campaign were disastrous: Sugar, used in the production of moonshine, disappeared from the shelves as bootlegging became epidemic; the government lost nearly 2 billion rubles in revenues; and the population became angered by the abrupt unavailability of alcohol. Within three years, Gorbachev retreated from his antialcohol policy, but it took nearly four years for the government to recover lost revenues. Moreover, there is no evidence to suggest the campaign had any effect on the level of alcohol consumption. By 1993, Russia was ranked ahead of France as the world's heaviest drinking nation.

Kate Transchel

See also: Alcohol, Consumption of (Russia); Russia, Imperial, Temperance in; Vodka

References

Philips, Laura. *The Bolsheviks and the Bottle: Drink and Worker Culture in St. Petersburg, 1900–1929.* DeKalb: Northern Illinois University Press, 2000.

Simpura, Jussi, and Boris M. Levin, eds. *Demystifying Russian Drinking: Comparative Studies from the 1990s.* Helsinki: Stakes, 1997.

Stone, Helena. "The Soviet Government and Moonshine, 1917–1929" in *Cahiers du monde russe et sovietique.* 27 (July–December 1986): 359–379.

Transchel, Kate. "Liquid Assets: Vodka and Drinking in Early Soviet Factories" in *The Human Tradition in Modern Russia,* edited by William Husband. Wilmington, DE: Scholarly Resources, 2000.

———. "Staggering towards Socialism: The Soviet Anti-Alcohol Campaign, 1928–1932" in *The Soviet and Post Soviet Review.* 23, no. 2 (1998): 191–202.

Treml, V. *Alcohol in the USSR.* Durham, NC: Duke University Press, 1982.

White, Stephen. *Russia Goes Dry: Alcohol, State, and Society.* Cambridge: Cambridge University Press, 1996.

Spain

Spanish drinking patterns have traditionally manifested the characteristics of a Mediterranean wine-drinking culture in which wine, routinely consumed with meals, has played the largest part in overall alcohol consumption. Consequently, Spanish society before the late nineteenth century showed little interest in temperance reform. By the late twentieth century, however, Spain had begun to adopt measures to control various adverse effects of drinking.

Alcohol Production and Trade

Spain is among the top wine-producing nations in the world. In the years 1995–1998, Spain ranked third in wine production worldwide and also among wine-producing countries in Europe, just after Italy and France. The majority of Spain's wine production, over 50 percent, focuses on table wines. Red wines make up about 50 percent of annual consumption in Spain. Quality wines are produced in 38 specific regions, where regulatory boards control a variety of factors: types of grapes grown, pruning, production quantities, processing, and aging. In the 1990s, wine's share in the production of alcoholic beverages in Spain fell. It declined to 46 percent. The growing popularity of beer among Spaniards—production rose to 29 percent—accounted for the change in habits. Spirits comprise the remaining production—about 25 percent. Between 1980 and 1994, total wine production declined by 53 percent. This trend of falling production and consumption of wine reverses the historical importance of wine in Spain, where, between 1400 and 1800, wine was a mainstay of its cultural life.

Sherries are produced only in Spain; in 1996, the European Union (EU) ruled that the name "sherry" was restricted to wine grown and made in the Spanish region of Jerez. This decision ended a 30-year-long legal battle waged by the country and its sherry producers to secure to Spain the exclusive use of this trademark name. Distilling alcohol to fortify sherry is a centuries-old practice that in Jerez dates to at least 900 C.E. After 1492, spirits were exported on a large scale to colonial Spanish America. The most popular exports were brandy, anise, and aguardente, a strong liquor made from either sugarcane or mashed grapes. Sherries are categorized as *finos* (dry wine) or *olorosos* (sweet wine). Finos are classified into three groups: Mananilla, Fino, or Amontillado. The olorosos are described as cream or brown.

Although beer was imported into Spain before World War I, because of its high cost relative to domestically produced wine, it remained a drink of the affluent. In recent decades, as the Spanish economy experienced a dramatic improvement, consumption of beer has increased. The increase in consumption prompted a rise in production. In the 1950s, Spain produced only 603,770 hectoliters (21.7 million U.S. gallons), but by the mid-1990s, it produced over 25 million hectoliters (900 million U.S. gallons). The brewing of beer in Spain is quite recent, but by 1995, fifteen independent brewing companies operated twenty-five breweries in Spain. Of these, the four largest breweries accounted for three-quarters of all production.

Alcohol Consumption

Typically consumption of alcohol for Spain, on a per capita basis, appears quite high, but the actual level for Spaniards is considered to be much lower. This disparity is due to Spain's position as probably the most popular tourist destination in the world. Since 1989, its population of about 40 million receives over 50 million tourists annually. In 1991, it was estimated that tourists accounted for 15 to 25 percent of alcohol consumption. Based on figures from the mid-1990s, the per capita consumption of spirits peaked in 1981 at 3.2 liters (1.1 U.S. gallons). By 1995, it had fallen to 2.5 liters per person (0.9 U.S. gallon), where it has stayed. The consumption of wine declined 51 percent between 1974 and 1989 and has remained around this level, with 38 liters per person recorded for 1995 (13.7 U.S. gallons). Since 1990, beer consumption per capita has risen steadily, and in 1995, it was almost 67 liters (24.1 U.S. gallons). The decline in the consumption of wine and spirits has offset the rise in beer consumption, putting overall consumption slightly lower than it was in the 1970s.

Drinking Habits

Patterns of drinking in Spain reflect Mediterranean wine-drinking culture, giving alcoholic beverages a key role in social and dietary habits. In 1984, a study of consumption habits of the adult Spanish population found that almost 85 percent of adults in Spain consume alcoholic beverages, and only 8 percent drink only occasionally. About 52 percent reported that they drank daily. Typically drinking is accompanied by food and is done in a social but also work setting, especially around the lunch hour and later. But increasingly, Spain reflects the drinking habits of other members of the EU and alcohol is less available in factories and office canteens during the working day, now considered to be from 8:30 A.M. to 6:00 P.M. In 1987, the results of another study of consumption habits in the city of Madrid concluded that drinking patterns in the city could be summarized as habitual consumption in moderate amounts. Researchers in 1992 found that beer is more popular with younger consumers, while those over forty prefer wine, and youthful drinkers partake on the weekends while those over fifty consume on a daily basis. While accounting for the influx of tourists in overall consumption levels, Spaniards rank high in their consumption of total alcohol, a 1995 study discovered; the country ranks ninth out of 31 countries assessed. It is eighth in spirits consumption, seventh in wine, and seventeenth for beer.

Spain has severe alcohol-related problems, a fact identified in a 1978 study. It has a comparatively high rate of liver cirrhosis mortality. In 1991, it was about 20 per 100,000 population. Although this is lower than some industrialized countries, it is higher than Canada, which has a rate of 7.5, or the Netherlands rate of 5 per 100,000. This rate has stayed stable from the late 1960s through the early 1990s. Alcohol consumption currently is the main cause in 4.2 percent of total deaths, making it the sixth leading cause of death in Spain.

One factor in alcohol consumption among Spaniards is that consumption levels of at least 75cc of alcohol per day (about two-thirds of a liter bottle of wine) has traditionally not been considered excessive. Throughout the nation, two areas of concern are trends that developed among youth during the early 1980s and that were confirmed by a study in 1985. Spanish youth now begin to drink heavily at a much earlier age and they prefer to consume beer and not the wine of previous generations. Also youth now gather in the streets, usually at a public plaza, late at night and drink liter bottles of beer, a fashion referred to as "cultura de la litrone" (Karlsson 1997). In 1995, a study found that of habitual drinkers age 14 to 18, some 24 percent get drunk at least once a month. Youth drinking is viewed as a pervasive national problem. In 2002, the government announced that legislation would be introduced to increase the age for drinking alcohol from 16 to 18 and to ban drinking on the streets. Spain's Interior Minister, Mariano Rajoy, said that 76 percent of Spanish teenagers between 14 and 18 years of age drink alcohol.

Drinking is also a leading cause of 30 to 50 percent of road accidents. Between 1990 and 1992, authorities estimated that alcohol-related accident rates in urban areas increased by 35 percent and in rural areas by over 50 percent, although overall traffic accidents that caused injuries had declined.

Temperance Reform Comes to Spain

Temperance reform in an American mold was introduced into Spain in 1892 by Alice Gordon Gulick (1847–1903), wife of the Reverend William H. Gulick (1835–1922), an American who directed the Congregational effort to reintroduce Protestantism to Spain. The Congregationalists initiated their effort in 1872 after Spain adopted the liberal Constitution of 1868 with its provision for religious freedom. Alice Gulick was also the sister of Anna Adams Gordon (1853–1931), who served as the personal assistant to Frances Willard (1839–1898), the dynamic leader of the Woman's Christian Temperance Union (WCTU) and founder of an international organization, the World's WCTU or WWCTU, created to carry the temperance message to every part of the globe and the largest social reform organization in the United States in the late nineteenth and early twentieth centuries. In 1891, under Willard's leadership, the WCTU approved the creation of an international organization, the World's WCTU or WWCTU, to carry the temperance message to every part of the globe. In April 1891, Gulick was designated by Willard and the WWCTU as the vice president for the organization in Spain.

In 1892, Gulick began her work in that capacity, although she acknowledged that during thirty years of work in Spain she had seen only one prostrate drunkard, a statement confirmed by impressions recorded by travelers in northern Spain. In 1890, however, another visitor, Felix L. Oswald, M.D., recounted in an article on alcoholism in Spain published in the *Union Signal,* the WCTU newspaper, that the opposite condition was common in the south of Spain. Oswald agreed that in northern Spain drunkenness was not a problem among the people, whom he termed "industrious." But among the impoverished peasants of Andalucia, Oswald faulted not their miserable economic condition but their supposed laziness.

Gulick focused her organizational efforts for the WWCTU on two groups. One was supervised by her husband, the Spanish Protestant ministers, and their wives, who staffed a chain of fourteen missions and preaching stations across northern Spain. The second consisted of her very able female students enrolled in the boarding and day school that she and her husband had established in 1881 in San Sebastian, the stylish international resort city on Spain's northern coast, where she and the Reverend Gulick made their home.

The school, incorporated in the United States in 1892 as the International Institute for Girls in Spain, offered a rigorous curriculum patterned on that of Mt. Holyoke College, Mrs. Gulick's alma mater. Advanced students were prepared by Gulick and her staff to take examinations at the province's government institute. After passing those examinations, the girls were awarded government diplomas, which opened to them the possibility of a career in teaching—and made them the first girls to be prepared by female instructors to pass such examinations in the country's history. Gulick's approximately 150 students were drawn from local families, both Protestant and Roman Catholic, but the boarders consisted of girls from Spanish Protestant families from the provinces in northern Spain reached by the Reverend Gulick's mission system and also from the provinces of Andalucia and Huelva.

Anna Gordon visited the San Sebastian school in 1893 and inaugurated its WWCTU chapter, administering the temperance oath to the older students and welcoming them as white ribboners. The enthusiasm and commitment of the girls to the WWCTU cause reflected their evangelical backgrounds and bode well for the future of the WWCTU in Spain. After receiving their diplomas, the school's graduates followed one of two career paths. Most graduates returned to their families and taught in local Protestant day schools or state-run elementary schools. Some married Spanish Protestant ministers. The young couple would then move to a new site in Spain and open or staff a Protestant mission, with the wife organizing a day school and leading the activities for women and children.

In 1896, Gulick encouraged one of her most promising graduates, Julia Campos, who moved to the capital to begin work for the British-sponsored Young Women's Christian Association, to begin temperance reform work in the capital city.

Campos's temperance activities consisted of speaking at small meetings of like-minded Spanish women and distributing temperance pamphlets directed to a Spanish audience. When the Gulicks moved their school to Madrid in 1903, the older girls continued to be educated in the principles of temperance as part of their evangelical training. When Alice Gulick died in 1903, however, the momentum and fund-raising skills behind the school's successful academic thrust and evangelical and temperance reform training was lost. In 1907, the school's mission was divided, and the evangelical-training component moved to Barcelona, where it faded gradually and closed in the 1920s. Moreover, as secularism increased in Spain—and in the United States—interest in organized religion and church-related programs such as temperance waned. The WWCTU and the WCTU suffered the same fate.

In the late 1800s, a revolutionary working-class political movement that flourished only in Spain, anarchism, also advocated temperance reform. Centered in the impoverished industrial slums of Barcelona and the equally poor rural areas of Andalucia, its other strongholds were in Aragon, the Levante, and Galicia. Anarchists espoused self-improvement, efforts to revive traditional society, a belief in the moral transformation of the world, trade unionism, and a few, terrorism. Committed anarchists, in addition to practicing vegetarianism, also abstained from liquor. The movement, in its political form, did not survive the Spanish Civil War of 1936 to 1939.

Modern Drinking Patterns and Control Measures

From 1923 through 1945, various national governments did little to study alcoholism, to offer solutions to alcohol-related problems, or to address the social problems stemming from alcohol abuse. This lack of data makes it difficult to determine if Spain, like other countries in Western Europe experienced changes in drinking patterns and behavior during years of urbanization and industrialization. After World War II, the regime led by the dictator Francisco Franco from 1939 to 1975 was intent on molding Spanish behavior to traditional models. In terms of alcoholism, this meant upholding the notion that Spain had no problem with alcoholism because, in keeping with Spanish cultural norms, visible signs of drunkenness continued to be absent from public view in daily life.

From 1953 through the late 1960s, the era of the Catholic technocrats, this inattention to socially challenging problems was especially typical. Beginning in 1953, the members of Opus Dei, the elitist and secret lay order founded in 1928 by the Spaniard Jose Maria Escriva de Balaguer, dominated the government. Although the policies and guidance provided by the Opus Dei technocrats dramatically altered Spain's economy, social problems were to a large measure ignored. According to the English historian Frances Lannon, this approach dominated because Opus Dei was "elitist, male, individualist, hostile to sexuality, silent on economic and social problems" (Lannon 1987: 226).

By the time of Franco's death in 1975, an array of welfare-state programs and services such as free public education, health care, and social security were available throughout Spain. Spain had also experienced a rise in consumerism that had triggered a substantial decline in rural life and an internationalization of Spanish culture. After Franco's death, the country experienced growing political and economic stability and joined the European Common Market. In the 1990s, prosperity and rapid change affected Spanish culture, especially the youth, in much the same way as in other industrialized countries.

Alcohol Control Measures

In 1984, the government established the Bureau of Alcoholism and Drug Addiction in the Ministry of Health. It developed educational programs aimed at school children on the consumption of alcohol and drugs and it established hospital facilities for treatment of alcoholics. Regional governments of Navarre and Basque also sponsored advertising campaigns against alcohol misuse directed mainly at youth and pregnant women. In 1993, the alcoholic beverage industry created a new social responsibility group, the Asociación De Distribuidores de Grandes Marcas de Bebidas (ADIGRAM). The organization's purpose was to link the Spanish and local government efforts. Its 1995 major advertising campaign slogan informed consumers that "To Drink with Moderation Is Your Responsibility." In 1995, a national road safety plan was initiated with the goal of reducing the number of drunken driving incidents. Checkpoints and surveillance equipment were increased and more up-to-date breathalyzers were introduced. In 1995, the government's national campaign against drunk driving emphasized the cost to society at large of alcohol abuse on public health.

Following traditional practices, Spain has a limited government approach to the control and sale of alcoholic beverages; unlike other countries, there is no government monopoly controlling the sale and distribution of alcoholic beverages. Alcoholic beverage licensing requirements are not imposed on individual cafes, restaurants, grocery stores, or like establishments.

Although advertising of alcoholic beverages is allowed, there are some restrictions. In 1988 and 1990, legislation was passed that governed advertising. Advertising of beverages over 20 percent alcohol by volume is banned from state or private broadcasting media. Alcohol of a lesser level is barred from state television before 9:30 P.M. Some regions imposed bans independently on all television advertising of alcoholic beverages, and the autonomous regions of the Basque country, Catalonia, and Galicia implemented their own advertising regulations. When cinemas show films directed at youth, sales of alcohol are banned. Outdoor advertising is permitted for beer and wine, but voluntary restrictions are in force for spirits advertising. In 1995, the Spanish Brewers Association adopted a self-regulatory code that required members to include educational statements on the hazards of alcohol in all advertising and statements on abuse, drinking, youth, and consumption in the workplace. In 1996, the Spanish Federation of Drinks Producers adopted the Spanish Advertisers' Association's code governing advertising of alcohol. The code acknowledges the benefits of moderate drinking but notes the potential for misuse. Its guidelines include alcoholic beverages and youth, driving, health, work, social behavior, alcoholic strength, and sport. Its goal was to harmonize with new legislation.

In terms of pricing of alcoholic beverages in Spain, there is no regulation. In cases of apparent alcohol-impaired driving, breathalyzer testing is authorized and the standard for driving while drunk is 0.08 percent. In the event of a failed breathalyzer test, a blood test can be requested by the individual or ordered by a judge in accident cases, and sizable fines are applicable.

Alcohol Taxation

In 1986 and 1993, Spain revised its tax system to achieve compatibility first with the European Economic Community (EEC) and then with EU requirements. This affected rates for alcohol. In 1986, an excise tax was placed on alcoholic beverages, and when it was revised in 1993, all types of wine were included in the excise system at a zero tax rate. There are now two types of taxes for alcohol—special taxes and other taxes. Special taxes for alcohol are excise taxes and this reflects EU policy. When alcohol products are sold in Spain they are taxed and the record of the sale is the control document for determining the tax. Special taxes increased between 1986 and 1989, and then stabilized. They increased again in 1993, and since Spain entered the EU, the excise rates have increased annually. Regarding other taxes, in 1986, on entering the EEC, Spain was required to adopt two taxes and put in place a national value-added tax. In 1993, the value-added tax rate on alcoholic beverages was set at 15 percent and increased in 1995 to 16 percent. In 1993, in accordance with EU policy, Spain eliminated its customs duties on imports from EU member countries, while simultaneously adjusting its duties on products from non-EU countries to conform to EU rates. Except for wine, these changes had only slight impact on the level of taxation between beverages and on the tax content of the products.

Carol Scally-Grigas

See also: Gordon, Anna Adams; World's Woman's Christian Temperance Union (WWCTU)

References

Carr, Raymond. *Spain, 1808–1973.* Oxford: Oxford University Press, 1982.

Gordon, Elizabeth Putnam. *Alice Gordon Gulick: Her Life and Work in Spain.* New York: Fleming H. Revell, 1917.

Karlsson, Thomas. "Spain: Pattern of Drinking and Beverages Consumed" in *International Survey: Alcoholic Beverage Taxation and Control Policies,* edited by W. Hurst, E. Gregory, and T. Gussman, 399–417. Ottawa: Brewers Association of Canada, 1997.

Karlsson, Thomas, and Esa Österberg. "Spain" *in Alcohol Policies in EU Member States and Norway: A Collection of Country Reports,* edited by Esa Österberg and Thomas Karlsson, 270–286. Helsinki: Stakes, 2000.

Lannon, Frances. *Privilege, Persecution and Prophecy: The Catholic Church in Spain, 1875–1975.* Oxford: Oxford University Press, 1987.

"News in a Nutshell" in *Union Signal.* 14 May 1891, 18.

Oswald, Felix L., M.D. "International Temperance Studies" in *Union Signal.* October 2, 1890, 3.

Shubert, Adrian. *A Social History of Modern Spain.* London: Unwin Hyman, 1990.

"Spain: Higher Drinking Age" in *New York Times.* 13 February 2002, Section A, 10.

Tyrrell, Ian. *Woman's World, Woman's Empire: The Woman's Christian Temperance Union in International Perspective, 1880–1930.* Chapel Hill: University of North Carolina Press, 1991.

World Drink Trends. 1996 Edition. Oxfordshire, United Kingdom: NTC Publications, 1996.

Spence, Francis Stephens (1850–1917)

Francis Stephens Spence (1850–1917), teacher, journalist, and politician, was one of the driving forces of the Canadian temperance campaigns of the late nineteenth century. An ardent Methodist and proponent of the Social Gospel movement, Francis and his younger brother, Benjamin H., were organizers of the Dominion Alliance for the Total Suppression of the Liquor Traffic, the leading lobby group for the prohibition of alcoholic beverages in Canada. Spence's magnum opus was *The Facts of the Case: A Summary of the Most Important Evidence and Argument Presented in the Report of the Royal Commission on the Liquor Traffic . . .* (1896). The alliance dissented from the findings of the commission his group forced the government to strike and pushed for the prohibition plebiscite of 1898, which turned out to be politically divisive and ultimately failed because the margin of voters in favor of prohibition was so small.

Francis Spence was born in Donegal, Ireland, on 29 March 1850, one of twelve children of a Methodist minister, Jacob Spence, and Elizabeth Stephens, a female suffragist. The family immigrated to Toronto, Upper Canada, in 1861. Jacob, who had been active in Father Theobald Mathew's abstinence movement in Ireland, became secretary of the Ontario Temperance and Prohibitory League. He took his children on the lecture circuit, and producing temperance pamphlets became a cottage industry for the family. Francis and Benjamin took over their father's mission. Francis married Sara Violet Norris in 1879 and had two daughters.

Francis was the headmaster of a public school until 1882, when he abandoned teaching for journalism. He edited a number of moral reform journals, including *The Canadian Citizen,* the *Vanguard,* and the *Ontario Good Templar,* and in 1902 he founded the *Pioneer,* the official journal of the Ontario temperance movement. He served as secretary for the Central Committee of the plebiscite campaign in Ontario in 1894 as well as for the Toronto Prohibitive Union in 1895 and the Dominion Alliance for the Suppression of the Liquor Traffic in 1897. He was elected on the municipal Liberal ticket as an alderman in 1896–1897, 1899–1900, 1902–1905, and 1914. He served as acting mayor, city controller, and chairman of the Toronto Harbour Commission and ran unsuccessfully for mayor twice on a temperance platform. He opposed Sunday streetcars, supported female suffrage and the playground movement, and was one of the founders of Ontario Hydro, the provincially owned electricity generator and distributor.

His major work, *The Facts of the Case,* was a painstakingly researched document. Its topics ranged widely, covering the extent of the liquor business; physical and social effects of intemperance; prohibitory legislation and liquor licensing laws in various Canadian provinces, the United States, and other jurisdictions; and the beneficial consequences that prohibitory legislation would have for Canada's social conditions and economic sectors.

Spence died suddenly of pneumonia in 1917, just as federal prohibition was becoming a reality as part of the War Measures Act. He did not live to see the eventual disintegration of both national and provincial prohibitory laws after the Great War.

Cheryl Krasnick Warsh

See also: Dominion Alliance for the Total Suppression of the Liquor Traffic; Federal Prohibition (Canada); Laurier, Sir Wilfrid; Mathew, Father Theobald; Royal Commission on the Liquor Traffic (Canada)

References

Decarie, Graeme. "Spence, Francis Stephens" in *Dictionary of Canadian Biography,* vol. 14, *1911–1920.* Toronto: University of Toronto Press, 1998.

Morgan, Henry J. *The Canadian Men and Women of the Time: A Handbook of Canadian Biography.* Toronto: W. Briggs, 1898.

Spence, Francis. *The Facts of the Case: A Summary of the Most Important Evidence and Argument Presented in the Report of the Royal Commission on the Liquor Traffic Compiled under the Direction of the Dominion Alliance for the Total Suppression of the Liquor Traffic.* Toronto: Newton & Treloar, 1896; Facsimile Reprint, Toronto: Cole's, 1973.

Spence, Ruth Elizabeth. *Prohibition in Canada: A Memorial to Francis Stephens Spence.* Toronto: Ontario Branch of the Dominion Alliance, 1919.

Sporting Events, as Drinking Sites

The history of sporting events as sites of alcohol consumption goes back to the very earliest days of sport and has persisted, not without controversy, up to the present day. In the English-speaking world of the eighteenth and nineteenth centuries, sporting events were often hosted and organized by owners of pubs. These activities included blood sports such as cock-fighting, bear-baiting, and bull-running as well as running and prize-fighting contests. These events served to attract customers to the pub and played a role in establishing the pub as an important center of community life. They were also associated with drunkenness and gambling, however.

Middle-class reformers were appalled by the drinking at these primarily working-class events. In the nineteenth century, the muscular Christianity movement promoted competitive sport and physical activity as a Christian alternative to the saloon, dance hall, and other undesirable leisure activities. The growth of muscular Christianity prompted the development of alcohol-free sport, especially in the United States. Despite the growth of the Young Men's Christian Association (YMCA) and temperate sporting clubs, however, sporting competition continued to be linked with alcohol consumption. From the mid-nineteenth century, intercollegiate sporting competition in the United States was associated with drunken revelries by athletes and fans. In Britain, pubs continued to host football, darts, pigeon shooting, rabbit coursing, and a variety of other sports, and both fans and athletes continued to imbibe.

Commercial Sport

Large-scale commercial sport took off with American baseball and British football ("soccer" in the United States) in the 1880s and 1890s. In the United States, the first two professional leagues were divided over what role alcohol would play in the game. The first professional league, the National League, did not permit beer sales or the playing of Sunday games. In 1882, a new league, the American Association, started up. Unlike the sabbatarian National League, the American Association sold beer at games. It became known as the "beer and whiskey" league. Four of its owners were brewers. Although the American Association had a quick demise, this would not mark the end of the important role brewers would play in baseball. In 1915, Jacob Ruppert, owner of the Ruppert Brewery, purchased the New York Yankees. He acquired Babe Ruth, Lou Gehrig, and Joe DiMaggio, and the Yankees won seven World Series championships during his tenure as owner.

In Britain, football clubs were often attached to local pubs, although unlike earlier sports, football was not organized by the pub itself. The football clubs were independent organizations that rented fields from the pubs and made sponsorship arrangements that would include items such as under-the-table payments for players and free beer after matches. As in the United States, there was a split over the role alcohol should play in football. Teetotalers helped to found the Football League in 1888; nonetheless, brewers such as John Davies were successful in their efforts to commercialize the game, and by the turn of the century newspapers were complaining that football matches were the site of considerable drunken behavior.

Since World War II, both the alcohol industry and the sporting industry have become increasingly centralized and commercialized. There has been enormous growth in the number of commercial sporting events around the world, and much of this growth has been fueled by the advertising sponsorship of alcohol companies, accompanied by alcohol sales in the stadium. In the 1950s, August Busch bought the St. Louis Cardinals and acquired the rights to sell his beer to fans and to place Budweiser signs throughout the stadium. In the 1970s, Miller and Anheuser-Busch poured money into sports-related marketing, turning sporting events into lengthy beer commercials.

In recent years, there has been growing controversy about the sponsorship of sporting events by companies selling alcohol beverages. Nongovernmental organizations such as AlkoKutt in Norway and Utah's Alcohol Policy Coalition have argued that sporting events, which are attended by many young people, should not be sponsored by liquor companies. In 1991, France introduced a ban on direct and indirect alcohol advertising in association with sport. In Malaysia in 1998, the government banned Carlsberg's sponsorship of the Commonwealth Games just two months before they began.

Alcohol and Hooliganism

In the 1980s and 1990s, many spectators complained that violence, obscenities, and other loutish behaviors were due to the high rates of alcohol consumption at professional sporting events. As a result, in stadiums around the world owners are increasingly providing alcohol-free seating areas, providing training to servers in dealing with drunken customers, and placing limits on the amount of alcohol and the times at which alcohol can be sold.

Governments and the media have paid particular attention to football hooliganism in Europe, a phenomenon frequently linked to excessive alcohol consumption by fans, especially in England. Although it is often referred to as the "English disease," there has also been serious football violence in Italy, Germany, Belgium, and the Netherlands. The British are the best known for causing havoc at international matches, however, and although there is little research on the matter, alcohol is often implicated in the violence. Riots in Brussels in 1985, Marseille in 1998, Charleroi in 2000, and Munich in 2001 were blamed, at least in part, on excessive alcohol consumption.

There have been efforts across Europe to restrict the sale of alcohol at football matches. In 1985, Britain prohibited the possession of alcohol on the way to matches on football special coaches, trains, and minibuses and made it an offense to be drunk on these conveyances. It also made it illegal to enter a ground already drunk or in possession of alcohol, to possess or consume alcohol within view of the pitch, or to be drunk inside the ground during the match.

British soccer fans shout and drink beer in front of Spiller Cafe in Munich on 1 September 2001. (Reuters NewMedia Inc./Corbis)

The Fédération Internationale de Football Association (FIFA) banned alcohol sales at Italia 1990 and France 1998 and had partial restrictions during USA 1994. In order to honor an agreement with sponsor Anheuser-Busch, there were limited alcohol sales in Korea/Japan 2002.

2002 Winter Olympics

Historical conflicts over alcohol sponsorship and sales at sporting events came to a head at the 2002 Winter Games in Salt Lake City, Utah, where alcohol other than low-alcohol beer is normally available only at private clubs. However, under pressure from sponsor Anheuser-Busch, alcohol was widely available at the games despite considerable community opposition. This accommodation reflected a connection between sporting events and alcohol consumption that has become an enduring feature of modern life.

Catherine Carstairs

References

Collins, Tony, and Wray Vamplew. *Mud, Sweat and Beers: A Cultural History of Sport and Alcohol.* Oxford: Berg, 2002.

Guttman, Allen. *Sports Spectators.* New York: Columbia University Press, 1986.

Johnson, William Oscar. "Sports and Suds: The Beer Business and the Sports World Have Brewed Up a Potent Partnership" in *Sports Illustrated.* 69, no. 6 (8 August 1988): 69–82.

March, P., K. Fox, G. Carnibella, J. McCann, and J. March. *March Football Violence in Europe.* Amsterdam: Amsterdam Group, 1996.

Revke, Dag. "Counteracting Sports Marketing by the Drinks Industry" in *Towards a Global Alcohol Policy: The Proceedings of the Global Alcohol Policy Advocacy Conference, Syracuse, New York, 2000,* edited by Andrew Varley. London: Institute of Alcohol Studies, 2001.

St. Germain en Laye Convention, 1919

The convention signed at St. Germain en Laye in 1919 gave new direction to the attempts by European colonial powers to control the supply of distilled liquor to Africans. These attempts had begun with the Brussels Act of 1890, itself an offshoot of the Berlin Congress of 1885; together, the bundle of obligations thus assumed by the signatory powers (by no means all of which related to alcohol) came to be known as the "Congo Basin Treaties." The restrictions on spirits agreed upon at St. Germain—one of three postwar conventions that revised the agreements originally made at Berlin and Brussels—were supported by the League of Nations. The league thus added to the rather vague requirement imposed upon "Mandatory" powers, who had taken over the former German colonies, that they "exercise a strict control" over the supply of spirituous liquors in these territories. Overall, the restrictions under the convention maintained the implicitly racist rhetoric of a European imperial mission of trusteeship that involved protecting Africans from alcohol, which they were allegedly unable to manage. The restrictions endured, in theory, until around 1950, but they were rapidly and quietly forgotten thereafter, along with the other provisions of the Congo Basin Treaties.

The 1919 Agreement

Although the Brussels Act had revolved around the idea that certain parts of Africa were as yet free of spirits, and should remain so, the St. Germain Convention focused rather on the undesirability of certain kinds of spirituous liquors. These were known in English as "trade spirits"—cheap beverage spirits produced specifically for the African market by European distillers, mostly in the Netherlands—and the convention alleged that they were particularly dangerous to human health, either "by the nature of their products entering into their composition or by the opportunities which a low price gives for their extended use." These were prohibited entirely from the area covered by the convention: the whole of Africa, with the exception of Algiers, Tunis, Morocco, Libya, Egypt, and South Africa. In contrast to the Brussels Act, the convention included no provision for importation for European use. Other spirits were to be subject to a minimum duty of 800 francs per hectoliter (about 26 U.S. gallons) of absolute alcohol. Overall, the effect was to replace a control policy based upon protecting certain areas from all kinds of spirits with one that relied on making spirits so expensive that they would be unaffordable for most Africans; in essence, the convention ensured that spirits would cost at least 8 francs per liter of absolute alcohol. By 1930, the convention had been ratified by all the remaining colonial powers: Britain and its Dominions, France, Italy, Belgium, and Portugal. An international bureau to collect and disseminate information on the alcohol trade and the working of the convention was established at Geneva.

The effects of the convention were inevitably limited by the uncertain definition of the term "trade spirits." An attempt to insert a precise definition—that these were beverage spirits that used industrial alcohol or cost less than 2 francs per liter of absolute alcohol in their country of origin—was defeated by the resistance of the Italian government. Once the convention was ratified, there was considerable debate in British-ruled parts of West Africa, for example, as to which imported spirits should be prohibited, and it was decided—for a while—that "Dutch gin" should not be classed as a trade spirit; this led the French government to adopt a similar attitude in territories neighboring British ones. The later imposition of new restrictions on "Dutch gin" in the British-ruled Gold Coast owed more to local African debates over temperance than to the St. Germain Convention. In most of eastern, central, and southern Africa, the convention seems not to have led to any change in the effect of existing colonial laws, which remained cast in racial terms: In British territories, Africans were still forbidden to drink any bottled beer or wine, as well as spirits; in French and Belgian territories, they were forbidden all spirits. British officials in the colonies and the metropole evidently had a rather vague grasp of the details of beverage production and of the wording of the convention; they viewed it simply as a reaffirmation of the principle that Africans should not be permitted to consume "European" beverages.

Undoing the Convention

After World War II, there was a strong sense among some colonial officials of the need to encourage the emergence of a new group of Africans who would share power and responsibility with their colonial mentors. African access to "European" drink became an important issue: How could cooperation be encouraged when social events were segregated by beverage laws? Initial experiments with allowing a small elite of African society to drink would not necessarily have violated the convention, which was, after all, technically phrased in terms of forbidding cheap spirits; postwar inflation had anyway weakened the effects of even that restriction. But it was symptomatic of the general lack of knowledge of the terms of the convention that, among British officials at least, there was much concern over whether cautious schemes of giving "permits" to drink spirits to African politicians or professionals might violate the St. Germain agreement. In the event, the rapid spread of illicit distillation offered a cheap and subversive alternative to the official cocktail party, and by the early 1950s officials in some colonial territories (notably the Gold Coast and Uganda) were anxious both to end racial restrictions on drinking and to create local distilling industries that would allow the suppression of illicit distillation by providing cheap spirits for an African market. In these circumstances, the St. Germain Convention was simply allowed to lapse, its disappearance unnoticed in an international climate that no longer favored the rhetoric of colonial paternalism.

Justin Willis

See also: Africa, Central; Brussels Act and Conventions, 1890–1912; Indigenous Peoples and the Liquor Traffic Controversy

References

Akyeampong, E. *Drink, Power and Cultural Change: A Social History of Alcohol in Ghana, c. 1800 to Recent Times.* Portsmouth, NH and Oxford: Heinemann and James Currey, 1996.

Pan, Lyn. *Alcohol in Colonial Africa.* Helsinki and Uppsala: Finnish Institute for Alcohol Studies, 1975.

Willis, Justin. "Demoralised Natives, Black-Coated Consumers and Clean Spirit: European Liquor in East Africa, 1890–1955" in *Journal of Imperial and Commonwealth History.* 29 (2001): 55–74.

St. Patrick's Day

There are longstanding and widespread links between St. Patrick's Day, alcoholic drink, and temperance. Across the world, wherever the Irish diaspora has settled, 17 March is celebrated with great enthusiasm through public parades, church services and dinners, and a multitude of gatherings in homes, restaurants, and bars. For many people, whether of Irish descent or otherwise, the celebration of 17 March is synonymous with consumption of "patriotic" drinks such as green-dyed beer, Irish stouts and ales, or one of a host of Irish whiskeys. Nonetheless, in historical terms the day has featured differences between those who wished to celebrate it through carnival and drink and others who saw the anniversary of Ireland's patron saint as a religious event, thus favoring contemplation via church services, prayers, and temperate behavior.

A key factor in understanding the relationship between St. Patrick's Day and drink is the position of the day in the Christian religious calendar: 17 March usually falls within the period of the Lenten fast. Despite this restriction, a church-sanctioned tradition was established in the sixteenth century that freed the Irish from their Lenten duties of abstinence on that date. They could now celebrate the memory of their patron saint in a full and festive manner—drinking, dancing, and singing.

The shamrock became an important symbol in St. Patrick's Day celebrations. According to legend, St. Patrick had used the shamrock to explain to Ireland's pagan kings the concept of the Trinity. Subsequently, Ireland's Christians wore the shamrock on 17 March as a symbol honoring the memory of their patron saint. Significantly, the shamrock also became part of the drinking culture of St. Patrick's Day. Historical documents from the seventeenth and eighteenth centuries are full of references to the "wetting" or "drowning" of the shamrock. According to custom, toward the end of the big day the shamrock was placed in a glass then covered with punch, whiskey, poteen, or any other available alcohol, and ingested along with the drink itself. This ritual of closure put drink at the very heart of the 17 March celebration, though it must be admitted that the drowning of the shamrock was preceded by widespread consumption of alcohol throughout St. Patrick's Day.

These drinking traditions were carried with the Irish when they immigrated to North America, Australia, Britain, and elsewhere in the late eighteenth and early to mid-nineteenth centuries. During the same period, St. Patrick's Day celebrations slowly developed into two distinct and largely separate forms of observance. The elite classes, especially in the United States, chose to celebrate their links with Ireland and the patron saint by staging sumptuous private dinners. Although drink was served at such events, the tenor of the gathering was respectable. The main themes were social improvement for the Irish community, religious observance, middle-class respectability, and patriotic support for the Irish at "home." In this socially "proper" setting, drink, or rather excessive drinking, had no place. But an alternative form of celebration among the lower classes of Irish expatriates offered drink and merriment in abundance. The nonelite groups, especially those who had recently arrived from Ireland, founded public forms of celebration such as parades and sporting events on 17 March. These gatherings sought to openly demonstrate Irish pride and presence in a new land, but they also attracted men and women whose main aim was to celebrate indulgently under the influence of alcohol. Examples of drunkenness on St. Patrick's Day helped to foster negative stereotypes of the Irish, and images of inebriated St. Patrick's Day revelers were seized upon by critics who preferred temperate and contemplative behavior on 17 March.

During the mid-nineteenth century, the temperance movement entered the St. Patrick's Day arena and tried to rid the parades of drink. Father Theobald Mathew's temperance crusade in Ireland during the 1830s and 1840s had a quick impact: The first "modern" St. Patrick's Day parade through the streets of Dublin was actually staged by members of Irish temperance societies. In 1847, the *Illustrated London News* attempted to sum up the stark choice facing the Irish. A double-sided cartoon depicted, on one half, a happy, healthy, and prosperous Irish family overseen by Father Mathew. Those ignoring temperance were choosing the other half of the image: a drunken, fighting, fornicating mass of inhumanity that was poor and in ill health. The temperate group in the cartoon was being watched over by a benevolent St. Patrick, while he turned his back on those who had not forsaken drink. The patron saint was now firmly part of the antidrink crusade—at least among temperance advocates.

Father Mathew took his campaign to America in 1849–1851. There is evidence to show that numerous temperance societies, such as the Albany Catholic Total Abstinence Association and the Pennsylvania Catholic Total Abstinence Society, joined St. Patrick's Day parades across the country. However, the temperance movement failed to capture control of 17 March celebrations. During the second half of the nineteenth century, various nationalist movements and campaigners seeking to free Ireland from British rule took precedence on the holiday, and temperance took a back seat.

The "victory" of drink as a key feature of St. Patrick's Day celebrations was secured during the twentieth century. Once the Irish Free State had secured independence, at least for a large part of Ireland, the day became a focus abroad for conceiving and venerating Irish culture. Given this changing context, the politics of celebrating St. Patrick's Day in the diaspora became less important. Concurrently, though, commercial forces became more influential, with companies seeing opportunities to link their products to the celebration of Irish culture. One of the key beneficiaries of this initiative was alcoholic drink manufacturers. From the 1920s onward, U.S. beer companies advertised their products specifically for consumption on St. Patrick's Day as a key part of the celebration. After World

A woman drinks from a large beer glass during the St. Patrick's Day parade in downtown Chicago, Illinois. (Sandy Felsenthal/Corbis)

War II, Irish products such as Guinness and Harp joined them, and a global market that linked drink with the celebration of Ireland's patron saint grew. Indeed, during the past twenty years the phenomenon of the Irish bar has spread across the world, and these venues, whether in Boston, Lagos, or Beijing, have become a major focus for 17 March celebrations.

The last "official" stand of temperance on St. Patrick's Day took place, strangely enough, in Ireland. The Irish Free State, when formed in 1922, took the view that St. Patrick's Day was a ritual of religious observance and sober contemplation; hence, the government closed all licensed premises on 17 March. This ban stayed in place until 1961, and so for forty years the only place to legally buy a drink on St. Patrick's Day in the Irish homeland was at the annual 17 March Dublin Dog Show (which, surprisingly, was allowed to serve drink). Since the ban was lifted, Ireland has marketed itself as a place to enjoy 17 March in a full and festive manner. The Irish government now embraces a four-day St. Patrick's Festival in Dublin, sponsored by Guinness, and bars and hotels have extended opening hours to facilitate the drowning of the shamrock.

Contemporary St. Patrick's Day celebrations are more in tune with the spirit of the inaugural festivities. Of course, new factors, notably globalization and commercialization, have ensured that Irish drink brands can now be consumed in most corners of the world on 17 March. The spirit of those who first drowned the shamrock lives on, both in Ireland and abroad.

Daryl Adair and Mike Cronin

See also: Ireland

References

Cronin, Mike, and Daryl Adair. *Wearing of the Green: A History of St. Patrick's Day.* London and New York: Routledge, 2002.

Kerrigan, Colm. *Father Mathew and the Irish Temperance Movement, 1838–49.* Cork: Cork University Press, 1992.

Malcolm, Elizabeth. *Ireland Sober, Ireland Free: Drink and Temperance in 19th Century Ireland.* Dublin: Gill and Macmillan, 1986.

Standard Encyclopedia of the Alcohol Problem

The Anti-Saloon League of America published the *Standard Encyclopedia of the Alcohol Problem* in six illustrated volumes in the years 1925–1930. They contained a total of nearly 3,000 large pages. Appearing during the heyday of National Prohibition in the United States, the *Standard Encyclopedia* was a celebration of the temperance movement and of prohibition as its natural goal and presented the temperance point of view.

The encyclopedia was international in scope and strong on biography, covering alcoholic drink as well as the temperance movement. Because of the level of detail that it achieved, it is unlikely to ever be completely superseded. If the league had not been short on money, there would have been additional volumes and even greater detail. In contrast with the more compressed later volumes, the first volume covered only the letter A and most of the letter B. When it was published, the editors promised annual supplements that in fact never appeared.

In 1900, John G. Woolley originated the idea for such an encyclopedia, to be published by the New Voice Company. William Eugene ("Pussyfoot") Johnson joined him in the project. Beginning in 1903, they collected materials for what they first called the *Standard Encyclopedia on Temperance and Prohibition.* In 1904, when the scope was broadened to include alcoholic drinks, they renamed their work the *Standard Encyclopedia of the Alcohol Problem.* Financial difficulties forced the temporary abandonment of the project. Woolley and Johnson then took jobs with the Anti-Saloon League. This helps explain why it was the league that revived the project. In addition, the league was the most important U.S. temperance organization at the time of the enactment of the Eighteenth Amendment establishing National Prohibition in the United States. One of the senior officials in the Anti-Saloon League, Ernest Hurst Cherrington, headed the project as general editor, while Dr. Albert Porter, who had edited other encyclopedias, was in charge of the day-to-day work as managing editor. "Pussyfoot" Johnson and Cora Frances Stoddard served as associate editors. Woolley died in 1922 so was unable to serve as one of the editors. Some of the articles that he had collected, however, appeared in the league's *Standard Encyclopedia.*

What is most impressive about the *Standard Encyclopedia* is its comprehensiveness. It is more than a reference work for the United States. In its global scope, the encyclopedia shared in the same ambitious spirit as the World League against Alcoholism, founded by the Anti-Saloon League in 1919. The preface to the first volume declared: "One of the important reasons for the production and publication of the Encyclopedia at this particular time is the purpose of assisting the present active worldwide movement against alcoholism." The editors collected information from around the world with the help of questionnaires written in several languages. The same preface acknowledges the cooperation of government officials in various countries, U.S. consuls, and "officials of temperance organizations in America and throughout the world, who have willingly cooperated in securing valuable historical data relating to their several societies."

What can readers today expect to find in the *Standard Encyclopedia?* Probably most people will read biographical articles and those on individual temperance societies. Although uneven in quality, they are generally factually accurate: The biographies were often based on questionnaires or on obituaries that appeared in temperance periodicals, and the historical sketches of temperance societies were often written by their own officials. The *Standard Encyclopedia* made a special effort to include articles on women's organizations against drink. Readers today would make a mistake if they overlooked the lengthy articles on countries, ancient and modern, and American states, which provide handy compilations of information that is hard to find elsewhere. As might be expected in books written in the 1920s and presented as a scholarly work, the *Standard Encyclopedia* included many articles on the medical and scientific aspects of the alcohol problem. Some topics, such as the use of alcohol in Communion wine, were controversial even among temperance reformers. In these articles, the editors tried to provide balanced accounts and to avoid partisanship. It is unfortunate that there is no index. Information for many topics is scattered in miscellaneous entries. Moreover, the editors acknowledged that no more than one individual in ten and one society in five, among those mentioned in the *Standard Encyclopedia,* is the subject of a separate article.

The Anti-Saloon League donated copies of the *Standard Encyclopedia* to many school and college libraries. As a result, it is widely available in the United States. Unfortunately, relatively few copies found their way to other countries, and the *Standard Encyclopedia* cannot be found outside the United States at great national libraries such as the British Library in London. Fortunately, it is available on microfilm.

David M. Fahey

See also: Cherrington, Ernest Hurst; Johnson, William Eugene "Pussyfoot"; Stoddard, Cora Frances

References

Anti-Saloon League of America website, Westerville Public Library, Ohio. http://www.wpl.lib.oh.us/AntiSaloon/print/encyclop.html.

Higley, Mary S. "Series XII, Standard Encyclopedia of the Alcohol Problem 1904–1930," in *Guide to the Microfilm Edition of Temperance and Prohibition Papers,* edited by Randall C. Jimerson et al., 221–225. Ann Arbor: University of Michigan, 1977.

"Preface" in *Standard Encyclopedia of the Alcohol Problem,* vol. 1, edited by Ernest Hurst Cherrington et al. Westerville, OH: American Issue Publishing, 1925.

State Prohibition (United States)

Using the flexibility permitted by the U.S. federal system, many states experimented with prohibition of the manufacture or sale of alcoholic beverages during the period from the 1850s to the 1920s. Statewide prohibition laws varied in their constitutional or statutory basis, in the activities they banned, and in their procedures for enforcement. At some point during this period, every state but Maryland adopted a statewide prohibition law, although some of the laws adopted after ratification of the Eighteenth Amendment consisted simply of an enforcement code for National Prohibition. The movement for state prohibition began in the Northeast with the famous Maine Law (1851). By the early twentieth century, however, the movement's momentum had shifted to the South and West, while the Northeast had become the nation's antiprohibition stronghold. A few state prohibition laws were repealed during National Prohibition, and many more followed in the aftermath of its demise. Some, however, survived for much longer. The last, in Mississippi, was repealed only in 1966.

State prohibition laws came in four waves, during the 1850s, the 1880s, the 1900s and 1910s, and following the ratification in 1919 of the Eighteenth Amendment. In the last wave, the impetus came from adoption of National Prohibition. The first three waves, in contrast, were pushed forward by developments within the states. Common to these state-level initiatives was disenchantment among temperance supporters with the ability of less coercive measures, such as pledge-signing campaigns, direct appeals to liquor dealers, and local option, to reduce drinking and curb drinking problems.

The successful movement to adopt a statewide prohibition law in Maine was led by Neal Dow, the mayor of Portland. The Maine Law, enacted in 1851 to replace an ineffective 1846 statute, forbade the manufacture of intoxicating liquor and allowed sale only by municipal agents and only for industrial and medicinal purposes. Police were empowered to search premises in which violations of the law were suspected and to seize illegal liquor, whose owners were to be deemed guilty until proven innocent. First offenders received heavy fines, and repeat offenders were jailed. The Maine Law provided a model that was adopted by twelve other states and territories during the four years that followed its enactment. In addition, other states passed restrictive laws that fell short of full prohibition. By 1855, however, this first wave began quickly to recede. Perceiving prohibition as an attack on their culture, ethnic groups mobilized in opposition, and the liquor industry also organized in self-defense. The antiprohibition backlash succeeded in bringing about the repeal or judicial invalidation of most of the state prohibition laws.

The second wave of campaigns for statewide prohibition sprang primarily from political motives. Support for the Prohibition Party had been growing since its founding in 1869, and the party's strengthening appeal threatened the electoral base of the Republican Party, where prohibition sentiment was strong. Within the Woman's Christian Temperance Union (WCTU), national president Frances E. Willard was moving toward support of the Prohibition Party. Judith Ellen Foster, a leading Republican WCTU activist, saw in campaigns for state referenda a way to deflect prohibitionist activism into a channel that did not lead to the Prohibition Party. Her reasoning found support among other major-party leaders—most of whom were Republicans in the northern states, where the Prohibition Party appeal was most potent—and the result was a series of referenda on the question of amending state constitutions to incorporate prohibition. The example was provided by Kansas, where a victorious prohibition referendum was held in 1880 under the leadership of Republican governor John P. St. John. Seventeen other states and territories held a total of nineteen referenda. As in the 1850s, the prohibitionist cause advanced at the same time in states that did not hold prohibition referenda. Although prohibitionists counted 1.6 million votes cast for their cause in the referenda, the net result was disappointing, as only four western states were added to Maine in the "dry" column. The Prohibition Party's advance was slowed, and this contributed to the divisive internal battles that would eventually fracture the party in 1896.

During the early years of the twentieth century, a new series of campaigns for statewide prohibition began. A recently formed national prohibitionist organization with state branches, the Anti-Saloon League of America, often took the lead in running the campaigns. Many of these used popular referenda to force state legislatures to act; a few went forward as traditional lobbying and petition campaigns. Some states in the northeastern quadrant submitted the prohibition question to their voters, but the arrival of massive numbers of antiprohibitionist immigrants into the region generally rendered such efforts moot for the dry forces. In southern and western states, in contrast, state prohibition spread rapidly. The previous two waves of statewide prohibition campaigns had risen against a background of stable or declining per capita alcohol consumption. The early-twentieth-century wave, however, arrived in a context of rising consumption. Although the new state prohibition laws banned manufacture and sale of alcoholic beverages for personal use (other than medicinally), they sometimes allowed importation from jurisdictions that still tolerated liquor production, and, of course, implicitly or explicitly permitted consumption of the imported products. Such provisions had two results favorable to prohibition's adoption: They gave voters who wished to continue to drink themselves, and who could afford to order their own supplies, but who wished to suppress others' drinking, reason to support prohibition; and they divided the liquor industry between producers and sellers who operated within the affected state and those who saw new opportunities in shipping liquor into prohibition jurisdictions. By the time state legislatures began to consider the proposed Eighteenth Amendment in 1917–1919, thirty-one states had adopted statewide prohibition.

The question of jurisdiction over liquor imported into dry states involved the federal government, through its interstate commerce and taxation powers, in the enforcement of state prohibition laws. Through a series of acts beginning in 1890, Congress ceded some degree of control over interstate commerce in liquor to state governments of prohibition states. In 1913, the Webb-Kenyon Act allowed states to prohibit liquor shipments from other states. This process culminated in the Reed Amendment to the Postal Act of 1917, which banned liquor shipments into any dry state, whether its laws forbade such consignments or not.

The Eighteenth Amendment provided for "concurrent" enforcement of National Prohibition by the federal government and the states. In accordance with this provision, states—with the exception of Maryland—modified their existing prohibition statutes or passed new laws so as to harmonize state with federal procedures. Within a short time after National Prohibition began, however, domestic political pressures within some states brought about repeal of the state prohibition laws, thereby seriously undercutting National Prohibition's chances of success, since the federal enforcement effort was underfunded and poorly organized. The leader in repealing its state prohibition law was New York in 1923. After the repeal of National Prohibition in 1933, the other states gradually repealed their prohibition statutes. This did not mean, however, that the states withdrew from regulation of the manufacture, sale, or consumption of alcoholic beverages. Instead, the states generally adopted far-reaching liquor codes, which, according to some authorities, imposed nearly as extensive a set of controls over drinking as had been attempted by National Prohibition.

Jack S. Blocker Jr.

See also: Anti-Saloon League of America (ASLA); Dow, Neal; Federal Regulation of Alcohol before 1920 (United States); Foster, Judith Ellen Horton; Maine Law; National Prohibition (United States); Prohibition Party (United States); Willard, Frances Elizabeth Caroline

References

Bader, Robert S. *Prohibition in Kansas.* Lawrence: University Press of Kansas, 1986.

Blocker, Jack S., Jr. *American Temperance Movements: Cycles of Reform.* Boston: Twayne, 1989.

Clark, Norman H. *The Dry Years: Prohibition and Social Change in Washington.* Seattle: University of Washington Press, 1965.

Engelmann, Larry. "O Whiskey: The History of Prohibition in Michigan" (Ph.D. dissertation, University of Michigan, 1971).

Hamm, Richard H. *Shaping the 18th Amendment: Temperance Reform, Legal Culture, and the Polity, 1880–1920.* Chapel Hill: University of North Carolina Press, 1995.

Isaac, Paul E. *Prohibition and Politics: Turbulent Decades in Tennessee, 1885–1920.* Knoxville: University of Tennessee Press, 1965.

Levine, Harry Gene. "The Birth of American Alcohol Control: Prohibition, the Power Elite, and the Problem of Lawlessness" in *Contemporary Drug Problems.* 12 (1985): 63–115.

Pearson, C. C., and J. Edwin Hendricks. *Liquor and Anti-Liquor in Virginia, 1619–1919.* Durham, NC: Duke University Press, 1967.

Pegram, Thomas R. *Battling Demon Rum: The Struggle for a Dry America, 1800–1933.* Chicago: Ivan R. Dee, 1998.

Stegh, Leslie. "Wet and Dry Battles in the Cradle State of Prohibition" (Ph.D. dissertation, Kent State University, 1975).

Tyrrell, Ian R. *Sobering Up: From Temperance to Prohibition in Antebellum America.* Westport, CT: Greenwood, 1979.

West, William Elliott. "Dry Crusade: The Prohibition Movement in Colorado, 1858–1933" (Ph.D. dissertation, University of Colorado, 1971).

State Stores (United States)

State stores operate in jurisdictions in the United States that control the sale of distilled spirits, and, in some cases, wine, through government agencies at the wholesale level. These same jurisdictions may also regulate on-premise alcohol sales, though most do not. The system of state stores was one form of alcoholic beverage regulation instituted by individual states when National Prohibition was repealed in 1933. The federal Bureau of Alcohol, Tobacco, and Firearms recognizes eighteen states (Alabama, Idaho, Iowa, Maine, Michigan, Mississippi, Montana, New Hampshire, North Carolina, Ohio, Oregon, Pennsylvania, Utah, Vermont, Virginia, Washington, West Virginia, and Wyoming) and Montgomery County in Maryland as "control jurisdictions" operating state stores.

The repeal of National Prohibition in 1933 returned the regulation of alcoholic beverages to the states. In the two years following repeal, fifteen states set up wholesale monopolies. These states were joined by three additional states after they, too, repealed state-level prohibition laws. The legislatures in these states were strongly influenced by a study financed by John D. Rockefeller Jr. that argued a monopoly system would be more efficient and less corrupt than a system of licensed private enterprise.

According to supporters of the monopoly system, the control jurisdictions are able to provide more choices for customers while also promoting moderation and reducing alcohol abuse. Some opponents of control argue that the government monopoly is "un-American" and use economic arguments to illustrate that the jurisdictions could increase revenue by licensing alcohol distribution. Others, taking a moral perspective, abhor the fact that government is directly profiting from the sale of distilled spirits and wine. Both positions can be found in the regularly occurring debates to privatize alcohol sales in the control jurisdictions. Most of the policy debates have centered on economic issues and not on the public health issue of controlling alcohol consumption.

Since 1970, six states (Idaho, Iowa, Maine, Virginia, West Virginia, and Washington) have eliminated public monopolies on sales of wine, imported wine, or fortified wine. According to researchers, the result of these changes has been an increase in the number of outlets for wine sales, longer hours for customers to purchase wine, and lower prices caused by increased competition. By 2000, only two states (Pennsylvania and Utah) had complete monopolies, and the remaining control jurisdictions had relinquished some state control. In 1981, for example, West Virginia ended the state monopoly on the retail sale of wine and implemented a system of licensing private retailers, including supermarkets.

The National Alcohol Beverage Control Association (NABCA) is the trade association representing the control jurisdictions. Established in 1937, the association serves as an information clearinghouse collecting data to help the boards and commissions in each state counter the privatization movement. NABCA also works as a liaison for the control jurisdictions in their relationships with the federal government, the thirty-two license states, and the alcoholic beverage industry. In addition, the organization conducts public education and awareness campaigns. NABCA maintains a website at www.nabca.org.

John David Rausch Jr.

See also: Gothenburg Schemes/Disinterested Management; Provincial Stores (Canada)

References

Fosdick, Raymond B., and Albert L. Scott. *Toward Liquor Control.* New York: Harper & Brothers, 1933.

Holder, Harold D. "Changes in Access to and Availability of Alcohol in the United States: Research and Policy Implications" in *Addiction.* 88/Supplement (January 1993): 67S-74S.

Wagenaar, Alexander C., and Harold D. Holder. "A Change from Public to Private Sale of Wine: Results from Natural Experiments in Iowa and West Virginia" in *Journal of Studies on Alcohol.* 2, no. 2 (March 1991): 162–173.

Stayton, William H. (1861–1942)

Founder of the Association Against the Prohibition Amendment (AAPA) and a leader in the campaign to overturn the Eighteenth Amendment to the U.S. Constitution, William H. Stayton (1861–1942) was influential in developing arguments against National Prohibition based on a philosophy of limited government rather than the desirability of alcoholic beverages. He thereby drew into his organization business, legal, and political leaders who in turn propelled the repeal movement to success in 1933. Writing in the *Baltimore Sun* (30 October 1932) H. L. Mencken declared Stayton "The Man Who Really Busted Prohibition."

Stayton grew up on a farm near Leipsic, Delaware, in an old-stock Delaware family of Swedish origin. He graduated from the U.S. Naval Academy in 1881, married the commandant's daughter, and served on ships assigned to the North Atlantic and the Pacific Coast. While on duty in Washington, D.C., he earned a law degree from Columbian (now George Washington) University, after which he resigned from the navy in 1891 to practice law, first in New York and later in Baltimore. During a brief return to active duty during the Spanish-American War, Stayton earned the rank of captain and proudly used it for the rest of his life. He became active in the Navy League of the United States, a naval power advocacy group, serving as its executive secretary and spokesman from 1916 to 1918. Through his presidency of the Naval Academy Alumni Association, his law practice, and involvement in the shipping industry as well as in the Navy League, Stayton developed an extensive social network that he exploited once he became interested in the prohibition issue.

To Stayton, National Prohibition represented an inappropriate and alarming expansion of federal power at the expense of local government. His conservative constitutional views, he maintained, had developed during his naval career. States' rights constitutionalism lay at the heart of his initial 1918 appeal for navy friends and Baltimore neighbors to join in an effort to oppose National Prohibition and to support the 1920 incorporation of the AAPA. For several years, he was the AAPA's prime financial supporter. His contribution of $1,000 per month was supplemented only by voluntary membership dues of $1 per person and a few larger gifts. Until 1923, he refused to accept contributions from anyone associated with the liquor industry. By the mid-1920s, Stayton's arguments had persuaded a number of wealthy individuals whom he had met through the Navy League and other contacts to enlist in the AAPA. Thereafter, the financial burden and the active leadership of the organization shifted away from Stayton.

Stayton remained a visible presence in the AAPA, serving as chairman of the board of directors, making public speeches, and maintaining contact with members. He testified at several congressional hearings on Prohibition and appeared before the platform committees of both the Republican and Democratic national conventions in 1924, 1928, and 1932 to advocate repeal. Cheered by the adoption of the Twenty-First Amendment, Stayton for the rest of his life headed a small organization, Repeal Associates, opposed to any renewed effort at federal liquor regulation.

David E. Kyvig

See also: Association Against the Prohibition Amendment (AAPA); National Prohibition (United States)

Reference

Kyvig, David E. *Repealing National Prohibition.* 2d ed. Kent, OH: Kent State University Press, 2000.

Mencken, H. L. "Man Who Really Busted Prohibition Gives All Credit to Opposite Sex" in *Baltimore Sun* (30 October 1932).

Stevens, Lillian Marion Norton Ames (1844–1914)

Lillian Marion Norton Ames Stevens (1844–1914), temperance leader and suffragist, was a tireless lecturer for prohibition and actively involved in the Woman's Christian Temperance Union (WCTU) on the state, national, and international level. She became the third president of the national WCTU in 1898 and served in that capacity until her death on 6 April 1914.

Stevens was born 1 March 1844 in Dover, Maine, where her father was a teacher. After her graduation from Westbrook Seminary, she taught school for several years. In 1865, she married Michael T. Stevens, a wealthy grain and salt wholesaler. They had one child. In 1874, she joined the Women's Temperance Crusade against saloons that swept through the Midwest and the East. She helped found the Maine WCTU in 1875 and became its president in 1878. She held this position until her death.

At about the same time, she became active in the national WCTU, serving as assistant recording secretary from 1880 to 1892 and recording secretary in 1893. She formed a close relationship with Frances Willard, the national WCTU president. Willard, who, in effect, designated her own successor, appointed Stevens to the new post of executive vice president in 1894. After Willard's death in 1898, Stevens became acting president and was elected president at the convention that year. In 1903, she also became the vice president of the World's WCTU (WWCTU).

Stevens gradually de-emphasized Willard's "Do Everything" policy that had called on the membership to attack a host of social issues that were seen as related to alcohol use. These included political corruption, women's suffrage, lynching, the eight-hour workday, and social insurance. Instead, during Stevens's tenure the WCTU concentrated largely on temperance and prohibition. Stevens's more conservative vision helped stem the decline in membership that the organization had seen in the early 1890s. Under her leadership, new local unions were formed, and from 1900 to 1910 membership in the WCTU rose from 168,000 to 248,000.

Beginning in 1907, six states adopted statewide prohibition in rapid succession. This breakthrough was in part attributable to the efforts of both the WCTU and the Anti-Saloon League of America, a competing organization founded in 1893. Encouraged, Stevens predicted in 1911 that National Prohibition would be attained within a decade. With this goal in mind, she urged a shift away from local prohibition efforts. She spent more time lobbying in Washington and participating in prohibition demonstrations. Stevens's prediction proved to be correct when the Eighteenth Amendment (which prohibited the manufacture, sale, transportation, importation, or exportation of intoxicating liquors) went into effect in 1920. Unfortunately, she did not live to see it.

Stevens was involved in other reform efforts as well. Reflecting her interest in women's suffrage, she served as treasurer of the National Council of Women from 1891 to 1895. She represented Maine at the National Conference of Charities and Correction for several years and campaigned for a state women's reformatory. Stevens also was one of the founders of the Maine Industrial School for Girls and a manager of the Maine contribution to the World's Columbian Exposition (1892–1893).

Jana Brubaker

See also: National Prohibition (United States); Willard, Frances Elizabeth Caroline; Woman's Christian Temperance Union (WCTU) (United States); World's Woman's Christian Temperance Union (WWCTU)

References

Bordin, Ruth. *Woman and Temperance: The Quest for Power and Liberty, 1873–1900.* Philadelphia: Temple University Press, 1981.

Byrne, Frank L. "Stevens, Lillian Marion Norton Ames" in *Notable American Women, 1607–1950: A Biographical Dictionary,* edited by Edward T. James, vol. 3, 370–372. Cambridge: Belknap Press of Harvard University Press, 1971.

Stoddard, Cora Frances (1872–1936)

A founder and executive secretary of the Scientific Temperance Federation, Cora Frances Stoddard (1872–1936) and her organization published and distributed statistical information concerning the effect of alcohol on health and social problems in additional to alcohol education material. Stoddard in her later career took a leadership role in the Woman's Christian Temperance Union (WCTU) and was associate editor of the *Standard Encyclopedia of the Alcohol Problem* from 1918 to 1930.

Born in Irvington, Nebraska, on 17 September 1872, Stoddard was the daughter of farmers. Her family moved to Massachusetts, where her mother was active in the East Brookfield WCTU, during her childhood. Stoddard attended the public schools and graduated from Wellesley College in 1896. She taught school for a year in Middletown, Connecticut, and then worked in business from 1897 to 1898. In 1899, she moved to Boston and became private secretary until 1904 to Mary H. Hunt, director of the department of scientific temperance instruction for the national WCTU. She also became involved with Hunt's Scientific Temperance Association, which had been founded to study and endorse temperance education texts. In this position, she educated teachers and administrators on the dangers of alcohol and evaluated temperance textbooks. She resigned from this job on account of illness and worked in an administrative position in Courtland, New York, but returned to Boston two years later upon Hunt's death. Because Hunt had received royalties from publishers, the WCTU refused to incorporate the association. In 1906, Stoddard helped reorganize the association as the Scientific Temperance Federation (STF) and became its executive secretary, a position she held for the rest of her life. She also began to edit the *Scientific Temperance Journal,* which served as a forum for research reports on the negative effects of alcohol.

In addition, Stoddard was secretary of the executive committee of the National Temperance Council and a member of the executive committee of the World League Against Alcoholism. From these positions she conducted surveys and issued reports. She attributed the passage of National Prohibition to the effort to educate a generation of youth concerning temperance and the dangers of alcohol. She was a delegate of the U.S. government to International Congresses on Alcoholism and other meetings between 1909 and 1923 and secretary of the United Committee on Temperance Activities from 1917 to 1920.

In 1913, after the federation ran into financial difficulties, the American Issue Publishing Company, operated by the Anti-Saloon League of America (ASLA), took over publication of the *Scientific Temperance Journal.* Stoddard, however, remained editor and was allowed to run her organization independently. Her editorship, many writings, and association with the Anti-Saloon League brought her national recognition among temperance workers. Stoddard's work with the ASLA also returned her to the WCTU fold. In 1918, the WCTU appointed her head of the Bureau of Scientific Temperance Investigating, and in 1922 she became director of the Department of Scientific Temperance Instruction for both the national and World's WCTUs.

After passage of the Eighteenth Amendment, Stoddard produced several publications that purported to show the positive effects of National Prohibition on social and health problems. One of her most noted books was *Fifteen Years of the Drink Question in Massachusetts* (1929), coauthored by Stoddard and Amy Woods, the STF's social welfare secretary. Her work as associate editor of the six-volume *Standard Encyclopedia of the Alcohol Problem,* which was published from 1925 to 1930, helped to produce a significant work on alcohol and the temperance movement still consulted by historians today. Stoddard wrote many publications that supported antialcohol education. Owing to severe arthritis in her later years, she was forced to curtail her activities; she resigned from her posts with the WCTU in 1933, but she remained active in the Scientific Temperance Federation until her death. She never married and died at her brother's home in Oxford, Connecticut, close to where she had been living in Seymour.

Ruth Clifford Engs

See also: Anti-Saloon League of America (ASLA); Cherrington, Ernest Hurst; Hunt, Mary Hannah Hanchett; International Congresses on Alcoholism; Scientific Temperance Federation; *Standard Encyclopedia of the Alcohol Problem;* Woman's Christian Temperance Union (WCTU) (United States); World League Against Alcoholism (WLAA); World's Woman's Christian Temperance Union (WWCTU)

References

Cherrington, Ernest Hurst, et al., eds. *Standard Encyclopedia of the Alcohol Problem.* 6 vols. Westerville, OH: American Issue Publishing Co., 1925–1930, vol. 6: 2325–2536.

Perry, Marilyn Elizabeth. "Stoddard, Cora Frances" in *American National Biography,* edited by John A. Garraty and Marc C. Carnes. 24 vols. New York: Oxford University Press, 1999, vol. 20: 816–817.

Stoddard, Cora Frances, and Amy Woods. *Fifteen Years of the Drink Question in Massachusetts.* Westerville, OH: American Issue Publishing, 1929.

"Stoddard, Cora Frances" in *Who Was Who in America: A Companion Volume to Who's Who in America, 1897–1942.* 6 vols. Chicago: A. N. Marquis Company, 1942–1976, vol. 1: 1190.

Zimmerman, Jonathan. *Distilling Democracy: Alcohol Education in America's Public Schools, 1880–1925.* Lawrence: University Press of Kansas, 1999.

Strength of Britain

Strength of Britain, an organization formed in 1916 in the midst of World War I, was established to bring pressure on the British government to initiate a complete ban on both the consumption and manufacture of alcohol. It attracted support from people not associated with the peacetime temperance movement. The organization argued for the need to conserve the cereals used by brewers and distillers for consumption as food during the wartime crisis. Strength of Britain was the brainchild of Arthur Mee (1875–1943), a Nonconformist reporter and newspaper editor who also edited the extremely successful *Children's Encyclopedia* (sold in the United States as Grolier's *Book of Knowledge*). Sir Arthur Booth, chairman of the Cunard Steamship

Line, served as the first president, but Mee dominated the organization in its earliest days. He used his tremendous energy and journalistic skill to draw the British public's attention to the wastefulness of alcohol production and consumption in wartime.

The first activity undertaken by Strength of Britain was to organize a petition presented to David Lloyd George upon his return to Downing Street at the close of his first day as prime minister. There were 2,448 signatories, including more than 100 members of Parliament (M.P.s), numerous military leaders, and other prominent persons, such as Lloyd George's own wife. The petition called for the government to prohibit the drink trade for the duration of the war. Apparently not understanding Strength of Britain's intentions, Lloyd George assured the petitioners that he was willing to accept state purchase of the liquor trade and allow localities to decide whether to remain "wet" or "dry." In response, Strength of Britain made it clear to the prime minister that they insisted on total prohibition for the United Kingdom and would not accept half measures.

Strength of Britain sent speakers throughout the nation to argue for complete wartime prohibition. There was public interest in the cause: A mass meeting held on 21 March 1917 at Queens Hall, London, was filled to capacity, so an overflow meeting was held concurrently at another hall. More than 4,000 people attended in all. According to the temperance authorities Gwymor Prys Williams and George Thompson Brake, those attending included "representatives of 20 British universities, 243 headmasters of schools, 100 admirals and generals, 250 directors of munitions factories, 100 members of parliament and privy councillors, 200 baronets and knights, 90 members of the Royal Society and the Royal Academy, 200 leaders of finance and industry, 100 persons distinguished in the arts and in literature, 450 doctors, 500 magistrates and 100 municipal leaders," as well as 9 Victoria Cross recipients (Williams and Brake 1980: 58).

The most important work undertaken by Strength of Britain came in publishing a series of pamphlets, most of which Mee wrote. His first major tract was *Defeat? The Truth behind the Betrayal of Britain.* In all, his twenty pamphlets sold some 1.5 million copies. In the Dominions, there was sometimes a hostile reception to Mee's tracts. They were banned in Australia and South Africa; in Canada, possession of a copy was an offense punishable by five years in jail. What produced this negative response was Mee's claim that soldiers from the empire were contracting venereal diseases and committing murders under the influence of alcohol after drinking it for the first time upon their arrival in France. Attitudes soon changed in Australia, where in 1918 a parallel Strength of the Empire movement was organized.

Although Strength of Britain does not deserve all the credit, it did play a role in convincing the British government to call for a reduction in annual beer production, from a prewar 36 million barrels to 10 million barrels, while also mandating the end of alcoholic beverage distilling for the duration of the war. Strength of Britain ultimately failed to convince the government of the need to impose complete prohibition, however, and the tensions over this failure led to dissent within the organization.

Shortly before the end of the war, in October 1918, Strength of Britain was reorganized as a limited liability company, with Frederick George Creed as chairman of a newly formed executive committee. After the war, the organization shifted its focus. It wanted distillers to manufacture alcohol for industrial use only and pubs to be converted to nonalcoholic gathering places. Strength of Britain no longer had the benefit of Mee's services. By the time that it reorganized in 1918, Mee had grown tired of the internecine fights within the prohibitionist organization. As a result of the postwar problems of National Prohibition in the United States, he began to rethink his advocacy of prohibition. His work for Strength of Britain is barely mentioned by his biographer Sir John Hammerton, a sign of how rapidly the prohibition movement declined in the United Kingdom.

Strength of Britain continued as an independent entity until 1923, when it merged with the National Commercial Temperance League.

Kenneth Pearl

See also: Lloyd George, David

References

Hammerton, Sir John. *Child of Wonder: An Intimate Biography of Arthur Mee.* London: Hodder and Stoughton, 1946.

Longmate, Norman. *The Waterdrinkers: A History of Temperance.* London: Hamilton, 1968.

Williams, Gwylmor Prys, and George Thompson Brake. *Drink in Great Britain, 1900–1979.* London: Edsall, 1980.

Stroh Brewing Company

The Stroh Brewing Company began in Detroit in 1850 as the Brewery of B. Stroh. B. Stroh was Johann Bernhard Stroh (1821–1882), a German immigrant who had learned brewing in his family's brew house in the town of Kirn in the German Palatinate. The Stroh family had begun brewing in 1775. Johann Bernhard learned the trade in his father's brewery, but his older brother would inherit the business. Thus, Johann Bernhard immigrated to the United States during the political turmoil of 1848 to seek his fortune.

Stroh initially settled in Harrisburg, Pennsylvania, and worked for Barnitz Brewery before relocating to Detroit. His brewery prospered as the German community in Detroit grew. In 1867, the year his first wife died, Stroh moved into a new brewery building, and in 1875 the firm was renamed the Lion Brewing Company. Two of his sons, Bernhard Jr. and Julius, joined him in the firm, which was incorporated for a ten-year period. In 1870, Stroh remarried and the brewery lost sales owing to problems with the quality of its product. Konrad Koppitz joined the brewery as brewmaster in 1878, and by 1880 sales exceeded 30,000 barrels a year.

Stroh's death in 1882 led to a crisis for the firm. Connecticut Mutual Insurance Company had assumed ownership of the brewery because of a default, and Stroh left the majority of the firm's stock to his widow, Clothilde. In 1883, she and her stepsons clashed over using the brewery's surplus to repurchase the plant versus paying a dividend. As the majority stockholder, Clothilde Stroh got the dividend she wanted, but the brothers bought the plant back themselves a short time later. They also began buying other family members' stock, and when the Lion Brewery's charter expired they established the B. Stroh Brewing Company.

The new firm prospered, shipping beer as far east as New England and establishing a branch in Cleveland. In 1890, brewmaster Koppitz left the firm and was replaced by Otto Rosenbusch, who remained with Stroh for thirty-five years. In 1893, the company won a Gold Medal at the Chicago World's Columbian Exposition, and by 1900 Stroh was the leader among Detroit's brewers. Its main label was Bohemian, a lager. Success led to expansion, and capacity increased from 300,000 to 500,000 barrels a year. The firm was reincorporated as the Stroh Brewing Company in 1902.

In 1908, Julius Stroh became head of the firm when Bernhard Jr. retired because of health problems. Julius wanted to expand production and in 1911 sent Rosenbusch to Europe. While there, Rosenbusch observed the use of direct fire in brewing in Pilsen, Bohemia. This "fire-brewing" method was adopted by Stroh, and a new, eight-story brew house was built in Detroit to use the technique, with production beginning in 1914.

Unfortunately, Prohibition began in Michigan in 1917, and the investment in the new brew house stretched the firm's resources. Julius bought out the interest of his brother's heirs and renamed the firm Stroh Products Company. The new firm produced a number of products, including near beer under several labels, birch beer, several ginger ales, a cola, an orange soda, and Alaska brand ice cream. The best-selling product was a hopped malt syrup that could be used for home brewing. Julius Stroh also built an eighteen-story office building in Detroit called the Stroh Building. The company continued operating during Prohibition and was in a good position to resume operations quickly when repeal came in 1933.

Stroh beer delivery truck, ca. 1910 (Bettmann/Corbis)

All of the Prohibition products, except the ice cream, renamed Stroh's, were dropped after repeal. The renamed Stroh Brewing Company had a large stock of beer awaiting de-alcoholization and thus was able to get it to market on the day repeal took effect. Sales rose rapidly, from 440,000 barrels in 1934 to more than 700,000 in 1936. Expanded sales led to replacing the old brewery buildings with large new facilities. The company had three labels during this period: Stroh Bohemian Style; Bock, a seasonal beer; and a premium Imperial Pilsner with limited distribution.

In 1939, Julius Stroh died and leadership of the firm passed to his son, Gari, who led Stroh through World War II, when demand was strong but raw materials were rationed. Stroh limited production rather than dilute its product. This maintained the beer's reputation for quality but left it out of step with changing taste as consumers adapted to the lighter beers produced by most breweries. Stroh sold 877,000 barrels in 1947 and expanded its packaging facility. Gari Stroh died in 1950 and was succeeded by his brother John.

John Stroh took charge at a difficult time. Changing consumer tastes led to a steep decline in sales, which fell to 500,000 barrels in 1950. At the same time, Stroh's local rivals, Goebel and Pfeiffer, saw their sales surge to more than a million barrels each. After a long run as the leading brewer in Detroit, Stroh found itself number three. The company lightened its formula and rebounded, reclaiming first place among Detroit brewers in 1956 with sales of 2.7 million barrels.

During the spring of 1958, all of Detroit's brewers were hit by a prolonged strike that brought out-of-state beers into the Detroit market and depressed local label sales for a long time afterward. Goebel was especially hard hit and at the end of the 1959 baseball season gave up its radio sponsorship of the Tigers baseball games. Stroh took over the sponsorship, decided that longtime play-by-play man Van Patrick was too closely associated with Goebel, and brought in Ernie Harwell from Baltimore to team up with George Kell. Thus began one of the longest and most successful radio broadcast teams in baseball history. In 1964, Goebel filed for bankruptcy and Stroh acquired its plant and product line.

John Stroh became chairman and chief executive officer and his nephew, Peter Stroh, Gari's son, became president in 1967. Sales surged, reaching 3 million barrels in 1970 and 4 million just two years later. Stroh moved into new territories and expanded its facilities, and sales passed 6 million barrels by 1977. This growth taxed Stroh's capacity; at the same time, the niche occupied by regional brewers began to shift. Regional and local brewers like Stroh faced increasing competition from national brands, a competition in which few were doing well. Stroh's price made it a premium beer—good for prestige, but not for market share. As the 1970s ended, Stroh found itself in difficulty, more because of the changing nature of the industry than owing to anything the company had done or not done.

Stroh adopted the same strategy as several other regional brewers—expansion by acquisition. In 1980, Stroh acquired the Schaefer Brewery of New York, which had recently relocated its brewing operation to Allentown, Pennsylvania. This move brought Stroh into the East Coast market. The company bought the larger but deeply troubled Jos. Schlitz Brewing Company of Milwaukee in 1982. This acquisition made it the third largest brewer in the country. It now had a large number of labels in all areas of the industry, both regionally and nationally. The acquisition also saddled Stroh with a substantial debt, however.

When John Stroh died in 1984, the company he left behind was not only larger; because of the several mergers, it was also a mix of a number of corporate cultures. The year 1985 saw Stroh's sales reach 24 million barrels. Capacity, however, was 31 million. The brewery in Detroit, where the company had begun, was the oldest of its breweries and was closed. By the end of the decade, Stroh's sales were down substantially. A deal was made in 1989 to sell Stroh's breweries to Coors, but the deal fell apart before completion.

Following the collapse of the Coors sale, the Stroh family became more active in the firm. One of the breweries acquired with Schlitz was closed and another sold to Coors. Stroh's stake in Mexican brewer Cruzcampo was also sold, allowing the company to retire its debt and increase advertising. Stroh also began doing contract brewing. By 1994, the company was able to mount an unsuccessful campaign to acquire bankrupt rival G. Heileman. When the successful bidders failed to turn the company around, Stroh acquired it in 1996. The following year saw the beginning of an industry-wide price war that continued into 1998 and severely stressed Stroh with a new debt load. Chairman Peter Stroh retired in 1997 and was replaced by John W. Stroh III. The company continued to struggle, and in April 1999 the firm sold its brewing operations and labels to Pabst and Miller. Pabst acquired the Stroh label along with other well-established labels Stroh had acquired: Schlitz, Schaeffer, and Old Style. This acquisition ended the independent existence of Stroh Brewing after nearly 150 years.

William H. Mulligan Jr.

See also: Coors, Adolph, Brewing Company; Heileman, G., Brewing Company; Miller Brewing Company; Pabst Brewing Company; Schlitz, Joseph, Brewing Company

References

Baron, Stanley. *Brewed in America: The History of Beer and Ale in the United States.* Boston: Little, Brown, 1962.

Blum, Peter H. *Brewed in Detroit: Breweries and Beers since 1830.* Detroit: Wayne State University Press, 1999.

Greer, Douglas F. "The Causes and Consequences of Concentration in the US Brewing Industry" in *Quarterly Review of Economics and Business.* 21, no. 4 (1981): 87–106.

McGahan, A. M. "The Emergence of the National Brewing Oligopoly: Competition in the American Market, 1933–58" in *Business History Review.* 65, no. 2 (1991): 229–284.

Stuart, Wilson (1873–1934)

Wilson Stuart (1873–1934) was a Wesleyan Methodist minister and United Kingdom Alliance official known for his opposition to the Carlisle scheme of public ownership of the drink trade. A minor figure, he has interest as one of the few twentieth-century temperance advocates for whom a book-length biography is available. Based on personal knowledge, the recollections of Stuart's associates, and Stuart's correspondence, Joseph Malins's memoir, published in 1935, provides insights into the career of this otherwise obscure temperance official.

Stuart was born in Staffordshire, the son of a Wesleyan Methodist minister and staunch temperance man. A friend said about both father and son: "Compromise and opportunism did not exist in their scheme of things" (Malins 1935: 17). Wilson Stuart received bachelor's and master's degrees at Yorkshire College, Leeds, a component of what became the Victoria University of Manchester. He later studied at St. John's College, Cambridge.

He first won a reputation as a champion of temperance while a minister on the Douglas circuit, Isle of Man, from 1910 to 1913. He was outraged in 1916 when the Central Control Board (Liquor Traffic), created during the wartime emergency, experimented with public ownership of the manufacture and sale of alcoholic beverages in Carlisle and several other industrial districts. The board recommended state purchase of the liquor trade throughout the United Kingdom but opposed wartime prohibition. In the fight against the Carlisle scheme, Stuart came into conflict with Henry Carter, a member of the Central Control Board and secretary of the Wesleyan Methodist temperance committee. In 1918, Stuart became so disillusioned by Wesleyan Methodist support for Carter and insinuations that he "sniped" at his work that he offered his resignation as a minister. The pleas of many prominent Wesleyans, including Carter, persuaded Stuart to withdraw his resignation.

Stuart became organizing secretary of the UKA, a job that included responsibility for business and financial matters and forming local-option unions to promote prohibition, in September 1919 and served in that capacity until 1921. During the Essex Hall riot in November 1919, which cost W. E. "Pussyfoot" Johnson one of his eyes, Stuart tried to protect Johnson against the toughs and students who attacked the U.S. prohibitionist.

As part of his fight against public ownership of the drink trade, Stuart visited Carlisle in 1917, 1919, 1925, and 1930 to collect evidence. He was a witness before the Southborough committee on the disinterested management of public houses in 1926 and before the Royal Commission on Licensing in 1930. His critique of the latter's report appeared in the *Methodist Times* on 11 February 1932.

Joseph Malins, the author of the Stuart memoir, also wrote a biography of his own father, the elder Joseph Malins, who had been an ally of Stuart's in the fight against public ownership of the drink trade. In 1932, Stuart had reviewed the biography of Joseph Malins the elder in two articles in *Joyful News.* In February 1934, heart trouble made Stuart an invalid; he died six months later.

David M. Fahey

See also: Carlisle System; Carter, Henry; Johnson, William Eugene "Pussyfoot"; Malins, Joseph

Reference

Malins, Joseph. *Wilson Stuart: A Memoir.* London: James Clarke, 1935.

Students Against Destructive Decisions (SADD)

High-school hockey coach Robert Anastas and his students created Students Against Destructive Decisions (SADD; formerly known as Students Against Driving Drunk) in Wayland, Massachusetts, in 1981 after two of the school's players died in separate car crashes in less than a week. SADD has chapters in all fifty states and in Canada. The name was changed in 1997 to reflect the fact that there are many dangerous decisions facing youth, including drinking, drugs, and violence. Today, SADD's mandate goes beyond prevention of drunk driving to encompass information about HIV and AIDS, smoking, blood-alcohol poisoning, suicide, and seatbelt use. With its goal of reducing teen deaths caused by drunk driving accidents to 2,000 per year by the year 2000, SADD lobbied governments to implement graduated licensing for beginning drivers, stricter seatbelt laws, laws that would strip people of their driver's licenses for using false identity documents to purchase alcohol, and laws to reduce the maximum allowable blood-alcohol content for drivers to 0.08 percent. Partly because of a substantial increase in the size of the youth cohort, SADD's goal was not achieved, but the organization continues its efforts at raising awareness of dangers to teens on the road. SADD has taken a strong position against underage drinking, and its national policies state that SADD groups should not sponsor Safe Ride or Designated Driver programs that might enable teen drinking.

The centerpiece of SADD is the Contract for Life, a document to be signed by a young person and his or her parent, or another caring adult, that was developed by Anastas and his students. By signing the contract, the young person commits to doing everything in his or her power to avoid making decisions that jeopardize her or his health, safety, and overall well-being or violate the adult's trust. He or she pledges to remain alcohol- and drug-free, never to accept a ride from someone who is impaired, and always to wear a seatbelt. The adult commits to doing everything he or she can to understand and communicate with the young person about the decisions he or she faces, to provide safe transportation home if ever the young person is in a situation that threatens his or her safety, and to defer discussion about the situation until a later time. Finally, the adult pledges that he or she will never drive under the influence of alcohol or drugs.

The adult influence in SADD is strong. The national organization has an all-adult board of directors, and its newsletter is largely written by and for adults. The high-school population is transient, and school-based chapters of SADD tend to be most successful where there is a staff member committed to the organization. Activities vary widely by school, but they commonly include mock-up car crashes, safe-prom campaigns, and grim-reaper days. SADD's success in reducing teenage drunk driving and other destructive decisions is not known.

Catherine Carstairs

See also: Drunk Driving Laws (United States); Mothers Against Drunk Driving (MADD)

Reference

SADD website. http://www.saddonline.com.

Sunday, William Ashley (1862–1935)

William Ashley "Billy" Sunday (1862–1935) was one of the most popular temperance evangelists of the twentieth century. A former professional baseball player for the Chicago White Stockings and Pittsburgh Pirates from 1883 to 1891, Sunday had a flamboyant preaching style in which he often used baseball terms and stories as he railed against the evils of alcohol. There was no middle road in life, according to Billy Sunday. His grass-roots revivals in the Midwest, particularly from 1910 to 1920, emphasized an absolute view of moral behavior: no drinking, no smoking, no idle amusements, and a determined sense of discipline in the home. His antialcohol sermons also included a strong nativist point of view that blamed foreigners and immigrants for America's problems.

Sunday was born on 18 November 1862 in Ames, Iowa. His father was killed shortly thereafter while fighting in the Civil War; subsequently, his mother could only sporadically care for him. His youth was spent in poverty as he was shuttled from one orphan home to another.

In 1883, while working in Marshalltown, Iowa, and playing baseball for a local team, Sunday came to the attention of Cap Anson, a Marshalltown native and captain of the major league White Stockings. Sunday was signed to the team, and he made his mark as a fleet outfielder, if only a mediocre batter. As part of the ballplayer fraternity, he often accompanied his teammates on their evening excursions to saloons, where heavy drinking and rowdy behavior were the norm. Sunday would occasionally have a drink, but he remained aloof from the carousing. According to his later revival stories, he underwent a conversion to the evangelical life in 1886 when he and his friends heard the singing and testimony of a "gospel wagon" mission band.

Sunday left baseball in 1891. After working with the Chicago Young Men's Christian Association (YMCA) and evangelist J. Wilbur Chapman, he was licensed to preach in 1898 and ordained as a minister in 1903. His career as an evangelist began modestly in small-town churches and rural tent revivals, his preaching marked by a fire-and-brimstone style that combined conservative politics with a strong moral fiber as the means to salvation.

William Ashley "Billy" Sunday (1862–1935) (Library of Congress)

Sunday was in some measure responsible for the passage of the Eighteenth Amendment prohibiting the manufacture, sale, and use of alcohol in 1917. As his popularity grew, he staged huge revivals in large cities such as Chicago, Pittsburgh, Cincinnati, and Indianapolis, quickly building "tabernacles" that held thousands of worshipers, choirs, and musicians, and his "free will" offerings brought in tens of thousands of dollars. At one revival, Sunday symbolically killed and buried "John Barleycorn" in an elaborate funeral service. His most famous sermon, reprinted many times, was "Get on the Water Wagon," a dramatic invective against the liquor interests in which he challenged the proponents of temperance and Prohibition to save the young men of the nation by rejecting the use of spirits and voting for their taxes to be used for pro-family platforms.

At his death in 1935, Billy Sunday's influence had waned with the repeal of Prohibition, though his method of conducting urban religious revivals became a mainstay of evangelism. In the decades that followed, Sunday became a part of pop culture, with theatrical productions, poems, and novels centered on his life.

Kevin Grace

References

Bruns, Roger A. *Preacher: Billy Sunday and Big-Time American Evangelism.* New York: W. W. Norton, 1992.

Calkins, David L. "Billy Sunday's Cincinnati Crusade" in *The Cincinnati Historical Society Bulletin.* 27, no. 4 (1969): 292–303.

Rodeheaver, Homer. *Twenty Years with Billy Sunday.* Nashville: Cokesbury Press, 1936.
Sunday, Billy. *Burning Truths from Billy's Bat.* N.p., 1914.

Sweden

Sweden is an interesting case in the history of alcohol and temperance: Although it has had a strong popular and legislative tradition of temperance that produced innovations attracting international attention, public drunkenness and drink-related violence, the result of "spree" drinking, have also been a part of Swedish history. The Swedes have experienced a recent convergence with West European behavior (notably, the shift from spirits to beer). Unlike most other Scandinavian countries—Norway, Finland, and Iceland—the Swedes did not adopt prohibition but instead experimented with other highly restrictive alcohol-control measures. An important factor historically for Sweden was the influence of Anglo-American evangelical Protestantism, which facilitated the reception of U.S. temperance organizations such as the Good Templars.

Traditional Drinking Cultures

Until the seventeenth century, beer was the Swedish national drink, but since the 1700s Swedes have preferred distilled liquor known as aquavit, the national equivalent of German schnapps and Russian vodka. Especially in the countryside, home distilling was widespread, especially in the eighteenth and the first half of the nineteenth century. Though drinking was not normally a daily occurrence (as in wine-drinking societies), a pattern of occasional heavy drinking of spirits led to considerable drunkenness. In the 1820s, consumption per capita was almost 46 liters (12.15 U.S. gallons) of distilled spirits per capita (compared with 9 liters [2.4 U.S. gallons] at the end of the 1900s). In 1825, more than 171,000 stills paid taxes. In order to concentrate distilling in fewer and larger distilleries (and increase government revenue), home distilling was prohibited in 1855. High-alcohol beer did not contribute much as a means of intoxication in the second part of the nineteenth century, as was the case in Denmark. When Swedes worried about alcohol, they were concerned about spirits.

First Measures to Control Alcohol Consumption

In 1837, a Lutheran minister named Peter Wieselgren organized the Swedish Temperance Society (Svenska Nykterhetssällskapet), which campaigned against rural drunkenness. The society demanded total abstinence from spirits, and members pledged not to drink distilled liquor. Besides holding meetings and taking pledges, the society published pamphlets and lobbied for legal action against home distilling. It enrolled about 100,000 men and women in the 1840s. Although the society mostly has been viewed as an elite-dominated organization similar to the Swedish Mission Society and the Evangelical Society, revisionist historians emphasize the connection with Anglo-American revivalism in the countryside at a time when the Lutheran state church forbade any such lay religious activity. Indeed, Sweden showed a remarkable capacity for adopting aspects of American activity. Letters from emigrants who had settled in North America and returned emigrants encouraged religious revivalism and temperance agitation. The Swedish Temperance Society disappeared after the 1855 law responded to its agenda. Moreover, an inward-looking pietism came to dominate Lutheran religious life, and the state church lost interest in temperance and other social reforms.

A New Temperance Movement

The growing strength of new nonconformist evangelical religious movements outside the state church spurred the next phase of temperance reform. Methodists, Baptists, and the Svenska Missionsforbundet gained influence in the second half of the nineteenth century and provided the basis for the development of more radical temperance activities. After the turn of the century, the Blue Ribbon societies had about 100,000 members, while the Independent Order of Good Templars (IOGT) International boasted 150,000 members. The Swedish temperance organizations had enrolled about 20 percent of the adult population in all. The temperance movement, especially the IOGT, developed as a secular self-help movement with a strong cultural, folkloric orientation. The IOGT established study circles, sporting events, and health activities. Therefore, the IOGT can be viewed as one of the basic organizations that developed civil society in Sweden on a local level after 1870. (Disagreement about the role of religion in lodge ritual and other disputes divided the Good Templars among several rival societies until the reunion of the major organizations in the mid-twentieth century.)

These temperance activities combined with the workers' movement to shape the Swedish national identity. The concept *Den skötsamme arbetaren* ("the conscientious worker") characterized the responsible, dutiful, helpful, careful, temperate worker fostered through the temperance movement. This ideal type emerged as a precursor of—and an ideological icon for—the development of the Swedish welfare citizen.

Pressure from temperance reformers helped bring about the introduction of the first major modern policy initiative in 1865, the so-called Gothenburg system established by S. A. Hedlund in the Swedish town of Göteborg (Gothenburg in English). The principal targets were working-class restaurants in the towns. This system originally provided for municipal ownership of taverns but eventually was extended to retail shops that sold spirits for off-premise consumption. The shopkeeper was restricted to a 5 percent profit. The municipality, the county, and the state divided any additional profit. In the late nineteenth and early twentieth centuries, the Gothenburg system attracted attention in English-speaking and Nordic countries as an alternative to prohibition. Even

Woman serving aquavit and mumma in Fjaderholmen, Sweden. Mumma (the dark drink) is a traditional Christmastime drink made from beer, porter beer, sweetened soda, and sherry and usually served with aquavit, also a Swedish tradition. (Bo Zaunders/Corbis)

some British and American prohibitionists became interested in the Gothenburg approach as a method of weakening the profit motive for drink sellers.

Modern Swedish Alcohol-Control Policy

The Gothenburg system was never a national policy. The innovative national policy that Sweden adopted in 1917 responded to the growing strength of the temperance movement without outlawing alcoholic drink. This compromise was called the Bratt system, named after the physician Ivan Bratt. Bratt and others who were concerned about drink but disliked prohibition were under pressure. By the late nineteenth century, northern Sweden, including Lapland, already was "dry." In 1909, temperance societies organized a private referendum to facilitate prohibition in Sweden. Although 55.6 percent of the voters had cast ballots for prohibition in this referendum, the Liberal government then in office did not follow its advice.

Instead, in 1911 the government created a committee to consider whether prohibition was the appropriate answer to the drink question. Although eight of the eleven members of the committee were identified with the temperance movement, one of its members—Bratt—succeeded in persuading the committee to endorse an alternative to prohibition. He approached the drink problem as a scientist, seeing it as a health and social question and not a moral issue. He had the support of the medical profession, the newspapers, and the leadership of the Social Democratic and Liberal parties (most of whose members were prohibitionists). He proposed two measures, namely, control over the individual's purchase of alcohol and elimination of the profit motive in the alcohol trade. Following Bratt's and the committee's suggestions, Stockholm experimented with ration books as early as 1914. In 1917, they were introduced nationally. Sweden adopted a ration-book system *(motbok)* to manage restrictions at the individual level. Under that system, the motbok was used to register and ration the amount of alcohol each individual purchased.

Local temperance boards, whose members knew the individuals of their jurisdictions, were empowered to make decisions that varied according to the social and moral behavior of those who applied. A criminal, drunkard, or morally incompetent person had difficulty obtaining a ration book and so could not buy alcohol. In line with prevailing assumptions about women's capacity for drinking and theories about the likely social impact of excessive women's drinking, women in general received lower rations. Ration books were mostly allocated to the male head of each household. In 1938, 72 percent of the adult male population had a ration book, whereas just 8 percent of the females possessed one. The system therefore implied that women normally could not purchase alcohol on their own. Women could get their own ration books if they were unmarried, and working women were also more liable to get them. The moral standard for unmarried women, however, had in general to be higher than for unmarried men in order to get a ration book. The ration-book system was not abolished until 1955. It was most harshly enforced in the 1930s when there were home inspections based on anonymous informants.

To manage this restriction system, Bratt proposed to concentrate all wholesale trade in one big company, a state-controlled central wine and spirits company. The retail stores were inspired by the existing Gothenburg system. Each municipality controlled retail outlets, including private restaurants. In 1919, bar guests could drink 15 centiliters (0.31 U.S. pints) with a meal. In 1921, the regulations only allowed a bar guest to drink 7.5 centiliters (0.16 U.S. pints; but 15 centiliters for men during the afternoon). The state guaranteed that the sale and consumption took place under acceptable circumstances. Bratt kept prices low in working-class restaurants despite a temperance faction that favored high prices as prevailed in Denmark.

The Bratt and Gothenburg systems help explain why Sweden—in spite of strong temperance influence—never adopted prohibition and in the late 1930s abandoned local option. The crucial moment was a new state-organized referendum in 1922 that showed 51 percent of the voters against and 49 percent for prohibition. There had been a debate about how to organize the referendum. Everyone knew, from the referendum in 1909, that women were more inclined to temperance values and that prohibition would probably be introduced if it were a purely female decision. Women had just got their right to vote in 1921. The referendum was thus organized so it was possible to see the gender distribution in the votes cast: As it turned out, 58.5 percent of female voters and 40.9 percent of the male voters cast ballots in favor of prohibition. Prohibition found strong support mostly among the older, agrarian population and in the traditional religious areas. These were about the same areas and groups that opposed joining the European Union (EU) eighty years later.

The failure of the Swedish referendum meant that Swedish alcohol consumption between the two wars was relatively higher per capita than the Danish average consumption despite the restrictive Swedish alcohol policies. High taxes on spirits had turned the Danes into beer drinkers. Because it helped to defeat prohibition, the Bratt system came to be unpopular with supporters of prohibition, but it survived the Great Depression and World War II. Even after the end of the ration books in 1955, the monopoly state stores remained.

Contemporary Developments

Since Sweden joined the EU in 2001, the Swedish government has been forced to end some of the alcohol restrictions that have characterized Swedish society for many decades. The monopoly stores seem doomed, but at the beginning of the twenty-first century the Systembolaget stores, with their distinctive green and yellow signs, cling to their monopoly for the sale of spirits, wine, and strong beer and offer only limited hours (and days) when they are open. Alcohol prices have declined, and partly as a consequence, alcohol consumption is rising. The restriction on high-alcohol beer has ended. Swedes are still not allowed to buy most kinds of alcohol in ordinary stores. Like Americans and unlike almost any other Europeans, Swedes under the age of eighteen cannot purchase alcohol. The blood-alcohol content permitted to drive a car legally is very low, less than .02 percent. These strict laws are reminiscent of the Bratt system's insistence on strong institutional controls and the supervision of drinking by individuals.

In recent years, Swedes have changed their preferences in alcoholic drink. Like most Europeans, they now prefer beer to spirits. But there is much continuity with past practices. In a study emphasizing international and comparative perspectives, Robin Room has reported in a convenient summary form key Swedish concerns about alcohol policy, noting distinctive patterns. Social concerns about the problems of drinking have, Room argued, been oriented primarily to "matters of welfare, safety and order. This way of thinking about alcohol problems reflects the longstanding forms and habits of Swedish drinking customs." Swedish drinkers imbibe, according to Room, "relatively infrequently, compared to other western Europeans" (Room 2001). Official statistics support these observations. As the (British) Institute of Alcohol Studies has stated, "Sweden ranks among the lowest consumers of alcohol, at twenty-seventh of 31 countries in 1995" (Institute of Alcohol Studies, n.d.). The same institute, however, reported that actual consumption is probably higher as a result of home brewing, which does not come within the net of census collecting. Moreover, as Room reported, when they do drink, Swedes tend to consume "at least five or six drinks at a time" (Room 2001). This persistence of heavy episodic drinking does not mean that Swedish drinking habits "have remained totally unchanged" (ibid.). There are now fewer abstainers. Although drinking in restaurants and taverns has increased since the 1960s, "it still accounts for only about 20 percent of total consumption" (ibid.). Yet despite this "relatively infrequent drinking," a pattern of intoxication as a "frequent result" of such drinking persists in Sweden as a legacy of older institutional structures (ibid.).

Sidsel Eriksen

See also: Denmark; Finland; Norway

References

Ambjörnsson, Ronny. *Den skötsamme arbetaren: Idéer och ideal i et norrländskt sågverkssamhälle, 1880–1930* (The conscientious worker: Ideas and ideals in a sawmill society in northern Sweden, 1880–1930). Stockholm: Carlssons, 1988.

Bruun, Kettil, and Per Frånberg. *Den svenska supen: En historia om brännvin, Bratt og byråkrati* (Swedish drink: A story on boozing, Bratt and bureaucracy). Stockholm: Prisma, 1985.

———. "The Swedish Schnapps: A History of Booze, Bratt, and Bureaucracy" in *State Monopolies and Alcohol Prevention: Report and Working Papers of a Collaborative International Study,* edited by Timo Kortteinen, 257–309. Social Research Institute of Alcohol Studies, Report No. 181. Helsinki: Alko, 1989.

Eriksen, Sidsel. "Drunken Danes and Sober Swedes? Religious Revivalism and the Temperance Movements As Keys to Danish and Swedish Folk Cultures" in *Language and the Construction of Class Identities: The Struggle for Discursive Power in Social Organisation, Scandinavia and Germany after 1800,* edited by Bo Straath, 55–94. Gothenburg: Department of History, Gothenburg University, 1990.

Eurocare website. http://www.eurocare.org/profiles/sweden/consumption.htm.

Frånberg, Per. *Umeasystemet: En Studie I Alternative Nykerhetspolitik, 1915–1945.* Umeå Studies in the Humanities, 50. Stockholm: Almqvist & Wiksell International, 1983.

Hurd, Madeleine. "Liberals, Socialists, and Sobriety: The Rhetoric of Citizenship in Turn-of-the-Century Sweden" in *International Labor and Working-Class History.* 45 (Spring 1993): 44–62.

Institute of Alcohol Studies website. http://www.ias.org.uk/.

Jansson, Torkel. *Adertonhundratalets associationer: Forskning och problem kring ett sprängfullt tomrum eller sammanslutningsprinciper och föreningsformer mellan två samhällsformationer, ca. 1800–1870.* Almqvist & Wiksell International, 1985.

Johansson, Lennart. *Systemet lagom: Rusdrycker, interesseorganisationer och politisk kultur under förbudsdebattens tidevarv, 1900–1922* (The middle system: Alcohol, interest organizations, and political culture in the days of the debate about prohibition, 1900–1922). Lund: Lund University Press, 1998.

Knobblock, Inger. *Systemets långa arm: En studie av kvinnor, alkohol och kontroll i Sverige, 1919–1955* (The long arm of the system: A study of women, alcohol and control in Sweden, 1919–1955). Stockholm: Carlssons, 1995.

Lundqvist, Torbjorn. "The Making of a National Cartel in the Swedish Brewing Industry, 1885–1908" in *Scandinavian Economic History Review.* 46, no. 3 (1998): 42–63.

Nillsson, Tom. "Alcohol in Sweden: A Country Profile" in *State Monopolies and Alcohol Prevention: Report and Working Papers of a Collaborative International Study,* edited by Timo Kortteinen, 311–354. Social Research Institute of Alcohol Studies, Report No. 181. Helsinki: Alko, 1989.

Room, Robin. *CAN 100 År: 1901–2001.* Jubilee number of *Alkohol och Narkotika.* 95, no. 6 (2001): 139–142. Summarized in English as "Sweden in an International Perspective: Alcohol Policy and Drinking Habits" at http://www.bks.no/sweden.htm.

Rosenqvist, Pia. "The Physicians and the Swedish Alcohol Question in the Early Twentieth Century" in *Contemporary Drug Problems.* 13, no. 3 (Fall 1986): 503–525.

Synanon

Synanon was a controversial offshoot of Alcoholics Anonymous established by Chuck Dederich in 1958 that expanded rehabilitation efforts to drug addicts. It then evolved into a planned community that sought a higher state of actualization for its members through intense encounter groups called the "Game." At its apex in 1972, membership numbered 1,700. Eventually, Synanon discontinued being a community based on abstinence and allowed members to "earn" the right to drink by reaching a certain level of development as judged by senior members. Synanon engaged in unconventional lifestyle activities, such as sponsoring vasectomy parties, during which numerous members would undergo the surgery, and limiting marriages to three-year stints, after which members would be randomly assigned a new spouse. The group became embroiled in many legal conflicts, from filing libel and slander suits against journalists to defending themselves against charges that members physically attacked an individual who left the group. The highest level of notoriety occurred in 1978 when members attempted to murder a lawyer who had won a judgment against them by placing a rattlesnake in his mailbox. Synanon eventually disintegrated. Many of its early members began work in the substance-abuse treatment field, where its influence is still felt today in techniques that employ highly confrontational and shame-based counseling tactics.

Historian Roy Janzen views Synanon's history as falling into two distinct phases. The first phase (or Synanon I, 1958–1969) was the era when the group rehabilitated hardened drug addicts through a combination of approaches modeled upon Alcoholics Anonymous (AA), encounter-group tactics, and "milieu therapy" (milieu therapy attempts to infuse psychotherapeutic norms and behaviors designed to foster personal growth through constant, everyday social interaction). During this time, Synanon received accolades from the press as well as from politicians and celebrities for its success in rehabilitating the most intractable of addicts. The second phase (Synanon II, 1969–1991) was one in which the group moved beyond its focus on alcoholics and drug addicts and instead began portraying its therapeutic techniques as a way for all people to reach a higher level of self-awareness and development. This goal was to be accomplished primarily through the Game, a highly confrontational, intense form of encounter-group therapy in which individuals would attempt to gain personal insight through the use of explicit honesty, humor, and graphic profanity. Synanon even attempted to apply this model to juvenile delinquents whom the state remanded to Synanon for rehabilitation.

There were several key turning points at which this idealistic community transformed into what some called a dangerous cult. The decision to expand beyond rehabilitating alcoholics and addicts was certainly one of these. Synanon, in a manner similar to an earlier temperance movement, the Washingtonians, eventually failed because it tried to expand beyond its original mission, began to politicize its activities, experienced excessive domination by its founder, and became involved in controversies. Another factor that contributed to its demise was a shift away from a nonviolent stance to one in which physical violence was used against former members and juvenile delinquents who were entrusted to their care. Finally, Chuck Dederich's declining mental health undoubtedly played a role in Synanon's turn for the worse. He apparently had an undiagnosed bipolar affective disorder. An alcoholic who returned to destructive binge drinking, he was immobilized by the death of his wife, who had been an early member and steadying influence.

Synanon offers an excellent foil with which to contrast AA's success. Alcoholics Anonymous has thrived in part because of its adherence to the Twelve Traditions, which have provided AA with a "singleness of purpose" and kept it from political controversies and cults of personality.

Synanon disbanded in 1991 following a judicial hearing that denied its status as a tax-exempt religion. Dederich died in 1997.

Paul E. Priester

See also: Alcoholics Anonymous (AA); Washingtonians

References

Alcoholics Anonymous World Services. *Twelve Steps and Twelve Traditions.* New York: AAWS, 2001.

Janzen, Roy. *The Rise and Fall of Synanon: A California Utopia.* Baltimore: Johns Hopkins University Press, 2001.

T

Taxation of Liquor (United States)

The current federal system of taxation of alcoholic beverages in the United States dates back to 1862. At that time, the government instituted excise taxes on liquor, tobacco, and other items as well as imposing an income tax. Most of these Civil War taxes were short lived; the liquor and tobacco taxes were permanent. Until the imposition of the federal income tax in 1913, liquor taxes generated a significant portion of the nation's internal revenue and played an important part in maintaining the economic health of the country.

Prior to the Civil War, the federal government had sought income from liquor taxation in 1791 and 1813. The first attempt, part of an ambitious plan developed by Alexander Hamilton, secretary of the treasury, resulted in the Whiskey Rebellion of 1794. When President Thomas Jefferson took office in 1801, he abolished this troublesome system of internal revenue. Although liquor once again provided a prime source for taxation in 1813, tensions similar to those of the 1790s did not develop, and taxes on distilled spirits remained in force until 1817. Conceiving of the internal revenue measures as temporary, Congress abolished the collection bureaucracy when it rescinded the taxes.

Until the passage of the Volstead Act establishing enforcement procedures for National Prohibition, the federal government had a limited view of its proper role in the regulation of the liquor industry. From 1862 on, officials conceived of liquor taxation as an easy, painless, and morally expedient way to raise revenue. High excise rates led to speculation, corruption, and illegal distilling, significantly reducing the amount of money the government received. The Internal Revenue Act of 1862 created many new patronage positions and opportunities for spoils.

Despite reform efforts, the combined forces of speculators and government spoilsmen dominated the federal tax policy and its administration. In the generally lax atmosphere of the Ulysses S. Grant administration, corruption reached new heights. Using the need for funds for Grant's reelection as a pretext, mid-level revenue officials in St. Louis and other Midwestern cities set up a collection ring that cost the federal government millions in revenue from St. Louis alone. Following the breakup of the Whiskey Ring, the administration of the Bureau of Internal Revenue stabilized. In March 1875, Congress raised the liquor tax to ninety cents a gallon. During the 1880s, there were several additional attempts to raise the tax. Linked to efforts to reduce tariffs, none of the legislation succeeded.

The federal system of liquor taxation had come into being because of the need of the state for, at first, an emergency supply of revenue, and after the Civil War, an ongoing secure source of finance. In 1894, the exigencies of the depression required additional revenue; Congress raised the distilled spirits tax to $1.10. The tax did not change again until the beginning of World War I. In 1898, however, the United States again went to war and sought financial assistance from the liquor industry. For the first time in more than thirty years, the federal government raised the tax rate on beer from $1 to $2 per barrel (31 U.S. gallons), leaving distilled spirits alone. The war tax of an additional dollar a barrel was temporary; in 1902, Congress restored the $1 tax.

The Income Tax

In 1909, after much political maneuvering and at a great political cost to President William Howard Taft, and, ultimately, the Republican Party, Congress submitted an income-tax amendment to the states. Congress also passed a 1 percent corporate income tax. In 1913, the United States ratified the Sixteenth Amendment and established the federal income tax, marking the beginning of our present income-tax system.

The significance of these two actions was not readily apparent to most observers. Less than 2 percent of the labor force filed returns from 1913 to 1915; before World War I, over 90 percent of federal revenue continued to come from excise taxes and customs. In 1914, the Wilson administration faced a revenue shortfall caused by the European conflict. The Emergency Revenue Act of 22 October 1914 raised the beer tax to $1.50 a barrel, increased the rates for wine, tobacco dealers, and manufacturers, and instituted a stamp tax. The legislation represented a continuation of the tax policies of the Spanish-American War and did not reflect a new importance for the income tax. In 1916, the Bureau of Internal Revenue collected almost $250 million from the liquor industry; excise taxes and customs duties still accounted for 74.8 percent of the government's income.

During World War I the income tax gradually replaced liquor taxes as the country's primary source of revenue. As part of this process, prohibitory and financial legislation, beginning in 1917, eroded the liquor industry's ability to operate. Perceiving food as an essential element in the nation's mobilization, in the summer of 1917 Congress considered legislation to create a Food Administration that would oversee all aspects of food production and distribution. Prohibitionists sought to ban the use of grains for distilling and brewing. Their stated goal was conservation. Although Herbert Hoover, head of the new Food Administration, and others believed that the use of barley was a more significant drain on food resources than the use of corn, brewers were successful in exempting beer from the Lever Bill. The final legislation prohibited the use of grains in distilling and granted President Woodrow Wilson latitude in determining how much barley and hops the brewing industry could use.

Congress passed the War Revenue Act of 1917 on 3 October 1917 following U.S. entry into war with Germany. Although the Lever Bill sharply limited production, distillers still paid a tax of $3.20 a gallon, whereas brewers contributed $3.00 a barrel, twice the 1914 rate. Congress passed several revenue bills during the course of the war; all expanded the role of the income tax in generating money for the federal government. By 1920, the income tax accounted for 58.6 percent of revenue; the tariff had ceased to have any significance either as a political issue or a source of revenue. Income-tax revenues now occupied the principal place in the internal revenue system that excise and customs had held for so long.

During the years 1913–1920, the internal revenue system of the United States changed dramatically. Simultaneously, more than thirty-six states ratified the Eighteenth Amendment outlawing the manufacture and sale of alcohol. The Eighteenth Amendment and the Volstead Act did not eliminate the relationship between the federal government and the liquor industry. Congress placed the responsibility for enforcement of National Prohibition in the hands of the Bureau of Internal Revenue. The enforcement strategy of the Anti-Saloon League of America relied heavily on state and local cooperation, a policy that the bureau had pursued for many years concerning tax compliance and evasion. The bureau also had existing personnel; any other agency would require a new staff. Throughout the 1920s, the country seemingly enjoyed economic prosperity. The federal government had a surplus of almost $1 billion in 1924. Secretary of the Treasury Andrew Mellon cut taxes five times.

Repeal and After

On 29 October 1929, the stock market crashed, bursting the bubble of the Roaring Twenties and very quickly bringing into question the claims of the Anti-Saloon League and other prohibitionists regarding the positive economic benefits of National Prohibition. The debate over the enactment of the Eighteenth Amendment had presented Prohibition as a panacea for much of industrial society's ills. Repeal became a similar "quick fix" for depression America.

With the economic downturn, the Association Against the Prohibition Amendment and the Women's Organization for National Prohibition Reform, the two major anti-Prohibition organizations, were able to use the prohibitionists' economic arguments against them. Proponents of repeal touted the economic benefits legalization of alcohol would bring. The liquor industry had always argued that it made an irreplaceable contribution to federal and state government. In 1932, most Americans agreed, electing Franklin D. Roosevelt as president. The Democratic platform promised legalization of near beer and repeal of the Eighteenth Amendment.

Because of the linkage of repeal with increased revenue, liquor taxes were inevitable. In April 1933, a special session of Congress legalized beer with an alcohol content of 3.2 percent; a week later legislators passed a tax bill. Echoing their Civil War predecessors, congressmen sought the highest possible rate from the beer tax that would not cause fraud and corruption. Legal brewers paid $5 a barrel; following passage of the Twenty-First Amendment, distillers paid $2 a gallon. Because of these measures and other aspects of the New Deal, tax revenues, which had fallen to $1.5 billion in 1932—the lowest collection since 1917—began to rise. Both the government and the liquor industry were quite comfortable reestablishing their old relationship, particularly since officials were willing to limit tax increases, citing concern over the continued presence of bootleggers. Following repeal, liquor taxes grew in strength; by 1936, excise taxes on alcohol contributed 13 percent of federal tax revenues.

There were tax increases for both beer and distilled spirits during World War II; then, in 1951, beer was taxed at $9 per barrel (31 U.S. gallons), distilled spirits at $10.50 per proof gallon, and wine (unfortified) at 17 cents per gallon. Beer did not receive another tax increase for forty years; the current rates, which were set in 1991, are beer at $18 a barrel, distilled spirits $13.10 per proof gallon, and wine (unfortified) at $1.07 per gallon. Wine is taxed at a higher rate per gallon than beer.

The states also tax alcoholic beverages and have done so since the colonial period. The oldest legislation dealt with licenses for taverns and other drinking establishments. Some colonies, later states, also had excise taxes. Prior to Prohibition and repeal, the states focused on licensing distillers, brewers, and saloon keepers. Following repeal, states have followed the lead of the federal government, taxing alcoholic beverages at the point of finished production. The rate for beer is usually lower than that for distilled spirits. On average, states tax beer at $8 a gallon, wine at $11, and distilled spirits at $14. Presently, neither state governments nor the federal government receive a major portion of their revenue from taxes on alcoholic beverages.

Amy Mittelman

See also: Anti-Saloon League of America (ASLA); Association Against the Prohibition Amendment (AAPA); Bureau of Alcohol, Tobacco, and Firearms (BATF); Federal Regulation of Alcohol before 1920 (United States); National Prohibition (United States); Prohibition Bureau; Whiskey Rebellion; Women's Organization for National Prohibition Reform (WONPR)

References

Chaloupka, Frank, et al. "Economic Perspectives on Alcohol Taxation." http: //tigger.uic.edu/~fjc/Presentations/Slides/madd-niaaa.pdf.

Downward, William L. *Dictionary of the History of the American Brewing and Distilling Industry.* Westport, CT: Greenwood, 1980.
Hu, Tun Yuan. *The Liquor Tax in the United States: A History of the Internal Revenue Taxes Imposed on Distilled Spirits by the Federal Government.* New York: Columbia University Graduate School of Business, 1950.
Kenkel, Donald, and Willard Manning. "Perspectives on Alcohol Taxation" in *Alcohol Health and Research World.* 20 (1996): 230–235.
McGowan, Richard. *Government Regulation of the Alcohol Industry.* Westport, CT: Quorum, 1997.
Mittelman, Amy H. "The Politics of Alcohol Production: The Liquor Industry and the Federal Government, 1862–1900" (Ph.D. dissertation, Columbia University, 1986).
———. "Who Will Pay the Tax? The Federal Government and the Liquor Industry, 1890–1933" in *The Social History of Alcohol Review.* 25 (Spring 1992): 28–39.

Temperance Bars in East Lancashire

Temperance bars provided working-class customers with nonalcoholic drinks in parts of northern England from the 1870s until after World War II. Their stronghold was East Lancashire, where herbalists created and sold temperance beverages. At one time, most mill towns there supported several temperance bars. Today, the only surviving temperance bar is Fitzpatrick's in Rawtenstall.

Industrialization led to the growth of mill towns in nineteenth-century East Lancashire. Living in small, terraced houses, mill workers generally had large families and a hard life. Many of them escaped through drink. In Salford, it was estimated that there was a public house or alehouse for every 160 people. Against this background the temperance movement flourished, too.

What came to be called "temperance bars" appeared in numbers in around 1875. Their origins are unclear; although one would expect a link with the temperance movement, this is hard to prove. Even nonalcoholic alternatives to the public house went against the objectives of temperance reformers, who believed that men should spend more time at home with their families. Yet the National Temperance League did establish the British Workman Public Houses, and a number of other temperance-influenced coffee, tea, and cocoa houses provided nonalcoholic hot drinks.

What does clearly link temperance bars with the temperance movement was the search for acceptable soft drinks. For these, a better, purer water supply was required, and from about 1820 the water supply in England began to improve. In the 1840s, such drinks as ginger beer and nettle beer appeared. Discussing temperance cordials, the historian Brian Harrison (1994) mentioned a temperance bar at Yeovil in Devon in 1860.

By the late 1870s, the climate seems to have been right for a large-scale development of temperance bars. Many of these new businesses resulted from the entrepreneurship of herbalists, who made their living selling patent remedies to poor people who could not afford doctors' fees. The fact that herbalists were known as "medical botanists" suggests that they had expertise in the use of plants, herbs, and the like to make medicinal remedies. This skill, in turn, equipped them to extend their activity to the preparation of nonalcoholic drinks. The herbalists knew where to obtain herbs, roots, spices, and barks as well as colorings and flavorings for such drinks. To create the temperance bar, the herbalist simply set out tables and chairs in the open space in front of the counter in his or her shop. The only time temperance bars were not connected to an herbalist's shop was when an herbalist worked from home. In such cases, the front room of a terraced house became the bar; the brewing was done in the only other downstairs room, the kitchen.

A typical day in the life of a herbalist who was a temperance bar owner started at 6 A.M., when the shop was opened in order to catch those mill-hands, on their way to work, who wanted to fill their bottles with a soft drink; mill work was a thirsty business. Herbalists would advise later customers on appropriate medicines, such as Beecham's powders for flu, Jubes lozenges for sore throats, Bile Beans for constipation, and yeast tablets as a pick-me-up. In between serving patent medicines, the herbalist would brew drinks such as herb beer, bitters, sarsaparilla, ginger beer, and hop stout and prepare the raspberry, lemon, orange, and pineapple concentrates for fruit drinks. All of these would be sold in the temperance bar. Medicinal sweets were also made, as was ice cream from about 1930.

Temperance bars played an important role in the social life of these hard times. Children started in the mills at eleven years of age, working long hours. Six days a week was the norm for a long time, until Saturday working was reduced to mornings only. There was therefore very little time for recreation, little money to spend, and few facilities where boys and girls could meet. There were youth clubs, but public houses were not an option. And so a tradition developed where youth would walk up and down the main street on Sunday evenings after church. This was known variously as the "monkey run," the "chicken run," or the "rabbit run." Various acts of Parliament forbidding unlawful assembly were used to prevent young people from standing around in groups. If policemen saw a group forming, they would move the young people on. In turn, they would go to the only places open on a Sunday evening, the temperance bars.

Young people also met at temperance bars during the week, and the bars were popular with adults as well. Apart from the early morning customers, there was a regular clientele throughout the day. Nonworking mothers and the unemployed would drop in for an inexpensive drink and a chat, using the temperance bar in much the same way as coffee bars came to be used later. Some of the more prestigious temperance bars in the town centers also served theatergoers both before and after shows. As a consequence, temperance bars grew in number, and most mill towns would have at least three. At various times, the author's grandfather, John Joseph Fitzpatrick, and his brothers and sisters owned some twenty-four different temperance bars.

These temperance bars flourished until about 1950, when the implementation of the 1948 National Health Act, with its free medicine, removed the need for herbalists. Together with greater affluence and a more relaxed attitude toward young people in pubs, the welfare state meant that there was no longer a role for the combined herbalist and temperance bar.

Kevin Hardman

See also: National Temperance League (NTL); Pubs and Beerhouses (United Kingdom)

References

Burnett, John. *Liquid Pleasures: A Social History of Drinks in Modern Britain.* London and New York: Routledge, 1999.

Harrison, Brian. *Drink and the Victorians: The Temperance Question in England, 1815–1872.* London: Faber & Faber, 1971; 2d ed., Keele: Keele University Press, 1994.

Temperance Commission of the Federal Council of Churches

The Federal Council of Churches of Christ in America (FCCCA)—a moderate group of representatives from Protestant denominations—founded its Temperance Commission in 1913 to stimulate support for the cause. The Temperance Commission was the immediate successor to the Temperance Committee, a permanent committee established at the time of the founding of the FCCCA in 1908. The commission conducted a campaign of agitation and propaganda, publishing a large quantity of literature and organizing speakers around the country. It also coordinated the work of existing denominational temperance committees and encouraged the creation of additional committees among its constituent church bodies.

The first chairperson was the Reverend Rufus W. Miller. The most active members of the Temperance Commission included Daniel A. Poling, Joshua Levering, and Luther Wilson. The thirty constituents of the FCCCA included Baptist, Congregationalist, Methodist Episcopal, African Methodist Episcopal, Methodist Protestant, Presbyterian, Evangelical, Mennonite, Society of Friends, and Lutheran representatives.

The creation of a Temperance Commission was consistent with the council's interpretation of its religious and ethical responsibilities. Activism on behalf of temperance was also popular because it assuaged tensions between liberal and conservative constituents: Conservatives favored the agenda because of its long tradition among American Christians, and liberals embraced its place within the Social Gospel movement.

Strategy and Program of the Temperance Commission

The Temperance Commission used education and moral suasion to promote temperance. Exemplifying this strategy, the commission initiated a very successful nationwide campaign for total abstinence in 1915. It organized the National Temperance Union for this purpose and sponsored mass meetings in Philadelphia, New York, Pittsburgh, and other major cities to encourage thousands of men to sign the pledge. The National Temperance Union also distributed literature that emphasized the scientific and economic rationale for abstinence. This information was made available to schoolchildren, immigrant populations, and social welfare departments.

After the U.S. entry into World War I, the Temperance Commission launched an intense "Strengthen America" campaign to insist upon prohibition as a wartime measure. It also demanded the elimination of the army canteen. The Temperance Commission cooperated with the United Committee on Temperance War Activities in the army and navy to assure the availability of practical and scientific information about alcohol and temperance to military personnel.

Other examples of the Temperance Commission's creative outreach programs included the placement of advertisements in labor publications to encourage the removal of union meetings from saloons, campaigns among university alumni associations and college students to discontinue liquor service at school functions, and the publication of a children's journal entitled *The Water Lily.* The Temperance Commission also stimulated denominational support for a National Prohibition amendment by merging with the National Temperance Society and Publication House, the oldest and most significant publication agency in the country. Between 1916 and 1920, the Temperance Commission spent more than $150,000 on publicity and educational materials about Prohibition.

The FCCCA supported the Eighteenth Amendment because of four assumptions: that individual liberty must be controlled in the interest of the public welfare; that liquor was evil; that no other means of eliminating alcohol consumption was viable; and that Prohibition—for all of its problems—worked better than any other program. Neither extreme fundamentalism nor nativism seems to have weighed significantly in the council's pro-Prohibition position.

The Decline of the Temperance Commission

The FCCCA remained formally pro-Prohibition, but its activism began to wane after legislative victory. After 1920, the Temperance Commission appears to have had no budget and no paid staff. It was maintained almost exclusively through the enthusiasm of Carl A. Millikan and soon fell under the direct control of the administrative committee of the council instead of retaining the status of a commission. The agreement between the National Temperance Society and the Temperance Commission expired in 1920 and was not renewed.

In 1923, F. Ernest Johnson, head of the research department of the Commission on the Church and Social Service, another arm of the council embarked on a tour of the eastern cities where Prohibition was seemingly the least effective. His conclusion—that the churches needed to withdraw from legal enforcement, leaving this task to government officials—suggested that the churches should exclusively pursue an educational agenda. The Anti-Saloon League of America's continuing emphasis on law enforcement prompted the FCCCA to challenge the Anti-Saloon League's leadership of the temperance movement. The council envisioned a council-sponsored group that would conduct a "dry" educational campaign as a successor to the old strategy and leadership. It called a Conference on Prohibition and Law Observance in the spring of 1927 to discuss a renewal of educational strategies. However, the council's attempt to lead the movement was quickly rebuffed by the Anti-Saloon League, many of whose members sat within the council itself as denominational officials.

After the Conference on Prohibition and Law Observance, the Temperance Commission completely ceased to function. Nonetheless, other commissions, such as the Christian Education Commission, continued some of the Temperance Commission's work. The Federal Council of Churches of Christ in America was reestablished as the National Council of the Churches of Christ in the USA in 1950.

Lisa Andersen

See also: Anti-Saloon League of America (ASLA); National Temperance Society and Publication House (NTSPH)

References

Cavert, Samuel McCrea, ed. *The Churches Allied for Common Tasks: Report of the Third Quadrennium of the Federal Council of Churches of Christ in America, 1916–1920.* New York: Federal Council of Churches of Christ in America, 1921.

Hutchison, John A. *We Are Not Divided: A Critical and Historical Study of the Federal Council of the Churches of Christ in America.* New York: Round Table Press, 1941.

Timberlake, James H. *Prohibition and the Progressive Movement, 1900–1920.* Cambridge: Harvard University Press, 1966.

Temperance Council of the Christian Churches (TCCC)

The Temperance Council of the Christian Churches (TCCC) existed in Britain from 1915 to 1979, when a new body, the Churches Council on Alcohol and Drugs (CCOAD), was formed to replace it. During World War I, the British government imposed severe restrictions on the retail sale of alcohol under the Defence of the Realm Act. The temperance movement was split over the issue of public ownership and membership on the government's Central Control Board (Liquor Traffic). Henry Carter of the Wesleyan Methodist Church was in favor of public ownership and serving on the board; the prohibitionist United Kingdom Alliance (UKA) was implacably opposed. Carter was the prime mover in establishing the TCCC with its aim "to secure legislative reform." The TCCC's greatest and lasting success in the field of legislation was in its advocacy of raising the public drinking age. In 1922, Parliament set the legal drinking age at eighteen (effective in 1923). Previously, there had been no statutory minimum age for purchasing and consuming alcoholic beverages on licensed premises. (Currently, the age remains at eighteen, except in restaurants, where it is sixteen.)

When the issue of public ownership came to the fore again in the early 1930s, it almost split the temperance movement. As a member of the Royal Commission on Licensing (1929–1931), Carter was in favor of disinterested management by means of public ownership. For this he had the support of William Temple, then the archbishop of York as well as chairman of the TCCC. Both railed against the stance of the United Kingdom Alliance over the issue. A split in temperance ranks was averted by a concordat (9 November 1933) reached between the Methodist Church Temperance and Social Welfare Department and the UKA that set out a program of temperance reform focusing on areas of agreement. For all practical purposes, this document ended the quarrel between the supporters of public ownership and local veto that had weakened the temperance movement since the turn of the century.

The temperance movement was able to come to this agreement because reformers were frustrated by the failure of the Conservative-dominated government to act upon the findings of the Royal Commission on Licensing. Responding to their own decline, the Nonconformist denominations, the chief supporters of temperance reform, showed a new spirit of unity and cooperation. For instance, in 1932 Carter's Wesleyan Methodist Church united with other Methodist denominations that held more advanced temperance views. It may not be a coincidence that he never published the second volume of his controversial history of the English temperance movement. Most important, as early as August 1933 temperance leaders had learned about a private speech by the director of the Brewers' Society that proposed an advertising campaign to encourage young nondrinkers to drink beer. Outraged, temperance reformers were disposed to cooperate against a common enemy.

The Brewers' Society launched its "Beer Is Best" advertising campaign in order "to get the beer drinking habit instilled into thousands, almost millions, of young men who do not at present know the taste of beer" (Sir Edgar Sanders, director of the Brewers' Society, quoted in *Times,* 18 November 1933). The TCCC played an important role in mounting opposition by the churches to the brewers' strategy.

In 1952, the work of the TCCC was divided between the Temperance Council of the Christian Churches General and the Temperance Council of the Christian Churches Educational Limited. The former concentrated on legislation and the latter on education, having become registered as an educational charity in order to gain tax relief. The educational work constituted more than 90 percent of TCCC activities.

The postwar period saw two dynamic secretaries of the TCCC: Dr. T. G. Dunning and Arthur C. Davies, both Baptist ministers. Dunning was president of the Baptist Union in 1958 and in the same year wrote *Alcohol and the Christian Ethic.* On retiring from the position of general secretary in 1961, Dunning was succeeded by Davies. He proved to be an effective advocate in the London liquor-licensing courts and played an important role in briefing members of Parliament against the 1976 Private Members Bill to liberalize the licensing laws, which was defeated.

In 1975, the council celebrated its diamond jubilee. Davies retired in 1978 and was succeeded by Kenneth Lawton, a Methodist minister. Lawton immediately set about the task of creating a new body—the Churches Council on Alcohol and Drugs. Writing in *Alliance News*, Lawton stated: "A new, revitalised, contemporary, practical ecumenical council had been formed . . . gaining a new commitment by the whole Church" (Lawton 1979: 7). This proved not to be the case. The new council existed for only a decade. The Christian denominations that made up the TCCC had their own social responsibility departments, and there was no commitment from the "whole church." Lacking finance and leadership, it ended (to borrow the words of the poet T. S. Eliot) "not with a bang but a whimper."

Derek Rutherford

See also: Advertising (United Kingdom); Carter, Henry; Davies, Arthur C.; Gothenburg Schemes/Disinterested Management; United Kingdom Alliance (UKA); Wilson Black, Robert

References

Brake, George Thompson. *Drink: Ups and Downs of Methodist Attitudes to Temperance.* London: Oliphants, 1974.

Lawton, Kenneth. *Alliance News.* May/June 1979.

Urwin, E. C. *Henry Carter, C.B.E.: A Memoir.* London: Epworth, 1955.

Temperance Standard Bearers

Temperance Standard Bearers of the Nineteenth Century: A Biographical and Statistical Temperance Dictionary was compiled and published by the Liverpool temperance historian Peter Turner Winskill. Its two volumes, published in 1897 and 1898, contain more than 1,000 pages. A portrait of Winskill precedes the first volume and his autobiographical memoir the second. Winskill's temperance biographical dictionary provides brief sketches of more than 7,000 temperance reformers, mostly United Kingdom residents. The entries vary greatly in length and detail. For instance, Winskill used no more space than decency required to present his fellow historian James Dawson Burns. Burns had accused his rival of copying passages from a *Temperance Biography* that Burns published serially in 1863 (reaching only "Bur" in the alphabet) for use in Winskill's own *Temperance History* (1881).

Winskill started to collect information for the biographical dictionary when he had to eliminate biographical information intended for his earlier book, *The Temperance Movemènt and Its Workers* (1891–1892), including, he said in his *Standard Bearers* preface, "much that was actually in type" (Winskill 1897–1898, 1: 5). In addition to providing a biographical work of reference, Winskill wanted to collect data for statistical tables to show the "advantages arising from the faithful practice of total abstinence from all kinds of intoxicating liquors and tobacco" (ibid., 1: 6). In particular, he wanted to prove that teetotalers lived longer than drinkers. Having difficulty in finding dates and ages, he sometimes had to visit graveyards to obtain the information for the early letters of the alphabet. Beginning with the letter "F," Winskill excluded most people from his biographical dictionary for whom such dates were unavailable, explaining, "We have no space to spare for those not available for the statistical statements" (ibid.). He also abandoned his original plan to restrict the book to persons over the age of thirty who had been abstainers for twenty years or more. In the end, he arranged his entries so family members appeared together.

Winskill admired the men and women whose lives he summarized. Many, "in addition to being faithful teetotalers for forty, fifty, sixty, seventy or more years, were also non-consumers of tobacco, many of them vegetarians, and almost without exception devout Christians" (ibid., 1: 16). One exception to the last characteristic was William Henry Quilliam, a third-generation teetotaler who was so impressed by Muslim rejection of alcoholic drink that he embraced Islam, established a mosque in Liverpool, and edited two Islamic magazines.

Winskill provided six statistical tables. He was most careful about Table I, showing ages and life spans; Table II, showing years of abstinence; and Table III, for life abstainers. At least two-sevenths of the people for whom he provided biographies were life abstainers. He acknowledged that information for the other tables was incomplete so they were only suggestive. Table IV identified trades and professions of teetotalers; for example, there were 26 estate agents, 43 lawyers, 564 Nonconformist ministers, 5 shepherds, and 2 town criers. Table V reported membership in temperance organizations (the largest being 930 Good Templars, followed by 891 Rechabites, 533 members of the United Kingdom Alliance, and 433 Bands of Hope). Table VI counted members of religious organizations (385 Quakers, followed by 335 Congregationalists, 324 Wesleyan Methodists, 228 members of the Church of England, 222 Presbyterians outside Scotland, 209 Baptists, 197 Primitive Methodists, and 179 United Methodist Free Church members).

A standard library database identifies only ten copies of *Temperance Standard Bearers* still extant in major libraries, but it is available on microform.

David M. Fahey

See also: Burns, (James) Dawson; Good Templars (IOGT); Islam as Cultural Influence; Nonconformists (United Kingdom); Rechabite Friendly Society; United Kingdom Alliance (UKA); Winskill, Peter Turner

Reference

Winskill, Peter Turner. *Temperance Standard Bearers of the Nineteenth Century: A Biographical and Statistical Temperance Dictionary.* Liverpool: the author, 1897–1898.

Temple, Frederick (1831–1902)

Frederick Temple (1831–1902) was remembered by the *Temperance Chronicle* as a "giant . . . of temperance warfare, comparable with Wellington, Nelson and Drake" (CETS 1914: 22). This most famous of teetotal prelates led several antidrink organizations, including the Church of England Temperance Society (CETS). He was known as a theological radical (*Essays and Reviews,* 1860), democrat, social reformer, teetotaler, and prohibitionist, but his high ecclesiastical office forced him to use restraint.

As a nonabstaining Rugby headmaster (1857–1869), Temple was hounded by teetotalers and prohibitionists. In 1862, echoing the Huxley-Wilberforce evolution debate, the Clerkenwell teetotal parson Robert Maguire answered Temple's argument that an abstainer, trained rather than self-directed, resembled a monkey, with the assertion that, since a drunkard had no more inner restraint than a monkey, he required the pledge. As bishop of Exeter (1869–1885), Temple patronized the moral suasionist National Temperance League, the teetotal fraternal Order of Rechabites, and the prohibitionist United Kingdom Alliance (UKA) and sat on the House of Lords' Committee on Intemperance (1876–1879). Victorian temperance hagiography recounted rowdy drinkers invading a platform and covering Temple and UKA President Sir Wilfrid Lawson with flour.

The founding of CETS in 1873 and Temple's promotions, to bishop of London in 1885 and to archbishop of Canterbury in 1896, advanced the temperance cause among the Anglican establishment but also imposed limits. Extreme teetotal clergymen applauded the moderate CETS chairman Henry Ellison's 1891 resignation and his replacement by Temple but were disappointed when, as chairman, Temple emphasized the "dual basis" for teetotalers and nonteetotalers and recommended licensing restriction rather than prohibition.

Temple refused to condemn wine drinkers but urged "better-class" teetotal Christians to set an example for working people who suffered from poverty and adulterated drinks. At the Canterbury Convocation and at the 1888 Lambeth world bishops' conference, Temple defended fermented Communion wine, possibly watered-down. He advised Archbishop E. W. Benson of Canterbury not to endorse the UKA and suggested arguments for compensating dispossessed publicans.

Although he served as bishop of London, Temple, surprisingly, presided over almost all CETS meetings, introduced its bills in the House of Lords, and helped persuade Prime Minister Salisbury to initiate the the Royal Commission on Liquor Licensing Laws, also known as the Peel Commission (1896–1899). As archbishop of Canterbury, Temple resigned the CETS chairmanship as too controversial and stopped regularly attending Peel Commission meetings. The teetotal bishop of Winchester, Randall Davidson, advised him not to support a sober welcome of Boer War veterans, lest he be "too closely identified with definite temperance work." His negotiations for a national temperance Sunday and for a teetotal coronation of Edward VII failed.

Yet Temple endorsed Davidson's successful efforts to enact consensus Peel Commission recommendations, and when the commission failed to advocate local prohibition, he formally protested, supporting his UKA associates. His Canterbury Cathedral memorial acknowledges his temperance efforts.

Gerald Wayne Olsen

See also: Church of England Temperance Society (CETS); Davidson, Randall; Ellison, Henry; Lawson, Sir Wilfrid; National Temperance League (NTL); United Kingdom Alliance (UKA)

References

Church of England Temperance Society (CETS). *Temperance Chronicle.* 42, no. 9 (January 1914): 22.

Dant, Charles H. *Archbishop Temple.* London: Walter Scott, 1903.

Olsen, G. W. *Drink and Temperance in the Church of England, 1829–1875.* Unpublished book manuscript.

———. *Drink and the British Establishment: The Church of England Temperance Society, 1873–1914.* Unpublished book manuscript.

———. "From Parish to Palace: Working-Class Influences on Anglican Temperance Movements, 1835–1914" in *Journal of Ecclesiastical History.* 40 (1989): 239–252.

Temple, Frederick. *Bishop Temple on the Permissive Bill.* Manchester: United Kingdom Alliance, 1870.

———. *National Aspects of the Temperance Reformation.* London: National Temperance League, 1884.

———. *Occasions of Stumbling.* London: Church of England Temperance Society, 1894.

Tennent, Hugh (1863–1890)

Hugh Tennent (1863–1890) was a pioneer of lager brewing in the United Kingdom and played a short but significant role as head of J&R Tennent of Wellpark Brewery, Glasgow. This enterprise, which had origins in the seventeenth century, expanded rapidly with the industrialization of Glasgow and its hinterland, becoming by the mid-nineteenth century the largest brewery outside Edinburgh and Alloa, the major centers of brewing in Scotland. Tennent's predecessors developed a large business specializing in ales and stouts. In 1855, it was valued at £220,000 and was said in the 1850s and early 1860s to be making "large and increasing profits" under the management of Tennent's grandfather, Hugh, and his father, Charles. Unfortunately, both died in the same year, 1864, and for the next twenty years trustees ran the firm. They must have been highly enterprising, for they built the business to a value of more than £300,000 in 1870, greatly extended the brewery, modernized the plant, and promoted the international reputation of the firm's products. The famous red "T" trademark was registered in 1876.

Tennent evidently assumed a management role in his late teens, and by 1884, when he reached twenty-one, he had bought out his brother's share and taken sole control of the firm. Highly ambitious and clever enough to see the consumer trend to light, bright beers, he took a close interest in the German industry and its rapid expansion into export markets. In 1881, Tennent, accompanied by an associate, Wyllie Clarke, who later became managing director, paid what was possibly the first of several visits to Bavaria to familiarize themselves with lager brewing, and presumably, the product. One thing they must have acknowledged was the similarity in the process to Scottish brewing, which often used low-temperature fermentation. It is believed that Tennent began lager brewing at Wellpark in 1885, and sufficiently encouraged by the response, decided in 1889 to build a dedicated lager brewery. Accordingly, the engineers, with the firm L. A. Riedinger of Augsburg, which had previously set up plants in locations as far apart as Tiflis, Yokohama, Cuba, and Buenos Aires, were commissioned to construct a model brewery in Glasgow.

After considerable delay, the new brewery was completed in 1891. Under the initial direction of German brewers, Tennent's lager became an immediate success in Scotland and overseas. Sadly, Hugh Tennent did not live long enough to see his vision become a reality, for he died, unmarried and childless, in 1890 at age twenty-seven. According to the entry in the register of death, the cause of death was acute fever and an (undeveloped) fatty heart. Above his share of the firm, his personal estate was worth at least a quarter of a million pounds. A second Tennent trust was set up to run the business, but Tennent was the last member of the family in direct control of this important and dynamic enterprise. Since lager ultimately became its most famous product, Hugh Tennent could reasonably claim to have influenced the development and prosperity of the firm long after his untimely demise.

Ian Donnachie

See also: Beer; Scotland

References

Donnachie, Ian. *A History of the Brewing Industry in Scotland.* 2d ed. Edinburgh: John Donald, 1998 [1979].

———. "Tennent, Hugh" in *Dictionary of Scottish Business Biography,* edited by Anthony Slaven and Sydney Checkland. 2 vols. Aberdeen: University of Aberdeen Press, 1986–1990.

Tenskwatawa (1775–1826)

Tenskwatawa, the Shawnee "Prophet" (1775–1826), younger brother of the Shawnee war chief Tecumseh, was briefly the spiritual leader of a revival of traditional Shawnee religious practices but added a significant anti-American, anti-Christian, and temperance message to his teachings. The Prophet's early childhood was illustrative of the pattern of destruction wrought by American westward expansion on the Native Americans of the old Northwest. Abandoned by their mother, the Prophet and his siblings were raised by relatives amidst the violence and scarce resources of the Shawnees, who had embraced alcohol just as their primary economic and subsistence resource, wild game, was fading. Losing an eye in a hunting accident, the Prophet was thought of as a marginal figure within the tribe, a poor husband and father, and a drain on tribal resources until 1805, when he underwent a transformative experience after studying traditional practices with a tribal medicine man.

Awakening from a trance, the Prophet, taking the name Tenskwatawa ("open door"), announced that the Shawnee, and all Native Americans, could regain the favor of their Great Spirit by rejecting American technology, dress, religion, and especially alcohol. He quickly gathered followers who began a purge of these influences among themselves and neighboring tribes. His settlement at Prophetstown attracted large numbers of followers, which strained his abilities to provide food, and which alarmed William Henry Harrison, the governor of Ohio. In conjunction with his brother, Tecumseh, who was organizing for a military and political solution to white encroachment, the Prophet was perceived by whites as a major threat. Harrison marched on Prophetstown in 1811 while Tecumseh was on a mission to the southern tribes. Foolishly, the Prophet ordered an attack on Harrison, which became the Battle of Tippecanoe, and the Shawnee defeat there ended the Prophet's hold on his followers.

For the rest of the Prophet's career, he was in the shadow of Tecumseh, eventually becoming a "government" chief under the British in Canada, a position he despised. Lured back to the United States by Indian Agents in 1824 to lead the Shawnee to a reservation in Kansas, he was generally disregarded by his former converts and his teachings ignored. Although he continued to protest the missionary efforts of the Baptists among the Shawnee in Kansas, he died in 1826 without restoring the traditional practices of the Shawnee or freeing them from the insidious influence of American whiskey.

Margaret Sankey

See also: Native Americans: Drinking Patterns and Temperance Reform

References

Edmunds, R. David. *The Shawnee Prophet.* Lincoln: University of Nebraska Press, 1983.

Eggleston, Edward, and Elizabeth Eggleston Seelye. *Tecumseh and the Shawnee Prophet.* New York: Dodd, Mead, 1880.

Sugden, John. *Tecumseh: A Life.* New York: Henry Holt, 1998.

Tequila

Tequila is a beverage distilled from the juice of the *pias,* the heart of the blue agave plant. The name comes from the small town in a valley in Jalisco state, Mexico, where the Spanish reportedly first distilled the beverage. Tequila is similar to mescal, since both are distilled from the agave. By Mexican law, tequila must be distilled from the blue agave, while mescal can be made from any one of five different varieties. For most of the history of tequila, the two beverages were essentially the same.

The Aztec people of Mexico made a fermented beverage called *pulque* from the agave (in Mexico, called *maguey*) plant. This plant was considered sacred, and pulque was used in Aztec religious ceremonies. In the early 1500s, when the newly arrived Spanish Conquistadores ran out of brandy, they fermented agave juice. The spirit was called mescal (or mezcal) brandy,

A laborer collects the sap of the maguey plant, one of the agave family of flowering plants, to produce an alcoholic drink called pulque, which is distilled to make a colorless liquor, mescal, and also tequila. (Hulton/Archive)

agave wine, mescal tequila, or tequila. Eager to maintain the market for Spanish products, Philip II banned new vineyard planting in 1595. The first tequila factory was established in 1600, when Don Pedro Sanchez de Tagle began to cultivate agave and distill tequila in the town of the same name. In 1608, the governor of New Galicia imposed the first taxes on mescal wine, and by 1621 "wines of mescal" were being supplied to nearby Guadalajara. In 1636, governor Don Juan Canseco y Quiñones authorized the distillation of mescal wines and taxed the product.

Mescal wines became an important export through the eighteenth century, though they did face numerous restrictions. In 1785, the viceroy Don Matias de Galvez outlawed eighty-five types of alcoholic beverages, including five brands of mescal, to promote the importation of Spanish wines and liquors. Production went underground for ten years, until King Ferdinand IV lifted the ban around 1795. During the prohibition, the process of baking the agave underground may have been developed. The War of Independence redirected trade routes, and tequila declined in importance, but after Mexican Independence in 1821 Spanish products became more difficult to acquire and tequila's popularity grew.

Notable tequila distillers trace their beginnings back as early as the mid-eighteenth century. Jose Antonio Cuervo became the first licensed manufacturer of tequila in 1758 when he obtained the right to cultivate a parcel of land that included a private distillery. In 1795, Cuervo's son José Maria acquired a license to produce mescal wine and founded the first official Mexican distillery. By the middle of the nineteenth century, Cuervo's fields had more than 3 million agave plants. In 1873, Don Cenobio Sauza purchased La Antigua Cruz; he changed the name to La Perservancia in 1888 and began making mescal wine. In 1906, he purchased thirteen more distilleries and numerous agave fields.

Railway expansion in the 1880s helped to spread tequila across North America. In 1893, "mezcal brandy" won an award at the Columbian Exposition in Chicago. During the last quarter of the nineteenth century, mescal wine was becoming popularly known as tequila. Tequila gained national importance during the Mexican Revolution in the early 1900s when it became associated with the rebels and heroes of the period.

After the revolution, the government embarked on a process of land redistribution that significantly hurt the tequila manufacturers, who saw acres of their agave plantations given back to peasants. These distillers were further crippled by the Great Depression, and the fact that some postrevolutionary leaders, such as Victoriano Huerta, preferred French cognacs. Tequila's fortunes were boosted, however, by actions north of the border. National Prohibition in the United States resulted in tequila smuggling, and the beverage grew in popularity. The growth in U.S. consumption was aided by the decision among distillers to use non-agave sugars in production, which resulted in a lighter product that was more palatable to U.S. tastes, though an aberration to aficionados.

Tequila's popularity in the United States grew again during World War II, when liquor imports from Europe were disrupted. Agave fields expanded 110 percent between 1940 and 1950, although after 1948 exports declined while domestic production grew. The increase in revenue enabled distilleries to modernize and upgrade their facilities. Tequila's popularity had a further boost when, sometime between 1930 and 1955, the margarita was invented, becoming one of the most popular mixed drinks in North America.

Until the 1940s, the tequila industry was relatively unregulated. In 1944, the Mexican government decided that to be called "tequila" a product had to be distilled in the state of Jalisco. In 1949, the Normas Oficial Mexicana Tequila stipulated that only certain plants could be used to make tequila. Tequila gained international recognition, and by 1977 it was defined by international convention as a product originating only in Mexico. The Normas Oficial Mexicana of 1978 regulated the agricultural, industrial, and commercial processes related to tequila, and in 1994 the Tequila Regulatory Council (Consejo Regulado de Tequila) formed to oversee production, quality, and industry standards. That same year, the government determined that tequila must be made from at least 51 percent blue agave juice.

In 1996, Mexico signed an international agreement for all countries to recognize tequila as a product from a specific region of Mexico. This treaty has been challenged by manufacturers in South Africa, Spain, and Japan, all of which have made or have threatened to make tequila. By the end of the twentieth century, tequila could be manufactured in five regions of Mexico, all within 100 miles of Guadalajara. In 1997, the agave crop was attacked first by a freak frost, then a fungus. Since agave takes eight years to mature, the result was devastating. Pias rose in price from 1,000 pesos to 15,000. A number of distilleries failed, and it would take years for the industry to recover.

Daniel J. Malleck

See also: Mexico

References

Chadwick, Ian. "In Search of the Blue Agave." http://www.ianchadwick.com/tequila.

"Drought in Margaritaville" in *Economist.* 360, no. 8239 (September 15, 2001): 33.

"The History of Tequila." http://www.lasmargaritas.com/tequila/history.html.

Luxner, Larry. "From Blue Agave to Good Spirits" in *Americas.* 47, no. 4 (July–August 1995): 16–23.

Miller, Summer. "Tequila." http://www.mixed-drink.com/tequilainfo.html.

Tilley, Samuel Leonard (1818–1896)

Samuel Leonard Tilley (1818–1896) was a New Brunswick premier who promoted the confederation of Canada and enacted the first Canadian prohibition law. Born in Gagetown, New Brunswick, on 8 May 1818, Tilley quit school at the age of thirteen and moved to the city of Saint John, where he worked in a drugstore for the next seven years. He joined the Young Men's Debating Society, developed his public-speaking skills, and honed his debating skills on the temperance issue. He became the owner of "Tilley's Drug Store" at the age of twenty, joined a temperance society in 1832, and became a total abstainer from alcohol five years later. He soon began to focus his energies on politics and helped organize the New Brunswick railway league. In June 1850, he was elected to a seat in the House of Assembly as a Liberal candidate from Saint John. Tilley and two others resigned soon after, but he was reelected in 1854. In November of that year, he became part of the cabinet as provincial secretary, and except for a period of a few months, he would be a minister or premier until 1888.

Tilley's dedication to temperance almost cost him his political career when in 1856 he used his influence to gain passage of a prohibition statute, the first such law in any province in Canada. Based on the Maine Law of 1851, it outlawed the manufacture and sale of alcohol. Enforcement of the law proved difficult. The public at large was not in favor of the measure, and the Liberals were beaten at the polls. The new government moved quickly to repeal the law. Within a few months, a dissolution occurred and Tilley's party regained power. He quickly ascended to the premiership in June 1857, a position he would maintain until 1865. Although remaining privately committed to temperance, he would never again try to put it on the public agenda.

Tilley's main historic importance lies in his advocacy of and political maneuvering for a united Canada. In 1864, he attended a maritime parliamentary meeting of representatives from Nova Scotia, New Brunswick, and Prince Edward Island at Charlottetown, Prince Edward Island, to discuss a legislative union between the provinces. Leaders of other provinces, upon hearing of the meeting, expressed a desire to attend. Abandoning the smaller scheme, Tilley pushed for discussion of a larger Canadian union. Meeting in Quebec on 10 through 27 October, the representatives developed the so-called "Quebec scheme" for unification. On 5 March 1865, he placed the issue before the voters of New Brunswick. Even though he remained personally popular, the anticonfederates won three to one, with Tilley going down to defeat. The new government resigned the next year and he came back into power. Hopes for a confederation revived, and he attended a conference in London to finalize plans. The British North America Act of 1867 brought union to Canada. Tilley then resigned his position and accepted John A. Macdonald's invitation to serve as minister of customs in the first federal government. Resigning from the federal government in 1873, he became lieutenant governor of New Brunswick, then reentered Parliament in 1878 as the minister of finance. Retiring in 1893 because of ill health, he died on 25 June 1896 in St. John. Although his dreams of a prohibition law could never be sustained, he was able to maintain his teetotal attitudes for his entire life and inspired thousands. He also attended temperance gatherings throughout North America even into the last decade of his life.

T. Jason Soderstrum

See also: Provincial Prohibition (Canada)

Reference

Hannay, James. *Wilmot and Tilley.* Toronto: Morang, 1910.

Wallace, Carl Murray. *Sir Leonard Tilley: A Political Biography.* Edmonton: University of Alberta, 1972.

Toasts

In making a "toast," two or more people drink to the good health of each other or of an eminent person or institution that they admire. The consumption of toasts at formal dinners reached its greatest popularity in both Britain and North America during the late eighteenth and early nineteenth centuries, when frequent toasting led many people to consume a great deal more wine than they might otherwise have done, and temperance campaigners called for the practice to be abandoned.

Toasting appears to have derived from the earlier practice of "pledging," which served in Anglo-Saxon and medieval Britain as a means of demonstrating one's loyalty to a particular tribe and willingness to fight on its behalf. Since a man was unable to draw his sword while drinking, he would ask his neighbor at table to pledge him: to guarantee his safety by drawing his sword (or knife) to guard the first man while he drank. At this period, it was common practice for everyone at table to drink from the same vessel. When individual glasses replaced communal goblets, diners took to clinking their glasses together to show that they were still sharing the same liquid. Glasses grew in popularity from the sixteenth century onward partly because of their ring: the sound they made when tapped. As men were now carrying arms much less frequently, the practice of pledging changed. Now a man showed his loyalty to another simply by drinking to his health.

During the sixteenth century, the English acquired an international reputation as a nation of heavy drinkers. Some observers attributed this to the popularity of pledging. The importance and frequency of this practice increased in the late seventeenth century, when loyalties were divided first between king and Parliament, and then between the restored Protestant monarchy and the Catholic king in exile. The practice of drinking to someone's health in polite English and Scottish society attracted derision from some foreign visitors in the seventeenth and eighteenth centuries, suggesting that it was practiced to a far greater extent in Britain than elsewhere in Europe.

Toasting was also popular in colonial America. Puritans disapproved of the practice, considering it profane, and in 1639 the General Court in Boston declared it illegal, but this prohibition proved impossible to enforce and was repealed after only six years.

Toasting eventually fell into a gradual decline in both North America and Britain during the nineteenth century under the influence of the temperance movement. It survives today at celebratory meals and official banquets as a consciously antique custom.

Andrew Barr

See also: Germany

References

French, Richard Valpy. *The History of Toasting.* London: National Temperance Publication Depot, 1882.

———. *Nineteen Centuries of Drink in England.* London: Longmans, Green, 1884.

Hackwood, Frederick W. *Inns, Ales, and Drinking Customs of Old England.* London: T. Fisher Unwin, 1909.

Treatment Institutions

Throughout temperance history, a variety of institutions have been established to "redeem," "reform," "rehabilitate," or "treat" individuals experiencing problems in their relationship with alcohol and/or other drugs. Historically, the combined use of alcohol and other drugs has been very common; institutions established to treat obsessive drinkers rapidly found themselves dealing with habitual users of opium, morphine, heroin, cocaine, and, in later years, a variety of more exotic substances.

Therapeutic Temperance

Pleas from physicians and social reformers for the creation of specialized institutions for the care and control of habitual drunkards came on the heels of a tripling of annual per capita alcohol consumption in the decades following U.S. independence. It was in this context of widespread heavy drinking and related problems that medical leaders such as Dr. Benjamin Rush and Dr. Samuel Woodward conceptualized chronic drunkenness as a disease and called for its treatment.

The first such institutions were established by temperance organizations shortly before and after the Civil War. Although the temperance movement would become increasingly associated with the goal of alcohol prohibition, groups such as the Woman's Christian Temperance Union (WCTU), founded in 1874, typically supported efforts to sober up and rehabilitate obsessive drinkers. Most temperance leaders believed that although prohibition would prevent the creation of drunkards and make treatment measures unnecessary at some point, in the meantime treatment was an important element in the battle against the Demon Rum.

The temperance movement's most important contribution to the history of treatment was a fellowship-based approach drawn from Protestant religious practices (most notably early Methodism). Put simply, "therapeutic temperance," as practiced by the Washingtonians in the 1840s, fraternal temperance societies, and reform clubs, relied on collective measures to exhort drunkards to pledge their abstinence and keep their pledges. Sobriety was to be achieved within a network of like-minded others who provided support and maintained surveillance. Usually, this process was undertaken without the use of any segregation in a formal treatment facility. The affected individuals, typically men, attended temperance meetings and other "elevating" activities held in local temperance halls and tried to limit their social partners to others pledged to abstinence. Visiting committees looked in on recovering people and their families. However, some temperance groups established formal residences for recovering people. Often, these were called "homes" to convey their simultaneously supportive and controlling character.

The first "inebriate homes" based on these principles were established in Boston (1857), San Francisco (1859), and Chicago (1863), with many others following their lead. These homes shared several important characteristics: Their residents were present on a legally voluntary basis rather than treated by force of law; they were private organizations (although some received public funds); they employed recovering people as staff; they were relatively small, housing fewer than fifty residents at a time; and they were located in cities so that family, friends, and the members of temperance fellowships would be available to each resident. Finally, they relied on a very short term of residence, usually just long enough for residents to get through withdrawal symptoms and be restored to reasonable health. The real work of achieving sobriety was to be accomplished in fellowship outside of the institution.

It is impossible to know how effectively such institutions and their related fellowships restored alcoholics to sobriety. For every testimonial to their success there is a condemnation of their methods. The homes' critics focused mainly on the voluntary nature of the treatment and its short duration. They developed an alternate view of treatment derived from institutions for the treatment of people with mental illness. In time, the asylum model prevailed, but inebriate homes never disappeared entirely. Though transformed in significant ways, the philosophy of therapeutic temperance remains influential.

Nineteenth-Century Inebriate Asylums

In some part, the conflict between the supporters and critics of inebriate homes was about understandings of human nature. Although many supporters of inebriate homes were physicians, and most used the language of disease to characterize habitual drunkenness, they emphatically denied that inebriety could be reduced to an involuntary state created by changes in the brain or nervous system. Their logic was religious: Human beings had immortal souls that represented the spirit of God; thus, habitual drunkards had a residual self-control that could never be entirely extinguished. Although treating the physiological symptoms of inebriety, the homes' methods spoke to matters of human purpose and community in ways that were often frankly spiritual.

Their critics tended to be younger physicians trained more rigorously in scientific medicine and enormously influenced by the neurological research emerging from Europe. They took a decidedly material approach to inebriety: It was a disease of the brain and nervous system, often incurable, and always requiring lengthy treatment in settings distinctly segregated from insalubrious influences, including those of friends and family. Asylum enthusiasts had little regard for the methods of therapeutic temperance, and to achieve treatment of the sort they admired, they turned to the model of the insane asylum.

Important reasons quite apart from therapeutic ideology inclined these men to admire the asylum. In the late nineteenth century, public insane asylums—or mental hospitals, as they would begin to be called early in the twentieth century—represented the single largest annual expenditure of state governments. They were grand, castellated affairs, and their superintendents were men of great professional and political power. The Association of Medical Superintendents of American Institutions for the Insane (AMSAII), the forerunner of the American Psychiatric Association, was the model for any professional group seeking power and influence. The promoters of inebriate asylums were attempting to create a new medical specialty, and the AMSAII's success was not lost on them. A specialty needed an institutional base. The American Association for the Cure of Inebriates (AACI) was formed in 1870 to do for medical specialists in inebriety what AMSAII had done for "alienists," physicians now known as psychiatrists.

In addition to the prestige and power associated with the control of public institutions, the inebriety doctors sought financial stability. The cyclical depressions that followed the Civil War caused many inebriate homes to fold, especially those that depended on payments from patients. Early in its career, the AACI took up the cause of creating public inebriate asylums on substantially the same political and financial footing as asylums for the insane.

The asylum model offered another advantage that was both therapeutic and political: The force of legal commitment could be brought to bear most easily on the patients of public institutions designed to provide some measure of secure custody. Legal commitment would permit the lengthy detention of patients, thus allowing the AACI to portray the inebriate asylum as a potential solution to the endemic homelessness of the late nineteenth and early twentieth centuries. Just as the insane asylum had to some extent allowed local poorhouses to transfer the care of the insane to state institutions, inebriate asylum promoters envisioned a similar transfer of tramps and habitual drunkards who turned up in local police courts over and over again. Indeed, as the inebriate asylum idea was elaborated over time, it became two institutions in one: a treatment facility for "recent and hopeful cases," as the asylum rhetoric often put it, and a custodial facility for the castoffs of poorhouses and jails.

The strategy failed. Very few public inebriate asylums were ever opened, and even the best-run and most long-lived example, in Foxborough, Massachusetts, closed with the advent of Prohibition after only twenty-seven years in operation (1893–1920). Ironically, the seed of the public inebriate asylum movement's failure was contained in its therapeutic approach. Although the methods of therapeutic temperance were derided as sentimental and unscientific in an era increasingly enamored of hardheadedness, therapeutic temperance at least had methods appropriate to its philosophy. The asylum approach, in contrast, had no therapeutic methods consistent with its claims about the nature of inebriety. Simple custody, healthy diet, exercise, the routine of institutional work—these were not medical interventions. Moreover, such methods were the stock-in-trade of a variety of institutions that managed inebriates at far less cost. In the end, the inebriate asylum was perceived in most jurisdictions as a costly and redundant enterprise. In Toronto, Ontario, the only North American jurisdiction in which public support for an inebriate asylum was put to a vote (in 1889), it failed in every ward, usually by a wide margin.

Other responses to the treatment of inebriety in the nineteenth century included private, for-profit addiction cure institutes, bottled home cures offered by the same patent medicine industry that was distributing alcohol-, morphine-, and cocaine-laced patent medicines, and religiously oriented urban rescue missions and rural inebriate colonies. The most culturally visible and controversial of the nineteenth-century treatments promised brief, low-cost treatment usually involving some medicinal specific that was promised to destroy all craving for one's pet poison. Most of these cures bore the names of their founding entrepreneurs: Keeley, Neal, Gatlin, Key, and Oppenheimer were among the most prominent.

The New York State Inebriate Asylum at Binghamton (Corbis)

The Influence of the Mental Hygiene Movement

In 1875, the AMSAII grudgingly approved the creation of public inebriate asylums. The superintendents were reluctant to create political competitors, but this was outweighed by their intense desire to rid their institutions of patients whom they bluntly characterized as "nuisances." Indeed, by the 1870s many well-established private mental hospitals had banned the admission of inebriates. Others hoped for the day when their finances would permit them to do the same. But as the years went by and few inebriate asylums materialized, state insane asylums remained the principal sites of public treatment. In large states with several hospitals, it was common for one to be designated mainly for inebriates. From the superintendents' point of view, this concentrated the evil in one location.

In the decade before World War I, however, what we now call "deinstitutionalization" began to take hold in several states under the influence of what is customarily called the "mental hygiene movement." Mental hospitals were scandalously overcrowded, and their therapeutic intent had been, in most places, reduced to professional pieties. Involuntary commitment resulted in many infamous abuses of civil liberties. At the same time, office practice had become a more common method among psychiatrists and neurologists, who devoted themselves increasingly to the treatment of mental distress that fell short of psychosis. In this context, the treatment of what was by now frequently called "alcoholism" was recast, particularly if the patient was employed or had a family to support. In Massachusetts, the Foxborough State Hospital was reorganized in 1908 to emphasize brief, voluntary inpatient treatment combined with systematic aftercare in local outpatient clinics. Here, the older methods of therapeutic temperance were reworked in the service of building a coherent system for the treatment of inebriates that linked hospitals and community care. Although the "deinstitutionalization" of inebriates would not occur until the 1960s and 1970s, the methods employed by Foxborough during its last decade were a striking anticipation of treatment as it developed after World War II as the result of community psychiatry and the rapid growth of Alcoholics Anonymous.

Early and Mid-Twentieth Century Treatment

The number of inebriate homes, inebriate asylums, and private addiction cure institutes diminished dramatically during the first two decades of the twentieth century as physicians and psychiatrists in the United States sought to resolve problems related to alcohol and other drug abuse mainly by prohibiting or aggressively controlling the manufacture and distribution of these drugs.

Four different types of institutions filled the continuing need for treatment: (1) outpatient clinics that utilized recovered alcoholics as lay psychotherapists; (2) private sanatoria and hospitals, such as the Towns Hospital in New York City, that provided detoxification for the affluent; (3) public hospitals that treated narcotic addiction (such as Riverside Hospital in New York City); and (4) outpatient narcotic maintenance clinics, most of which operated only briefly between 1919 and 1924. The brunt of care for the impoverished inebriate fell upon the large public hospitals, the "drunk tanks" of city jails, county work farms—many of which functioned as inebriate colonies—and state psychiatric hospitals.

Beyond private hospitals and sanatoria, there was very little specialized institutional treatment for alcohol and drug addiction during the 1920s and early 1930s. Only California funded a specialized facility (the State Narcotic Hospital at Spadra, 1929–1941) for the treatment of narcotic addiction, and few state-funded alcoholism treatment units existed. This situation began to change in 1935 with the opening of the first of two U.S. Public Health Hospitals for the treatment of narcotic addiction and the founding of Alcoholics Anonymous (AA).

Through much of the 1930s and 1940s, the only addiction treatment facilities were these federal hospitals, located in Lexington, Kentucky, and Fort Worth, Texas. During this period, a growing number of hospitals began to collaborate with AA, however. The first two were Rockland State Hospital, a psychiatric facility in Orangeburg, New York, and Blythewood Sanitarium in Greenwich, Connecticut. To detoxify and stabilize the large number of "late-stage" alcoholics entering the AA fellowship, members pioneered a model of brief detoxification and treatment at St. Thomas Hospital in Akron, Ohio; St. Vincent's Hospital in Cleveland, Ohio; and Knickerbocker Hospital in New York City. So-called "AA wards" spread across the United States in tandem with AA's growth. AA "retreats," "farms," and "rest homes" were also started by AA members to meet the posthospitalization needs of alcoholics. Many of these small institutions, such as Alina Lodge (Kenvil, New Jersey), High Watch Farm (Kent, Connecticut), and Beech Hill Farm (Dublin, New Hampshire), later evolved into formal alcoholism treatment programs. In 1939, having worked with alcoholics along the lines of therapeutic temperance since the 1880s, the Salvation Army opened its first alcoholism treatment facility. The army subsequently became one of the largest providers of alcoholism treatment services in the United States.

During the 1940s, several new models of alcoholism treatment gained prominence. First, an inpatient psychiatric model of addiction treatment was promoted by private psychiatric hospitals such as the Menninger Clinic in Topeka, Kansas. This approach provided medical detoxification and treatment of the primary psychiatric illnesses of which alcoholism was thought to be a symptom. The second approach was an outpatient clinic model pioneered at the Georgian Clinic and Rehabilitation Center for Alcoholics (Atlanta, Georgia), the Yale Plan Clinics (New Haven and Hartford, Connecticut), the Institute of the Pennsylvania Hospital (Philadelphia, Pennsylvania), and Johns Hopkins Hospital (Baltimore, Maryland). These clinics viewed alcoholism psychodynamically as an escape from life's travails and, like the Menninger Clinic, sought to resolve underlying problems. The third approach was a residential model of alcoholism treatment developed within three Minnesota institutions: Pioneer House, Willmar State Hospital, and Hazelden. The major components of the "Minnesota model" included conceptualization of alcoholism as a progressive, primary disease (not as merely symptomatic of other disorders); the use of a multidisciplinary treatment team that incorporated recovered alcoholics as primary counselors (a practice that recalled the era of therapeutic temperance); the infusion of AA philosophy and AA "step work"; the focus on abstinence from all mood-altering drugs; and reliance on continued support from AA following treatment. This approach became closely aligned with a "halfway house" movement in the 1950s that provided a structured transition from institutional treatment to a sustained recovery lifestyle in the community.

Although there were other residential models of alcoholism treatment during this period (notably, Bridge House in New York City, Portal House in Chicago, Brighton Hospital for Alcoholism in Brighton, Michigan), the Minnesota model evolved into the dominant approach in the second half of the twentieth century. A rise in juvenile narcotic addiction led to the reopening of New York City's Riverside Hospital as a juvenile treatment facility and the creation of addiction wards in such hospitals as the Detroit Receiving Hospital, Chicago's Bridewell Hospital, and Bellevue, Kings County, Manhattan General, and Metropolitan hospitals in New York City. Local religious organizations also sponsored new counseling agencies aimed at juvenile addiction. Some of the more notable were St. Mark's Clinic in Chicago, the Addict's Rehabilitation Center in Manhattan, and Exodus House in East Harlem. During this period, many states organized alcoholism treatment units within their state psychiatric hospitals, and a few states organized hospitals that specialized in alcoholism treatment (Blue Hills Hospital in Connecticut and Avon Park in Florida, for example).

Treatment Comes of Age

The National Council on Alcoholism and a joint committee of the American Medical Association and the American Bar Association were at the forefront of advocacy for the expansion of treatment for alcoholism and drug abuse during the 1950s and 1960s. To be successful, this movement needed models of addiction treatment that could be widely replicated. Added to the outpatient clinic, detoxification, and residential treatment and halfway house models were three new approaches to narcotic addiction and "polydrug abuse."

Ex-addict-directed therapeutic communities (TCs), representing a long-term, residential model for the treatment of drug addiction, began with the opening of Synanon in 1958. TCs viewed drug addiction as a problem of immaturity and poor socialization that required a reconstruction of personality and character. By 1975, there were more than 500 TCs in the United States modeled after Synanon in its early form. In 1964, Drs. Vincent Dole and Marie Nyswander conceptualized heroin addiction as a metabolic disease and introduced the daily oral administration of methadone as a means of stabilizing the addict's disordered metabolism so that social rehabilitation could begin. By 1973, more than 80,000 heroin addicts were maintained on methadone in licensed treatment programs in the United States. Growing concerns about youthful alcohol and polydrug use during this same period generated an outpatient clinic model that provided individual, group, and family counseling for young people experiencing problems with drugs other than narcotics. Outpatient drug-free treatment quickly became the most frequently utilized treatment modality in the United States.

Federal support for community-based treatment of alcoholism and other drug addiction increased through the 1960s and culminated in the passage of landmark legislation in the early 1970s. The Comprehensive Alcoholism Prevention and Treatment Act (Hughes Act) of 1970 and the Drug Abuse Treatment Act of 1972 created a federal, state, and local partnership to treat alcoholism, drug addiction, and drug abuse. The major elements of this partnership were two federal institutes (the National Institute on Alcohol Abuse and Alcoholism and the National Institute on Drug Abuse) designated as treatment planning authorities within each U.S. state and territory, and community-based treatment agencies. Together, these partners planned, built, staffed, operated, and evaluated treatment programs across the United States. The remote federal narcotic hospitals and alcoholism wards in state psychiatric hospitals gave way to community-based treatment agencies.

The emerging field of addiction treatment was marked by expansion (from fewer than 200 programs in the 1960s to more than 500 in 1973, 2,400 by 1977, and 6,800 by 1987), increased regulation (development of accreditation and program licensure standards), and professionalization (preparatory training and worker certification/licensure). The field also reorganized itself from what had essentially been two separate fields (one treating alcohol problems, the other treating "drug" problems) to a single field that addressed all alcohol- and other drug-related problems within an integrated framework. This very contentious integration process was nearly complete at the state and local levels by the mid-1980s, leaving in its wake new language such as "chemical dependency" and "substance abuse."

The 1980s witnessed significant growth in for-profit and hospital-based addiction treatment programs and an expansion of programs for special populations of clients: adolescents, women, ethnic and cultural minorities, and those with co-occurring psychiatric illness. The service missions of many treatment institutions also expanded to include early intervention with alcohol- and other drug-impaired employees, students, and drivers.

The growth of residential treatment programs was reversed in the 1990s when ethical concerns about the field's business and clinical practices led to an aggressive scheme of managed behavioral health care that significantly reduced inpatient treatment admissions and lengths of stays. This shift led to the closure of many for-profit and hospital-based treatment programs and, in all programs, a greater emphasis on outpatient, brief therapies.

Current Status of Addiction Treatment Institutions

Some 15,239 institutions participated in the latest (1999–2000) national survey of alcoholism and addiction treatment facilities in the United States, 45 percent of which were concentrated in eight states. This national network of facilities is made up of private nonprofit agencies (60 percent), private for-profit organizations (26 percent), and state/local government-operated facilities (11 percent). In 65 percent of these facilities, treating addiction is the primary organizational mission. Ninety-six percent of the facilities treat both alcohol and other drug-related problems. Types of care provided by these agencies include outpatient rehabilitation services (82 percent of facilities), residential rehabilitation services (25 percent), partial hospitalization (19 percent), outpatient counseling (13 percent), and residential detoxification (5 percent).

Seventy percent of all clients admitted to U.S. treatment institutions are men. The racial/ethnic composition of these clients is 60 percent non-Hispanic white, 25 percent non-Hispanic black, 10 percent Hispanic, and 5 percent other. The primary drug choices of clients being admitted to these facilities are alcohol only (26 percent), alcohol with a secondary drug (20 percent), opiates (16 percent), cocaine (14 percent), marijuana (14 percent), and other stimulants (5 percent). More than 1,200 facilities (8 percent of all facilities) dispense methadone or LAAM (levo-alpha-acetylmethadol) for the treatment of narcotic addiction. Two-thirds of the facilities provide both treatment and prevention services.

Addiction treatment programs in the United States are today funded by a combination of federal, state, and local grants and contracts; public (Medicare and Medicaid) and private health insurance; and client self-payment. Costs of treatment vary widely by modality and by type of provider organization (public versus private). Most programs meet the accreditation standards of the Joint Commission for the Accreditation of Healthcare Organizations or the Council on Accreditation of Rehabilitation Facilities and/or state program licensure standards. The programs are staffed by interdisciplinary teams of physicians, nurses, social workers, counselors, counselor assistants, and outreach workers.

Individuals with alcohol and other drug problems get to these programs by self-referral or referrals from physicians, community service agencies, the courts, employee assistance programs, schools, and alumni and members of recovery support groups such as AA and Narcotics Anonymous. Most treatment consists of a combination of one or more of the following: outreach/engagement services, detoxification; individual, group, and family counseling; pharmacotherapy (for example, methadone, LAAM, antabuse, naltrexone); relapse prevention training; linkage to community mutual aid groups; and a structured program of follow-up counseling. Nearly all addiction treatment programs in the United States provide treatment that is based on the goal of complete abstinence, and the majority provide treatment based on AA's Twelve Steps.

The U.S. government spends more than $3.1 billion per year on addiction treatment and treatment-related research, and more than 1.5 million people each year are admitted to the nation's treatment institutions. Today's field of addiction treatment has achieved partial ownership of the nation's alcohol and other drug problems. The field has attained a high level of professional organization, and it is supported by multiple federal and state addiction-focused agencies. Its interests are promoted by public advocacy organizations (such as the National Council on Alcoholism and Drug Dependence) and numerous trade organizations (for example, the American Society of Addiction Medicine, the National Association of Addiction Treatment Providers, and the National Association of Addiction Treatment Professionals). Major philanthropic foundations (including the Smithers Foundation and the Robert Wood Johnson Foundation) are supporting the field's development.

Jim Baumohl and William White

See also: Alcoholics Anonymous (AA); Alcoholism; Christopher D. Smithers Foundation; Dashaway Association; Hazelden Foundation; Homewood Retreat; Inebriate Institutions (Australia); Keeley Institutes; National Council on Alcoholism and Drug Dependence (NCADD); Patent Medicines; Synanon; Turner, Joseph Edward; Washingtonians

References

Baumohl, J. "Maintaining Orthodoxy: The Depression-Era Struggle over Morphine Maintenance in California" in *Contemporary Drug Problems.* 27 (2000): 17–75.

Baumohl, J., and R. Room. "Inebriety, Doctors, and the State: Alcoholism Treatment Institutions before 1940" in *Recent Developments in Alcoholism,* edited by M. Galanter, vol. 5, 135–174. New York: Plenum, 1987.

Substance Abuse and Mental Health Services Administration. *Treatment Episode Data Sets (TEDS). 1994–1999. National Admission to Substance Abuse Treatment.* SAMHSA, Office of Applied Studies, n.d.

———. *Uniform Facility Data Set (UFDS): 1999 (Data on Substance Abuse Treatment Facilities).* Rockville, MD: SAMHSA, Office of Applied Studies, 2001.

White, W. *Slaying the Dragon: The History of Addiction Treatment and Recovery in America.* Bloomington, IL: Chestnut Health Systems, 1998.

Trotter, Thomas (1760–1832)

Thomas Trotter (1760–1832), a Scottish physician, pioneered a comprehensive approach to the physical, social, and mental components of habitual drunkenness that provided a road map for later investigators. His 200-page *An Essay Medical, Philo-*

sophical, and Chemical on Drunkenness and Its Effects on the Human Body (London, 1804) was reprinted several times (including a U.S. edition in 1813) and translated into German and Swedish. The *Essay,* a much-expanded version of his 1788 M.D. thesis at the University of Edinburgh, was the first book-length analysis of drunkenness by a British doctor. Trotter's explicit claim that the drunkard suffered from a type of mental illness—and therefore physicians, not clergymen, knew the best remedies—was novel for its clarity, but not for its fundamental insights. Trotter's *Essay* offered an eclectic mix of physiological theory, moralizing remarks, quotations from famous poets, and colorful anecdotes but managed to omit any reference to important predecessors in the field.

Trotter believed that alcohol was a stimulant like other narcotics, such as opium and hashish. The torpor that followed was the result of exhaustion from overstimulation. The *Essay* cataloged the harmful effects of alcohol on the body, beginning with redness of the eyes and nose, progressing to inflammation of the stomach, bowels, and liver, and culminating in disorders of the brain ranging from melancholy to outright madness. Trotter argued that alcohol frequently precipitated deaths ascribed to other causes: Drunkenness made accidents more likely, lowered resistance to diseases such as scurvy and tropical fevers, and exacerbated existing medical conditions such as dropsy, apoplexy, hysteria, and epilepsy. Spontaneous combustion was another hazard. Trotter devoted almost thirty pages of the *Essay* to accounts of old sots who were so marinated in liquor that they suddenly burned up, leaving only ashes and a few charred bones behind.

Trotter saw habitual drunkenness as part of a larger pattern of unhealthy indulgence in a civilization gone soft. The never-ending search for new excitements had its origins in childhood; early weaning was followed by a diet of rich foods, tea, and alcoholic beverages. Such luxurious fare predisposed the young to later excesses. Social conventions that encouraged drinking, such as serving wine with meals or following them with cordials, placed temptations in the path of the weak.

Trotter offered no sweeping program for freeing society from the disease of drunkenness, focusing instead on the personal relationship between physician and patient. The watchful doctor could spot the signs of an alcohol problem, reason with the patient, and extract a pledge of immediate and total abstinence. Helping a drunkard on the path to recovery was a tricky business: Patients might break their pledge and then lie about it, and family members could intervene with violent condemnations of the drunkard's vices, provoking despair and suicide rather than speeding the cure. Instead, Trotter prescribed a simple diet and positive thinking under the vigilant care of a physician well schooled in human nature.

Isaac Land

See also: Alcoholism

References

Lloyd, Christopher, and Jack L. S. Coulter. *Medicine and the Navy, 1200–1900.* 3 vols. Edinburgh: Livingstone, 1961.

Trotter, Thomas. *An Essay Medical, Philosophical, and Chemical on Drunkenness and Its Effects on the Human Body.* Edited with an Introduction by Roy Porter. London: Routledge, 1988 [1804].

Turner, Joseph Edward (1822–1889)

Joseph Edward Turner (1822–1889) was, in the words of the title of his one 1888 book, the founder of "the first inebriate asylum in the world." Although others in the United States, such as the Washingtonians and the Dashaways, had established institutions providing care to drunkards based on the belief that drunkenness was a disease, Turner was the first individual to combine the disease concept with the rise of the asylum movement. A tireless and charismatic promoter, Turner opened the New York State Inebriate Asylum at Binghamton in 1864 using both public and private funds. Unfortunately, as inept an administrator as he was zealous in promotion, he was unable to meet setbacks and alienated his own board of trustees. In 1867, the board fired Turner and sold the asylum for $1 to the state, which turned it into a mental institution. Later, Turner would attempt to establish the Women's National Hospital for Inebriates and Opium Eaters in Connecticut but was blocked by his old enemies. He died in 1889 before he could overcome these obstacles.

Joseph Edward Turner was born in Bath, Maine. Growing up with the upheaval surrounding temperance issues that eventually led to the Maine Laws in 1851, Turner became interested in alcohol problems in part from taking care of an uncle prone to drinking sprees. After attending medical lectures in Philadelphia, he became a physician in Paterson, New Jersey. Convinced of his calling to cure inebriates, he traveled throughout Europe gathering information on the subject. Turner's conclusions differed little from those of Benjamin Rush, surgeon general during the American Revolution more than half a century earlier. Like many physicians, Turner insisted that drunkenness had a physiological basis and that drunkards possessed a constitutionally weakened system.

Turner's contribution lay in his combination of this disease concept with an extension of the asylum movement to the disease of inebriety. Before this time, inebriates had been placed in almshouses, jails, workhouses, and lunatic asylums. None of these had worked, and the asylums actually began to refuse drunkards as disturbing to the regular population. Nevertheless, Turner insisted that an asylum-type setting was necessary to keep inebriates away from alcohol. The key to these asylums lay in the total control of a gentle but authoritarian superintendent with specific rules enforced equally for all individuals. Since Turner, like others, believed that no cure could be effected in less than a year, involuntary treatment was preferred. When the Binghamton asylum was opened, Turner demonstrated the strength of his convictions by treating each patient as a potential suicide with an attendant in constant watch, the heavy use of restraint and physical discipline, the prohibition of money and stamps, and a prohibition on speaking to workingmen within the asylum. Also, although he required twice-daily compulsory prayers, Turner refused admittance to ministers and even to families and friends of patients.

Although Turner quickly faded from the scene, his admirers, such as Thomas Trotter, helped found the American Association for the Study and Cure of Inebriety (AASCI) in 1870. The AASCI added the study of inebriety within asylums to Turner's basic model, and reflected the shift to private ownership of almost all the asylums, but otherwise shared Turner's vision.

Mark C. Smith

See also: Alcoholism; American Association for the Study and Cure of Inebriety (AASCI); Dashaway Association; Treatment Institutions; Washingtonians

References

Brown, Edward M. "What Shall We Do with the Inebriate? Asylum Treatment and the Disease Concept of Alcoholism in the Late Nineteenth Century" in *Journal of the History of the Behavioral Sciences.* 21 (January 1983): 48–59.

Lender, Mark Edward. "Jellinek's Typology of Alcoholism: Some Historical Antecedents" in *Journal of Studies on Alcohol.* 40 (1979): 361–375.

Rypiens, Senta. "Joseph Turner and the First Inebriate Asylum" in *Quarterly Journal of Studies on Alcohol.* 10 (1949): 127–134.

Turner, Joweph Edward. *The History of the First Inebriate Asylum in the World.* New York: Arno, 1981 [1888].

White, William L. *Slaying the Dragon: The History of Addiction Treatment and Recovery in America.* Bloomington, IL: Chestnut Health Systems, 1998.

U

United Kingdom, Alcohol Consumption in

See Alcohol, Consumption of (United Kingdom)

United Kingdom Alliance (UKA)

The United Kingdom Alliance (UKA) was the leading prohibitionist pressure group in the United Kingdom throughout the late nineteenth and early twentieth centuries. It supported the policy of local veto that gave local inhabitants the power to vote for their areas to go "dry." It was a well-organized, populist pressure group drawing support from Nonconformist liberals and reached its apogee of influence in the 1880s. It had declined as a political force by the turn of the century.

Policy and History

In 1852, a Manchester Quaker cotton manufacturer, Nathaniel Card, was much impressed by the effect of prohibition in the state of Maine in the United States. Following his initiative, the United Kingdom Alliance for the Suppression of the Traffic in All Intoxicating Liquors was established in 1853 to work for a "Maine Law" in the United Kingdom. This step marked a departure for the British temperance movement, which hitherto had concentrated on preaching teetotalism to individuals. The UKA regarded alcoholic beverages as inherently addictive, like narcotic drugs; hence, the state had a duty to suppress the liquor trade. In the pursuit of this objective, the UKA welcomed nonabstainers to its membership. In 1857, the alliance pursued a strategic shift in policy, pushing for a permissive prohibition bill that would allow a majority of two-thirds of the rate-payers in a locality to suppress the drink trade in their area. This policy later became known as "local veto" or "direct local veto" and was doggedly pursued by the UKA for the next fifty years. After 1870, the UKA gained increasing influence in the Liberal Party, along with other similar pressure groups that appealed to Nonconformist radicals. A resolution was passed in the House of Commons in 1883 favoring local option, and in 1892–1895 the Liberal government introduced local-veto bills but failed to press them through with any vigor. Local veto for Scotland was passed in 1913, coming into force in 1920, although with disappointing results.

The UKA proved singularly uninterested in other reforms of the licensing laws and found it difficult to work with more moderate reformers. The alliance saw schemes for reduction in numbers of licenses or for broader forms of popular control as blocking the way for local veto. By 1895, most politicians, including the Liberal leadership, increasingly regarded the UKA as an obstructive force. The growing interest in ideas for more moderate reforms, as well noncommercial, disinterested public control of the drink trade, also meant that after 1900 the UKA could no longer control the discourse of public debate.

During World War I, in a mood where drinking was perceived as harming the nation's war effort and national efficiency, national prohibition came briefly on the political agenda and was promoted by the UKA. However, the work of the Central Control Board (Liquor Traffic) in restricting drink sales weakened such moves. The UKA took encouragement from the advance of prohibition in the United States, but in the United Kingdom both the cause and the UKA itself entered a sharp decline after 1919 as the Nonconformists, who had been the core of the support for prohibition, dwindled in numbers. An expensive libel suit about medicated wines also weakened the UKA. After World War II, the UKA concentrated more on educational work concerning alcohol than on a legislative agenda and cooperated closely with the Christian Economic and Social Research Foundation.

Political Strategy and Ethos

The UKA self-consciously modeled itself on the Anti–Corn Law League, which had succeeded in forcing the repeal of the Corn Laws in 1846. UKA leaders believed that political progress came "from below." According to the UKA, there were three stages in any political struggle: the first was agitational, the second parliamentary, and the third governmental. The political class only acted in the interests of the people when coerced by the organized force of public opinion. The UKA saw itself as a democratic force championing the cause of the people, who yearned to be set free from the thralls of liquordom.

Another feature of the strategy was that the UKA consistently claimed to be above party battles and refused to endorse any political group or party. Membership in the alliance was open to men and women of all creeds and doctrines. These ideas helped the UKA in its formative years, when there was a rising tide of popular radicalism directed against the narrow governing elite. However, they were more of a handicap after 1870, as party polarization on the drink issue developed. More comfortable as a provincial force on the outside of politics, the UKA found it difficult to offer constructive support to those

Liberal politicians who were prepared, for their own ends, to co-operate.

Support

Membership was open to all who paid a shilling, and the UKA retained a mass membership throughout the nineteenth century. In 1856, there were 30,000 members. The annual income ran consistently at around £15,000. The bedrock of support came from the Nonconformist middle class in the north of England. In the early years, support and income came from wealthy manufacturers, but this funding fell off after 1870. In addition, a few prominent figures that graced the alliance's platforms: for example, Cardinal Henry Edward Manning, F. W. Newman, Samuel Pope, and Lady Rosalind Carlisle. T. H. Green, the liberal political philosopher, was another supporter. The UKA's parliamentary champion was a radical member of the landed gentry, Sir Wilfrid Lawson, who championed local veto with wit and lifelong commitment. He remained, however, somewhat aloof from the organization. The UKA never succeeded in attracting any political heavyweights such as John Bright or Joseph Chamberlain who possessed a clearer grasp of political realities.

Organization and Campaigning

The UKA was well organized with a central executive of around twenty-five members. There was a strong sense of collective responsibility. For most of its existence, the alliance was based in Manchester. After 1880, when the policy of political neutrality came under strain, there were tensions, particularly with the small number of local "auxiliary" branches that had been established in the larger cities. There were also divisions over policy after 1899, following the report of the Royal Commission on Liquor Licensing Laws. The organization employed district agents who did much to further the electoral and campaigning work of the UKA during the nineteenth century. There were close links with a number of other temperance bodies, particularly prohibitionist or Sunday closing organizations operating in Scotland, Ireland, and Wales and other temperance groups concentrating on nonpolitical aspects. The staple methods of campaigning were the production of pamphlets and tracts, a newspaper called the *Alliance News*, public meetings, and, at times of major legislation, large rallies in Hyde Park or elsewhere.

Many UKA officials were long serving. For instance, Thomas Holliday Barker was secretary from 1853 to 1884; James Whyte, 1884–1903; William Williams, 1903–1909; and George Bailey Wilson, 1909–1919. Henry Septimus Sutton edited the *Alliance News* from 1854 to 1898; Robert Arthur Jameson, 1898–1914; and George Bailey Wilson, 1914–1943.

In 1942, the United Kingdom Temperance Alliance, an educational organization, was formed and, in effect, superseded the UKA, a political organization that had become irrelevant in the mid-twentieth century. UKA records can be found at Alliance House in London, the headquarters of the United Kingdom Temperance Alliance.

John Greenaway

See also: Carlisle, Countess of (Rosalind Howard); Carlisle System; Chamberlain, Joseph; Gothenburg Schemes/Disinterested Management; Hayler, Guy; Lawson, Sir Wilfrid; Liberal Party (United Kingdom); Local Option; Maine Law; Manning, Henry Edward; National Temperance League (NTL); Nonconformists (United Kingdom); Pope, Samuel; United Kingdom Temperance Alliance (UKTA)

References

Dingle, A. E. *The Campaign for Prohibition in Victorian England: The United Kingdom Alliance, 1872–1895.* London: Croom Helm, 1980.

Hamer, D. A. *The Politics of Electoral Pressure: A Study in the History of Victorian Reform Agitations.* Hassocks: Harvester Press, 1977.

Harrison, Brian. *Drink and the Victorians: The Temperance Question in England, 1815–1872.* London: Faber & Faber, 1971; 2d ed., Keele: Keele University Press, 1994.

Hayler, Mark H.C. *The Vision of a Century, 1853–1953: The United Kingdom Alliance in Historical Retrospect.* London: United Kingdom Alliance, 1953.

United Kingdom Temperance Alliance (UKTA)

The United Kingdom Temperance Alliance (UKTA) is an educational charity founded by leaders of the United Kingdom Alliance (UKA). In 1937, Robert Wilson Black outlined to the UKA executive a scheme to ensure the alliance's future financial viability and organizational vitality by constructing a headquarters building that would provide a secure rental income. Wilson Black's six-story commercial office block in Westminster was built at a cost of around £75,000. Shortly before his death in 1952, he was able to announce that the mortgage for Alliance House had been paid off. In 2002, the building was valued at around £6 million.

Alliance House was officially opened on 26 November 1938. Nine months later, Britain entered World War II, and London soon was to experience the Blitz. Despite the trying circumstances, the building by 1941 was fully occupied, producing an income of some £10,000 per annum. Wartime conditions, with a hefty burden of income tax, motivated the executive in 1941 to seek the status of an incorporated charitable society. As an educational charity, the United Kingdom Temperance Alliance Limited would not pay taxes. The primary objective of the new body, according to UKA executive minutes (24 April 1941), was "to acquire and take over as a going concern and carry on the educational and charitable work of temperance reform hitherto carried on by the present unincorporated body called or known by the name of 'The United Kingdom Alliance' with the exception of the political activities therefor." The Board of Trade certified the new constitution in February 1942.

Twenty years later, the need to separate the educational work from overt political lobbying had a critical impact on the relationship between the two bodies. From the outset, the UKTA was independent of the UKA; it was not possible to include in the UKTA's articles a clause that would have given the UKA power to nominate a majority of the members of the new body. Such a clause would have deprived the new body of its autonomy. Relationships between the two bodies became strained when, in the mid-1960s, the Charity Commission cautioned charities against involvement in political activities. The two bodies drew further and further apart until 1985, when the UKA ceased occupying an Alliance House office.

Without the United Kingdom Temperance Alliance, the voice of temperance would have been lost in the United Kingdom. Postwar Britain saw little need for the temperance movement. In 1951, the government refused to pay for a representative to a World Health Organization meeting on alcoholism on the grounds that the problem did not exist in Britain. This claim was in one sense a tribute to the work of the temperance movement; however, it also revealed the temperance movement's decline. The spirit of the 1960s was freedom, summed up by Sir Jocelyn Simon's peroration at the final reading of the Liquor Licensing Act of 1962: "Now is the time to dance. Now is the time to stamp the floor with the feet of freedom."

Against this backdrop, the UKTA enabled temperance leaders, through a variety of initiatives, to meet the challenges presented by the alcohol problem in the second half of the twentieth century. The UKTA has been responsible for such initiatives as, in the 1950s, the establishment of the Economic Research Council (which in 1966 became the Christian Economic Research Foundation) and, in 1969, the Teachers Advisory Council on Alcohol and Drug Education (TACADE). The UKTA was largely responsible for financing both of these bodies until 1983. At that time, the UKTA decided to establish and fund the work of the Institute of Alcohol Studies. From 1985 to 1990, the UKTA also sponsored an alcohol-free educational campaign in cooperation with the National Union of Students. In addition, the UKTA has been responsible for a wide range of publications, such as the *Alliance News.*

In May 2003, the UKTA was restructured as the Alliance House Foundation. It is home to a large temperance history library that makes available materials deposited by various British temperance societies. UKA and UKTA executive minutes are available, as well as past issues of the *Alliance News.*

Derek Rutherford

See also: Heath, H. Cecil; Institute of Alcohol Studies (IAS); Rutherford, Derek; United Kingdom Alliance (UKA); Wilson, George Bailey; Wilson Black, Robert

Reference

Hayler, Mark H.C. *The Vision of a Century, 1853–1953: The United Kingdom Alliance in Historical Retrospect.* London: United Kingdom Alliance, 1953.

United States, Alcohol Consumption in

See Alcohol, Consumption of, by Indices (United States); Alcohol, Consumption of, per Capita (United States).

United States Brewers' Association (USBA)

The United States Brewers' Association (USBA) was the trade organization representing the majority of U.S. brewers until the adoption of the Eighteenth Amendment. Combining the resources of its members, it vigorously opposed prohibition at the state and local levels in the nineteenth century and fought a heated but eventually fruitless battle against National Prohibition.

Origin and Growth

The Internal Revenue Act of 1862 levied new taxes on beverage alcohol. In addition to licensing fees, the law imposed a $1 per barrel tax on beer and ale and a $0.20 per gallon tax on distilled spirits. In response, New York brewers formed a local association and called for a national meeting of the brewing industry. The organization did not adopt the name United States Brewers' Association until two years later, but the meeting, conducted primarily in German in November 1862, established the structure of the organization. Member brewers paid dues ranging from $25 to $1,000 based on production. This stratification helped to ensure the dominant influence within the association of larger producers.

The USBA was remarkably successful in its early years. Very quickly, industry representatives established close relations with government officials. The association secretary became a ubiquitous lobbying presence in Washington. In 1865, the U.S. commissioner of internal revenue addressed the USBA's annual meeting, promising cooperation with the brewers. These efforts paid off. In 1863, Congress reduced the tax on beer to $0.60 per barrel and imposed a tariff on imported beer to benefit U.S. producers. The liquor distillers, who did not yet have a national organization, did not fare so well. In 1864, the tax on distilled liquor was raised to $2 per gallon.

The rise of the USBA coincided with an increase in the consumption of beer in the United States. In 1870, U.S. brewers produced more than 200 million U.S. gallons (757,000 hectoliters) of beer. Ten years later, they produced half a billion U.S. gallons (1.9 million hectoliters) in one year. However, these decades also witnessed an increase in prohibition sentiment. The USBA and affiliated state brewers' associations vigorously opposed any threat to their livelihood. Brewers pressured or bribed local politicians to oppose prohibition measures where they were being pushed forward and to avoid too scrupulously enforcing them where the prohibitionists had succeeded. They also turned to the courts for relief. The USBA funded suits and appeals on behalf of brewers in a number of states to question the constitutionality of prohibition laws.

In 1881, the organization formulated a strategy of challenging prohibition laws in the U.S. Supreme Court, arguing that prohibition violated the due process clause of the Fourteenth Amendment by depriving brewers of property without due compensation, an issue that the Court earlier had indicated it would be willing to consider. The trials and appeals through the courts stretched on for six years, and when the Court handed down its decision in *Mugler v. Kansas* and *Ziebold v. Kansas* in 1887, the result was a defeat for the brewers. States could prohibit the sale of alcohol and did not have to compensate producers or sellers. The USBA would have to find relief in the political arena.

Political Struggle and Defeat

The political landscape was quickly changing, however. In addition to increasing efforts in the states to enact prohibition, the brewers perceived a threat in the growing demands for women's suffrage. In large part owing to the efforts of the Woman's Christian Temperance Union (WCTU), female political activism was linked to prohibition. USBA Secretary Hugh F. Cox warned brewers in 1913 that women's suffrage threatened their livelihood if they did not work harder to clean up the saloons. The more direct response was to fight women's access to the ballot. The USBA secretly provided financial support for antisuffrage campaigns at the state level, particularly in the western states, where the saloon had been a target of female anger.

Opposing female suffrage would not be sufficient to stem the rising tide of prohibition. After the formation of the Anti-Saloon League of America, the USBA faced a determined and well-organized foe whose members already could vote. Working covertly with sympathetic local organizers, the USBA financed "Personal Liberty Leagues" to oppose local-option and state prohibition campaigns. Efforts to bribe local and state officials and newspaper editors continued, often with quite discouraging results when zealous prosecutors uncovered the practices. In Pennsylvania, a federal grand jury indicted the USBA and a number of other industry-related organizations and businesses on charges of conspiracy after it was discovered that county commissioners had been bribed to provide lists of registered voters to the brewers. State and local officials had been paid according to their rank and influence, and poll workers routinely received stipends to encourage "wet" voters. USBA officials pled guilty and were assessed a stiff fine. The USBA was also implicated in scandals in New York and Texas. In the Texas case, seven brewers paid $281,000 in fines plus court costs after their extensive efforts to influence the outcome of local-option elections were revealed. The Anti-Saloon League later published the subpoenaed documents as a two-volume campaign handbook.

The USBA attempted to blacklist businesses that they felt were hostile to their interests. Brewers were asked to avoid doing business with the Pennsylvania Railroad and with U.S. Steel because of their strict sobriety rules for employees. Other firms were blacklisted because of suspected contributions to the Anti-Saloon League. H. J. Heinz was on the list because he served as president of the Pennsylvania Sunday School Association.

In 1913, the USBA levied a special assessment of $0.03 per barrel of production on members to finance its campaign against prohibition. In addition to funding questionable payments to politicians and journalists, these contributions were used in a massive campaign publishing effort. In 1915, the brewers underwrote the production and distribution of nearly half a billion pages of campaign materials, including the *Anti-Prohibition Handbook—A Textbook of True Temperance,* intended to help the association's local allies. In an unusual move, the USBA joined forces with the United Brewery Workers in 1914 to form state Trade Union Liberty Leagues. Within two years, there were leagues in nineteen states.

Despite these enormous efforts, the USBA failed to stop National Prohibition. Observers were appalled by the organization's tactics and blatant disregard for the law. The brewers were frequently outmaneuvered by the more disciplined Anti-Saloon League. In addition, divisions within the USBA between wealthy brewers and smaller operators over the burden of financing the campaigns caused problems. The USBA never was able to work effectively with the distilled liquor producers and dealers to combat their common enemy. Efforts to modernize and streamline the association's structure were haphazard and came too late to be effective. The organization struggled for its existence throughout Prohibition, dividing into competing associations after repeal. Renamed the Brewing Industry Foundation in 1942, it later rejoined its rivals to form the United States Brewing Foundation in 1944. In 1961, the organization again adopted its original name.

In 1970, under the leadership of President Henry B. King, the USBA moved its headquarters from New York to Washington, D.C. During that decade, membership dropped off as major producers withdrew their support from the organization, pursuing lobbying efforts on their own. In the mid-1980s, the remaining members reconciled with the former members and agreed to form a new trade organization. In 1986, the USBA dissolved and was replaced by the Beer Institute. Donald Shay, the last president of the USBA, became the first president of the new association. The Beer Institute eliminated the state offices of its predecessor and focused efforts on national policy. Representing both domestic producers and importers, the Beer Institute continues its activities of information gathering, education, and lobbying on behalf of the brewing industry.

James D. Ivy

See also: Anti-Saloon League of America (ASLA); Beer; Federal Regulation of Alcohol before 1920 (United States); Labor in the U.S. Liquor Industry; Lauer, Frederick; Personal Liberty League; Taxation of Liquor (United States)

References

Asbury, Herbert. *The Great Illusion: An Informal History of Prohibition.* New York: Greenwood, 1968.

Baron, Stanley. *Brewed in America: A History of Beer and Ale in the United States.* Boston: Little, Brown, 1962.

Cochran, Thomas C. *The Pabst Brewing Company: The History of an American Business.* New York: New York University Press, 1948.

Downard, William L. *The Cincinnati Brewing Industry: A Social and Economic History.* Athens: Ohio University Press, 1973.

Gordon, Ernest. *When the Brewer Had the Stranglehold.* New York: Alcohol Information Committee, 1930.

Hamm, Richard F. *Shaping the Eighteenth Amendment: Temperance Reform, Legal Culture, and the Polity, 1880–1920.* Chapel Hill: University of North Carolina Press, 1995.

Kerr, K. Austin. *Organized for Prohibition: A New History of the Anti-Saloon League.* New Haven: Yale University Press, 1985.

Mittelman, Amy H. "The Politics of Alcohol Production: The Liquor Industry and the Federal Government, 1862–1900" (Ph.D. dissertation, Columbia University, 1986).

Rose, Kenneth D. *American Women and the Repeal of Prohibition.* New York: New York University Press, 1996.

V

Vegetarianism and the Anti-Tobacco Movement (United Kingdom)

In Victorian Britain vegetarianism was a logical progression for many extreme teetotalers, who saw it as ultra-temperance. Having rejected alcoholic drink, they found similar physiological, social, and moral reasons for rejecting meat. For instance, the temperance authority Dr. F. R. Lees, in his prize essay *An Argument on Behalf of the Primitive Diet of Man* (1857), asserted that the wisest men will seek the ideal practices, vegetarianism as well as abstinence from alcoholic drink.

Teetotalism attracted perhaps a million adults in Britain in the 1860s. In contrast, vegetarianism attracted at most in the Victorian period as a whole some 6,000 men and women and their children. They were organized in the Vegetarian Society (founded in 1847) and in a network of often short-lived metropolitan and provincial societies. Several journals, such as the *Vegetarian Messenger,* published in Manchester, and an important journal of the late Victorian movement, *The Vegetarian,* published in London, acted as organs for the movement. The temperance affiliations of rank-and-file vegetarians are obvious. Few vegetarians were not also teetotal.

Vegetarianism's relations with the temperance movement were strained at times. Early teetotalers were expected to eat meat, and vegetarianism was attacked for bringing ridicule to the temperance cause. Silence on the subject of vegetarianism in Samuel Couling's (1862) early history of the British temperance movement probably indicates his embarrassment about the association. His entry for the Vegetarian Society's first president, James Simpson (1812–1859), ignored Simpson's vegetarianism. The temperance historian P. T. Winskill felt required to emphasize that vegetarianism was by no means a majority choice for teetotalers despite its espousal by many teetotal leaders. In spite of this wariness, the temperance press understood that vegetarianism, along with other concerns, such as opposition to capital punishment, was a useful topic to attract a wider reformist readership.

One early, influential examination of vegetarianism and teetotalism, an (unsigned) article by Samuel Brown in the *Westminster Review,* placed these reforms in the context of a "physical puritanism" that also included homeopathy, hydropathy, and mesmerism. According to Brown, vegetarianism was a physiological heresy. Yet he identified vegetarianism only in part with people whom he condemned as fanatics: supporters of Cowherdism (a vegetarian Swedenborgian sect, based in the northern English industrial centers of Manchester and Salford, with an offshoot in Philadelphia) and mystics such as J. P. Greaves (1777–1842) and his disciples. Brown acknowledged that vegetarianism was also spreading among phrenologists, natural-religionists, and general reformers (Anon. [Brown] 1852).

Alcohol-related concerns were important motives for vegetarians. Supposedly, vegetarianism protected against alcoholism and diseases such as cholera and tuberculosis. Vegetarians called for cooperation and consistency from teetotalers in the same way that they asked for support from societies opposed to vivisection and animal cruelty. Vegetarian connections with temperance and a range of other physiological, moral, and social reforms continued throughout the period, as recognized by the London correspondent of a provincial newspaper, the *Morpeth Herald:* "I fear most of them, if not all, are water-drinkers. I suspect few of them smoke. I am penetrated with the conviction that all of them are connected with the Spelling Reform or some other dismal association for the improvement of humanity" (cited in *Herald of Health* 1881).

Many vegetarians were involved in the still smaller antitobacco movement. Like teetotalers and vegetarians, opponents of tobacco marshaled a broad array of medical, economic, social, and moral arguments. Smoking was denounced as the source of much of profligacy, penury, and crime. Physiologically, it was claimed to cause racial degeneracy, impotence, insanity, and paralysis. (Only after World War II did the association between tobacco and cancer become generally accepted.) Few medical professionals supported the movement. Although the tobacco question gained coverage in journals such as *Lancet* and the *British Medical Journal,* most physicians and popular medical books did not condemn the use of tobacco. Tobacco continued to figure in folk medicine.

Nonconformists and temperance reformers were prominent among antitobacco activists. Very few women were publicly identified with the antitobacco movement. Most teetotalers were not automatically opposed to tobacco, but the Good Templars applied an antismoking pledge to their juvenile auxiliary, and some Band of Hope branches prohibited tobacco. The Rechabite order opposed smoking at public meetings. Antitobacco literature was distributed at the annual Crystal Palace temperance fetes. Important temperance opponents of tobacco included Joseph Livesey (1794–1884), Peter Spence (1807–1883), and F. W. Newman (1805–1897). Some nonteetotalers also opposed smoking, such as Albert, prince consort of Queen Victoria from 1840 to 1861, W. E. Gladstone, and John Ruskin.

An Anti-Nicotine Society was established in 1838 at Congleton, Cheshire, by the temperance worker and future vegetarian William Horsell (1807–1863). Thomas Cook of Leicester (1808–1892), pioneer of working- and middle-class tourism, published the *Anti-Smoker and Progressive Temperance Reformer* in 1841 to 1842. In 1844, there was also a National Philanthropic Anti-Tobacco Association based in London. This association, like Cook's journal and Horsell's society, were short lived, but two national societies proved more durable. The Anti-Tobacco Society, established in London in 1853, had some illustrious supporters, including the bishop of Ripon, a physician to the queen, and temperance leaders such as Samuel Morley (1809–1886) and Sir Walter Trevelyan (1797–1879). Its most important personality was its secretary, Thomas Reynolds (d. 1872), who also edited the *Anti-Tobacco Journal,* which was continued after his death in 1872 by his daughter Frances Emma Reynolds (c. 1825–1900). There were a few local branches, and a parallel Scottish Anti-Tobacco Society was established in 1869. The second national society began in November 1867 as a local organization, the Manchester and Salford Anti-Tobacco Society. However, it grew in national support, changed its name to the English Anti-Tobacco Society in 1872, produced pamphlets, and published an annual letter to the newspapers. It became affiliated with the Anti-Narcotic League, which promoted antitobaccoism in its journal *Beacon Light* (a monthly from 1896).

In the Victorian period, the British antitobacco movement failed to secure any legislation to regulate the sale of tobacco. The passage in 1908 of the Children's Act, which made it an offense to sell cigarettes to children under the age of sixteen and banned juvenile smoking in public, was mainly due to the efforts of unrelated figures such as Winston Churchill and Robert Baden Powell, who supported an Anti-Cigarette League. Cigarettes were controversial even among cigar and pipe smokers.

After having started on the fringe of nineteenth-century "physical puritanism," the vegetarian and antitobacco movements during the second half of the twentieth century became large-scale and even mainstream campaigns in Britain and other Western countries. By the interwar period, British vegetarianism was associated with pacifism, various internationalist movements (such as Esperanto, an invented language meant to supersede national ones), and natural healing. An amalgamation of earlier organizations created the Vegetarian Society of the United Kingdom Limited in 1969. This registered charity has its own periodical, cooking school, and research center; licenses its logo for an increasing range of vegetarian products; and is affiliated with international and European vegetarian unions. The former dates from 1908, while the latter was founded much more recently, in 1985. The popularity of contemporary vegetarianism had been strengthened by concern over animal welfare, fears about the health risks of consuming meat (most recently, anxieties stimulated by "mad cow disease," bovine spongiform encephalopathy, first identified in 1986), and the rise of environmentalism that questions the efficiency of providing meat-based diets. It has been estimated that 5 percent of British households include at least one vegetarian, with females slightly more likely than males to abstain from meat and people in the middle class more likely than working-class people to follow vegetarian diets. There also are regional variations. For instance, support for vegetarianism is stronger in England than in Northern Ireland. In addition to several million vegetarians, many other British people who eat meat have curtailed their consumption of it. More radical vegetarians (vegans) oppose the eating of animal-derived products such as eggs and milk as well as meat. The Vegan Society, founded in Britain in 1944, has promoted veganism with increasing success.

Cigarettes became popular in Britain during World War I and were consumed in increasing quantity in the first half of the twentieth century. At the end of World War II, 65 percent of adult men and 41 percent of adult women smoked cigarettes. Yet medical research increasingly revealed the health risks: nicotine addiction, cancers, and vascular and respiratory diseases. In 1954, a British health minister acknowledged a causal link between lung cancer and smoking. Recently, there have been worries over "passive smoking," nonsmokers unwillingly breathing other people's smoke. Government funds began to support propaganda organizations such as ASH (Action on Smoking and Health, established in 1971). Government has tightened the regulation of the retail sale of cigarettes and restricted advertising and smoking in public places. Although smoking rates have declined considerably, the reduction has varied by social class, with manual and unskilled workers smoking the most. Tobacco in the twenty-first century, like alcoholic drink in the nineteenth century, has become an important concern of public policy.

James Gregory

See also: Barker, Thomas Halliday; Churchill, Winston; Gladstone, William Ewart; Good Templars (IOGT); Hills, Arnold Frank; Hoyle, William; Lees, Frederic Richard; Livesey, Joseph; Rechabite Friendly Society; Winskill, Peter Turner

References

Anonymous. [W. Brown]. "Physical Puritanism" in *Westminster Review.* (1852): 405–442.

The Anti-Tobacco Journal. London. 1858–1900.

Antrobus, Derek. *A Guiltless Feast: The Salford Bible-Christian Church and the Rise of the Modern Vegetarian Movement.* Salford: Salford City Council, 1997.

Couling, Samuel. *History of the Temperance Movement in Great Britain and Ireland from the Earliest Date to the Present Time, with Biographical Notices of Departed Temperance Worthies.* London: W. Tweedie, 1862.

Forward, C. W. *Fifty Years of Food Reform. A History of the Vegetarian Movement in England.* London: Ideal Publishing Union, 1898.

Gregory, James. "The Vegetarian Movement in Victorian Britain and Its Relationship with Temperance" in *Social History of Alcohol Review.* 16, nos. 1–4 (Fall 2001–Spring 2002): 14–34.

The Herald of Health. (April 1881): 41.

Hilton, Matthew. *Perfect Pleasures. Smoking in British Popular Culture, 1800–2000.* Manchester: Manchester University Press, 2000.

Hilton, Matthew, and Simon Nightingale. "A Microbe of the Devil's Own Make: Religion and Science in the British Anti-Tobacco Movement, 1853–1908" in *Ashes to Ashes: The History of Smoking and Health,* edited by S. Lock, L. A. Reynolds, and E. M. Tansey. Amsterdam: Rodopi, 1998.

House of Commons. "Second Report of the Health Committee, 2000." Session 1999–2000, (5 June 2000.)

Spencer, Colin. *The Heretic's Feast: A History of Vegetarianism.* London: Fourth Estate, 1993.

Twigg, J. M. "The Vegetarian Movement in England from 1847–1981: A Study of the Structure of Its Ideology" (Ph.D. dissertation, London School of Economics, 1982).

Walker, R. B. "Medical Aspects of Tobacco Smoking and the Anti-Tobacco Movement in Britain in the Nineteenth Century" in *Medical History.* 24 (1980): 391–402.

Victoria Wine Company

Victoria Wine Company was the name given to the oldest British chain of specialized wine and spirits stores. It was established in 1865 by William Winch Hughes, whose first store was located in Mark Lane in the City of London. Within five years, Hughes had opened nine other stores, most of them in London's East End. Ten years later, there were sixty-three, and when Hughes died in 1886 there were ninety-eight stores in locations as far apart and as nonmetropolitan as Birmingham, Brighton, and Bristol. Hughes, who dealt in ales, tea, and later tobacco as well as wines and spirits, popularized wine consumption at a time when the British wine trade, highly personalized, was catering largely to the aristocracy, the prosperous middle classes, and traditional institutions. The secret of his success, he maintained, was a policy of cheap prices for cash sales, small profits, and quick returns. He concentrated on customers with known preferences who had no servants and no cellars. He advertised in local newspapers but also reached more affluent customers by advertising in the *Illustrated London News* and by maintaining wine lists of considerable range. His obituaries paid tribute to his Victorian values: He was said to have offered a striking example of what "untiring energy and unswerving honesty of purpose" might achieve (Ridley 1886).

Hughes's chain of stores was very much a family business, and when he died, leaving no male heir, it was not a subordinate, but his widow, Emma Susan Durrant Hughes, who succeeded him. She married again and lived until 1911. Six years before her death, she appointed as her sole manager Frank Wood, who had joined the firm in the 1870s, and it was he who after buyouts became sole proprietor of the enterprise upon her death. Unlike Hughes, Wood purchased properties for his stores rather than leasing them, and in 1920 he converted the business into a private limited liability company. When he died in 1921, his widow took charge, but in 1924 the business was acquired by Charles Edward Cottier, who dealt in bread as well as in wines and spirits. Cottier converted the business into a public company. Its 106 freehold and leasehold properties were then valued at £329,547 and its initial loan and share capital was £600,000. "Our Branches Are Your Wine Cellar" was its first motto. Some of the branches were managed by women. In addition to a warehousing department and a small office staff, there were seven branch supervisors, those in London traveling by bicycle.

The pattern of the wine and spirits lists did not change significantly between 1886 and 1939, although exclusive rights were acquired in 1911 in a sweet Dutch liqueur, Advocaat, and in 1939 cocktails headed the lists. What did change—and continued to change—was the structure and control of the business. In 1929, one year after Cottier's death, a brewery firm that recently had become a public company, Taylor, Walker and Company, became the major shareholder in the business (54.6 percent) after acquiring his shares. Following the difficult years of the Great Depression, priority was given to the modernization of warehousing, transport, and premises. There were also vigorous and expensive advertising campaigns. In 1937, Commander Redmond McGrath, a businessman with wide social contacts, became managing director as well as chairman, a post he retained until 1958 when the wine and spirits business was reviving after the restraints and tribulations of World War II.

A year later, the firm of Taylor, Walker and Company was taken over by Ind Coope, a large brewery concern with interests in wines and spirits. In 1961, it merged into a large breweries group that came to be called Allied Breweries. Through these moves, Victoria Wine, which retained its identity, acquired a new structure with four regional divisions—London, Northern, Western, and Scottish. The wine and spirits subsidiary of Ind Coope, B. Grant and Company was converted under Allied control into Grants of St. James, which dealt in wine and spirits, largely a wholesale company but partly retail. It sold to some of Victoria Wine's competitors. The decentralized structure of Allied Breweries encouraged both businesses to expand, and by 1965 Victoria Wine, through a series of local mergers with other wine chains, had as many as 630 high-street shops. Ten years later, the number peaked at 990. Meanwhile, Allied merged in 1968 with Showerings, West of England producers of British drinks, including Babycham, a popular pear-based sparkling wine. Allied also took over an old traditional Bristol wine firm, Harveys, that specialized in sherries, which had been founded long before Victoria Wine in 1796. Other mergers followed, the first of them in 1978 when Allied merged with J. Lyons and Company, famous for its cakes and tearooms, and changed its name to Allied-Lyons.

Social change favored Victoria Wine as its management structure and style changed. British consumption of wine increased, and the media grew more interested in the range and quality of wines on offer. Between 1967 and 1980, sales rose each year, increasing a total of tenfold. At the same time, the company faced increasing competition from supermarkets, which, guided by skilled wine advisers, challenged the liquor shops through clearly marked prices, a competitive range of products, and quality. Tempting displays and self-service arrangements attracted new customers. So, too, did ease of access to private transportation and the convenience of buying wine and spirits along with other purchases.

High-street competition itself increased during the 1990s, and although Victoria Wine and a leading rival, Threshers, both became part of an Allied Domecq Whitbread group, all its wine chains, including Wine Rack, were disposed of to the Japanese investment bank Nomura International in October 2000. Under the name First Quench, they were incurring losses of £30 million a year. A new manager appointed three months later, who had run Burger King's operations in Europe, the Middle East, and Africa, described the business as he found it as "a legacy business . . . crying out for change" (*Times,* 26 March 2002). The days of Victoria Wine seemed almost over.

Asa (Lord) Briggs

See also: Allied Breweries; Wine

Reference

Briggs, Asa. *Wine for Sale: Victoria Wine and the Liquor Trade, 1860–1984.* London: Batesford; Chicago: University of Chicago Press, 1985.

Ridley and Company's Monthly Wine and Spirit Trade Circular. 11 September 1886.

Vodka

Vodka is a colorless, odorless, distilled beverage of common usage in Russia. Although there are Polish and Scandinavian vodkas, the term is believed to be derived from the Russian word for water *(voda)* with an attached diminutive suffix *(-ka);* the literal translation is "little water." The term is etymologically similar to other European descriptions of strong, clear grain liquors such as *eau de vie,* and the Scandinavian *aquavit* ("living water" or "water of life") but the exact date of its appearance in Russian territories and its provenance are continuing subjects of debate. Before the fourteenth century, Russians consumed mead *(mёd),* beer *(pivo),* ale *(braga; ol or olus),* wine *(vinogradnoe vino),* and *kvass*—a lightly alcoholic beverage made, variously, by the fermentation of oatmeal, malted barley, wheat, rye, wheat or buckwheat flour, or pastry, bread, or rusks. For centuries after its introduction, vodka was simply referred to in the literature as *vin;* it was not until the nineteenth century that the use of the term *vodka* became fairly widespread in the literature; thus, the historian must be careful to distinguish it from fermented wine, or *vinogradnoe vino.*

Origins

Vodka's point of origin is a problematic and contentious issue. Some historians have argued that it was introduced into the territories now occupied by the Russian Republic and the Ukraine after 1240 C.E. by the Livonian knights. Others point to the introduction of alcohol discovered by Muslims during the late medieval period via a southwesterly route through Tartar territories by Genoese merchants operating out of their colony at Cafta, and its spread northward and eastward into Russian territories after the fifteenth century. Post-Soviet authors have argued that neither of the above scenarios is accurate, claiming that vodka grew out of an indigenous tradition of manufacture in early modern Russia known as "pot-distillation," and that it is therefore a quintessentially Russian beverage, not something of Polish or Scandinavian origins.

This pot-distillation process involved the "setting out" of beer or mead by pouring the fermented beer wort or honey solution into pots, which were then placed in a stove and covered with other pots, so that the wort was heated up. A wooden tub was usually placed beneath the pots in order to avoid wasting wort in case the pots boiled over. During a long period in the stove under an even temperature, the brewing produced a form of distillation, albeit a primitive one. One problem with the process was that the alcoholic products and the wort were not sufficiently separated. Most of the liquid remained in the vessels. At any rate, the heating concentrated the product by driving off the excess water and the ethyl alcohol—the chief ingredient in distilled spirits. What remained were the highly poisonous and foul-smelling congeners, the amyl and butyl alcohols known as "fusel oil." The process was not only slow and expensive but potentially fatal for consumers. Inconsumable as it was, the product of this pot distillation, when combined with advances in processing and refining introduced from the southwest, it is argued, eventually led to the distillation of a purer and more powerful alcoholic product—vodka.

Early Problems in Distillation

For more than 300 years, the continued unreliability of the equipment for refining and processing grain alcohol compelled Russian distillers to use complicated methods to avoid poisoning potential consumers. At this point, distilling was small in scale and the methods were primitive. By the seventeenth century, a "simple wine" *(prostoe vino)* could be created by fermenting a mixture of grain and water (mash), distilling this product into an intermediary stage called *raka,* and then distilling it again. This simple wine could be used to create several types of vodka and grain spirits—one of the most widely distributed types of which was *polugar,* obtained by diluting three "buckets" (sing. *vedro;* pl. *vedra*—approximately 50 liters [13.2 U.S. gallons]) of simple wine with one bucket of pure, cold water.

By the 1700s, both simple wine and polugar came to be seen as unfinished and even intermediate products contaminated by trace elements of residual fusel oils, thereby making them still dangerous for widespread consumption. This did not, of course, prevent its consumption, although vodka's deleterious effects were increasingly denounced by the Russian Orthodox Church. Several distilling methods were used in attempts to remedy this problem. The state exacerbated the situation, however, by heavily taxing vodka consumption without regard for either quality or abuse and without making any distinctions between the many forms the product took between the sixteenth and the late nineteenth centuries. Such taxation encouraged "bootlegging" *(korchma, shinkarstvo)* by private individuals eager to profit through the illegal sale of "homemade" *(samogon)* vodka. These bootleggers frequently were none too cautious in their manufacturing methods; their low-quality drinks aggravated the effects of vodka, producing sometimes fatal consequences. The methods of distillation that were employed in an attempt to eliminate the residual fusel oils included, inter alia, further distillations of varying dilutions. Triple-tested "wine" *(trekhprobnoe vino)* was obtained by distilling 100 buckets of simple wine with 33 buckets of water; quadruple-tested "wine" *(cheterykhprobnoe vino)* was a distillation of 100 buckets of simple wine with 50 buckets of water. Double-tested "wine" *(dvukhprobnoe vino)* consisted of 100 buckets each of simple wine and water. Both double-and triple-tested vodka were sold in the taverns *(kabaki)* of Russia in the first half of the 1800s, but the latter, naturally, was the stronger and hence the subject of increasing concern because of its devastating effects on drinkers.

The eighteenth century also saw the development of double wine *(dvoinoe vino)*. This was grain spirit distilled from "simple wine," in effect involving a triple-distillation process from mash to raka to simple wine to double wine. The distillation of triple wine—a further distillation of double wine (quadruple-distilled spirit) and quadruple wine (quintuple-distilled spirit) followed. In this manner, the alcoholic content was increased and the contaminants reduced—although many still remained, as subsequent temperance advocates would point out. The resulting liquor was usually no stronger than 40 percent alcohol. To give the product added "kick," some distillers added spices and aromatic ingredients. Ordinarily, if the vodka was flavored with fruits or berries, the product was termed *nalivki*, and if with herbs, *nastoiki*. During the notorious vodka tax-farm system *(otkup)* in effect during most of the nineteenth century, the operators of *kabaki* (usually the employees of the tax-farmers [*otkupshchiki*] themselves) would add pepper, eau de cologne, nicotine, and even arsenic as well as other foreign substances to achieve the same or greater effects.

Impact of Factory Distillation

Beginning with the mid-nineteenth century, the introduction and use of increasingly sophisticated filtration systems and the evolution of distilling equipment produced a transition from gentry distillation as an exclusive prerogative of the landowning class to factory distillation. Indeed, the very size and complexity of the equipment necessitated this step. Temperance critics bemoaned this shift for several reasons: First, the increased productivity of industrial distillation made more vodka available to hard-drinking Russians, and second, the improved availability and quality of the vodka encouraged Russians to visit the notorious kabaki more often and thus increased drunkenness. Moreover, although the official strength of the vodka was to be between 38 and 40 percent alcohol, instances of strengths approaching 47 percent were found. These considerations were underscored by the prevalence since early times of binge drinking *(zapoi)* among Russians of all social categories, a practice that was further exacerbated by the strength of the vodka. The issue of quality aside, the *quantity* of vodka available for consumption was difficult to measure before the advent of industrial distillation.

Social Uses

Vodka was the center of many social occasions for upper-class Russians, as described by such writers as Mikhail Lermontov, Leo Tolstoy, and Fyodor Dostoevski. But it was also a social lubricant among the peasants. Weddings, baptisms, christenings, betrothals, marriages, and, finally, wakes were normally celebrated with the free flow of vodka. It was also a medium of exchange. Numerous would-be reformers commented on the practice of *kabak* operators under the *otkup* system selling vodka to peasants "on credit" in exchange for part of their following year's crop. Moreover, the consumption of several glasses of vodka as a means of "sealing" agreements between peasants was a standard feature of country life. And in the cities of Russia's industrial era, it was likewise common for "new men" just arrived in the factories and mills to "treat" their more experienced workmates to one or several rounds of vodka upon receiving their first pay as a rite of passage. Finally, well into the second decade of the twentieth century, both the Russian army and navy continued to provide daily and special "celebratory" vodka rations to the men in order to "strengthen" and "warm" them.

These social practices did not end with the introduction of the state vodka monopoly in the 1890s. The exclusive sale of vodka in sealed bottles of state-mandated volume for off-premises consumption in fact only drove urban and rural people into the streets, where their subsequent behavior irritated both middle-class Russians and police authorities. Along with increased street drunkenness *(ulichnoe p'ianstvo)* and hooliganism, the direct state monopoly of the sale of vodka brought in its wake considerable income—estimated at anywhere between one-third to one-quarter of annual state revenues by 1914. This included the sale of vodka from state shops, the retail sale of vodka in hotels and restaurants, and sales at select railway stations that purchased special licenses.

World War I

With the outbreak of war in August 1914, prohibition of the manufacture and sale of vodka was declared "for the duration" of the war; this law was subsequently extended "permanently." However, as the authority of first the tsarist government and then the Provisional Government waned, bootlegging, moonshining, and the abuse of vodka surrogates multiplied until the Bolsheviks reinforced prohibition during the Russian Civil War (1918–1921).

The Bolsheviks and the Soviet Era

Before the October Revolution of 1917 the consumption of vodka, and, indeed, inebriation, was viewed by committed workers and revolutionaries as inappropriate; abstention, in contrast, was viewed as "heroic" and a sign of worker solidarity and discipline. Lenin was personally abstinent, and from 1921 until 1923 the distillation and sale of vodka was prohibited on Soviet soil. Beginning in 1923, however, vodka production was resumed. The Soviet state monopolized both its manufacture and sale in order to increase revenues and to eliminate the illicit manufacture and sale of samogon, and the official strength of vodka was maintained at 40 percent. This situation continued

Bottles of vodka in a Saint Petersburg, Russia, kiosk include the brand Rasputin, named for the controversial monk whose influence reached to Tsarina Alexandra Feodorovna and Tsar Nicholas II on the eve of the Bolshevik Revolution. (Steve Raymer/Corbis)

until World War II, when the curve of consumption officially fell. However, there is written record of vodka being served to frontline troops both to "warm" and "embolden" Red Army soldiers in battle.

The postwar period witnessed a return to and a surpassing of prewar levels of consumption. Attempts by the Leonid Brezhnev regime (1964–1983) to curb vodka consumption resulted only in the virtual disappearance of sugar from the shelves of state stores during the attempt as enterprising Russians resorted once again to the distillation of samogon. Russian physicians and psychiatrists reported increasing instances of abuse and subsequent societal, public health, and familial problems as well as physical and mental illnesses during the period. "Temperance" efforts by Yuri Andropov in 1984, and the better-publicized efforts by Mikhail Gorbachev in the mid-1980s, likewise failed, although in these instances an elaborate but superficial system of "temperance organizations" was set up to attempt to curb consumption. Similar efforts to prevent the abuse of vodka through stringent measures of control in the war in Afghanistan produced numerous cases of fatalities through the consumption of dangerous surrogates. With the collapse of the Soviet Union in 1991, vodka consumption rose to such levels that, by the last years of the 1990s, alcohol abuse—with vodka being the chief culprit—was deemed to be one of the leading causes of the decline in Russian male longevity to Third World levels.

George Snow

See also: Alcohol, Consumption of (Russia); Russia, Imperial, Temperance in; Soviet Union and Russia since 1917, Alcohol and Temperance in

References

Christian, David. *Living Water: Vodka and Russian Society on the Eve of Emancipation.* New York: Oxford University Press, 1990.

Herlihy, Patricia. *The Alcoholic Empire: Vodka and Politics in Late Imperial Russia.* New York: Oxford University Press, 2002.

Phillips, Laura L. *Bolsheviks and the Bottle.* DeKalb: Northern Illinois University Press, 2000.

Pokhlebkin, William. *A History of Vodka.* New York: Verso, 1992.

Romanov, Sergei. *Istoriia russkoi vodki* (A history of Russian vodka). Moscow: Veche, 1998.

Segal, Boris. *The Drunken Society: Alcohol Abuse and Alcoholism in the Soviet Union.* New York: Hippocrene, 1990.

———. *Russian Drinking: Use and Abuse of Alcohol in Pre-Revolutionary Russia.* New York: Rutgers Center for Alcohol Studies, 1988.

Smith, R. E. F., and David Christian. *Bread and Salt: A Social History of Food and Drink in Russia.* Cambridge: Cambridge University Press, 1984.

White, Stephen. *Russia Goes Dry: Alcohol State and Sobriety.* Cambridge: Cambridge University Press, 1996.

Voluntary Committee of Lawyers (VCL)

The Voluntary Committee of Lawyers (VCL) was a New York–based nationwide network of attorneys opposed to National Prohibition that played an influential role in negotiating the distinctive technical process by which the Twenty-First Amendment to the U.S. Constitution was adopted in 1933. It began in 1927 as an informal group of prominent New York attorneys drawn together by their growing discontent with the Eighteenth Amendment. Not only did National Prohibition upset the Constitution's federal balance, they believed, it created disrespect for law, corrupted policing and the judicial system, and infringed due process of law. Opening an office in October 1928 and formally incorporating in January 1929, the VCL was led by two prominent New York attorneys, Joseph H. Choate Jr. and Harrison Tweed. Although the VCL never acquired a large membership, peaking at 3,626, the organization's influence was manifested by its success in obtaining an American Bar Association (ABA) referendum on Prohibition. The autumn 1930 poll revealed that, with three-fourths voting, the ABA membership favored repeal two to one.

Many VCL members were also affiliated with the Association Against the Prohibition Amendment (AAPA) and, in effect, the VCL served as the legal arm of the larger repeal organization. From its earliest days, the AAPA had expressed doubts about the legitimacy of the Eighteenth Amendment because of its ratification by state legislatures, which were often rural-dominated, rather than by directly elected state conventions, an alternative method specifically provided by Article V of the U.S. Constitution. Legal challenges to Eighteenth Amendment ratification were thwarted by the U.S. Supreme Court in *Hawke v. Smith* (1920) and *U.S. v. Sprague* (1931). As support for repeal grew, the VCL devoted its attention to assuring that a new amendment would be submitted to convention ratification.

Unexpectedly swift congressional adoption in February 1933 of a repeal amendment according to the terms for which the AAPA and the Women's Organization for National Prohibition Reform had lobbied strenuously produced legislative confusion. State convention ratification had not been employed since 1788. VCL leaders quickly drafted a model convention bill; its provision for slates pledged to repeal or retention of Prohibition gave delegate elections the character of a referendum. VCL members in every state where the legislature was in session arranged for the bill's prompt introduction. The availability of a well-drafted bill, one that prominent attorneys urged them to adopt, led thirty-nine state legislatures to act within four months. Twenty states embraced the VCL measure with, at most, minor modifications; many other states employed its fundamental ideas and at least some of its language. Eight states allowed for a slate of unpledged delegates, but only Wyoming failed to provide for any pledged delegates. Contrary to most expectations, prompt scheduling and conduct of elections and conventions proved possible. Voting results mirrored public opinion polls showing 73 percent of Americans in favor of repeal. By 5 December 1933, fewer than ten months after Congress had acted, the states had completed the repeal process, thanks in no small part to the VCL placing in their hands an instrument for effective expression of their preference.

David E. Kyvig

See also: Association Against the Prohibition Amendment (AAPA); National Prohibition (United States); Women's Organization for National Prohibition Reform (WONPR)

References

Brown, Everett S., ed. *Ratification of the Twenty-first Amendment to the Constitution of the United States: State Convention Records and Laws.* Ann Arbor: University of Michigan Press, 1938.

Kyvig, David E. *Repealing National Prohibition.* 2d ed. Kent, OH: Kent State University Press, 2000.

Vose, Clement E. "Repeal As a Political Achievement" in *Law, Alcohol, and Order: Perspectives on National Prohibition,* edited by David E. Kyvig, 97–121. Westport, CT: Greenwood, 1985.

W

Wales

Wales had a distinctive history in the growth and development of its trade in alcoholic beverages, patterns of regulation of alcoholic sales, and temperance discourse and organization. The context for the history of alcohol and temperance in Wales was the shift from largely rural community- and work-based drinking customs in the eighteenth century to the problems associated with rapid industrialization and urban dwelling in nineteenth-century South Wales when escalating levels of drunkenness stimulated intensive temperance activity. The history of commercial brewing begins with small, independent brewing companies in the nineteenth century. Their almost complete absorption by national and international breweries took place by the end of the twentieth century.

Tradition and Custom

Throughout the eighteenth century, Wales was a predominantly agricultural country. Brewing was a cottage-based industry, and beer was the drink of choice. Beer (*cwrw* in Welsh) was, and still is, considered to be the national drink of Wales. At a time when water supplies were unreliable and very often contaminated and when ale was cheaper than tea, beer fulfilled the dual functions of thirst quencher and a basis of the social framework underpinning Welsh cultural practices. Consumption of ale was closely associated with rituals bound up in agricultural practices, rites of passage, and traditional leisure pursuits. Home-brewed beer was an important feature of wedding custom. The traditional practice of "bidding" *(cwrw bach),* that is, of selling beer considerably above its market price, helped to raise money for newly engaged couples to help them set up their new homes. A similar collective exercise in self-help was the *cwrw gwadd,* a feast held among the poorer members of the community. When a member suffered from prolonged illness or other misfortune, local communities organized a feast of beer, bread, and cheese. A charge was made for the food to avoid being prosecuted for selling beer, and the money handed over to the person concerned. Ale was also liberally distributed both before and after funerals; mourners spent three or four hours drinking beer in the house of the deceased before forming the funeral cortege.

Before the growth in influence of Nonconformity in Wales later in the nineteenth century, every parish celebrated its *gwylmabsant* festival, or saint's day, which began on a Saturday and continued until the following Tuesday and was associated with varying degrees of intoxication and rough behavior. Communal drinking of beer also consolidated contracts between local tradesmen. Home-brewed beer was provided at events like pig slaughtering *(cwrw bwtshwr),* house building *(cwrw cwple),* and the wheelwright's work of putting a tire onto a wheel *(cwrw bando).* As the temperance movement gained pace in the nineteenth century, the use of beer on such occasions began to die out, with tea and cakes being provided instead.

The consumption of beer was also important in the political arena. Alcohol was liberally distributed at contested and noncontested elections, a custom that disappeared after the passage of the Corrupt Practices Act in 1883. Voters in all constituencies, at parliamentary and municipal elections, looked forward to handouts of free beer in return for a promised vote.

Many drinking customs gradually came to an end as the influence of Nonconformity, and, in particular, the aggressive line taken by Methodism on alcohol, began to spread throughout Wales during the nineteenth century. Traditional pastimes and practices associated with a culture of drinking were condemned as harmful, foolish, and ultimately sinful. Celebrations marking the important landmarks in a person's life were stripped of much of their traditional character. The replacement of rural custom and traditional drinking practices with sober, rational leisure activities, however, was by no means uniform. A new influence in the form of the public house and the drinking club, particularly in the rapidly expanding industrial communities of South Wales, worked to temper the hegemonic influence of Nonconformist discourse.

Public Houses, Beer Shops, and Drunkenness

During the course of the nineteenth century, the rapid growth in numbers and popularity of public houses in Wales occasioned much introspective debate. In 1850, 305 drinking places were recorded in the towns of Merthyr Tydfil and Dowlais. By 1854, the numbers had risen to 504. Temperance reformers believed that the Beer Act of 1830 had compounded the problem. The legislation aimed to wean people off harder drinks such as gin and to reduce drunkenness by making beer accessible and cheap. Beer was sold from licensed private houses for consumption on or off the premises. Conditions of tenure were similar to public houses, but closing hours in beerhouses were more strictly defined. According to many contemporaries, the overall result of the beerhouse license in Wales was to increase drunkenness, particularly in the newly industrialized areas.

During the nineteenth century, Wales acquired a reputation for high levels of drunkenness. A panel of commissioners charged with the task of examining the state of education in Wales in 1847 caused a furor when they complained of widespread drunkenness and immorality among the Welsh. Throughout the century the areas of greatest drunkenness coincided with the coalfields. Theories put forward as an explanation for high levels of drunkenness included the observation that modes of production concentrated drinking opportunities into specific and short periods of time, that overcrowding and inadequate sanitation in industrial communities adversely affected the health and morals of the laboring classes, and that intemperance was a direct result. In economic terms, the mono-industrial base of coalfield communities, the dangerous nature of the work, the unstable security inherent in coalfield employment, and the male-dominated nature of the work produced a predominantly male, working-class migrant population. Industrialization was largely considered to be inimical to sobriety.

In the newly established mining communities, counterattractions to the pub were largely absent. Beer-drinking was strongly associated with contemporary notions of masculinity; a belief in the health-giving benefits of beer as a reviver of male strength resulted in the pub and the beerhouse becoming the focus of working-class masculine leisure. Convictions for drunkenness in the county of Glamorgan, the area most affected by the dramatic increase in the trade and extraction of coal, increased from 179 in 1842–1843 to 3,202 by 1887–1888. At the end of the nineteenth century, the summary offense of drunkenness occupied much of the work of local police and magistrates. Fines for drunkenness remained static throughout the period—most individuals prosecuted at magistrates' courts were fined five shillings. The gradual increase in prosecutions for drunkenness throughout the century should be understood in the context of more efficient policing and the proliferation of magistrates' courts as much as in terms of increasing intemperance. These factors, along with a strong condemnatory temperance discourse, led to Wales's reputation as a land needing to be purged of impurity and immorality.

Apart from the emphasis on beer rather than spirits, Wales does not appear to have possessed an identifiable drinking culture in the nineteenth century. Moreover, statistics confirm that the Welsh were no more intemperate than their neighbors in England and Scotland. What was distinctive about the Welsh experience was the strength of a largely Nonconformist temperance discourse and the zeal with which reformers pursued their aim of reforming the drinking habits of the country's citizens.

Temperance and Reform

At the heart of the temperance movement in Wales, as elsewhere, was the belief that economic and social success could be combined with moral purity. The campaigning zeal of temperance reformers was targeted at elevating the moral, social, and cultural character of the individual as a necessary first step toward ultimately raising the profile of the nation as a whole. The first temperance society established in Wales was a branch of the British Foreign and Temperance Society, which formed in Holywell in North Wales in 1832. The society advocated moderate drinking; members were allowed to consume beer, porter (strong beer), and wine. The first total abstinence society was formed at Llanerchymedd on Angelsey in 1835; all members were expected to sign "the pledge." Throughout the nineteenth century, both moderation in drinking and total abstinence were advocated by temperance societies. As the temperance movement gained pace and spread from North to South Wales, a raft of campaigning literature emerged, much of it in Welsh. *Y Dirwestydd* (The abstainer), 1836; *Y Dirwestydd Deheuol* (The southern abstainer), 1838; and *Y Dirwestwr Deheuol* (The temperance advocate), 1840 all vigorously promoted the cause. In the eyes of temperance reformers, drink and drunkenness were portrayed as being at the root of all social ills; the public house led to the pawnshop, the police court, and the prison. Intemperance led to the destruction of family life, bringing poverty, disease, and premature death.

Both sexes were actively involved in the promotion of temperance, although Welsh women were not to come to the fore until the 1890s. In 1892, Undeb Dirwestol Merched Gogledd Cymru (The North Wales Women's Temperance Union) formed to unite disparate women's campaigning groups spread throughout Wales, and in 1901 Undeb Dirwestol Merched y De (The South Wales Women's Temperance Union) was formed. By 1916, some 140 branches, run by women, worked to promote the cause of temperance in South Wales. The membership reflected the middle-class, Nonconformist base of male temperance societies. A formative figure in the women's temperance movement was Sarah Jane ("Cranogwen") Rees (1839–1916).

In Welsh temperance literature, the problem of drink and drunkenness were articulated broadly within a masculinist framework. Despite the rhetoric, reports by social commentators and local newspapers testify to significant levels of female inebriety. In the context of a strengthening discourse that emphasized women's maternal duties to family and home, the female drunk was represented as the antithesis of contemporary Welsh femininity. In such a reading, gender as well as class played a significant role in all debates concerned with alcohol and temperance.

The temperance campaigns of the nineteenth century reached their apogee with the passage of the Welsh Sunday Closing Act (1881). The growing number of Welsh, Nonconformist Liberal members of Parliament gave their support to campaigners for Sunday closing in Wales. In an unprecedented move, Welsh drinking establishments were ordered to close their doors to all customers on Sundays. Exceptions to this rule were "bona fide" travelers who were able to prove that they had traveled more than three miles. The clause led to much subterfuge as those determined to buy a drink used great ingenuity to achieve their goal. A direct consequence of the legislation was to increase the numbers of private drinking clubs, or *shebeens.* Despite a royal commission in 1889 set up to investigate the workings of the act, with specific reference to reports of increasing levels of drunkenness, the legislation remained on the statute book until 1961, when a new licensing act allowed every area to hold a referendum, bringing to an end the eighty-year-old tradition of Sunday closing in Wales.

Beer and Brewers

The first commercial breweries appeared in Wales at the end of the eighteenth century. The inward migration of workers into the newly created iron and tin-plate industries created a market for mass-produced beer. The Cambrian Porter Brewery in Swansea (founded in 1792) and the Pontycapel Brewery in Cefncoed near Merthyr Tydfil (founded in 1840) supplied beer to thirsty workers from nearby works. The mountainous topography of Wales presented serious difficulties for brewers looking to extend their markets beyond local boundaries, but the rapid development of an extensive rail network from 1850 opened up markets further afield in Wales. A significant and persistent theme in the development of the brewing trade in Wales was the failure of brewers to penetrate English markets. This was in contrast to Scottish and Irish brewers, who successfully marketed indigenous beers. The failure of Welsh brewers to capitalize on English markets might in part be understood as having its roots in the reluctance of independent brewers to relinquish their separate identities and work together.

From the middle of the nineteenth century, independent breweries capitalized on the rapidly growing markets in the coal-producing valleys of South Wales. Buchans of Pontypool, the Rhondda Valley breweries of Treherbert and Pontypridd (who were eventually taken over by the Cardiff-based Ely Brewery in 1920), and Brain's Brewery, also of Cardiff, were well established by 1900. Their strength lay in the coal-producing valleys; miners were often heavy drinkers. Sales figures, however, were dramatically affected by the economic vagaries of coal production. The reputation of Brain's Beer has long been associated with the city of Cardiff. The "Old Brewery," established in Cardiff in 1713, traded under a number of owners before it was taken over by Samuel Brain in 1882. By 1900, Brain's owned the leasehold of more than eighty public houses in Cardiff. Advertisements in and around the city vigorously promoted the company's products, proclaiming loudly, "Its Brains You Want," a slogan that endured throughout the twentieth century. Brain's was one of the first companies in Wales to bottle its own beers and the beers of companies such as Guinness and Bass.

In 1900, despite the reputation of Brain's Beer, Hancock's Brewery, with breweries in Newport, Cardiff, and Swansea, was Wales's most prolific brewer and the country's largest wine and spirit merchant. Also worthy of note in South Wales were Buckley's Brewery in Llanelli and the Felinfoel Brewery, also of Llanelli, which pioneered the development of canned beer.

In North Wales, Wrexham dominated the brewing trade. The town was ideally placed to exploit natural springs ideal for brewing beer. The first commercial brewery, Charles Bates' Union Brewery, opened in 1799, followed by Soames' Wrexham Brewery and Island Green Brewery (1856), the two latter companies combining to form the Border Brewery in 1931. In 1984, Border's was incorporated with the Burton brewers, Marstons, ending a long tradition of independent brewing in the town.

Wrexham was also famous for the production of lager. The Wrexham Lager Beer Company (1881) achieved limited success; local drinkers continued to prefer stouts, milds, and pale ales. Lager was even promoted as a temperance drink or as a "tonic" to aid digestion in an effort to boost sales. To overcome the problem of limited enthusiasm for the product, the company found a small market in private hotels and restaurants. The firm went into liquidation in 1892, but new life was breathed into the company when Robert Graesser, an industrialist, installed more efficient refrigeration in the plant and established a specialist export market. It would be some forty years later, just before World War II, when lager began to become popular among Welsh drinkers. In 1949, the brewery was sold to Coope Allsopp, and in 1963 a state-of-the-art brew house was built, transforming the company into the largest modern brewery in Europe.

As the twentieth century progressed, many of the largest breweries in Wales were taken over by English and Scottish brewing giants. Bass Charrington incorporated Hancock's in 1968, and Buckley's Brewery fell to Guinness in 1990. (Buckley's independent status was revived in 1993.) At the end of the twentieth century, Brain's Brewery of Cardiff was the only large-scale independent brewery remaining in Wales.

In the 1970s and 1980s, the Campaign for Real Ale (CAMRA) encouraged small-scale Welsh brewers to capitalize on a growing demand for specialist beers. Beers such as Master Blaster (Afan Brewery), Druid's Ale (Gwent Ales), and Snowden Strong (Gwynedd Brewers) achieved limited success, but by 1991 only two small producers remained in business—specialist beers would always be a niche market and economies of scale meant that small-scale brewers struggled to compete with multinational companies.

A strong sense of independence, an insularity that thrived on brand loyalty, and a culture of beer drinking underpin the history of alcohol and brewing in Wales. As the heavy extractive industries began to decline and the Welsh developed a taste for a wider range of alcoholic drinks, independent brewers found it increasingly hard to sustain their profit margins and sold their shares to large English and Scottish conglomerates. It is significant that Welsh brewers did not look to each other to form a Welsh brewing company as brewers did in Scotland—the preference always appears to have been to look outside Wales for help. Welsh brewers, unlike their English and Scottish counterparts, also had to battle against a vigorous temperance movement. As late as the 1930s, temperance reformers succeeded in banning all drinks advertising Brain's Beer from the sides of buses and trams.

Deborah James

See also: Bass; Beer; Campaign for Real Ale (CAMRA); Central Sunday Closing Association; Guinness Brewery; Liberal Party (United Kingdom); Nonconformists (United Kingdom); Pubs and Beerhouses (United Kingdom); Shebeens

References

Glover, Brian. *Prince of Ales: The History of Brewing in Wales.* Stroud, UK: Alan Sutton, 1993.

Harrison, Brian. *Drink and the Victorians: The Temperance Question in England, 1815–1872.* 2d ed. Staffordshire, UK: Keele University Press, 1994.

Lambert, W. R. *Drink and Sobriety in Victorian Wales, c. 1820–c. 1895.* Cardiff, Wales: University of Wales Press, 1983.

Lloyd-Morgan, Ceridwen. "From Temperance to Suffrage?" in *Our Mothers' Land: Chapters in Welsh Women's History, 1830–1939,* edited by Angela John, 135–158. Cardiff, Wales: University of Wales Press, 1991.

Washingtonians

The Washingtonian movement, or the Washingtonian revival, as it is sometimes known, represented a groundswell of working- and lower-middle-class teetotal sentiment that profoundly democratized the U.S. temperance movement during the 1840s. Regenerate hard drinkers, unlettered men, and some women were prominent among the Washingtonians, although skilled craftsmen composed the movement's core. In some groups, self-confessed habitual drunkards may have numbered half the membership. Although the movement dissipated by the 1850s, its philosophy and practice of mutual aid and the rehabilitation of habitual drunkards influenced subsequent efforts. These stand at the head of a tradition of "therapeutic temperance" that includes urban rescue missions such as the Salvation Army, the fellowships of Alcoholics Anonymous and its derivatives, and the modern therapeutic community.

The movement spread rapidly on the East Coast following the initiation in May 1840 of the Washington Temperance Society of Baltimore by a group of heavy drinkers who resolved to abstain from alcohol. They seem to have taken Washington's name because of its association with the country's liberation from English taxation, an analogy similar to those of later temperance advocates who compared sobriety to escape from chattel or wage slavery. By 1843, the Washingtonians numbered their membership in the millions, though this likely was exaggerated.

The Washingtonians differed considerably from older temperance organizations made up largely of elite evangelicals. The movement's local associations tapped a population of artisans awash in alcoholic beverages (the per capita alcohol consumption of the era was staggering by today's standards) and caught up in the uncertainties and demands of a nascent industrial economy. Traditional work routines and hierarchies were being rationalized in many trades, if not yet widely subjected to the authority and discipline of the factory; precarious, waged employment was becoming common. Indeed, the movement took off in the midst of a depression that extended from 1837 to 1843. Although Washingtonian membership was heterogeneous and included some from the old elite, the movement appealed most to workingmen for whom careful (though not ascetic) self-discipline seemed essential to a decent "competence" (livelihood). Small employers who believed that their prospects would be greatly improved by more sober, calculating management and more reliable workers joined them. Women, not "slaves to drink" as commonly as men, but along with children often the victims of hard-drinking men, created auxiliaries called Martha Washington Societies. The Washingtonians helped make temperance the important "women's issue" that it would remain.

Washingtonian Practices

Washingtonian practices were grounded in religious and secular aspects of working-class life. Spurning the formalism and hierarchy of elite Christianity, Washingtonians took inspiration from the democratic and sentimental core of Methodist revivalism, an important religious influence of the era. The "experience meeting" featured melodramatic "drunkard's tales," stereotyped stories of degradation and redemption that inspired a literary genre. These powerful, oral narratives, still common in the recovery movement, introduced the drunkard's tortured inner life to large audiences and made palpable the possibility and benefits of reform. Especially when addressed to drunkards, the conversion formula of Washingtonianism was right out of revivalism: The testimony of the saved insinuated a conviction of sin that invited redemption, often through the visceral suffering of renunciation (in the form of withdrawal symptoms). Moreover, local associations functioned much like religious congregations. Members (often women) visited the newly sober, bolstered their resolve, nursed them and their families through sickness, and sometimes found work for them. Always, members were encouraged to seek out others to spread the news of reform, an evangelical tactic recognizable today in the Twelve Step work of Alcoholics Anonymous fellowships.

Washingtonian mutual aid derived from older workingmen's associations, some of which allied themselves with the movement en masse. Similarly, Washingtonians drew on existing practices to provide sober alternatives to the alcohol-saturated environments of working-class life. Probably borrowing from the Odd Fellows, to compete with the taverns the Washingtonians established free reading rooms that stocked newspapers and literature of the day (especially temperance literature). They rented or constructed halls to mount more lively entertainments—theatrical productions and group singing, in particular—that could compete with the theaters, bowling alleys, and other decidedly "wet" amusements of city life and the all-purpose taverns of America's myriad small towns. The Washingtonians were the first to employ the alternative social activities that became so important to environmentally attuned, Progressive-era temperance campaigners such as the Woman's Christian Temperance Union and the settlement house movement.

The Decline and Legacy of the Washingtonians

In the movement's early years, the resources of elite, old-guard temperance adherents were crucial to its spread. However, this was an uneasy cooperation. Elites regarded their support as a kind of tutelage aimed at taming and co-opting the new phenomenon, whereas many Washingtonian leaders insisted on the movement's autonomy.

At bottom, the class distance between the old guard and the newcomers was the most important source of the conflict that quickly splintered the Washingtonians. Old-line temperance groups were dominated by the formal, learned piety of elite denominations such as Congregationalism; Washingtonians, when affiliated at all, were more likely to be steeped in the "heart religion" of the Methodists and Baptists, and they were skeptical of sectarian claims to righteousness and highbrow approaches to religion. Similarly, Washingtonian amusements borrowed freely from popular theater and tavern culture and thus seemed vulgar to many among the old guard who preferred sobriety to be accompanied by a refinement of manners and mores uncongenial to working-class sensibilities. Finally, and in part as the result of backsliding among some notable Washingtonians, the temperance elite developed a faith in coercive methods, especially the prohibition of the manufacture and sale of alcoholic beverages and the institutionalization of drunkards. This shift in thinking alienated many Washingtonians for whom moral suasion was the sine qua non of democratic practice.

The Washingtonian movement did not decline so much as shatter. Some members were captured by the older organizations, some left the temperance cause to participate solely in the labor movement or other working-class organizations that promoted mutual aid, and still others temporized with elite criticism and created numerous "neo-Washingtonian" organizations. The best known of these is the Sons of Temperance, founded in New York City in 1842 by former Washingtonian artisans. The Sons of Temperance preserved for a time Washingtonian mutual aid practices and instituted a more formal cash benefit system based on the insurance schemes of workingmen's associations of the period. They sanitized alternative entertainments, supported prohibition, and by the 1860s endorsed and raised money for the institutional treatment of drunkards (by court order if necessary). Yet they, too, were beset by internal conflicts over such policies, along with conflicts over the admission of women and African Americans. The Good Templars, established in 1851, had a similar early history, although the Good Templars admitted women from the beginning.

The Washingtonians also seeded many smaller, regional temperance organizations whose history is little known. For example, in San Francisco, a city that could not be reached by train from the East until 1869, the Dashaway Association was founded in January 1859 by firemen who hailed from mid-Atlantic and New England hotbeds of earlier Washingtonian fervor. The association soon had thousands of adherents in chapters around the state of California. In direct imitation of the Washingtonians, the Dashaways adopted experience meetings and the practices of mutual aid, founded a Home for Inebriates (where residence was voluntary), built a large hall for reading, study, theater, and music, rejected prohibition, and adopted a relentlessly nonsectarian style (they even admitted Catholics and Jews). But the organization soon was rent by the same conflicts that had splintered its predecessor, particularly the issue of prohibition. Although the Dashaway Association survived until 1892, it was in decline by 1870, many of its members defecting to more militant organizations such as the Good Templars and the Sons of Temperance.

A more durable legacy of the Washingtonian movement was its influence on the urban rescue missions that proliferated in U.S. slums beginning in the 1870s. Jerry McAuley, former drunkard, thief, and prison inmate ("a dirty rag shop of a man," he called himself), founded New York City's Water Street Mission in 1872 with money from a wealthy temperance adherent. With his wife (a former prostitute), McAuley and his successors operated the mission along Washingtonian lines. Indeed, the whole panoply of Protestant city missions, including the transplanted Salvation Army, owe much to the Washingtonian tradition and its affinity with a practical religion suffused with powerful sentiment and a regard for the wretched and outcast.

Jim Baumohl

See also: Alcoholics Anonymous (AA); Dashaway Association; Good Templars (IOGT); Moral Suasion; Sons of Temperance; Treatment Institutions; Martha Washington Societies

References

Baumohl, Jim. "On Asylums, Homes, and Moral Treatment: The Case of the San Francisco Home for the Care of the Inebriate, 1859–1870" in *Contemporary Drug Problems.* 13 (Fall 1986): 395–445.

———. "Inebriate Institutions in North America, 1820–1920" in *Drink in Canada: Historical Essays,* edited by Cheryl Krasnick Warsh, 92–114. Montreal: McGill-Queens University Press, 1993.

Tyrrell, Ian R. *Sobering Up: From Temperance to Prohibition in Antebellum America, 1800–1860.* Westport, CT: Greenwood, 1979.

Wilentz, Sean. *Chants Democratic: New York City and the Rise of the American Working Class, 1788–1850.* New York: Oxford University Press, 1985.

Watneys

Watneys, a fast-growing London ale brewer in the nineteenth century, merged with two other metropolitan firms in the 1890s and, surviving early financial difficulties, became the United

Kingdom's most adventurous brewer by the 1930s. The undervalued property assets of the company attracted a takeover bid from an outsider to the industry in 1959. Although defeated, this unwelcome approach helped precipitate a spate of mergers that led to rapid concentration of the United Kingdom industry. Expansion was vigorously pursued in the 1960s, turning Watneys into a national brewer. The company became part of a conglomerate in 1972 when acquired by Grand Metropolitan Hotels, which had purchased London brewers Truman, Hanbury, Buxton and Company Limited in the previous year. The 1970s exposed the company's strategy of overdependence on one brand, Red Barrel, as consumers abandoned heavily promoted mediocre keg bitter for its lager equivalent. In the 1990s, the holding company swapped its breweries with another brewer in exchange for pubs and subsequently sold the entire enlarged estate. The Watney name is now extinct on both beers and pubs in the United Kingdom.

In 1837, James Watney bought a quarter share of the long-established Stag Brewery, Pimlico, and by 1858 the firm was trading as Watney and Company. Increasing output more than threefold to more than 500,000 hectoliters (13.2 million U.S. gallons) per year, both James and his namesake son were to die millionaires in the 1880s. Floated as a limited liability company in 1885, the firm continued to prosper, acquiring a brewery in Mortlake in 1889 before a merger in July 1898 created Watney, Combe, Reid and Company Limited. Reid's brewery was closed within a year, Combe's in 1905, and production switched to the Stag and Mortlake breweries. The new company, with Combe's influential Cosmo Bonsor as chairman, issued £8.71 million of shares and £6 million of fixed-interest debentures. These amounts, inflated by overvalued properties, made the new firm easily the most highly capitalized brewing company in the United Kingdom. Loans made to publicans for buying leases, and thus tying pubs to the brewery, produced mounting bad debts as property values dipped. Heavy interest charges and falling sales further hamstrung the company, and profits collapsed. The managing director was sacked in 1902. In 1905, the firm's deferred ordinary stock was written down by 75 percent, but still no dividend was paid between 1908 and 1916. In the 1920s, the company recovered, restoring the capital by 1925; by taking over eight breweries and their tied estates in the next thirty years, it consolidated its base in London and southeastern England.

Under Threat

In 1958, the company purchased Mann, Crossman and Paulin Limited, whose Whitechapel brewery replaced the by then cramped Stag Brewery, which closed in April 1959, leaving a site ripe for redevelopment. In May 1959, the new company, Watney Mann Limited, which owned 3,670 pubs, attracted a bid from Charles Clore, a financier of humble origins who specialized in targeting companies with underexploited assets. Clore's move, although ultimately unsuccessful, affronted Simon Combe, the patrician chairman of Watneys, and startled the insular United Kingdom brewing industry. The ensuing defensive consolidation created the "Big Six" pub-owning brewers, who by the late 1960s controlled the lion's share of the industry. Having escaped Clore's clutches, Watney Mann rapidly expanded, buying six sizable brewers across the United Kingdom by 1965 in moves that doubled the company's tied estate. The firm went on to acquire brewing interests in Belgium and in 1972 bought both the conglomerate drinks company International Distillers and Vintners and the Halifax brewer Samuel Webster before itself being swallowed, after a struggle, by Maxwell Joseph's Grand Metropolitan Hotels group. Joseph, who had bought Trumans in 1971 against acrimonious competition from Watneys, merged the two companies in 1973. Truman's Black Eagle Brewery in Spitalfields had been one of the great London porter breweries. It was the largest brewery in the world in 1860, producing more than 750,000 hectoliters (19.8 million U.S. gallons) per year. Truman's also brewed ale in Burton from 1873 to 1971 and was completing an extensive modernization program when taken over.

Over a Barrel

Watneys embraced the new with an enthusiasm not shown by any other United Kingdom brewer. In the period 1918–1940, the firm was preeminent in improving pubs, spending £2.78 million on rebuilding 286 properties. It pioneered chilled and filtered keg bitter in the 1930s and vigorously and successfully promoted it from the late 1950s. It ran a first-rate laboratory and skillfully operated the tricky technology of production-scale continuous brewing earlier and longer than any other Northern Hemisphere brewer from 1960 to 1975. But in the 1970s, the company became a prime target for a newly formed consumer group, the Campaign for Real Ale, that was vehemently opposed to keg beers, the replacement of tenants with managers, and closure of country pubs and breweries. Watneys was in the vanguard in all these areas and was vilified for it. Crucially, the firm had become too closely associated with Red Barrel, its archetypal premium keg bitter. A massive marketing campaign to relaunch and extend the brand "The Red Revolution" was poorly received and the company went into terminal decline.

Watneys had little to fall back on. It had already spurned traditional cask beer and was slow to brew draft lager. By the time a new joint venture plant with Carlsberg started brewing in Northampton in late August 1973, United Kingdom lager sales had been growing at an average 30 percent per annum for a decade. From the mid-1970s, the firm made frantic attempts to recapture lost ground by re-creating a local image for its breweries. But even painting the pubs any color other than the corporate red, removing all traces of the Watney name, and reintroducing cask beer did not restore credibility. In 1990, Grand Metropolitan exchanged its breweries for the pubs owned by Elders IXL. Grand Metropolitan had brewed Foster's brand for Elders IXL since 1981, and Elders IXL had acquired Courage in 1986. In 1997, the pubs were sold to the Japanese banking group Nomura, and Grand Metropolitan merged with Diageo. Anheuser-Busch now operates the Mortlake brewery, where Budweiser had been brewed under license since 1985.

Raymond G. Anderson

See also: Anheuser-Busch; Beer; Campaign for Real Ale (CAMRA); Courage Brewery; National Trade Defence Association (NTDA)

References

Boston, Richard. *Beer and Skittles.* Glasgow: Collins, 1976.

Cornwell, Martyn. "Three Men Holding Me Down: A History of Truman, Hanbury, Buxton & Co. Ltd." in *Journal of the Brewery History Society.* 57 (1989): 4–12.

Gourvish, Terry R., and Richard G. Wilson. *The British Brewing Industry, 1830–1980.* Cambridge: Cambridge University Press, 1994.

Janes, Hurford. *The Red Barrel: A History of Watney Mann.* London: John Murray, 1963.

Serocold, Walter Pearce. *The Story of Watneys.* St Albans: Watney Combe Reid, 1949.

Weston, Agnes (1840–1918)

Agnes Weston (1840–1918) was an English moral reformer who devoted fifty years of her life to improving the physical and spiritual welfare of naval seamen. She was called the "Mother of the Navy." Her work in naval temperance followed in the tradition of earlier maritime missions and was part of a larger evangelical Christian temperance movement that considered temperance a vital step toward salvation. Although Weston herself was Anglican, her many organizations, such as her sailors' rests, claimed to be nonsectarian.

During Weston's childhood, her barrister father moved the family from London to Bath. In her youth, she was active in church missionary work there: playing the church organ, teaching Sunday school to children, and instructing Bible classes for workingmen. Her philanthropic interest in sailors developed from her initial missionary work on behalf of soldiers stationed in Bath. Weston's missionary work with sailors began in 1868 in the form of written correspondence to naval men overseas. With the guidance and financial support of the National Temperance League, she helped to establish the Royal Naval Temperance Society (RNTS) in 1873 with the goal of creating a sober Christian navy. Admiral Sir William King-Hall, an ardent teetotaler, was appointed president, and Weston, who had already gained a reputation as a naval philanthropist, served as superintendent. By the early 1870s, she had also established the Royal Naval Christian Union. By the late 1880s, Weston could boast nearly 200 Christian and temperance branches throughout the fleet. Apart from recruiting individual men to temperance, Weston and the RNTS lobbied both the Admiralty and Parliament to increase the monetary allotment for naval abstainers in place of the rum ration and to abolish the daily rum ration entirely. However, the Admiralty did not abolish the rum ration as a daily allowance until 1970.

Weston also established sailors' rests to provide lodging for naval seamen ashore at the main naval ports of Devonport in 1876 and Portsmouth in 1882. Organized on temperance and Christian principles, these rests acted as the headquarters for her naval temperance movement. Her friend Sophia Wintz, the daughter of a middle-class Anglo-Swiss family from Devonport, aided Weston through her years of work. Wintz particularly helped Weston in writing and editing the many publications of her naval temperance campaign. Weston's monthly magazine *Miss Weston's Ashore and Afloat* was sent to sailors throughout the fleet and to subscribers in Britain.

In 1892, her rests received the Royal Warrant and were renamed the Royal Sailors' Rests. Weston became the first woman to receive an honorary LL.B., Bachelor of Laws, from the University of Glasgow, in 1901. In 1917, she and Wintz received royal endorsement of their work when they both were awarded the Dame Grand Cross of the Order of the British Empire. Weston continued to preside over her various organizations until her death at the age of seventy-eight in October 1918, when she became the first woman honored by the Admiralty with a naval burial.

Mary A. Conley

See also: National Temperance League (NTL); Robinson, Sarah

References

Conley, Mary. "'You Don't Make a Torpedo Gunner out of a Drunkard:' Agnes Weston, Temperance, and the British Navy" in *Northern Mariner/Le Marin du nord.* 9, no. 1 (January 1999): 1–22.

Gulliver, Doris. *Dame Agnes Weston.* London: Phillimore, 1971.

Shiman, Lilian Lewis. *Crusade against Drink in Victorian England.* New York: St. Martin's, 1988.

Weston, Agnes. *My Life among the Bluejackets.* London: Nisbet, 1911.

Wintz, Sophia. *Our Bluejackets: Miss Weston's Life and Work among Our Sailors.* London: Hodder and Stoughton, 1890.

Wheeler, Wayne Bidwell (1869–1927)

Wayne Bidwell Wheeler (1869–1927) was a prominent leader of the Anti-Saloon League of America (ASLA). He sought and obtained personal publicity, earning a reputation as "America's Dry Boss," during the 1920s. A leader of enormous energy and an unsurpassed dedication to prohibition, Wheeler served as general counsel of the league after 1916 and as legislative superintendent after 1919, when he moved his office to Washington, D.C. From that post, Wheeler advocated strong prohibition statutes, strictly enforced. His publicity-seeking was in part a successful attempt to cultivate an aura of political invincibility, although Wheeler was unable to obtain the strict enforcement practices he sought.

Wheeler was newly graduated from Oberlin College (A.B., 1893; A.M., 1894) when Howard Russell recruited him for a career in temperance work. His first position with the fledgling Ohio Anti-Saloon League was as "agitator and organizer," a job description he filled very well indeed. Wheeler's first success occurred quickly, when he rode his bicycle long distances to meet with voters and persuade them to unseat a "wet" state senator. Early in his career with the league, Wheeler realized the importance of acquiring legal skills; he graduated from the Law School of Western Reserve University in 1898. Thereafter, he devoted himself to political maneuvering and legal action.

Promoted to superintendent of the Ohio league in 1903, for a time Wheeler proved politically adept. The "dry" strategy was to work on local-option laws, measures that allowed voters in a local district to declare the liquor trades illegal. There was much resistance to local option, except for the smallest civil divisions, and the fight to achieve expansion of local option was difficult. When Governor Myron T. Herrick, a Republican, opposed the league in 1905, Wheeler led a successful effort to replace him with John M. Pattison, a Democrat who promised to help the drys. The now powerful Ohio league lobbied successfully for higher taxes on liquor businesses and in 1908 achieved a huge victory, a law allowing county option. The next step was state prohibition.

After 1908, however, Wheeler was much less successful. Ohio's brewers launched a self-reform movement that reduced the league's appeal and managed to defeat the league's annual campaigns for state prohibition until 1918. These defeats eventually led Purley Baker, the national superintendent, to appoint Wheeler to his national duties, removing him from the day-to-day fray of Ohio politics. Such were Wheeler's talents, however, that failure in one arena led to a higher, more prominent office.

While serving the league in Washington, Wheeler studied prohibition statutes carefully and worked with sympathetic politicians, especially Andrew Volstead, a congressman from Minnesota, to devise effective national enforcement laws for use in punishing the liquor trades. Wheeler failed to achieve a federal law outlawing the personal possession of liquor, however. He was also frustrated when the Warren G. Harding and Calvin Coolidge administrations were much more restrained than he had hoped, and when Republican congressional majorities refused to appropriate substantial funds for enforcement. Thus, Wheeler's reputation as a dry boss, although sought, was inaccurate; he proved much less powerful in reality than his reputation claimed. He did, however, command a faction within the national Anti-Saloon League that sought to emphasize strict enforcement as the organization's principal strategy following the enactment of Prohibition.

K. Austin Kerr

See also: Anti-Saloon League of America (ASLA); Baker, Purley Albert; National Prohibition (United States)

References

Chalfant, Harry Malcolm. *These Agitators and Their Idea.* Nashville, TN: Cokesbury, 1931.

Hogan, Charles Marshall. "Wayne B. Wheeler, Single Issue Exponent" (Ph.D. dissertation, University of Cincinnati, 1986).

Kerr, K. Austin. *Organized for Prohibition: A New History of the Anti-Saloon League.* New Haven, CT: Yale University Press, 1985.

Steuart, Justin. *Wayne Wheeler, Dry Boss: An Uncensored Biography of Wayne B. Wheeler.* New York: Fleming H. Revell, 1928.

Whiskey Rebellion

The Whiskey Rebellion of 1794 was an armed revolt against the federal government of the United States. Its direct cause was a 1791 federal excise tax imposed on domestic whiskey distillers. The rebellion was centered on the four westernmost counties of Pennsylvania. Accounts of the Whiskey Rebellion's ultimate causes, meaning, and long-term consequences have varied. The events of the rebellion can be traced with better certainty, stretching back to the winter of 1791 and earlier.

On 3 March 1791, Congress passed a law to collect taxes on spirits distilled in the United States. Similar laws existed, but were largely ignored, in Pennsylvania and other states. The new law showed a new resolve on the part of the national government. The whiskey tax was seen by Treasury Secretary Alexander Hamilton as an important source of revenue complementing his fiscal policy. Hamilton had proposed in his "Report on the Public Credit" that the federal government assume all state debts resulting from the American Revolutionary War. The cost to the national government of defending the frontier against hostile Indians also required expenditures, which a whiskey tax could help offset. But passage of the whiskey tax sparked protests in western sections of the country, especially in Pennsylvania.

Concerned citizens, many of whom were of Scots or Ulster-Scots backgrounds, protested in various ways during the summer and autumn of 1791. Some protested on newspaper pages and by petition. Others met together, at Redstone Old Fort (Brownsville) and Pittsburgh, for instance, to discuss ways to proceed against the tax. Others reacted unlawfully and violently. Tax collectors, such as Robert Johnson, the excise agent for Allegheny and Washington counties in Pennsylvania, were harassed, robbed, beaten, and, in some cases, tarred and feathered. Tensions were particularly high in the four westernmost counties of Pennsylvania: Allegheny, Westmoreland, Fayette, and Washington. But as recent historical studies have shown, it is a mistake to see the unrest as exclusive to Pennsylvania. The Carolinas, Kentucky, and Virginia all experienced protests. Distillers across the west were disgruntled by a whiskey excise tax. Most refused to pay it.

Explaining the virulence of that protest is not easy. Contemporaries at the time, and modern historians since, have differed in their assessments. Some then and now have argued that the tax was unfair because for frontier farmers whiskey production

A 1794 wood engraving shows a large mob with a tarred-and-feathered tax collector riding on a rail. (Library of Congress)

was a commercial necessity. The Mississippi River was closed to American commerce. Transportation costs to haul commercial goods across the mountains to eastern markets were high. Many farmers therefore turned their bulky grains into compact whiskey, a product that was less costly to transport. Much whiskey was also consumed locally. It even functioned as a currency for barter exchange. Another explanation emphasizes that the whiskey tax would have drained hard currency away from the west, where it was already in short supply.

A further point of contention was that violators of the whiskey excise were to be tried a long way away from their homes and livelihoods, at the U.S. district court in Philadelphia. Others argued that government inspectors aimed to spy on the "mysteries" of whiskey production, that the law required too much onerous paperwork, or that distilling whiskey was a personal right with which the government ought not interfere. The whiskey excise was seen by some as being in the same vein as the Stamp Act of 1765. The significance of that parallel was that the Stamp Act was commonly thought to have been an unfair tax leveled on the colonists by an overbearing British government. It was one of the precipitating events of the Revolutionary War. All in all, the whiskey excise was widely portrayed as a ploy whereby privileged business interests of the "east" sought to control the underprivileged "west."

These protests, whatever their root, were sufficient to dissuade full-scale enforcement. Violent reactions to attempted enforcement continued in 1792 and 1793. In early August 1792, William Faulkner of Washington County was threatened by a man with a drawn knife for allowing a building Faulkner owned to be used as a tax office. Days later a group of rioters broke into Faulkner's house, which was damaged in part by gunfire. Luckily for him, he was not home at the time. Benjamin Wells, collector of the excise for Fayette County, was harassed on 23 November 1793. By 1794, Hamilton had determined to enforce the whiskey excise more sternly in Pennsylvania.

In June 1794, David Lenox, a U.S. marshal, rode west to serve processes on several dozen distillers who had evaded their taxes. Accompanied by General John Neville, inspector of excise for the fourth survey, Lenox served process on Samuel Miller on 15 July, but not without a confrontation with some 35 to 40 local farmers. On 16 July, approximately 100 men from the Mingo Creek militia aimed to capture Lenox, who was staying with Neville. In that confrontation, a militiaman, Oliver Miller, was shot and killed. The next day, 500 to 700 militiamen returned to find the house defended by a handful of soldiers from Fort Pitt under the leadership of Major James Kirkpatrick. A battle followed in which Neville's house and outbuildings were destroyed. Men were killed on both sides, including James McFarlane, the militia commander. During the last week of July and the first week of August, protests escalated, culminating in a march of approximately 7,000 rural people to Pittsburgh on 2 August.

Hearing details of the July unrest, President George Washington issued a proclamation calling up a militia army of 13,000 men from eastern Pennsylvania, Maryland, New Jersey, and Virginia. This army marched west under Governor Henry Lee of Virginia, accompanied by Hamilton. Washington also sent in advance three peace commissioners: William Bradford, James Ross, and Jasper Yeates. These three met with a rebel committee of twelve on 20–23 August. Before the army arrived in Pittsburgh on 24 October, the rebellion had come to an end and the insurgents had dispersed. However, several dozen suspected rebels were apprehended and taken to Pittsburgh. Twenty of these were marched back to Philadelphia. Two, John Mitchell and Philip Vigol, were charged with treason and given death sentences; Washington later pardoned them, however. Coverage of the Whiskey Rebellion has waned in modern U.S. history textbooks published after the Civil War, against which it pales. But that is an anachronistic measure. Contemporaries interpreted the Whiskey Rebellion as an event of national significance, which it was. Historians will continue to debate its origins, events, and consequences, not least because it illustrates the many tensions that existed in the late-eighteenth-century United States.

Mark G. Spencer

See also: Taxation of Liquor (United States)

References

Baldwin, Leland D. *Whiskey Rebels: The Story of a Frontier Uprising.* Pittsburgh: University of Pittsburgh Press, 1939; rev. ed., 1968.

Boyd, Steven R., ed. *The Whiskey Rebellion: Past and Present Perspectives.* Westport, CT: Greenwood, 1985.

Slaughter, Thomas P. *The Whiskey Rebellion: Frontier Epilogue to the American Revolution.* New York: Oxford University Press, 1986.

Whiskies

Whiskies are spirits distilled from grains. Legend states that St. Patrick introduced distilling to Ireland in the fifth century, though other historians have suggested it was brought from the east by Irish monks. At any rate, when the English first invaded Ireland in the twelfth century, distilling was well established there, and by then the Scots also possessed the art. "Whisky" (the English, Scottish, and Canadian spelling), or "whiskey" (the Irish and American spelling) is a modern derivation from the Gaelic *uiscea beatha, Uisge beatha,* or *usquebaugh,* which means "water of life." It may also be called aqua vitae, though that term also may refer to wine or brandy. Samuel Johnson recognized the anglicized term in his 1755 dictionary. Different whiskies are distinguished by ingredients, equipment used, and place of manufacture.

Method of Production

The grain in whiskey may be malted or unmalted. In malting, grain is soaked in water, allowed to begin to germinate, then dried in a kiln to stop germination and impart certain flavor characteristics. Unmalted grains are ground and cooked. Once cooking is completed, some barley malt is always added, since the enzymes (diastase) in malted barley enable starches to convert to fermentable sugars. The grains are "mashed" to complete this conversion, then mixed with warm water in a vessel with a false bottom that allows liquid (the wort) to drain off. The wort is mixed with yeast in a fermentation vessel, and the yeast converts the sugars to alcohol. The resulting liquid, "wash" or "beer," is ready for distillation.

To distill, the wash must be heated to the boiling point of alcohol, which is lower than that of water. This allows the alcohol to vaporize and separate from the wort. Distillation is done in one of two ways. Traditional distilling took place in a pot still, a sealed pot with a neck through which the vapor traveled through a coiled copper tube immersed in cold water to cause condensation. This liquid, the "low wine," is not potable and must be distilled at least once more. In the patented column or Coffey still, which was designed by Robert Stein in the 1820s and modified by Aeneas Coffey in 1831, the wash enters the top of the column-shaped still and filters down through a series of perforated plates, where it mixes with steam. The alcohol vaporizes and collects in a separate vessel. The resulting spirit is lighter in flavor than that produced by the pot still and does not need to be distilled a second time.

The final step is maturation. Whiskey is stored in wooden casks to allow oils and impurities to leech away, thereby mellowing the flavor. The process varies by length of time and type of barrel, depending on brand, style, law, and tradition. Some manufacturers then mix different grades and qualities of whiskies, to achieve blended whiskey.

Scotch Whiskey

The earliest evidence of Scottish whiskey is from 1494–1495, when the Scottish Exchequer Roll included an entry of malt delivered to Friar John Car to make aqua vitae. In 1497, a barber brought a gift of whiskey to King James IV at Dundee. In 1505, the Guild of Surgeon Barbers in Edinburgh was given the monopoly of distilling and selling aqua vitae within the city boundaries. This link with medicine is not an accident of historical records. Generally, whiskey was used for medicinal purposes. By 1550, however, increasing prosecutions for infringements of the barbers' privilege suggest that whiskey was becoming more popular generally as a beverage.

In 1644, the Scottish Parliament imposed an excise tax on spirits. Oliver Cromwell lowered the tax, which lapsed after the Restoration in 1660. It was reimposed in 1693, and after the 1707 union between England and Scotland, the British Parliament created the Board of Excise, which continued to collect the whiskey duty. Parliament extended the English malt tax to Scotland in 1713 but kept it at half the English tax. Nevertheless, the measure caused riots in Edinburgh and Glasgow and drove the creation of illicit stills across the country.

Excise taxes were widely evaded, especially in the Highlands, which were difficult to access, though illicit distillation was not confined to the Highlands. By 1777, Edinburgh excise officers claimed there were eight licensed and more than 400 illicit stills in that city. Protests from English gin distillers resulted in the Wash Act of 1784, which drew a formal legal distinction between Lowland and Highland whiskey and subjected the latter to a license duty on the gallon content of the still rather than on the amount of whiskey distilled. The rationale was that since the distillation process took a fixed amount of time, officials could predict how much whiskey each still produced. Consequently, distillers developed ways of speeding up distillation, thereby reducing the effect of the license duty. Over the next few decades, the duty was repeatedly increased. Between 1786 and 1800 it rose 1000 percent. Smuggling increased, quality diminished, and the exciseman's job became increasingly difficult. Parliament continued to attempt to eliminate illicit distillation when, in the Government Act of 1814, it prohibited all stills of fewer than 500 imperial gallons (600 U.S. gallons) capacity in the Highlands. By 1823, officials had discovered 14,000 illicit stills.

The 1823 Act to Eliminate Illicit Distilling made significant changes in the whiskey industry. The duties outlined in the act were based upon more sophisticated gauging techniques using the saccharometer, which measured the specific gravity of the wash, and the Sykes hydrometer, which measured proof spirit. Excise collectors could base their assessments upon a more precise indication of whiskey's alcohol content. The act placed a flat license fee of £10 on stills of 40 imperial gallons (48 U.S. gallons) capacity—down from the minimum capacity of 500 imperial gallons (600 U.S. gallons)—and placed a duty on every gallon of proof spirit distilled. The act permitted stock to be warehoused in bond without duty being paid. Warehousing helped distillers deal with problems of supply, thereby enabling them to have better control over market conditions.

After the 1823 act, the whiskey industry began to grow and change. Coffey's patented still instituted a revolution in whiskey distilling. Patent stills took hold in the Lowlands, which had seen considerable distilling of lighter whiskies from a variety of grains for years. Between 1853 and 1855, spirit duties were equalized throughout the United Kingdom, bringing Scotish duties up to the level English distillers were paying. Patent distillers reacted by attempting to share markets and fix prices, and patent distillation soon began to outstrip traditional pot-still whiskey in Scotland. In 1850, 59.8 percent of the Scotch produced was from pot stills, and 40.2 percent from patent stills; by 1860, the ratio was 42.1 percent pot still and 57.9 percent patent. Between 1850 and 1860, the number of operating Scottish distilleries declined from 161 to 125. When pot distilleries closed, other businesses took over their stock and facilities, using their stocks in blending operations.

Blending and vatting offered significant competition to the traditional whiskies and opened up the market. Vatting is the mixing of whiskies from different distillations of various years at the same distillery and was permitted under the provisions of the law of 1853. Blending is the mixing of different brands of malt and grain whiskies to produce a lighter product. It was not a new process, as some distillers had blended their whiskies for decades. But in the latter half of the nineteenth century, blended whiskey grew in popularity. Pot distillers felt blending lowered the quality of their industry, but they also benefited as many of them became dependent upon the blenders to buy the pot-still whiskey for inclusion in blends.

Blending was officially recognized in the 1860 Spirits Act, making it possible for individuals such as Andrew Usher, who had pioneered the practice of vatting, to expand their blending operations. Blending created a lighter product that was more palatable to the English market. In the late 1860s, the decimation of French vineyards by the *Phylloxera vastatrix* pest and the subsequent collapse of the French wine industry caused the price of brandy, the English drink of choice, to skyrocket. Soon whiskey and soda replaced brandy and soda as the Englishman's drink.

The growing popularity of blends in the last quarter of the nineteenth century saw a whiskey boom that ended in a bust. In 1877, a new combination emerged that drastically changed the distilling landscape when six Lowland distilleries formed the Distillers Company Limited (DCL). DCL quickly became one of the largest distilleries. Other enterprises were less successful, yet equally notorious. Robert and Walter Pattison's distillery was the result of successful marketing gone awry. The immediate result was public investment in Pattison's Limited and tremendous wealth for the brothers. More distilleries were built, and public investment increased, but in 1898 the company collapsed, sending shockwaves throughout the industry. The Pattison brothers were convicted of fraud, many investors lost their savings, and the whiskey boom came to an abrupt halt.

The biggest challenge to the industry came in the early 1900s with the question of whether blended whiskey violated the Food and Drugs Act. In 1904, the Islington Borough Council took several publicans to court for deceptively selling blended brandy. The council's success in this case encouraged it to turn to whiskey. The initial hearing resulted in a decision that patent-still whiskey was not true whiskey; upon appeal the bench was divided and the original decision stood. The patent distillers demanded and received a Royal Commission into Whisky and Other Potable Spirits. The commission's 1909 report defined whiskey as "a spirit obtained by distillation from a mash of cereal grains saccharified by the diastase of malt." Patent-still whiskey was whiskey.

World War I caused considerable upheaval in whiskey production. In 1915, the Immature Spirits (Restriction) Act required that all spirits had to be held in bond for two years, later extended to three years. It was designed to limit the potential damage caused by perceived impurities in immature spirits. The effects were a rise in prices and the failure of many small distilleries. The larger distilleries survived. In 1917, distilling was restricted to those patent-still distillers producing industrial alcohol, a measure that lasted until November 1919. Patent distilleries, notably DCL and James Calder and Company, were also increasingly important suppliers of baker's yeast, which previously had come from Germany and Belgium.

In the interwar period, the market became more competitive as recession kept the demand low, North American interests expanded into Britain, and U.S. National Prohibition limited access to that important market. Combinations resulted. DCL merged with the big five during this time, beginning with Haig in 1919 and finishing with Mackie in 1927. Between 1919 and 1926, pot-still distillers discussed creating a protective association, resulting in the Pot-Still Malt Distillers Association of Scotland. By 1933, only fifteen distilleries were still in operation in Scotland, in contrast to 150 in 1900. The industry faced collapse.

It was saved by the end in 1933 of U.S. National Prohibition, which in its duration over the previous fourteen years had presented both challenges and promise to the whiskey industry. Americans remained illicit purchasers of Scotch. Exports increased to Canada, the British West Indies, and the Bahamas, and much of that liquor found its way to the United States. When Prohibition ended in 1933, the market was already primed for a major marketing push by the Scotch whiskey industry.

During World War II, production plummeted and the government encouraged export of whiskey to boost national income. The distillers redirected their efforts to export. Patent distilling was prohibited in 1940, but limited malt whiskey distilling continued until the end of 1941. From 1942 to 1944, distilling was not permitted in order to ensure adequate levels of grain and produce for home consumption. In the fiscal year 1944–1945, malt distilling began again, but full-scale distilling (with some restrictions) did not return until 1949.

By the 1950s, North American companies had begun to compete much more directly with the Scotch distillers. Seagram bought Chivas Brothers just after the war, and Strathisla in 1950; Hiram Walker bought Scapa and Glencadam in 1954; Schenley Industries bought Seager Evans in 1956; and Inver House set up grain and malt distilleries in 1959. By the 1960s, a rapidly expanding market in North America, combined with the elimination of the last wartime rationing measures, had encouraged more distillation. Between 1959 and 1967, Scottish malt distillers increased their number of stills by more than half. Numerous malt distilleries that had closed in the early years of the century were reopened or rebuilt and reconditioned. This expansion continued into the 1970s as demand grew in the North American, European, and Japanese markets.

The last quarter of the twentieth century saw increased consolidation, takeovers, and amalgamation of the liquor industry generally, and the whiskey industry was affected. Among its many purchases in the last half of the twentieth century, the Seagram Company acquired Glenlivet, possibly the most famous malt distillery in the world, in 1977. In 1985, Guinness Brewery acquired the Distillers Company, which some industry watchers called the greatest takeover of all time.

Irish Whiskey

Irish and Scotch whiskies are distinct in several ways. Unlike Scotch, Irish whiskey is not made with malted barley dried in kilns stoked with peat. Also, unlike most pot-still Scotch, Irish whiskey is usually distilled three times rather than two, making Irish whiskey smoother than Scotch. Like single-malt Scotch, Irish whiskey is distilled only in pot stills, though Irish distillers use unmalted barley and occasionally other cereal grains in their stills along with malted barley. Some still argue about what is real Irish whiskey: whether it must come from only barley, malted and unmalted, or whether other grains may be used. However, historically, any whiskey distilled in Ireland from malted and unmalted cereal grains in pot stills was Irish whiskey. In 1950, the Irish government confirmed this definition. This essay does not discuss "poteen" (whiskey made illegally).

Evidence of the popularity of Irish whiskey (or simply "Irish") in the early modern period comes from royal proclamations and laws. A 1556 law drawn up to restrict the distillation of aqua vitae owing to perceived negative social effects from drunkenness required distillers to acquire a license. Whiskey was often connected to political activity. When martial law was declared in Munster in 1580, it was applied to, among others, "makers of aqua vitae." Irish was first taxed in the 1660s under Charles II. The excise tax of 4 pence per gallon became law on 24 December 1661, establishing two other notable Irish institutions: the Excise Office in Dublin, and illicit distillation across the country.

Enforcement was a challenge. The excise law required the distillers to declare the amounts they produced in the course of a year. The excise officers had little means of confirming the distillers' declarations, however. In 1717, the law was revised to include more stringent gauging methods. A 1731 act required stills to be erected only in premises in or within two miles of market towns. In 1757, commercial stills were restricted to those over 200 imperial gallons (240 U.S. gallons), and stills for personal consumption had to be 12 imperial gallons (14.4 U.S. gallons) or below. It was not until 1761 that a law passed giving excise officers power to enter premises without first obtaining the consent of the distiller. Yet the penalties for evading the law remained minimal.

The majority of legal distillation took place in Dublin, a port city that had easy access to imported fuel and grains and a considerable local market. By the middle of the eighteenth century, Dublin's distillation was fully four to five times greater than that of Coleraine, the next largest center.

The first major upheaval in the industry came in 1779, when the government introduced a license duty based upon the estimated number of times a still could be worked off in twenty-eight days. The immediate result was a dramatic decline in the number of legal distilleries, from 1,000 in 1779 to 250 in 1790. As happened in Scotland, the distillers responded by developing ways of increasing their output, while legislation increased the duty. In 1780, the law was amended to increase duties and distinguish and tax stills based upon their size. The output of legal distilleries increased from 1.2 million imperial gallons (1.4 million U.S. gallons) in 1780 to 3 million imperial gallons (3.6 million U.S. gallons) in 1790.

The late 1700s also saw a growing concern about intemperance. In 1791, a member of Parliament (MP) noted that though there were 8,000 spirits licenses in Ireland, census returns showed that about 80,000 houses sold whiskey. Some suggested that it would be best to wean the public off spirits in favor of beer and ale. In 1796, a duty on beer was repealed and the duty on malt increased. Distillers paid both the malt tax and the still duties.

The Act of Union that merged Ireland and the United Kingdom in 1800 did not directly affect the fate of Irish whiskey. Like Scotland, Ireland saw its products treated differently from those in England until legislation in the middle of the century harmonized most spirits regulations. Legislation continued to increase the still duties, and otherwise the laws remained the same.

The effects of British rule were more subtle. For example, when distilling was prohibited because of food scarcity, some Irish MPs charged that the restriction was really concerned with guaranteeing Irish grain imports to England rather than with the welfare of the Irish people. Although Irish whiskey imported to Scotland was subject to full import duties, Scottish distillers received a drawback that actually exceeded the amount of duty they paid. Fortunately for the Irish distillers, Scotch was not very popular in Ireland. An 1817 law restricted the sale of raw spirit in England and considered Irish whiskey to be raw spirit. The law was changed in 1820. In 1823, the Distillery Act (discussed above) established uniformity in the regulations and collection of duty in Ireland and Scotland. Two years later, the Distilling Act extended the principles of the 1823 act to English distillers. The act was an amalgamation of facets from the Irish and the Scottish distilling laws. One of the components that persisted from the Irish system was the warehousing provision, which had existed in Ireland from 1804.

Between 1823 and 1900, the number of Irish distilleries declined but the amount being distilled rose dramatically, from about 3 million imperial gallons (3.6 million U.S. gallons) to 14.5 million (17.4 million U.S. gallons). This number reflects the popularity of Irish whiskey in foreign markets in spite of increasing competition from the patent stills and blends. In 1840, 320,000 imperial gallons (384,000 U.S. gallons) of Irish whiskey were shipped to England; two decades later, this number had risen to more than 1 million imperial gallons (1.2 million U.S. gallons).

The nineteenth century was also the beginning of industry consolidation and cooperation. In 1867, the Cork Distillers company formed from a group of five distillers in that county. In 1886, two Ulster blending companies, Kirk, Greer and Company and Mitchell and Company, joined with a distiller to build a large distillery at Connswater in Belfast. In 1902, United Distillers formed from firms in Belfast and Londonderry.

Competition from the blenders caused distilleries to band together temporarily in a public relations campaign. In a series of newspaper articles, the big four distilleries (John Jameson, William Jameson, John Power, and George Coe) published "Truths about Whisky" to explain how some blenders were diluting pot-still whiskey with patent-still spirit. They urged the consumer to be vigilant and pressed for legislation to deal with the problem.

In the early twentieth century, Irish whiskey faced several challenges from which it never recovered. World War I placed restrictions on distilling. Patent-still whiskey could be used in manufacturing munitions; pot-still whiskey was not prohibited, though it was not encouraged. The 1915 Immature Spirits (Restriction) Act placed a further burden on distillers. The 1916 Easter Uprising exacerbated relations between Ireland and the United Kingdom, and the 1919–1921 Anglo-Irish War restricted trade. After the creation of the Free State, many markets of the empire were closed or restricted to Irish distillers. U.S. Prohibition and the growing strength of lighter blended Scotch whiskey further undermined the industry. It also suffered from a lack of foresight. Irish distillers appeared to view the growth of popularity of blends as simply a trend that would die out. Exports from the Irish Free State contracted from more than 1 million imperial gallons (1.2 million U.S. gallons) in 1923 to slightly more than 390,000 imperial gallons (468,000 U.S. gallons) in 1933. Many distilleries closed.

In an attempt to increase the reputation of Irish whiskey, the Free State raised the compulsory bond period from three to five years in 1926. The measure kept much foreign whiskey off the Irish market, since it was rarely bonded for more than three years. However, it also limited the competitiveness of Irish whiskey abroad, as many countries forbade the sale of whiskey that was not eligible for sale in its country of origin. These conditions made Irish whiskey more expensive than others on foreign markets. Five-year-old Irish whiskey competed with three-year-old Scotch, and lost. Finally, during World War II, while Scottish distillers voluntarily restricted domestic sales in order to encourage export sales, the Irish government legislated the opposite, restricting exports to maintain the domestic excise revenue. Irish whiskey virtually disappeared from the international market.

Despite extreme constriction of the market, the industry persisted. In 1950, the Irish Whiskey Act defined Irish as pot-still grain spirit distilled in Ireland. The Irish Export Board, Coras Trachtala, began to develop an overseas marketing plan for Irish whiskey in 1953. In 1966, the remaining Irish distillers, Jameson, Power, and Cork, combined to form the United Distillers of Ireland, later the Irish Distillers Group (IDG), expanding to include Old Bushmills in the early 1970s. In 1969, the compulsory bonding period was dropped to three years. Despite these efforts, sales of Irish whiskey continued to be low. The IDG remained weak internationally. In the 1980s, French group Pernod-Ricard purchased IDG and began to market Jameson and Bushmills internationally, leaving the other IDG brands to home consumption. As the century came to a close, the Irish whiskey industry appeared again to be growing, with new brands and smaller distilleries competing successfully for market share.

American Whiskey

Historically there have been three main types of American whiskey in the United States: rye, bourbon, and Tennessee. American distillation used indigenous grains, and the preponderance of corn (or maize) in North America led to a preponderance of corn in whiskey production. Even rye has a large proportion of corn, although by law it has to include a majority of rye in the mash. The whiskey distillers typically blended grains in the mash and in the final product. The term "straight" in the name means that the whiskey is not blended with neutral spirits.

Though it likely began earlier, American distilling expanded in the early 1700s with a rush of Scottish and "Scotch-Irish" immigrants into Pennsylvania, Maryland, and western Virginia. They grew rye, because it was a hearty crop, and distilled it and corn into whiskey. When the U.S. government's 1791 excise tax on spirituous liquors spurred the 1794 Whiskey Rebellion in western Pennsylvania, many former whiskey rebels joined a migration west, eventually settling in Kentucky and Tennessee. These states had land that was ideal for corn, and the whiskies that developed were primarily corn-based. "Bourbon" is named after the original Bourbon County, Kentucky, which included all or part of thirty-four present-day counties, though some "bourbons" were distilled elsewhere in the state. Corn whiskey is the unaged product of corn distillation. Bourbon is corn whiskey aged in charred white oak casks. The technique is credited to the Reverend Elijah Craig of Lexington, who is said to have casked his whiskey in barrels he accidentally had charred. Charcoal naturally filters out some whiskey impurities, and the process imparted a distinctive flavor that other distillers soon emulated. In neighboring Tennessee, distillers developed the method of filtering their corn whiskey through thick beds of sugar maple charcoal, the process that marks the difference between bourbon and Tennessee whiskey.

Thomas Jefferson rescinded the liquor excise in 1802, and, except for three years after the war of 1812, spirits remained untaxed until 1862. In that year, Congress instituted a new excise to help pay for the Civil War. In 1868, Congress created a stamp tax on tobacco and liquor but permitted a bonding period of one year, during which time the duties did not need to be paid. This bonding period was increased to three years in 1879, and in 1894 it rose to eight years, and then to a maximum of twenty in 1958. The 1868 act led to the creation of the notorious Whiskey Ring, a conspiracy between distillers and Republican tax assessors to redirect excise duties into Republican Party funds.

Between Reconstruction and World War I, the industry saw considerable changes. The 1862 excise tax forced a number of distilleries out of business. The advent of the Coffey still further exacerbated the situation for smaller distillers, who could not compete with the deep pockets of larger distilleries. The introduction of hinged metal bottle molds at the end of the 1860s allowed bottle sales, making whiskey more portable than it had been when it was sold in casks, from which customers would fill their own receptacles, or in bulkier ceramic jugs. Kentucky distillers bottled 400,000 U.S. gallons in 1903, and 9 million U.S. gallons by 1913.

In the 1880s, the Distillers' and Cattle Feeders' Trust of Peoria, Illinois, began buying small-scale distilleries with an eye to controlling the market. Its tactics were suspect, and in 1893 the "Whiskey Trust" was investigated under the Sherman Anti-Trust Act. It reconfigured itself into the Distilling and Cattle Feeding Company and fell in and out of receivership and suspicion into the 1900s. It reemerged during National Prohibition as the U.S. Food Products Corporation, making yeast, vinegar, and cereal products, and then became the National Distillers Products Corporation in 1924. It bought up stocks of many closed bourbon distilleries, and when Prohibition ended, National Distillers, now a reputable company, owned over half the aged whiskey in the country.

The definition of whiskey itself was a major point of debate at the beginning of the twentieth century, especially between straight whiskey distillers and blenders. When the Pure Food Act of 1906 left the definition relatively loose, Dr. Harvey W. Wiley, chief of the bureau of chemistry in the Department of Agriculture, who was charged with enforcing the act, began to investigate various cases of deception and adulteration. He stopped Canadian Club whiskey at the border because it was not labeled a blend. In 1909, after pressure from the liquor industry, President William Howard Taft announced that since all potable liquor distilled from grain had been known as whiskey for centuries, that definition would stand.

During World War I, Congress passed the War Revenue Act of 1917, which prohibited the use of food to manufacture spirits. Later that year, Congress adopted a Prohibition resolution that became the Eighteenth Amendment. National Prohibition fundamentally reshaped the face of the distilling industry. Smaller distilleries that could not redirect their production into legal products closed or were bought by larger companies, such as National Distillers. Moreover, foreign companies such as Seagram, Hiram Walker, and DCL had stocks ready when Prohibition was repealed in 1933. To combat some abuses, the Federal Alcohol Administration Act of 1936 banned bulk liquor sales and placed enforcement authority in the Federal Alcohol Control Administration. It also established more precise classifications for different types of liquor.

During World War II, the distilleries turned their attention to industrial alcohol production. Beverage whiskey was rationed, and rum began a resurgence in the market. Whiskey production remained low; full production of aged whiskies did not return until the early 1950s. Yet American whiskey languished, and it was not until the renewed popularity of single-malt Scotch in the 1980s that rebranded and newly marketed bourbons, ryes, and others gained a broader public awareness.

Canadian Whiskey

A simple definition of Canadian whiskey is that it is spirit distilled from cereal grains in Canada. In Canada it is often called "rye," but this is technically a misnomer because generally the majority of the grain used in the distillation of Canadian whiskey is corn or wheat, not rye. Canadian whiskies are always blended, usually from malted and unmalted grains mixed with lighter neutral spirit. From the middle of the nineteenth century, Canadian whiskey was distilled in patent stills or rectified and filtered through charcoal or charred bone dust to remove impurities.

As in the United States, distilling in Canada has its roots in early colonial times. Spirits were being taxed in 1758 in Halifax. Rum distilling was in place in Quebec in 1769. In 1792, John Graves Simcoe, the newly appointed governor of the new colony of Upper Canada (the southern part of the modern-day province of Ontario), reported that fifty-one still licenses had been issued that year. Many of the earliest colonists of Upper Canada were migrants from the former American colonies, where the distillation of surplus grain was standard practice.

Early Canadian distilleries were often tied to brewing or grain-milling operations. Many members of the Molson family had, by the middle of the century, considerable interest in brewing and distilling, though the Molsons sold their distilling assets by 1867. James Morton founded a brewery in Kingston with Robert Drummond and, after Drummond's death in 1834, expanded to distilling. His distillery operated until 1900. Ezekiel Gooderham and James Worts formed a partnership in York (later Toronto) in 1832 to establish a flour mill, which included distilling in 1837. The mash was predominantly wheat, followed by barley and rye. In 1861, G&W built a distillery that was at the time the largest in Canada. Theirs was one of more than 200 distilleries in Upper and Lower Canada by the middle of the nineteenth century.

The 1860s were a period of growth for distilling in Canada, notably because the U.S. Civil War seriously interrupted the industry in that country. J. P. Wiser of Prescott and Hiram Walker in Windsor were both close to the border. Hiram Walker's main competitor was Joseph E. Seagram, who acquired full ownership in 1883 of the distillery that was to bear his name. In 1888, Henry Reifel immigrated to British Columbia and established his British Columbia Breweries in Nanaimo, Victoria, and Vancouver, which he renamed the British Columbia Distillery Company.

The standard by which Canadian whiskey became known internationally was established by Hiram Walker and followed soon thereafter by Seagram. When Walker introduced his "Club Whisky" in 1884, he aimed it specifically at an upscale U.S. market. At a time when most whiskies were sold from barrels and often carried in buckets or jugs, Walker's was bottled and labeled to guarantee its source. The term "Canadian" was added after U.S. distillers convinced their government that imported liquor should note its country of origin. The tactic backfired, since its Canadian origins raised the status of Walker's "Canadian Club." Soon, competitors were trying to market other "club" whiskies with the name "Canadian."

In 1890, the Dominion government introduced a law requiring distillers to age their product in oak barrels of 150 imperial gallons (180 U.S. gallons) or less for at least two years. The law caused some distilleries to fold and required larger distilleries to build substantial warehouses, though several were already aging their products. The two-year regulation increased to three in 1974.

The twentieth century saw threats and promises to the industry. American pure food legislation (discussed above) challenged the veracity of Canadian blends in the U.S. market. World War I and National Prohibition presented more formidable challenges. During the war, all provincial governments, followed by the federal government, instituted some form of prohibition, including austerity measures to ensure that grain would be used for food and that workers would not waste their energies on drink. Since alcohol was essential to munitions, some companies, such as the Bronfmans' Distillers Corporation, managed to maintain and even expand their businesses. During National Prohibition, several large Canadian distilleries began bootlegging operations into the "dry" United States. Many smaller distillers were unable to survive Prohibition, which had ended in many provinces in the 1920s but continued in the United States until the end of 1933. When the U.S. market reopened, Seagram, Hiram Walker, Wiser, and others flooded the market with Canadian whiskey.

During World War II, distilleries turned their attention to making industrial alcohol, though the government permitted beverage alcohol production on "distillers' holidays." By the middle of the century, many brands of Canadian whiskey were symbols of upscale quality, with notable entries such as Hiram Walker's Canadian Club and Seagram's Crown Royal and V.O. brands. By the 1960s, sales of Canadian whiskey were increasing by about 10 percent each year, and Canadian distillers began shipping whiskey to the United States in bulk and building distilleries in the west to supply the West Coast and Asian markets.

Into the Twenty-First Century

Recent years have seen distilling increase tremendously. Single-malt Scotches are a fashion item, and bourbons are following a similar trend. Other countries, notably Japan, have developed a significant distillation industry. Japanese whiskey follows the Scotch tradition. It began in 1926 and gained considerable quality by the 1960s. Little Japanese whiskey is exported. Meanwhile, some of the bigger players, such as Seagram, have been amalgamated into nonexistence.

Daniel J. Malleck

See also: Bronfman Family; Corby Distilleries Limited; Gooderham and Worts Distillery; Hiram Walker; Ireland; Japan; Moonshine; Scotland; Seagram; Taxation of Liquor (United States); Whiskey Rebellion; Wiser Distillery

References

Brander, Michael. *The Original Scotch: A History of Scotch Whisky from the Earliest Days.* London: Hutchinson, 1974.

Brown, Lorraine. *Two Hundred Years of Tradition: The Story of Canadian Whisky.* Markham, Ontario: Fitzhenry & Whiteside, 1994.
Carson, Gerald. *The Social History of Bourbon: An Unhurried Account of Our Star-Spangled American Drink.* New York: Dodd, Mead, 1963.
Cooper, Derek. *The Century Companion to Whiskies.* London: Century Publishing, 1983.
Crowgey, Henry G. *Kentucky Bourbon: The Early Years of Whiskeymaking.* Lexington: University Press of Kentucky, 1971.
Dabney, Joseph Earl. *Mountain Spirits: A Chronicle of Corn Whiskey from King James' Ulster Plantation to America's Appalachians and the Moonshine Life.* New York: Charles Scribner's Sons, 1974.
Daiches, David. *Scotch Whisky: Its Past and Present.* London: Macmillan, 1969.
Downard, William L. *Dictionary of the History of the American Brewing and Distilling Industries.* Westport, CT: Greenwood, 1980.
Getz, Oscar. *Whiskey: An American Pictorial History.* New York: David McKay Co., 1978.
History of Whisky website. http://www.history-of-whisky.com.
Jackson, Michael. *The World Guide to Whisky.* London: Dorling Kindersley, 1987.
Journal of the Beverage Tasting Institute website. http://www.tastings.com/spirits/index.html.
Magee, Malachy. *1000 Years of Irish Whiskey.* Dublin: O'Brien, 1980.
McDowall, R. J. S. *The Whiskies of Scotland.* London: John Murray, 1975.
Mcguire, E. B. *Irish Whiskey: A History of Distilling, the Spirit Trade and Excise Controls in Ireland.* Dublin: Gill and Macmillan, 1973.
Moss, Michael S., and John R. Hume. *The Making of Scotch Whisky: A History of the Scotch Whisky Distilling Industry.* Edinburgh: James and James, 1981.
Murray, Jim. *Jim Murray's Irish Whiskey Almanac.* Glasgow: Neil Wilson, 1994.
Rannie, William F. *Canadian Whisky: The Product and the Industry.* Lincoln, Ontario: W. F. Rannie, 1976.
Regan, Gary, and Mardee Haidin Regan. *The Book of Bourbon and Other Fine American Whiskeys.* Shelburne, VT: Chapters, 1995.
Ross, James. *Whisky.* London: Routledge and Kegan Paul, 1970.
Wilson, Ross. *Scotch: The Formative Years.* London: Constable, 1970.

Whitbread

Samuel Whitbread's London porter brewery, founded in 1742, was the biggest in the world by the 1780s. After his death, the firm regressed under a lackluster partnership before new blood brought stability. In common with other London brewers, the company struggled in the first decade of the twentieth century as bad debts incurred by overextended loan-tied publicans accumulated. At Whitbread, however, the effects were ameliorated by the success of the company's pioneering national bottling network. Sydney Nevile, one of the UK brewing industry's most influential figures, became a managing director in 1919 and guided the company through the difficult interwar years and beyond. Whitbread became almost reluctant acquirers of other breweries in the 1960s as its relatively benign symbiotic "umbrella" strategy of expansion was overtaken by events. Increasing diversification culminated in the shedding of all involvement in brewing and pub ownership in 2000–2001 as the company became a hotel-owning purveyor of fast food and fitness.

Apprenticed as a brewer at the age of sixteen in 1736, Samuel Whitbread (1720–1796) entered into a partnership in a brewery six years later. The partners moved to Chiswell Street, then on the eastern rim of London, in 1750 and built a new brewery to exploit the trade in the cheap, brown, heavily hopped beer known as porter. Whitbread was sole proprietor from 1761. For twenty years after 1780, the much-extended brewery, with its huge storage vats and cisterns, had the greatest output of any in London and therefore the world. Samuel's namesake son had little interest in the brewery, and in 1799, three years after his father's death, formed a partnership, freeing himself as far as he could from direct involvement in the business. Within five years, in the hands of an inattentive, quarrelsome partnership, output had fallen by half, and the firm to fourth place among London's brewers. Later partnerships, which included the established brewer John Martineau and his son Joseph, achieved a partial revival of the company, but it was not until the 1860s that output consistently returned to the levels reached in the 1790s. In the ninety years prior to incorporation in 1889, Whitbread was controlled by a total of eight different partnerships comprising thirty individuals. Until 1834, when ale production began, the company made only porter. From 1866, pale ale was also brewed, and by 1875 Whitbread was producing more ale than porter.

The company began bottling beer in 1868 and from 1891 opened a string of forty-eight depots and stores in the United Kingdom and abroad. At the turn of the century, Whitbread was once again the capital's biggest brewer, largely on the strength of this initiative. By 1914, over half of the company's output of 1.4 million hectoliters (37 million U.S. gallons) was bottled at a time when bottled beer accounted for less than 10 percent of the beer drunk in the United Kingdom. The success of the bottling operations helped to insulate the company from the full effects of the problems that plagued London brewers after 1900, when loans made to publicans in order to protect trade turned sour as property prices collapsed. Not that Whitbread was immune; in 1908 and 1913, capital was written down, and between 1900 and 1914 profits were halved and dividends cut from 12 percent to 0.5 percent. In the interwar years, Sydney Nevile pushed the company to the forefront in the "fewer and better" reformed pub movement and through acquisitions and relatively heavy advertising sought to minimize the effects of the loss of free trade as competitors increasingly went into bottling. In 1948, the ordinary stock of the company was sold to the public for the first time. In the same year, Nevile retired as a managing director, but for another twenty years he continued to hold a seat on a board still brimming with descendants of the former partners, with his protégé Colonel W. H. "Bill" Whitbread as chairman.

In the mid-1950s, Whitbread began a policy of offering assistance to smaller brewers who feared unwelcome takeover bids that threatened their independence. Whitbread took a protective equity shareholding and a directorship in the client company, which in turn sold selected beers such as Mackeson Stout, Whitbread's bestseller, in its pubs. This tactic seems to have been genuinely appreciated, at least initially, by the majority of the more than twenty regional and local companies accommodated under what became known as the "Whitbread Umbrella." Matters took on a different character when the emergence of the Canadian E. P. Taylor as a powerful outside predator caused Whitbread to increase its shareholdings and then mop up its smaller brethren in self-defense. Twenty-three brewing companies were acquired in the period 1961–1971. Output went from 2.1 to 7.4 million hectoliters (55.4 to 195 million U.S. gallons) as Whitbread became Britain's third largest brewer.

With acquisition came rationalization. Thirteen breweries were closed between 1966 and 1969, and new ones were built at Luton in 1969, Samlesbury near Preston in 1972, and Magor in South Wales in 1979. Chiswell Street stopped brewing in 1976. Between 1981 and 1984, eight of the remaining sixteen breweries were closed, including the new strike-prone Luton brewery. Whitbread, the protector of independence, ended up closing more breweries than any other of Britain's "Big Six" brewers. The company's response to the British lager boom was to brew Heineken under license from 1968, followed by Stella Artois in the 1970s. Diversification into wines and spirits, soft drinks, and off-licenses began seriously in the 1960s, with increasing investment in restaurants and hotels in the 1980s. The wines and spirits business was sold in 1990 and a chain of health clubs purchased in 1995. In 1999, an attempt to double pub ownership by purchasing Allied Domecq's 3,500 strong chain ended in defeat after an acrimonious battle. Whitbread embarked on a U-turn, selling its breweries in May 2000, its off-licenses in September 2000, and its pubs in May 2001. The price Interbrew paid for the breweries was £400 million; the same company paid £2.3 billion for Bass a month later. Bass had nearly twice the barrelage but the huge price differential also reflected the weakness of Whitbread's position as a licensor rather than owner of its lager brands.

After Whitbread sold its breweries, the Whitbread Brewing Archive was dispersed. Business records were deposited at London or county records offices, according to their geographical origin, while books went to the new National Brewing Library at Oxford Brookes University.

Raymond G. Anderson

See also: Allied Breweries; Bass; Beer; Interbrew; Pubs and Beerhouses (United Kingdom)

References

Knox, Diana M. "The Development of the London Brewing Industry, 1830–1914, with Special Reference to Whitbread and Company" (B.Litt. thesis, University of Oxford, 1956).

Mathias, Peter. *The Brewing Industry in England, 1700–1830.* Cambridge: Cambridge University Press, 1959.

Nevile, Sydney O. *Seventy Rolling Years.* London: Faber & Faber, 1958.

Redman, Nicholas Barritt. *The Story of Whitbread plc, 1742–1990.* London: Whitbread, n.d.

Ritchie, Berry. *An Uncommon Brewer: The Story of Whitbread, 1742–1992.* London: James & James, 1992.

White, Mary (1827–1903)

Mary White (1827–1903) was a Quaker woman prominent in the temperance movement in the central region of Scotland, especially Glasgow and Edinburgh. She was a founding member of the Scottish Christian Union (SCU) and sat on the executive committees of both the Glasgow District Union and the Edinburgh Central Branch. A respected member of the Religious Society of Friends, she helped to found two rescue homes in Glasgow, the Whitevale Mission Shelter and an orphanage.

She was born just outside Glasgow on 3 November 1827. The youngest child of William White, a merchant, and his wife Jane, she attended Wigton School for three years. She has been described as well read and a lover of natural history, and her mother has been credited with influencing her keen sense of social responsibility. As a girl, she accompanied her mother to meetings with women interested in the antislavery cause and social reform.

White's temperance work began in 1860 when she helped the wife of the governor of the Duke Street Reformatory distribute temperance tracts in Bluevale, outside Glasgow. In 1871, she aided in the resettlement of ninety orphan boys in Canada. During her time in Canada, White became interested in Eliza "Mother" Stewart's Women's Temperance Crusade in Ohio. When White returned to Scotland, she contacted female Good Templars in Glasgow, and in 1875 she helped to establish the Glasgow District Union (originally the Glasgow Prayer Union and the Glasgow branch of the SCU). Mary White was secretary of the Glasgow District Union from 1881 until about 1898 and president from 1893 to 1902. As president, she also was a vice president of the Edinburgh Central Branch, the headquarters of the SCU.

White was well known in the Society of Friends and was a member of the Edinburgh monthly meeting. She was an eloquent preacher and was recorded as a minister in 1878. In cooperation with her close friend and fellow Quaker Agnes Ann Bryson, White drew on religious networks to establish the Whitevale Mission Shelter in 1878 (also known as the Glasgow Mission Shelter and, originally, the Prison Gate Mission). The

Whitevale Mission Shelter was committed to the "rescue" of "inebriate" and "fallen" women.

White was also involved in the resettlement of "waifs" in the white settler colonies. She collaborated with Annie Macpherson, herself well-known for her work with London's orphans. In the period 1871–1873, White twice escorted destitute children to Canada. By the mid-1870s, she had contacted Glasgow's famous evangelical reformer William Quarrier. Aided by Quarrier, she helped set up the Emigration Home for Glasgow Waifs.

Megan Smitley

See also: Good Templars (IOGT); Scotland; Scottish Christian Union (SCU)

References

"Mary White" in *Annual Monitor.* 63 (1905): 145–156.

Smitley, Megan. "'Inebriates,' 'Heathens,' Templars and Suffragists: Scotland and Imperial Feminism, c. 1870–1914" in *Women's History Review.* 11, no. 3 (2002): 455–480.

———. "'Woman's Mission': The Temperance and Women's Suffrage Movements in Scotland, c. 1870–1914" (Ph.D. dissertation, University of Glasgow, 2002).

White, Mary. "Recollections of My Temperance Work" in *Scottish Women's Temperance News.* 2, no. 2 (February 1898): 21–22.

"White, Mary" in *Dictionary of Quaker Biography.* Typescript at Library of the Society of Friends, London.

Whittaker, Sir Thomas Palmer (1850–1919)

Sir Thomas Palmer Whittaker (1850–1919), member of Parliament for Spen Valley from 1892 until his death, was a leading English prohibitionist who served on the Royal Commission on Liquor Licensing Laws from 1896–1899. From that time he began to favor more general local control of drink and finally the Gothenburg system of disinterested management. He was a leading figure in the Temperance Legislation League (TLL) and ended up a bitter critic of the prohibitionists.

Whittaker was the son of an eminent English temperance pioneer, Thomas Whittaker. The younger Whittaker worked in the hardware and iron trade and later was managing director (from 1898) and then chairman (from 1908) of the United Temperance and General Provident Institution. Earlier, he had been a loyal supporter of the United Kingdom Alliance (UKA) and had published articles in favor of its policy of local veto. He had also written on the economic effects of drinking.

Whittaker was one of eight representatives of the temperance movement on the Royal Commission on Liquor Licensing Laws, which produced divided reports. The chairman, Lord Peel, advocated in his minority report a broad program of licensing reform based around a scheme of reducing numbers of licenses, the abolition of grocers' retail licenses, and various other restrictive reforms along with a time limit before the state acquired the monopoly value of all licenses with opportunity for local veto at the end. Whittaker set about rallying support for the Peel program with the aim of establishing it as Liberal Party policy. In the process, he quarreled with Sir Wilfrid Lawson and many other supporters of the orthodox policy of local veto in the UKA. Tactical differences soon turned into strategic conflicts. By 1903, Whittaker and his allies had come to an agreement with Joseph Rowntree and Arthur Sherwell to work for a broad licensing reform that included provision for disinterested management, a policy the latter had been championing. The TLL was formed in 1904 and henceforth became the main vehicle for reformers who sought to promote disinterested management as a solution to the British temperance problem. Such reformers found their policy enshrined in the work of the Central Control Board (Liquor Traffic) in World War I, and in particular in the Carlisle scheme. The divisions from 1899 to 1904 marked the start of a bitter internecine war within the British temperance movement that was to last until the 1930s. Whittaker and the TLL worked hard to support Herbert Asquith's Licensing Bill of 1908 and, later, David Lloyd George's scheme for state purchase of the drink trade in World War I, an idea that prohibitionists opposed.

Although temperance reform was his major interest, Whittaker was also an active supporter of free trade and a defender of trade unions' rights. He was knighted and appointed a privy councillor in 1906, and in 1916 he chaired a royal commission on the importation of paper. He remained a supporter of Lloyd George's after the Liberal Party split of 1916. Contemporaries paid tribute to his combative qualities and his blunt Yorkshire manner, which he combined with effective organizational talents.

John Greenaway

See also: Caine, William Sproston; Carlisle System; Carter; Henry; D'Abernon, Lord (Sir Edgar Vincent); Gothenburg Schemes/Disinterested Management; Lawson, Sir Wilfrid; Liberal Party (United Kingdom); Lloyd George, David; Rowntree, Joseph; United Kingdom Alliance (UKA); Whittaker, Thomas

References

Fahey, David M, "Temperance and the Liberal Party—Lord Peel's Report, 1899," in *Journal of British Studies.* 10 (1971): 132–159.

———. "T. P. Whittaker" in *Biographical Dictionary of Modern British Radicals,* vol. 3, *1870–1974,* edited by Joseph O. Baylen and Norbert J. Gossman, 866–868. Hemel Hempstead, UK: Harvester Wheatsheaf, 1988.

Whittaker, T. P. "Practical Temperance Reform" in *Twentieth Century Quarterly.* (August 1906): 48–51.

Whittaker, Thomas (1813–1899)

Thomas Whittaker (1813–1899) was a pioneering North of England temperance advocate and the author of one of the few Victorian temperance autobiographies, *Life's Battles in Temperance Armour* (1884). As a young man, he undertook the hard work of street evangelization, generally traveling from town to town on foot. Whittaker said that on his travels, when he could not find a teetotaler, he would locate a Methodist. Once, after a town crier refused to announce a temperance meeting, Whittaker brought out a rattle to attract a crowd. The rattle became his trademark.

Whittaker's early campaigns took place when the temperance movement had yet to take its familiar shape. At first, many middle-class temperance reformers opposed only the drinking of spirits, and well-off teetotalers quarreled over the question of serving alcoholic beverages to nonabstaining friends and guests. Working-class abstainers, such as Whittaker, played a more prominent role as teetotal missionaries and organizers than they did in the late Victorian period. Until the founding of the United Kingdom Alliance in 1853, teetotalers aimed only at voluntary pledges, not prohibition.

Several temperance organizations hired Whittaker to spread the word. For instance, in 1836 the British Association for the Promotion of Temperance; in 1838, the New British and Foreign Temperance Society; and, much later, in 1874, the National Temperance League. In the mid-1870s, he lectured in the United States. He also helped convert J. H. Raper and Sir Wilfrid Lawson's father to teetotalism.

Whittaker criticized many aspects of the multisided temperance movement. In his autobiography, he complained about the waste of resources on things peripheral to reform. He considered temperance halls, processions, and musical bands a waste of money, the Good Templar fraternal society an expensive distraction, and gospel temperance empty excitement barren of enduring results. Although Whittaker eventually became a prohibitionist, his heart always remained with the pioneer "moral suasion" temperance reformers. He dedicated his autobiography to Joseph Livesey. In other ways, too, Whittaker was a traditionalist: He opposed women speaking from temperance platforms and complained about the indifference that the church exhibited toward the drink evil.

In his later years, Whittaker rose to middle-class status as an agent of the United Kingdom Temperance and General Provident Institution. In 1880, he was elected mayor of Scarborough in Yorkshire and later was appointed a justice of the peace. In 1897, a local paper, the *Scarborough Gazette,* of which he had become part owner, published reminiscences that supplemented his 1884 autobiography. At the turn of the century, his son, Sir Thomas Palmer Whittaker, acquired prominence as a temperance politician.

David M. Fahey

See also: British Temperance League (BTL); Good Templars (IOGT); Lawson, Sir Wilfrid; Livesey, Joseph; National Temperance League (NTL); Raper, James Hayes; United Kingdom Alliance (UKA); Whittaker, Sir Thomas Palmer

References

Harrison, Brian. *Dictionary of British Temperance Biography.* Aids to Research, no. 1, *Bulletin Supplement.* Coventry, UK: Society for the Study of Labour History, 1973: 135.

Whittaker, Thomas. *Life's Battles in Temperance Armour.* London: Hodder and Stoughton, 1884.

Wightman, Julia Bainbrigge (1817–1890)

Julia Bainbrigge Wightman (1817–1890), wife of the Shrewsbury vicar Charles Wightman (1816–1896), wrote fifteen works promoting Christian teetotalism. Her best-selling *Haste to the Rescue or Work While It Is Day* (1858), with British, U.S., and Dutch editions, was credited with the 1862 founding of the Church of England Total Abstinence Society. A volunteer in an era when well-bred women were denied paid employment, Julia Wightman joined male "improvers," helping working people adopt more utilitarian and industrial attitudes toward drink and temperance, expenditure and savings, recreation and work.

The Calvinist, evangelical Wightmans actually performed their duties at the St. Alkmond's vicarage, a sinecure Charles had taken over from his father. There, they had initiated several unsuccessful programs to attract working people. Despite fourteen years' exposure to teetotalism, Mrs. Wightman first condemned the abstinence of some parishioners. She considered drink medically necessary and feared teetotalism's association with Chartism and its competition with religion. She reversed her opposition when Richard Stedman, a "devout" parishioner and secret drunkard, obeying his dying wife, signed the pledge; an increasing number of fellow working people also did so.

As president of the resultant parish teetotal society and night school, Julia Wightman instituted Christian principles but relied on working-class missionaries to distribute tracts, raise funds, and judge pledge violations. She became a life abstainer despite continuing sickliness, shunning Bath Abbey church after a curate, preaching there, counted how much champagne, beer, porter, brandy, or whiskey a moderate could drink.

Julia Wightman and her Shrewsbury parish teetotal council opened the Shrewsbury Working Men's Hall in 1863. Financed by donations and book profits, the hall included facilities for religious services and benefit society meetings as well as bathrooms, a "ragged" school for poor children, and a library and offered concerts by a brass band, lectures, and inexpensive meals. By 1872, 4,500 working people, mostly abstainers, used the hall.

Julia Wightman's role in founding the Church of England Total Abstinence Society has been exaggerated. She did not attend the all-male, organizational London coffeehouse meeting of fifty teetotal clergymen in May 1862, although her bishop, John Lonsdale of Lichfield, read the letter that she had written. The meeting was inspired, however, partly by the distribution of 10,300 copies of *Haste to the Rescue* by the National Temperance League (NTL). Several Anglican clergymen had embraced teetotalism after reading her book, as did the Baptist prohibitionist W. S. Caine. The church teetotal society founding resulted from teetotalism's greater acceptability—medically, socially, and religiously—by the mid-1850s, and constant pressure on parishes by the largely Nonconformist NTL and United Kingdom Alliance, to which many Anglican teetotal clergymen belonged.

With the formation of the "dual-basis" Church of England Temperance Society (CETS), Julia Wightman, considered the more intelligent of the Wightmans, led the parish teetotal section, leaving the "general section" to her husband Charles. Nationally, she supported the CETS Women's Union.

Fearing secularism and gambling, she later abandoned her support for the Working Men's Hall when laymen assumed control.

Gerald Wayne Olsen

See also: Caine, William Sproston; Chartism; Church of England Temperance Society (CETS); National Temperance League (NTL); United Kingdom Alliance (UKA)

References

Fletcher, J. M. J. *Mrs. Wightman of Shrewsbury.* London: Longmans, Green, 1906.

Olsen, G. W. *Drink and the British Establishment: The Church of England Temperance Society, 1873–1914.* Unpublished book manuscript.

———. *Drink and the Temperance in the Church of England, 1829–1875.* Unpublished book manuscript.

———. "From Parish to Palace: Working Class Influences on Anglican Temperance Movements, 1835–1914" in *Journal of Ecclesiastical History.* 40 (1989): 239–252.

Shiman, Lilian L. *Women and Leadership in Nineteenth-Century England.* New York: St. Martin's, 1992; London: Macmillan, 1992.

Wightman, Julia. *Haste to the Rescue or Work While It Is Day.* London: Nisbet, 1858.

Wilberforce, Basil (1841–1916) and Wilberforce, Ernest (1840–1907)

Basil Wilberforce (1841–1916) and Ernest Wilberforce (1840–1907) were grandsons of William Wilberforce (1759–1833), the evangelical slave abolitionist, and sons of "Soapy Sam" Wilberforce (1805–1873), the high church bishop of Oxford and Thomas Huxley's antagonist on evolution. A teetotal prohibitionist radical clergyman, Basil quarreled with his brother Ernest, bishop of Newcastle and later of Chichester, who was the moderate third chairman of the Church of England Temperance Society (CETS).

The Wilberforces show the progress of teetotalism in three generations. In keeping with the social and biblical principles of the Cambridge evangelical Charles Simeon, William encouraged his son Samuel, when an Oxford student, to drink "the fruit of the vine" at evening parties. As bishop of Oxford (1845–1870), when drinking still cheered university life, Sam advised the lemonade-drinking vicar and CETS founder Henry Ellison "to take a little wine for his stomach's sake" (quoted in Sherlock 1910: 16). Sam's sons, Basil and Ernest, took the total abstinence pledge the year he died and implored the Anglican establishment to help working people through temperance, demonstrating the church's usefulness in a disestablishment-minded era.

At age thirty-one, Basil became an abstainer in the first year of the CETS, when he also initiated a CETS parish branch for teetotalers only at Southampton. A CETS vice president, he also joined the prohibitionist United Kingdom Alliance (UKA) and the militant, U.S.-inspired, teetotal Blue Ribbon Army. As vicar of Seaforth, Ernest publicly declared his teetotalism in 1873, established a CETS parish society for both abstainers and moderate drinkers, and became a Chester CETS diocesan officer. His wife supported the temperance efforts of the CETS Women's Union among women, children, and cabmen.

Basil rejected both his grandfather's evangelicalism and father's high church allegiance for a radicalism that was at once political, social, and religious. At first, he emphasized the analogy of drunkenness with slavery as the objects of sacred church crusades. Disillusioned, he later claimed that the church had neglected both emancipation and temperance. Like the Scottish American robber-baron/philanthropist Andrew Carnegie, and extreme British Nonconformists, Basil and other Anglican temperance militants exaggerated the Church of England's connections with drink, especially the promotion of pubs on its extensive lands.

Condoning Prime Minister W. E. Gladstone's exclusion of Basil from episcopal appointment for teetotal excesses, Queen Victoria condemned Basil's attacks on the ecclesiastical commissioners for allowing pubs on church property. When Basil was appointed canon of Westminster, she warned him through Gladstone not to preach militant teetotal sermons at Westminster Abbey. As Prince of Wales, the future Edward VII snubbed Basil for criticizing his liberality toward drink. Dean George H. Connors of Windsor, Ernest's teetotal father-in-law, condemned Basil's "excess of zeal" (Browne to Gladstone 1883). Basil's convert to teetotalism, an earlier canon of Westminster, the Broad Churchman F. W. Farrar, chairman of the convocation of Canterbury's Committee on Intemperance, was also a UKA officer, but as a more prudent temperance advocate, he won promotion as dean of Canterbury.

The way that Basil advocated teetotalism and prohibitionism exceeded the tolerance of churchmen, aristocrats, and royals. Appearing on public platforms with prohibitionist UKA members and other temperance militants, he offended CETS officials, including Ernest, who as bishop and CETS chairman advanced temperance principles more cautiously. Basil and other CETS radical priests believed not only in teetotalism and prohibition but in the abolition of the sacramental, medical, and social uses of intoxicants and in termination of any investment in what they derided as the "drink traffic." Ernest and moderate CETS clergymen favored a more balanced position: the CETS dual-basis, with sections for proletarian teetotalers and for privileged nonabstainers who attacked intemperance's deeper social, cultural, and legislative causes; restriction rather than prohibition of drinking places; retention of fermented Communion wine; and appropriate social drinking for the responsible classes. They did not condemn all investment in the drink trade.

In 1890, Basil helped lead a CETS rebellion of almost 300 members, including Lady Henry Somerset, the British Women's Temperance Association president, and the Anglo-Catholic Charles Gore of Pusey House, Oxford, against CETS support for a Conservative government's license purchase scheme, which they thought too friendly to the drink trade. Basil temporarily resigned membership in the CETS. This revolt helped cause the retirement, for health reasons, of the first CETS chairman, the moderate Henry Ellison. When Frederick Temple, Ellison's successor, in turn resigned, Ernest, recently made bishop of Chichester, was elected chairman. The two brothers became increasingly distant.

Ernest's moderate compromise won greater church preferment but also influenced the CETS, so closely identified with the establishment. Ernest steadily advanced in the church, while his brother, previously considered more promising, became only archdeacon of Westminster. In 1878, Ernest was appointed canon of Winchester. The first of several abstaining bishops, he next headed the Newcastle diocese, created in 1882 for the mostly working-class population, many of them teetotalers. From 1896 to 1907, he was bishop of Chichester and CETS chairman. Unlike Basil, Ernest helped the CETS, divided between extremists and moderates, find compromises for church and nation. Ernest's, not Basil's, temperance program shaped ecclesiastical and governmental policy.

During Ernest's CETS chairmanship, the Royal Commission on Liquor Licensing Laws (1896–1899), also known as the Peel Commission, set the agenda for licensing proposals for Conservative and Liberal governments during the next decade. Seeking legislation based on the Peel Commission reports, a wide cross-section of temperance reformers, led by the CETS and some relatively pragmatic members of the UKA, founded the Temperance Legislation League in November 1899, with the brothers Wilberforce for a time in political collaboration. This cooperation helped pass the Conservative 1902 Licensing Act. Family relations, nonetheless, remained strained. The *CETS Temperance Chronicle* mentioned Basil's presence at Ernest's 1907 funeral, without indicating his temperance activities. Yet, when Basil died in 1916, the CETS *War Issue,* ignoring his occasional bolting from the society, stressed his continuing service as CETS vice president.

Wilberforce papers are at the Bodleian Library, Oxford University, in the Gladstone Papers at the British Library, and in the Archbishops of Canterbury papers at Lambeth Palace.

Stephanie Olsen and Gerald Wayne Olsen

See also: Christian Socialism; Church of England Temperance Society (CETS); Church of England Temperance Society (CETS) Inebriate Homes; Church of England Temperance Society (CETS) Police Court Mission; Close, Francis; Conservative Party (United Kingdom); Cruikshank, George; Davidson, Randall; Ellison, Henry; Evangelical Temperance (United Kingdom); Gore, Charles; Moule, Handley Carr Glyn; Wightman, Julia Bainbrigge

References

Atlay, J. B. *The Life of the Right Reverend Ernest Roland Wilberforce.* London: Smith Elder, 1912.

Browne, Bishop E. H., to W. E. Gladstone, 30 September 1883. Gladstone Papers, Add. 4415 ff. 196–197, British Library.

Olsen, G. W. *Drink and the British Establishment: The Church of England Temperance Society, 1873–1914.* Unpublished book manuscript.

———. *Pub and Parish: Drink and Temperance in the Church of England, 1829–1875.* Unpublished book manuscript.

Russell, G. W. E. *Basil Wilberforce: A Memoir.* London: Murray, 1917.

Sherlock, Frederick. *Henry John Ellison: Founder of the Church of England Temperance Society.* London: Wells Gardner, 1910.

Willard, Frances Elizabeth Caroline (1839–1898)

Frances Elizabeth Caroline Willard (1839–1898) was the leading women's temperance reformer of the nineteenth century in the United States—and indeed anywhere. Most closely associated with the Woman's Christian Temperance Union (WCTU), she was also an advocate of suffrage and other causes in the aid of women's emancipation.

Upbringing and Early Career

Born in Churchville in western New York State on 28 September 1839, Willard was one of five children, but two died in infancy. She was raised on a thousand-acre Wisconsin farm in Janesville by her Methodist parents, Josiah and Mary Willard, who had migrated there in 1846. She was close to her brothers and sisters, but her family life was several times marked by tragedy, particularly with the death of her sister Mary and later her older brother, Oliver. Willard was the only surviving offspring by the late 1880s.

After a tomboyish (by her own admission) upbringing, Willard attended college in Milwaukee in 1857 and then in Evanston, Illinois, which she retained after 1860 as her home base because her parents had moved there. She served upon graduation from the Northwestern Female College in 1859 as a schoolteacher in Pittsburgh and other locations across the northern states until 1868, when, after the death of her father, she toured Europe with a friend, Kate Jackson. When she returned to Evanston in 1870, she was appointed president of her alma mater, but she resigned in 1874 after a stormy tenure. Willard had clashed badly with Charles Fowler, the president of Northwestern University, the men's institution to which her college had become affiliated. It did not help matters that she had for a short time in 1861 been Fowler's fiance and that the engagement had been broken off in murky and acrimonious circumstances.

Frances Elizabeth Caroline Willard (1839–1898) (Library of Congress)

Temperance Sympathies

Having left academia, she worked first for Chicago-based evangelist Dwight Moody in 1877, establishing her reputation by preaching to crowds of thousands of women at a time. She then turned to temperance as an alternative career. Willard had learned from her Methodist youth of the dangers of alcohol use and abuse. Her parents were stern temperance supporters. She accepted the conventional bourgeois belief that alcohol constituted a threat to the home and to women's place within it. Because of alcohol, she believed, women were subject to domestic violence, not to mention loss of income, and husbands were tempted to spend time in the saloon rather than the home. By drinking, men could also fall prey to such evils as gambling and prostitution. The saloon would undermine political, social, and domestic tranquility unless church people and reformers united in a campaign against drink.

This much she shared with other women of her middle-class status and Protestant, temperance persuasion in the mid-nineteenth century. The Women's Temperance Crusade of 1873–1874 in the Midwestern states indicated widespread middle-class concern over the threats that drinking appeared to pose to the stability of family life. Willard knew of women picketing saloons and praying for the souls of the drinkers, though she did not engage in such direct action herself and always preferred reform to militancy. The death of her brother in 1878, and a nephew's subsequent alcohol abuse, later reinforced her conviction of drink's dangers, but by that time she was already a leading temperance reformer.

Joining and Leading the WCTU

The WCTU that emerged after the upheaval of the Women's Temperance Crusade not only offered Willard a new career but also gave her a clear moral purpose in life. As president, she had no salary until 1886 and thereafter received only modest remuneration at most. Instead, she survived mainly on lecture fees. Upon the founding of the national WCTU, Willard became national corresponding secretary in 1874. Her rise thereafter within the organization was mercurial. Elected as Illinois state president in 1878, and national president the following year, she was a vigorous leader after the conservatism of President Annie Wittenmeyer. Willard began to call for what she termed "Home Protection." In 1878, she conceived the novel idea of a huge petition of the women of Illinois to the legislature to gain the vote on liquor questions. The Home Protection petition was a successful publicity stunt and aided recruitment of supporters, though women did not achieve the ballot in the state until much later. However, in 1882 Willard established a Franchise Department within the WCTU to distribute pro-suffrage material. Home Protection showed how she used the conservative rhetoric of women's domestic obligations as mothers, wives, sisters, and daughters to draw women into wider social roles.

Willard was a strong as well as an innovative leader, remaining as national president for nineteen years until her death. She altered the organizational structure of the WCTU, replacing committees of supervision over reform work with departmental superintendents in a system modeled on U.S. business methods. She thought of the temperance movement as a kind of moral bureaucracy specializing in a variety of causes but run by women whom she termed "Protestant nuns" (Tyrrell 1991: 123). As she remained single herself, the tag fit. Under her influence, the WCTU also adopted the "Do Everything" policy. Thus the list of issues the organization would tackle included all conceivable moral reforms that could be connected, directly or indirectly, to temperance. Thus antiprostitution work, scientific temperance instruction, foreign and home missionary work, kindergartens, flower missions, labor concerns, campaigns for peace and for the Sabbath, antitobacco and antinarcotic campaigns, and women's suffrage were among the causes she championed, and she worked to establish departments of work for these within the WCTU.

In the late 1880s and early 1890s, Willard's reformism flowed into more radical causes. She came to believe that wage reform and trade unions were important parts of the temperance reform agenda. She developed in 1886 a friendship with Terence Powderly, the Knights of Labor leader and temperance advocate. In the 1890s, she declared herself a Christian Socialist. This stance sprang from her commitment to the Social Gospel, justified in terms of the biblical "golden rule." But it was reinforced by her observation of poverty in large cities and her exposure to labor and working-class politics in England.

Personal Appeal and Abilities

Frances Willard had a compelling platform presence, and her rank-and-file followers adored her. But even though she was a convincing speaker, she preferred the pen as a means of communication. As a journalist, she contributed to the *Union Signal,* the WCTU's national paper, edited that paper from 1892, and edited other papers earlier in her career, including a daily newspaper in Chicago. She also supported Matilda Carse's Woman's Temple and the associated establishment of the Woman's Temperance Publishing Association. This business venture centered around the temple was designed to show women's capacity for economic initiative and independence. Though backed by the WCTU, it failed in the depression of the 1890s.

Under Willard's leadership, the WCTU membership rose dramatically in the 1880s but then slumped in the economic troubles of the early 1890s and in the wake of controversies over policies and leadership, including dissension from Willard's own radical turn. Though Willard was immensely successful as a publicist and leader, several of her policy stances provoked internal opposition and controversy. She sought an alliance with the Prohibition Party as a third force in U.S. politics in the 1880s, and thereby offended Republican Party loyalists. As a result, she made an enemy of Judith Ellen Foster, who took the nonpartisan forces out of the WCTU. In 1892, Willard moved close to the Populist Party but was unable to effect a full alliance.

Willard addressed hundreds of conventions and other meetings and gave innumerable interviews to the press. She crisscrossed the United States in the 1880s to speak in support of Home Protection. The WCTU was established in the South during an 1881 trip, and Willard took the temperance message to every western state in 1883. She recorded her experiences on these trips, writing for her journalistic outlets on trains and in hotels. Though from 1892 to 1896 she spent most of her time living in England, she did not—apart from this period and visits to Canada—travel outside the United States during her presidency. She suffered badly from seasickness and was too much in demand in her home country.

Nevertheless, Willard was a great supporter of temperance internationalism. In addition to adding peace reform to the WCTU's program and lobbying for the protection of Armenian refugees in 1895, she founded the World's WCTU after visiting San Francisco in 1883 and observing the impact of opium upon Chinese immigrants. She became second world's president of the WCTU in 1891, a position she held until her death.

Willard's hectic schedule gave her little time for reflection. She wrote much as a journalist but little as a more serious and reflective writer. Her *Woman and Temperance* (1884) and *Glimpses of Fifty Years* (1889) illustrate her temperance worldview and her Victorian cultural values. Although she was not an original thinker, she could grasp the significance of new ideas and used them to propel women forward into social reform and political activity. She was able to inspire vast audiences and gained an intensely loyal following that verged on love. Outside the movement itself, she became one of the best-known women in America.

Willard formed strong female friendships. Emotionally she remained very close to her mother, with whom she lived until the latter's death in 1892. From 1877, her closest companion and adviser was Anna Adams Gordon, her private secretary and a WCTU official, but she also developed close bonds with Lady Henry Somerset, the English aristocrat and British Women's Temperance Association president in the 1890s, and lived with Somerset while in England, where she sought to project the WCTU upon the world stage.

Advocate of Women's Political and Religious Emancipation

A Methodist by upbringing and conviction, Willard was, however, tolerant of other faiths. She supported the Parliament of Religions at Chicago, held in 1893 as part of the Chicago Columbian Exposition, and at first backed Elizabeth Cady Stanton's *Woman's Bible* in 1888, a text that exposed the sexism of conventional religious beliefs. Willard also battled her own Methodist Episcopal church for the right for women to be delegates to the church's general conferences. Of this problem she had personal experience. After being elected as a woman delegate in 1887, she had been refused seating.

Described by some subsequent historians as a feminist, she was certainly a prominent if pragmatic advocate of women's suffrage. In 1873, she was elected an inaugural vice president of the Association for the Advancement of Women sponsored by Sorosis, a club for women writers and jounalists, to promote moderate feminism. In the 1890s, she embraced the bicycle for women as a sign of emancipation and for exercise. More important, she was also influential in the 1888 organization of the International Council of Women in Washington, D.C., served as first president of the National Council of Women, and was on friendly terms with leaders of the women's rights movement, such as Susan B. Anthony.

Years of Dissension and Decline

In the mid-1890s, Willard's advocacy, along with that of Lady Henry Somerset, of modification of prohibition through her reputed support for "high license" policies caused concern among temperance women, most of whom, by this time, were, as Willard had been, committed to total prohibition. Equally damaging was her support for Lady Henry when the latter condoned keeping women in brothels for the benefit of the British troops in India. Worn out by overwork, internal political struggles, and travel, Willard was in ill health for the last few years of her life. She died on 17 February 1898 in New York City of influenza and "pernicious anemia." Greatly mourned by community and temperance leaders alike, she was buried in Chicago, near the family home. In 1905, a statue of her was placed in the National Capitol by the State of Illinois in recognition of her pioneering leadership of temperance reform and women's emancipation.

Ian Tyrrell

See also: British Women's Temperance Association (BWTA); Gordon, Anna Adams; Home Protection; Non-Partisan Woman's Christian Temperance Union; Prohibition Party (United States);

Somerset, Isabella Caroline (Lady Henry Somerset); Wittenmeyer, Annie Turner; Woman's Christian Temperance Union (United States); Women's Temperance Crusade; World's Woman's Christian Temperance Union (WWCTU)

References

Bordin, Ruth. *Frances Willard: A Biography.* Chapel Hill: University of North Carolina Press, 1986.

———. *Woman and Temperance: The Quest for Power and Liberty, 1873–1900.* Philadelphia: Temple University Press, 1980.

Earhart, Mary. *Frances Willard: From Prayers to Politics.* Chicago: University of Chicago Press, 1944.

Tyrrell, Ian. *Woman's World/Woman's Empire: The Woman's Christian Temperance Union in International Perspective, 1880–1930.* Chapel Hill: University of North Carolina Press, 1991.

Willard, Frances. *Glimpses of Fifty Years: The Autobiography of an American Woman.* New York: Source Book, 1970 [1889].

———. *Woman and Temperance or, The Work and Workers of the Woman's Christian Temperance Union.* New York: Arno, 1972 [1883].

Wilson, George Bailey (1863–1952)

George Bailey Wilson (1863–1952), for many years general secretary to the United Kingdom Alliance (UKA), was best known as a statistician for alcoholic drink. He was a lifelong total abstainer, having signed a Band of Hope pledge in 1873. He practiced law as a solicitor in Birmingham. In 1901, he was persuaded to join the executive of the Birmingham Auxiliary of the UKA. While in Birmingham, he came to prominence through his attack on the number of licensed premises in a notorious slum area; he opposed the renewal of forty-three drink licenses for a population of only 7,000. After the passage of the Licensing Act of 1904, the magistrates were able to slightly reduce the numbers. Wilson's articles in the *Daily News* (20 April 1904) on the finances of the drinks industry brought him national publicity.

In 1909, Wilson was invited to become the general secretary of the UKA and in the following year launched the publication of the *Alliance Year Book,* which he edited for many years. He changed offices within the UKA in 1919 to become political and literary secretary. In the same year, he joined the executive of the U.S.-based World League against Alcoholism. He edited the *Alliance News* from 1914 to 1943.

In 1910, Wilson competed with a "drink trade" rival for the Howard Medal of the Royal Statistical Society, contributing a statistical review of the variations in the consumption of intoxicating drinks and convictions for drunkenness over two decades. It amused and surprised the prize committee when it discovered the award had been made to an official of a temperance society. That same year, the *Times* invited him to compile the annual letter reporting what was called the National Drink Bill, which he did for thirty years. (William Hoyle had begun this series, and after his death Dawson Burns had continued it until his own death.) The letter described the cost to the United Kingdom of consumption of alcoholic beverages. Wilson's first "National Drink Bill" letter appeared in the *Times* on 31 March 1910. Wilson's lifetime research in this field is on permanent record in his book *Alcohol and the Nation,* for which the University of London awarded him with a doctorate in 1940. It was published in the same year.

As a statistician, Wilson had a reputation for scrupulous accuracy that even his opponents admired; it was a great satisfaction to him that the drinks industry members of the Royal Commission on Liquor Licensing, 1929–1932, accepted his figures. He gave testimony to the royal commission for a day and a half and missed attending only one of its sixty sittings. The editor of *Ridley's Wine and Spirit Trade Circular,* when reviewing *Alcohol and the Nation,* paid further tribute to his accuracy and honesty. Although he warned his readers that Wilson was an official of a prohibitionist society, he then described him as "the only flower which has sprung from that 'dry' ground" (Wilson, "Looking Back" 1947: 74).

Wilson also wrote a biography of Leif Jones, for many years UKA president and a Liberal member of Parliament.

Derek Rutherford

See also: Band of Hope; Burns, (James) Dawson; Hoyle, William; Jones, Leif; United Kingdom Alliance (UKA); World League Against Alcoholism (WLAA)

References

Wilson, George B. *Alcohol and the Nation: A Contribution to the Study of the Liquor Problem in the United Kingdom from 1800 to 1935.* London: Nicholson and Watson, 1940.

———. "Looking Back" in *Alliance Year Book and Temperance Reformers' Handbook for 1947,* 46–77. London: United Kingdom Alliance, 1947.

Wilson, William Griffith (1895–1971)

William Griffith Wilson (1895–1971), cofounder of Alcoholics Anonymous, was born on 26 November 1895 in his family's inn in East Dorset, Vermont, the only son of Gilman and Emily Griffith Wilson. His parents' marriage was not a happy one, and after their separation, Bill lived with his grandfather, Gilman Griffith.

After attending Burr and Burton Academy in Manchester, Wilson entered Norwich University, a military college in Northfield, in the fall of 1914. At the outbreak of World War I in the spring of 1917, he was mustered to Fort Monroe in Virginia; two months later, he was commissioned as second lieutenant and stationed at Fort Rodman, Rhode Island. It was at a party thrown by a patriotic family in New Bedford, Massachusetts, that Bill Wilson first discovered the positive mood-changing effect of beverage alcohol. From then on, Wilson drank at virtually every opportunity, rarely stopping while there was more alcohol available and he was still standing.

On 24 January 1918, Wilson married Lois Burnham of Brooklyn, New York. Six months later, his unit shipped for England, then France, but the armistice was signed before they saw military action. Back in the United States, Wilson had difficulty finding a job commensurate with his own opinion of his skills. After taking some night courses at Brooklyn Law School, he found a position with the first of several brokerage firms. His responsibilities were primarily investigative, but he immediately fell in love with Wall Street. By 1929, Wilson was an obvious alcoholic and could no longer find employment. Shortly after the October crash, in which he lost heavily, he found employment with a brokerage firm in Montreal. He was fired in less than a year because of his heavy drinking.

Returning to Brooklyn, the Wilsons moved in with Lois's father. Four times in 1933 and 1934, Wilson was admitted to Manhattan's Towns Hospital for detoxification. There he met Dr. William Duncan Silkworth, who understood alcoholism as a kind of allergy that involved obsession and compulsion. Silkworth explained to Wilson and his wife that Wilson's choices were abstinence, insanity, or death. Despite this knowledge, Wilson drank again, beginning his last drunk on Armistice Day (November 11, now called Veterans Day) in 1934.

Near the end of the month, one of Wilson's old drinking buddies, Ebby Thatcher, dropped by for a visit. Offered a drink by Wilson, Thatcher declined, explaining that he did not need it any more. Thatcher went on to tell Wilson of his involvement in the Oxford Group, but Wilson shied at the mention of "religion." Still, days later, Wilson decided to investigate what Thatcher had told him and made his way to Towns Hospital to be detoxified yet again. During his hospital stay, Wilson had a classic "spiritual experience" like those described by William James in *Varieties of Religious Experience.* Leaving the hospital, he attempted to carry the message of his conversion to other alcoholics. He found none interested.

In late April 1935, Wilson traveled to Akron, Ohio, as a representative in a corporate proxy fight stemming from his continuing work as a broker. On 11 May, for the first time since his Towns experience of December 1934, Wilson experienced a craving for alcohol. Recognizing that he had not had that experience while trying to work with drunks, Wilson began a series of telephone calls seeking help; these led him to Dr. Robert Holbrook Smith, who would become the other cofounder of Alcoholics Anonymous.

From this point on, Bill Wilson's story is so closely intertwined with that of Alcoholics Anonymous that it is best followed by reading that entry in this encyclopedia. Wilson "led" Alcoholics Anonymous largely by recognizing that alcoholics cannot be "led." He was the main author of its book literature—the Big Book, *Alcoholics Anonymous* (1939); *Twelve Steps and Twelve Traditions* (1953); and *Alcoholics Anonymous Comes of Age* (1957). In the 1960s, Wilson attempted to separate himself from his AA role and eventually did return briefly to a position on Wall Street. But by this time a lifetime of cigarette smoking had caught up with him.

William Griffith Wilson died of emphysema from smoking on 24 January 1971. His death was marked by a front-page *New York Times* obituary that revealed his full name for the first time since AA's tradition of anonymity had been adopted in 1950.

Ernest Kurtz

See also: Alcoholics Anonymous (AA); Smith, Robert Holbrook

References

Anonymous. *Alcoholics Anonymous.* New York: AAWS, 1939, 1955, 1976, 2002.

Kurtz, Ernest. *Not-God: A History of Alcoholics Anonymous.* Center City, MN: Hazelden, 1979.

Thomsen, Robert. *Bill W.* New York: Harper & Row, 1975.

Wilson Black, Robert (1871–1951)

Robert Wilson Black (1871–1951) served as president of the United Kingdom Alliance (UKA) from 1932 until his death in 1951. He was responsible for the building of Alliance House and the creation of the United Kingdom Temperance Alliance Limited.

His boyhood was greatly influenced by his parents and elder brother, Sydney, an inspiring evangelist with a deep social conscience. Educated at Bishop Stortford College, Wilson Black left school at age fifteen and was articled to a firm of estate agents. At the age of nineteen, a firm of builders and estate agents, Knight and Company, engaged him, and on his insistence he was not paid a salary until he proved that it was to their advantage to employ him. Wilson Black was soon offered a partnership. He later bought out the other partners and remained with the business for fifty years. Tribute to his integrity in business is seen in the confidence of David Lloyd George, a former British prime minister, who when considering purchases of property would do nothing until he had consulted Wilson Black. Wilson Black's entrepreneurial spirit and acumen were to benefit the causes he espoused, including buildings constructed under his supervision for the National Free Church Federal Council, the UKA, and the Band of Hope Union.

Of his work for the UKA, his biographer Henry Townsend said: "No history of Temperance reform in Britain during the last fifty years could possibly neglect the influence of his leadership, he became recognised as one of the foremost—many thought of him as the foremost exponent of Temperance principles. His annual [UKA] addresses were regarded as the authoritative utterance of Temperance statesmanship in the country" (Townsend 1954: 169).

In the year that Wilson Black became UKA president, the Royal Commission on Liquor Licensing published its findings. The temperance movement welcomed its recommendation regarding the need for alcohol education for all schoolchildren, but for the most part the UKA opposed its view "that a prima facie case of considerable strength has been made out in favour of public ownership" (ibid.: 174). Dr. Henry Carter, secretary of the Methodist Temperance Department and a member of the royal commission, was an advocate of state control. A split in temperance ranks seemed likely. Although Wilson Black defended the UKA's stance on state control, the last thing he wished to do was to split the movement. He felt a concordat had to be found, and in November 1933 the Methodist Church's Temperance and Social Welfare Department and the United Kingdom Alliance signed a temperance program. It recognized their difference on the question of public ownership and public house improvement but outlined clear areas of agreement in the advocacy of personal abstinence; temperance instruction in schools; rapid reduction of liquor licenses; prohibition of alcohol advertising; Sunday closing, and a number of other areas. The manifesto called for an immediate closing of temperance ranks.

In 1940, Wilson Black was elected vice president of the Baptist Union and, as was the custom, president in the following year. He helped raise more than £1 million to build new churches as the commissioner of the Baptist Forward Movement. He also served as president of the London Free Church Federation and as chairman of the Temperance Committee of the National Free Church Council.

The most lasting impact of his presidency of the alliance was the establishment of the United Kingdom Temperance Alliance Limited in the early 1940s. Wilson Black died, as he would have wished, in "active service." The day before his death, he had spent an hour in his office discussing the future of the alliance with its general secretary, and that evening he attended a deacons' meeting at his church, Twynholm Hall, where he was taken ill.

Derek Rutherford

See also: Carter, Henry; Temperance Council of the Christian Churches (TCCC); United Kingdom Alliance (UKA); United Kingdom Temperance Alliance (UKTA)

References

Townsend, Henry. *Robert Wilson Black.* London: Carey Kingsgate, 1954.

Williams, Gwylmor Prys, and George Thompson Brake. *Drink in Great Britain, 1900 to 1979.* London: Edsall, 1980.

Wine

Wine is an alcoholic beverage made by fermenting the juice of grapes. Although wine can also be made from other fruit, these products are usually called "fruit wines" (and, more specifically, "apple wine," "blueberry wine," and so on), leaving the simple term "wine" for the grape-based product. The process for making wine is to crush grapes and promote fermentation by introducing yeast to the liquid (called "must"). Some winemakers use natural yeasts that occur in vineyards and wineries, but more often commercial winemakers employ cultured yeasts. Fermentation occurs when the yeast consumes the sugars in the grape juice and produces alcohol and carbon dioxide.

Types of Wine

Wine comes in many styles, with the most common distinction being between white and red. White wine can be made from green or dark grapes because the flesh of both is generally light in color and produces a clear or yellowish liquid. In order to make red wine, dark grapes (with red, purple, blue, or black skins) are used, and the skins are left in the liquid to allow the pigments to dye the otherwise clear grape juice.

In addition to "still" types, there are sparkling wines with bubbles that emerge when carbon dioxide has been dissolved in the wine. In some wines, especially those made according to the "Champagne method" (the method used to make the highly prestigious Champagne from the region of that name in France), the carbon dioxide produced during fermentation is dissolved in the wine rather than being allowed to dissipate (as it is in the production of still wine). In other cases, gas may be pumped into the wine.

Yet other styles of wine include fortified varieties (such as port and sherry), where grape spirits such as brandy are added to wine at one point in the fermentation process so as to contribute alcoholic strength and sweetness. Finally, there has historically been a wide range of wine-based beverages in which wine is mixed with other liquids or substances.

Wine in Ancient Societies

The origins of wine are unknown, and it is likely that wine was first made accidentally, perhaps when grapes that were collected for consumption as fresh fruit were crushed and began to ferment spontaneously as wild yeasts acted upon the sugars in the juice. Historians speculate that wine was initially made from wild grapes and that, like other food plants, vines were first cultivated by settled societies during the Neolithic period (8500–4000 B.C.E.). The earliest clear evidence of wine dates to about 5000 B.C.E. in the Middle East, where archaeologists have discovered earthenware jars and other vessels containing grape seeds, stems, and other evidence of wine. The earliest known wine jar, dating to 5000 B.C.E., was found in the Zagros Mountains of modern western Iran.

Wine played a part in the diet and culture of all ancient societies from the Neolithic period onward. For the most part, it was a privileged beverage consumed only by the elites, whereas beer was the drink of the masses. One reason for the special status of wine was its scarcity. Grain grew far more widely and easily than grapes, and beer (which can be thought of as liquid bread) could be made year-round as long as stocks of grain were available. But grapes grew only in certain localities and ripened only once a year, so that there was limited scope for winemaking. Moreover, each year's wine had to last until the next vintage was ready for drinking. In regions where grapes did not grow, wine had to be imported, thus adding to its cost. Wine trade routes emerged in ancient societies. One, which lasted for millennia from about 3000 B.C.E., ran 1,000 miles down the Tigris and Euphrates rivers from the vine-clad mountains of northern Mesopotamia to southern Mesopotamian cities such as Ur, Babylon, and Sumer.

Egypt provides the most coherent evidence of an ancient wine culture. Hundreds of clay jars of wine (with a total volume of some 4,500 liters [about 1,200 U.S. gallons]) were buried with one of the first Egyptian kings, Scorpion I, in about 3150 B.C.E. It was most probably imported from the modern Israel-Palestine region. Between 3000 and 2500 B.C.E., viticulture was established in Egypt, mainly in the Nile Delta, where the earth was fertile and the heat was moderated by the Mediterranean. Royalty, great officials, and priests owned the vines. A census taken in about 1000 B.C.E. showed 513 vineyards that were associated with temples. The elite status of wine is indicated by its prominence in the burial chambers of the kings. Thirty-six jars of wine were buried with the young King Tutankhamen. Wine played an important role in Egyptian religion, as it did in religions in other parts of the ancient world, and it was poured as a libation or offering to the gods as prayers were said. Wine was also used for medical purposes. Physicians prescribed it to increase the appetite, purge the body of worms, and treat asthma. It could also be applied externally to bring down swelling and to treat wounds.

From Egypt, viticulture and wine production spread to Europe, first to Greece and then to Italy. Along with grain and olive oil, it became one of the three staple products of the Mediterranean region. Greek wine could be found in locations as diverse as France, Egypt, the area around the Black Sea, and in the Danube region. Moreover, the Greeks introduced viticulture to France (with limited plantings near modern Marseilles), southern Italy, and Sicily.

Greek males of the upper social strata developed a specific institution for consuming wine: the *symposium,* which literally meant "drinking together." A dozen or more men, all wearing garlands on their heads, reclined on couches and drank diluted wine while conversing, being entertained by young men and women, and playing games that often involved wine. Symposia were idealized as occasions for elevated discussion and cultural activities, but often they were merely boisterous drinking sessions. Greek wine cups are often decorated with scenes of drunkenness and sexual activities at symposia. Women were excluded from such events (except as servers, entertainers, and prostitutes), and there is evidence in Greek writings of male anxiety about women drinking wine.

The Greeks paid serious attention to viticulture and winemaking. They adopted techniques of growing vines along trellises and up stakes to make the grapes more accessible during harvest. But it was the Romans, whose empire superseded the Greeks', who left us the most coherent documentation on wine in the classical period. A host of writers, including Cicero, Pliny, and Cato, described viticultural and winemaking practices and wrote extensively about the wines available to them. Roman writers also focused on winemaking and gave recipes for wines that would appeal to Roman tastes. Unlike modern winemaking methods, where additives are minimal, Roman wine was a grape-based concoction that might include seawater, honey, and all kinds of herbs and spices. Additional flavors might be contributed by the pitch and resin sometimes used to seal the inside of earthenware jars, and sweetness could be added by boiling the grape juice in a lead vessel. Lead not only sweetened wine but also preserved it by killing some bacteria. (Its potential toxicity was recognized but was largely ignored until the seventeenth century.)

The engine of the Roman wine industry was Rome itself, which grew from 300,000 to more than 1 million inhabitants between 300 B.C.E. and the beginning of the Christian era. By that time, Romans were consuming an estimated 1.8 million hectoliters (47 million U.S. gallons) of wine a year, about half a liter a day for every inhabitant. This massive volume of wine came not only from the region around Rome itself but from other parts of the Italian peninsula. The Romans extended viticulture throughout Europe as their empire expanded. By the first century C.E., most of the famous French wine regions (including Bordeaux, the Rhône, and Burgundy) had been planted, as had areas in England, Germany, Hungary, and other parts of southeastern Europe. The Romans were thus responsible for the beginnings of the European wine industry.

Roman attitudes about wine were complex. On the one hand, Roman commentators broadly condemned drunkenness, whether on the part of women or men. Cicero frequently labeled his opponents drunkards and alleged that his main rival, Mark Antony, started drinking early each morning. Others cautioned against excessive drinking for a variety of physical and mental reasons. Lucretius argued that wine could disturb the soul and weaken the body; Seneca wrote that wine revealed and magnified character defects; and Pliny the Elder praised quality wines but warned that many of the truths spoken under the influence of wine were better not expressed. On the other hand, classical medical opinion generally held that wine, alone or with other substances, had curative properties, particularly for gastric and urological ailments. Cato recommended certain flowers soaked in wine as effective for snakebite, constipation, gout, indigestion, and diarrhea.

If wine had achieved a privileged status at the center of the Roman Empire, some non-Roman populations on the margins of Roman control carved out their own relationship with the beverage. For Jews, wine was a powerful expression of divine power. When Moses sent out scouts to survey the promised land, they returned with a bunch of grapes so massive that it took two men to carry it. Grapes and wine were such important signs of the bounty provided by God to the Jews that the Old Testament frequently threatens that God will make the vines barren if Jews disobey God's word.

Wine in Christianity

This intense symbolism of wine carried over to Christianity. The first miracle performed by Christ was to turn water into wine at the wedding at Cana. Wine became integral to Christian theology, ritual, and tradition. In the Eucharist, wine represents the blood of Christ, and there are many representations in art of "Christ in the wine press," where Christ's blood, flowing from wounds inflicted during the crucifixion, mixes with the red juice flowing from the grapes as they are crushed.

Because the Eucharist required wine, Christianity and wine became so intimately connected that in the first centuries of the Christian era conversion from beer to wine became a sign of conversion from paganism to the new religion. Many religious houses had their own vineyards. Monasteries were centers of learning; not only in theology but also in the practical sciences, and for hundreds of years religious orders were at the forefront in developing new techniques in viticulture and winemaking.

The invasion of the western region of the Roman Empire by tribes from central and eastern Europe from the fifth century did not affect European viticulture as dramatically as once thought. It is possible that some vineyards were abandoned, but overall it seems that Europe's new rulers were as interested in protecting viticulture as the Romans had been. What did suffer was the wine trade as the single Roman Empire was broken up into smaller political units, each dominated by one of the invading tribes.

It is a mistake, then, to think of a Dark Ages of wine, and certainly misleading to suggest, as some scholars have done, that viticulture survived only because of the vineyards owned by the Christian church and various religious houses. They were undoubtedly important; some were extensive, and the church sponsored the expansion of vineyards in the important Rhine region and in Austria and Switzerland. Even so, many vineyards had secular owners, and viticulture and winemaking were not particularly threatened in this period. But with decline in trade, many regions began to cultivate their own grapes.

The Impact of Islam

The real threat to wine (and alcoholic beverages generally) emerged not in Europe, but in the Middle East, the birthplace of wine. There, the Islamic religion took hold in the seventh century, and within a hundred years it had extended its control across northern Africa, the Iberian Peninsula, and, for a short time, parts of southwestern France. The Prophet Muhammad forbade his followers the consumption of alcohol. Although he acknowledged that wine could make people happy and sociable, he believed that its threats to social order and morality were so great that it should be banned. Wine production practically dried up in many parts of the Islamic empire, but in others (Spain, for example) it was generally tolerated and even acknowledged insofar as it was taxed by Muslim authorities.

Expansion of Wine Culture in the Middle Ages

Wine production and trade in Christian Europe began to boom around the year 1000 C.E. One reason was the creation of a large political unit in Europe under the Emperor Charlemagne; another was the growth of population, cities, and trade that took place in Europe between 1000 and 1300. In northern Europe, northern Italy, and elsewhere, new urban middle classes of entrepreneurs and merchants emerged, all with a thirst for wine. Wine regions close to these new urban markets (such as those in Tuscany and other regions of northern Italy) prospered. But many of the new cities were in areas unsuitable for viticulture, and wine trade routes developed to serve them. Among the most important were the sea route from southwestern France (now the Bordeaux region) to England and the northern European ports and the route down the Rhine River from the vineyards of central and southern Germany to the North Sea and Baltic ports.

This boom period for the medieval wine industry ended with the Black Death that struck Europe from the mid-1300s. Europe's population declined by as much as a third, and as markets contracted and vineyard workers died or fled the plague, many vineyards were abandoned. Production and trade began to recover as population and markets grew again in the 1500s. In this period, wine was part of the daily diet in many parts of Europe. Reliable statistics on per capita consumption are hard to come by (because the information was not collected), but common estimates suggest per capita consumption rates of between 0.5 and 3 liters (between 0.13 and 0.79 U.S. gallons) a day. The impact of these volumes depended on the alcohol content of the wine. Wine was also part of some people's income or entitlement.

Introduction of Wine to the Americas and Other European Colonies

It was in this period that Europeans extended viticulture beyond Europe itself. Vines were planted in Mexico in the 1520s, and viticulture rapidly spread down the west coast of South America in the wake of the invading Spanish armies and Jesuit missionaries. As mission stations were established, vineyards were planted, and the connection was so strong that the grape commonly planted became known as the Mission variety. By the 1550s, major vineyards had been established in Peru, Chile, and Argentina. During the mid-1600s, the Dutch established vineyards in what is now South Africa, and in 1788 the first vines were planted in Australia.

Viticulture in North America was far less successful. Settlers tried to make wine from native grapes from the 1600s and later tried unsuccessfully to grow European varieties. A combination of climate and disease condemned most of these attempts to failure, and even though Franciscan missionaries established vineyards in California in the 1700s, it was not until the nineteenth century that wine was produced in America in meaningful volumes. In the first half of the nineteenth century, the heart of the American wine industry lay not on the West Coast, as it does today, but in states such as Indiana and Ohio, which provided adequate growing conditions relatively close to the major concentrations of population that were the main wine markets.

It was only after the end of the California Gold Rush in the late 1840s that vines were planted extensively near the Pacific Coast. The influx of population that accompanied the Gold

Rush, good conditions for viticulture, and later the extension of the railroad (which gave California wine producers easy and less expensive access to eastern markets) were all-important in the rapid development of the California wine industry.

The Setback from *Phylloxera*

The first two thirds of the nineteenth century saw a general expansion of the wine industry throughout much of the world, but from the 1860s onward, many vineyards were devastated by a North American aphid called *Phylloxera vastatrix.* Unwittingly brought to Europe on the roots of American vines that were planted in France for experimental purposes, *P. vastatrix* spread to nearly all French wine regions and later to other parts of Europe, including the massive Spanish and Italian vineyards, and to wine regions as distant as parts of Australia and South Africa.

Unable to eradicate the pest, growers began to graft their vines onto the roots of native American vines that were tolerant of the aphid. Most European wine regions were replanted with grafted vines (and most European vines remain grafted in this way), but vineyards in marginal regions (such as parts of northern France) were abandoned. At the same time, regions such as Languedoc and Roussillon in southern France expanded, not only because vines grew easily there but also because rail transportation provided access to population centers in the north.

P. vastatrix had far-reaching effects on wine. It altered the geographical distribution of viticulture by providing opportunities for the wine industries in such diverse regions as California and Algeria. Moreover, it forced the French wine industry to fight to recover the markets it had lost when production was crippled and counterfeit wine (made from dried raisins rather than fresh grapes) flooded the market. The Bordeaux region responded to the crisis by popularizing the *château,* an icon of the grand vineyard estate that promoted the image of Bordeaux wines as ancient, aristocratic, and timeless—something that the new vines, grafted onto American stock, were not. French winemakers began to advertise their wines as "château-bottled" (*mis en bouteilles au château*) as a guarantee against fraud or blending by merchants, and French legislators began to introduce "appellation" legislation to guarantee that wine came from a specific region, be it Champagne, Bordeaux, or Burgundy.

Although the *P. vastatrix* disaster affected European wine production for several decades, it gave a boost to production elsewhere. California vineyards grew even more rapidly as producers happily watched the disaster in Europe and imagined that *P. vastatrix* would destroy the European wine industry, allowing California to replace Europe as the source of most of the world's wine.

On the negative side, throughout this same period a wave of antialcohol sentiment swept across many countries. Mass movements such as the Woman's Christian Temperance Union argued for limitations on, if not the prohibition of, the production, sale, and consumption of alcoholic beverages. The temperance and abstinence movements were quite varied in their composition and aims. Some sought total prohibition of all beverage alcohol; others focused on distilled alcohol and were tolerant of moderate consumption of fermented beverages such as beer and wine.

Doctors and Wine

Wine was frequently exempted as a special beverage because of its religious and therapeutic traditions. It was frequently mentioned in positive contexts in the Bible, was an element in the Christian Eucharist, and had long been regarded as having valuable medicinal properties. A number of doctors who supported the temperance cause, among them Benjamin Rush, U.S. surgeon general during the American Revolution, and Henry Lindeman (the founder of Australia's most famous early winery, and still a large company today) in Australia, believed so much in the therapeutic value of wine that they promoted its production. Several of the leading temperance movements in France called for an increase in wine production so that people had an alternative to distilled spirits, which they believed to be far more dangerous to health and social order.

The Impact of Temperance Movements

Despite support for wine in the medical community and among some temperance campaigners, many others in the temperance movement considered wine to be just as pernicious as other forms of alcohol. They argued that doctors contributed to alcoholism by prescribing wine for their patients and that the positive references to wine in the Bible were in fact references to unfermented grape juice. This view became known as the "two-wine" theory, in which only the negative references to wine were believed to designate the alcoholic beverage.

The antialcohol lobby had varying success in having alcohol laws tightened. During the last three decades of the nineteenth century and the first two of the twentieth, some two-thirds of the American states introduced some form of prohibition. In 1919, Congress amended the Constitution, and from 1920 Prohibition became national policy. Between then and 1933, when National Prohibition was repealed, wine production declined dramatically. Many vineyards continued to produce grapes for unfermented juice and for consumption as table grapes, for raisin production, and for religious and medical purposes that were categorized as exemptions from Prohibition.

Prohibition in the United States echoed policies that had been adopted in a number of other countries a few years earlier during World War I. Governments in many belligerent countries had halted or limited alcohol production to conserve food resources (especially grain) and to maximize war-related industrial productivity by reducing intoxication. Russia, for example, banned the production of vodka for the duration of the war, and in Britain and Germany regulations lowered the alcoholic content of beer.

In France, however, wine was regarded as a healthy national beverage. Although the wartime government banned absinthe and restricted bar hours, it went to lengths to ensure that soldiers on the frontlines had regular rations of wine. The government requisitioned wine at fixed prices (which was good for bulk producers and hard on premium wine producers) and shipped it in hundreds of rail tankers to the troops. By 1917, French soldiers were receiving up to a liter and a half of wine a day, and in that year French forces consumed some 12 million hectoliters (317 million U.S. gallons) of wine. At the end of the war, wine was given credit for its role in French military success. One military paper wrote, tongue partly in cheek: "No doubt our brilliant generals and heroic soldiers were the immortal artisans of victory. But would they have been without the plonk that kept them going to the end, that endowed them with spirit, courage, tenacity, and scorn for danger, and made them repeat with unbreakable conviction, 'We will prevail'."

The French government's promotion of wine among its troops was an exception, however. Most of the belligerents took measures to restrict alcohol consumption in the military and on the home front. And in general, restrictions on alcohol that were adopted during the war for military purposes were maintained after hostilities had ended. In Great Britain, for example, rules limiting pub opening hours to lunchtime and the evening were kept in place until the 1990s. During the war, moreover, several countries adopted forms of prohibition that lasted beyond the war. In Canada, all the provinces except Quebec adopted prohibition at some time during the war, but by the end of the 1920s they had all abandoned it. There, as elsewhere, wine was often treated differently from other alcoholic beverages. In Ontario, which had a small wine industry, prohibition regulations permitted the production and consumption of wine but restricted its sale. Consumers could purchase wine only from the winery itself and in bulk quantities.

The Modern Organization of the Wine Industry

Throughout much of the twentieth century, France maintained its reputation as virtually the only producer of quality wine, although Spain and Portugal exported their well-known fortified wines, sherry and port, respectively. Italy and Spain were known as sources of inexpensive wines of mediocre quality. Most German wine was also considered mediocre, apart from a cluster of highly regarded Rieslings.

In the 1930s, the French codified their appellation regulations into a comprehensive system that classified wine into three main categories: table wine (the lowest quality and least regulated), regional wine, and quality wine from an appellation, or specified region. Only wine made from grapes grown in these appellations could carry the appellation name on the label. The French appellation law, the first comprehensive national wine law of this sort in the world, became a model for laws in many other countries, particularly those that joined the European Economic Community (the forerunner of the European Union) in the 1950s and 1960s.

From the 1960s onward, almost all wine-producing countries adopted wine laws regulating such questions as appellations, the percentage of specific grape varieties in varietally named wines, and alcohol content. Depending on the country, wine laws could also regulate the entire process of growing grapes (such as permitted varieties, irrigation, and harvesting), making wine (fermentation, aging, chaptalization—the addition of sugar to increase potential alcohol level—and acidification), and even bottle shape and color.

Changes in Consumption since the 1960s

A major change from the 1960s on was the emergence of New World wines, especially from the United States (notably California), Australia, Chile, South Africa, New Zealand, and Argentina. The 1970s and 1980s saw the growth of a global wine market, and in the 1990s the New World began to challenge European wines for reputation and dominance. For example, European (and especially French) wines had historically dominated the important English wine market, but by 2002 Australia had surpassed France as the single largest exporter of wines to England. At the same time, Old World producers such as Italy and Germany lost ground to exports from Chile and California.

One of the motors driving the international commerce in wine from the 1960s was the decline in wine consumption in countries that historically had high per capita rates. In France, Italy, and Spain, per capita consumption of wine fell by half between the 1960s and the 1990s, and the same was true of high New World consumers, such as Chile and Argentina. The reasons included changes in patterns of sociability, growing awareness of the health dangers of heavy consumption, and generational shifts in beverage preferences. Faced with contracting domestic markets, some wine producers participated in government-sponsored programs to pull out vines and others began to look more aggressively for export markets.

Toward the end of the twentieth century, several countervailing trends affected wine and some other alcoholic beverages. One was the discovery of the so-called "French paradox." This was a much-contested assertion that, despite a diet that should have resulted in high rates of heart disease, the French enjoyed relatively low rates because they consumed significant quantities of red wine. Scientists argued that wine, especially red wine, contained *resveratrol,* a substance that protected against heart disease when consumed in moderate quantities. This health argument is often credited with much responsibility for a consumer shift toward red wine, a decline in wine consumption in high-consumption societies, and an increase elsewhere. It is worth noting, however, that the decline in wine consumption in France, Italy, and other such countries preceded the discovery of the French paradox.

At the beginning of the twenty-first century, wine was established in many Western countries as a "lifestyle" alcohol. It was

French villagers working in a vineyard, ca. 1948 (Hulton/Archive)

the center of a vast consumer industry that included publications, wine bars, wine websites, wine tourism, and wine appreciation courses. With the possible exception of Scotch whiskey, no other beverage alcohol had achieved this cultural status, and whiskey was far more limited in its consumption. The modern status of wine reflects a historical continuity that can be traced back thousands of years. In this tradition, wine was associated with divinity, health, and sociability and was accorded privileged status among beverage alcohols.

Rod Phillips

See also: Absinthe; Archaeological Approaches to Drinking and Temperance; Art, Representations of Drinking in; Australia; Banana Wine; Beer; Drinking Establishments (France); Drinking Sites and Culture (Australia); France, Production and Consumption of Alcohol in; France, Temperance in; Gallo; Islam as Cultural Influence; Islamic World; Italy; National Prohibition (United States); New Zealand; Palm Wine; Spain

References

Johnson, Hugh. *The Story of Wine.* London: Mitchell Beazley, 1989.

Phillips, Rod. *A Short History of Wine.* New York: HarperCollins, 2001.

Pinney, Thomas. *A History of Wine in America from the Beginnings to Prohibition.* Berkeley: University of California Press, 1989.

Unwin, Tim. *Wine and the Vine: A Historical Geography of Viticulture and the Wine Trade.* London: Routledge, 1991.

Winskill, Peter Turner (1834–1912)

Peter Turner Winskill (1834–1912), English temperance reformer, wrote several histories of the temperance movement and compiled a biographical dictionary of temperance reformers. Among eyewitness temperance historians in Victorian England, he ranks second to Dawson Burns. Winskill's historical work was mostly biographical, and there is considerable overlap among his books. Some were self-published by the author with the aid of subscriptions. His works include *A Comprehensive History of the Rise and Progress of the Temperance Reformation* (1881), *History of the Temperance Movement in Liverpool and District* (1887), *The Temperance Movement and Its Workers* (four volumes, 1891–1892), and *Temperance Standard Bearers of the Nineteenth Century* (two volumes, 1897–1898). The latter briefly described more than 7,000 reformers, mostly men. Winskill generally neglected women reformers.

Late in life, he published a short undated memoir, *A Key to the Mystery; or, A Short Sketch of the Life and Labours of Peter T. Winskill, Temperance Historian, Advocate, Poet, Songster, Reciter . . . with a Selection from his Latest Temperance Songs.* As the title indicates, Winskill served the temperance movement in many ways. The memoir also appears at the beginning of the second volume of his *Temperance Standard Bearers.* In the introduction to the first volume, he made clear his hostility to "the stinking fumes of tobacco" (Winskill 1897–1898, 1: 27), and although "not in practice a strict vegetarian," his sympathy to the argument that vegetarianism was the most healthy diet (ibid., 1: 16).

In the early 1870s, Winskill was an organizer for the Good Templars, a temperance fraternal society then recently introduced into England. He soon quarreled with its leader Joseph Malins. When the international organization split in 1876, he backed Malins's rival, F. R. Lees. In 1882, Winskill moved to Liverpool, a stronghold of Lees's faction. Lees contributed an introduction to *The Temperance Movement and Its Workers.* The newspaper of Malins's organization criticized Winskill's histories as prejudiced and inaccurate in their account of Good Templar history.

Winskill's life was typical of antidrink reformers who rose from working-class beginnings to a precarious lower-middle-class status as a result of their service to the temperance movement. Born at Newcastle-on-Tyne, Winskill was a frail child until the age of six. His mother was an army officer's daughter, and his father was a disabled soldier turned shoemaker and Methodist local preacher. Although Winskill's father had been an early North of England temperance advocate, he later was ruined by drink. Young Winskill's embittered mother brought him up in the temperance movement. At the age of ten, he became a juvenile Rechabite. He received his scanty education in Houghton-le-Spring, Durham. Until he was in his mid-thirties, Winskill often changed employment. He was successively a pupil-teacher (that is, a monitor assisting a teacher), construction worker, iron molder, book and insurance agent, and auctioneer. Beginning in 1871, he made his living in the temperance movement as organizer, speaker, and writer. In 1857, Winskill married his wife Elizabeth. They had fourteen children, several of whom died in early adulthood.

David M. Fahey

See also: Burns, (James) Dawson; Good Templars (IOGT); Lees, Frederic Richard; Malins, Joseph; Rechabite Friendly Society; *Temperance Standard Bearers;* Vegetarianism and the Antitobacco Movement (United Kingdom)

References

Cherrington, Ernest Hurst, et al., eds. *Standard Encyclopedia of the Alcohol Problem.* 6 vols. Westerville, OH: American Issue Publishing, 1925–1930. 2880.

Winskill, Peter Turner. *Temperance Standard Bearers of the Nineteenth Century: A Biographical and Statistical Temperance Dictionary.* Liverpool: the author, 1897–1898.

Winterton, Ernest (1873–1942) and Winterton, Wilfrid (1885–1976)

Ernest Winterton (1873–1942) and Wilfrid Winterton (1885–1976) were brothers who were prominent in the English temperance movement in the early and mid-twentieth century. Born in Leicester, they were brought up by parents who became teetotal through the influence of the Blue Ribbon movement in the 1880s.

A teacher by training, Ernest left the profession to become secretary of the Leicester and District Temperance Society in 1896. In 1906, he moved to Manchester to become secretary of the Temperance Union in that area. During World War I, he came into national prominence as a leader in the Strength of Britain movement, which advocated wartime prohibition. In 1929, he was elected to the House of Commons as a Labour member. He lost his seat in the 1931 general election, when Labour was routed. During this period, he took on the editorship of the *White Ribbon,* the official publication of the National British Women's Total Abstinence Union. In the year of his death, 1942, he compiled a book, *People I Have Met and Places I Have Seen: Some Memories of Agnes E. Slack.* A memorial lecture series by the United Kingdom Temperance Alliance honored the name of Ernest Winterton.

On leaving school, Wilfrid Winterton joined Parr's Bank, Leicester, in 1902. At that time, junior clerks were paid a pittance and an employee had to remain single until he was thirty. Not afraid to challenge the meanness of the banks, he set out to better the lot of bank employees, but it was not until 1918 that a union for bank employees was established. Wilfrid was one of the earliest of the fellows of the Banker's Institute.

In 1936, Wilfrid Winterton's temperance advocacy came to national attention. As a delegate from the Leicester Temperance Society to the annual meeting of the United Kingdom Alliance (UKA), he moved a resolution that called for a blood-alcohol test to be taken in all cases when drivers were involved in road accidents involving personal injury. Professor Erik Widmark's test for measuring the amount of alcohol in the blood influenced Winterton. Although the resolution carried, there was not unanimity over the blood test among temperance workers. In 1937, when the House of Lords appointed a Select Committee on the Prevention of Road Accidents, the quarrel among temperance advocates surfaced. Dr. Courtney Weeks, medical director of the National Temperance League, and James Hudson, secretary of the National Temperance Federation, refused to include the blood test proposal in their testimony. Weeks feared that the test had questionable scientific accuracy; Hudson felt it was a violation of personal liberty. Wilfrid Winterton persevered. He persuaded William Eccles, consultant surgeon at St. Bartholomew Hospital, London, to give evidence in favor of the blood test. In 1939, the Select Committee on the Prevention of Road Accidents recommended that scientific tests should at once be carried out on a voluntary basis and expressed the hope that in the future the public would accept obligatory testing.

World War II intervened. The idea of compulsory blood testing remained the stumbling block for identification of drunk drivers until the Breathalyzer came along in the 1960s.

To further the campaign, Wilfrid Winterton, together with his brother Ernest, financed and published the Research Student Service beginning in January 1937. In 1946, the UKA took over the financing of the Research Student Service until it ceased publication in October 1972. Wilfrid Winterton saw the culmination of his fight in 1967, when the Road Safety Act was enacted. The act established an alcohol breath test for motorists and set 80 milligrams per liter as the criterion for drunk driving. The penalty for violations was a minimum twelve-months license suspension. In 1968, Winterton wrote and privately published *Breath-Taking History: Britain's War with Drinking Drivers.*

Derek Rutherford

See also: National Temperance Federation (NTF); National Temperance League (NTL); Strength of Britain; United Kingdom Alliance (UKA)

Reference

Winterton, Wilfrid. *Harvest of the Years: Autobiography.* Birmingham: Templar, 1969.

Wiser Distillery

The Wiser distillery in Prescott, Ontario, was one of Canada's largest distilleries during the late nineteenth and early twentieth centuries, and presently the Wiser brand name lives on as an important player in the Canadian rye whiskey market. According to some accounts, when J. P. Wiser, a native of upstate New York, bought the distillery at Prescott in 1857, he intended to use it as a "value-added" feed station in his cattle-exporting business. (Spent distilling grains make for ideal cattle food, and the added value comes from the grain's initial use in distilling.) This plan was an intriguing reversal of the usual relationship between distilling and animal-fattening—distillers traditionally graduated into stock-raisers and not vice versa—but it stands to reason, despite its exceptionalism. Wiser succeeded in both ends of the business. Possessing knowledge of the U.S. commercial world, and quickly learning about the Canadian one, he found a favorable international reception for his product. The distiller took immediate advantage of the U.S. Civil War's disruption of the U.S. distilling industry to develop a market for his Canadian whiskey south of the border. By 1862, production at Prescott had grown fivefold to nearly 116,500 imperial gallons (139,800 U.S. gallons) from the 22,746 (27,295 U.S. gallons) registered in 1857. Not all sales were export-driven; Wiser also enjoyed rising domestic demand. Indeed, provincial sales were sufficiently brisk that he ably recovered from a fire, which virtually destroyed his plant, in 1864.

Above all, Wiser was a distiller of that most Canadian of whiskeys: rye (traditionally one part rye to several parts corn or other grain). His two principal early brands were the domestic Wiser's Red Letter Rye and the export Wiser's Canada Whiskey. The latter brand intentionally assumed the Irish spelling for the drink ("Whiskey"), since that spelling had become the norm in the United States, rather than the Scottish form of the word ("Whisky"), which had taken root in Canada. He also crafted a quality product aged for up to ten years and cultivated a wide network of retail sales agencies. His efforts did not go unrewarded. From the late 1860s through the late 1880s, with annual output hovering between 350,000 and 500,000 imperial gallons (420,000 to 600,000 U.S. gallons), Wiser's distillery was the third largest in Canada, behind Hiram Walker and Gooderham and Worts but ahead of Seagram and Corby. Over these years, Wiser improved both plant and machinery. His competitive advantage was such that the spread of local-option prohibition under the Canada Temperance Act during the 1880s did not harm his production figures. Indeed, in 1887, amidst the Canada Temperance Act's pinnacle of geographic force in Ontario, the strength of Wiser's whiskey business encouraged him to build a new tankhouse. In the late 1880s, the Prescott distiller admitted his sons into a partnership styled J. P. Wiser and Sons. Incorporation of the J. P. Wiser Distillery Limited followed in 1893.

Cattle-fattening was the other half of Wiser's distilling operation and, as stated above, may have been his motivation for purchasing the distillery in the first place. His cattle business was truly international in scope. Most of his stock was raised on his 150,000-acre cattle ranch in Kansas, then shipped by rail to Prescott for fattening on spent distillery grain, which, during routine peaks in the distilling season at the Wiser distillery amounted to 900 bushels per day. Locally grown hay was used as a supplemental feed. Once fattened, the live cattle went to markets in Canada, the United States, and Great Britain. Some of Wiser's overseas shipments numbered as many as 1,000 head of cattle. The entire enterprise revolved around several giant livestock barns in Prescott, which could collectively house thousands of cattle at a time. Wiser, like so many other wealthy distillers, also participated in raising, racing, and selling prize horses. Some of his finest specimens fetched tens of thousands of dollars on the international horse market. Profit in whiskey and livestock permitted Wiser, his sons, and their families the good life. They built grand mansions on Prescott's most fashionable streets, traveled extensively, donated to local charities and public causes, and participated in civic life. J. P. Wiser, for instance, sat for the Liberal Party as Prescott-Grenville County's member of Parliament in 1878. His local popularity is reflected in the fact that he was elected in absentia while traveling overseas. Not all of his community involvement was purely altruistic, however. Wiser and his sons, Harlow and Frank, subscribed stock in the Prescott Elevator Company Limited not simply for reasons of local boosterism but also to have a say in the area's grain traffic for the benefit of their distillery.

Toward the end of the nineteenth century, strong competition and a general price depression saw output at Wiser's dip somewhat, although the distillery continued as one of Canada's largest into the new century. World War I interrupted whiskey making in favor of producing industrial alcohol for the manufacture of cordite, an explosive. The consequences of province-wide prohibition under the Ontario Temperance Act of 1916 encouraged the Wisers to sell their distillery to Mortimer Davis's holding company, the Canadian Industrial Alcohol Company, in 1918. This move affiliated the Wiser name with the Corby distillery, Davis's other major drink property. Under the Volstead Act, Wiser whiskey enjoyed a renewed popularity in the U.S. market, particularly in the northeastern states. An interesting legacy of Wiser's sustained niche presence in the United States after Volstead Prohibition is that Wiser was one of only six whiskey brand references that George Thorogood and the Delaware Destroyers chose for the lyrics to their hit blues-rock anthem of 1985, *I Drink Alone,* a song used in the U.S. "Don't Drink and Drive" educational campaign of the late 1980s. In the early 1930s, when Harry Hatch bought out Davis, Wiser became a member of the Hiram Walker–Gooderham and Worts family. The Wiser distillery in Prescott closed in 1932. Today, Wiser's De Luxe Whisky, Wiser's Special Blend, and Wiser's Very Old are components of Allied Domecq's whiskey portfolio through its Canadian subsidiaries Hiram Walker and Sons Limited and Corby Distilleries Limited.

Glen C. Phillips

See also: Canada Temperance Act; Corby Distilleries Limited; Gooderham and Worts Distillery; Hiram Walker; Provincial Prohibition (Canada); Whiskeys

References

Morris, John Alfred. *Prescott, 1810–1967.* Prescott, Ontario: Prescott Journal, 1967.

Prescott Telegraph. Various issues. 1886–1890.

Rannie, W. F. *Canadian Whisky: The Product and the Industry.* Lincoln, Ontario: W. F. Rannie, 1976.

Wittenmeyer, Annie Turner (1827–1900)

Annie Turner Wittenmeyer (1827–1900) served as the first president of the national Woman's Christian Temperance Union (WCTU) in the United States. She was selected to lead the organization at its founding in 1874 and continued as president until her defeat by Frances Willard in an 1879 reelection bid. Under Wittenmeyer's leadership, the WCTU became the largest women's organization in the United States and a major force for temperance reform.

Wittenmeyer began her reform work before the Civil War. She led efforts to found a school for poor children in Keokuk, Iowa, before the establishment of public schools in that town, and she was active in local religious work. After the onset of the Civil War, she helped to organize the Keokuk Ladies' Soldiers' Aid Society and was appointed its first corresponding secretary. Assisting in the formation of similar organizations across Iowa, Wittenmeyer developed a network of relief and provision for Iowa's Union troops that proved more effective than the state sanitary commission established later by the state's governor. After the two organizations were merged into the Iowa Sanitary Commission, Wittenmeyer moved on to other projects. She proposed the establishment of an orphanage for the children of Civil War casualties, an effort that resulted in the founding of two orphans' homes in Iowa. She also was appointed supervisory agent for the Diet Kitchens by the United States Christian Commission and worked to provide more nutritious meals at Union field hospitals.

After the war, Wittenmeyer helped to form the Methodist Home Missionary Society and served as that organization's first corresponding secretary. She edited a religious newspaper in Philadelphia, the *Christian Woman,* and published her first book, *Woman's Work for Jesus,* in 1873. That year also marked the beginning of the Women's Temperance Crusade, when hundreds of women took to the streets to protest and disrupt the effects of saloons in their communities. The spontaneous demonstrations spread throughout the Midwest and Northeast and led to a call for the establishment of a national organization. Consequently, delegates from sixteen states gathered in Cleveland in November 1874 to organize the Woman's Christian Temperance Union. Wittenmeyer was chosen president after a tie vote on the first ballot.

Wittenmeyer worked tirelessly organizing local unions and publicizing the organization's goals. Although there was considerable sentiment for women's suffrage, particularly among western delegates, she tried to keep the organization from taking an official position on the issue. Personally opposed to suffrage for women, Wittenmeyer did believe that women should exercise their right to petition government. In February 1875, she presented Congress with more than 40,000 signatures petitioning for congressional investigation of the liquor traffic. Her petition was received, but no action was taken.

Reelected unanimously at the second annual national WCTU convention in 1875, Wittenmeyer faced growing opposition from more radical forces in the organization in following years. In 1879, she was defeated for reelection by Frances Willard. She remained active in the leadership of the national WCTU until the organization officially endorsed the ballot for women in 1881. She left the WCTU, along with a number of other conservative members, and later helped to found the Non-Partisan Woman's Christian Temperance Union when disputes over partisanship led to another secession from the WCTU.

Wittenmeyer remained active in reform and charitable work after her break with the national WCTU. She also continued to write and to edit. In 1882, she published *The History of the Woman's Temperance Crusade* about the uprisings that energized the movement.

James D. Ivy

See also: Non-Partisan Woman's Christian Temperance Union; Willard, Frances Elizabeth Caroline; Women's Temperance Crusade

References

Bordin, Ruth. *Woman and Temperance: The Quest for Power and Liberty, 1873–1900.* New Brunswick, NJ: Rutgers University Press, 1990.

Leonard, Elizabeth D. *Yankee Women: Gender Battles in the Civil War.* New York: W. W. Norton, 1994.

Willard, Frances. *Woman and Temperance, or the Work and Workers of the Woman's Christian Temperance Union.* Hartford, CT: Park Publishing, 1883.

Woman's Christian Temperance Union (WCTU) (Australia)

The first Australian branch of the Woman's Christian Temperance Union (WCTU) was formed in 1882 in Sydney by Eli Johnson, husband of Mary Johnson, the recording secretary of the U.S. national WCTU. In 1885, Mary Leavitt, the WCTU world missionary, visited Australia and formed WCTU branches in other colonies. It was Jessie Ackermann, traveling to Australia on behalf of the World's WCTU in 1889, however, who helped the branches organize effectively into colonial unions. In 1891, Ackermann confederated the colonial branches into a quasi-national body, the WCTU of Australasia (renamed the Australasian WCTU in 1903 and the National WCTU of Australia in 1933).

Structure and Policies

The national body had its own executive, triennial conventions, which were widely attended, and, from 1898–1903, a magazine called *Our Federation.* The colonial branches largely worked independently of each other, however. The Australasian WCTU was in turn affiliated with the World's WCTU and regularly sent delegates to the world conventions. A number of Australian WCTU members served as World's WCTU superintendents of departments. Two Australians, Isabel McCorkindale and Millicent Harry, were to serve as World's WCTU presidents, in 1959–1962 and 1977–1986, respectively. By 1894, the WCTU had 7,400 members in Australia. Per head of population it was as large as the U.S. national WCTU. The numbers, however, fluctuated throughout the next century, declining in the 1900s and rising during World War I, when membership reached 10,000.

Although the WCTU was not nearly as large in membership as other temperance societies, such as the Rechabites, its members did play an important part in temperance campaigns and conventions. The Australian temperance movement of the late nineteenth century, like its British counterpart, focused on securing the right of local districts to prohibit public houses ("local option"). Within a year of the founding of the local WCTU in Victoria, its members were campaigning for a reduction of licenses at local-option polls throughout the colony. The WCTU was also active in the campaign for six o'clock closing, which consumed temperance energies during World War I. The long-term goal of the WCTU of Australasia, from its first convention, was nationwide prohibition. At this time, the WCTU was more ambitious than most Australian temperance groups, which concentrated on local option.

The WCTU devoted special attention to the effects of liquor on women and children. It was specifically opposed to licensed grocers, from whom women bought liquor to drink at home. The branches also waged a special campaign against the employment of women as barmaids. Soon after it was formed, the Sydney union organized a petition to oppose what it saw as this pernicious form of employment. Raising the minimum drinking age and introducing compulsory temperance education in public schools were also WCTU concerns.

The Australian WCTU branches embraced Frances Willard's "Do Everything" policy from their very earliest days. Like the U.S. national WCTU, the Australasian WCTU and the colonial unions incorporated a number of different departments led by a supervisor. In addition to tackling matters directly related to temperance, the WCTU in Australia set up departments for combating various vices and addictions such as gambling, narcotics, smoking, and prostitution. The Victorian WCTU magazine *White Ribbon Signal,* and the New South Wales WCTU magazine of the same name, carried articles on all of these subjects. There was also a Department of Peace and Arbitration, which protested against compulsory military conscription in World War I. In the 1930s, the subject of Aboriginal rights gained WCTU attention, too.

The WCTU had a particular interest in introducing laws and policies for the protection of women and children. It was the women of the WCTU who were responsible for raising the age of consent in each colony in the 1890s. They were also instrumental in the introduction of maternity allowances in 1912. They pioneered kindergartens, playgrounds, and juvenile courts and set up hostels and recreation centers for young people in cities.

Role in Suffrage and Politics

The WCTU was crucial in winning the vote for Australian women in 1902. When Mary Leavitt arrived in 1885, only Victoria had a women's suffrage organization. Within a few years, however, the WCTU was agitating for women's right to vote in every Australian colony. The women organized petitions, wrote and distributed pamphlets and newspaper articles, and conducted public meetings on women's suffrage. They were sometimes assisted by other suffrage bodies, but in Western Australia and Tasmania they were virtually the only force fighting for women's right to vote. Indeed, it was Elizabeth Nicholls, the president of the WCTU of Australasia, who in 1895 proposed that a clause be inserted in the draft constitution for the new Commonwealth of Australia granting adult women the right to vote on the same terms as men. The WCTU was the only suffrage organization to cover each Australian colony, and thus was in the best position to campaign for the federal vote.

In the 1880s, very few political organizations were run by women. The women of the United States had been speaking in public since the days of the abolitionists campaigning against slavery, but Australian women had had no such political initiation. The coming of the WCTU fulfilled this function. After the vote had been won, the temperance women directed their energies to encouraging women's political participation, and in the 1930s they began a campaign to procure more women parliamentary candidates. Many of the first women politicians in Australia were WCTU members.

After World War II, the WCTU focused on preventing the extension of hotel opening hours and the increase of licenses. Its other main activity has been public education on the health and social effects of alcohol.

Anna E. Blainey

See also: Australia; Barmaids (Australia); Home Protection; Liquor Licensing (Australia); Willard, Frances Elizabeth Caroline; Women Publicans (Australia); World's Woman's Christian Temperance Union (WWCTU)

References

Oldfield, A. *Woman Suffrage in Australia: A Gift or a Struggle.* Melbourne: Cambridge University Press, 1992.

Pargeter, J. *For God, Home and Humanity: National Woman's Christian Temperance Union of Australia, Centenary History, 1891–1991.* Golden Grove, South Australia: National Woman's Christian Temperance Union of Australia, 1995.

Tyrrell, I. "International Aspects of the Woman's Temperance Movement in Australia: The Influence of the American WCTU, 1882–1914" in *Journal of Religious History.* 12, no. 3 (June 1983): 284–304.

Woman's Christian Temperance Union (WCTU) (Dominion of Canada)

The Dominion Woman's Christian Temperance Union (WCTU) was the largest nondenominational women's temperance organization in Canada. Founded to counter what temperance reformers perceived to be the evils of alcohol, the Dominion WCTU rapidly grew into a multifaceted organization that championed a variety of forms of childhood and adult education; homes for abandoned and "fallen" women as well as poor, abandoned, or orphaned children; humane care of the indigent aged; residences and "Travelers' Aid" for single working women; women's hospitals; coffeehouses; reading rooms; and support for traveling lecturers and missionaries. The first local organization, established in Picton, Ontario, in 1874, was followed by the founding of the Ontario provincial WCTU in 1877 (the strongest provincial union throughout the organization's long history) and the Dominion WCTU, which held its first conference in 1888. Long after other temperance organizations collapsed, the Dominion WCTU survived and remained active in the cause of temperance education.

An important reason for its high level of productivity was a decentralized organization that allowed its members to pursue a wide range of goals through energetic lobbying for social legislation that would benefit women, children, and the family, including female suffrage. It must be granted, however, that the Dominion annual conventions of the WCTU were more inclined to place their confidence in the value of the franchise to expedite social reform than were women at the local or even provincial levels of the organization. Perhaps most important, the organization insisted on a public role for women and provided them with a forum through which the skills necessary for such a role could be developed.

Social Appeal

Most, but by no means all, of the Canadian women holding membership in the WCTU were white, middle-class Protestants. The culture of WCTU women, particularly in the late nineteenth and early twentieth centuries in Ontario and at the Dominion organizational level, was based on an evangelical vision for society. Characterized by a view of salvation as personal and experiential, and as dependent on a spiritual "awakening," evangelicalism created a sense of moral authority for its middle-class female members, who sought expression and action on a more prominent stage than in their own homes.

The last quarter of the nineteenth century in Canada was a period of profound social change and uncertainty characterized by the dislocating effects of industrialization, immigration, urbanization, British imperial decline, and secularization. One reaction was a deep and persistent concern about the survival of the family unit as it had been idealized in nineteenth-century literature. Most worrisome was the "intemperate" behavior of men. The latter included men's tendency to enjoy violence in sports, a double standard of sexual morality, their indulgence in personally destructive pastimes, such as gambling and tobacco use, and especially their violent behavior toward wives and children, often as a result of drunkenness. Within this context, the WCTU was formed and thrived.

Organizational Structure

The Dominion WCTU was remarkably democratic in its functioning. The executive officers at all levels were usually chosen by election. As a general rule, the executive did not set policy, although it did establish the agenda to be debated at each convention. Representatives to annual conventions delivered reports and considered resolutions drafted by the executive to determine that level's official policies. The executive then had the task of implementing these decisions and reporting any difficulties that arose in the course of its work to the unions through written reports. The local groups operated in a loose confederation to form the next level of the hierarchy, deciding the degree to which they would support the provincial or Dominion structures financially and ideologically. This virtual autonomy may help to account for the longevity and vitality of the local unions in otherwise difficult circumstances.

The organization operated at all levels through Departments of Work. In this way, it attacked a broad range of social problems, all of which were associated in some way with alcohol abuse. The Departments of Work provided for self-education, both of information and strategies to effect change, curricular and social supports for children and mothers, fund-raising and mobilization of other resources to support social services, and internal group support to reinforce resolve in the face of disappointing public response.

The Dominion WCTU functioned as a "mother organization" to a wide range of subgroups, many of which were established to carry through its program of "values instruction" through childhood education. The vibrancy and evangelical fervor of these satellite groups help to explain the organization's long-term vitality. As the mother organization waned in energy and membership, many of the subgroups carried on almost independently and with great resolve.

Both middle- and working-class children under the age of seven were welcomed into the Little White Ribboners, along with their mothers. Similarly, after-school clubs such as the Bands of Hope and Loyal Temperance Legions (LTL) attempted to attract both boys and girls, though the WCTU was consistently more successful in its efforts with girls than with boys. Members were expected to sign the "triple pledge" in which they foreswore alcohol, tobacco, and foul language. Both groups used an array of colorful room decorations and uniforms to attract children. Band of Hope members, for example, wore blue neck scarves with "Band of Hope" embroidered in white, and badges made with red, white, and blue ribbons were looped into buttonholes. Members could further deck their outfits with temperance medals, which they could win for making speeches, creating posters for the community, or performing songs. The Dominion WCTU encouraged these group activities by publishing Medal Contest Books with suggestions of appropriate selections for recitations. Working-class youths who were thought to be particularly vulnerable to the allure of alcohol and tobacco were also served in specialized groups where skills such as literacy, carpentry, needlecraft, or homemaking were combined with Bible and temperance study.

Pious single and young women joined the Young Woman's Christian Temperance Union (YWCTU), the group from which the leaders of the Bands of Hope and the LTL were usually drawn. Although members in the young women's sector were expected to join the WCTU one day, many stayed on in the YWCTU "farm team" well into adulthood, carrying on the many projects the local groups had initiated. Along with the Bands of Hope, the YWCTU was the most active youth group created by the WCTU. In fact, it often rivaled the mother organization in the range of activities it undertook and the successes it enjoyed, especially in childhood education and evangelical proselytizing. The YWCTU offered a role for single and often intensely evangelical women to contribute to their communities in a wide variety of ways.

By 1891, the Dominion WCTU reported a membership of almost 10,000 women. By 1914, 16,838 members were on the rolls, making it still among the largest nondenominational women's organizations in Canada in this period. In 1927, it reported 30,043 members across the Dominion. Hereafter, however, membership fell off, eventually sliding to its present figure of around 2,000. The official WCTU figures should be supplemented by the membership of the subgroups. In 1914, for example, when the Dominion WCTU boasted a membership of 16,838, there were also 11,535 children enrolled in the Bands of Hope and the LTLs and 1,596 young women in the YWCTU.

Sharon Anne Cook

See also: Evangelical Temperance (United Kingdom); Loyal Temperance Legion (LTL); Woman's Christian Temperance Union (United States); Wright, Bertha; Youmans, Letitia

References

Cook, Sharon Anne. "The Canadian Woman's Christian Temperance Union" in *Canadian History in Multimedia, 1867 to the Present,* edited by Chris Hackett and Bob Hesketh. CD-ROM. Edmonton: Chinook Multimedia, 2001.

———. "The Ontario Young Woman's Christian Temperance Union: A Study in Female Evangelicalism, 1874–1930" in *Changing Roles of Women within the Christian Church in Canada,* edited by M. Fardig Whiteley and E. Muir, 299–320. Toronto: University of Toronto Press, 1995.

———. "'Sowing Seed for the Master': The Ontario W.C.T.U. and Evangelical Feminism, 1874–1930" in *Journal of Canadian Studies.* 30, no. 3 (1995): 175–194.

———. *"Through Sunshine and Shadow": The Woman's Christian Temperance Union, Evangelicalism, and Reform in Ontario, 1874–1930.* Montreal and Kingston: McGill-Queen's University Press, 1995.

Malleck, Daniel J. "Priorities of Development in Four Local Woman's Christian Temperance Unions in Ontario, 1877–1895" in *The Changing Face of Drink: Substance, Imagery, and Behavior,* ed. Jack S. Blocker Jr. and Cheryl Kraswick Warsh, 189–208. Ottawa: Histoire Social/Social History, 1997.

Mitchinson, Wendy. "The WCTU: For God and Home and Native Land" in *A Not Unreasonable Claim: Women and Reform in Canada, 1880s–1920s,* ed. Linda Kealy, 151–167. Toronto: Women's Press, 1979.

Sheehan, Nancy. "The WCTU on the Prairies, 1886–1930: An Alberta-Saskatchewan Comparison" in *Prairie Forum.* 6, no. 1 (1981): 17–33.

Youmans, Letitia. *Campaign Echoes.* Toronto: William Briggs, 1893.

Woman's Christian Temperance Union (WCTU) (Japan)

From its inception to the present, the Japanese Woman's Christian Temperance Union (WCTU) has made it a priority to correct Japan's sexual double standards. Convincing people to confine sexual intercourse within a monogamous relationship has thus been a key issue. The World WCTU's first round-the-world missionary, Mary Clement Leavitt, inspired the formation of the

Japanese WCTU in 1886 when she toured Japan. After her visit, local male and female temperance unions, including the Tokyo Fujin Kyofu Kai (Tokyo Woman's Reform Society), were established. As its name infers, this union in Tokyo aimed to reform Japanese society by borrowing U.S. middle-class Protestant values and discourse. By the early 1890s, the Tokyo union had begun petitioning drives to "uplift" Japan's sexual moral standard to that of their vision. Their goals included the establishment of civil and criminal codes to punish not only a wife's but also a husband's extramarital relationship as adultery and the abolition of prostitution systems licensed by local governments. At the same time, members of the Tokyo union started efforts to build a women's home to "rescue" and "rehabilitate" prostitutes. In 1893, the local female temperance unions that emerged from Leavitt's organizing tour coalesced into a national organization, the Japanese WCTU (Nihon Fujin Kyofu Kai). Kajiko Yajima, a Japanese teacher of a Presbyterian female mission school in Tokyo, assumed the presidency.

Although members of the Japanese WCTU were originally unenthusiastic about promoting temperance on liquor consumption, World's WCTU missionaries such as Clara Parrish and Kara Smart, who worked in Japan during the 1890s and 1900s, strongly advocated the temperance cause. Parrish's efforts led to the formation of the Japan Temperance League in 1897, which united Christian and pro-Christian male temperance unions. Smart promoted temperance education at Japanese schools and popularized the temperance cause among Japanese youths. Her efforts were funded by the World's WCTU and continued by Japanese women such as Moriya Azuma, who presided over the Japanese WCTU's Little Temperance League (LTL). At the same time, the Japanese WCTU's foreign auxiliary members, mainly missionary women of the Anglo-American Protestant churches, led a movement against tobacco consumption. These efforts established Japan's national laws prohibiting minors from smoking in 1900 and from drinking in 1922.

Wider Reform Agenda

Similar to the national WCTU in the United States, the Japanese WCTU worked for a wide range of causes, including expansion of women's political rights. The Japanese WCTU cooperated with other women's organizations in petitioning the government to lift a ban on women joining and forming political organizations and participating in political meetings. After failing in 1917 to prevent the Osaka prefectural government from granting a new plot of land for the reconstruction of the region's pleasure quarter that had been consumed by fire, Japanese WCTU members came to recognize the need for women's suffrage to achieve their goals. Urged by Tsuneko Gauntlett, a Japanese WCTU officer married to a British national, who had recently returned from her European tour, Japanese WCTU members formed the Japanese Woman's Suffrage Association (Nihon Fujin Sanseiken Kyokai) in 1921 and established affiliation with the International Woman Suffrage Association headquartered in London. In 1924, the Japanese Woman's Suffrage Association merged with the New Woman's Society (Shin Fujin Kyokai), another women's organization working for women's suffrage, to become the League for the Realization of Women's Suffrage (Fujin Sanseiken Kakutoku Kisei Domei), one of the key groups in the history of Japan's women's suffrage movement.

Establishing itself as an acceptable and respectable women's organization in Japan when the country was aspiring to become an imperialist power, the Japanese WCTU led reform efforts commensurate with Japan's imperialistic expansion in Asia and the Pacific. Although world peace was one of the Japanese WCTU's causes, it cooperated with the Japanese government in its imperialistic wars: the Sino-Japanese War (1894–1895), the Russo-Japanese War (1904–1905), and the Fifteen-Years' War (1931–1945). Furthermore, starting in 1905, the Japanese WCTU extended its influence over the emerging Japanese communities overseas by creating branch unions on the American West Coast; in Japan's colonies, namely Taiwan and Korea; and in its puppet state, Manchuria.

Post-1945 Activities

During the U.S. occupation era after World War II, several of the Japanese WCTU's causes were achieved. Women obtained suffrage in 1945, the licensed prostitution system was abolished in 1946, and the sexual double standards in civil and criminal laws were eliminated by legal revisions whereby neither a wife's nor a husband's extramarital relationship would be punished as adultery.

In postwar Japan, the Japanese WCTU resumed its broad spectrum of activities under the slogan of "Peace, Purity, and Elimination of Liquor." Japanese WCTU members, in repenting of their past support of Japan's militarism, became keen supporters of Japan's new constitution renouncing war and engaged in movements to eliminate nuclear and hydrogen bombs around the world. When the postwar economic devastation and the arrival of U.S. soldiers stationed in Japan contributed to the continuing visibility of the prostitution business, the Japanese WCTU, in cooperation with other women's organizations, campaigned against prostitution. Their efforts led to the establishment of the Prostitution Prevention Law in 1956, whereby prostitution became a punishable offense. Today, the Japanese WCTU's women's home in Tokyo provides shelter to women of various nationalities for "protection" and "rehabilitation." At the same time, the organization's efforts for the "elimination of liquor" aspire to eradicate public nuisances caused by drinking, such as drunk driving and violence committed under the influence of alcohol.

Rumi Yasutake

See also: Japan; Leavitt, Mary Greenleaf Clement; Sake; World's Woman's Christian Temperance Union (WWCTU)

References

Bbior, Sharman Lark. *Women of a Tokyo Shelter: Domestic Violence and Sexual Exploitation in Japan* (Ph.D. dissertation, UCLA, 1993).

Cherrington, Ernest Hurst, et al., eds. *Standard Encyclopedia of the Alcohol Problem.* Westerville, OH: American Issue Publishing Co., 1926–1930: 1384–1391.

Nihon Kirisutokyo Fujin Kyofukai. *Nihon Kirisutokyo Fujin Kyofukai Hyakunenshi* (Centennial history of the Japanese WCTU). Tokyo: Domesu Shuppan, 1986.

Tyrrell, Ian. *Woman's World/Woman's Empire: The Woman's Christian Temperance Union in International Perspective, 1880–1930.* Chapel Hill: University of North Carolina Press, 1991.

Yasutake, Rumi. "Transnational Women's Activism: The Woman's Christian Temperance Union in Japan and the United States" in *Women and Twentieth-Century Protestantism,* edited by Margaret Lamberts Bendroth and Virginia Lieson Brereton. 93–112. Urbana: University of Illinois Press, 2002.

Woman's Christian Temperance Union (WCTU) (United States)

Founded in the wake of the Women's Temperance Crusade against saloons that began in Fredonia, New York, in 1873, the Woman's Christian Temperance Union (WCTU) was organized the following August at Lake Chautauqua, New York. It went on to become the most important temperance organization in the nineteenth century. As significant, it was the largest women's organization in the world by the end of the century.

Women, Temperance, and Organization

The first national convention of the WCTU was held in November 1874 in Cleveland, Ohio, with representatives from sixteen states in attendance. Annie Wittenmeyer, who had been prominent in reform circles since her relief work during the Civil War, was selected as the union's first president, and Frances Willard became corresponding secretary.

Voting membership was limited to women, a decision that ensured that control of the organization would remain in female hands. Unions were organized locally, and states were represented at the national convention based on congressional district apportionment, a feature that benefited the eastern, more conservative wing of the union, until 1881, when representation became based on dues-paying members. The 1874 convention committed its membership to the principle of total abstinence and resolved to support women's efforts to continue the fight against saloons with mass meetings and publications. The union called on Congress to investigate the liquor traffic and asked public officials to refrain from serving alcohol at public functions.

Reflecting the middle-class, evangelical origins of its early leadership, the WCTU in its early years emphasized neither militant action nor broad reform. Wittenmeyer and other conservative leaders opposed the vote for women and feared that political endorsements would thwart efforts at organization. More radical members, such as Willard, hoped to expand the agenda of the WCTU but were persuaded, and outvoted, by leaders who believed that women more properly should utilize the petition rather than the ballot.

In a contentious battle at the national convention in 1879, Frances Willard defeated Annie Wittenmeyer in the vote for president. Although a few women were put off by Willard's more radical views, more were attracted by her enthusiasm and vision. Willard was also a tireless worker. Even before her election to the presidency, Willard had traveled extensively to organize local unions. After her election, she spent the next decade traveling and writing, ceaselessly cultivating the organization's reach and visibility. Willard's election also signaled a shift in the organization's policy on women's suffrage.

Before her ascent to the presidency of the WCTU, Willard had endorsed the notion that women should have at least a limited access to the ballot, and in 1881 the WCTU endorsed the "home protection" ballot for women. Ingeniously taking a term from the debates over tariff policy, Willard argued that in matters affecting the home, women should have greater influence. The justification was not the egalitarian argument of more radical suffragists who believed that sex should not be a barrier to political power. Instead, Willard and her colleagues were able to place themselves closer to the mainstream of nineteenth-century views of the role of women in society. If a woman's sphere is the home, they reasoned, then a logical extension of her role would be to grant her the ballot to protect the home from the evils of the saloon. But even this more cautious attempt to gain political power was controversial. A few of the WCTU's leaders, including Annie Wittenmeyer, walked out of the 1881 convention in protest, and later helped to create a rival organization.

Willard's efforts drew considerable national attention. The consolidation of the WCTU newspapers in the formation of the *Union Signal* increased the organization's visibility, but much of the WCTU's most important work was accomplished at the state and local levels. Local and state unions organized petition drives and sponsored meetings, often including speeches by traveling national members. Particularly in the western states, where the Women's Temperance Crusade had started so much of the agitation, local unions supported local-option campaigns and efforts to close illegal saloons. Women set up booths at polling places and handed out ice water, sandwiches, and prohibition ballots. Members raised money for shelters, visited jail and prison inmates, organized social events for young adults, and wrote columns for local newspapers.

The WCTU provided many women their first experience in social reform. Many of the union's leaders had gained considerable experience in church and relief organizations before joining the cause, but for much of the membership the WCTU was the first institution, outside of the church, that allowed them to have a public voice. In rural areas and small towns, the WCTU provided a sympathetic network that eased the isolation of women's lives. African American women participated, both across racial lines in alliances with white members and in local unions of their own.

Frances Willard and other leaders of the WCTU continued their efforts to broaden the agenda of the movement. At the same 1881 convention that endorsed women's suffrage, Willard proposed a "Do Everything" policy that would link temperance reform to a variety of causes. Departments were established to work among immigrants, industrial workers, and miners. In the pages of the *Union Signal,* the leadership of the WCTU endorsed higher wages for workers, prison reform, and exercise and more sensible clothing for women. A Social Purity Department was established to minister to prostitutes, an effort that led to the establishment of women's shelters and eventually to successful lobbying of state legislatures to raise the age of consent to at least fourteen years. Eventually, thirty-five different departments were established to deal with a variety of issues, most of which had little direct relation to temperance. In 1884, Willard and the WCTU launched the World's Woman's Christian Temperance Union, a pioneering effort to organize women internationally, to pursue on a broader scale the national WCTU's wide-ranging program of reform.

The Department of Scientific Temperance Education was one of the most successful efforts of the WCTU. The union from its inception had been preaching temperance to children, both in organized efforts at the local level and in publications designed for Sunday schools. The Department of Scientific Temperance Education expanded those efforts to lobbying state legislatures to adopt compulsory "scientific temperance" curricula for public schools. Such efforts bore fruit across the United States although implementation of scientific temperance laws raised issues that the WCTU never satisfactorily resolved.

Women, Temperance, and Politics

On the national level, the WCTU's political influence was less effective. Although the Republican Party was more open to temperance sentiment, and the Democratic Party more openly hostile, leaders in both major parties generally tried to keep the issue of prohibition from complicating the political mix, particularly in states where majorities were razor thin. Within the union itself, political loyalty, so far as it can be gauged among women who could not vote, was more often a function of class and region than of devotion to the temperance cause. Nevertheless, the WCTU leadership, particularly Frances Willard, felt that the organization should have a presence on the national political stage. Under Willard's leadership, the WCTU had endorsed women's suffrage and nationwide prohibition, but neither of the major political parties expressed much interest in either issue.

Willard's solution lay in an alliance with the Prohibition Party, which posed a threat to Republican majorities in parts of the Midwest and Northeast and embraced both prohibition and women's suffrage. Willard failed in attempts at the 1882 and 1883 WCTU conventions to pass resolutions endorsing the Prohibition Party, but she was successful in 1884, although endorsement drove out of the organization a number of Republican women, who organized the Non-Partisan WCTU under the leadership of Judith Foster. As the representative of tens of thousands of potential voters, Willard was invited to speak at the Prohibition Party National Convention. She and other leaders of the WCTU campaigned vigorously for the party that fall. After Democrat Grover Cleveland won the election, the *Union Signal* claimed that the WCTU shared credit with the Prohibition Party for the defeat of the insufficiently prohibitionist Republicans. Four years later, the Prohibition Party fared less well, in part because the Republicans began to move closer to an endorsement of the issue. In 1892, Willard tried to forge a union of the Prohibition Party and the Populists, but neither side was willing to make the concessions necessary for fusion.

Despite Willard's failure to find an influential national political party that would include both the goals and the membership of the WCTU, the union continued to grow through the last decades of the century. WCTU speakers toured the nation organizing local unions. Willard herself traveled constantly through the early 1880s, visiting every city and large town in the nation. Her 1883–1884 tour took her and her assistant and companion Anna Gordon to every state in the union. This tireless organizational effort paid off. In 1890, the WCTU reported a membership of nearly 150,000, but the rate of growth subsequently began to fall. The union continued its successful lobbying and campaigning efforts at the state and local levels, but its visibility in national politics declined. Frances Willard remained president until her death in 1898, but her failing health and long absences from the country left the organization without a strong, visible leader much of the time.

The WCTU also faced financial difficulties in the 1890s. In 1890, construction began on the Temple Office Building, a project developed by Matilda Carse, president of both the Chicago WCTU and the Woman's Temperance Publishing Association, the publishing arm of the WCTU. The WCTU was not the controlling owner of the project, but the impressive building in downtown Chicago was intended to house the national headquarters and provide income for the organization. The building was completed in 1892, the year before the Panic of 1893 and ensuing depression swept the U.S. economy. The value of the WCTU's share in the project fell, and other investors began to pull out. Despite support from Chicago retailer Marshall Field and a year-long fund-raising effort by Frances Willard, the WCTU's ties to the building proved disastrous. Heated debates over the wisdom of the project raged at national WCTU conventions and divided the union's leadership. Willard died in February 1898, and later that year the WCTU abandoned its affiliation with the Temple project.

Eclipse of Women's Efforts

After the turn of the century, the WCTU remained an important part of the temperance movement, but by the time National Prohibition moved to the top of the national agenda it was no longer leading the fight. The Anti-Saloon League of America (ASLA), a male-dominated, nonpartisan organiza-

Members of the Woman's Christian Temperance Union (WCTU) march in Washington, D.C., to present a petition supporting prohibition, ca. 1921. (Hulton/Archive)

tion, successfully supported state and local prohibition campaigns and lobbied for a federal constitutional amendment. The ASLA was also a single-issue advocate, unencumbered by association with other reform goals. Its leaders could avoid partisan acrimony, and they could deliver votes. The WCTU had done neither.

The WCTU also lost its preeminence among women's organizations. After Willard's death, fewer new members were recruited, and during the presidency of Lillian Stevens (1898–1914) the broad range of reform issues was curtailed. Many reform-minded women found other outlets for their energies, and other organizations appeared to provide for sororal association. The General Federation of Women's Clubs surpassed the WCTU's membership numbers early in the century. The WCTU continued to be a presence in prohibition campaigns, but it became more and more an auxiliary to the real contest.

Toward the end of National Prohibition, the WCTU rallied to fight the repeal efforts, but again the organization found itself in the minority. In 1945, the union became a charter member of the United Nations Non-Governmental Organizations. Today, the WCTU maintains headquarters in the Willard House, Frances Willard's home at Evanston. The *Union Signal* is still published, now as a quarterly journal. On its official website (www.wctu.org), the Woman's Christian Temperance Union declares its opposition to alcohol, tobacco, illegal drugs, gambling, and pornography. It continues to lobby politicians and sponsors an annual essay contest for schoolchildren on temperance-related topics. It also notes the role it has played historically in providing a venue for women to run their own organization and have a public voice.

James D. Ivy

See also: Anti-Saloon League of America (ASLA); Boole, Ella Alexander; Democratic Party (United States); Foster, Judith Ellen Horton; Gordon, Anna Adams; Home Protection; Hunt, Mary Hannah Hanchett; Loyal Temperance Legion (LTL); Non-Partisan Woman's Christian Temperance Union; Prohibition Party (United States); Republican Party (United States); Scientific Temperance Federation; Stevens, Lillian Marion Norton Ames; Stoddard, Cora Frances; Willard, Frances Elizabeth Caroline; Wittenmeyer, Annie Turner; Women's Temperance Crusade; World's Woman's Christian Temperance Union (WWCTU)

References

Blocker, Jack S., Jr. *American Temperance Movements: Cycles of Reform.* Boston: Twayne, 1989.

Bordin, Ruth. *Woman and Temperance: The Quest for Power and Liberty, 1873–1900.* New Brunswick, NJ: Rutgers University Press, 1990.

Epstein, Barbara Leslie. *The Politics of Domesticity: Women, Evangelism, and Temperance in Nineteenth-Century America.* Middletown, CT: Wesleyan University Press, 1981.

Pegram, Thomas R. *Battling Demon Rum: The Struggle for a Dry America, 1800–1933.* Chicago: Ivan R. Dee, 1998.

Tyrrell, Ian R. *Woman's World/Woman's Empire: The Woman's Christian Temperance Union in International Perspective.* Chapel Hill: University of North Carolina Press, 1991.

Willard, Frances E. *Glimpses of Fifty Years: The Autobiography of an American Woman.* Chicago: Woman's Temperance Publication Association, 1889.

———. *Woman and Temperance: The Work and Workers of the Woman's Christian Temperance Union.* Hartford, CT: Park Publishing, 1883.

Woman's Christian Temperance Union website. http://www.wctu.org.

Zimmerman, Jonathan. *Distilling Democracy: Alcohol Education in America's Public Schools, 1880–1925.* Lawrence: University Press of Kansas, 1999.

Women Publicans (Australia)

The tradition of women holding liquor licenses in Australia dates back to the earliest years of European colonization. The first woman to hold a publican's license in New South Wales, the ex-convict Sarah Bird, received her license for the Three Jolly Settlers in 1797. Female licensees remained central to the Australian hotel industry throughout the nineteenth and twentieth centuries.

Publicans' licenses were first issued in the penal colony in 1792, but legislation regulating the licensing system was not enacted until 1825. Prospective licensees were subject to a certification process designed to ensure that the applicant was a person of "good fame and reputation" and "fit and proper" to keep a public house (Wright 2000: 50). In the official reckoning of "fitness," gender, marital status, and convict stain were of no consequence—so long as applicants could show that they had not obtained the money for their license or house by immoral means. Many accounts of Australia's early colonial period point to the success of women in obtaining the necessary certificates and recognizances to be issued a license to retail alcohol from their own homes as either pardoned convicts, convicts' wives, widows, or, from the 1820s, single and married free women.

In the first half of the nineteenth century, the proportion of female licensees was between 5 and 25 percent of all publicans' licenses. Colonial legislators turned their backs on the English custom of only granting widows the privilege of selling alcohol—a form of communal charity. Eschewing the traditional association between women, alcohol, and disorder, authorities saw women as part of the solution to the "liquor problem" in a disproportionately male, transient population. Female publicans were represented by politicians and customers alike as industrious, capable, and deserving women exhibiting the feminine qualities of dignity, hospitality, and maternal self-restraint.

By 1901, when the Australian colonies federated, at least a quarter to a half of all publicans' licenses were held by women. In the state of Victoria, 50 percent of hotels were under the management of female licensees until the mid-twentieth century. This predominance of female hotelkeepers provides a stark contrast to other Western nations; in the United States, for example, late-nineteenth-century legislation banished women to the fringes of informal, illegal "kitchen-selling." Australian women, by contrast, were at the apex of the hotel industry at precisely the time that the industry was at its most socially, culturally, and financially influential.

Complex laws prohibiting multiple holdings and the use of "dummy" nominees ensured that female licensees were genuinely the proprietors of their own independent businesses. Women also had to prove that the profits of their enterprise flowed directly to their own separate estates. Upon marriage, a female publican retained the rights and privileges of her liquor license, a situation that contravened the customary laws of coverture whereby a wife's property became vested in her husband. Female publicans in Australia thus held significant property rights long before the passage of the Married Women's Property Acts in the 1870s.

Some conservative judges in the late nineteenth century challenged the legal autonomy of the female publican, but legislators were less willing to revoke long-standing free-trade entitlements. Female publicans often had the backing of powerful industry supporters such as the Licensed Victuallers' Associations and, more important, the politically persuasive brewing lobby. Brewers had a commercial interest in promoting the rights of female licensees; since the early 1800s, brewers had been lending money to women to go into their "tied houses," an important factor in women's access to the hotel industry in the first place.

Although temperance advocates campaigned against the employment of barmaids in the late nineteenth and early twentieth centuries, they never similarly pursued the female publican as either a source of immorality or a victim of male degeneracy. Temperance agitation did lead to some restrictions on female license-holding—in some states, single women were prohibited from becoming licensees; in others, women had to be over the age of twenty-five—but this did not affect the overall proportion of women pursuing hotel-keeping as a vocation.

Women were attracted to the hotel industry for many reasons: financial independence, public prominence, social mobility, and emotional satisfaction. As colonial Australia witnessed a particularly high incidence of wife desertion and early widowhood, hotel-keeping provided a respectable means by which a single woman could support her family. Licensing laws that required the publican to reside on-premises, as well as provide accommodation and meals to travelers (until the 1960s in some states), determined the character of the Australian pub as an essentially domestic industry. This feature was enhanced by the cultural archetype of the "matriarch behind the bar," an image perpetuated in bush ballads, folklore, and popular literature.

Rates of female license-holding declined after World War II, when alternative employment opportunities for women expanded at the same time that business management in the hotel industry became more corporatized and less family oriented. Female publicans are still a prominent and widely respected facet of the Australian hotel industry.

Clare Wright

See also: Australia; Barmaids (Australia); Woman's Christian Temperance Union (WCTU) (Australia)

References

Alford, Katrina. *Production or Reproduction? An Economic History of Women in Australia, 1788–1850.* Melbourne: Oxford University Press, 1984.

Atkinson, Alan. "Women Publicans in 1838" in *The Push from the Bush.* No. 8 (1980).

Kirkby, Diane. *Barmaids: A History of Women's Work in Pubs.* Cambridge: Cambridge University Press, 1997.

Nichols, G. R. *The Licensed Victuallers' Consolidation Act.* Sydney: William Moffitt, 1838.

Wright, Clare. *Beyond the Ladies Lounge: A History of Female Publicans in Australia.* Melbourne: Melbourne University Press, 2003.

———. "Beyond the Ladies Lounge: A History of Female Publicans, Victoria, 1875–1945" (Ph.D. thesis, University of Melbourne, 2001).

———. "Of Public Houses and Private Lives: Female Hotelkeepers as Domestic Entrepreneurs" in *Australian Historical Studies* no. 116 (April 2001): 57–75.

Women's Organization for National Prohibition Reform (WONPR)

The Women's Organization for National Prohibition Reform (WONPR) was a short-lived but large national membership organization that played a significant role in achieving the repeal of U.S. constitutional Prohibition. Its emergence in 1929 shattered the prevailing image that U.S. women universally favored Prohibition and raised prospects for the achievement of a constitutional consensus in favor of repeal previously thought to be unreachable. Aided by the onset of the Great Depression, the WONPR proved able to work with other anti-Prohibition organizations, in particular the Association Against the Prohibition Amendment (AAPA), in mounting a politically effective campaign for the adoption of a new constitutional amendment to overturn the Eighteenth Amendment.

Female temperance advocates played a prominent role throughout the quest for prohibition in the United States. Women such as Frances Willard and Carry Nation became the most well-known temperance crusaders. With their critique of alcoholic beverages centered on damage to women, children, and family life caused by male drinking, organizations such as the Woman's Christian Temperance Union (WCTU) fostered the impression that women were united in their opposition to alcoholic beverages. Following simultaneous campaigns for National Prohibition and women's suffrage amendments in which support for these reforms was often linked, the adoption of both constitutional provisions within eighteen months of one another reinforced the view that women naturally favored a liquor ban.

Women, however, had never been universally hostile to alcoholic beverages. In the late nineteenth century, many saloons featured a separate "ladies entrance," and women were primary consumers of high-alcohol patent medicines. For instance, Lydia Pinkham's Vegetable Compound, a popular tonic, contained 20 percent alcohol. More important, gender distinctions in alcohol use began to disappear as it became more common for men and women to drink wine and cocktails together at gatherings in middle-class homes. With the arrival of Prohibition and the closing of male-oriented saloons, drinking shifted toward mixed-sex speakeasies and particularly toward the home, where in some circumstances alcohol use remained legal, and where in any case Volstead Act enforcement did not intrude. In a domestic environment where alcoholic beverages were becoming more familiar and were largely being used moderately, female doubts about Prohibition increased.

The WONPR emerged in the late 1920s from discussions among upper-class women concerned with the impact of National Prohibition upon social stability and regard for law. These women had come to perceive federal and state efforts to enforce National Prohibition as both heavy-handed and ineffective. Even though overall alcohol consumption may have declined compared to its pre-Prohibition level, their attention, as well as that of the general public, focused on highly visible violations of the "dry law." Public discussion began to shift from the matter of alcohol abuse to concerns about declining respect for law and government.

In the spring of 1929, Pauline Morton Sabin began discussing her concerns about Prohibition with other wealthy and socially prominent women, first in New York and then in Chicago. On 28 May, a group of twenty-four women from eleven states, many of whom were married to men active in the AAPA, formed the WONPR and immediately began enlisting additional members. Aided initially by the AAPA in opening a small New York office, the WONPR quickly became self-supporting. Within a year, it had acquired more than 10,000 members and was effectively recruiting middle- and upper-class women in most states outside of the Deep South.

In April 1930 at its first national convention in Cleveland, the WONPR set forth what would remain its basic position. Declaring itself in favor of temperance and protection of the home, the

WONPR held that National Prohibition had reversed a trend toward moderate drinking; stimulated misuse of alcohol; encouraged crime, political corruption, and disrespect for law; and functioned contrary to constitutional principles of local self-government and individual rights. All of these developments threatened the home and the welfare of children. WONPR members favored Prohibition repeal, Sabin asserted, "because they don't want their babies to grow up in the hip-flask, speakeasy atmosphere that has polluted their own youth" (Kyvig 2000: 122).

From the moment the WONPR's founding was announced, a flood of press reports began to call attention to its activities. Newspaper and magazine articles and feature stories aimed at a female readership not only publicized the WONPR argument but also gave it an image of respectability that had previously attached only to temperance advocates. WONPR women were characterized as sophisticated, modern, fashionable, and as concerned with home and family protection as their dry rivals. A *Time* magazine cover story about the WONPR in July 1932, just as the national political parties were debating how to deal with Prohibition in their campaign platforms, also aided the repeal cause.

Favorable press notice, together with the WONPR's own active recruitment efforts, helped its membership grow rapidly. By April 1931, the organization claimed 300,000 members. A year later, 600,000 were said to be enrolled; by the 1932 election, 1.1 million; and ultimately 1.5 million in 1933. After it first exceeded the size of the WCTU upon reaching 400,000 members in December 1931, the WONPR began touting its membership figures as a powerful sign that the political balance was tipping in favor of repeal.

The appearance of a large women's organization opposed to Prohibition raised prospects for Prohibition repeal. WONPR lobbying of congressional candidates and grassroots canvassing contributed to the repeal movement's rising political appeal. Throughout the repeal campaign, the WONPR was an important presence, reminding politicians and voters that many women favored an end to the liquor ban. WONPR leaders addressed both 1932 national political conventions, and, after the Democratic Party embraced repeal, endorsed Franklin D. Roosevelt for president. After the election, they urged Congress to adopt the repeal amendment, joined slates of pro-repeal states ratifying convention delegates, and engaged in get-out-the-vote campaigns for state elections. Two days after the ratification of the Twenty-First Amendment was complete, the WONPR held a nonalcoholic celebratory dinner at the Mayflower Hotel in Washington, D.C., and disbanded, depositing its remaining $30,000 in a Barnard College scholarship fund for female graduate students in political science.

David E. Kyvig

See also: Association Against the Prohibition Amendment (AAPA); National Prohibition (United States); Sabin, Pauline Morton; Willard, Frances Elizabeth Caroline

References

Kyvig, David E. *Repealing National Prohibition.* 2d ed. Kent, OH: Kent State University Press, 2000.

Murdock, Catherine G. *Domesticating Drink: Women, Men, and Alcohol, 1870–1940.* Baltimore: Johns Hopkins University Press, 1998.

Root, Grace C. *Women and Repeal: The Story of the Women's Organization for National Prohibition Reform.* New York: Harper, 1934.

Rose, Kenneth D. *American Women and the Repeal of Prohibition.* New York: New York University Press, 1996.

Women's Temperance Crusade

The Women's Temperance Crusade of 1873–1874 was the largest women's direct-action movement of the nineteenth century in the United States. Empowered by their religious faith and social convictions, tens of thousands of women, in more than 900 communities in thirty-one states and the District of Columbia, publicly protested the sale and importation of intoxicating liquors in an era of excessive drinking.

In the 1870s, liquor was a rapidly growing $95-million industry with more than 200,000 retail outlets serving the United States. From 1866 to 1873, per capita consumption of beer increased from 4.4 to 7 U.S. gallons and production almost doubled, and distilled spirits were consumed at a rate of over 2 U.S. gallons per person. In a context of alcohol abuse and lack of enforcement of liquor laws and ordinances, women and children suffered from the effects of poverty, domestic violence, and broken homes.

The Crusade was most prominent in the Midwest, especially in Ohio, where more than 32,000 women marched in over 300 Crusades throughout the winter. Extant sources reported 93 Crusades in Indiana, 80 in Illinois, 69 in Michigan, 60 in New York, 39 in Iowa, 32 in Pennsylvania, and 30 in Wisconsin. Most Crusades took place in rural areas and small towns with access to railroads and various forms of communication; however, Crusades were also organized in many cities, including Chicago, Pittsburgh, Dayton, Cincinnati, and Cleveland.

This grassroots social movement organized by women to protect their homes, children, communities, and nation began in Fredonia, New York, on 15 December 1873 following a lecture, "The Duty of Christian Women in the Temperance Work," delivered by Diocletian Lewis, a Boston medical doctor. Lewis outlined a plan for women to shut down liquor establishments by marching through city streets, praying, singing, invading saloons, and demanding that liquor dealers quit their trade. Lewis related how his mother had used these nonviolent techniques to rid her town of four saloons forty years earlier in Clarksville, New York. Although Lewis had given his temperance speech hundreds of times over the previous twenty years, very few communities had experienced public demonstrations, and only two had been successful in shutting down liquor outlets. The

127 marching women of Fredonia were responsible for giving birth to the national Women's Temperance Crusade. They believed they were part of a holy crusade sanctioned by God to save the bodies and souls of men.

On 16 December 1873, Lewis gave his stock temperance lecture in nearby Jamestown, New York, precipitating women's demonstrations, and then traveled to Hillsboro, Ohio, on 22 December and Washington Court House, Ohio, on 25 December. On Christmas Eve, Eliza Jane Trimble Thompson, the "mother" of the Crusade in Hillsboro, led seventy-five women into drugstores to secure signatures from three proprietors, who promised not to sell liquor for beverage purposes. Women then began marching in Washington Court House under the leadership of Matilda Gilruth Carpenter on 26 December; within eight days they shut down all saloons and elicited temperance pledges from the druggists. Although Lewis returned to the East Coast and stopped lecturing for six weeks, the women's movement quickly gained national recognition and adherents by word of mouth, newspaper articles, temperance "missionaries," and personal letters.

Eliza Daniel "Mother" Stewart of Springfield, Ohio, who rose to national prominence for her organizing efforts, speaking ability, liquor sting operations, and dramatic court appearances, exemplified the female Crusader. Such women transcended traditional nineteenth-century gender and religious conventions when they boldly marched in streets and held prayer meetings in saloons. In an era in which women were not allowed to speak in mainline churches or take a public role in politics, Crusaders led public prayers, sang hymns, preached, lectured, marched, picketed, staged sit-ins, testified in courts, lobbied voters, petitioned local and state governing bodies, and used moral suasion while men assumed a more passive role, supporting the women Crusaders financially and spiritually. Men prayed in churches, gathered wood for fires to keep the Crusaders warm, provided transportation, took care of domestic responsibilities, and built shelters—"tabernacles"—to protect women during outdoor prayer services. The Crusade was an ecumenical, privileged, largely white women's movement; however, some marches did include women of color and women of lower social and economic classes.

Faced with the onslaught of Crusaders, some druggists, saloon keepers, and physicians signed temperance pledges and treated the women with decorum; others organized and fought against the Crusaders through court injunctions, physical and verbal violence, and boycotts of the businesses owned by the male relatives of the Crusaders. One of the most notorious saloon keepers, John Calvin Van Pelt of New Vienna, Ohio, displayed a blood-covered ax in his saloon and threatened to "hang, draw, and quarter" Crusaders who dared to enter. (He later surrendered to the Crusaders and reportedly transformed himself into a temperance lecturer.) Women were also subjected to violence by drunken and unruly crowds, especially in major urban centers. Crusaders were targets of bricks, stale beer, dirty water, food, and rotten eggs, and a few were seriously beaten or arrested. In Bucyrus, Ohio, an elderly temperance woman suffered a deep laceration when she was dragged through the street and thrown down cellar stairs.

The Crusade, which lasted through the winter of 1873–1874, was more effective in small towns and less effective in urban areas. By 26 February 1874, at least 241 saloon keepers in Ohio and Indiana had quit their trade and 128 towns had become "dry." In over 20 percent of the towns in which the results were recorded in newspapers and other public documents, Crusaders successfully limited the number of liquor outlets; fewer than 10 percent of the towns that experienced Crusades reported complete failure. Throughout the spring and summer of 1874, women organized local, county, and state temperance meetings and unions. Women gathered for a National Temperance Convention in Cleveland, Ohio, in November 1874, and elected Annie Turner Wittenmeyer, a Methodist laywoman, as its first president. The convention gave birth to the national Woman's Christian Temperance Union (WCTU), which evolved into the largest and most powerful women's organization in the nineteenth century.

Michelle J. Stecker

See also: Wittenmeyer, Annie Turner; Woman's Christian Temperance Union (United States)

References

Blocker, Jack S., Jr. *"Give to the Winds Thy Fears": The Women's Temperance Crusade, 1873–1874.* Westport, CT: Greenwood, 1985.

Bordin, Ruth. *Woman and Temperance: The Quest for Power and Liberty, 1873–1900.* Philadelphia: Temple University Press, 1981.

Mattingly, Carol. *Well-Tempered Women: Nineteenth-Century Temperance Rhetoric.* Carbondale: Southern Illinois University Press, 1998.

Murdock, Catherine Gilbert. *Domesticating Drink: Women, Men, and Alcohol in America, 1870–1940.* Baltimore: Johns Hopkins University Press, 1998.

Stewart, Eliza Daniel. *Memories of the Crusade: A Thrilling Account of the Great Uprising of the Women of Ohio in 1873, Against the Liquor Crime.* 2d ed. Columbus, OH: William G. Hubbard, 1889.

Wittenmeyer, Annie. *History of the Woman's Temperance Crusade.* Philadelphia: Office of the Christian Woman, 1878.

Women's Total Abstinence Union (WTAU)

The Women's Total Abstinence Union (WTAU) was a British single-issue organization. It rejected the women's suffrage policy that Lady Henry Somerset had borrowed from the leader of the national WCTU in the United States, Frances Willard, and that the majority of British Women's Temperance Association (BWTA) members favored.

A New Temperance Organization

When the majority of the executive committee of the BWTA left that organization in 1893, they immediately founded the Women's Total Abstinence Union. They reorganized along the lines of the BWTA's original constitution and formed a provisional committee composed of members of the old BWTA executive and influential representatives of the provincial and metropolitan branches. They elected former BWTA officers Martha Holland, Lucy Brooks, Mary E. Docwra, and, as chair, Louisa Stewart. The officers were experienced and capable temperance women who also had the advantage of retaining the journal *Wings.* The executive committee owned the majority of the shares in the publication.

Many of the members of the new organization had been long-standing members of the BWTA. They included Charlotte Cowen (Mrs. Henry J.) Wilson, a veteran temperance advocate who became the first president of the WTAU. In common with many other temperance women she was a supporter of a number of causes. She organized women's rescue homes in the Sheffield area and assisted Josephine Butler in her campaign against the Contagious Diseases Acts (1864; repealed 1886). Wilson brought to the presidency a calm disposition and personal links to the Liberal Party. Her husband was a prominent member of Parliament (M.P.). Her guiding principle was that temperance work could be improved under Christian influence. In general, the WTAU appointed presidents for one year only, but Wilson led the organization for two years.

The WTAU published a manifesto in November 1893 highlighting its priorities. These included a demand for a local veto, ending the sale of intoxicating drinks to children, repeal of grocers' and shopkeepers' licenses, and legislation for treatment of inebriates. The first annual meeting, convened the following March, was well attended. The WTAU believed that by emphasizing its nonpolitical role and subscribing to no political party it would appeal to many women who had reservations about Somerset's overt political stance. Despite this, most members were well aware of the reality of the politics of temperance. The wives of many temperance M.P.s (for example, W. S. Caine and T. P. Whittaker) joined the WTAU.

Accomplishments and Frustrations

There is no doubt that the WTAU was active within the temperance movement. It urged members to provide them with evidence of the evils of alcohol, such as the increase of drinking and drunkenness in specific areas as a result of licenses being granted to grocers and other shops. Anne Hawkes, a Good Templar and Band of Hope activist, gave evidence to the Royal Commission on Liquor Licensing Laws (1896–1899) on behalf of the WTAU. She reported on shopkeepers' licenses, beer cart vendors, and prohibition as well as the situation in Ireland. Hawkes's evidence on prohibition differed radically from Lady Henry Somerset's. Hawkes wished to see intoxicating liquors removed from public houses and clubs and the abolition of the manufacture of alcohol.

The union continued to be solid in its desire to be a single-issue organization. After the retirement of Lady Henry Somerset, her successor, Rosalind Howard, Countess of Carlisle, suggested that the two societies should merge in order to fight Prime Minister Arthur James Balfour's Licensing Bill in 1904. The WTAU declined the offer. After some deliberation, the WTAU agreed to share a platform with Carlisle's society, now called the National British Women's Temperance Association (NBWTA), on the condition that the WTAU could use its own name and nominate its own speakers. Both organizations condemned Balfour's compensation scheme. The WTAU printed a powerful editorial in *Wings* arguing that Balfour's legislation would give the drink trade the scope for claiming "audacious" sums of money for compensation. Other attempts to merge the two women's societies in 1905–1906 came to nothing as the WTAU was determined to maintain its single-issue status.

The WTAU continued to encourage links with other societies that supported the same aims and objectives, including prohibition by local veto. They cooperated in order to assist protemperance candidates who supported the local veto, abolition of grocers' licenses, Sunday closing, and the closing of public houses on election day. Women could vote and stand as candidates in local elections. For instance, in 1907 temperance women were encouraged to stand as candidates for the London County Council elections. WTAU members who stood for election did so as individuals, as the society continued to be politically neutral. The WTAU did praise the London County Council for allowing liquor licenses to lapse in connection with their street improvement schemes.

The WTAU was frustrated by the tardiness of the Liberal government to introduce a local veto bill, so much so that executives began to develop an interest in the M.P.s of the new Labour Party. The WTAU proved pragmatic when the government finally introduced a Licensing Bill in 1908. It passed a resolution accepting the bill and approving its main features. It encouraged members to write to their M.P.s and participate in demonstrations, but the House of Lords rejected the bill. The temperance movement was fading.

During the early days of World War I, the WTAU provided water carts for the troops and continued to agitate for prohibition. The temperance movement had little influence on wartime government policies such as the introduction of state management of the liquor trade, generally referred to as the Carlisle experiment, which encompassed the largest munitions factory in the country on the borders between England and Scotland. To fight drunkenness, other legislation was enacted. The opening hours were cut dramatically, all-day drinking was abolished, and by 1917 40 percent of the pubs in the Carlisle district area had closed and off-sales licenses had been revoked. These restrictions continued in place until 1921.

When the war ended, the WTAU and other temperance organizations campaigned for wartime restrictions such as those on opening hours to be made permanent. The group of temperance M.P.s sent a letter to Prime Minister David Lloyd George requesting this. The WTAU became increasingly active in this campaign, and at its annual meeting in 1921, a political dimension was added to the union in an effort to retain the wartime status quo. The WTAU was determined not to go back to prewar conditions. When relaxations looked probable, the WTAU formulated resolutions and branches began to participate in national politics. They were encouraged by the fact that two temperance women, Nancy Astor and Margaret Wintringham, had been elected to the House of Commons and began working with them to prohibit the sale of alcohol to under-eighteens. Other gains were eroded. Although the Licensing Act of 1921 retained the shortened opening hours, Prime Minister Stanley Baldwin reduced drink taxes in 1923.

The two women's temperance organizations, the WTAU and the NBWTA, became closer following the death of the controversial Rosalind Howard, Countess of Carlisle, and reunited in 1926 under a new name, the National British Women's Total Abstinence Union (NBWTAU). The WTAU took the initiative in this reunion. By this time, women's suffrage was no longer a divisive issue.

Many WTAU records survive: Executive committee meeting minutes, 1893–1907, general committee minutes, 1900–1926, and annual reports, 1894–1925, are on file at the headquarters of the White Ribbon Association, the current name of the NBWTAU.

Margaret Barrow

See also: Band of Hope; British Women's Temperance Association (BWTA); Carlisle, Countess of (Rosalind Howard); Carlisle System; National British Women's Temperance Association (NBWTA); Somerset, Isabella Caroline (Lady Henry Somerset); Willard, Frances Elizabeth Caroline; Woman's Christian Temperance Union (United States)

References

Barrow, Margaret. "Temperate Feminist: The British Women's Temperance Association" (Ph.D. dissertation, University of Manchester, 1999).

Wings. Selected issues. 1893–1926.

Working Men's Clubs Breweries (United Kingdom)

In the first decades of the twentieth century, as Britain's working men's club movement grew in confidence, workingmen challenged the power of the commercial brewers through a series of experiments in collective brewing. Local clubs were able to harness the traditional attachment of the industrial classes, particularly in the north and midlands, to both beer drinking and the principles of cooperative enterprise. Though some schemes proved vulnerable, others survived to display a business acumen and rigor comparable to that of any brewery undertaking.

The Working Men's Club and Institute Union (CIU) was founded in 1862 with the intention of freeing the laboring classes from the public house by establishing clubs free from intoxicating drinks. In 1875, these restrictions on alcoholic drinks were removed, and thenceforth clubs competed directly with the public houses. The Industrial and Provident Societies Acts passed in 1852 and 1862 aided the growth of working men's clubs, and CIU's membership grew sixfold in the four decades after 1890.

Growing in strength, the clubs resolved to form brewing cooperatives. Some steps were taken in the early twentieth century: The Sowerby and District Clubs Brewery and the Burnley Clubs Brewery were registered in about 1900 and a Leeds and District Clubs Brewery was formed in 1911. During World War I, the 60,000-circulation *CIU Journal* stoked up a general disenchantment about beer supplies and mistreatment by brewers, so clubs were determined to act once hostilities ended. The largest and most successful of the clubs breweries, the Northern Clubs Federation Brewery Limited (henceforth "the Federation"), was registered in 1919. Clubs became shareholding members of the Federation on the basis of individual club membership, received interest on share capital, and qualified for a share of profits distributed in proportion to purchases. After a false start with an unworkable, redundant brewery in Alnwick, Northumberland, the Federation began brewing in Newcastle in 1921. By then, the Medway Federation of Clubs Brewery of Kent, the South Wales and Monmouthshire United Clubs Brewery, and the Northants and Leicestershire Cooperative Clubs Brewery had opened.

For the clubs brewery movement, the 1920s were a time of mixed fortunes. New breweries were opened: the Clubs Breweries (Wakefield) Limited in 1923, the York and District Clubs Brewery in 1924, the Labour Clubs Brewery Society in Preston in 1927, and the Shamrock Working Men's Brewery, Cheshire, in 1929. In the same period, earlier proposals in Derbyshire, Coventry, and London fell by the wayside and the United Clubs Brewery (Fleetwood) Limited and the Amalgamated Clubs Brewery Company of Tamworth went into liquidation before reaching the brewing stage. More significantly, some schemes that had graduated to brewing closed down: The Sowerby District Clubs Brewery was wound up and the Gwent Union Clubs Brewery went into liquidation in 1929. Another failure was the brewery at Wakefield, which brewed for only three years.

While other brewery schemes were struggling to transform their best intentions into viable levels of operation, the Federation was increasing its capacity. By the mid-1930s, it was producing 78,000 barrels annually. Its nearest rival, the United Clubs Brewery in South Wales, had an annual output of fewer than 30,000 barrels. Most clubs breweries survived the 1930s intact, although they did so very much as the poor relations of their local rivals, the traditional brewers. But there were losses: the West Midlands Working Men's Brewery closed around 1931 and the Leeds and District Clubs Brewery absorbed its neighbor at York and subsequently changed its name to the Yorkshire Clubs Brewery. Two new brewing cooperatives had been formed in Lancashire—one at Bolton in 1933 and the Moor Park Clubs Brewery in 1936—but neither brewed.

After World War II, the difficult prewar conditions that afflicted the brewing trade persisted, especially the long-term decline in beer consumption and the growing popularity of counterattractions. The relatively weak position of clubs breweries meant some casualties: During the war itself, the Rochdale and District Clubs Brewery in 1942, the Medway Federation Brewery in 1946, and the Burnley Clubs Brewery in 1950 all failed. By the early 1950s, the combined annual output of the remaining seven clubs breweries was around 300,000 barrels, a little over 1 percent of United Kingdom beer production, of which the Federation was responsible for around 210,000. The clubs brewery in South Wales built a new brewery in the mid-1950s, but others were approaching the end. In 1960, the Lancashire Clubs Federation Brewery stopped trading and the Walsall and District Clubs Brewery was taken over by a conventional brewer. The Preston Clubs Brewery closed in 1962, and the Midland Clubs Brewery (formerly the Northants and Leicestershire) ceased in 1969.

The clubs brewery movement was reduced to two participants when the Federation took over the Yorkshire Clubs Brewery in 1973. For a while, the Federation had cooperated with, and provided financial assistance to, the other survivor, the United Clubs Brewery of South Wales. By the mid-1970s, the close relationship between the two cooperatives had faltered, leaving the Federation with an expanding market and an enhanced reputation, but consigning the Welsh brewery to a troublesome period of uncertainty and change that ended with its demise in the late 1980s. The Federation continues as a major regional brewer.

Brian Bennison

See also: Beer; Working-Class Social Clubs (United Kingdom)

References

Bennison, Brian. "A Happy Band of Brewers? The Federation and the Clubs Brewery Movement" in *North East History.* 31 (1997): 75–91.

Elkins, T. *So They Brewed Their Own Beer.* Newcastle: Northern Clubs Federation Brewery, 1970.

Glover, Brian. *Loyalty Pays. The History of the United Clubs Brewery.* Stroud: Allan Sutton, 1995.

Tremlett, George. *The First Century.* London: CIU, 1962.

Working-Class Social Clubs (United Kingdom)

Working-class social clubs evolved out of the movement started by the Working Men's Club and Institute Union (CIU) of Great Britain, established in 1862. Philanthropic and religious reformers wanted to provide workingmen with an appropriate environment to attain a lifestyle dedicated to moral and self-improvement. Alcohol was forbidden, for the clubs were conceived as a temperance alternative to the thousands of beerhouses that had opened throughout the country after the 1830 Beer Act. Licensing justices lacked the power to deny beerhouse licenses. Within a few years after the inauguration of the club movement, it became apparent to all involved that the CIU would fold through lack of patronage unless beer was made available. Subsequently, clubs affiliated to the union became another outlet for breweries.

The Working Men's Club and Institute Union expanded slowly. By 1895, only 404 clubs were affiliated. However, the policy of restricted licensing in the country, reintroduced with the Wine and Beerhouse Act of 1869, made it difficult for would-be licensees to obtain full on-premise licenses. As a result, the drink industry turned to the club movement and opened working-class social clubs. The majority of these clubs were dedicated to a specific sporting pastime, such as golf, bowling, or fishing, and were not merely drinking clubs. By 1904, 6,462 working-class social clubs were registered in England and Wales. By 1930, 2,660 such clubs were affiliated with the CIU.

The emphasis in the clubs was social interaction, with or without the purchase of drink, whereas the emphasis in public houses was obtaining drink, with or without social interaction. In contrast to the patrons of public houses, club members could abstain from alcohol while enjoying all the amenities that the clubs had to offer. Also, club members came together to pursue a common interest other than consuming alcohol.

As working-class social clubs grew in number, they began to attract the attention of licensing justices. Justices feared that the worst elements of pub culture—drunkenness, violence, and idleness—would naturally migrate from public houses to clubs, where they were free from supervision. It was not until the 1902 Licensing Act that liquor-supplying clubs were required to register with the clerk to the justices of petty sessional divisions. A club could be struck off the register if, upon proven complaint, it was found to have fewer than twenty-five members or was improperly conducted. Nevertheless, the majority of working-class social clubs were law-abiding, well-run groups offering workers a recreational alternative to the drinking culture found inside public houses.

Nigel Ansell-Roberts

See also: Working Men's Clubs Breweries (United Kingdom)

References

Rowntree, Joseph, and Arthur Sherwell. *The Taxation of the Liquor Trade,* vol. 1. London: Macmillan, 1906.

Tremlett, George. *Clubmen: The History of the Working Men's Club and Institute Union.* London: Secker and Warburg, 1987.

Wilson, George B. *Alcohol and the Nation.* London: Nicholson and Watson, 1940.

World Health Organization (WHO)

The World Health Organization (WHO) is the most important international institution in the field of disease control and prevention. It was founded in 1946 and officially established in

1948 as a Specialized Agency of the United Nations, the successor of the League of Nations. In addition to the headquarters in Geneva there are largely autonomous regional offices (local headquarters) and committees (assemblies of the member states) for Africa, the Americas (formerly Pan American Sanitary Organization), Europe, the Far East, the Near East, and the Western Pacific.

In 1948, the WHO constitutional assembly approved a far-reaching definition of health: "a state of complete physical, mental and social well-being." Accordingly, the WHO—unlike the League of Nations—assumed responsibility for alcoholism. In the 1950s, there was a short but important phase of research in this field. Alcohol policy, however, ranked low among WHO activities. There were other challenges in the aftermath of World War II, and the health problems in the colonies and underdeveloped countries due to poverty and overpopulation took precedence. From the 1970s onward, alcohol policy gradually gained momentum. The most comprehensive approach was the action plan by the regional office for Europe (WHO-EURO). According to the "total consumption model," its main objective was to cut down consumption by 25 percent by the year 2000. Started with great expectations, the plan had no great impact on the national policies except in the Nordic countries. Moreover, it came under attack, not only from alcohol interest groups but also from a growing number of scholars. Finally, in 1999 a new plan was launched that omitted the 25-percent goal.

International Agreements before World War II

The first multinational convention on alcohol control was a byproduct of the antislavery congress on Africa held in Brussels in 1889–1890. The signatory states of the so-called General Act declared the prohibition of the trade and manufacture of distilled liquors in an area south of the Sahara and north of the English and German colonies in southern Africa. The regulations were aimed only at the African and other nonwhite populations; thus their practical value was limited, as white colonists, traders, and officials were the main liquor consumers. Nonetheless, the General Act was a blow against the Russian and German distillers who had flooded the world market with spirits.

As a lesson of World War I, the League of Nations was established in 1919–1920. From the outset standing on shaky ground, it proved to be helpless when Japan, Italy, and finally Germany started their wars of aggression. In 1939–1940, the league virtually came to an end. However, in the meantime some progress was made in alcohol control: The League of Nations widened and tightened the measures of the General Act on Africa and settled agreements curbing illicit liquor traffic in other parts of the world, especially on the Baltic Sea. (Because the United States, the biggest "dry" country, stayed outside the League of Nations, it had to negotiate on a bilateral basis with numerous countries.) Moreover, efforts were made to install an Advisory Committee on Alcoholism analogous to the league's Narcotics Control Board to provide for investigations and—in the long run—for control measures as well. Fostered, among others, by the Finnish delegates, this initiative was fought, among others, by the French ones. As "alcoholism" could mean every kind of negative consequence of drinking—even drinking as such—the opposition was too strong. One side wanted to label alcohol a "dangerous drug"; the opponents successfully insisted on their cultural identity and pointed to health benefits from wine drinking. These arguments are not settled to this day—in the heated atmosphere of global quarrels among "wet" and "dry" experts a compromise was impossible to find. After 1930, with the United States on the road to repeal of National Prohibition, the debate petered out.

Activities of the WHO

By 1948, the dry experts had fallen silent and the WHO general assembly could pass a decision to claim alcoholism—now regarded solely a mental disorder—as part of its work (namely, of the Mental Health Unit at WHO's headquarters). Initially, this work mainly consisted in the activities of E. M. Jellinek, who had been invited to work as a consultant in Geneva. From 1950 to 1955, in many conferences, study groups, and seminars, he helped to define and refine the disease model of alcoholism. Although this concept had in principle been established in medical thought and everyday knowledge by around 1900, at least in Europe, it was the WHO that definitively defined "alcohol addiction." Transferred into the International Classification of Diseases, the WHO definition became standard in diagnostics and therapy. Despite substantial changes in wording ("dependence" instead of "addiction," for example) and a growing scholarly criticism, it still serves as the prevailing paradigm in clinical practice and in lay perception of heavy drinking.

The WHO, however, was hardly interested in alcohol policy. Against the background of the failure of Prohibition, experts and laymen alike regarded alcoholism as not primarily caused by alcohol but instead by inner malfunctions and defects. Except in Scandinavia, alcohol control bore the stigma of zealotry. Jellinek made clear that it lay outside his sphere of interest. Although he later admitted that socioeconomic factors also played an important role in the etiology of alcoholism, he was less concerned with control and policy than with diagnostics and therapy. When Jellinek left Geneva, alcoholism ranked low on WHO's agenda. This was in accordance with the general trend in medicine and the media that depicted alcohol as an integral part of the fashionable life.

Around 1970, the tide turned again when a series of meetings was held at the WHO head office. The renewed interest came from the developed countries, mainly from Europe, and had several sources. First, improved life expectancy focused attention on noncommunicable "civilization diseases," such as coronary heart disease and cancer, and in turn on lifestyles, especially with regard to eating, smoking, and drinking. Second, the Nordic countries faced fundamental challenges to their restrictive alcohol policies owing to attempts at rapprochement toward the European Community (EC) and its successor since 1995, the European Union (EU), as well as to cultural changes in the urban middle classes. Third, the growing staff and budget of the WHO called for new fields of activity (between 1951 and 1971 the staff at the European regional office—which in 1957 had moved from Geneva to Copenhagen—grew from 13 to 218). Under the auspices of WHO-EURO, broad epidemiological and sociological research was carried out. The study group—headed by Kettil Bruun and supported by Finnish and Canadian institutions—published a pioneering report on "Alcohol Control Policies" in 1975. In a nutshell, the report concluded that reducing per capita consumption was the key to reducing health hazards in general and alcoholism in particular. The latter was assumed again as primarily induced by alcohol and was regarded as only one negative consequence of drinking among others. In other words, the focus shifted from secondary prevention of alcoholism back to primary prevention of alcohol-related problems.

This approach matched with WHO's efforts toward a "new" public health. In a series of declarations culminating in the Ottawa Charter (1986), the claim was laid to shaping national policies by drawing upon the 1948 definition of health. In numerous conventions, this all-embracing—meanwhile nearly forgotten—objective of "well-being" was broken down into individual "targets" to be reached by limited "plans." Of course, among the lifestyle-influenced health hazards, alcohol was taken into account as well. The phase of indulgence was coming to an end. However, the WHO headquarters and regional offices, including the one for the Americas, concentrated more on other targets, such as tobacco. Only WHO-EURO made alcohol a field of highest priority.

The European Alcohol Action Plan

In 1984, the regional committee for Europe passed a "Health for All" resolution containing 38 targets. No. 17 ("decreasing health-damaging behavior") aimed inter alia at alcohol-related problems. In its explanation of the document, the regional office called for lowering consumption by 25 percent. Thus, for the first time since the interwar period, an international organization was demanding a reduction in the consumption (and indirectly the production) of alcoholic beverages. This attempt was the fruit of debates that had occurred in the wake of the study group's report in 1975. In the 1980s, this approach was transferred into a coherent theory called the "total consumption model."

The theory assumed a close relationship between average consumption and perils from drinking in any given population (with consumption distributed according to the so-called Ledermann-curve, which the Frenchman Sully Ledermann had calculated in 1956). Owing to a kind of "infection," the rate of "heavy drinkers," or alcoholics, was considered to be a function of the per capita consumption; cutting back that figure meant cutting back alcoholism. Likewise—and even more important—the model stated that the lion's share of alcohol-related problems (death, diseases, accidents, and violence) was suffered and caused by the "moderate drinkers," who were consuming most of the alcohol. Although moderate drinkers might have a low likelihood of dying from liver cirrhoses, widespread moderate consumption induces greater harm from alcohol-related problems than that induced by heavy drinkers. Alcohol policy, therefore, had to concentrate on the "normal drinkers"; it had to de-normalize traditional drinking patterns by means of high taxation, reduction of availability, bans on advertising, and other related measures.

Except in the Nordic countries, target No. 17 had almost no impact on national policies. The Soviet Union made a futile attempt to curb vodka consumption, but in the EC, by far the most important addressee, the public and even experts took virtually no notice of it. Experts from Norway, Sweden, and Finland tried all the more to promote the total consumption approach. Because they were considering membership in the EC, these countries faced a fundamental threat to their tough control policies (Finland and Sweden joined the EU in 1995; Norway did not). Finally, after several resolutions similar to target No. 17, the European Alcohol Action Plan (EAAP) was worked out by WHO-EURO and passed by the committee in 1992.

The official resolution as such did not go very far. The member states were urged, inter alia, to help to reduce alcohol *abuse* and the problems related to it. Again, WHO-EURO added a lengthy explanation that had a more far-reaching message. Most alcohol-related problems, it read, cropped up in connection with moderate drinking, and there was a close relationship between average consumption and the number of heavy drinkers. So the "primary target" of the EAAP was to reduce the average *consumption* between 1980 and 2000 by 25 percent in all member states. Therefore, the main target group was to be the "normal drinker." Secondary prevention, which aims at risk groups, was, although necessary, of "limited value."

WHO-EURO harshly claimed the leading role in alcohol policy, but the EAAP received broader attention only when a new study group under the auspices of the WHO had gathered data that provided an epidemiological basis for the plan. The report of the research group—headed by Griffith Edwards and cofinanced by institutions in Toronto, London, Helsinki, and Oslo—was published only after the EAAP, namely in 1994. The report underlined the manifold physical and social harm alcohol does to the "public good," especially as manifested in the majority of "moderate consumers." Confirming the strategy of the first WHO report from 1975, the report pleaded for an all-embracing alcohol policy in accordance with the total consumption model.

A debate on the EAAP was set in train. Because WHO-EURO had blurred the boundaries between the official resolution and its own explanation, the debate focused on lowering average consumption, although this objective was not part of the document approved by the committee. Unsurprisingly, alcohol interest groups opposed this goal, whereas those in the temperance and health sphere supported it (in many countries adopting the 25-percent goal into their national programs). In science, too, reception was mixed. After an overwhelmingly positive initial response, gradually a critical stance gained ground.

It became obvious that the approach merely reflected the views and problems of temperance cultures. Critics argued that the European Alcohol Action Plan would be more accurately called a Scandinavian American one and that Edwards's study group had used it as a means of gaining cultural hegemony. Indeed, among the seventeen researchers of the group, none were from Russia, Germany, or France, whereas 41 percent were from Scandinavia and 35 percent from North America. Culturally biased as it was, the approach tended to turn a blind eye to unsuitable evidence such as health benefits from moderate drinking or data indicating a weak relationship between "problems" and per capita consumption. The approach ignored the fact that de-normalizing integrated drinking patterns may risk increasing hazardous consumption. Bruun's notion that drinking is part of a "meaningful whole" had gone out of sight. Certainly, the EAAP mentioned that alcohol is part of life and can have some health benefits, but this idea was opposed to the inner logic of the approach faded out in later analysis. Because at any consumption level "problems" will occur, the demand for further lowering is always obvious—hypothetically until alcohol is abolished (as Ledermann had already clearly shown). Critics therefore called the approach "crypto-prohibitionism." Finally, the total consumption model became contentious among Scandinavian researchers, too. Labeled a "doctrine," it was being replaced by other strategies, in particular—once again—by secondary prevention.

Probably as a result of this growing criticism, the plan was dropped. After passing several resolutions (for example, an "Alcohol Charter"), in 1999 WHO-EURO launched a new EAAP. The official document hinted at the "different cultures" among member states and stated its recommendations in a different way than they had been presented before. No longer did it speak of lowering the average consumption, but a justification for this reorientation was missing, confirming Bruun's former observation that policy making within the framework of the WHO is a matter of "insiders" and not the fruit of an open scholarly discussion. Instead, the "primary target" of the old EAAP—reducing consumption by 25 percent—was reworded to emphasize reducing "adverse health effects." WHO-EURO insisted that the old plan should still function as a guideline, however, and thus the new document failed to correct the flaws in the old one. These ambiguities suggest that the controversy over the total consumption model is anything but settled at the turn of the millennium.

Hasso Spode (English translation by Jennifer Hosek)

See also: Alcoholism; Brussels Act and Conventions, 1890–1912; Jellinek, Elvin Morton; League of Nations Health Organisation; St. Germain en Laye Convention, 1919

References

Brunn, Kettil, et al. *Alcohol Control Policies in Public Health Perspective.* Helsinki: Finnish Foundation for Alcohol Studies, 1975.

———. *The Gentlemen's Club: International Control of Drugs and Alcohol.* Chicago: University of Chicago Press, 1975.

Edwards, Griffith, ed. *Alcohol Policy and the Public Good.* Oxford: Oxford University Press, 1994.

Room, Robin. "The World Health Organization and Alcohol Control" in *British Journal of Addiction.* 79 (1984): 85–92.

Spode, Hasso. "Präventionskonzepte in Geschichte und Gegenwart" in *Strategien und Projekte zur Reduktion alkoholbezogener Störungen,* edited by Gerhard Bühringer. Lengerich: Pabst Science, 2002.

Sulkunen, Pekka, et al., eds. *Broken Spirits. Power and Ideas in Nordic Alcohol Control.* Helsinki: NAD, 2000.

WHO Regional Office for Europe. *European Alcohol Action Plan.* Copenhagen: WHO, 1993; German version, Gamburg: Conrad, 1993.

———. *European Alcohol Action Plan, 2000–2005.* www.who.dk/adt/aaction.html.

World League Against Alcoholism (WLAA)

The World League against Alcoholism (WLAA) was established in Washington, D.C., in 1919 to spread information worldwide about the benefits of prohibition. Its founder, Ernest Hurst Cherrington, secretary of the Anti-Saloon League of America (ASLA), had first advocated the global adoption of prohibition in a speech to the ASLA in 1913. Coming after the Eighteenth Amendment had made the sale of liquor in the United States unconstitutional, the WLAA had a partly utilitarian purpose. A "dry" world would make it easier to defend Prohibition against liquor smugglers from abroad. However, the WLAA also had an idealistic side. It resembled a moral form of Wilsonian internationalism in remaking the world in the aftermath of World War I. It was strongly evangelical in its roots and reformist in its motivation to cleanse the world of the stain of alcohol and its attendant evils.

The World League established a general council of representatives of member organizations but worked through an executive committee. It initially appointed four joint presidents, the ASLA's Howard H. Russell, Woman's Christian Temperance Union (WCTU) president Anna Gordon, Dr. Robert Hercod of Switzerland, and Leif Jones, a Welsh temperance reformer. The real power was General Secretary Cherrington, who coordinated league activities from his Washington offices. The organization was heavily dependent on the parent ASLA for funding.

International offices were established in Oslo, Norway; in Lausanne, Switzerland, through the International Bureau against Alcoholism; and in London, where Henry Beach Carré, a Tennessee clergyman, was appointed as office manager. The WLAA had, as affiliates at home, the WCTU, the Intercollegiate

Prohibition Association (founded in 1892), and the Scientific Temperance Federation (1906). Apart from the initial organizing meetings in 1919, two major international conventions were held—in 1922 in Toronto, Canada, and in 1927 at Winona Lake, Illinois. Over 1,000 delegates attended each conference; fifty-eight countries were represented at the second and sixty-six at the first. The World League's everyday activities were reported in the WCTU's *Union Signal* and the ASLA's *American Issue.* Ultimately, the World League worked in 185 countries, including European and American colonies. It was most active in pushing Prohibition in the early 1920s, but the peak of activity in terms of circulation of its propaganda came from 1927 to 1930 as Prohibition in the United States went on the defensive.

Continuing the nonpartisan strategy of the ASLA, the World League professed to diverge from the political lobbying that the ASLA practiced at home. It would leave political activity to local groups in each country, instead serving as a clearinghouse for information to combat antiprohibitionist sentiment. Nonetheless, the WLAA and its officials repeatedly traveled to Europe as well as occasionally to Africa, Australia, and Asia to take an active part in electioneering whenever prohibition votes or strict licensing laws were being decided. In these political campaigns, William E. "Pussyfoot" Johnson, as international organizing secretary of the WLAA from 1919 to 1924, was most active. He undertook extensive and sometimes controversial tours of England (1919), Scotland (1920), India (1921), and Sweden (1922). Other leading ASLA and WLAA figures traveled in the interests of temperance reform. These included Bishop James Cannon, who went to England in 1920 and South Africa in 1922; Anna Gordon, who toured South America and Britain; and Howard H. Russell, who went to Britain in 1920.

International collaborators impressed by U.S. symbols of modernity and Prohibition encouraged the World League's foreign activities. Allies included European teetotalers such as Michel Légrain, who headed the World Prohibition Federation's European bureau; American and British missionaries in such places as China, Japan, and India, and pro-American reformers such as Australia's Robert B. S. Hammond, editor of the prohibitionist newspaper *Grit.* These overseas admirers equated Prohibition with U.S. material prosperity and moral superiority. The WLAA was encouraged by these allies to intervene in foreign prohibition debates in Scotland in 1920 and in Sweden in 1922 and gained support in Scandinavia, Australia, and New Zealand as well as in Britain. The WLAA took advantage of expatriate American clergy and businessmen for its lobbyists and lecturers, used immigrants who had lived in the United States and returned to their home countries, and enlisted the help of foreign nationals normally based in the United States. Ten Swedish American pastors headed by the Reverend David Ostlund were sent to lead the Swedish prohibition referendum, which narrowly lost in 1922.

The WLAA was not the first international temperance organization, nor the first to advocate prohibition. That honor belonged with the World's WCTU. Moreover, the World Prohibition Federation and International Reform Bureau all preceded the WLAA. These were dwarfed, however, by the WLAA in the 1920s because the latter was an extension of the work of the ASLA, whose prestige was at its highest after passage of the Eighteenth Amendment.

Even among its friends, the WLAA always had its detractors. Leading ASLA official Wayne Wheeler saw the WLAA as "too grand a scheme" (Tyrrell 1994), and abroad WLAA lecturers were often ridiculed in the popular press or subject to pickets and even violence. They also incurred the wrath of authorities, as when World League official Johnson gave rhetorical support for anticolonialism in India when he visited there in 1921. He was forced to curtail the tour and leave.

No real or lasting success came from any of these overseas campaigns. Though prohibition was in force briefly in Finland and Iceland, this stemmed from local sentiment. Norway and Sweden both narrowly rejected prohibition, as did New Zealand. Australia adopted six o'clock closing, not full prohibition. And in India, M. K. Gandhi favored prohibition but did not think the United States had the necessary moral traditions and political strength to maintain its own Prohibition.

The World League remained active until the fall of National Prohibition in the United States in 1933. After that defeat, the ASLA withdrew financial support and its international operations were effectively stymied, though Cherrington (and Hercod) continued to work for prohibition and temperance legislation until Cherrington's death in 1950.

Ian Tyrrell

See also: Anti-Saloon League of America (ASLA); Australia; Cherrington, Ernest Hurst; China; Finland; Gandhi, Mohandas Karamchand; Gordon, Anna Adams; India; Intercollegiate Prohibition Association (IPA); International Reform Federation, Inc.; Japan; Johnson, William Eugene "Pussyfoot"; New Zealand; Norway; Russell, Howard Hyde; Scotland; Sweden; Wheeler, Wayne Bidwell; World Prohibition Federation; World's Woman's Christian Temperance Union (WWCTU)

References

Brook, Susan. "The World League against Alcoholism: The Attempt to Export an American Experience" (M.A. thesis, University of Western Ontario, 1972).

Higley, Mary. "World League against Alcoholism, 1900–1937" in *Guide to the Microfilm Edition of Temperance and Prohibition Papers,* edited by Randall Jimerson, Francis Blouin, and Charles A. Isetts, 205–208. Ann Arbor: University of Michigan Press, 1977.

Roberts, Kenneth. *Why Europe Leaves Home.* New York: Arno, 1977 [1922].

Tyrrell, Ian. "Prohibition, American Cultural Expansion, and the New Hegemony in the 1920s: An Interpretation" in *Histoire Sociale/Social History.* 27 (November 1994): 413–445.

World Prohibition Federation

The movement for worldwide prohibition began organizationally in 1909 with the founding of the International Prohibition

Confederation (IPC), which took a new name, the World Prohibition Federation (WPF), in 1919. This propaganda organization harbored a small but impassioned group of reformers who endeavored to unite prohibition societies the world over in a crusade for international prohibition. In the waning years of the prohibition agitation in Britain, Guy Hayler (1850–1943), a tireless English temperance advocate, became interested relatively late in life in prohibition as an international movement. Devoting himself to spreading prohibition worldwide, he was the founding president of the IPC and continued as honorary president of the WPF until he retired in 1939 at the age of eighty-nine.

Hayler's interest in international prohibition developed gradually. He had joined the Independent Order of Good Templars (IOGT) in 1870 and had gained an administrative position in its international organization by 1905. He traveled to the United States in 1893, 1901, and 1905 to attend Good Templar international functions and to conduct prohibition-related research. His travels abroad, and membership in the Good Templars, ultimately fostered Hayler's internationalism to the point that he readily accepted a leading role in the IPC upon its creation.

The international propaganda society was the invention of a group of Anglo-American reformers led by Edward Page Gaston, European manager for Funk and Wagnalls. Gaston organized the first meeting of the IPC in London, at which Hayler was elected honorary president and Gaston honorary secretary. Reformers from nineteen countries endorsed the IPC's first conference, and delegates from thirty-nine countries approved its next conference in the Netherlands in 1911. When the WPF finally established official headquarters in London in 1925, Hayler's son Mark (1887–1986) became executive secretary.

The IPC operated on a slim budget. In its first two years, it spent just £118 of the mere £123 it had collected. Its modest income included £10 from Hayler and around £20 from Gaston. Isaac Funk from the United States, honorary treasurer, contributed £2, and his publishing company (Funk and Wagnalls) added another £20. In the early years, two of Hayler's principal supporters in the IPC were fellow Good Templars Tom Honeyman, leader of the Scottish Grand Lodge, and Joseph Malins, head of the English Grand Lodge. In 1918, at least half of the IPC's executive committee were members of the Good Templar order.

The IOGT remained a dominant influence on the IPC/WPF. In his 1911 report for the Good Templars on the "World's Fight against Alcoholism," Hayler resolved that the principles of world prohibition were essentially the same as those established by the Good Templars in 1851: The first principle was total abstinence from all intoxicating liquors, followed by the elimination of the liquor traffic, as well as an absolute prohibition not only on the sale but also on the production and importation of intoxicating liquors. Prohibition activists were to work toward the creation of a healthy public opinion through temperance education, help to elect upstanding men to make and administer laws, and continue their efforts to save individuals and communities from drunkenness and the liquor traffic.

As propaganda organizations, the IPC and the WPF championed international prohibition chiefly through their literature. In less than two decades, the organizations distributed more than 5 million leaflets, pamphlets, and other publications. Hayler, who was highly regarded as both a student of and propagandist for the movement, published countless pamphlets, reports, and books that espoused the prohibition cause. His publications consisted of detailed reports on the status of prohibition initiatives around the world, excerpts from government documents and other contemporary sources, as well as historical lessons imbued with temperance messages.

In 1917, Hayler became the founding editor of the IPC's *International Record*, a quarterly magazine of international news about prohibition designed to alert the world to the movement's progress. Since the movement had its genesis among a host of reform-minded intellectuals around the world, its chief publication was distributed free of charge to politicians, intellectuals, and other world leaders in anticipation that international reformers would see to it that postwar reconstruction included prohibition. Hayler edited the magazine until his death; it was the official organ of both the IPC and the WPF.

Hayler referred repeatedly to the advantages of worldwide prohibition. During his extensive travels, he observed that where nations had adopted prohibition, drunkenness had been practically eliminated, crime reduced, and a host of other social evils virtually overcome. "Dry" nations experienced improvements in moral, educational, industrial, commercial, political, and religious life. Total abstainers, insisted Hayler, were more thrifty, better able to accumulate savings and own their own homes, and noticeably more prosperous and well-to-do than drinkers.

Despite having vice presidents from a variety of countries, the WPF was essentially an Anglo-American organization. Although based in London, the organization balanced its British and American identity by maintaining U.S.-born honorary secretaries and treasurers and by intermittently holding its conferences in the United States, which it praised as a model for worldwide prohibition. In 1948, Dr. David Leigh Colvin became the first American to serve as president of the WPF. He headed the organization until his death in 1959.

Although the IPC had remained for its first ten years the principal international organization for prohibition propaganda, the WPF was challenged in 1919 by a better-financed U.S.-based organization, the World League against Alcoholism, an offshoot of the Anti-Saloon League of America (ASLA). After ASLA achieved its goal of National Prohibition, many of its leaders turned their focus to the abolition of the liquor traffic around the world, evidently dissatisfied with the WPF. By the 1930s, financial problems fatally weakened both organizations.

Although many temperance and prohibition organizations lost their impetus in the interwar years, the WPF survived until after World War II—nominally until at least 1968, the date of the last issue of its organ, the *International Record,* which Mark Hayler edited after his father's death. To a large extent, the Haylers and a few other activists were the organization.

Matthew K. McKean

See also: Anti-Saloon League of America (ASLA); Good Templars (IOGT); Hayler, Guy; Hayler, Mark; Honeyman, Tom; International Reform Federation, Inc.; Malins, Joseph; World League Against Alcoholism (WLAA)

References

Fahey, David M. "Guy Hayler (1850–1943) and Moral Reform Internationalism: England's Publicist for World Prohibition." Unpublished conference paper, Ohio Academy of History, 20 April 2002.

Hayler, Guy. *Prohibition Advance in All Lands: A Study of the World-Wide Character of the Drink Question.* London: International Prohibition Confederation, 1914 [1913].

World's Woman's Christian Temperance Union (WWCTU)

The World's Woman's Christian Temperance Union (WWCTU) was one of the most significant international women's organizations of the late nineteenth and early twentieth centuries and a forerunner of many transnational organizations interested in social justice, peace, and global governance as well as the narrower concerns of prohibition and Protestant morality. Established in 1884, the WWCTU grew out of the national Woman's Christian Temperance Union (WCTU) in the United States, founded in 1874. The drive for external expansion began as part of transatlantic reform movements that included antislavery, women's suffrage, socialism, and campaigns against the British Empire's contagious diseases acts, but its ideological basis was in the evangelical urge to missionary activity, an impulse that would take temperance women from the United States far beyond the familiar routes across the Atlantic. Prominent was a millennial vision of a whole world transformed by religiously inspired reform movements.

Founding and Organization

The WWCTU had an ephemeral predecessor in the International Woman's Christian Temperance Union, which met in Philadelphia during the 1876 U.S. centennial celebrations, but the International WCTU was inactive. Over the next few years, WCTU officials often discussed the need for international outreach. Indispensable to this international work was the spread of the WCTU to Canada in 1876, through the work of Ontario's Letitia Youmans, and to Britain, where a visit by Eliza "Mother" Stewart that same year was crucial. One of the leaders of the Women's Temperance Crusade against saloons that took place in the American Midwest in 1873–1874, Stewart's tales of the Crusade stirred great interest in Britain. Margaret Parker of Dundee, Scotland, who had toured the United States in 1875 to learn about the methods of the WCTU, also sparked interest in Britain. Parker and Stewart both took part in the founding of the British Women's Temperance Association (BWTA), with Parker as first president in 1876. Women's temperance unions inspired by the WCTU were soon founded in South Africa (1878), Calcutta, India (1878), and Sydney, Australia (1882), but no coherent organization lay behind these efforts, which were the work of individual women.

A serious effort to develop affiliated WCTUs across the world began with the work of Frances Willard, national president of the WCTU in the United States after 1879. She conceived of a global effort as a result of an 1883 organizing trip to the Pacific Coast, where she witnessed the opium problem among the Chinese immigrants of California. Concerned that alcohol and drug problems affecting women (and men) crossed national borders, Willard proposed a new organization devoted to a worldwide attack on all "brain poisons." This experience and contact with missionary efforts made her think of the need not just for transatlantic but wider action that would encompass the colonial world of India, Africa, and the Pacific Islands as well as China and Japan. The idea of a World's WCTU was officially sanctioned by the national WCTU and marked by the embarkation of Mary Leavitt in November 1884 on a round-the-world missionary trip. As a related, major initiative, Willard started the "Polyglot Petition" of women against alcohol and other drugs in 1885. Over the next ten years, 7 million "signatures" were obtained. Though many of these were contributed through Christian societies such as the Salvation Army and missionary groups signing on behalf of members, the Polyglot was an important organizing tactic. It stirred much interest in the evangelical churches, the press, and the general public and gave women in WCTU affiliates concrete work to do. Meanwhile, Margaret Bright Lucas of Britain was appointed first World's WCTU president in 1885. The first meeting of international delegates came at the U.S. national convention in Minneapolis in 1886. Under Willard's insistent vision, the organization held its first World's WCTU convention in 1891 in Boston; well-publicized and impressive conventions were later held in 1893 (Chicago), 1895 (London), 1897 (Toronto), 1900 (Edinburgh), 1903 (Geneva), 1906 (Boston), 1910 (Glasgow), and 1913 (Brooklyn). After a break during World War I, they resumed in 1920 (London), then continued in 1922 (Philadelphia), 1925 (Edinburgh), 1928 (Lausanne), 1931 (Toronto), and 1934 (Stockholm). The conventions were interrupted by World War II but continued into the 1980s.

Upon Lucas's death in 1890, Frances Willard herself took the position of World's WCTU president until her death in 1898. Willard was succeeded by her close confidante Lady Henry Somerset, a wealthy, aristocratic English woman with connections to the British Liberal Party. Somerset's tenure of office proved stormy, and she officially resigned in 1906 to be succeeded by another stalwart English Liberal aristocrat, Rosalind Stanley Howard, Countess of Carlisle. After Carlisle's death in

1921, Anna Gordon, Willard's devoted secretary and close friend, took charge while still national WCTU president, and in turn another national WCTU president from the U.S. organization, Ella Boole, took over; she served a record term, from 1931 to 1947. Boole was the last of the WCTU leaders to exert international political influence. She campaigned for world peace and against the international narcotics trade, securing, after World War II, United Nations recognition for the WWCTU as an official international organization. After the end of U.S. National Prohibition in 1933, the national WCTU suffered a marked decline in interest and influence. The World's WCTU also declined in importance, though world conventions continued to be held.

Organizationally, the WWCTU used a departmental structure to enable women to specialize in various branches of temperance, religious, political, and social work. They were given budgets and able to operate as "superintendents" as part of an elaborate "moral bureaucracy" linked to a worldwide chain of specialized agents, each reporting to her own local, state, or national convention. Delegates to the world conventions were chosen by the national affiliates. From 1895 through the 1940s, the administration of the WWCTU rested largely in the capable hands of Agnes Slack, corresponding secretary, an English Liberal, middle-class woman. Financially, the WWCTU was partly dependent on the dues of national affiliates, but much of its work had to be accomplished with the aid of special appeals for mission support and donations from interested and wealthy individuals such as Somerset.

Support

Membership peaked at 766,000 dues-paying members in 1927, though the organization from the early twentieth century had regularly claimed a following of more than a million women. Of the affiliates, Britain was the second largest group, with 157,000 members in 1910 compared to the United States' 235,000. Significant unions were also founded in Australia, Canada, New Zealand, and South Africa, and in Scotland, a separate Scottish Christian Union affiliated with the WWCTU in 1904 after secession from the BWTA. In a number of countries, for example Australia, the WCTU was the largest single women's organization and proved an effective vehicle for advancing claims for women's equality, especially demands for voting rights, and was active within the National Council of Women. Local and national affiliates typically established their own annual or biennial conventions, built headquarters, and established newspapers such as the Australian *White Ribbon Signal,* Canada's *White Ribbon Tidings,* and Scotland's *Scottish Women's Temperance News.*

These strong unions were either based in Britain or were part of the British Empire's major so-called settlement colonies, but the union staked a claim in forty-two countries by the 1920s. Mary Leavitt traveled for seven years from 1884 to 1891, starting vigorous organizations in Hawaii and New Zealand in 1885 before moving on to Asia, Africa, Europe, and South America. Following Leavitt's travels, the organization sent out many missionaries, recruiting its cadre not only from the United States but from many of its affiliates, and had notable work done by English women (such as Flora Strout and Christine Tinling); Scottish women (Mary Campbell and Mary Lochhead), New Zealanders (Mrs. Anderson Hughes and Bessie Cowie, based originally in Australia), South Africans (Emilie Solomon), and a host of Australians (Emily Cummins). The much publicized round-the-world missionary program begun by Leavitt continued with a journey to Hawaii, East Asia, and the southwest Pacific by Jessie Ackermann beginning in 1889. Ackermann was to be the most prolific of the travelers, circumnavigating the globe several times, though her impact was most strongly felt in Australia, where she helped organize the Australasian WCTU in 1891 and served as its first president. Her major missionary endeavors for the WWCTU occurred between 1889 and 1895. After this, another thirty-three round-the-world missionaries, later termed "organizers" traveled to various countries (for example, Alice Palmer to South Africa), and at least another thirty-four ad hoc missionaries served between 1884 and 1928. Including other individuals recruited for specific purposes, the WWCTU's cadre of international organizers totaled at least 108 by 1928.

Much of the missionary effort concerned the non-Western world, particularly India, China, and Japan. The WWCTU campaigned against the child bride practices of high-caste Hindus, supporting the work of Pundita Ramabai, a WCTU representative in India. After the United States acquired colonies from Spain in 1898, the WCTU also sent missionaries there, notably to the Philippines and Puerto Rico. Mexico was another major target, and South America was extensively covered by Elma Gowen, Hardynia Norville, and others after 1900.

The damaging conflict from 1884 over nonpartisanship in the U.S. national WCTU—after Willard had dabbled in third-party politics against Republican Party interests—spilled over into the WWCTU. Judith Ellen Foster, who helped to found the Non-Partisan Woman's Christian Temperance Union in 1889, did not support the WWCTU, but significantly, she visited Britain in 1890 and courted a breakaway of the BWTA. Numerically, however, the British Women's Total Abstinence Union (BWTAU) did not stop the growth of the BWTA in the 1890s, and the Non-Partisan WCTU did not develop into a viable alternative to Willard's creation in the United States. Eventually, the BWTAU and the BWTA were reunited, but not until the 1920s.

Europe was not a WCTU stronghold, with the exception of the Scandinavian countries Finland, Sweden, and Norway. The WCTU also developed a strong union in Iceland, and several of its international missionaries came from Iceland. In the rest of Europe, the union was strongest in Germany, aided by the efforts of Ottilie Hoffmann and the Deutscher Bund Abstinenter Frauen (Federation of German Women Abstainers), but in France and Switzerland (as well as Italy) it was very small indeed and its work was associated with the Blue Cross League, an organization that did not even practice total abstinence. Thus the WWCTU's strength was inversely proportionate to the influence of European wine cultures and closely mirrored the Protestant, evangelical geography of Europe; moreover, its popularity in Scandinavia can be accounted for by the fact that in those countries women's individual rights to express their views were given at least modest support in the political culture. The WWCTU was fundamentally middle class in its support base, though it sometimes received elite patronage and garnered some working-class and interracial allegiances.

Policies

Among policy issues, the defining feature was the "Do Everything" policy first enunciated within the U.S. national WCTU by Frances Willard. National unions were not obliged to adopt the same elements of policy as the WCTU in the United States, nor was the WWCTU; in principle, however, the WWCTU took on all of the major activities associated with the U.S. branch, though it did stress some more than others. Key issues involved support for missionary work; opposition to prostitution and support for other elements of "purity" reform; antinarcotic legislation; treaties to ban alcohol from colonial possessions of the European powers; peace reform; social welfare; and social justice. Only a minority adopted the socialist sympathies of the leaders surrounding Frances Willard, but a number of national affiliates opened Temperance and Labor Departments as a means of advancing the condition of working people, especially the wages of working women. Racially, the World's WCTU proclaimed and to some extent practiced ideas of racial equality and ecumenicalism in matters of faith. Humanitarianism was a strong theme, with financial, political, and emotional support given to the cause of international refugees, especially on behalf of Armenians after the massacres of 1895 in the Ottoman Empire. Yet the WWCTU's reliance on European dominance of the colonial world to gain access to indigenous audiences, together with its strong Christian faith and its belief in the equal treatment of women, tended to produce conflicts over religion and culture and meant that the organization was tinged with "cultural imperialism."

Suffrage was a major issue almost everywhere the WCTU flourished. In Australia and New Zealand, the WCTU was deeply involved in the agitation to give women the right to vote, an aim achieved in New Zealand in 1893, in South Australia in 1894, and nationally in Australia in 1902, far ahead of the United States, where the federal suffrage was achieved in 1920. Though suffrage was to be used to promote temperance legislation, arguments based on the "justice" of giving women voting rights were also used.

Only slightly less important was peace work, led by Hannah Clark Bailey of Maine, who tapped Quaker support at home and in Britain in the 1890s. Though compromised by World War I, the peace agitation was revived in the 1920s with the support of Carrie Chapman Catt's American Committee on the Cause and Cure of War after 1925, the World Court, and the Kellogg-Briand Pact of 1928. In Australia, the WCTU campaigned successfully for women to be appointed as alternates on the Australian delegation to the League of Nations.

Like its U.S. national counterpart, the World's WCTU always stood officially against any alcohol use, but it was not always in favor of prohibition. In the mid-1890s, Willard, Somerset, and some others favored a high-license system called the Raines Law and investigated the Gothenburg plan and other systems of government or municipal ownership of hotels. The Norwegian WCTU supported for a time the modified Samlag system, which was a variant of many "disinterested management" schemes proposed or introduced in Scandinavian countries.

The organization was split by the controversial opinions of Vice President Somerset from 1895 to 1898, not only on prohibition but also over her advocacy of licensed brothels in India as a protection for British troops against venereal infection in the army cantonments. This put Somerset on a collision course with the influential World's WCTU Purity Department superintendent, Josephine Butler. Though Butler had already resigned from the WCTU over differences of opinion with Willard, she still had much support in the WWCTU affiliates, particularly in such countries as South Africa and parts of Australia, where the Contagious Diseases Acts or similar laws regulating prostitution in the interests of the colonial state and its armies had been in force. Under severe political pressure, Somerset was forced to recant in January 1898, but the WWCTU was badly affected by internecine rivalries between the pro-Somerset and pro-Butler camps through 1902, a period in which organizational tasks took back stage. Somerset finally resigned from the BWTA presidency in 1903, though she remained titular WWCTU leader until 1906.

In the 1920s, the need to defend National Prohibition consumed more and more of the energies of the U.S. branch, and the World's WCTU joined in the drive to extend prohibition globally. Its leaders campaigned vigorously in many countries for prohibition, but only in Iceland, Finland, and Norway was a measure of (temporary) success achieved. At the same time, the WWCTU continued to campaign for peace, social welfare, and other laws affecting women; moreover, it continued to exercise influence within politics and the women's movement internationally and in individual countries throughout the 1920s. Not until the repeal of U.S. National Prohibition in 1933 did the organization seriously begin to lose its influence internationally.

Ian Tyrrell

See also: Boole, Ella Alexander; British Women's Temperance Association (BWTA); Carlisle, Countess of (Rosalind Howard); Foster, Judith Ellen Horton; Gordon, Anna Adams; Gothenburg Schemes/Disinterested Management; Leavitt, Mary Greenleaf Clement; Lucas, Margaret Bright; Non-Partisan Woman's Christian Temperance Union; Parker, Margaret Eleanor; Scottish Christian Union (SCU); Slack, Agnes Elizabeth; Somerset, Isabella Caroline (Lady Henry Somerset); Willard, Frances Elizabeth

Caroline; Woman's Christian Temperance Union (WCTU) (Australia); Woman's Christian Temperance Union (WCTU) (United States); Women's Temperance Crusade

Reference

Tyrrell, Ian. *Woman's World/Woman's Empire: The Woman's Christian Temperance Union in International Perspective.* Chapel Hill: University of North Carolina Press, 1991.

Wright, Bertha (1863–1949)

Hannah Bertha Wright (Carr-Harris) (1863–1949) was an evangelist, Sunday school teacher, founder of the Home for Friendless Women, active temperance reformer, and author of six books. From the 1870s, many middle-class Canadian women, inspired by the demands of their evangelical faith and by the challenges of urbanization, materialism, and intemperance, organized groups such as the Woman's Christian Temperance Union (WCTU). Its subgroup for single women, the Young Woman's Christian Temperance Union (YWCTU), undertook campaigns to attest to their faith and temperance and to evangelize among those who had ignored the call to purity and piety, typically the imbibing working poor. One especially active YWCTU group was led in the Ottawa region by Bertha Wright, born Hannah Bertha Carr-Harris in 1863.

As a member of Ottawa's social and political elite, Bertha Wright could easily have spent her life in blissful self-absorption. However, as a serious-minded young woman intent on personal and societal reform, she embodied many of the qualities of the middle-class evangelical woman. Intelligent, articulate, and deeply religious, Bertha Wright took charge of almost any situation in which she involved herself. As a member of the Ottawa YWCTU during the 1880s, she organized a night school for working girls to improve literacy, a series of sewing schools and "Kitchen Garden Programs" where girls learned homemaking and employment skills, calisthenics classes for working-class boys and girls, and, of course, Bible study groups for all of these, with a liberal dose of temperance instruction. Occasionally, gospel temperance meetings or revivals were arranged and opened to the public. She provided "training classes" in evangelism for young women to arm them with the necessary abilities so that they might distribute tracts, lead Bible and temperance tract study, or engage in door-to-door visitations without rebuff. She wrote a column for the WCTU periodical, the *White Ribbon Tidings,* to help other youth-group leaders cope with their challenging (working-class) students. But two special projects commanded much of her attention during this time: Bands of Hope and the Home for Friendless Women.

The Bands of Hope each accommodated up to one hundred children and met on Friday evenings or Saturday afternoons to hear lectures, read the Bible, and sing hymns and temperance songs. Through her development of pedagogical skills to teach these often-unpopular topics to children inclined to be restive at the best of times, she developed a considerable reputation in temperance circles as an education authority, self-taught and indisputably expert.

In 1887, Wright led about a dozen members of the local YWCTU group in establishing and maintaining a refuge for "fallen" and abandoned women. The Home for Friendless Women took in women who had been released from the local jail or found to be inebriated, along with their children. On admission, women were urged to sign a pledge saying that they would never again use alcohol themselves and that they would raise temperate children in the shelter of the home for a full year. They would agree to work in the home's commercial laundry, and eventually to become self-supporting, in order to avoid the old patterns of a dissolute life with a man who drinks. Through her experiences directing this moral rescue project, Wright became more aware of social inequities that drove women to alcohol and vagrancy and gradually developed a broader societal critique to explain individual sin. However, this did not prevent her from engaging in spirited proselytizing.

In 1890, she and a "brave little band" of temperance workers from the Ottawa YWCTU decided that the time was ripe to establish a mission in the Catholic stronghold of Hull, Quebec. The ensuing riot reverberated through Ottawa society, reaching even to the floor of the House of Commons, where both Prime Minister John A. Macdonald and Wilfrid Laurier defended the young women's evangelism and deplored the "brutal and cowardly" violence directed toward them (*Ottawa Evening Journal,* 10 February 1890).

Wright's group had intended to lure men and women from the local bars in Hull to the mission hall, where temperance leaflets and evangelical speakers would encourage listeners to begin the long process of moral reform through abstaining from drink. The YWCTU also hoped to direct intemperate women with children, as well as abused or abandoned mothers living with men made violent through alcohol, to the Home for Friendless Women.

The Ottawa YWCTU began its campaign in the spring of 1888 to save the men and women of Hull by distributing handbills announcing regular meetings at the mission hall. Very quickly, however, the meeting turned violent. A week later, the YWCTU found itself in the midst of a second, and much more serious, riot. This time, a crowd estimated at 400 "toughs and sluggers" directly attacked the young women, knocking them down and forcibly driving them from the hall. Although Wright insisted that the YWCTU would carry on "by His Grace" in evangelizing the dangers of alcohol, the mission soon after was quietly closed (*Ottawa Evening Journal,* 25 February 1890). In no sense, however, did this deter the Ottawa Y and other members across the country from their evangelism. Many branches of the YWCTU carried on with a daunting program of social activism, visiting and supporting the elderly, infirm, and orphaned, mounting antitobacco campaigns in schools and youth groups, teaching and nurturing poor children, and teaching "domestic science" to young women, among many other initiatives.

Sharon Anne Cook

See also: Band of Hope; Woman's Christian Temperance Union (WCTU) (Dominion of Canada)

References

Cook, Sharon Anne. "'A Gallant Little Band': Bertha Wright and the Late Nineteenth-Century Evangelical Woman" in *Journal of the Canadian Church Historical Society.* 37, no. 1 (1995): 3–21.

———. "The Ontario Young Woman's Christian Temperance Union: A Study in Female Evangelicalism, 1874–1930" in *Changing Roles of Women within the Christian Church in Canada,* edited by M. Fardig Whiteley and E. Muir, 299–320. Toronto: University of Toronto Press, 1995.

———. "'Sowing Seed for the Master': The Ontario W.C.T.U. and Evangelical Feminism, 1874–1930" in *Journal of Canadian Studies.* 30, no. 3 (1995): 175–194.

———. *"Through Sunshine and Shadow": The Woman's Christian Temperance Union, Evangelicalism, and Reform in Ontario, 1874–1930.* Montreal and Kingston: McGill-Queen's University Press, 1995.

Ottawa Evening Journal, 10 February 1890.

Ottawa Evening Journal, 25 February 1890.

Y

Yale Center of Alcohol Studies

The Yale Center of Alcohol Studies was the first research institution focusing on alcohol problems to emerge in the United States following the repeal of National Prohibition in 1933. Established in 1943 as the Section of Studies on Alcohol in the Laboratory of Applied Physiology at Yale University, the team of scientists and social scientists developed a major multidisciplinary research program dealing with physiological, psychological, social, and epidemiological aspects of alcohol use. The Yale Center was home to the *Quarterly Journal of Studies on Alcohol,* founded in 1940 as a forum for international alcohol-related research. During the 1940s, the center expanded its programs to include professional education and training through the Summer School of Alcohol Studies, outpatient treatment for alcoholics through the Yale Plan Clinics, and consultation services for businesses through the Yale Plan for Business and Industry. The Yale Center also helped establish the forerunner of the National Council on Alcoholism and Drug Dependence and became a strong supporter of the disease concept of alcoholism. The center's various programs became models for the development and growth of similar activities around the world.

Origins and Early Initiatives

The "founding father" of the Yale Center was Dr. Howard W. Haggard, a noted physiologist and director of the Laboratory of Applied Physiology. Haggard had done some experiments on alcohol metabolism in the 1930s, and in 1937 he joined a group of scientists and physicians in the Research Council on Problems of Alcohol (RCPA). The goal of the RCPA was to raise money to fund research on the effects of alcohol on the body and to publish the results of this research. In 1940, Haggard founded the *Quarterly Journal of Studies on Alcohol* to serve as an organ for the work of the RCPA and to publish other significant research on alcohol problems. Believing alcohol problems should be studied from a wide range of perspectives, he began to build a team of researchers in his laboratory that included physiologists, psychologists, and biochemists as well as a sociologist, a jurisprudent, and an economist. In 1943, he formally designated a Section of Studies on Alcohol to bring together the various alcohol research projects. As director, he appointed E. M. (Elvin Morton) Jellinek, a fellow member of the RCPA who had directed the council's project to review and evaluate the existing research literature on alcohol's effects on the body. Jellinek had published in the *Quarterly Journal* in 1942 a detailed research outline for alcohol problems, in which he prioritized those areas most in need of investigation. By giving first priority to the study of alcohol addiction and its treatment, he set the stage for a narrowing of the research focus from the broader alcohol problems perspective to an emphasis on alcoholism.

The center's researchers embarked on an ambitious research program that included a model of the phases of drinking in alcoholism, a formula for estimating numbers of alcoholics, and the development of the first Breathalyzer to measure blood-alcohol content. The results of these efforts were published for the research community in the *Quarterly Journal,* but Haggard and Jellinek wanted also to educate health professionals, social service workers, clergy, and law enforcement personnel about the nature and consequences of alcohol problems. In 1943, they inaugurated the first Summer School of Alcohol Studies with the goal of building a core of trained professionals to provide education and treatment services for alcohol problems. This was the first such professional education program of its kind, and it served as a model for many similar programs over the next several decades. The school has continued annually since its inception.

Alcoholism: Disease and Treatment

In 1944, the Yale Center collaborated with Marty Mann to reach out to the broader public. Mann was the first woman alcoholic to recover through Alcoholics Anonymous, and she wanted to bring the message to the public that alcoholism was a disease that could be treated. She established the National Committee for Education on Alcoholism, which was headquartered and subsidized by the Yale Center. Marty Mann proved to be an effective spokesperson for the disease concept, and the National Committee quickly spawned affiliates in many cities across the country. In 1950, the renamed National Council on Alcoholism became an independent organization and moved its headquarters to New York City.

The Yale Center recorded another milestone in 1944 when it opened two Yale Plan Clinics, the first outpatient clinics in Connecticut (and the United States) for treating alcoholics. In addition to helping alcoholics, the clinics were designed to provide training for social service workers and to serve as research bases for clinical studies. The clinics provided free services, and as the volume of patient traffic increased, the number of clinics expanded to five. By 1949, the clinics had become a major financial burden, and the Yale Center handed over control to the Connecticut Commission on Alcoholism (established in 1945 with help from Dr. Selden Bacon of the Yale Center).

The Yale Center also developed a plan to help businesses deal with the ramifications of alcoholism among employees. The first programs to identify and treat alcoholic employees began in the 1940s at several large companies, including DuPont and Eastman Kodak. In 1951, Yale researchers published a study on alcoholism and social stability that included occupational characteristics, and in 1953 the center formally announced the Yale Plan for Business and Industry. It was designed to help businesses assess the extent and costs of alcoholism in their workforce and to establish treatment and referral services for employees. A major component of the plan was the need to educate employees and management about the disease nature of alcoholism and to establish company policy for treatment. The number of workplace alcoholism programs grew through the next few decades, gradually evolving into employee assistance programs dealing with a broad range of substance abuse, health, financial, and family issues.

By 1950, the programs of the Section of Studies on Alcohol dominated the work of the Laboratory of Applied Physiology, and the section was renamed in that year as the Center of Alcohol Studies. When Jellinek moved on to the World Health Organization in 1951, Bacon was appointed as director. Bacon, a sociologist and a graduate of Yale, had joined the Laboratory of Applied Physiology in 1943 to do research on sociological aspects of alcohol use. His strong personal and family connections to Yale University were a major factor in helping the center to withstand efforts by Yale through the 1950s to withdraw university support. The Yale administration had supported creation of the original Section of Studies on Alcohol, but when Whitney Griswold became president in 1950 he had a different vision for the university—one that focused on a classical education and did not include applied projects such as the Center of Alcohol Studies. After a ten-year battle, he finally succeeded in persuading the board of governors to pass a resolution that the center should find a more appropriate home. Bacon negotiated with several universities, and in 1962, with financial help from the National Institute of Mental Health and from philanthropist R. Brinkley Smithers, the center moved its operations to Rutgers University, where it continued to expand its research, training, and information programs.

Penny B. Page

See also: Alcoholism; Bacon, Selden Daskam; Christopher D. Smithers Foundation; Employee Assistance Programs (EAPs); Haggard, Howard Wilcox; Jellinek, Elvin Morton; *Journal of Studies on Alcohol;* Keller, Mark; Mann, Marty; National Council on Alcoholism and Drug Dependence (NCADD); Research Council on Problems of Alcohol (RCPA); Rutgers Center of Alcohol Studies; World Health Organization (WHO)

References

Keller, Mark. "Mark Keller's History of the Alcohol Problems Field" in *The Drinking and Drug Practices Surveyor.* No. 14 (March 1979): 1, 22–28.

———. "The Origins of Modern Research and Responses Relevant to Problems of Alcohol: A Brief History of the First Center of Alcohol Studies" in *Research Advances in Alcohol and Drug Problems,* vol. 10, edited by Lynn T. Kozlowski et al., 157–170. New York: Plenum, 1990.

Page, Penny Booth. "E. M. Jellinek and the Evolution of Alcohol Studies: A Critical Essay" in *Addiction.* 97, no. 12 (1997): 1619–1637.

Youmans, Letitia (1827–1896)

Letitia Youmans (1827–1896), Sunday school teacher, youth organizer with the Independent Order of Good Templars (IOGT), editorial writer for the Templars' *Temperance Union,* teacher at the Picton Ladies' Academy, and stepmother of eight, was the daughter of a farmer, born Letitia Creighton near Cobourg, Ontario, in 1827. She first organized the Woman's Christian Temperance Union (WCTU) in Canada. Despite being the offspring of the U.S. national WCTU and founded in the same year (1874), the Canadian organization was both more evangelical and more decentralized than the U.S. one. This unique character was largely due to Letitia Youmans's distinctive personality and societal critique.

In 1874, Youmans attended an international Sunday School Conference at Chautauqua, New York, where the idea of forming a women's temperance organization in the United States was conceived. Later that same year, she traveled to Cleveland for the founding of the U.S. national WCTU; soon afterward, she established the first local union of a Canadian WCTU in Picton, Ontario. Youmans would go on to create the Ontario provincial WCTU, which remained the strongest provincial union throughout the organization's long history, in 1877, and in 1885 she became president of the recently formed Dominion WCTU, which held its first conference in 1888. Youmans also acted as the WCTU's organizer throughout Canada, encouraging women to establish local and county unions. She had been raised in an intensely evangelical Methodist household. As a student, and later as a teacher at several ladies' seminaries in Cobourg and Picton, she experienced the passion and intellectual fervor of revivalism. Ever after, she held to an evangelical analysis. Every project she undertook, including the founding of the WCTU, attested to her evangelical principles. As well, she chose to root the Canadian women's temperance movement in the local community rather than to organize it from the national level, as had been the pattern adopted in the United States, and to engage less in legislative battles and more in educational campaigns. The result was an organization with a strong local base that was organically connected to the community that benefited from her leadership—but that was never overshadowed by that leadership, as Frances Willard was accused of doing with the WCTU in the United States.

Youmans was the main proponent in Ontario of the successful juvenile Bands of Hope youth organization sponsored by many temperance lodges and, eventually, closely associated with the Canadian WCTU. The Bands of Hope (and later the Loyal Temperance Legions) met weekly to study temperance literature, to practice "declaiming" in public meetings in the style of the day, and to engage in public entertainments for family and friends. Most of the activities emphasized and developed children's oratorical powers, providing the upwardly mobile with useful skills. At the same time, they encouraged children to develop serious moral habits to last a lifetime.

Steeped as she was in the evangelical tradition, Youmans made effective use in her writings and lectures of shocking portraits and compelling stories of drunkards dragging themselves and their families into the mire. Although she drew on a long narrative tradition from temperance literature and meetings, she molded contemporary ideas into a new temperance critique that emphasized the plight of women and children at the mercy of irresponsible, indeed, self-centered and violent men. Her solutions were also well situated within evangelicalism: education, and especially education of women and children, would reform the Canadian Christian family, and thereafter, Canadian society.

Sharon Anne Cook

See also: Band of Hope; Willard, Frances Elizabeth Caroline; Woman's Christian Temperance Union (WCTU) (Dominion of Canada); Woman's Christian Temperance Union (WCTU) (United States)

References

Cook, Sharon Anne. "Letitia Youmans: Ontario's Nineteenth-Century Temperance Educator" in *Ontario History.* 84, no. 4 (1992): 329–342.

———. *"Through Sunshine and Shadow": The Woman's Christian Temperance Union, Evangelicalism, and Reform in Ontario, 1874–1930.* Montreal and Kingston: McGill-Queen's University Press, 1995.

Youmans, Letitia. *Campaign Echoes.* Toronto: William Briggs, 1893.

Young Women's Christian Association (YWCA)

The Young Women's Christian Association (YWCA) is a worldwide organization initially established in the nineteenth century to provide Christian leadership and a homelike environment to young women who had left home to find employment in industrialized cities. The YWCA stressed traditional Christian values of sexual propriety, temperance, and proper female behavior.

The YWCA traces its roots to the explosion of Protestant evangelism that spread throughout Great Britain and the United States in the middle decades of the nineteenth century. Women, encouraged by male religious leaders committed to the concept of the social gospel, searched for ways to put their Christian principles into practice. Two British women, Emma Robarts and Mary Jane Kinnaird (Mrs. Arthur and later Lady Kinnaird), followed different paths toward Christian activism but almost simultaneously organized groups of women who later would claim to be the first YWCAs. Robarts, the daughter of a small-town merchant, began in 1855 a prayer union among her friends whose purpose was to pray each Saturday night for young female friends and relatives who had left home for employment and education. This spiritual activity was popular among fervent churchwomen, and prayer unions were soon established in surrounding villages and towns. In 1859, borrowing from the previously established Young Men's Christian Association (YMCA), Robarts named her organization the Young Women's Christian Association.

In the same year that this first prayer union came into being, a wealthy and prominent London matron, Mary Jane Kinnaird, concerned for the safety of young women who had enlisted as nurses in the Crimean War, established the London Home on Charlotte Street. At the end of the war, Kinnaird transformed this temporary shelter into a permanent home for young women who had come from England's countryside in search of employment. Soon the one home grew into a network of working girls' homes located throughout London that offered more than lodging. The staff of the homes taught Bible classes, offered vocational and educational instruction, and chaperoned social activities for its residents. As the movement grew in scope and spread beyond London, it, too, adopted the name of Young Women's Christian Association. This YWCA established employment agencies, lunchrooms, travelers' aid bureaus, and working girls' clubs. In 1877, Kinnaird's homes for young women and Robarts's prayer unions combined to form one Young Women's Christian Association.

The same conditions that prompted the formation of British YWCAs—the Industrial Revolution, war, and religious revivalism—also combined to form the impetus of the movement in the United States. The so-called "Businessmen's Revival" of 1857–1858 that stressed efficiency in Christian charity also promoted the creation of interdenominational organizations and institutions. Added to this religious movement were concerns for young women who had left the protected environment of their Christian homes for the potentially corruptive influence of the impersonal, industrialized city amid the general societal upheaval caused by the Civil War. Although no direct connection can be made between the development of the British and U.S. YWCAs, the causes and timing were similar.

Mrs. Marshall Roberts organized thirty-five of her friends into a prayer union in 1858 in New York City whose special mission was to pray for the "temporal, moral, and religious welfare" of young working women (Sims 1950, 3). Within a few years, the group moved beyond prayer and established a recreation center

A 1917 poster shows a young woman reaching out from behind a blue triangle, the symbol of War Work Council relief efforts. (Library of Congress)

in midtown Manhattan, changing its name to the Young Women's Christian Association. Similarly, a group of middle-class women established a supervised home for young working women in Boston in 1866. They also called their organization the Young Women's Christian Association. By 1875, twenty-eight U.S. cities supported YWCAs, and the movement had spread to Canada as well.

The YWCA took root in continental Europe by the end of the nineteenth century for many of the same reasons. France, first to establish a YWCA, was the slowest to develop because of its overwhelming Catholic population. Germany used a leadership structure developed through churchwomen's organizations and created a national association in 1893. The YWCAs in Norway and Denmark were closely linked to Lutheranism and the Young Men's Christian Association. The Swedish YWCA grew out of working girls' meetings held in the 1880s and thrived under the leadership of Beatrice Dickson, who had visited local YWCAs in Great Britain. However, it was the British YWCA, through its Colonial Division, that initially spread the movement throughout the world and introduced not only evangelical Protestantism but a need for gender-specific social services to combat the injustices of male-dominated traditional societies. By the late 1880s, U.S. YWCAs were competing with the British for "spheres of influence" in missionary work around the globe. This led to the formation of the World's YWCA, headquartered in London beginning in 1894. At its first international conference held four years later, the World's YWCA stressed its evangelical interdenominational roots but looked to a future of interracial and international cooperation on social issues involving women and girls.

The original YWCAs in Great Britain and the United States had been strongly influenced by evangelism and existed to protect the pure Christian character of the young women whom they served. Early documents of both national and local associations reveal a strong concern for the Christian morality of its members and often contain extensive rules dictating personal behavior. Initially, membership was extended to any woman who was an upstanding member of a Protestant church. This designation implied support of temperance and disapproval of smoking, card playing, gambling, dancing, or unchaperoned activities. Over time, many restrictions concerning leisure activities were dropped. By the 1920s, most YWCAs allowed smoking and card playing, except on Sundays, as well as mixed dances and parties. However, the use of alcohol at YWCA-sponsored events or by YWCA residents was strictly forbidden. Although the YWCA was not an active temperance organization, its very nature suggested support for the movement. By the twentieth century, the organization had decreased its emphasis on individual character and societal temptation and had begun to investigate the industrialized environment. The YWCA, through its Industrial Departments, reported on working conditions, hours, and wages, empowered working women through the formation of clubs, and supported trade unionism. Desiring that women should be healthy in mind and body, YWCAs pursued a universal program of physical fitness, building gymnasiums, swimming pools, and recreation centers throughout the world. Because of their work with immigrants and as overseas "emissaries," YWCA members were confronted with issues surrounding ethnicity, race, and imperialism. The YWCA also aided women during the upheavals of two world wars and engaged in such social reform activities as the civil rights movement in the United States and the campaign to end apartheid in South Africa.

Margaret A. Spratt

References

Boyd, Nancy. *Emissaries: The Overseas Work of the American YWCA, 1895–1970.* New York: Woman's Press, 1970.

Mjagkij, Nina, and Margaret Spratt, eds. *Men and Women Adrift: The YMCA and the YWCA in the City.* New York: New York University Press, 1997.

Rice, Anna V. *A History of the World's Young Women's Christian Association.* New York: Woman's Press, 1947.

Seymour-Jones, Carole. *Journey of Faith: The History of the World YWCA, 1945–1994.* London: Allison & Busby, 1994.

Sims, Mary. *The YWCA: An Unfolding Purpose.* New York: Women's Press, 1950.

Appendix A: Documents

Petition for Interdiction and Curatorship (1890)

Various modern societies have created legal processes for limiting the damage an habitual drunkard can inflict upon his or her family. Such procedures necessarily raise the issue of individual autonomy versus familial or communal welfare. In the Canadian province of Quebec during the late nineteenth century, the procedure, called interdiction and curatorship, was initiated by a petition such as this. See Interdiction and Curatorship.

Province of Quebec

Superior Court

District of Montreal.

To any of the Judges of the Superior Court for the Province of Quebec.

The petition of Dame Agnes M. of the City of Montreal, wife, separate as to property by contract of marriage, of William H. of the same place without profession,

Respectfully showeth:—

That for a number of Years past the said William H. the petitioner's husband has been an habitual drunkard indulging in intoxicating liquors to great excess, has become and is now very violent and dangerous as well to himself your petitioner and others who come in contact with him, that he uses the most insulting language to your petitioner and the servants, and others and ill treats and insults your petitioner, and by reason of his said drunkenness he squanders and mismanages any property coming into his hands, and places his family in trouble and distress, and it is therefore desirable in virtue of the law that the said William H. be interdicted as an habitual drunkard and that if found necessary he be ordered to be confined in an establishment for habitual drunkards for such space of time as may be deemed necessary. Wherefore your petitioner prays that the interdiction of said William H. as an habitual drunkard be pronounced in accordance with the law and that if found advisable he be ordered to be confined in an establishment for habitual drunkards for such space of time as may be deemed necessary and further also that for the purposes of the present demand all authority which may be deemed necessary may be granted to your petitioner.

Montreal 22nd December 1890

Selkirk C.

Atty. for petitioner

Dame Agnes M. of the City of Montreal the petitioner named in the foregoing petition being duly sworn on the Holy Evangelist doth depose and say the facts alleged in the foregoing petition are true and that the said petition has not been made through malice nor with a view to oppress, and she hath signed after the same hath duly read to her.

Sworn before me at the City of Montreal this twenty-second day of December A.D. 1890

Agnes H.

Selkirk C. a commissioner Superior Court District of Montreal

Source: Tutorships and curatorships (CC 601); 2 January 1891; no. 1, Quebec National Archives, Montreal, Quebec, Canada.

The Wilson Act (1890)

The nature of the United States federal system of government meant that although efforts to regulate or prohibit the manufacture and sale of beverage alcohol originated at the local or state level, state prohibition involved the federal government either as hindrance or accessory because of the federal tax and commerce powers. Prohibitionists therefore campaigned to enlist federal power in aid of prohibitory acts by the states. One of their first victories was the Wilson Act, which allowed dry states to intercept liquor shipments originating in other states. See Federal Regulation of Alcohol before 1920.

CHAP. 728.—An act to limit the effect of the regulations of commerce between the several States and with foreign countries in certain cases.

Be it enacted by the Senate and House of Representatives of the United States of America in Congress assembled, That all fermented, distilled, or other intoxicating liquors or liquids transported into any State, or Territory or remaining therein for use, consumption, sale or storage therein, shall upon arrival in such State or Territory be subject to the operation and effect of the laws of such State or Territory enacted in the exercise of its police powers, to the same extent and in the same manner as though such liquids or liquors had been produced in such State or Territory, and shall not be exempt therefrom by reason of being introduced therein in original, packages or otherwise.

Approved, August 8, 1890.

Source: *Congressional Record, 51st Congress, 1st Session:* 313.

The Prohibition Party Platform (1892)

The Prohibition party had played a key role in influencing the outcome of the presidential election of 1884, but since then the initiative in American reform had been seized by the People's, or Populist, party. As party prohibitionists met in their 1892 national convention, they sought to regain their lost impetus. To do so, they adopted the most radical platform in the party's history, borrowing from the Populists on some issues and adopting more extreme positions on others. Although the Prohibition campaign in 1892 attracted the party's largest vote total ever, the radicalism of the platform repelled some party supporters and provoked an internal debate that was to rupture the party four years later, opening the way for the emergence of the Anti-Saloon League of America and the latter's eventual takeover of the leadership of the prohibition movement. See Anti-Saloon League of America; Prohibition Party (United States).

THE PLATFORM OF 1892

The Prohibition Party, in National Convention assembled, acknowledging Almighty God as the source of all true government, and

His law as the standard to which human enactments must conform to secure the blessings of peace and prosperity, presents the following declaration of principles:

1. The liquor traffic is a foe to civilization, the arch enemy of popular government, and a public nuisance. It is the citadel of the forces that corrupt politics, promote poverty and crime, degrade the nation's home life, thwart the will of the people, and deliver our country into the hands of rapacious class interests. All laws that under the guise of regulation legalize and protect this traffic or make the Government share in its ill-gotten gains, are "vicious in principle and powerless as a remedy." We declare anew for the entire suppression of the manufacture, sale, importation, exportation, and transportation of alcoholic liquors as a beverage by Federal and State legislation. The full powers of Government should be exerted to secure this result. No party that fails to recognize the dominant nature of this issue in American politics deserves the support of the people.

2. No citizen should be denied the right to vote on account of sex, and equal labor should receive equal wages, without regard to sex.

3. The money of the country should consist of gold, silver, and paper, and be issued by the general Government only, and in sufficient quantity to meet the demands of business and give full opportunity for the employment of labor. To this end an increase in the volume of money is demanded. No individual or corporation should be allowed to make any profit through its issue. It should be fixed at a definite sum per capita, and made to increase with our increase in population.

4. Tariff should be levied only as defense against foreign governments which levy tariff upon or bar out our products from their markets, revenue being incidental. The residue of means necessary to an economical administration of the Government should be raised by levying the burden on what the people possess, instead of upon what they consume.

5. Railroad, telegraph, and other public corporations should be controlled by the Government in the interest of the people, and no higher charges allowed than necessary to give fair interest on the capital actually invested.

6. Foreign immigration has become a burden upon industry, one of the factors in depressing wages and causing discontent; therefore, our immigration laws should be revised and strictly enforced. The time of residence for naturalization should be extended, and no naturalized person should be allowed to vote until one year after he becomes a citizen.

7. Non-resident aliens should not be allowed to acquire land in this country, and we favor the limitation of individual and corporate ownership of land. All unearned grants of lands to railroad companies or other corporations should be reclaimed.

8. Years of inaction and treachery on the part of the Republican and Democratic parties have resulted in the present reign of mob law, and we demand that every citizen be protected in the right of trial by constitutional tribunals.

9. All men should be protected by law in their right to one day of rest in seven.

10. Arbitration is the wisest and most economical and humane way of settling national differences.

11. Speculations in margins, the cornering of grain, money and products, and the formation of pools, trusts, and combinations for the arbitrary advancement of prices should be suppressed.

12. We pledge that the Prohibition Party, if elected to power, will ever grant just pensions to disabled veterans of the Union army and navy, their widows and orphans.

13. We stand unequivocally for the American public school and opposed to any appropriation of public moneys for sectarian schools. We declare that only by united support of such common schools, taught in the English language, can we hope to become and remain a homogeneous and harmonious people.

14. We arraign the Republican and Democratic parties as false to the standards reared by their founders; as faithless to the principles of the illustrious leaders of the past to whom they do homage with the lips, as recreant to the higher law, which is as inflexible in political affairs as in personal life; and as no longer embodying the aspirations of the American people, or inviting the confidence of enlightened, progressive patriotism. Their protest against the admission of "moral issues" into politics is a confession of their own moral degeneracy. The declaration of an eminent authority that municipal misrule is "the one conspicuous failure of American politics," follows as a natural consequence of such degeneracy, and is true alike of cities under Republican and Democratic control. Each accuses the other of extravagance in Congressional appropriations, and both are alike guilty. Each protests, when out of power, against infraction of the Civil Service laws, and each when in power violates those laws in letter and in spirit. Each professes fealty to the interests of the toiling masses, but both covertly truckle to the money power in their administration of public affairs. Even the tariff issue, as represented in the Democratic Mills bill and the Republican McKinley bill, is no longer treated by them as an issue between great and divergent principles of government, but is a mere catering to sectional and class interests. The attempt in many states to wrest the Australian ballot system from its true purpose, and so to deform it as to render it extremely difficult for new parties to exercise the right of suffrage, is an outrage upon popular government. The competition of both old parties for the vote of the slums, and their assiduous courting of the liquor power and subserviency to the money power, have resulted in placing those powers in the position of practical arbiters of the destinies of the nation. We renew our protest against these perilous tendencies, and invite all citizens to join us in the upbuilding of a party that has shown, in five national campaigns, that it prefers temporary defeat to an abandonment of the claims of justice, sobriety, personal rights and the protection of American homes.

15. Recognizing and declaring that Prohibition of the liquor traffic has become the dominant issue in national politics, we invite to full party fellowship all those who, on this one dominant issue, are with us agreed, in the full belief that this party can and will remove sectional differences, promote national unity, and insure the best welfare of our entire land.

Source: David Leigh Colvin, *Prohibition in the United States: A History of the Prohibition Party and of the Prohibition Movement* (New York: George H. Doran Co., 1926), 247–249.

The Webb-Kenyon Act (1913)

American prohibitionists won one of their greatest victories before the Eighteenth Amendment when Congress passed the Webb-Kenyon Act over the veto of President William Howard Taft. The act tightened the strictures around interstate liquor shipments created by the Wilson Act of 1890. *See Federal Regulation of Alcohol before 1920 (United States).*

CHAP. 90.—An Act Divesting intoxicating liquors of their interstate character in certain cases.

Be it enacted by the Senate and House of Representatives of the United States of America in Congress assembled, The shipment or transportation, in any manner or by any means whatsoever, of any spirituous, vinous, malted, fermented, or other intoxicating liquor of any kind, from one State, Territory, or District of the United States, or place noncontiguous to but subject to the jurisdiction thereof, into any other State, Territory, or District of the United States, or place noncontiguous to but subject to the jurisdiction thereof, or from any foreign country into any State, Territory, or District of the United States, or place noncontiguous to but subject to the jurisdiction thereof, which said spirituous, vinous, malted, fermented, or other intoxicating liquor is intended, by any person interested therein, to be received, possessed, sold, or in any manner used, either in the original package or otherwise, in violation of any law of such State, Territory, or District of the United States, or place noncontiguous to but subject to the jurisdiction thereof, is hereby prohibited.

Source: *Congressional Record, 62d Congress, Third Session:* 699–700.

The Volstead Act (1919)

The Eighteenth Amendment to the United States Constitution established the principle of national prohibition, but the implementation of that principle required federal legislation. Such legislation was forthcoming in the National Prohibition Act, which became known as the Volstead Act after its sponsor in the House of Representatives, Andrew J. Volstead, a Republican from Minnesota. The chief author of the act, however, was the General Counsel and Washington lobbyist for the Anti-Saloon League of America, Wayne B. Wheeler. It was the Volstead Act, not the Eighteenth Amendment, that created the severe definition of "intoxicating liquor" as containing 0.5 percent alcohol by volume. The Volstead Act was amended once, in 1929, and that was to stiffen the penalties for violating the act. See Jones, Wesley Livsey; National Prohibition (United States); Wheeler, Wayne Bidwell.

CHAP. 85.—An Act to prohibit intoxicating beverages, and to regulate the manufacture, production, use and sale of high-proof spirits for other than beverage purposes, and to insure an ample supply of alcohol and promote its use in scientific research and in the development of fuel, dye and other lawful industries.

Be it enacted by the Senate and House of Representatives of the United States of America in Congress assembled, That the short title of this Act shall be the "National Prohibition Act."

TITLE I

TO PROVIDE FOR THE ENFORCEMENT OF WAR PROHIBITION

The term "War Prohibition Act" used in this Act shall mean the provisions of any Act or Acts prohibiting the sale and manufacture of intoxicating liquors until the conclusion of the present war and thereafter until the termination of demobilization, the date of which shall be determined and proclaimed by the President of the United States. The words "beer, wine, or other intoxicating malt or vinous liquors" in the War Prohibition Act shall be hereafter construed to mean any such beverages which contain one-half of 1 per centum or more of alcohol by volume: *Provided,* That the foregoing definition shall not extend to dealcoholized wine nor to any beverage or liquid produced by the process by which beer, ale, porter or wine is produced, if it contains less than one-half of 1 per centum of alcohol by volume, and is made as prescribed in section 37 of Title II of this Act, and is otherwise denominated than as beer, ale, or porter, and is contained and sold in, or from, such sealed and labeled bottles, casks, or containers as the commissioner may by regulation prescribe.

SEC. 2. The Commissioner of Internal Revenue, his assistants, agents and inspectors, shall investigate and report violations of the War Prohibition Act to the United States attorney for the district in which committed, who shall be charged with the duty of prosecuting, subject to the direction of the Attorney General, the offenders as in the case of other offenses against laws of the United States; and such Commissioner of Internal Revenue, his assistants, agents, and inspectors may swear out warrants before United States commissioners or other officers or courts authorized to issue the same for the apprehension of such offenders, and may, subject to the control of the said United States attorney, conduct the prosecution at the committing trial for the purpose of having the offenders held for the action of a grand jury.

SEC. 3. Any room, house, building, boat, vehicle, structure, or place of any kind where intoxicating liquor is sold, manufactured, kept for sale, or bartered in violation of the War Prohibition Act, and all intoxicating liquor and all property kept and used in maintaining such a place, is hereby declared to be a public and common nuisance, and any person who maintains or assists in maintaining such public and common nuisance shall be guilty of a misdemeanor, and upon conviction thereof shall be fined not less than $100 nor more than $1,000, or be imprisoned for not less than thirty days or more than one year, or both. If a person has knowledge that his property is occupied or used in violation of the provisions of the War Prohibition Act and suffers the same to be so used, such property shall be subject to a lien for, and may be sold to pay, all fines and costs assessed against the occupant of such building or property for any violation of the War Prohibition Act occurring after the passage hereof, which said lien shall attach from the time of the filing of notice of the commencement of the suit in the office where the records of the transfer of real estate are kept; and any such lien may be established and enforced by legal action instituted for that purpose in any court having jurisdiction. Any violation of this title upon any leased premises by the lessee or occupant thereof shall, at the option of the lessor, work a forfeiture of the lease.

SEC. 4. The United States attorney for the district where such nuisance as is defined in this Act exists, or any officer designated by him or the Attorney General of the United States, may prosecute a suit in equity in the name of the United States to abate and enjoin the same. Actions in equity to enjoin and abate such nuisances may be brought in any court having jurisdiction to hear and determine equity causes. The jurisdiction of the courts of the United States under this section shall be concurrent with that of the courts of the several States.

If it be made to appear by affidavit, or other evidence under oath, to the satisfaction of the court, or judge in vacation, that the nuisance complained of exists, a temporary writ of injunction shall forthwith issue restraining the defendant or defendants from conducting or permitting the continuance of such nuisance until the conclusion of the trial. Where a temporary injunction is prayed for, the court may issue an order restraining the defendants and all other persons from removing or in any way interfering with the liquor or fixtures, or other things used in connection with the violation constituting the nuisance. No bond shall be required as a condition for making any order or issuing any writ of injunction under this Act. If the court shall find the property involved was being unlawfully used as aforesaid at or about the time alleged in the petition, the court shall order that no liquors shall be manufactured, sold, bartered, or stored in such room, house, building, boat, vehicle, structure, or places of any kind, for a period of not exceeding one year, or during the war and the period of demobilization. Whenever an action to enjoin a nuisance shall have been brought pursuant to the provisions of this Act, if the owner, lessee, tenant, or occupant appears and pays all costs of the proceedings and files a bond, with sureties to be approved by the clerk of the court in which the action is brought, in the liquidated sum of not less than $500 nor more than $1,000, conditioned that he will immediately abate said nuisance and prevent the same from being established or kept therein a period of one year thereafter, or during the war and period of demobilization, the court, or in vacation the judge, may, if satisfied of his good faith, direct by appropriate order that the property, if already closed or held under the order of abatement, be delivered to said owner, and said order of abatement canceled, so far as the same may relate to said property; or if said bond be given and costs therein paid before judgment on an order of abatement, the action shall be thereby abated as to said room, house, building, boat, vehicle, structure, or place only. The release of the property under the provisions of this section shall not release it from any judgment, lien, penalty or liability to which it may be subject by law.

In the case of the violation of an injunction, temporary or permanent, granted pursuant to the provisions of this Title, the court, or in vacation a judge thereof, may summarily try and punish the defendant. The proceedings for punishment for contempt shall be commenced by filing with the clerk of the court from which such injunction issued information under oath setting out the alleged facts constituting the violation, whereupon the court or judge shall forthwith cause a warrant to issue under which the defendant shall be arrested. The trial may be had upon affidavits, or either party may demand the production and oral examination of the witnesses. Any person found guilty of contempt under the provisions of this section shall be punished by a fine of not less than $500 nor more than $1,000, or by imprisonment of not less than thirty days nor more than twelve months, or by both fine and imprisonment.

SEC. 5. The Commissioner of Internal Revenue, his assistants, agents and inspectors, and all other officers of the United States whose duty it is to enforce criminal laws, shall have all the power for the enforcement of the War Prohibition Act or any provisions thereof which is conferred by law for the enforcement of existing laws relating to the manufacture or sale of intoxicating liquors under the laws of the United States.

SEC. 6. If any section or provisions of this Act shall be held to be invalid, it is hereby provided that all other provisions of this Act which are not expressly held to be invalid shall continue in full force and effect.

SEC. 7. None of the provisions of this Act shall be construed to repeal any of the provisions of the "War Prohibition Act," or to limit or annul any order or regulation prohibiting the manufacture, sale, or disposition of intoxicating liquors within certain prescribed zones or districts, nor shall the provisions of this Act be construed to prohibit the use of the power of the military or naval authorities to enforce the regulations of the President or Secretary of War or Navy issued in pursuance of law, prohibiting the manufacture, use, possession, sale, or other disposition of intoxicating liquors during the period of the war and demobilization thereafter.

TITLE II

PROHIBITION OF INTOXICATING BEVERAGES

SEC. 1. When used in Title II and Title III of this Act

(1) The word "liquor" or the phrase "intoxicating liquor" shall be construed to include alcohol, brandy, whisky, rum, gin, beer, ale, porter, and wine, and in addition thereto any spirituous, vinous, malt or fermented liquor, liquids, and compounds, whether medicated, proprietary, patented, or not, and by whatever name called, containing one-half of 1 per centum or more of alcohol by volume which are fit for use for beverage purposes: *Provided,* That the foregoing definition shall not extend to dealcoholized wine nor any beverage or liquid produced by the process by which beer, ale, porter or wine is produced, if it contains less than one-half of 1 per centum of alcohol by volume, and is made as prescribed in section 37 of this title, and is otherwise denominated than as beer, ale, or porter, and is contained in, or from such sealed and labeled bottles, casks, or containers as the commissioner may by regulation prescribe.

(2) The word "person" shall mean and include natural persons, associations, copartnerships, and corporations.

(3) The word "commissioner" shall mean Commissioner of Internal Revenue.

(4) The term "application" shall mean a formal written request supported by a verified statement of facts showing that the commissioner may grant the request.

(5) The term "permit" shall mean a formal written authorization by the commissioner setting forth specifically therein the things that are authorized.

(6) The term "Bond" shall mean an obligation authorized or required by or under this Act or any regulation, executed in such form and for such a penal sum as may be required by a court, the commissioner or prescribed by regulation.

(7) The term "regulation" shall mean any regulation prescribed by the commissioner with the approval of the Secretary of the Treasury for carrying out the provisions of this Act, and the commissioner is authorized to make such regulations.

Any act authorized to be done by the commissioner may be performed by any assistant or agent designated by him for that purpose. Records required to be filed with the commissioner may be filed with an assistant commissioner or other person designated by the commissioner to receive such records.

SEC. 2. The Commissioner of Internal Revenue, his assistants, agents, and inspectors shall investigate and report violations of this Act to the United States attorney for the district in which committed, who is hereby charged with the duty of prosecuting the offenders, subject to the direction of the Attorney General, as in the case of other offenses against the laws of the United States; and such Commissioner of Internal Revenue, his assistants, agents, and inspectors may swear out warrants before United States commissioners or other officers or courts authorized to issue the same for the apprehension of such offenders, and may, subject to the control of the said United States attorney, conduct the prosecution at the committing trial for the purpose of having the offenders held for the action of a grand jury. Section 1014 of the Revised Statutes of the United States is hereby made applicable in the enforcement of this Act. Officers mentioned in said section 1014 are authorized to issue search warrants under the limitations provided in Title XI of the Act approved June 15, 1917 (Fortieth Statutes at Large, page 217, et seq.).

SEC. 3. No person shall on or after the date when the eighteenth amendment to the Constitution of the United States goes into effect, manufacture, sell, barter, transport, import, export, deliver, furnish or possess any intoxicating liquor except as authorized in this Act, and all of the provisions of this Act shall be liberally construed to the end that the use of intoxicating liquor as a beverage may be prevented.

Liquor for nonbeverage purposes and wine for sacramental purposes may be manufactured, purchased, sold, bartered, transported, imported, exported, delivered, furnished and possessed, but only as herein provided, and the commissioner may, upon application, issue permits therefor: *Provided,* That nothing in this Act shall prohibit the purchase and sale of warehouse receipts covering distilled spirits on deposit in Government bonded warehouses, and no special tax liability shall attach to the business of purchasing and selling such warehouse receipts.

SEC. 4. The articles enumerated in this section shall not, after having been manufactured and prepared for the market, be subject to the provisions of this Act if they correspond with the following descriptions and limitations, namely:

(a) Denatured alcohol or denatured rum produced and used as provided by the laws and regulations now or hereafter in force.

(b) Medicinal preparations manufactured in accordance with formulas prescribed by the American Institute of Homeopathy that are unfit for use for beverage purposes.

(c) Patented, patent, and proprietary medicines that are unfit for use for beverage purposes.

(d) Toilet, medicinal, and antiseptic preparations and solutions that are unfit for use for beverage purposes.

(e) Flavoring extracts and syrups that are unfit for use as a beverage, or for intoxicating beverage purposes.

(f) Vinegar and preserved sweet cider.

A person who manufactures any of the articles mentioned in this section may purchase and possess liquor for that purpose, but he shall secure permits to manufacture such articles and to purchase such liquor, give the bonds, keep the records, and make the reports specified in this Act and as directed by the commissioner. No such manufacturer shall sell, use, or dispose of any liquor otherwise than as an ingredient of the articles authorized to be manufactured therefrom. No more alcohol shall be used in the manufacture of any extract, sirup, or the articles named in paragraphs b, c, and d of this section which may be used for beverage purposes than the quantity necessary for extraction or solution of the elements contained therein and for the preservation of the article.

Any person who shall knowingly sell any of the articles mentioned in paragraphs a, b, c, and d of this section for beverage purposes, or any extract or sirup for intoxicating beverage purposes, or who shall sell any of the same under circumstances from which the seller might reasonably deduce the intention of the purchaser or use them for such purposes, or shall sell any beverage containing one-half of 1 per centum or more of alcohol by volume in which any extract, sirup, or other article is used as an ingredient, shall be subject to the penalties provided in section 29 of this Title. If the commissioner shall find, after notice and hearing as provided for in section 5 of this Title, that any person has sold any flavoring extract, sirup, or beverage in violation of this paragraph, he shall notify such person, and any known principal for whom the sale was made, to desist from selling such article; and it shall thereupon be unlawful for a period of one year thereafter for any person so notified to sell any such extract, sirup, or beverage without making an application for, giving a bond, and obtaining a permit so to do, which permit may be issued upon such conditions as the commissioner may deem necessary to prevent such illegal sales, and in addition the commissioner shall require a record and report of sales.

SEC. 5. Whenever the commissioner has reason to believe that any article mentioned in section 4 does not correspond with the descriptions and limitations therein provided, he shall cause an analysis of said article to be made, and if, upon such analysis, the commissioner shall find that said article does not so correspond, he shall give not less than fifteen days' notice in writing to the person who is the manufacturer thereof to show cause why said article should not be dealt with as an intoxicating liquor, such notice to be served personally or by registered mail, as the commissioner may determine, and shall specify the time when, the place where, and the name of the agent or official before whom such person is required to appear.

If the manufacturer of said article fails to show to the satisfaction of the commissioner that the article corresponds to the descriptions and limitations provided in section 4 of this title, his permit to manufacture and sell such article shall be revoked. The manufacturer may by appropriate proceeding in a court of equity have the action of the commissioner reviewed, and the court may affirm, modify, or reverse the finding of the commissioner as the facts and law of the case may warrant, and during the pendency of such proceedings may restrain the manufacture, sale, or other disposition of such article.

SEC. 6. No one shall manufacture, sell, purchase, transport, or prescribe any liquor without first obtaining a permit from the commissioner so to do, except that a person may, without a permit, purchase and use liquor for medicinal purposes when prescribed by a physician as herein provided, and except that any person who in the opinion of the commissioner is conducting a bona fide hospital or sanatorium engaged in the treatment of persons suffering from alcoholism, may, under such rules, regulations and conditions as the commissioner shall prescribe, purchase and use, in accordance with the methods in use in such institution, liquor, to be administered to the patients of such institution under the direction of a duly qualified physician employed by such institution.

All permits to manufacture, prescribe, sell, or transport liquor, may be issued for one year, and shall expire on the 31st day of December next succeeding the issuance thereof: *Provided,* That the commissioner may without formal application or new bond extend any permit granted under this Act or laws now in force after August 31 in any year to December 31 of the succeeding year: *Provided further,* That permits to purchase liquor for the purpose of manufacturing or selling as provided in this Act shall not be in force to exceed ninety days from the day of issuance. A permit to purchase liquor for any other purpose shall not be in force to exceed thirty days. Permits to purchase liquor shall specify the quantity and kind to be purchased and the purpose for which it is to be used. No permit shall be issued to any person who within one year prior to the application therefore or issuance thereof shall have violated the terms of any permit issued under this Title or any law of the United States or of any State regulating traffic in liquor. No permit shall be issued to anyone to sell liquor at retail, unless the sale is to be made through a pharmacist designated in the permit and duly licensed under the laws of his State to compound and dispense medicine prescribed by a duly licensed physician. No one shall be given a permit to prescribe liquor unless he is a physician duly licensed to practice medicine and actively engaged in the practice of such profession. Every permit shall be in writing, dated when issued, and signed by the commissioner or his authorized agent. It shall designate and limit the acts that are permitted and the time when and place where such acts may be performed. No permits shall be issued until verified, written application shall have been made therefore, setting forth the qualification of the applicant and the purpose for which the liquor is to be used.

The commissioner may prescribe the form of all permits and applications and the facts to be set forth therein. Before any permit is granted the commissioner may require a bond in such form and amount as he may prescribe to insure compliance with the terms of the permit and the provisions of this title. In the event of the refusal by the commissioner of any application for a permit, the applicant may have a review of his decision before a court of equity in the manner provided in section 5 hereof.

Nothing in this title shall be held to apply to the manufacture, sale, transportation, importation, possession, or distribution of wine for sacramental purposes, or like religious rites, except section 6 (save as the same requires a permit to purchase) and section 10 hereof, and the provisions of this Act prescribing penalties for the violation of either of said sections. No person to whom a permit may be issued to manufacture, transport, import, or sell wines for sacramental purposes or like religious rites shall sell, barter, exchange, or furnish any such to any person not a rabbi, minister of the gospel, priest, or an officer duly authorized for the purpose by any church or congregation, nor to any such except upon an application duly subscribed by him, which application, authenticated as regulations may prescribe, shall be filed and preserved by the seller. The head of any conference or diocese or other ecclesiastical jurisdiction may designate any rabbi, minister, or priest to supervise the manufacture of wine to be used for the purposes and rites of this section mentioned, and the person so designated may, in the discretion of the commissioner, be granted a permit to supervise such manufacture.

Sec. 7. No one but a physician holding a permit to prescribe liquor shall issue any prescription for liquor. And no physician shall prescribe liquor unless after careful physical examination of the person for whose use such prescription is sought, or if such examination is found impracticable, then upon the best information obtainable, he in good faith believes that the use of such liquor as a medicine by such person is necessary and will afford relief to him from some known ailment. Not more than a pint of spirituous liquor to be taken internally shall be prescribed for use by the same person within any period of ten days and no prescription shall be filled more than once. Any pharmacist filling a prescription shall at the time indorse upon it over his own signature the word "canceled," together with the date when the liquor was delivered, and then make the same a part of the record that he is required to keep as herein provided.

Every physician who issues a prescription for liquor shall keep a record, alphabetically arranged in a book prescribed by the commissioner, which shall show the date of issue, amount prescribed, to whom issued, the purpose or ailment for which it is to be used and directions for use, stating the amount and frequency of the dose.

Sec. 8. The commissioner shall cause to be printed blanks for the prescriptions herein required, and he shall furnish the same, free of cost, to physicians holding permits to prescribe. The prescription blanks shall be printed in book form and shall be numbered consecutively from one to one hundred, and each book shall be given a number, and the stubs in each book shall carry the same numbers as and be copies of the prescriptions. The books containing such stubs shall be returned to the commissioner when the prescription blanks have been used, or sooner, if directed by the commissioner. All unused, mutilated, or defaced blanks shall be returned with the book. No physician shall prescribe and no pharmacist shall fill any prescription for liquor except on blanks so provided, except in cases of emergency, in which event a record and report shall be made and kept as in other cases.

Sec. 9. If at any time there shall be filed with the commissioner a complaint under oath setting forth facts showing, or if the commissioner has reason to believe, that any person who has a permit is not in good faith conforming to the provisions of this Act, or has violated the laws of any State relating to intoxicating liquor, the commissioner or his agent shall immediately issue an order citing such person to appear before him on a day named not more than thirty and not less than fifteen days from the date of service upon such permittee of a copy of the citation, which citation shall be accompanied by a copy of such complaint, or in the event that the proceedings be initiated by the commissioner with a statement of the facts constituting the violation charged, at which time a hearing shall be had unless continued for cause. Such hearings shall be held within the judicial district and within fifty miles of the place where the offense is alleged to have occurred, unless the parties agree on another place. If it be found that such person has been guilty of willfully violating any such laws, as charged, or has not in good faith conformed to the provisions of this Act, such permit shall be revoked, and no permit shall be granted to such person within one year thereafter. Should the permit be revoked by the commissioner, the permittee may have a review of his decision before a court of equity in the manner provided in section 5 hereof. During the pendency of such action such permit shall be temporarily revoked.

Sec. 10. No person shall manufacture, purchase for sale, sell, or transport any liquor without making at the time a permanent record thereof showing in detail the amount and kind of liquor manufactured, purchased, sold, or transported, together with the names and addresses of the persons to whom sold, in case of sale, and the consignor and consignee in case of transportation, and the time and place of such manufacture, sale, or transportation. The commissioner may prescribe the form of such record, which shall at all times be open to inspection as in the Act provided.

Sec. 11. All manufacturers and wholesale or retail druggists shall keep as part of the records required of them a copy of all permits to purchase on which a sale of any liquor is made, and no manufacturer or wholesale druggist shall sell or otherwise dispose of any liquor except at wholesale and only to persons having permits to purchase in such quantities.

Sec. 12. All persons manufacturing liquor for sale under the provisions of this title shall securely and permanently attach to every container thereof, as the same is manufactured, a label stating name of manufacturer, kind and quantity of liquor contained therein, and the date of its manufacture, together with the number of the permit authorizing the manufacture thereof; and all persons possessing such liquor in wholesale quantities shall securely keep and maintain such label thereon; and all persons selling at wholesale shall attach to every package of liquor, when sold, a label setting forth the kind and quantity of liquor contained therein, by whom manufactured, the date of sale, and the person to whom sold; which label shall likewise be kept and maintained thereon until the liquor is used for the purpose for which such sale was authorized.

SEC. 13. It shall be the duty of every carrier to make a record at the place of shipment of the receipt of any liquor transported, and he shall deliver liquor only to persons who present to the carrier a verified copy of a permit to purchase which shall be made a part of the carrier's permanent record at the office from which delivery is made.

The agent of the common carrier is hereby authorized to administer the oath to the consignee in verification of the copy of the permit presented, who, if not personally known to the agent, shall be identified before the delivery of the liquor to him. The name and address of the person identifying the consignee shall be included in the record.

SEC. 14. It shall be unlawful for a person to use or induce any carrier, or any agent or employee thereof, to carry or ship any package or receptacle containing liquor without notifying the carrier of the true nature and character of the shipment. No carrier shall transport nor shall any person receive liquor from a carrier unless there appears on the outside of the package containing such liquor the following information:

Name and address of the consignor or seller, name and address of the consignee, kind and quantity of liquor contained therein, and number of the permit to purchase or ship the same, together with the name and address of the person using the permit.

SEC. 15. It shall be unlawful for any consignee to accept or receive any package containing any liquor upon which appears a statement known to him to be false, or for any carrier or other person to consign, ship, transport, or deliver any such package, knowing such statement to be false.

SEC. 16. It shall be unlawful to give any carrier or any officer, agent or person acting or assuming to act for such carrier an order requiring the delivery to any person of any liquor or package containing liquor consigned to, or purporting or claimed to be consigned to a person, when the purpose of the order is to enable any person not an actual bona fide consignee to obtain such liquor.

SEC. 17. It shall be unlawful to advertise anywhere, or by any means or method, liquor, or the manufacture, sale, keeping for sale or furnishing of the same, or where, how, from whom, or at what price the same may be obtained. No one shall permit any sign or billboard containing such advertisement to remain upon one's premises. But nothing herein shall prohibit manufacturers and wholesale druggists holding permits to sell liquor from furnishing price lists, with description of liquor for sale, to persons permitted to purchase liquor, or from advertising alcohol in business publications or trade journals circulating generally among manufacturers of lawful alcoholic perfumes, toilet preparations, flavoring extracts, medicinal preparations, and like articles: *Provided, however,* That nothing in this Act or in the Act making appropriations for the Post Office Department, approved March 3, 1917 (Thirty-ninth Statutes at Large, Part 1, page 1058, et seq.), shall apply to newspapers published in foreign countries when mailed to this country.

SEC. 18. It shall be unlawful to advertise, manufacture, sell, or possess for sale any utensil, contrivance, machine, preparation, compound, tablet, substance, formula direction, or recipe advertised, designed, or intended for use in the unlawful manufacture of intoxicating liquor.

SEC. 19. No person shall solicit or receive, nor knowingly permit his employee to solicit or receive, from any person any order for liquor or give any information of how liquor may be obtained in violation of this Act.

SEC. 20. Any person who shall be injured in person, property, means of support, or otherwise by any intoxicated person, or by reason of the intoxication of any person, whether resulting in his death or not, shall have a right of action against any person who shall, by unlawfully selling to or unlawfully assisting in procuring liquor for such intoxicated person, have caused or contributed to such intoxication, and in any such action such person shall have a right to recover actual and exemplary damages. In case of the death of either party, the action or right of action given by this section shall survive to or against his or her executor or administrator, and the amount so recovered by either wife or child shall be his or her sole and separate property. Such action may be brought in any court of competent jurisdiction. In any case where parents shall be entitled to such damages, either the father or mother may sue alone therefor, but recovery by one of such parties shall be a bar to suit brought by the other.

SEC. 21. Any room, house, building, boat, vehicle, structure, or place where intoxicating liquor is manufactured, sold, kept or bartered in violation of this title, and all intoxicating liquor and property kept and used in maintaining the same, is hereby declared to be a common nuisance, and any such person who maintains such a common nuisance shall be guilty of a misdemeanor and upon conviction thereof shall be fined not more than $1,000 or be imprisoned for not more than one year, or both. If a person has knowledge or reason to believe that his room, house, building, boat, vehicle, structure, or place is occupied or used for the manufacture or sale of liquor contrary to the provision of this title, and suffers the same to be so occupied or used, such room, house, building, boat, vehicle, structure, or place shall be subject to a lien for and may be sold to pay all fines and costs assessed against the person guilty of such nuisance for such violation, and any such lien may be enforced by action in any court having jurisdiction.

SEC. 22. An action to enjoin any nuisance defined in this title may be brought in the name of the United States by the Attorney General of the United States or by any United States attorney or any prosecuting attorney of any State or any subdivision thereof or by the commissioner or his deputies or assistants. Such action shall be brought and tried as an action in equity and may be brought in any court having jurisdiction to hear and determine equity cases. If it is made to appear by affidavits or otherwise, to the satisfaction of the court, or judge in vacation, that such nuisance exists, a temporary writ of injunction shall forthwith issue restraining the defendant from conducting or permitting the continuance of such nuisance until the conclusion of the trial. If a temporary injunction is prayed for, the court may issue an order restraining the defendant from conducting or permitting the continuance of such nuisance until the conclusion of the trial. If a temporary injunction is prayed for, the court may issue an order restraining the defendant and all other persons from removing or in any way interfering with the liquor or fixtures, or other things used in connection with the violation of this Act constituting such nuisance. No bond shall be required in instituting such proceedings. It shall not be necessary for the court to find the property involved was being unlawfully used as aforesaid at the time of the hearing, but on finding that the material allegations of the petition are true, the court shall order that no liquors shall be manufactured, sold, bartered, or stored in such room, house, building, boat, vehicle, structure, or place, or any part thereof. And upon judgment of the court ordering such nuisance to be abated, the court may order that the room, house, building, structure, boat, vehicle, or place shall not be occupied or used for one year thereafter; but the court may, in its discretion, permit it to be occupied or used if the owner, lessee, tenant, or occupant thereof shall give bond with sufficient surety, to be approved by the court making the order, in the penal and liquidated sum of not less than $500 nor more than $1,000, payable to the United States, and conditioned that intoxicating liquor will not thereafter be manufactured, sold, bartered, kept, or otherwise disposed of therein or thereon, and that he will pay all fines, costs, and damages that may be assessed for any violation of this title upon said property.

SEC. 23. That any person who shall, with intent to effect a sale of liquor, by himself, his employee, servant, or agent, for himself or any person, company or corporation, keep or carry around on his person, or in a vehicle, or other conveyance whatever, or leave in a place for another to secure, any liquor, or who shall travel to solicit, or take, or accept orders for the sale, shipment, or delivery of liquor in violation of this title is guilty of a nuisance and may be restrained by injunction, temporary and permanent, from doing or continuing to do any of said acts or things.

In such proceedings it shall not be necessary to show any intention on the part of the accused to continue such violations if the action is brought within sixty days following any such violation of the law.

For removing and selling property in enforcing this Act the officer shall be entitled to charge and receive the same fee as the sheriff of the county would receive for levying upon and selling property under execution, and for closing the premises and keeping them closed a reasonable sum shall be allowed by the court.

Any violation of this title upon any leased premises by the lessee or occupant thereof shall, at the option of the lessor, work a forfeiture of the lease.

SEC. 24. In the case of the violation of any injunction, temporary or permanent, granted pursuant to the provisions of this title, the court, or in vacation a judge thereof, may summarily try and punish the defendant. The proceeding for punishment for contempt shall be commenced by filing with the clerk of the court from which such injunction issued information under oath setting out the alleged facts constituting the violation, whereupon the court or judge shall forthwith cause a warrant to issue under which the defendant shall be arrested. The trial may be had upon affidavits, or either party may demand the production and oral examination of the witnesses. Any person found guilty of contempt under the provisions of this section shall be punished by a fine of not less than $500 nor more than $1,000, or by imprisonment of not less than thirty days nor more than twelve months, or by both fine and imprisonment.

SEC. 25. It shall be unlawful to have or possess any liquor or property designed for the manufacture of liquor intended for use in violating this title or which has been so used, and no property rights shall exist in any such liquor or property. A search warrant may issue as provided in Title XI of public law number 24 of the Sixty-fifth Congress, approved June 15, 1917, and such liquor, the containers thereof, and such property so seized shall be subject to such disposition as the court may make thereof. If it is found that such liquor or property was so unlawfully held or possessed, or had been so unlawfully used, the liquor, and all property designed for the unlawful manufacture of liquor, shall be destroyed, unless the court shall otherwise order. No search warrant shall issue to search any private dwelling occupied as such unless it is being used for the unlawful sale of intoxicating liquor, or unless it is in part used for some business purpose such as a store, shop, saloon, restaurant, hotel, or boarding house. The term "private dwelling" shall be construed to include the room or rooms used and occupied not transiently but solely as a residence in an apartment house, hotel, or boarding house. The property seized on any such warrant shall not be taken from the officer seizing the same on any writ of replevin or other like process.

SEC. 26. When the commissioner, his assistants, inspectors, or any officer of the law shall discover any person in the act of transporting in violation of the law, intoxicating liquors in any wagon, buggy, automobile, water or air craft, or other vehicle, it shall be his duty to seize any and all intoxicating liquors found therein being transported contrary to law. Whenever intoxicating liquors transported or possessed illegally shall be seized by an officer he shall take possession of the vehicle and team or automobile, boat, air or water craft, or any other conveyance, and shall arrest any person in charge thereof. Such officer shall at once proceed against the person arrested under the provisions of this title in any court having competent jurisdiction; but the said vehicle or conveyance shall be returned to the owner upon execution by him of a good and valid bond, with sufficient sureties, in a sum double the value of the property, which said bond shall be approved by said officer and shall be conditioned to return said property to the custody of said officer on the day of trial to abide the judgment of the court. The court upon conviction of the person so arrested shall order the liquor destroyed, and unless good cause to the contrary is shown by the owner, shall order a sale by public auction of the property seized, and the officer making the sale, after deducting the expenses of keeping the property, the fee for the seizure, and the cost of the sale, shall pay all liens, according to their priorities, which are established, by intervention or otherwise at said hearing or in other proceeding brought for said purpose, as being bona fide and as having been created without the lienor having any notice that the carrying vehicle was being used or was to be used for illegal transportation of liquor, and shall pay the balance of the proceeds into the Treasury of the United States as miscellaneous receipts. All liens against property sold under the provisions of this section shall be transferred from the property to the proceeds of the sale of the property. If, however, no one shall be found claiming the team, vehicle, water or air craft, or automobile, the taking of the same, with a description thereof, shall be advertised in some newspaper published in such city or county, in a newspaper having circulation in the county, once a week for two weeks and by handbills posted in three public places near the place of seizure, and if no claimant shall appear within ten days after the last publication of the advertisement, the property shall be sold and the proceeds after deducting the expenses and costs shall be paid into the Treasury of the United States as miscellaneous receipts.

SEC. 27. In all cases in which intoxicating liquors may be subject to be destroyed under the provisions of this Act the court shall have jurisdiction upon the application of the United States attorney to order them delivered to any department or agency of the United States Government for medicinal, mechanical, or scientific uses, or to order the same sold at private sale for such purposes to any person having a permit to purchase liquor the proceeds to be covered into the Treasury of the United States to the credit of miscellaneous receipts, and all liquor heretofore seized in any suit or proceeding brought for violation of law may likewise be so disposed of, if not claimed within sixty days from the date this section takes effect.

SEC. 28. The commissioner, his assistants, agents, and inspectors, and all other officers of the United States, whose duty it is to enforce criminal laws, shall have all the power and protection in the enforcement of this Act or any provisions thereof which is conferred by law for the enforcement of existing laws relating to the manufacture or sale of intoxicating liquors under the law of the United States.

SEC. 29. Any person who manufactures or sells liquor in violation of this title shall for a first offense be fined not more than $1,000, or imprisoned not exceeding six months, and for a second or subsequent offense shall be fined not less than $200 nor more than $2,000 and be imprisoned not less than one month nor more than five years.

Any person violating the provisions of any permit, or who makes any false record, report, or affidavit required by this title, or violates any of the provisions of this title, for which offense a special penalty is not prescribed, shall be fined for a first offense not more than $500; for a second offense not less than $100 nor more than $1,000, or be imprisoned not more than ninety days; for any subsequent offense he shall be fined not less than $500 and be imprisoned not less than three months nor more than two years. It shall be the duty of the prosecuting officer to ascertain whether the defendant has been previously convicted and to plead the prior conviction in the affidavit, information, or indictment. The penalties provided in this Act against the manufacture of liquor without a permit shall not apply to a person for manufacturing nonintoxicating cider and fruit juices exclusively for use in his home, but such cider and fruit juices shall not be sold or delivered except to persons having permits to manufacture vinegar.

SEC. 30. No person shall be excused, on the ground that it may tend to incriminate him or subject him to a penalty or forfeiture, from attending and testifying, or producing books, papers, documents, and other evidence in obedience to a subpoena of any court in any suit or proceeding based upon or growing out of any alleged violation of this Act; but no natural person shall be prosecuted or subjected to any penalty or forfeiture for or on account of any transaction, matter, or thing as to which, in obedience to a subpoena and under oath, he may so testify or produce evidence, but no person shall be exempt from prosecution and punishment for perjury committed in so testifying.

SEC. 31. In case of a sale of liquor where the delivery thereof was made by a common or other carrier the sale and delivery shall be deemed to be made in the county or district wherein the delivery was made by such carrier to the consignee, his agent or employee, or in the county or district wherein the sale was made, or from which the shipment was made, and prosecution for such sale or delivery may be had in any such county or district.

SEC. 32. In any affidavit, information, or indictment for the violation of this Act, separate offenses may be united in separate counts and the defendant may be tried on all at one trial and the penalty for all offenses may be imposed. It shall not be necessary in any affidavit, information, or indictment to give the name of the purchaser or to include any defensive negative averments, but it shall be sufficient to state that the act complained of was then and there prohibited and unlawful, but this provision shall not be construed to preclude the trial court from directing the furnishing the defendant a bill of particulars when it deems it proper to do so.

SEC. 33. After February 1, 1920, the possession of liquors by any person not legally permitted under this title to possess liquor shall be prima facie evidence that such liquor is kept for the purpose of being sold, bartered, exchanged, given away, furnished, or otherwise disposed of in violation of the provisions of this title. Every person legally permitted under this title to have liquor shall report to the commissioner within ten days after the date when the eighteenth amendment of the Constitution of the United States goes into effect, the kind and amount of intoxicating liquors in his possession. But it shall not be unlawful to possess liquors in one's private dwelling while the same is occupied and used by him as his dwelling only and such liquor need not be reported, provided such liquors are for use only for the personal consumption of the owner thereof and his family residing in such dwelling and of his bona fide guests when entertained by him therein; and the burden of proof shall be upon the possessor in any action concerning the same to prove that such liquor was lawfully acquired, possessed, and used.

SEC. 34. All records and reports kept or filed under the provisions of this Act shall be subject to inspection at any reasonable hour by the commissioner or any of his agents or by any public prosecutor or by any person designated by him, or by any peace officer in the State where the record is kept, and copies of such records and reports duly certified by the person with whom kept or filed may be introduced in evidence with like effect as the originals thereof, and verified copies of such records shall be furnished to the commissioner when called for.

SEC. 35. All provisions of law that are inconsistent with this Act are repealed only to the extent of such inconsistency and the regulations herein provided for the manufacture or traffic in intoxicating liquor shall be construed as in addition to existing laws. This Act shall not relieve anyone from paying any taxes or other charges imposed upon the manufacture or traffic in such liquor. No liquor revenue stamps or tax receipts for any illegal manufacture or sale shall be issued in advance, but upon evidence of such illegal manufacture or sale a tax shall be assessed against, and collected from, the person responsible for such illegal manufacture or sale in double the amount now provided by law, with an additional penalty of $500 on retail dealers and $1,000 on manufacturers. The payment of such tax or penalty shall give no right to engage in the manufacture or sale of such liquor, or relieve anyone from criminal liability, nor shall this Act relieve any person from any liability, civil or criminal, heretofore or hereafter incurred under existing laws.

The commissioner, with the approval of the Secretary of the Treasury, may compromise any civil cause arising under this title before bringing action in court; and with the approval of the Attorney General he may compromise any such cause after action thereon has been commenced.

SEC. 36. If any provision of this Act shall be held invalid it shall not be construed to invalidate other provisions of the Act.

SEC. 37. Nothing herein shall prevent the storage in United States bonded warehouses of all liquor manufactured prior to the taking effect of this Act, or prevent the transportation of such liquor to such warehouses or to any wholesale druggist for sale to such druggist for purposes not prohibited when the tax is paid, and permits may be issued therefore.

A manufacturer of any beverage containing less than one-half of 1 per centum of alcohol by volume may, on making application and giving such bond as the commissioner shall prescribe, be given a permit to develop in the manufacture thereof by the usual methods of fermentation and fortification or otherwise a liquid such as beer, ale, porter, or wine, containing more than one-half of 1 per centum of alcohol by volume, but before any such liquid is withdrawn from the factory or otherwise disposed of the alcoholic contents thereof shall under such rules and regulations as the commissioner may prescribe be reduced below such one-half of 1 per centum of alcohol: *Provided,* That such liquid may be removed and transported under bond and under such regulations as the commissioner may prescribe, from one bonded plant or warehouse to another for the purpose of having the alcohol extracted therefrom. And such liquids may be developed, under permit, by persons other than the manufacturers of beverages containing less than one-half of 1 per centum of alcohol by volume, and sold to such manufacturers for conversion into such beverages. The alcohol removed from such liquid, if evaporated, and not condensed and saved, shall not be subject to tax; if saved, it shall be subject to the same law as other alcoholic liquors. Credit shall be allowed on the tax due on any alcohol so saved to the amount of any tax paid upon distilled spirits or brandy used in the fortification of the liquor from which the same is saved.

When fortified wines are made and used for the production of non-beverage alcohol, and dealcoholized wines containing less than one-half of 1 per centum of alcohol by volume, no tax shall be assessed or paid on the spirits used in such fortification, and such dealcoholized wines produced under the provisions of this Act, whether carbonated or not, shall not be subject to the tax on artificially carbonated or sparkling wines, but shall be subject to the tax on still wines only.

In any case where the manufacturer is charged with manufacturing or selling for beverage purposes any malt, vinous, or fermented liquids containing one-half of 1 per centum or more of alcohol by volume, or in any case where the manufacturer, having been permitted by the commissioner to develop a liquid such as ale, beer, porter, or wine containing more than one-half of 1 per centum of alcohol by volume in the manner and for the purpose herein provided, is charged with failure to reduce the alcoholic content of any such liquid below such one-half of 1 per centum before withdrawing the same from the factory, then in either such case the burden of proof shall be on such manufacturer to show that such liquid so manufactured, sold, or withdrawn contains less than one-half of 1 per centum of alcohol by volume. In any suit or proceeding involving the alcoholic content of any beverage, the reasonable expense of analysis of such beverage shall be taxed as costs in the case.

SEC. 38. The Commissioner of Internal Revenue and the Attorney General of the United States are hereby respectively authorized to appoint and employ such assistants, experts, clerks, and other employees in the District of Columbia or elsewhere, and to purchase such supplies and equipment as they may deem necessary for the enforcement of the provisions of this Act, but such assistants, experts, clerks, and other employees, except such executive officers as may be appointed by the Commissioner or the Attorney General to have immediate direction of the enforcement of the provisions of this Act, and persons authorized to issue permits, and agents and inspectors in the field service, shall be appointed under the rules and regulations prescribed by the Civil Service Act: *Provided,* That the commissioner and Attorney General in making such appointments shall give preference to those who have served in the military or naval service in the recent war, if otherwise qualified, and there is hereby authorized to be appropriated, out of any money in the Treasury not otherwise appropriated, such sum as may be required for the enforcement of this Act including personal services in the District of Columbia, and for the fiscal year ending June 30, 1920, there is hereby appropriated, out of any money in the Treasury not otherwise appropriated, the sum of $2,000,000 for the use of the Commissioner of Internal Revenue and $100,000, for the use of the Department of Justice for the enforcement of the provisions of this Act, including personal services in the District of Columbia and necessary printing and binding.

SEC. 39. In all cases wherein the property of any citizen is proceeded against or wherein a judgment affecting it might be rendered, and the citizen is not the one who in person violated the provisions of the law, summons must be issued in due form and served personally, if said person is found within the jurisdiction of the court.

TITLE III.

INDUSTRIAL ALCOHOL

SEC. 1. When used in this title—

The term "alcohol" means that substance known as ethyl alcohol, hydrated oxide of ethyl, or spirit of wine, from whatever source or whatever processes produced.

The term "container" includes any receptacle, vessel or form of package, tank, or conduit used or capable of use for holding, storing, transferring, or shipment of alcohol.

INDUSTRIAL ALCOHOL PLANTS AND WAREHOUSES

SEC. 2. Any person now producing alcohol shall, within thirty days after the passage of this Act, make application to the commissioner for registration of his industrial alcohol plant, and as soon thereafter as practicable the premises shall be bonded and permit may issue for the operation of such plant, and any person hereafter establishing a plant for the production of alcohol shall likewise before operation make application, file bond, and receive permit.

SEC. 3. Warehouses for the storage and distribution of alcohol to be used exclusively for other than beverage purposes may be established upon filing of application and bond, and issuance of permit at such places, either in connection with the manufacturing plant or elsewhere as the commissioner may determine; and the entry and storage of alcohol therein, and the withdrawals of alcohol therefrom shall be made in such containers and by such means as the commissioner by regulation may prescribe.

SEC. 4. Alcohol produced at any registered industrial alcohol plant or stored in any bonded warehouse may be transferred under regulations to any other registered industrial alcohol plant or bonded warehouse for any lawful purposes.

SEC. 5. Any tax imposed by law upon alcohol shall attach to such alcohol as soon as it is in existence as such, and all proprietors of industrial alcohol plants and bonded warehouses shall be jointly and severally liable for any and all taxes on any and all alcohol produced thereat or stored therein. Such taxes shall be a first lien on such alcohol and the premises and plant in which such alcohol is produced or stored, together with all improvements and appurtenances thereunto belonging or in any wise appertaining.

SEC. 6. Any distilled spirits produced and fit for beverage purposes remaining in any bonded warehouse on or before the date when the eighteenth amendment of the Constitution of the United States goes into effect, may, under regulations, be withdrawn therefrom either for denaturation at any bonded denaturing plant or for deposit in a bonded warehouse established under this Act; and when so withdrawn, if not suitable as to proof, purity, or quality for other than beverage purposes, such distilled spirits shall be redistilled, purified, and changed in proof so as to render such spirits suitable for other purposes, and having been so treated may thereafter be denatured or sold in accordance with the provisions of this Act.

SEC. 7. Any distillery or bonded warehouse heretofore legally established may, upon filing application and bond and the granting of permit, be operated as an industrial alcohol plant or bonded warehouse under the provisions of this title and regulations made thereunder.

SEC. 8. Alcohol may be produced at any industrial alcohol plant established under the provisions of this title, from any raw materials or by any processes suitable for the production of alcohol, and, under regulations, may be used at any industrial alcohol plant or bonded warehouse or sold or disposed of for any lawful purpose, as in this Act provided.

SEC. 9. Industrial alcohol plants and bonded warehouses established under the provisions of this title shall be exempt from the provisions of sections 3154, 3244, 3258, 3259, 3260, 3263, 3264, 3266, 3267, 3268, 3269, 3271, 3273, 3274, 3275, 3279, 3280, 3283, 3284, 3285, 3286, 3287, 3288, 3289, 3290, 3291, 3292, 3293, 3294, 3295, 3302, 3303, 3307, 3308, 3309, 3310, 3311, 3312, 3313, 3314, and 3327 of the Revised Statutes; sections 48 to 60, inclusive, and sections 62 and 67 of the Act of August 27, 1894 (Twenty-eighth Statutes, pages 563 to 568), and from such other provisions of existing laws relating to distilleries and bonded warehouses as may, by regulations, be declared inapplicable to industrial alcohol plants and bonded warehouses established under this Act.

Regulations may be made embodying any provision of the sections above enumerated.

TAX-FREE ALCOHOL

SEC. 10. Upon the filing of application and bond and issuance of permit denaturing plants may be established upon the premises of any industrial alcohol plant, or elsewhere, and shall be used exclusively for the denaturation of alcohol by the admixture of such denaturing materials as shall render the alcohol, or any compound in which it is authorized to be used, unfit for use as an intoxicating beverage.

Source: *Congressional Record, 66th Congress, 1st Session:* 305–319.

U.S. Supreme Court Decisions in the National Prohibition Cases (1920)

The Volstead Act to enforce National Prohibition was not secure in American law until the Supreme Court declared it constitutional. Opponents of prohibition quickly initiated challenges to the new law, and the Supreme Court was equally quick to render judgment. Bundling several cases into one decision, the Supreme Court spoke on 7 June 1920 to deny the challenges and thereby make National Prohibition unquestionably the law of the land. See National Prohibition.

STATE OF RHODE ISLAND v. PALMER,
Atty. Gen., et al.
No. 29, Original.
STATE OF NEW JERSEY v. SAME. No. 30, Original.

DEMPSEY v. BOYNTON,
U. S. Atty., et al.
No. 696.

KENTUCKY DISTILLERIES & WAREHOUSE CO. v. GREGORY,
U. S. Atty., et al.
No. 752.

CHRISTIAN FEIGENSPAN v. BODINE,
U. S. Atty., et al.
No. 788.

SAWYER, U. S. Atty., et al. v. MANITOWOC PRODUCTS CO.
No. 794.

ST. LOUIS BREWING ASS'N v. MOORE,
Collector, et al.
No. 837.

(Decided June 7, 1920.)

1. CONSTITUTIONAL LAW *10—RESOLUTION PROPOSING AMENDMENT NEED NOT CONTAIN DECLARATION THAT IT IS REGARDED AS ESSENTIAL.

A joint resolution proposing an amendment to the Constitution need not contain an express declaration that those voting for it regard it as essential; its adoption sufficiently showing that they deem it necessary.

2. CONSTITUTIONAL LAW *10—"TWO-THIRDS VOTE" OF MEMBERS PRESENT CONSTITUTING QUORUM MAY ADOPT RESOLUTION PROPOSING AMENDMENT.

The "two-thirds vote" in each house, which is required in proposing an amendment to the Constitution, is a vote of two-thirds of the members present, assuming the presence of a quorum, and not a vote of two-thirds of the entire membership.

3. CONSTITUTIONAL LAW *10—REFERENDUM PROVISIONS CANNOT BE APPLIED TO ADOPTION OF AMENDMENT TO FEDERAL CONSTITUTION.

The referendum provisions of state Constitutions and statutes cannot be applied, consistently with the Constitution of the United States, in the ratification or rejection of amendments to that Constitution.

4. CONSTITUTIONAL LAW *10—PROHIBITION AMENDMENT WITHIN POWER TO AMEND CONFERRED BY CONSTITUTION.

Const. Amend. 18, prohibiting the manufacture, sale, etc., of intoxicating liquors for beverage purposes, is within the power to amend reserved by article 5.

5. CONSTITUTIONAL LAW *10—PROHIBITION AMENDMENT HELD LAWFULLY PROPOSED AND RATIFIED.

Const. Amend. 18, prohibiting the manufacture, sale, etc., of intoxicating liquors for beverage purposes, has become, by lawful proposal and ratification, a part of the Constitution.

6. INTOXICATING LIQUORS *13—STATUTES AUTHORIZING WHAT PROHIBITION AMENDMENT PROHIBITS ARE INVALIDATED.

Const. Amend. 18, § 1, prohibiting the manufacture, sale, etc., of intoxicating liquors for beverage purposes, is operative throughout the entire territorial limits of the United States and of its own force invalidates every legislative act of Congress, state Legislatures, or territorial assemblies, authorizing or sanctioning what it prohibits.

7. INTOXICATING LIQUORS *13—PROHIBITION AMENDMENT ONLY AUTHORIZES STATUTES ENFORCING ITS PROVISIONS.

Const. Amend. 18, § 2, giving Congress and the states concurrent power to enforce such amendment by appropriate legislation, does not authorize Congress or the states to defeat or thwart the prohibition contained in section 1, but only to enforce it by appropriate means.

8. INTOXICATING LIQUORS *13—CONGRESSIONAL LEGISLATION UNDER PROHIBITION AMENDMENT NEED NOT BE JOINED IN OR SANCTIONED BY STATES; "CONCURRENT POWER."

The words "concurrent power," in Const. Amend. 18, § 2, giving concurrent power to Congress and the states to enforce that amendment, do not mean a joint power or require that legislation thereunder by Congress to be effective, shall be approved or sanctioned by the several states, or any of them.

9. INTOXICATING LIQUORS *13—POWER OF CONGRESS NOT LIMITED TO INTERSTATE TRANSACTIONS.

Const. Amend. 18, § 2, does not divide the power to enforce such amendment between Congress and the states along lines which separate or distinguish foreign and interstate commerce from intrastate affairs, but confides to Congress power territorially coextensive with the prohibition of the first section and embracing manufacture and other intrastate transactions as well as importation, exportation, and interstate traffic.

10. INTOXICATING LIQUORS *13—POWER OF CONGRESS NOT DEPENDENT ON ACTION OF THE STATES.

The power conferred on Congress by Const. Amend. 18, § 2, to enforce the prohibition contained in section 1, is in no wise dependent on, or affected by, action or inaction on the part of the states, or any of them.

11. INTOXICATING LIQUORS *13—CONGRESS MAY PROHIBIT DISPOSAL OF LIQUORS MANUFACTURED PRIOR TO PROHIBITION AMENDMENT.

Under Const. Amend. 18, Congress may prohibit the disposal, for beverage purposes, of liquors manufactured before such amendment became effective.

12. INTOXICATING LIQUORS *13—NATIONAL PROHIBITION ACT IS WITHIN POWERS OF CONGRESS.

The National Prohibition Act, which treats liquors containing one-half of 1 per cent. of alcohol by volume and fit for use for beverage purposes as within the powers of enforcement conferred on Congress by Const. Amend. 18, does not transcend the powers so conferred.

Mr. Justice McKenna and Mr. Justice Clarke dissenting in part.

No. 696: Appeal from the District Court of the United States for the District of Massachusetts.

No. 752: Appeal from the District Court of the United States for the Western District of Kentucky.

No. 788: Appeal from the District Court of the United States for the District of New Jersey.

No. 794: Appeal from the District Court of the United States for the Eastern District of Wisconsin.

No. 837: Appeal from the District Court of the United States for the Eastern District of Missouri.

Original suits by the State of Rhode Island and by the State of New Jersey against A. Mitchell Palmer, Attorney General, and others. Suits dismissed.

Suits by George C. Dempsey against Thomas J. Boynton, as United States Attorney, and others, by the Kentucky Distilleries & Warehouse Company against W. V. Gregory, as United States Attorney, and others, by Christian Feigenspan, a corporation, against Joseph L. Bodine, as United States attorney, and others, by the Manitowoc Products Company against Hiram A. Sawyer, as United States Attorney, and others, and by the St. Louis Brewing Association against George H. Moore, Collector, and others. From a decree in favor of plaintiff in the suit by the Manitowoc Products Company, defendants appeal, and from decrees for the defendants in the other suits, the plaintiffs appeal. Decree in the suit by the Manitowoc Products Company reversed, and decrees in the other suits affirmed.

For opinion below in Christian Feigenspan v. Bodine, see 264 Fed. 186.

See, also, State of Rhode Island v. Palmer, 40 Sup. Ct. 179, 64, L. Ed.—; State of New Jersey v. Palmer, 252 U.S. 570, 40 Sup. Ct. 345, 64 L. Ed.—

No. 29. Argued March 8 and 9, 1920:

Mr. Herbert A. Rice, of Providence, R.I., for complainant.

Mr. Solicitor General King and Mr. Assistant Attorney General Frierson, for respondents.

No. 30. Argued March 29, 1920:

Mr. Thomas F. McCran, of Paterson, N. J., for complainant.

Mr. Assistant Attorney General Frierson, for respondents.

No. 696. Argued March 9, 1920:

Mr. Patrick Henry Kelley, of Boston, Mass., for appellant.

Mr. Assistant Attorney General Frierson, for appellees.

No. 752. Argued March 9 and 10, 1920: Messrs. Levy Mayer, of Chicago, Ill., and William Marshall Bullitt, of Louisville, Ky., for appellant.

Mr. Solicitor General King and Mr. Assistant Attorney General Frierson, for appellees.

No. 788. Argued March 29 and 30, 1920:

Messrs. Elihu Root and William D. Guthrie, both of New York City, for appellant. Mr. Solicitor General King and Mr. Assistant Attorney General Frierson, for appellees.

No. 794. Argued March 30, 1920:

Mr. Solicitor General King and Mr. Assistant Attorney General Frierson, for appellants.

Mr. Ralph W. Jackman, of Madison, Wis., for appellee.

No. 837. Submitted March 29, 1920:

Messrs. Charles A. Bouts, John T. Fitzsimmons, and Edward C. Crow, all of St. Louis, Mo., for appellant.

Mr. Solicitor General King and Mr. Assistant Attorney General Frierson, for appellees.

Mr. Justice VAN DEVANTER announced the conclusions of the Court.

Power to amend the Constitution was reserved by article 5, which reads:

"The Congress, whenever two-thirds of both Houses shall deem it necessary, shall propose Amendments to this Constitution, or, on the Application of the Legislatures of two-thirds of the several States, shall call a Convention for proposing Amendments, which, in either case, shall be valid to all Intents and Purposes, as Part of this Constitution, when ratified by the Legislatures of three-fourths of the several States, or by Conventions in three-fourths thereof, as the one or the other Mode of Ratification may be proposed by the Congress; Provided that no Amendment which may be made prior to the Year One thousand eight hundred and eight shall in any Manner affect the first and fourth Clauses in the Ninth Section of the first Article; and that no State, without its Consent, shall be deprived of its equal Suffrage in the Senate."

The text of the Eighteenth Amendment, proposed by Congress in 1917 and proclaimed as ratified in 1919 (40 Stat. 1050, 1941), is as follows:

"Section 1. After one year from the ratification of this article the manufacture, sale, or transportation of intoxicating liquors within; the importation thereof into, or the exportation thereof from the United States and all territory subject to the jurisdiction thereof for beverage purposes is hereby prohibited.

"Sec. 2. The Congress and the several states shall have concurrent power to enforce this article by appropriate legislation."

We here are concerned with seven cases involving the validity of that amendment and of certain general features of the National Prohibition Law, known as the Volstead Act, c. 85, Acts 66th Cong., 1st Sess. (41 Stat. 305), which was adopted to enforce the amendment. The relief sought in each case is an injunction against the execution of that act. Two of the cases—Nos. 29 and 30, original,—were brought in this court, and the others in District Courts. Nos. 696, 752, 788, and 837 are here on appeals from decrees refusing injunctions, and No. 799 from a decree granting an injunction. The cases have been elaborately argued at the bar and in printed briefs; and the arguments have been attentively considered, with the result that we reach and announce the following conclusions on the questions involved:

1. The adoption by both houses of Congress, each by a two-thirds vote, of a joint resolution proposing an amendment to the Constitution sufficiently shows that the proposal was deemed necessary by all who voted for it. An express declaration that they regarded it as necessary is not essential. None of the resolutions whereby prior amendments were proposed contained such a declaration.

2. The two-thirds vote in each house which is required in proposing an amendment is a vote of two-thirds of the members present—assuming the presence of a quorum—and not a vote of two-thirds of the entire membership, present and absent. Missouri Pacific Ry. Co. v. Kansas, 248 U. S. 276, 39 Sup. Ct. 93, 63 L. Ed. 239, 2 A. L. R. 1589.

3. The referendum provisions of state Constitutions and statutes cannot be applied, consistently with the Constitution of the United States, in the ratification or rejection of amendments to it. Hawke v. Smith, 253 U. S. 221, 40 Sup. Ct. 495, 64 L. Ed.—, decided June 1, 1920.

4. The prohibition of the manufacture, sale, transportation, importation and exportation of intoxicating liquors for beverage purposes, as embodied in the Eighteenth Amendment, is within the power to amend reserved by article 5 of the Constitution.

5. That amendment, by lawful proposal and ratification, has become a part of the Constitution, and must be respected and given effect the same as other provisions of that instrument.

6. The first section of the amendment—the one embodying the prohibition—is operative throughout the entire territorial limits of the United States, binds all legislative bodies, courts, public officers and individuals within those limits, and of its own force invalidates every legislative act, whether by Congress, by a state Legislature, or by a territorial assembly, which authorizes or sanctions what the section prohibits.

7. The second section of the amendment—the one declaring "The Congress and the several states shall have concurrent power to enforce this article by appropriate legislation"—does not enable Congress or the several states to defeat or thwart the prohibition, but only to enforce it by appropriate means.

8. The words "concurrent power," in that section, do not mean joint power, or require that legislation thereunder by Congress, to be effective, shall be approved or sanctioned by the several states or any of them; nor do they mean that the power to enforce is divided between Congress and the several states along the lines which separate or distinguish foreign and interstate commerce from intrastate affairs.

9. The power confided to Congress by that section, while not exclusive, is territorially coextensive with the prohibition of the first section, embraces manufacture and other intrastate transactions as well as importation, exportation and interstate traffic, and is in no wise dependent on or affected by action or inaction on the part of the several states or any of them.

10. That power may be exerted against the disposal for beverage purposes of liquors manufactured before the amendment became effective just as it may be against subsequent manufacture for those purposes. In either case it is a constitutional mandate or prohibition that is being enforced.

11. While recognizing that there are limits beyond which Congress cannot go in treating beverages as within its power of enforcement, we think those limits are not transcended by the provision of the Volstead Act (title 2, § 1), wherein liquors containing as much as one-half of 1 per cent. of alcohol by volume and fit for use for beverage purposes are treated as within that power. Jacob Ruppert v. Caffey, 251 U. S. 264, 40 Sup. Ct. 141, 64 L. Ed.—

Giving effect to these conclusions, we dispose of the cases as follows:

In Nos. 29 and 30, original, the bills are dismissed.

In No. 794, the decree is reversed.

In Nos. 696, 752, 788 and 837, the decrees are affirmed.

Mr. Chief Justice WHITE concurring.

I profoundly regret that in a case of this magnitude, affecting as it does an amendment to the Constitution dealing with the powers and duties of the national and state governments, and intimately concerning the welfare of the whole people, the court has deemed it proper to state only ultimate conclusions without an exposition of the reasoning by which they have been reached.

I appreciate the difficulties which a solution of the cases involve and the solicitude with which the court has approached them, but it seems to my mind that the greater the perplexities the greater the duty devolving upon me to express the reasons which have led me to the conclusion that the amendment accomplishes and was intended to accomplish the purposes now attributed to it in the propositions concerning that subject which the court has just announced and in which I concur. Primarily in doing this I notice various contentions made concerning the proper construction of the provisions of the amendment which I have been unable to accept, in order that by contrast they may add cogency to the statement of the understanding I have of the amendment.

The amendment, which is reproduced in the announcement for the court, contains three numbered paragraphs or sections, two of which only need be noticed. The first prohibits—

"the manufacture, sale, or transportation of intoxicating liquors within, the importation thereof into, or the exportation thereof from the United States and all territory subject to the jurisdiction thereof for beverage purposes."

The second is as follows:

"Sec. 2. The Congress and the several states shall have concurrent power to enforce this article by appropriate legislation."

1. It is contended that the result of these provisions is to require concurrent action of Congress and the states in enforcing the prohibition of the first section and hence that in the absence of such concurrent action by Congress and the states no enforcing legislation can exist, and therefore until this takes place the prohibition of the first section is a dead letter. But in view of the manifest purpose of the first section to apply and make efficacious the prohibition, and of the second to deal with the methods of carrying out that purpose, I cannot accept this interpretation, since it would result simply in declaring that the provisions of the second section, avowedly enacted to provide means for carrying out the first, must be so interpreted as to practically nullify the first.

2. It is said, conceding that the concurrent power given to Congress and to the states does not as a prerequisite exact the concurrent action of both, it nevertheless contemplates the possibility of action by Congress and by the states and makes each action effective, but as under the Constitution the authority of Congress in enforcing the Constitution is paramount, when state legislation and congressional action conflict the state legislation yields to the action of Congress as controlling. But as the power of both Congress and the states in this instance is given by the Constitution in one and the same provision, I again find myself unable to accept the view urged because it ostensibly accepts the constitutional mandate as to the concurrence of the two powers and proceeds immediately by way of interpretation to destroy it by making one paramount over the other.

3. The proposition is that the concurrent Powers conferred upon Congress and the states are not subject to conflict because their exertion is authorized within different areas, that is, by Congress within the field of federal authority, and by the states within the sphere of state power, hence leaving the states free within their jurisdiction to determine separately for themselves what, within reasonable limits, is an intoxicating liquor, and to Congress the same right within the sphere of its jurisdiction. But the unsoundness of this more plausible contention seems to me at once exposed by directing attention to the fact that in a case where no state legislation was enacted there would be no prohibition, thus again frustrating the first section by a construction affixed to the second. It is no answer to suggest that a regulation by Congress would in such event be operative in such a state, since the basis of the distinction upon which the argument rests is that the concurrent power conferred upon Congress is confined to the area of its jurisdiction and therefore is not operative within a state.

Comprehensively looking at all these contentions, the confusion and contradiction to which they lead, serve in my judgment to make it certain that it cannot possibly be that Congress and the states entered into the great and important business of amending the Constitution in a matter so vitally concerning all the people solely in order to render governmental action impossible, or, if possible, to so define and limit it as to cause it to be productive of no results and to frustrate the obvious intent and general purpose contemplated. It is true indeed that the mere words of the second section tend to these results, but if they be read in the light of the cardinal rule which compels a consideration of the context in view of the situation and the subject with which the amendment dealt and the purpose which it was intended to accomplish, the confusion will be seen to be only apparent.

In the first place, it is indisputable, as I have stated, that the first section imposes a general prohibition which it was the purpose to make universally and uniformly operative and efficacious. In the second place, as the prohibition did not define the intoxicating beverages which it prohibited, in the absence of anything to the contrary, it clearly, from the very fact of its adoption, cast upon Congress the duty, not only of defining the prohibited beverages, but also of enacting such regulations and sanctions as were essential to make them operative when defined. In the third place, when the second section is considered with these truths in mind it becomes clear that it simply manifests a like purpose to adjust, as far as possible, the exercise of the new powers cast upon Congress by the amendment to the dual system of government existing under the Constitution. In other words, dealing with the new prohibition created by the Constitution, operating throughout the length and breadth of the United States, without reference to state lines or the distinctions between state and federal power and contemplating the exercise by Congress of the duty cast upon it to make the prohibition efficacious, it was sought by the second section to unite national and state administrative agencies in giving effect to the amendment and the legislation of Congress enacted to make it completely operative.

Mark the relation of the text to this view, since the power which it gives to state and nation is, not to construct or perfect or cause the amendment to be completely operative but as already made completely operative, to enforce it. Observe also the words of the grant which confines the concurrent power given to legislation appropriate to the purpose of enforcement.

I take it that if the second section of the article did not exist no one would gainsay that the first section in and of itself granted the power and imposed the duty upon Congress to legislate to the end that by definition and sanction the amendment would become fully operative. This being true it would follow, if the contentions under consideration were sustained, that the second section gave the states the power to nullify the first section, since a refusal of a state to define and sanction would again result in no amendment to be enforced in such refusing state.

Limiting the concurrent power to enforce given by the second section to the purposes which I have attributed to it, that is, to the subjects appropriate to execute the amendment as defined and sanctioned by Congress, I assume that it will not be denied that the effect of the grant of authority was to confer upon both Congress and the states power to do things which otherwise there would be no right to do. This being true, I submit that no reason exists for saying that a grant of concurrent power to Congress and the states to give effect to, that is, to carry out or enforce, the amendment as defined and sanctioned by Congress, should be interpreted to deprive Congress of the power to create, by definition and sanction, an enforceable amendment.

Mr. Justice McREYNOLDS concurring

I do not dissent from the disposition of these causes as ordered by the court, but confine my concurrence to that. It is impossible now to say with fair certainty what construction should be given to the Eighteenth Amendment. Because of the bewilderment which it creates, a multitude of questions will inevitably arise and demand solution here. In the circumstances I prefer to remain free to consider these questions when they arrive.

Mr. Justice McKENNA, dissenting.

This case is concerned with the Eighteenth Amendment of the Constitution of the United States, its validity and construction. In order to have it, and its scope in attention, I quote it:

"Section 1. After one year from the ratification of this article the manufacture, sale, or transportation of intoxicating liquors within, the importation thereof into, or the exportation thereof from the United States and all territory subject to the jurisdiction thereof for beverage purposes is hereby prohibited.

"Sec. 2. The Congress and the several states shall have concurrent power to enforce this article by appropriate legislation."

The court in applying it has dismissed certain of the bills, reversed the decree in one, and affirmed the decrees in four others. I am unable to agree with the judgment reversing No. 791 and affirming Nos. 752, 696, 788, and 837.

I am, however, at a loss how or to what extent to express the grounds for this action. The court declares conclusions only, without giving any reasons for them. The instance may be wise—establishing a precedent now, hereafter wisely to be imitated. It will undoubtedly decrease the literature of the court if it does not increase its lucidity. However, reasons for the conclusions have been omitted, and my comment upon them may come from a misunderstanding of them, their present import and ultimate purpose and force.

There are, however, clear declarations that the Eighteenth Amendment is part of the Constitution of the United States, made so in observance of the prescribed constitutional procedure, and has become part of the Constitution of the United States, to be respected and given effect like other provisions of that instrument. With these conclusions I agree.

Conclusions 4, 5, and 6 seem to assert the undisputed. I neither assent to them or dissent from them except so far as I shall presently express.

Conclusion 7 seems an unnecessary declaration. It may, however, be considered as supplementary to some other declaration. My only comment is that I know of no intimation in the case that section 2 in conferring concurrent power on Congress and the states to enforce the prohibition of the first section, conferred a power to defeat or obstruct prohibition. Of course, the power was conferred as a means to enforce the prohibition and was made concurrent to engage the resources and instrumentalities of the nation and the states. The power was conferred for use, not for abuse.

Conclusions 8 and 9, as I view them, are complements of each other, and express, with a certain verbal detail, the power of Congress and the states over the liquor traffic, using the word in its comprehensive sense as including the production of liquor, its transportation within the states, its exportation from them, and its importation into them. In a word, give power over the liquor business from producer to consumer, prescribe the quality of latter's beverage. Certain determining elements are expressed. It is said that the words "concurrent power" of section 2 do not mean joint power in Congress and the states, nor the approval by the states of congressional legislation, nor its dependency upon state action or inaction.

I cannot confidently measure the force of the declarations or the deductions that are, or can be made from them. They seem to be regarded as sufficient to impel the conclusion that the Volstead Act is legal legislation and operative throughout the United States. But are there no opposing considerations, no conditions upon its operation? And what of conflicts, and there are conflicts, and more there may be, between it and state legislation? The conclusions of the court do not answer the questions and yet they are submitted for decision; and their importance appeals for judgment upon them. It is to be remembered states are litigants as well as private citizens, the former presenting the rights of the states, the latter seeking protection against the asserted aggression of the act in controversy. And there is opposing state legislation, why not a decision upon it? Is it on account of the nature of the actions being civil and in equity, the proper forum being a criminal court investigating a criminal charge? There should be some way to avert the necessity or odium of either.

I cannot pause to enumerate the contentions in the case. Some of them present a question of joint action in Congress and the states, either collectively with all or severally with each. Others assert spheres of the powers, involving no collision, it is said, the powers of Congress and the states being supreme and exclusive within the spheres of their exercise—called by counsel "historical fields of jurisdiction." I submit again, they should have consideration and decision.

The government has felt and exhibited the necessity of such consideration and decision. It knows the conflicts that exist or impend. It desires to be able to meet them, silence them and bring the repose that will come from a distinct declaration and delimitation of the power of Congress and the states. The court, however, thinks otherwise and I pass to the question in the case. It is a simple one, it involves the meaning of a few English words—in what sense they shall be taken, whether in their ordinary sense, or have put upon them an unusual sense.

Recurring to the first section of the amendment, it will be seen to be a restriction upon state and congressional power, and the deduction from it is that neither the states nor Congress can enact legislation that contravenes its prohibition. And there is no room for controversy as to its requirement. Its prohibition of "intoxicating liquors" "for beverage purposes" is absolute. And, as accessory to that prohibition, is the further prohibition of their manufacture, sale or transportation within or their importation into or exportation "from the United States." Its prohibition, therefore, is national, and considered alone, the means of its enforcement might be such as Congress, the agency of national power might prescribe. But it does not stand alone. Section 2 associates Congress and the states in power to enforce it. Its words are:

"The Congress and the several states shall have concurrent power to enforce this article by appropriate legislation."

What, then, is meant by the words "concurrent power"? Do they mean united action, or separate and independent action, and, if the actions differ (there is no practical problem unless they differ), shall that of Congress be supreme?

The government answers that the words mean separate and independent action, and, in case of conflict, that of Congress is supreme, and asserts besides, that the answer is sustained by historical and legal precedents.[1] I contest the assertions and oppose to them the common usage of our language, and the definitions of our lexicons, general and legal.[2] Some of the definitions assign to the words "concurrent power" action in conjunction, contribution of effort, certainly bar, many of action, not antagonism. Opposing laws are not concurring laws, and to assert the supremacy of one over the other is to assert the exclusiveness of one over the other, not their concomitance. Such is the result of the government's contention. It does not satisfy the definitions, or the requirement of section 2—"a concurrent power excludes the idea of a dependent power." Mr. Justice McLean in the Passenger Cases, 7 How. 283, 399, 12 L. Ed. 702.

Other definitions assign to the words "existing or happening at the same time," "concurring together," "coexistent." These definitions are, as the others are, inconsistent with the government's contention. If coexistence of the power of legislation is given to Congress and the states by section 2, it is given to be coexistently exercised. It is to be remembered that the Eighteenth Amendment was intended to deal with a condition, not a theory, and one demanding something more than exhortation and precept. The habits of a people were to be changed, large business interests were to be disturbed, and it was considered that the change and disturbance could only be effected by punitive and repressive legislation, and it was naturally thought that legislation enacted by "the Congress and the several states," by its concurrence would better enforce prohibition and avail for its enforcement the two great divisions of our governmental system, the nation and the states, with their influences and instrumentalities.

From my standpoint, the exposition of the case is concluded by the definition of the words of section 2. There are, however, confirming considerations; and militating considerations are urged. Among the confirming considerations are the cases of Wedding v. Meyler, 192 U.S. 573, 24 Sup. Ct. 322, 48 L. Ed. 570, 66 L. R. A. 833, and Nielsen v. Oregon, 212 U. S. 315, 29 Sup. Ct. 383, 53 L. Ed. 528, in which "concurrent jurisdiction" was given respectively to Kentucky and Indiana over the Ohio river by the Virginia Compact, and respectively to Washington and Oregon over the Columbia river by act of Congress. And it was decided that it conferred equality of powers, "legislative, judicial and executive," and that neither state could override the legislation of the other. Other courts have given like definitions. 2 Words and Phrases Judicially Defined, p. 1391 et seq.; Bouvier's Dictionary, vol. 1, page 579. Analogy of the word "concurrent" in private instruments may also be invoked.

Those cases are examples of the elemental rule of construction that in the exposition of statutes and constitutions, every word "is to be expounded in its plain, obvious, and common sense, unless the context furnishes some ground to control, qualify or enlarge it," and there cannot be imposed upon the words "any recondite meaning or any extraordinary gloss." 1 Story, Const. § 451; Lake County v. Rollins, 130 U. S. 662, 9 Sup. Ct. 651, 32 L. Ed. 1060. And it is the rule of reason as well as of technicality, that if the words so expounded be "plain and clear, and the sense distinct and perfect arising on them" interpretation has nothing to do. This can be asserted of section 2. Its words express no "double sense," and should be accepted in their single sense. It has not yet been erected into a legal maxim of constitutional construction, that words were made to conceal thoughts. Besides, when we depart from the words, ambiguity comes. There are as many solutions as there are minds considering the section, and out of the conflict, I had almost said chaos, one despairs of finding an undisputed meaning. It may be said that the court, realizing this, by a declaration of conclusions only, has escaped the expression of antithetical views and considered it better not to blaze the trails, though it was believed that they all led to the same destination.

If it be conceded, however, that to the words "concurrent power" may be ascribed the meaning for which the government contends, it certainly cannot be asserted that such is their ordinary meaning, and I might leave section 2, and the presumptions that support it, to resist the precedents adduced by the government. I go farther, however, and deny the precedents. The Federalist and certain cases are cited as such. There is ready explanation of both, and neither supports the government's contention. The dual system of government contemplated by the Union encountered controversies, fears, and jealousies that had to be settled or appeased to achieve union, and the Federalist in good and timely sense explained to what extent the "alienation of state sovereignty" would be necessary to "national sovereignty," constituted by the "consolidation of the states," and the powers that would be surrendered, and those that would be retained. And the explanation composed the controversies and allayed the fears of the states that their local powers of government would not be displaced by the dominance of a centralized control. And this court after union had been achieved, fulfilled the assurances of the explanation and adopted its distribution of powers, designating them as follows: (1) Powers that were exclusive in the states reserved to them; (2) powers that were exclusive in Congress, conferred upon it; (3) powers that were not exclusive in either, and hence said to be "concurrent." And it was decided that, when exercised by Congress, they were supreme—"the authority of the states then retires" to inaction. To understand them, it must be especially observed that their emphasis was, as the fundamental principle of the new government was, that it had no powers that were not conferred upon it, and that all other powers were reserved to the states. And this necessarily must not be absent from our minds, whether construing old provisions of the Constitution or amendments to it or laws passed under the amendments.

The government nevertheless contends that the decisions (they need not be cited) constitute precedents for its construction of section 2 of the Eighteenth Amendment. In other words, the government contends (or must so contend for its reasoning must bear the test of the generalization) that it was decided that in all cases where the powers of Congress are concurrent with those of the states, they are supreme as incident to concurrence. The contention is not tenable; it overlooks the determining consideration. The powers of Congress were not decided to be supreme because they were concurrent with powers in the states, but because of their source, their source being the Constitution of the United States and the laws made in pursuance of the Constitution, as against the source of the powers of the states, their source being the Constitution and laws of the states, the Constitution and laws of the United States being made by article 6 the supreme law of the land, "any Thing in the Constitution or Laws of any State to the Contrary notwithstanding." McCulloch v. Maryland, 4 Wheat. 316, 426, 4 L. Ed. 579.

This has example in other powers of sovereignty that the states and Congress possess. In McCulloch v. Maryland, at pages 425, 430 of 4 Wheat. (4 L. Ed. 579), Chief Justice Marshall said that the power of taxation retained by the states was not abridged by the granting of a similar power to the government of the Union, and that it was to be concurrently exercised, and these truths, it was added, had never been denied, and that there was no "clashing sovereignty" from incompatibility of right. And, necessarily, a concurrence of power in the states and Congress excludes the idea of supremacy in either. Therefore, neither principle nor precedent sustains the contention that section 2 by giving concurrent power to Congress and the states, gave Congress supreme power over the states. I repeat the declaration of Mr. Justice McLean:

"A concurrent power excludes the idea of a dependent power."

It is, however, suggested (not by the government) that if Congress is not supreme upon the considerations urged by the government, it is made supreme by article 6 of the Constitution. The article is not applicable. It is not a declaration of the supremacy of one provision of the Constitution or laws of the United States over another, but of the supremacy of the Constitution and laws of the United States over the Constitutions and laws of the states. Gibbons v. Ogden, 9 Wheat. 1, 209, 6 L. Ed. 23, 211; Sec. 1838 et seq.; 2 Story, Const., 5th Ed.

The Eighteenth Amendment is part of the Constitution of the United States, therefore of as high sanction as article 6. There seems to be a denial of this, based on article 5. That article provides that the amendments proposed by either of the ways there expressed "shall be valid to all intents and purposes as part of this Constitution." Some undefinable power is attributed to this in connection with article 6, as if article 5 limits in some way, or defeats, an amendment to the Constitution inconsistent with a previously existing provision. Of course, the immediate answer is that an amendment is made to change a previously existing provision. What other purpose could an amendment have and it would be nullified by the mythical power attributed to article 5, either alone or in conjunction with article 6? A contention that ascribes such power to those articles is untenable. The Eighteenth Amendment is part of the Constitution and as potent as any other part of it. Section 2, therefore, is a new provision of power, power to the states as well as to Congress, and it is a contradiction to say that a power constitutionally concurrent in Congress and the states, in some way becomes constitutionally subordinate in the states to Congress.

If it be said that the states got no power over prohibition that they did not have before, it cannot be said that it was not preserved to them by the amendment, notwithstanding the policy of prohibition was made national, and besides, there was a gift of power to Congress that it did not have before, a gift of a right to be exercised within state lines, but with the limitation or condition that the powers of the states should remain with the states and be participated in by Congress only in concurrence with the states, and thereby preserved from abuse by either, or exercise to the detriment to prohibition. There was, however, a power given to the states, a power over importations. This power was subject to concurrence with Congress and had the same safeguards.

This construction of section 2 is enforced by other considerations. If the supremacy of Congress had been intended it would have been directly declared as in the Thirteenth, Fourteenth and Fifteenth Amendments. And such was the condition when the amendment left the Senate. The precedent of preceding amendments was followed, there was a single declaration of jurisdiction in Congress.

Section 2 was amended in the House upon recommendation of the Judiciary Committee and the provision giving concurrent power to Congress and to the states was necessarily estimated and intended to be additive of something. The government's contention makes it practically an addition of nothing but words, in fact denuding it of function, making it a gift of impotence, not one of power to be exercised independently of Congress or concurrently with Congress, or, indeed, at all. Of this there can be no contradiction, for what power is assigned to the states to legislate if the legislation be immediately superseded—indeed, as this case shows, is possibly forestalled and precluded by the power exercised in the Volstead Act. And meaningless is the difference the government suggests between concurrent power and concurrent legislation. A power is given to be exercised, and we are cast into helpless and groping bewilderment in trying to think of it apart from its exercise or the effect of its exercise. The addition to section 2 was a conscious adaptation of means to the purpose. It changed the relation between the states and the national government. The lines of exclusive power in one or the other were removed, and equality and community of powers substituted.

There is a suggestion, not made by the government, though assisting its contention, that section 2 was a gift of equal power to Congress and to the states, not, however, to be concurrently exercised, but to be separately exercised; conferred and to be exercised is the suggestion, to guard against neglect in either Congress or the states, the inactivity of the one being supplied by the activity of the other. But here again we encounter the word "concurrent" and its inexorable requirement of coincident or united action, not alternative or emergent action to safeguard against the delinquency of Congress or the states. If, however, such neglect was to be apprehended, it is strange that the framers of section 2, with the whole vocabulary of the language to draw upon, selected words that expressed the opposite of what the framers meant. In other words, expressed concurrent action instead of substitute action. I cannot assent. I believe they meant what they said and that they must be taken at their word.

The government with some consciousness that its contention requires indulgence or excuse, but at any rate in recognition of the insufficiency of its contention to satisfy the words of section 2, makes some concessions to the states. They are, however, not very tangible to measurement. They seem to yield a power of legislation to the states and a power of Jurisdiction to their courts, but almost at the very instant of concession, the power and jurisdiction are declared to be without effect.

I am not, therefore, disposed to regard the concessions seriously. They confuse—"make not light, but darkness, visible." Of what use is a concession of power to the states to enact laws which cannot be enforced? Of what use a concession of jurisdiction to the courts of the states when their judgments cannot be executed, indeed the very law upon which it is exercised may be declared void in an antagonistic jurisdiction exerted in execution of an antagonistic power?[3] And equally worthless is the analogy that the government assays between the power of the national government and the power of the states to criminally punish violations of their respective sovereignties, as, for instance, in counterfeiting cases. In such cases the exercises of sovereignty are not in antagonism. Each is inherently possessed and independently exercised, and can be enforced no matter what the other sovereignty may do or abstain from doing. On the other hand, under the government's construction of section 2, the legislation of Congress is supreme and exclusive. Whatever the states may do is abortive of effect.

The government seeking relief from the perturbation of mind and opinions produced by departure from the words of section 2, suggests a modification of its contention, that in case of conflict between state legislation and congressional legislation, that of Congress would prevail, by intimating that if state legislation be more drastic than congressional legislation, it might prevail, and in support of the suggestion, urges that section 1 is a command to prohibition, and that the purpose of section 2 is to enforce the command, and whatever legislation is the most prohibitive subserves best the command, displaces less restrictive legislation and becomes paramount. If a state, therefore, should define an intoxicating beverage to be one that has less than one-half of 1 per cent. of alcohol, it would supersede the Volstead Act and a state might even keep its legislation supreme by forestalling congressional retaliation by prohibiting all artificial beverages of themselves innocuous, the prohibition being accessory to the main purpose of power; adducing Purity Extract Co. v. Lynch, 226 U. S. 192, 33 Sup. Ct. 44, 57 L. Ed. 184, and Ruppert v. Caffey, 251 U. S. 264, 40 Sup.

Ct. 141. Of course this concession of the more drastic legislation destroys all that is urged for congressional supremacy, for necessarily supremacy cannot be transferred from the states to Congress or from Congress to the states as the quantity of alcohol may vary in the prohibited beverage. Section 2 is not quite so flexible to management. I may say, however, that one of the conclusions of the court has limited the range of retaliations. It recognizes "that there are limits beyond which Congress cannot go in treating beverages as within its power of enforcement" and declares "that those limits are not transcended by the provisions of the Volstead Act." Of course, necessarily, the same limitations apply to the power of the states as well.

From these premises the deduction seems inevitable that there must be united action between the states and Congress, or, at any rate, concordant and harmonious action; and will not such action promote better the purpose of the amendment—will it not bring to the enforcement of prohibition, the power of the states and the power of Congress, make all the instrumentalities of the states, its courts and officers, agencies of the enforcement, as well as the instrumentalities of the United States, its court and officers, agencies of the enforcement? Will it not bring to the states as well, or preserve to them, a partial autonomy, satisfying, if you will, their prejudices, or better say, their predilections; and it is not too much to say that our dual system of government is based upon them. And this predilection for self-government the Eighteenth Amendment regards and respects, and by doing so sacrifices nothing of, the policy of prohibition.

It is, however, urged that to require such concurrence is to practically nullify the prohibition of the amendment, for without legislation its prohibition would be ineffectual, and that it is impossible to secure the concurrence of Congress and the states in legislation. I cannot assent to the propositions. The conviction of the evils of intemperance—the eager and ardent sentiment that impelled the amendment, will impel its execution through Congress and the states. It may not be in such legislation as the Volstead Act with its 1/2 of 1 per cent. of alcohol or in such legislation as some of the states have enacted with their 2.75 per cent. of alcohol, but it will be in a law that will be prohibitive of intoxicating liquor for beverage purposes. It may require a little time to achieve, it may require some adjustments, but of its ultimate achievement there can be no doubt. However, whatever the difficulties of achievement in view of the requirement of section 2, it may be answered as this court answered in Wedding v. Meyler, supra:

"The conveniences and inconveniences of concurrent" power by the Congress and the states "are obvious and do not need to be stated. We have nothing to do with them when the lawmaking power has spoken."

I am, I think, therefore, justified in my dissent. I am alone in the grounds of it, but in the relief of the solitude of my position, I invoke the coincidence of my views with those entertained by the minority membership of the Judiciary Committee of the House of Representatives, and expressed in its report upon the Volstead Act.

Mr. Justice CLARKE dissents. See 253 U.S. 350, 40 Sup. Ct. 588, 64 L. Ed.—.

(253 U. S. 221)

HAWKE v. SMITH, Secretary of State of Ohio.

(Argued April 23, 1920. Decided June 1, 1920.) No. 582.

1. STATES *4—FEDERAL CONSTITUTION SUPREME LAW OF THE LAND.

The powers specifically conferred on the general government by the Constitution were surrendered by the states, and the Constitution and laws of the United States are the supreme law of the land.

2. CONSTITUTIONAL LAW *10—METHOD OF RATIFYING AMENDMENTS DETERMINABLE BY CONGRESS AND LIMITED TO METHODS SPECIFIED.

Under Const. art. 5, providing for the ratification of proposed amendments by the Legislatures of three-fourths of the states or by conventions in three-fourths thereof, as one or the other mode may be proposed by Congress, the power of determining the method of ratification is conferred upon Congress, and is limited to the two methods specified.

3. CONSTITUTIONAL LAW *10—COURTS OR LEGISLATIVE BODIES CANNOT ALTER METHODS OF RATIFYING AMENDMENTS.

It is not the function of courts or legislative bodies, national or state, to alter the method of ratifying proposed amendments to the federal Constitution, which the Constitution has fixed.

4. CONSTITUTIONAL LAW *10—"LEGISLATURES" EMPOWERED TO RATIFY AMENDMENTS DEFINED.

The word "legislatures," in Const. art. 5, relative to the ratification of the proposed amendments, has the same meaning as when the Constitution was adopted, and means the representative body which makes the laws of the people.

5. CONSTITUTIONAL LAW *10—RATIFICATION OF AMENDMENT NOT ACT OF "LEGISLATION."

Ratification by a state of a proposed amendment to the federal Constitution is not an act of "legislation," within the proper sense of the word, but the expression of the assent of the state to the proposed amendment.

6. CONSTITUTIONAL LAW *10—POWER TO RATIFY AMENDMENTS DERIVED FROM CONSTITUTION.

While the power of a state Legislature to legislate in the enactment of laws for the state is derived from the people of the state, the power to ratify a proposed amendment to the federal Constitution has its source in such Constitution.

7. CONSTITUTIONAL LAW *10—STATE CANNOT PROVIDE REFERENDUM ON RATIFICATION OF AMENDMENT TO FEDERAL CONSTITUTION.

A state has no authority to provide for the submission to a referendum under the state Constitution of the ratification of a proposed amendment to the federal Constitution, as is attempted by the amendment of 1918 to the Constitution of Ohio.

In Error to the Supreme Court of the State of Ohio.

Suit by George S. Hawke against Harvey C. Smith, as Secretary of State of Ohio. A Judgment sustaining a demurrer to the petition was affirmed by the Court of Appeals and Supreme Court of Ohio (126 N. E. 400), and plaintiff brings error. Reversed and remanded.

Mr. J. Frank Hanly, of Indianapolis, Ind., for plaintiff in error.

Mr. Lawrence Maxwell, of Cincinnati, Ohio, for defendant in error.

Mr. Justice DAY delivered the opinion of the Court.

Plaintiff in error (plaintiff below) filed a petition for an injunction in the court of common pleas of Franklin county, Ohio, seeking to enjoin the secretary of state of Ohio from spending the public money in preparing and printing forms of ballot for submission of a referendum to the electors of that state on the question of the ratification which the General Assembly had made of the proposed Eighteenth Amendment to the federal Constitution. A demurrer to the petition was sustained in the court of common pleas. Its judgment was affirmed by the Court of Appeals of Franklin County, which judgment was affirmed by the Supreme Court of Ohio, and the case was brought here.

A joint resolution proposing to the states this amendment to the Constitution of the United States was adopted on the 3d day of December, 1917. 40 Stat. 1050. The amendment prohibits the manufacture, sale or transportation of intoxicating liquors within, the importation thereof into, or the exportation thereof from, the United States and all territory subject to the jurisdiction thereof for beverage purposes. The several states were given concurrent power to enforce the amendment by appropriate legislation. The resolution provided that the amendment should be inoperative unless ratified as an amendment of the Constitution by the Legislatures of the several states, as provided in the Constitution, within seven years from the date of the submission thereof to the states. The Senate and House of Representatives of the state of Ohio adopted a resolution ratifying the proposed amendment by the General Assembly of the state of Ohio, and ordered that certified copies of the joint resolution of ratification be forwarded by the Governor to the Secretary of State at Washington and to the presiding officer of each House of Congress. This resolution was adopted on January 7, 1919; on January 27, 1919, the Governor of Ohio complied with the resolution. On January 29, 1919, the Secretary of State of the United States proclaimed the ratification of the amendment, naming 36 states as having ratified the same, among them the state of Ohio.

The question for our consideration is: Whether the provision of the Ohio Constitution, adopted at the general election, November, 1918, extending the referendum to the ratification by the General Assembly of proposed amendments to the federal Constitution is in conflict with article 5 of the Constitution of the United States. The amendment of 1918 provides:

"The people also reserve to themselves the legislative power of the referendum on the action of the General Assembly ratifying any proposed amendment to the Constitution of the United States."

Article 5 of the federal Constitution provides:

"The Congress, whenever two-thirds of both houses shall deem it necessary, shall propose amendments to this Constitution, or, on the application of the Legislatures of two-thirds of the several states, shall call a convention for proposing amendments, which, in either case, shall be valid to all intents and purposes, as part of this Constitution, when ratified by the Legislatures of three-fourths of the several states, or by conventions in three-fourths thereof, as the one or the other mode of ratification may be proposed by the Congress: Provided that no amendment which may be made prior to the year one thousand eight hundred and eight shall in any manner affect the first and fourth clauses in the ninth section of the first article; and that no state, without its consent, shall be deprived of its equal suffrage in the Senate."

[1] The Constitution of the United States was ordained by the people, and, when duly ratified, it became the Constitution of the people of the United States. McCulloch v. Maryland, 4 Wheat. 316, 402, 4 L. Ed. 579. The states surrendered to the general government the powers specifically conferred upon the nation, and the Constitution and the laws of the United States are the supreme law of the land.

The framers of the Constitution realized that it might in the progress of time and the development of new conditions require changes, and they intended to provide an orderly manner in which these could be accomplished; to that end they adopted the fifth article.

This article makes provision for the proposal of amendments either by two-thirds of both houses of Congress, or on application of the Legislatures of two-thirds of the states; thus securing deliberation and consideration before any change can be proposed. The proposed change can only become effective by the ratification of the Legislatures of three-fourths of the states, or by conventions in a like number of states. The method of ratification is left to the choice of Congress. Both methods of ratification, by Legislatures or conventions, call for action by deliberative assemblages representative of the people which it was assumed would voice the will of the people.

[2, 3] The fifth article is a grant of authority by the people to Congress. The determination of the method of ratification is the exercise of a national power specifically granted by the Constitution; that power is conferred upon Congress, and is limited to two methods, by action of the Legislatures of three-fourths of the states, or conventions in a like number of states. Dodge v. Woolsey, 18 How. 331, 348, 15 L. Ed. 401. The framers of the Constitution might have adopted a different method. Ratification might have been left to a vote of the people, or to some authority of government other than that selected. The language of the article is plain, and admits of no doubt in its interpretation. It is not the function of courts or legislative bodies, national or state, to alter the method which the Constitution has fixed.

All of the amendments to the Constitution have been submitted with a requirement for legislative ratification; by this method all of them have been adopted.

[4] The only question really for determination is: What did the framers of the Constitution mean in requiring ratification by "legislatures"? That was not a term of uncertain meaning when incorporated into the Constitution. What it meant when adopted it still means for the purpose of interpretation. A Legislature was then the representative body which made the laws of the people. The term is often used in the Constitution with this evident meaning. Article 1, section 2, prescribes the qualifications of electors of Congressmen as those "requisite for electors of the most numerous branch of the state Legislature." Article 1, section 3, provided that Senators shall be chosen in each state by the Legislature thereof, and this was the method of choosing senators until the adoption of the Seventeenth Amendment, which made provision for the election of Senators by vote of the people, the electors to have the qualifications requisite for electors of the most numerous branch of the state Legislature. That Congress and the states understood that this election by the people was entirely distinct from legislative action is shown by the provision of the amendment giving the Legislature of any state the power to authorize the executive to make temporary appointments until the people shall fill the vacancies by election. It was never suggested, so far as we are aware, that the purpose of making the office of Senator elective by the people could be accomplished by a referendum vote. The necessity of the amendment to accomplish the purpose of popular election is shown in the adoption of the amendment. In article 4 the United States is required to protect every state against domestic violence upon application of the Legislature, or of the executive when the Legislature cannot be convened. Article 6 requires the members of the several Legislatures to be bound by oath, or affirmation to support the Constitution of the United States. By article 1, section 8, Congress is given exclusive jurisdiction over all places purchased by the consent of the Legislature of the state in which the same shall be. Article 4, section 3, provides that no new states shall be carved out of old states without the consent of the Legislatures of the states concerned.

There can be no question that the framers of the Constitution clearly understood and carefully used the terms in which that instrument referred to the action of the Legislatures of the states. When they intended that direct action by the people should be had they were no less accurate in the use of apt phraseology to carry out such purpose. The members of the House of Representatives were required to be chosen by the people of the several states. Article 1, section 2. The Constitution of Ohio in its present form, although making provision for a referendum, vests the legislative power primarily in a General Assembly, consisting of a Senate and House of Representatives. Article 2, section 1, provides:

"The legislative power of the state shall be vested in a General Assembly consisting of a Senate and House of Representatives, but the people reserve to themselves the power to propose to the General Assembly laws and amendments to the Constitution, and to adopt or reject the same at the polls on a referendum vote as hereinafter provided."

[5] The argument to support the power of the state to require the approval by the people of the state of the ratification of amendments to the federal Constitution through the medium of a referendum rests upon the proposition that the federal Constitution requires ratification by the legislative action of the states through the medium provided at the time of the proposed approval of an amendment. This argument is fallacious in this—ratification by a state of a constitutional amendment is not an act of legislation within the proper sense of the word. It is but the expression of the assent of the state to a proposed amendment.

At an early day this court settled that the submission of a constitutional amendment did not require the action of the President. The question arose over the adoption of the Eleventh Amendment. Hollingsworth et al. v. Virginia, 3 Dall. 378, 1 L. Ed. 644. In that case it was contended that the amendment had not been proposed in the manner provided in the Constitution as an inspection of the original roll showed that it had never been submitted to the President for his approval in accordance with article 1, section 7, of the Constitution. The Attorney General answered that the case of amendments is a substantive act, unconnected with the ordinary business of legislation, and not within the policy or terms of the Constitution investing the President with a qualified negative on the acts and resolutions of Congress. In a footnote to this argument of the Attorney General, Justice Chase said:

"There can, surely, be no necessity to answer that argument. The negative of the President applies only to the ordinary cases of legislation. He has nothing to do with the proposition, or adoption, of amendments to the Constitution."

The court by a unanimous judgment held that the amendment was constitutionally adopted.

[6] It is true that the power to legislate in the enactment of the laws of a state is derived from the people of the state. But the power to ratify a proposed amendment to the federal Constitution has its source in the federal Constitution. The act of ratification by the state derives its authority from the federal Constitution to which the state and its people have alike assented.

This view of the amendment is confirmed in the history of its adoption found in 2 Watson on the Constitution, 1301 et seq. Any other view might lead to endless confusion in the manner of ratification of federal amendments. The choice of means of ratification was wisely withheld from conflicting action in the several states.

But it is said this view runs counter to the decision of this court in Davis v. Hildebrant, 241 U. S. 565, 36 S. Ct. 708, 60 L. Ed. 1172. But that case is inapposite. It dealt with article 1 section 4, of the Constitution, which provides that the times, places, and manners of holding elections for Senators and Representatives in each state shall be determined by the respective Legislatures thereof, but that Congress may at any time make or alter such regulations, except as to the place for choosing Senators. As shown in the opinion in that case, Congress had itself recognized the referendum as part of the legislative authority of the state for the purpose stated. It was held, affirming the Judgment of the Supreme Court of Ohio, that the referendum provision of the state Constitution, when applied to a law redistricting the state with a view to representation in Congress, was not unconstitutional. Article 1, section 4, plainly gives authority to the state to legislate within the limitations therein named. Such legislative action is entirely different from the requirement of the Constitution as to the expression of assent or dissent to a proposed amendment to the Constitution. In such expression no legislative action is authorized or required.

[7] It follows that the court erred in holding that the state had authority to require the submission of the ratification to a referendum under the state Constitution, and its judgment is reversed and the cause remanded for further proceedings not inconsistent with this opinion.

Reversed.

Source: *Supreme Court Reporter* 40 (1921): 486–499.

Notes

1. The following is the contention of the government which we give to accurately represent it: "It is true that the word 'concurrent' has various meanings, according to the connection in which it is used. It may undoubtedly be used to indicate that something is to be accomplished by two or more persons acting together. It is equally true that it means, in other connections, a right which two or more persons, acting separately and apart from each other, may exercise at the same time. It would be idle, however, to go into all the meanings which may attach to this word. In certain connections, it has a well-fixed and established meaning, which is controlled in this case."

   And again: "It is to be noted that section 2 does not say that legislation shall be concurrent, but that concurrent power to legislate shall exist. The concurrent power of the states and Congress to legislate is nothing new. And its meaning has been too long settled, historically and judicially, to now admit of question. The term has acquired a fixed meaning through its frequent use by this court and eminent statesmen and writers in referring to the concurrent power of Congress and the states to legislate."

   And after citing cases, the government says: "It will thus be seen that in legal nomenclature the concurrent power of the states and of Congress is clearly and unmistakably defined. It simply means the right of each to act with respect to a particular subject-matter separately and independently."
2. Definitions of the dictionaries are as follows: The Century: "Concurrent: . . ."
3. The government feels the inconsistency of its concessions and recessions. It asserts at one instant that the legislation of the states may be enforced in their courts, but in the next instant asserts that the conviction or acquittal of an offender there will not bar his prosecution in the federal courts for the same act as a violation of the federal law. From this situation the government hopes that there will be rescue by giving the Eighteenth Amendment "such meaning that a prosecution in the courts of one government may be held to bar a prosecution for the same offence in the courts of the other." The government considers, however, the question is not now presented.

Appendix B: Internet Resources on Temperance and Alcohol Studies

Dorothy A. Lander

This annotated bibliography is organized alphabetically and offers commentary on Internet resources in two areas of alcohol studies: (1) historical resources related to beverage alcohol and the temperance movement, including leaders and organizations; and (2) resources for contemporary alcohol research, treatment centers, public education, and policy development.

Alcohol and Temperance History Group (ATHG) (*The Social History of Alcohol Review*)
http://www.athg.org

The no-frills navigational links that appear on the entry screen for this website of the U.S.-based Alcohol and Temperance History Group deceptively suggest that the site will focus entirely on the group's quarterly journal, *The Social History of Alcohol Review.* The links to a searchable archive of the group's listserv beginning in 1995 are a rich resource in which seasoned social historians and researchers—including many contributors to this encyclopedia—converse with new scholars. Click on the Links icon to enter many of the websites listed in this appendix as well as full-text papers of alcohol historians and researchers, including Harry Gene Levine and Robin Room, and history websites related to the brewing industry and Alcoholics Anonymous. The collation of on-line bibliographies gleaned from the journal itself between 1995 and 2000 is a useful resource that can be manipulated through the "Edit" and "Find in this Page" command of the web browser. Recent issues of the journal include some full-text articles on-line to entice browsers to join the Alcohol and Temperance History Group and to subscribe to the journal and the listserv.

Anti-Saloon League, 1893–1933 (Westerville Public Library, Ohio)
http://www.wpl.lib.oh.us/AntiSaloon

The Westerville Public Library has produced an impressive digital archive of the Anti-Saloon League on this comprehensive and multimedia website. The library's work involves preserving the story of the Anti-Saloon League and organizing in digitized form documents, images, and publications—many of them from the American Issue Publishing Company. The main page index provides links to the league's history, bios of its leaders (Howard Hyde Russell, Purley Baker, Ernest Hurst Cherrington, Wayne Wheeler, and Francis McBride), printed material, related organizations (such as the Scientific Temperance Federation and the Lincoln-Lee Legion), and the Anti-Saloon League Museum, which is housed in the Westerville Public Library. The Classroom Activities section is pitched to young adults and provides educators with critical materials designed for students and information about alcohol and drug abuse in today's society. There are also links to the editorial cartoons of Frank Beard in *The Ram's Horn.*

Bacchus & Gamma Peer Education Network
http://www.bacchusgamma.org

On-line educational resources on issues related to alcohol, tobacco, and other drugs are offered on this website. There are also links to educational programs at colleges and universities throughout the United States. A link to the National Social Norms Resource Center explains the social-norms marketing approach to substance abuse and provides examples of programs and materials. The audience for these resources and materials is primarily student services departments in colleges and universities.

Center of Alcohol Studies, Rutgers University
http://www.rci.rutgers.edu/~cas2

The website for the Center of Alcohol Studies (CAS)—physically located at Rutgers University in Piscataway, New Jersey—offers a thorough introduction to CAS history and to its mandate as an interdisciplinary research center devoted to alcohol use and alcohol-related problems and treatment. Although there are several links to the clinical and educational programs at the center, the immediate value of the website for the Internet researcher are the links: Online Facts, CAS Databases, Online Resources, and Publication Catalogue, which includes the table of contents and abstracts of the CAS journal, *Journal of Studies on Alcohol.*

The Online Fact Sheets offer brief texts that focus on substance abuse as it relates to children, teens, gays and lesbians, women, and the elderly; domestic violence; and educational resources on the web. Click on Historical Resources under the Online Resources to access a full array of web resources, including many cited in this bibliography, and also American Brewery History, Images from the History of Medicine, and the Prohibition Party. The Internet researcher can go directly to CAS databases: The primary database contains citations to all materials collected since 1987. The index to the library's collection of more than 500 historical materials from the temperance and prohibition movements is included on the history database at http://www.scc.rutgers.edu/alcohol_history. The entry point to the history database is an mage credited to Charles Morris, *Broken Fetters: The Light of Ages on Intoxication* (Temperance Publishing Co., 1888). The search function is quite helpful. Keying in "John Newton Stearns," for example, returns five hits for titles between 1876 and 1893 related to this editor of the National Temperance Society and Publication House in New York. Keying in "autobiography" returns seven matching records, including titles by John Gough, Lyman Beecher, Luther Benson, and Thomas P. Hunt. A wealth of other information is also available. Author Portraits, under List of Illustrations, contains thirty-two matching records for texts in which portraits appear, including those of Henry William Blair, Timothy Shay Arthur, and Mrs. T. Narcisse Doutney. There are also twenty-five matching records for cartoons and seven matching records for charts. The "Select from List" organizer for the more than 60,000 citations in the primary database (alcohol studies) includes three fields—(1) physiological aspects, (2) social aspects, and (3) drug terms—and over ten populations. A search for "women and alcohol" calls up 2,773 matching records.

Drug Policy Alliance (formerly Lindesmith Center)
http://www.drugpolicy.org

The busy screen that is the entry to the website for the Drug Policy Alliance signals the reach of the alliance's work in broadening the public debate on drug policy in the United States. The name change in 2000 from the Lindesmith Center accords with the alliance's focus on policy reform and its guiding principle of harm reduction as opposed to drug prohibition. The Internet researcher will be drawn to the large on-line library of full-text drug- and drug policy–related documents, which can be searched in the Subject Index and narrowed to specific countries.

Not one of the subject areas refers to alcohol specifically. In general, this website, along with its links to other drug-related sites on the Internet, provides a useful resource for the alcohol researcher who is exploring alcohol issues in the wider context of drug use and drug policy. The link to Prohibitionists' Web Sites leads to organizations with ideologies that stand in contrast to harm reduction, such as the Norwegian League Against Intoxicants (http://www.fmr.no/eng/index.html). The catalog for the hard-copy library at the alliance's offices in New York can also be searched on-line. Keying in "alcohol" for the search engine for the entire website leads to a list of full-text documents on-line. This keyword search also led to a reprint of the 1977 "Selected Bibliography on Alcohol Control" compiled by Harry Levine and David Smith. The full-text research briefs and position papers provided on-line include information on cocaine, methadone maintenance, marijuana, and heroin, but not explicitly alcohol.

Goldsmiths' Library of Economic Literature
http://www.ull.ac.uk/goldsmiths

The on-line catalog of Goldsmiths' Library of Economic Literature, part of the University of London Library's historical collections, provides basic information and physical locations for Goldsmiths' holdings on temperance and moral reform as well as for the supplementary collection (500 volumes) on the temperance movement in the nineteenth century. The supplementary collection, acquired in 1930 from the collection of James Turner, a temperance advocate from Manchester, is oriented toward the temperance movement in the UK. There are many international records; the search database returned 811 hits for the keyword "temperance." Many documents in the special collection are marked "reference only"; the Internet offers the advantage of doing preliminary library research before of an in-person visit.

Institute of Alcohol Studies
http://www.ias.org.uk

The website of the Institute of Alcohol Studies, which is sponsored by the United Kingdom Temperance Alliance, successor to the United Kingdom Alliance fulfills its public education aims related to alcohol prevention by providing the table of contents for recent issues of its quarterly magazine, *Alcohol Alert,* along with a sample issue; downloadable (Adobe PDF) fact sheets on subjects such as alcohol and crime, youth, women, and the workplace; and an on-line catalog of publications on policy and educational activities (such as posters) that can be ordered on the Internet. The links to other alcohol or related websites are organized by area or country, from Australia to the United States, and offer a useful departure point for researchers conducting an international inquiry into contemporary organizations related to alcohol and drug education, treatment centers, and policy development. The Drinks Industry link also provides a useful consolidation of beverage- and brand-specific resources, including websites for the Beer Institute, Heineken (a pictorial history), and Southern Comfort (landmark events from its beginning in 1874 through Prohibition until today).

Lindesmith Center. *See* **Drug Policy Alliance**.

Livesey Collection, British National Temperance League Library
http://www.uclan.ac.uk/library/libspar1.htm

This website offers a searchable introduction to the special collection of temperance materials named after Joseph Livesey—founding member of the Preston Temperance Society (UK) in 1832—via the library catalog of the University of Central Lancashire Library in Preston, where this collection is housed. The special collection consists primarily of texts and journals related to the temperance movement in the UK and the British Empire from the early 1800s through the 1930s. A profile and pictures of Joseph Livesey, and mention of his writings as well as the work of other temperance workers, provide starting points for searching the library catalog on-line. The site provides, for example, the copy location of Agnes Weston's 1915 autobiography *My Life among the Bluejackets,* a record of her heavy involvement with the Royal Naval Temperance Society. Available in this collection are national temperance titles, including *The British Temperance Advocate* and *Alliance News,* as well as many obscure journals, and some 3,000 lantern slides targeting the temperance message to both children and adults. The few stunning visuals of journal illustrations and lantern slides provided on the website will lure readers to visit this collection in person.

PREVNET Network (European Professional Network Connecting Telematic Methods with the Prevention of Alcohol and Drug Dependencies)
http://www.prevnet.net

The website of PREVNET declares itself "a virtual portal for networking, training, and exploring prevention potentials through the use of telematics," and it is as this portal that this site will be most useful to the researcher or educator. The site offers an open public forum along with a closed forum restricted to members to develop on-line prevention methods. The Resources icon leads to many other websites related to substance abuse.

Temperance & Prohibition (Ohio State University, Department of History)
http://prohibition.history.ohio-state.edu

The table of contents on this site provides easy-to-navigate links to nineteenth- and early-twentieth-century texts, profiles of leaders, and still images—all related to temperance and Prohibition in the United States. The Roaring Twenties ragtime music that accompanies the welcome to this website prepares the reader for the dominance of texts related to National Prohibition during the 1920s. The close relationship between the brewing industry and Prohibition is documented in the texts and the twenty-one cartoons of the Prohibition Party, many by Russell Henderson.

This site provides a useful record of the anti-Prohibition movement, including testimony before the 1926 Senate Committee from Fiorella LaGuardia, a New York City politician, and Russell Lee Post, a Yale University student. It includes an address by Percy Andreae, closely associated with the brewing industry and a successful spokesman against Prohibition, in which he castigates the prohibitionists for their religious intolerance. Other historical profiles and pictures document the Women's Temperance Crusade of 1873–1874; the work of Frances Willard, early president of the Woman's Christian Temperance Union (with links to full texts of her speeches and her last interview); and details about old-time saloons in Illinois, Pennsylvania, and New York. The full text of "Woman and Organization," the 1891 address of Frances Willard, president of the Woman's National Conference at its first triennial meeting, touches on many issues related to the women's temperance movement, as revealed in the subheadings: societal purity, co-operative happiness, dress reform, scientific motherhood, and women and religion. This address makes reference to contemporary women activists from many disciplines, including religion, law, medicine, business, literature, journal publishing, and academia (such as Elizabeth Cady Stanton, the Reverend Juniata Breckenridge, Dr. Emily Blackwell, Mary F. Seymour, and Mary Clement Leavitt).

Temperance, Prohibition, and Alcoholism (Seagram Museum Library Collections, University of Waterloo)
http://www.lib.uwaterloo.ca/seagrams/temper/index.html

This website is a tantalizing exhibit that hints at the possibilities for historical research related to the social history of alcohol contained in the Seagram Museum Archives. These archives were donated to the University of Waterloo (Canada) and include records of the Seagram Company, information about the Seagram family, and documents related to the history of beverage alcohol and distillation. The only image on the website directly related to the Seagram Company is entitled "Bottling of Seagram's V.O. in Waterloo Plant, Ontario, 1928." The site provides a brief outline supported by still images in several areas. "Distillation and Distillation of Various Beverages," for example, includes photos from books on distillation, including life-size views of pages from the account and diagrams of Conrad Gesner (1516–1565), a Swiss physician, on distilling medicines and wines. "Cookery, Gastronomy, and Entertaining" provides information on books such as *A Treatise on the Art of Brewing* (second edition, 1821). "Temperance, Prohibition, and Alcoholism" refers to works that illustrate societal attitudes toward drinking: For example, a still image from the Reverend William Burgess's 1887 book *Land, Labor and Liquors* lays out in graph form the comparative expenditures in Canada on intoxicating liquors and on religious and educational work. "Trade and Industry, and Regulation" refers to titles in the collection related to the various trades involved in beverage alcohol production and offers photos related to the regulation of the alcohol industry dating from the seventeenth century. "Wine" is the link to a page that offers five images of titles related to viticulture and viniculture.

TimeLinks: Prohibition and Temperance (Historical website about Manitoba from 1910 to 1930), http://timelinks.merlin.mb.ca/referenc/db0012.htm

TimeLinks is a creation of River East School Division and the University of Manitoba and is a "historical website about Manitoba in the decade from 1910 to 1920." The one-page overview of "Prohibition and Temperance" found within the Politics reference section of this website provides numerous links to such topics as social gospel, the women's suffrage movement, nativism, maternal feminism, and allied institutions such as the Woman's Christian Temperance Union, the Trades and Labour Council, and the Political Equality League. The search engine for the larger website is a valuable tool providing further links to profiles and/or images of leaders associated with the politics of prohibition in Manitoba, including Nellie McClung, Dr. Amelia Yeomans, James S. Woodworth, Sir Rodmond Roblin, and the Reverend Charles W. Gordon. The Image Archive is a source of still images, including a 1910 Winnipeg liquor store, a WCTU float in a 1910 Winnipeg parade, and an 1890–1891 WCTU annual report signed by Nellie McClung. This site has plans that promises to activate more links and provide a selected bibliography by subject.

Virtual Clearinghouse on Alcohol, Tobacco, and Other Drugs
http://www.atod.org/english/home.asp

The welcome page of this website declares its emphasis as providing "access to the 'fugitive' or 'grey' literature . . . that are generally not included in existing indexing and abstracting services." Researchers can test this claim via the search engine for the on-line documents database, which allows a search across all fields and by country as well as the usual subject, author, title, and publisher. A search for "harm reduction," for example, returns four hits for full-text documents from the Alcohol and Other Drugs Council of Australia, the Netherlands Institute for Alcohol and Drugs, the Centre for Addiction and Mental Health (Canada), and the Canadian government. The Virtual Clearinghouse was initiated by the Canadian Centre on Substance Abuse and developed in partnership with representatives from mainly English-speaking countries. The format for downloading these documents varies. This website offers useful links categorized by organization, topic, and geographic area and also includes a form for web browsers to add their own links to related sites.

Woman's Christian Temperance Union
http://www.wctu.org

The white ribbon symbol of the Woman's Christian Temperance Union (WCTU) in the United States serves as a clickable icon for accessing the various sections of this website: News, Issues, History, Citizenship, Publications, Children and Youth, and Friends of WCTU. This attractive and easy-to-navigate site affords a comparison of the activities of today's WCTU with its nineteenth-century foremothers. For example, the history link provides a profile of the 1873 Women's Temperance Crusade and an illustration of the "Lady Crusaders" praying on snowy pavements outside the saloon doors; the News link lists the national winners in 2000 and 2001 of children's coloring and essay contests and itemizes the resolutions from the 128th Annual National WCTU Convention of 2001, which deal with issues such as binge drinking, fetal alcohol syndrome (FAS), and the Pray for the Children Weekend via the Internet (www.prayforthechildren.com). The pictures of the national presidents of the WCTU from 1874 (Annie Turner Wittenmeyer) to the present day (Sarah Ward) offer a visual text for tracing the history of the organization.

Appendix C: English Song Themes, 1600–1900

Table 1 Bacchanalian Song Themes

	17th–18th Centuries			*19th Century*			*Number of Songs*			
	Ale	*Wine*	*Distilled*	*Ale*	*Wine*	*Distilled*	*Ale*	*Wine*	*Distilled*	*Total*
Ale v. Wine	6	7	0	9	0	0	15	7	0	22
Hedonism	22	35	4	10	52	2	32	87	6	125
Health	18	17	0	7	12	2	25	29	2	56
Romance	7	23	3	1	42	1	8	65	4	77
Fellowship	9	8	0	5	19	1	14	27	1	42
Courage	8	4	0	2	2	1	10	6	1	27
Anti-aggression	3	2	0	1	8	0	4	10	0	14
Anti-world	1	17	1	0	21	0	1	38	1	40
Imagination	6	17	1	0	10	0	6	27	1	34
Drink > women	0	14	0	0	14	0	0	28	0	28
Woman drinks	6	4	0	1	1	0	7	5	0	12
Bad ends	7	2	0	0	2	0	7	4	0	11
Comic	4	11	0	2	8	0	6	19	0	25
Number of songs	54	45	9	84	2	36	138	66	45	249*

Note: Many songs contain more than one theme.

Table 2 Non-Bacchanalian Song Themes

	17th–18th Centuries			*19th Century*			*Number of Songs*			
	Ale	*Wine*	*Distilled*	*Ale*	*Wine*	*Distilled*	*Ale*	*Wine*	*Distilled*	*Total*
Ale v. Wine	5	0	0	5	0	0	10	0	0	10
Hedonism	10	7	1	18	4	6	28	11	7	56
Health	5	0	0	11	1	2	16	1	2	19
Fellowship	1	1	0	6	1	2	7	2	2	11
Romance	7	8	2	9	3	3	16	11	5	32
Woman critic	3	5	0	3	3	1	6	5	1	12
Woman drinkers	11	8	5	9	9	15	20	13	20	53
Bad ends	28	28	3	29	29	21	57	42	24	123
Comic	6	3	2	15	15	9	21	10	11	42
Work	3	1	0	15	15	2	18	1	2	21
Tax/prices	2	0	0	10	10	0	12	0	0	12
Number of songs	35	78	5	17	90	3	52	168	8	228

Note: Many songs contain more than one theme. The categories Imagination, Courage, Anti-World, Anti-Aggression, and Wine > Women (wine over women) have three or fewer songs each.

Table 3 Kinds of "Bad Ends" in Non-Bacchanalian Songs

	17th–18th Centuries	*19th Century*	*All*
Poverty	28	18	46
Fight	14	18	32
Health	10	15	25
Tavern con	7	9	16
Seduction	9	0	9
Abuse	2	4	6
Foolish	1	5	6
Jail	3	2	5
Whoring	4	0	4
Number of songs	56	64	120

Note: Some songs contain more than one "bad end."

Song Theme List

In order to concentrate on songs that depict the effects of drinking, many toasts and wassail songs have been excluded from this list. The latter are well reviewed by Marchant (1888) [need to provide full citation!], as are other songs about drinking on religious holidays.

1. Collections with abbreviations used in lists of song themes.

Amm *American Musical Miscellany,* edited by H. Wiley Hitchcock. Northampton, MA: Andrew Wright, 1798; Reprint, New York: Da Capo Press, 1972.

Bdw *Old English Songs,* edited by John Broadwood. London: Elliot Stock, 1843.

Bell *Ancient Poems, Ballads & Songs of the Peasantry of England,* edited by Robert Bell and John W. Parker. London, 1857.

Bgf *The Bagford Ballads,* edited by Joseph W. Ebsworth, 2 vols. New York: AMS Press, 1968.

Bkm *The Musical Entertainer,* edited by George Bickham, 2 vols. London, 1740.

Cent *A Century of Ballads Illustrating the Life, Manners and Habits of the English Nation during the Seventeenth Century,* edited by John Ashton. London: Elliot Stock, 1887, Reprint, Detroit: Singing Tree Press, 1968.

Chlt *Victorian Folk Songs,* edited by Charles Chilton. London: Essex Music, 1965.

Chp *Popular Music of the Olden Times,* edited by William Chappell. London, 1858–1859; Reprint, New York: Dover, 1965.

Drl *Merry Drollery Compleat: Jovial Poems, Merry Songs,* edited by Joseph W. Ebsworth. 1691; Reprint, Boston, England: R. Roberts, 1875.

Ebb *Everyman's Book of British Ballads,* edited by Roy Palmer. London: Dent, 1980.

Ebcs *Everyman's Book of English Country Songs,* edited by Roy Palmer. London: Dent, 1979.

Fgr *Folklore of Gloustershire and Rutland,* edited by Roy Palmer. Tiverton: Westcountry, 1985.

Fmr *Merry Songs & Ballads Prior to the Year l800,* edited by John S. Farmer, 5 vols. 1897; Reprint, New York: Cooper Square, 1964.

Fss *Folk Song in School,* edited by Roy Palmer. Cambridge: Cambridge University Press, 1987.

Full *Full and Bye.*

Glo *Folklore of Gloucestershire,* edited by Roy Palmer. Tiverton: Westcountry, 1994.

Harl *Ballads & Songs of Lancashire,* edited by J. Harland. London: J. Routledge, 1875.

Hol *Later English Broadside Ballads,* edited by John Holloway and Joan Black, 2 vols. London: Routledge and Keegan Paul, 1975, 1979.

Hrfd *Folklore of Hereford & Worcester,* edited by Roy Palmer. Almeley: Logaston, 1992.

Hugl *Shanties from the Seven Seas,* edited by Stan Hugill. London: Routledge & Kegan Paul, 1984.

Ken *Folksongs of Britain & Ireland,* edited by Peter Kennedy. London: Cassell, 1975.

Kids *English Peasant Songs,* edited by Frank Kidson. London: Ascherberg, Hopwood & Crews, 1929.

Lld *Folksongs in England,* edited by A. L. Lloyd. London: Lawrence & Wishart, 1967.

Mid *Songs of the Midlands,* edited by Roy Palmer. Wallefield: E. P. Publishing, 1972.

Msb *Modern Street Ballads,* edited by John Ashton. London: Chatto & Windus, 1888.

Oep *Old English Popular Music,* edited by William Chapell. London: Cahppell & Co., 1893; Reprint, New York: J. Brussel, 1961.

Oxf *Oxford Book of Sea Songs,* edited by Roy Palmer. Oxford: Oxford University Press, 1986.

Pain *The Painful Plough: A Portrait of the Agricultural Laborer in the Nineteenth Century,* edited by Roy Palmer. Cambridge: Cambridge University Press, 1973.

Ped *A Pedlar's Pack of Ballads and Songs,* edited by W. H. Logan. Edinburgh: W. Paterson, 1869.

Pils *Wit and Mirth: Or Pills to Purge Melancholy,* edited by Thomas D'Urfey, 6 vols. London: W. Pearson, 1719–1720.

Prs *Praise of Ale,* edited by W. T. Marchant. [ADD FULL INFO]

Purs *Marrowbones: English Folk Songs from the Hammond and Gardiner MS's,* edited by Frank Purslow. English Folk Dance & Song, 1965.

Rigs *Rigs of the Fair: Popular Sports and Pastimes in the Nineteenth Century,* edited by Roy Palmer and Jon Raven. Cambridge: Cambridge University Press, 1976.

Rits *Ancient Songs from the Time of K. Henry 3rd to the Revolution,* edited by Joseph Ritson. 1790.

Room *Room for Company: Folk Songs and Ballads,* edited by Roy Palmer. Cambridge: Cambridge University Press, 1974.

Rxb *The Roxburghe Ballads,* edited by J. Woodfall Ebsworth, 9 vols. London: Ballad Society, 1869–1899.

Shrp *Cecil Sharpe's Collection of English Folk Songs,* edited by Maud Karpels. Oxford: Oxford University Press, 1974.

Snd *The Sound of History: Songs and Social Comment,* edited by Roy Palmer. Oxford: Oxford University Press, 1988.

Str *Strike the Bell: Transport by Road, Canal, Rail and Sea in the Nineteenth Century,* edited by Roy Palmer. Cambridge: Cambridge University Press, 1978.

Tch *A Touch on the Times: Songs of Social Change (1770–1914),* edited by Roy Palmer. Cambridge: Cambridge University Press, 1974.

Thm *Folk-Songs of the Upper Thames,* edited by Alfred Williams. London: Duckworth, 1923.

Tka *A Tankard of Ale,* edited by Theodore Maynard. London: E. MacDonald, 1919.

Toa *A Tale of Ale* (recorded songs and notes). Free Reed, 1977.

Tst *A Taste of Ale,* edited by Roy Palmer. Gloucestershire: Green Branch Press, 2000.

U *The Universal Songster; or, Museum of Mirth,* edited by George Cruikshank, 3 vols. London: G. Routledge, 1878.

Vsb *Victorian Street Ballads,* edited by William Henderson. London: Country Life, 1937.

Wmb *A Williamsburg Song Book: Songs Convivial, Sporting, Amorous, & etc., from Eighteenth Century Collections,* edited by John Edmunds. New York: Holt, Rinehart & Winston, 1964.

Wts *The Musical Miscellany,* edited by James Watts, 6 vols. London, 1740.

2. Theme abbreviations:

A = ale, a-agr = antiaggression, a-miser = anti-miserly, a-wld = antiworldly, A>all other drinks = ale is better than all other drinks, A>W = ale over wine, bdy = brandy, b-end = bad end, com = comic, con = tavern con (swindle), crg = courage, eloq = eloquent, fel = fellowship, fgt = fight, fool = foolish, hed = hedonism, hth = health, img = imagination, mfg = manufacturer, mod = moderation, patr = patriotism, pov = poverty, rom = romance, sed = seduction, t-flies = time flies, w-crit = woman criticizes, w-drk = woman drinks, whor = whoring, wsk = whiskey, W>wom = wine over women.

Bacchanalian Songs (nineteenth-century titles with asterisk)

*Ale, ale, all ale [Prs-416] **A>W: hth**
Achreon in heaven (to) [U1-213] **W: rom, img**
Anacreontiques #2 [Tka-76, Prs-265] **W: hed, com**
Bacchanalian's wish (the) [Bkm 2-3, Wmb-46] **W: com**
Bacchus [Tka-79] **W: hth, img, a-wld**
Bacchus & time [U1-372] **W: fel, t-flies**
*Bacchus is a power divine [U2-271] **W: hed, a-wld**
*Bacchus swings out & I swing within [U2-322] **W: b-end (hth), com**
*Bacchus's speech in praise of wine [U?] **W: rom, img**
*Back & side go bare [Tka 24, Prs-214, Esb-72] **A: hed, hth, crg, w-drk**
Ballad on ale (a) [Prs-280, Tst-20] **A>all: hed, rom, crg, img**
Beauty & the bowl [U3-302] **Wom>W: img**
*Beauty, wit & wine [U2-319] **W: hed, rom, wit**
*Beer [Tka-124] **A: com**
Beer-drinking Briton (the) [Tka-116] **A>W: crg, com**
Birmingham beer [Tst-42] **A: hed, hth, fel**
*Bitter beer [Prs-277] **A>all other dks**
*Blann's beer [Pain-48] **A: hed, hth**
*Blessing on brandy & beer (a) [U3-272] **A: crg**
*Blooming sex, the charming sex (the) [U2-153] **W: rom**
*Bring the flask, the music bring [U2-133] **W: hed, fel, a-wld, t-flies**
*Bumper, a friend & the girl of your heart [U1-154] **W: hed, fel, rom**
*Bumper of good English ale (a) [U1-400] **A>W & others**
*Bumper of good liquor (a) [Prs-258] **W: a-agr**
*By drinking, drive dull care away [U1-413] **W: hed, a-agr**
Canary crowned [Drl-121] **W>A: rom**
*Careless gallant [Rxb 3-484] **W: hed, w-drk, crg, t-flies**
*Cast away care [Tka-62, Prs-257] **W: hed, hth, wit, a-wld**
Catch (a) [Prs-267] **W>wom**
Choice (the) [Amm-92] **Wom>W: rom**
*Churn supper (the) [Tka-149] **A: hth, hed, rom**
Claret-drinker's song (The) [Rxb 4-645] **W: hed, fel, a-wld**
*Come bring us wine in plenty [Pils 5-15] **W: hed, fel**
*Come fill me a glass, fill it high [U2-414] **W: hed, a-wld, a-int**
Come jolly Bacchus [Chp-658] **W: hed, a-wld**
Come, landlord, fill the flowing bowl [Chp-670, Tka-27] **A: hed, hth, t-flies**
Come let us drink about (= Jolly fellow) [Chp-670] **W: hth, hed, t-flies**
*Come my boys, fill a bumper of sherry [U2-373] **W: hed, rom**
Come, neighbors, now we've made our hay [Frm 3-246] **A: rom**
Come, ye friends of a social life [Thm-52] **A: hed, t-flies**
Contentment [Drl-124] **W: hed, a-wld**
*Crown me Bacchus, mighty god [U2-18] **W>wom: hed**
*Crown me with the branching vine [U3-270] **W: rom, a-agr**
Cupid among the bachelors [U2-100] **W>wom**
Delights of the bottle #1 [Rxb 4-42, Chp-193 & 598] **W: rom, img, w-drk**
Delights of the bottle #2 [Ward—ref. list] **W: hth, rom, wit, com,**
Delights of the bottle: a parody [Rxb 4-434] **W: rom, a-agr**
Down among the dead men [Tka-91] **W: rom**
*Dramatic morality [U2-82] **Wom>wine**
Drink, drink, drink the red wine [Tka-96] **W: hth, hed, patriot**
Drink today [Tka-62] **W: hed, img, t-flies**
*Drinking & kissing are pleasures divine [U1-254] **W: rom**
Drinking ode (a) [Rits-4] **W: hed, img**
Drinking song [Chl-68] **Wsk: hed, fel, img, a-wld, t-flies**
Drunk and sober [Frm 5-268] **W: rom**
Drunkard (the) [Drl-47] **W>A: hed (t-flies)**
Elysum on earth [U1-154] **W: hed, rom**
*England forever & a glass of good wine [U2-7] **W: hed, hth, fel**
Ex-Ale-tation of ale [Tst-16] **A: hed, hth, img**
Faith, I'll awa' to the bridal [U1-155] **W>wom: com**
Fathom the bowl = Punch ladle [Thm-88] **Bdy-Rum, W>wom**

*Feast of reason & the flow of soul [U1-406] **W: fel (modn)**
Fill me a bowl—a brimmer full [U1-324] **W>wom**
*Fill the bowl again [U1-300] **W: hed, rom**
Fill the goblet again [Tka-71] **W>wom: hed**
*Fill the sparkling bumper high [U2-121] **W: hed, rom**
For to be merry is to be wise [U1-344] **W: hed, img, t-flies**
*Friar's song (the) [Tka-162, Prs-161] **W: hed, com, rom**
From good liquor ne'er shrink [Wts-34] **W: hed, t-flies**
*Generous toper (the) [U3-264] **W: img**
*Gin palaces [Chlt-74] **Bdy: hth, hed, fel, fgt (jolly)**
*Give me but a friend & a glass [U-2-89] **W>wom**
*Give me but my bottle [U2-326] **W: rom, a-wld**
Glass of old English ale [Tka-118, Prs-264] **A>W: hed, fel**
*Glass of rich brown ale (a) [Prs-263] **A>W**
Good ale [Ebs-661, Tst-30] **A: hed, b-end (pov)**
Good ale for my money [Prs-229, Rxb 1-412] **A>W: hed, hth, crg, w-drk**
Good canary (on) [Drl-178] **W>A etc: img**
Good fellow (the) #2 [Pils 5-16] **W: hed, a-wld**
Good fellowes' best beloved [Rxb 3-248, Prs-220, Ash 2-188] **A: fel, a-agr, w-drk**
*Grand summum-bonum's a bumper of wine [U1-417] **W: hed, rom, a-wld, a-agr**
*Guzzling duck, the drunkard [U2-309] **W+A: hed, hth, com**
*Hail, Bacchus, hail—farewell to love [U1-307] **W>wom**
*Happy fellow (the) [Prs-298n] **W: fel, a-wld**
*He drinks & drinks forever [U3-64] **W: com**
*He loves best who drinks like me [U2-201] **W: rom**
Health to all good fellowes (a) [Rxb 1-447] **A: hed, hth, t-flies**
Healths (the) [Fmr 5-77] **W: rom**
*Here's a health to all honest men [Prs-180] **A: hed, a-agr**
*Here's richer juice from barley pressed [U3-267] **A>W: hth**
Here's to thee, my boy [Wts 4-58, Wmb-59] **W>wom: fel, hed**
*Hey, ho, hunt about [Cent-60] **W: fel, rom, w-drk**
High & mighty commendation of the vertue of a pot of good ale [Prs-234] **A>beer: hed, hth, img, crg, w-drk, b-end (fgt)**
*How happy is the man [Thm-43] **A: fel, hed**
*Humors of London [U1-218] **W: hth, com**
*Hunting, love & wine [U2-274] **W: hed, rom, a-wld**
I am the jolly prince of drinkers [Tka-47] **W: com**
*I'll give the land we live in [U2-119] **W: rom, (patriotic)**
In praise of ale [Tka-104, Prs-240] **A: hth, hed, crg, wit**
In praise of claret [Wsb-23] **W: com**
In praise of sack #1 [Drl-246] **W>A: hed (excess)**
In praise of the bottle [Tka-173] **W: a-wld (a-intel)**
*In vino veritas [U1-349n] **W>wom: a-wld**
Joke and push around the pitcher [Thm-97] **A: hed, hth, rom**
Jolly bacchanal (the) #1 [Tka-77] **W>wom: hed, w-crit, com**
Jolly bacchanal (the) #2 [Full-80] **W: hed, a-wld**
Jolly mortals [Wts 6-182, Wmb-94] **W>wom: crg**
Jolly old Bacchus, goodnight [U1-397] **W: b-end (hth)**
Jolly toper [Chp-680. Tka-169] **W>wom: a-agr, com**
Jolly town rakes (the) [Pils 5-123] **W: hed, b-end (fgt, sed)**
Jovial drinker (the) [Tka-65] **W: hth, crg, wit**
*Jovial souls [U1-427] **W: hed, rom**
*Joy & health be the lot of each worthy odd fellow [U1-363] **W: hed, fel, hth, a-agr**
*Joys of dear women & wine (the) [U3-252] **W: rom, hed**
*Joys of drinking (the) #1 [U2-242] **W>wom**
*Joys of drinking (the) #2 [U2-382] **A: rom, a-agr**
Joys of the bottle (the) [U1-376] **W: hth, fel**
*Joys that sparkle in the bowl (the) [U2-148] **W>wom**
Jug of this (a) [Oxf-73] (= Ye Mariners all) **Rum: hed, rom**
*Junction of Bacchus & Venus (the) [U1-134] **W: hed, rom**
King of good fellows (the) [Rxb 6-502] **W: hed (excess), hth**
*Let's be jovial, fill our glasses [U2-378] **W: hed, a-wld**
*Let's chorus the praises of gen'rous wine [U2-127] **W: hed, fel**
Let's drink & be merry [Pils-3-177] **W: rom, img, a-wld, t-flies**
*Liberty's in a flowering bowl [U2-271] **W>wom: hth**
Little barley corne [Rxb 2-29, Prs-312] **A: hed, hth, crg, img, b-end (fool)**
Loyal delights of the bottle [Fmr 4-214, Pils-1-709] **W: rom, a-wld**
Man in the moon drinks claret [Rxb 2-256] **W: hth, rom>A: b-end (hth)**
*Masonic drinking song (a) [U2-55] **W: hth, a-wld**
*May beauty crown our days & whisky our nights [U2-294] **Wsk: hed, rom**
*May odd fellows flourish forever [U1-294] **W: fel, a-wld**
*Mermaid Inn (the) [Tka-75] **W: fel**
Merry boys of Europe (the) [Rxb 5-85] **W>wom: hth, hed, fel**
Merry discourse (a) [Rxb 1-248] **A: w-drk, w-crit**
Merry fellows (the) [Tka-32, Thm-51] **A: hed, fel, rom, t-flies**
Merry gegs (the) [Bkm 2-64, Wsb-51] **W: fel**
Merry good fellowes (the) [Drl-326] **W: hed, a-wld**
Merry hostess (the) [Rxb 3-306, Cent-193, Tst-62] **A: hth, fel, w-drk**
Merry song (a) [Drl-268] **W: hed, a-wld**
Messe of good fellowes (a) [Rxb 2-142] **W: hed, fel**
*Mighty Bacchus, great thy wrongs [U2-341] **W: hth, (modern)**
Monday's worke [Rxb 2-149] **A: fel, w-crit**
Monks of the screw (the) [Tka-160] **W: hed, com**
Mr. Playstone's epithalium [Rxb 2-399] **W: hed, rom**
*My bottle & friend [U2-446] **W: fel, rom**
*My heart's in the bumper I pledge to my friend [U2-301] **W: fel**
*New Anacreonic song (a) [Amm-107] **W: rom, img**
*Nothing in life can sadden us [U2-150] **W: rom**
*Nothing like grog [Ful-47] **Rum: hed, crg**
Nottingham ale [Tka-194, Prs-414, Chp-573. Tst-43] **A: hth, com, (eloq)**
Nottingham ale (2nd song) [Prs-415] **A>spirits: hed, hth**
Now I'm resolved to love no more [Tka-171] **W>wom**
October brew [Toa] **A>all: hth**
*Odd fellows, drink & kiss the lasses [U1-330] **W: hed, rom, (a-wld)**
Old Simon the king [Frm 3-1, Pils 3-143, Chp-262] **A: hed, b-end (hth)**
*Our vicar still preaches that Peter & Paul [U1-124] **W: rom,**
*Port & good sherry will make us all merry [U1-231] **W>wom: fel**
*Potent delights of sparkling ale [U2-422] **A: hed, rom, fel**
Pot of porter, ho (a) [Prs-274] **A>W: hth**
Praise of Hull's ale [Prs-151] **A: hth, crg**
Prodigal's resolution (the) [Rxb 6-327, Chp-446] **W: hed, rom**
*Push around the brisk bowl [U2-174] **W: hed, rom, t-flies**
*Push around the pitcher [U2-191] **W: hed, a-wld**
Push the bowl around [Thm-89] **W: hed, rom, a-wld**
*Reason to fill my glass (a) [U2-259] **W: hed, rom, img**

*Rebel redcoat (the) [Drl-190] **W: hed, a-wld**
*Ring the bell, fill the bowl [U2-256] **W>wom: hed**
Round (a) [Tka-88] **W: hed, a-wld**
Sack for my money [Cent-185, Rxb 6-318] **W>A: hth, crg, img, rom, w-drk**
School for scandal (song from) [Tka-92] **W: rom (toast)**
Since Tom's in the chair [Pil-6-340] **W>cdr: hed, hth, rom, wit**
Song-1 [Drl-107] **W: hed, a-wld**
Song-2 [Drl-304] **W>wom**
Song-3 [Tka-60] **W: b-end (pov)**
Song in praise of ale [Tka-102] **A>W: hth**
Song in praise of punch (a) [Pils 5-138] **Wsk>W: hed, rom**
Song of a fallen angel over a bowl of rum punch [Tka-173] **Rum: hed**
Song of the mug [Tka-36] **W>wom: hed, img**
Song on Bacchus (a) [Tka-82] **W: com**
Strew sweet roses of pleasure between [Amm-270] **W: hth, wit**
Stroop of Rhenish [Tka-90] **W: hed, hth**
*Submit, bunch of grapes [Prs-288n] **A>W: fel, hth**
*Then drown dull care with sparkling wine [U1-355] **W: hed, rom**
*Then give me a friend with my glass [U2-314] **W: hed, fel, rom**
*Then glass after glass let me pursue [U1-31] **W>wom: hed, a-agr**
*This day I'll enjoy with wine in good store [U3-240] **W: hed, a-agr**
*Tho' Bacchus may boast of his care-killing bowl [Amm-268] **W: rom, img**
*Though all the world drink, not all the world loves [U2-253] **W>wom**
*Thousands or more [Ken-284] **W: hed, fel**
*Thousands or more [Ken-617] **W: hed, fel**
Three drunken maidens [Ped-240] **A: hed, rom, com, w-drk**
Tippling deities (the) [U3-375] **W: hed, hth**
Tippling philosophers [Wts-154, Tka-66, Wmb-79] **W: img, com**
*To Bacchus the libation pour [U2-93] **W: a-wld, t-flies**
*Tomorrow's a cheat [U2-95] **W: hed, fel, a-wld, t-flies**
Toper (the) [Tka-168] **W>wom: com**
*Toper's apology (the) [Full-101] **W: hed, rom, img**
*Toper's plea for drinking (the) [Hrl-429] **W: hed, com, t-flies**
*Toper's rant (the) [Tka-52, Tst-128] **A>W: hed, fel**
Toping song [Frm 5-18, Pils 6-200] **A: hed, crg, com, a-agr**
Tosse the pot [Tka-40] **A: hed, a-wld**
True use of the bottle (the) [Pils 6-367, Wmb-134] **W: rom**
Twankydillo [Pils-1-19, Ken #286] **A: fel, rom**
*Union of Cupid & Bacchus (the) [U2-43] **W: rom**
*Union of love & wine [U1-363] **W: a-agr**
Virtue of sack [Pils 3-327, Drl-293] **W>A: hth, rom, img, a-wld**
Virtue of wine (the) [Drl-218] **W: hed, hth, a-wld**
*Warrington Ale [Tka-195, Prs-412] **A>W: hed, hth**
Water drinker (the) [Vsb-98] **W: hed, hth**
What class in life [Wts 5-148, Wmb-121] **W: hed, a-wld**
*When Bacchus, jolly god, invites [U1-163] **W: rom**
*When betimes in the morn [Prs-251] **A: hed, fel (hunting)**
*When I drain the rosy bowl [U1-155] **W: hed, rom**
When Joan's ale was new [Prs-143, Pils-3-133] **A: fel, hed, b-end (pov)**
*When 'round the bowl we meet, boys [U2-298] **W: rom**
While Phyllis is drinking [Wts 1-40, Wmb-90] **W: hth, rom, w-drk**
*Wine, a toast & a chorus [U2-190] **W: hed, fel, crg**
*Wine & women we delight in [U1-335] **W: rom (a-intel)**
*Wit & mirth in wine are crowned [U1-422] **W: hth, rom, img**
*Wit sparkle like the bowl [U2-366] **W: hed, rom, wit**
*With an honest old friend [U1-68] **W: fel (a-miser)**
Women, love & wine [U2-45n] **W: rom (a-miser)**
World drowned in a glass (the) [Tka-49] **W: img**

Non-Bacchanalian Songs of the Seventeenth and Eighteenth Centuries

Advice to the beaus [Pils 2-10] **Tavern business**
Ale-wives' invitation to married men [Rxb 8-797] **A+W: b-end (pov)**
Answer (the) [Drl-306] **W: b-end (hth, fgt)**
Bad husband (a) [Rxb 2-414] **A: hed, w-crit, b-end (discord)**
Bad husband turned thrifty (the) [Rxb 6-484] **A: b-end (pov, fgt)**
Bad husband's experience of ill-husbandry [Rxb 7-820] **W: b-end (pov)**
Bad husband's folly (the) [Rxb 6-493] **A: b-end (abuse)**
Bad husband's reformation (the) [Rxb 8-796] **A: b-end (pov)**
Beer-drinking Briton (the) [Tka-116] **A>W: hth (patriotic)**
Blowzabella [Pils 1-194] **A: w-crit, b-end (pov, whor)**
Blue Lion (the) [Hol-31] **A+W: hed, w-drk**
Catalogue of contented cuckolds (the) [Rxb 3-481] **W>wom**
Catch (a) [Drl-69] **W: b-end (pov, hth, agr)**
Caveat for a spendthrift (a) [Rxb 8-802] **A: bad end (pov)**
Choice of intentions [Rxb 1-110] **A>W; W: b-end (pov)**
Come along, ye jolly watermen [Tka-145] **A: work**
Country farmers' vain glory [Tka-151,Rxb-?] **A: work, (patriotic)**
Country hostess's vindication [Rxb-3-380] **W: hostess cheats**
Craven churn-supper (the) [Bel-163] **A: hth, rom**
Cuckholds all in a row [Oep-342] **A: w-crit (bad wife)**
Cup of old stingo [Chp-308, Prs-328] **A: hed, rom, b-end (fool)**
Debauchery scared [Rxb 4-20] **W: rom (sed), com**
Disappointed taylor (the) [Pils 6-292, Fmr 4-227] **A: com**
Dropsied man (the) [Prs-262] **W: b-end (hth)**
Drunk was I last night [Pils 5-329] **A: w-critic**
Drunkard's legacy [Bel-100] **W: b-end (pov)**
English ale #1 [Prs-270] **A>W: a-tax**
Fairlop fair [Bel-191] **A: hed (excess) (fair—no fgt)**
Fayre warning [Rxb 1-372] **A: b-end (pov, fgt)**
Gilian of Croyden [Pls 1-707, Fmr 4-156] **A: w-drk**
Gondibert (On) [Drl-118] **A: b-end (fgt)**
Good ale [Ken-273, Chp-661,Tst-30] **A: hed, b-end (pov, fgt)**
Good fellow's consideration (the) [Rxb 6-339] **A: b-end (pov)**
Good fellow's counsel (the) [Rxb 6-499] **A: b-end (pov, whor)**
Good fellow's folly (the) [Rxb 6-346] **A: b-end, (abuse, pov)**
Good fellow's frolick (the) [Rxb 6-351] **A: b-end (pov)**
Good fellow's resolution (the) [Rxb 6-342] **A: b-end (pov)**
Groatsworth of good counsel for a penny [Rxb 6-480] **A: b-end (pov)**
Halfe a dozen of good wives [Rxb 1-454] **A: w-drk, b-end (pov)**
Heavy heart & a light purse (the) [Rxb 6-336] **W: b-end (pov)**
Here's to the bottle [Prs-507] **A: w-drk, b-end (pov, hth)**
Hop-planters' song [Tka-148] **A>W: rom, (patriotic)**
Horned miller (the) [Hol-111] **Gin: w-drk**
How 5 & 20 shillings were expended in 1 week [Tch-183] **Gin: w-drk**

I hate a fop at his glass [Pils 1-176] **W: w-crit**
In a cellar in Sodom [Fmr 3-214, Pils 2-297] **W: b-end, (sed, incest)**
In praise & dispraise of women & wine [Rxb 7-149] **W: w-drk, b-end (fgt, sed)**
Jack Had-Land's lamentation [Rxb 6-474] **A: b-end (pov)**
Jack Oakum in the suds [Hol-133] **Rum: b-end (death), com**
Jack Tar's frolick [Rxb 8-437/ 566] **A: b-end (con)**
John and Joan [Rxb 1-504] **A: w-drk, b-end (pov, fgt)**
John Barleycorn #1 [Rxb 2-273, Prs-320/338] **A: hed, hth**
John Barleycorn's a hero bold [Ken-627] **A>W: hed, hth**
Jolly gentleman's frolick (the) [Rxb 6-513] **W: b-end (jail)**
Jolly sailor (the) #1 [Amm-163] **W: hed**
Jolly sailor (the) #2 [Rxb 8-568] **Brandy: b-end (con)**
Jolly town rakes (the) [Pils 5-123] **W: hed, b-end (fgt, sed)**
Jolly Welsh-woman (the) [Rxb 7-724] **A: com, w-drk**
Leather bottell [Tka-134, Chp-514 , Rxb 6-468] **A: work**
London taylor's misfortune [Rxb 7-470] **W: w-drk, b-end (con)**
London's ordinaries [Rxb 2-24] **A: com**
Loyal delights of the bottle [Pils 1-709, Fmr-4-84] **W: rom, a-wld**
Mare & the foal (the) [Ebcs-64] **A: publican cheats**
Mark Noble's frolick [Rxb 6-510] **W: b-end (jail)**
Marriage it seems is for better or worse [Pils 5-272] **W: w-crit, rom**
Master Mault he is a gentleman [Rxb 2-379] **A: b-end (hth), w-drk, com**
Master's health (the) [Tka-155] **W: work**
Merry days of England (the) [Prs-331] **A: tax**
Missus Monday [Msb-135] **Bdy: com, w-drk, b-end (hth)**
Nick and froth [Rxb 6-486] **A: host cheats, w-drk**
Noble prodigal (the) [Rxb 6-489] **W: hed, rom**
Old man's vindication [Rxb 8-192] **W>wom, w-crit**
Old man's wish (the) [Rxb 6-506] **W: hed (mod)**
Old Sir Simon the king [Chp-264, Pils 3-144] **W: b-end, (fgt, sed)**
Pangyric on ale [Prs-492] **A: hed, fel, img>W: b-end (hth, fgt)**
Poor Tom the taylor [Rxb 7-472] **W: w-drk, b-end (con)**
Praise of ale [Prs-242] **A: hth, a-agr>W: b-end (fgt)**
Prodigal son converted [Rxb 4-48] **W: rom, b-end (pov, whor)**
Ranting rambler (the) [Bgf-205] **W: bad end (jail)**
Rap-a-tap-tap [Glo-77, Ebecs-77, Jm-4] **Gin: rom (sed)**
Reformed drinker [Rxb 6-317, Pils-2-42, Oep-269] **W: b-end (hth)**
Rioter's ruin (the) [Rxb 8-717] **A: b-end (pov), w-drk**
Roger's reknown [Rxb 7-236] **A: w-drk**
Room for a jovial tinker [Rxb 7-74] **A: rom, w-drk, com**
Salisbury plain [Glo-92] **W: rom (sed)**
Saturday night at sea [Amm-120] **Grog: hed, rom**
Stingo, oil of barley [Prs-328, Chp-308] **A: hth, rom, b-end (foolish)**
Tar's frolick (the) [Rxb 8-566, Hol-259] **A: b-end (con)**
Taylor's lamentation (the) [Rxb 7-474] **W: w-drk, b-end (con, fgt)**
Taylor's wanton wife of Wapping [Rxb-7-484] **A: b-end (con)**
There was a poor smith [Tka-137] **A: b-end (pov, fgt, sed)**Three buxome maids of Yoel [Rxb 8-647] **W: w-drk, b-end (pov)**
Three merry travellers [Ebb-38, Bgf-51] **A+W: hed, rom**
Time's abuses (the) [Rxb 2-577] **W: b-end (insult)**
Tippling John [Mm-66] **W: com**
Town rakes (the) [Pils 5-123] **W: hed, b-end (fgt, sed)**
Trooper watering his nag (the) [Pils 5-13] **A: rom (sed), com**
Turk in linen (the) [Fmr 3-5] **all drinks: com**
Two penny score (the) [Rxb 8-718] **A: b-end (pov)**
Upon the new inn [Pils 4-243] **W: b-end (con)**
Urserer and the spendthrift (the), pt. 2 [Rxb 1-110] **W: rom, b-end (pov)**
Warning to youth (a) [Rxb 3-36] **W: w-drk, b-end (sed)**

Non-Bacchanalian Songs of the Nineteenth Century

A very good hand at it [Chlt-154] **A+Gin: b-end (con), w-drk, com**
Adulterations [U2-269] **Gin: publican cheats**
All for the grog [Ebb-152, Kids-18] **Rum: b-end (pov, hth), w-drk**
All I want is my quantum of grog [U3-335] **Rum: hth, work**
All the folks would drink like fish [U3-265] **W: com**
Another cup & then [U2-135n] **W: b-end (reputn)**
Bacchus swings out & I swing within [U2-322] **W: com**
Ballad on ale [Prs-280, Tst-20] **A>W: hth, rom, crg, img**
Beautiful landlady (the) [Tst-67] **A: cheerful hostess, anti-gin**
Beer [Tka-124] **A>W: hed, hth, com**
Beer (the) [Prs-177] **A: anti tax**
Beer: a voice from the crowd [Prs-175] **A: b-end (abuse), com, cost**
Beer, boys, beer [Tka-115, Chlt-78, Prs-172, Tst-37] **A: hth, tax**
Ben Backstay [U2-276] **Rum: b-end (dies)**
Billy Yarn of Red-E-Riff [U1-447] **Gin: b-end (fgt), w-drk**
Bitter remonstrance (a) [Prs-173] **A: anti tax**
Black ram (the) [Ken-598, Barrrett-26] **A: hed, fel, work**
Bold cockney (the) [Ebb-162] **A: rom, fgt (over girl)**
Booze is there [Tka-54] **A: fel, w-drk, com**
Bottle of rum (the) [U3-298] **Rum: b-end (dies)**
Brown jug (the) [Prs-592, Tst-41] **A: com**
Bunch of grapes (the) [U2-93] **W: hed, b-end (fgt), w-drk, com**
Burton's ale [Prs-428] **A: hth, mfg cheats**
Choice (the) [Mm-92] **Wom>wine: rom**
Choice of intentions [Rxb-1-110] **W: b-end (pov)**
Cornish harvest-home song [Prs-192] **A: fel, work**
Country statues [Msb-199] **A: work**
Dan Dab & Dolly Deckswab [Un-3-201] **Gin: rom, com, w-drk**
Delights of a tea & supper party [U1-384] **W+Bdy: fel, rom**
Dockyard gate [Kids-22] **A: w-drk (happy)**
Dorset squire (the) [Prs-457] **A: a-agr**
Down by the dark arches [Ebb-160] **Bdy: b-end (con, hth)**
Dramatic morality [U2-82] **Drk?: b-end (fgt)**
Drinker (the) [U2-316] **Rum+Gin+Wine: hed, com, w-drk (gin)**
Drinking is a foolish thing [Chlt-76] **A: b-end (hth, pov)**
Drop of good beer (a) [Tst-25, Prs-476] **A: hed, hth, w-drk**
Drunkard and the pig (the) [Tst -118] **A: com**
Drunken bucher of Tideswell (the) [Prs-442] **A: b-end (hth)**
English ale #1 [Prs-251] **A>W: hed**
English ale #2 [Prs-408] **A>grog: hth**
English bright beer [Tka-117, Prs-271] **A: hed, crg**
Exciseman's disaster [U3-209] **A: com, tax**
Execution of John Hill and John Williams [Hrfd-194] **A: b-end (kill)**
Free & easy (the) [U1-350] **Bdy+Rum: hed, w-crit**
Friar (the) [Tka-164] **W: hed, com, b-end (hth)**
Friezland ale [Tst-33] **A: hed, hth**
Fuddling day [Tch-144] **A: w-crit , Gin: w-drk**
Gin shop (the) [U1-354] **Gin: w-drk, b-end (fgt, pov, jail)**
Gin shop (the) #2 [U2-331] **Gin: b-end (fgt)**

Good fellow's frolick [Prs-487] **A: hed, b-end (pov, foolish)**
Good old days of Adam & Eve (in the) [Prs-244, Chlt-47] **A: hth**
Gritten of Garway [Hrfd-192] **A: b-end (kill)**
Harvest home song [Ebcs-48, Prs-58] **A: hed, work**
Health to the barley mow (a) [Tka-152. Prs-343] **A: work**
Here, tapster [U2-394] **A>wsk: Wsk: w-drk, b-end (hth)**
Here's a health to all honest men [Prs-180] **A: fel (patriotic)**
Here's the bottle she loved [U2-383, Prs-507] **W: w-drk, b-end (hth)**
Here's to the grog [Kds-18, Ken-606] **Rum: bad end (pov), w-drk**
Hot coddlings [Msb-145] **Gin: w-drk, b-end (pov)**
How 5 & 20 shillings were expended [Tch-183, Msb-48] **Gin: w-drk, com**
Humors of Eccles' wake [Rigs-8] **A: b-end (fgt, sports dispute)**
I likes a drop of good beer [Tka-112, Prs-476, Tst-25] **A>others: hed, hth, tax**
I think the liquor's in me [U2-40] **Drk?: b-end (foolish)**
I'll never get drunk anymore [Tst-124] **A: b-end (abuse)**
Jack Junk [U2-101] **Rum: b-end (hth) blk-com**
Jealous husband (the) [Hol-141] **W: b-end (abuse), com**
John Appleby [U1-385] **A: b-end (quarrel), w-drk, com**
John Darm [U1-236] **W: b-end (fgt, jail), com**
John Strong [U2-173] **Drk?: w-drk, b-end (fgt)**
Jolly fat friar loved liquor (a) [U1-433] **W: b-end (fool), com**
Jolly Jerry of the Tarter frigate [U2-138] **Rum: b-end (jail)**
Jolly post-boy [Hol-147, Str-14] **A: hed, work**
Jolly wagonner (the) [Ebcs-16] **A: work**
Joys of drinking (the) [U2-382] **A: rom, a-agr**
Limbo [Ebb-147] **A: b-end (pov)**
London adulterations [Tch-175] **W-A-Gin: mfg cheat**
Lovers' farewell to pale ale (the) [Prs-171] **A: price, com**
Maggie May [Hugl-307) **W: w-drk, b-end (con)**
Mail coach guard [Tst-84] **A: rom, work**
Man that waters the workers' beer (the) [Tst-28] **A: mfg cheats**
Manchester races [U2-348] **A: hed, b-end (fgt)**
Mare & the foal (the) [Ebcs-64] **A: publican cheats**
Merry days of England (the) [Prs-331] **A: tax**
Miller & his wife (the) [Praise-480] **A: w-drk, a-agr, rom**
Modern blood of fashion [U2-2] **Drk?: hed, rom, (duel)**
Music and wine [Thm-42] **W: hed, hth**
My grandfather's days [Tch-83] **Gin: w-drk, b-end (hth)**
Navvy on the line [Tch-40, Str-42] **A: hed, rom, w-drk**
Nottingham ale #2 [Prs-415] **A>spirits: hed, hth**
Off to sea once more [Oxf-250] **A: b-end (con)**
Origin of grog (the) [U2-125] **Rum: com**
Outward bound [Shp 2-303, Kds-12] **Rum: hed, sailor, work**
Oyster girl (the) [Kds-108] **W: w-drk, b-end (con)**
Pace egging song [Rm-28] **A: b-end (fgt)**
Pangyric on ale (a) [Prs-492] **A: hed, fel, img>W: b-end (hth, fgt)**
Paris & London [U2-296] **A>W: (patriot)**
Peter McCawley's wife & the doctor [U2-321] **Gin: w-drk, b-end (fgt), com**
Pleasures of the fair (the) [U2-277] **A: b-end (fgt)**
Publican (the) [U2-156] **A: cheats**
Publican's new Sunday act (the) [Vsb-94] **A: b-end (pov)**
Quarry Bank mashers (the) [Ebb-125] **Wsk-Rum: hed, rom**
Rambling comber (the) [Tch-200] **A: b-end (pov)**
Ratcliffe highway [Ebb-159] **W: b-end (con)**
Rigs and fun of Nottingham goose fair [Tch-102] **A: hed, b-end (pov)**
Rise in ale (the): a bitter wish [Prs-171] **A: price, com**
Road-makers (the) [Str-7] **A: work**
Rum & milk [Tka-130] **Rum: hed, rom**
Sale of a wife [Tch-2] **A: b-end (pov), w-crit, com**
Salisbury plain [Ebcs-92] **A+W: rom (sed)**
Satire upon Derby ale [Prs-432] **A: hth**
Saturday night [Tch-92] **A: hed, work**
Sheep shearing (the) [Str-56, Harl-27] **A: hed, work**
Sheffield's a wonderful town (O) [Tch-94] **A: hed, work**
Snob & the bottle (the) [Msb-38] **Gin: b-end (pov)**
So was I [Tst-119] **A: b-end (jail)**
Soldiers three [JM-6] **A: com**
Song (a) [Tka-60] **W: b-end (pov)**
Song of John Barleycorn [Prs-354] **A: b-end (fool)**
Sot's paradise (the) [Prs-430] **A: b-end (hth)**
Stave for Bass & Company (a) [Prs-174] **A: anti-tax**
Success to old England & jolly old Tom [U2-10] **A: com**
Tarpaulin jacket (the) [Shp 2-228] **Rum: hed, fel**
Tea-drinking wives [Tch-166] **Gin: w-drk, b-end (pov)**
There's nothing goes wrong when the grog's mixed right [U1-67] **Rum: hed**
Thirsty family (the) [U2-154] **Gin: w-drk, com**
Thrasher (the) [Prs-475] **A: work**
Three physicians (the) [U2-219] **W: b-end (pov), com (black)**
Toper's calendar (the) [U2-95] **A: b-end (foolish)**
Very good hand at it (a) [Chlt-154] **Gin: w-drk, com, b-end (pov)**
Vicar & Moses (the) [U1-353] **A: b-end (fool), com**
Wedgefield wake (the) [Palmer-pers com] **A: b-end (fgt: sports dispute)**
Wednesbury cocking [Fss-77] **A: b-end (fgt: sports dispute)**
We're all jolly fellows that follow the plough [Vsb-117] **A: hed, fel, work**
When first I landed at Liverpool [Lld-284] **Rum: b-end (con)**
Wholesome advice for drunkards [Tst-127] **A: w-drk, b-end (pov)**
Wife for sale (a) [Tch-196] **A: w-crit, com**
Willie drunk again [Vsb-101] **A: b-end (pov, abuse)**
Wonderful effects on the Leister railroad [Tch-52] **A: rom, work, w-drk**
Word of advice (a) [Tch-172] **A: b-end (pov)**

Bibliographical Tools and Reference Works

Bibliographies and Bibliographical Essays

Alcohol and Temperance History Group/*Social History of Alcohol Review*. www.athg.org.

Fahey, David M. "'I'll Drink to That!': The Social History of Alcohol" in *Choice: Current Reviews for Academic Libraries.* 38, no. 4 (December 2000): 637–640, 642–645.

Gutzke, David W. *Alcohol in the British Isles from Roman Times to 1996: An Annotated Bibliography.* Westport, CT: Greenwood Press, 1996.

Jessup, Jacquie. "The Liquor Issue in American History: A Bibliography" in Jack S. Blocker Jr. (ed), *Alcohol, Reform and Society: The Liquor Issue in Social Context.* Westport, CT: Greenwood Press, 1979. 259–279.

Jimerson, Randall C., Francis X. Blouin, and Charles A. Isetts (ed). *Guide to the Microfilm Edition of Temperance and Prohibition Papers.* Ann Arbor: University of Michigan, 1977.

Social History of Alcohol Review. 1979–present.

Verhey, Jeffrey. "Sources for the Social History of Alcohol" in Susanna Barrows and Robin Room, eds., *Drinking: Behavior and Belief in Modern History.* Berkeley: University of California Press, 1991. 425–439.

Reference Works

Barber, Norman. *A Century of British Brewers, 1890–1990.* 2d ed. New Ash Green, Kent, UK: Brewery History Society, 1994.

Brewers Association of Canada. *Alcoholic Beverage Taxation and Control Policies.* Ottawa: Brewers Association of Canada, 1972–present.

Cherrington, Ernest Hurst et al., eds. *Standard Encyclopedia of the Alcohol Problem.* 6 vols. Westerville, OH: American Issue, 1925–1930.

Downard, William L. *Dictionary of the History of the American Brewing and Distilling Industries.* Westport, CT: Greenwood Press, 1980.

Harrison, Brian. *Dictionary of British Temperance Biography.* Coventry: Society for the Study of Labour History, Aids to Research, no. 1. *Bulletin Supplement,* 1973.

Heath, Dwight B., ed. *International Handbook on Alcohol and Culture.* Westport, CT: Greenwood Press, 1995.

Lender, Mark Edward. *Dictionary of American Temperance Biography: From Temperance Reform to Alcohol Research, the 1600s to the 1980s.* Westport, CT: Greenwood Press, 1984.

Österberg, Esa, and Thomas Karlsson, eds. *Alcohol Policies in EU Member States and Norway. A Collection of Country Reports.* Helsinki: Stakes, 2002.

Richmond, Lesley, and Alison Turton, eds. *The Brewing Industry: A Guide to Historical Records.* Manchester: Manchester University Press, 1990.

Simpura, Jussi, and Thomas Karlsson. *Trends in Drinking Patterns among Adult Population in 15 European Countries, 1950 to 2000. A Collection of Country Reports.* Helsinki: Stakes, 2001.

World Drink Trends 2002 Edition. Oxfordshire, UK: NTC Publications, 2002.

Index

Note: page numbers in bold type refer encyclopedia articles devoted to a given index topic heading.